Lecture Notes in Artificial Intelligence 8398

Subseries of Lecture Notes in Computer Science

Lecture Notes in Artificial Intelligence 8398

Subseries of Lecture Notes in Computer Science

Ngoc Thanh Nguyen Boonwat Attachoo
Bogdan Trawiński Kulwadee Somboonviwat (Eds.)

Intelligent Information and Database Systems

6th Asian Conference, ACIIDS 2014
Bangkok, Thailand, April 7-9, 2014
Proceedings, Part II

Volume Editors

Ngoc Thanh Nguyen
Bogdan Trawiński
Wrocław University of Technology, Poland
E-mail: ngoc-thanh.nguyen@pwr.edu.pl
E-mail: bogdan.trawinski@pwr.wroc.pl

Boonwat Attachoo
Kulwadee Somboonviwat
King Mongkut's Institute of Technology Ladkrabang
Bangkok, Thailand
E-mail: {boonwat, kskulwad}@kmitl.ac.th

ISSN 0302-9743 e-ISSN 1611-3349
ISBN 978-3-319-05457-5 e-ISBN 978-3-319-05458-2
DOI 10.1007/978-3-319-05458-2
Springer Cham Heidelberg New York Dordrecht London

Library of Congress Control Number: 2014933552

LNCS Sublibrary: SL 7 – Artificial Intelligence

Typesetting: Camera-ready by author, data conversion by Scientific Publishing Services, Chennai, India

Printed on acid-free paper

Springer is part of Springer Science+Business Media (www.springer.com)

Preface

ACIIDS 2014 was the sixth event in the series of international scientific conferences for research and applications in the field of intelligent information and database systems. The aim of ACIIDS 2014 was to provide an internationally respected forum for scientific research in the technologies and applications of intelligent information and database systems. ACIIDS 2014 was co-organized by King Mongkut's Institute of Technology Ladkrabang (Thailand) and Wrocław University of Technology (Poland) in co-operation with the IEEE SMC Technical Committee on Computational Collective Intelligence, Hue University (Vietnam), University of Information Technology UIT-HCM (Vietnam), and Quang Binh University (Vietnam) and took place in Bangkok (Thailand) during April 7-9, 2014. The first two events, ACIIDS 2009 and ACIIDS 2010, took place in Dong Hoi City and Hue City in Vietnam, respectively. The third event, ACIIDS 2011, took place in Daegu (Korea), while the fourth event, ACIIDS 2012, took place in Kaohsiung (Taiwan). The fifth event, ACIIDS 2013, was held in Kuala Lumpur in Malaysia.

We received almost 300 papers from over 30 countries from around the world. Each paper was peer reviewed by at least two members of the international Program Committee and International Reviewer Board. Only 124 papers with the highest quality were selected for oral presentation and publication in the two volumes of ACIIDS 2014 proceedings.

The papers included in the proceedings cover the following topics: natural language and text processing, intelligent information retrieval, Semantic Web, social networks and recommendation systems, intelligent database systems, technologies for intelligent information systems, decision support systems, computer vision techniques, machine learning and data mining, multiple model approach to machine learning, computational intelligence, engineering knowledge and semantic systems, innovations in intelligent computation and applications, modelling and optimization techniques in information systems, database systems and industrial systems, innovation via collective intelligences and globalization in business management, intelligent supply chains as well as human motion: acquisition, processing, analysis, synthesis, and visualization for massive datasets.

Accepted and presented papers highlight the new trends and challenges of intelligent information and database systems. The presenters showed how new research could lead to new and innovative applications. We hope you will find these results useful and inspiring for your future research.

We would like to express our sincere thanks to the honorary chairs, Prof. Tawil Paungma (Former President of King Mongkut's Institute of Technology Ladkrabang, Thailand), Prof. Tadeusz Więckowski (Rector of Wrocław University of Technology, Poland), and Prof. Andrzej Kasprzak, Vice-Rector of Wrocław University of Technology, Poland, for their support.

Our special thanks go to the program chairs and the members of the international Program Committee for their valuable efforts in the review process, which helped us to guarantee the highest quality of the selected papers for the conference. We cordially thank the organizers and chairs of special sessions who essentially contributed to the success of the conference.

We would also like to express our thanks to the keynote speakers (Prof. Hoai An Le Thi, Prof. Klaus-Robert Müller, Prof. Leszek Rutkowski, Prof. Vilas Wuwongse) for their interesting and informative talks of world-class standard.

We cordially thank our main sponsors, King Mongkut's Institute of Technology Ladkrabang (Thailand), Wrocław University of Technology (Poland), IEEE SMC Technical Committee on Computational Collective Intelligence, Hue University (Vietnam), University of Information Technology UIT-HCM (Vietnam), and Quang Binh University (Vietnam). Our special thanks are also due to Springer for publishing the proceedings, and to the other sponsors for their kind support.

We wish to thank the members of the Organizing Committee for their very substantial work and the members of the local Organizing Committee for their excellent work.

We cordially thank all the authors for their valuable contributions and all the other participants of this conference. The conference would not have been possible without them.

Thanks are also due to many experts who contributed to making the event a success.

April 2014

Ngoc Thanh Nguyen
Boonwat Attachoo
Bogdan Trawiński
Kulwadee Somboonviwat

Conference Organization

Honorary Chairs

Tawil Paungma	Former President of King Mongkut's Institute of Technology Ladkrabang, Thailand
Tadeusz Więckowski	Rector of Wrocław University of Technology, Poland
Andrzej Kasprzak	Vice-Rector of Wrocław University of Technology, Poland

General Chairs

Ngoc Thanh Nguyen	Wrocław University of Technology, Poland
Suphamit Chittayasothorn	King Mongkut's Institute of Technology Ladkrabang, Thailand

Program Chairs

Bogdan Trawiński	Wrocław University of Technology, Poland
Kulwadee Somboonviwat	King Mongkut's Institute of Technology Ladkrabang, Thailand
Tzung-Pei Hong	National University of Kaohsiung, Taiwan
Hamido Fujita	Iwate Prefectural University, Japan

Organizing Chairs

Boonwat Attachoo	King Mongkut's Institute of Technology Ladkrabang, Thailand
Adrianna Kozierkiewicz-Hetmańska	Wrocław University of Technology, Poland

Special Session Chairs

Janusz Sobecki	Wrocław University of Technology, Poland
Veera Boonjing	King Mongkut's Institute of Technology Ladkrabang, Thailand

Publicity Chairs

Kridsada Budsara	King Mongkut's Institute of Technology Ladkrabang, Thailand
Zbigniew Telec	Wrocław University of Technology, Poland

Conference Webmaster

Natthapong Jungteerapanich	King Mongkut's Institute of Technology Ladkrabang, Thailand

Local Organizing Committee

Visit Hirunkitti	King Mongkut's Institute of Technology Ladkrabang, Thailand
Natthapong Jungteerapanich	King Mongkut's Institute of Technology Ladkrabang, Thailand
Sutheera Puntheeranurak	King Mongkut's Institute of Technology Ladkrabang, Thailand
Pitak Thumwarin	King Mongkut's Institute of Technology Ladkrabang, Thailand
Somsak Walairacht	King Mongkut's Institute of Technology Ladkrabang, Thailand
Bernadetta Maleszka	Wrocław University of Technology, Poland
Marcin Maleszka	Wrocław University of Technology, Poland
Marcin Pietranik	Wrocław University of Technology, Poland

Steering Committee

Ngoc Thanh Nguyen (Chair)	Wrocław University of Technology, Poland
Longbing Cao	University of Technology Sydney, Australia
Tu Bao Ho	Japan Advanced Institute of Science and Technology, Japan
Tzung-Pei Hong	National University of Kaohsiung, Taiwan
Lakhmi C. Jain	University of South Australia, Australia
Geun-Sik Jo	Inha University, South Korea
Jason J. Jung	Yeungnam University, South Korea
Hoai An Le Thi	University Paul Verlaine - Metz, France
Toyoaki Nishida	Kyoto University, Japan
Leszek Rutkowski	Częstochowa University of Technology, Poland
Suphamit Chittayasothorn	King Mongkut's Institute of Technology Ladkrabang, Thailand
Ali Selamat	Universiti Teknologi Malaysia, Malyasia

Keynote Speakers

Hoai An Le Thi	University of Lorraine, France
Klaus-Robert Müller	Technische Universität Berlin, Germany
Leszek Rutkowski	Częstochowa University of Technology, Poland
Vilas Wuwongse	Asian Institute of Technology, Thailand

Special Sessions Organizers

1.*Multiple Model Approach to Machine Learning (MMAML 2014)*

Tomasz Kajdanowicz	Wrocław University of Technology, Poland
Edwin Lughofer	Johannes Kepler University Linz, Austria
Bogdan Trawiński	Wrocław University of Technology, Poland

2. *Computational Intelligence (CI 2014)*

Piotr Jędrzejowicz	Gdynia Maritime University, Poland
Urszula Boryczka	University of Silesia, Poland
Ireneusz Czarnowski	Gdynia Maritime University, Poland

3. *Engineering Knowledge and Semantic Systems (IWEKSS 2014)*

Jason J. Jung	Yeungnam University, South Korea
Dariusz Król	Bournemouth University, UK

4. *Innovations in Intelligent Computation and Applications (IICA 2014)*

Shyi-Ming Chen	National Taiwan University of Science and Technology, Taiwan
Shou-Hsiung Cheng	Cheinkuo Technology University, Taiwan

5. *Modelling and Optimization Techniques in Information Systems, Database Systems and Industrial Systems (MOT-ACIIDS 2014)*

Hoai An Le Thi	University of Lorraine, France
Tao Pham Dinh	National Institute for Applied Sciences - Rouen, France

6. *Innovation via Collective Intelligences and Globalization in Business Management (ICIGBM 2014)*

Yuh-Shy Chuang	Chien Hsin University, Taiwan
Chao-Fu Hong	Aletheia University, Taiwan
Pen-Choug Sun	Aletheia University, Taiwan

7. *Intelligent Supply Chains (ISC 2014)*

Arkadiusz Kawa	Poznań University of Economics, Poland
Milena Ratajczak-Mrozek	Poznań University of Economics, Poland
Konrad Fuks	Poznań University of Economics, Poland

8. *Human Motion: Acquisition, Processing, Analysis, Synthesis and Visualization for Massive Datasets (HMMD 2014)*

Konrad Wojciechowski	Polish-Japanese Institute of Information Technology, Poland
Marek Kulbacki	Polish-Japanese Institute of Information Technology, Poland
Jakub Segen	Gest3D, USA

International Program Committee

Ajith Abraham	Machine Intelligence Research Labs, USA
Muhammad Abulaish	King Saud University, Saudi Arabia
El-Houssaine Aghezzaf	Ghent University, Belgium
Jesús Alcalá-Fdez	University of Granada, Spain
Haider M. AlSabbagh	Basra University, Iraq
Troels Andreasen	Roskilde University, Denmark
Toni Anwar	Universiti Teknologi Malaysia, Malaysia
Giuliano Armano	University of Cagliari, Italy
Zeyar Aung	Masdar Institute of Science and Technology, United Arab Emirates
Ahmad-Taher Azar	Benha University, Egypt
Costin Bădică	University of Craiova, Romania
Emili Balaguer-Ballester	Bournemouth University, UK
Amar Balla	Ecole Superieure d'Informatique, France
Zbigniew Banaszak	Warsaw University of Technology, Poland
Dariusz Barbucha	Gdynia Maritime University, Poland
Ramazan Bayindir	Gazi University, Turkey
Maumita Bhattacharya	Charles Sturt University, Australia
Mária Bieliková	Slovak University of Technology in Bratislava, Slovakia
Jacek Błażewicz	Poznań University of Technology, Poland
Veera Boonjing	King Mongkut's Institute of Technology Ladkrabang, Thailand
Mariusz Boryczka	University of Silesia, Poland
Urszula Boryczka	University of Silesia, Poland
Abdelhamid Bouchachia	Bournemouth University, UK
Stephane Bressan	National University of Singapore, Singapore
Peter Brida	University of Žilina, Slovakia

Piotr Bródka	Wrocław University of Technology, Poland
Andrej Brodnik	University of Primorska, Slovenia
Grażyna Brzykcy	Poznań University of Technology, Poland
The Duy Bui	University of Engineering and Technology, Hanoi, Vietnam
Robert Burduk	Wrocław University of Technology, Poland
František Čapkovič	Slovak Academy of Sciences, Slovakia
Gladys Castillo	University of Aveiro, Portugal
Oscar Castillo	Tijuana Institute of Technology, Mexico
Dariusz Ceglarek	Poznań High School of Banking, Poland
Stephan Chalup	University of Newcastle, Australia
Boa Rong Chang	National University of Kaohsiung, Taiwan
Somchai Chatvichienchai	University of Nagasaki, Japan
Rung-Ching Chen	Chaoyang University of Technology, Taiwan
Shyi-Ming Chen	National Taiwan University of Science and Technology, Taiwan
Shou-Hsiung Cheng	Chein-Kuo University of Technology, Taiwan
Suphamit Chittayasothorn	King Mongkut's Institute of Technology Ladkrabang, Thailand
Tzu-Fu Chiu	Aletheia University, Taiwan
Amine Chohra	Paris-East University (UPEC), France
Kazimierz Choroś	Wrocław University of Technology, Poland
Young-Joo Chung	Rakuten, Inc., Japan
Robert Cierniak	Częstochowa University of Technology, Poland
Dorian Cojocaru	University of Craiova, Romania
Tina Comes	University of Agder, Norway
Phan Cong-Vinh	NTT University, Vietnam
José Alfredo F. Costa	Federal University of Rio Grande do Norte (UFRN), Brazil
Keeley Crockett	Manchester Metropolitan University, UK
Bogusław Cyganek	AGH University of Science and Technology, Poland
Ireneusz Czarnowski	Gdynia Maritime University, Poland
Piotr Czekalski	Silesian University of Technology, Poland
Tran Khanh Dang	HCMC University of Technology, Vietnam
Jerome Darmont	Université Lumiere Lyon 2, France
Paul Davidsson	Malmö University, Sweden
Roberto De Virgilio	Roma Tre University, Italy
Mahmood Depyir	Shiraz University, Iran
Phuc Do	Vietnam National University, HCMC, Vietnam
Tien V. Do	Budapest University of Technology and Economics, Hungary
Pietro Ducange	University of Pisa, Italy
El-Sayed El-Alfy	King Fahd University of Petroleum and Minerals, Saudi Arabia
Mourad Elloumi	University of Tunis-El Manar, Tunisia

Peter Erdi	Kalamazoo College, USA
Victor Felea	Alexandru Ioan Cuza University of Iasi, Romania
Thomas Fober	University of Marburg, Germany
Dariusz Frejlichowski	West Pomeranian University of Technology, Poland
Mohamed Gaber	University of Portsmouth, UK
Patrick Gallinari	Pierre and Marie Curie University, France
Dariusz Gąsior	Wrocław University of Technology, Poland
Andrey Gavrilov	Novosibirsk State Technical University, Russia
Deajan Gjorgjevikj	Ss. Cyril and Methodius University in Skopje, Macedonia
Daniela Godoy	UNICEN University, Argentina
Fernando Gomide	University of Campinas, Brazil
Vladimir I. Gorodetsky	St.Petersburg Institute for Informatics and Automation, Russia
Manuel Graña	University of Basque Country, Spain
Janis Grundspenkis	Riga Technical University, Latvia
Adam Grzech	Wrocław University of Technology, Poland
Sajjad Haider	Institute of Business Administration, Pakistan
Slimane Hammoudi	ESEO Institute of Science and Technology, France
Jin-Kao Hao	University of Angers, France
Habibollah Haron	Universiti Teknologi Malaysia, Malaysia
Tutut Herawan	Universiti Malaysia Pahang, Malaysia
Francisco Herrera	University of Granada, Spain
Huu Hanh Hoang	Hue University, Vietnam
Tzung-Pei Hong	National University of Kaohsiung, Taiwan
Wei-Chiang Hong	Oriental Institute of Technology, Taiwan
Timo Honkela	Helsinki University of Technology, Finland
Mong-Fong Horng	National Kaohsiung University of Applied Sciences, Taiwan
Dosam Hwang	Yeungnam University, South Korea
Natthakan Iam-On	Mae Fah Luang University, Thailand
Dmitry I. Ignatov	National Research University Higher School of Economics, Moscow, Russia
Lazaros Iliadis	Democritus University of Thrace, Greece
Khalid Jebari	LCS Rabat, Morocco
Joanna Jędrzejowicz	University of Gdańsk, Poland
Piotr Jędrzejowicz	Gdynia Maritime University, Poland
Janusz Jeżewski	Institute of Medical Technology and Equipment ITAM, Poland
Gordan Ježić	University of Zagreb, Croatia
Geun-Sik Jo	Inha University, South Korea
Kang-Hyun Jo	University of Ulsan, South Korea

Jason Jung	Yeungnam University, South Korea
Janusz Kacprzyk	Systems Research Institute of Polish Academy of Science, Poland
Tomasz Kajdanowicz	Wrocław University of Technology, Poland
Radosław Katarzyniak	Wrocław University of Technology, Poland
Tsungfei Khang	University of Malaya, Malaysia
Vladimir F. Khoroshevsky	Dorodnicyn Computing Centre of Russian Academy of Sciences, Russia
Muhammad Khurram Khan	King Saud University, Saudi Arabia
Pan-Koo Kim	Chosun University, South Korea
Yong Seog Kim	Utah State University, USA
Frank Klawonn	Ostfalia University of Applied Sciences, Germany
Joanna Kołodziej	University of Bielsko-Biała, Poland
Marek Kopel	Wrocław University of Technology, Poland
Józef Korbicz	University of Zielona Góra, Poland
Leszek Koszałka	Wrocław University of Technology, Poland
Adrianna Kozierkiewicz-Hetmańska	Wrocław University of Technology, Poland
Worapoj Kreesuradej	King Mongkut's Institute of Technology Ladkrabang, Thailand
Ondřej Krejcar	University of Hradec Králové, Czech Republic
Dalia Kriksciuniene	Vilnius University, Lithuania
Dariusz Król	Bournemouth University, UK
Marzena Kryszkiewicz	Warsaw University of Technology, Poland
Adam Krzyzak	Concordia University, Canada
Kazuhiro Kuwabara	Ritsumeikan University, Japan
Sergei O. Kuznetsov	National Research University Higher School of Economics, Moscow, Russia
Halina Kwaśnicka	Wrocław University of Technology, Poland
Pattarachai Lalitrojwong	King Mongkut's Institute of Technology Ladkrabang, Thailand
Helge Langseth	Norwegian University of Science and Technology, Norway
Henrik Legind Larsen	Aalborg University, Denmark
Mark Last	Ben-Gurion University of the Negev, Israel
Annabel Latham	Manchester Metropolitan University, UK
Nguyen-Thinh Le	Clausthal University of Technology, Germany
Hoai An Le Thi	University of Lorraine, France
Kun Chang Lee	Sungkyunkwan University, South Korea
Philippe Lenca	Telecom Bretagne, France
Thitiporn Lertrusdachakul	Thai-Nichi Institute of Technology, Thailand
Lin Li	Wuhan University of Technology, China
Horst Lichter	RWTH Aachen University, Germany

Kamol Limtanyakul	Sirindhorn International Thai-German Graduate School of Engineering, Thailand
Sebastian Link	University of Auckland, New Zealand
Heitor Silvério Lopes	Federal University of Technology - Parana (UTFPR), Brazil
Wojciech Lorkiewicz	Wrocław University of Technology, Poland
Edwin Lughofer	Johannes Kepler University Linz, Austria
Marcin Maleszka	Wrocław University of Technology, Poland
Urszula Markowska-Kaczmar	Wrocław University of Technology, Poland
Francesco Masulli	University of Genova, Italy
Mustafa Mat Deris	Universiti Tun Hussein Onn Malaysia, Malaysia
Jacek Mercik	Wrocław University of Technology, Poland
Saeid Nahavandi	Deakin University, Australia
Kazumi Nakamatsu	University of Hyogo, Japan
Grzegorz J. Nalepa	AGH University of Science and Technology, Poland
Prospero Naval	University of the Philippines, Philippines
Fulufhelo Vincent Nelwamondo	Council for Scientific and Industrial Research, South Africa
Ponrudee Netisopakul	King Mongkut's Institute of Technology Ladkrabang, Thailand
Linh Anh Nguyen	University of Warsaw, Poland
Ngoc-Thanh Nguyen	Wrocław University of Technology, Poland
Thanh Binh Nguyen	International Institute for Applied Systems Analysis, Austria
Adam Niewiadomski	Łódź University of Technology, Poland
Yusuke Nojima	Osaka Prefecture University, Japan
Mariusz Nowostawski	University of Otago, New Zealand
Manuel Núñez	Universidad Complutense de Madrid, Spain
Richard Jayadi Oentaryo	Singapore Management University, Singapore
Shingo Otsuka	Kanagawa Institute of Technology, Japan
Jeng-Shyang Pan	National Kaohsiung University of Applied Sciences, Taiwan
Tadeusz Pankowski	Poznań University of Technology, Poland
Marcin Paprzycki	Systems Research Institute of Polish Academy of Science, Poland
Jakub Peksiński	West Pomeranian University of Technology, Poland
Niels Pinkwart	Humboldt University of Berlin, Germany
Grzegorz Popek	Wrocław University of Technology, Poland
Elvira Popescu	University of Craiova, Romania
Piotr Porwik	University of Silesia, Poland
Bhanu Prasad	Florida A&M University, USA
Wenyu Qu	Dalian Maritime University, China

Christoph Quix	RWTH Aachen University, Germany
Preesan Rakwatin	Geo-Informatics and Space Technology Development Agency, Thailand
Ewa Ratajczak-Ropel	Gdynia Maritime University, Poland
Chotirat Ann Ratanamahatana	Chulalongkorn University, Thailand
Rajesh Reghunadhan	Central University of Bihar, India
Przemysław Różewski	West Pomeranian University of Technology, Poland
Miti Ruchanurucks	Kasetsart University, Thailand
Leszek Rutkowski	Częstochowa University of Technology, Poland
Henryk Rybiński	Warsaw University of Technology, Poland
Alexander Ryjov	Lomonosov Moscow State University, Russia
Virgilijus Sakalauskas	Vilnius University, Lithuania
Sakriani Sakti	Nara Institute of Science and Technology, Japan
Daniel Sánchez	University of Granada, Spain
Jürgen Schmidhuber	Swiss AI Lab IDSIA, Switzerland
Björn Schuller	Technical University of Munich, Germany
Ali Selamat	Universiti Teknologi Malaysia, Malaysia
S.M.N. Arosha Senanayake	University of Brunei Darussalam, Brunei
Alexei Sharpanskykh	Delft University of Technology, The Netherlands
Seema Shedole	M S Ramaiah Institute of Technology, India
Quan Z. Sheng	University of Adelaide, USA
Andrzej Siemiński	Wrocław University of Technology, Poland
Dragan Simić	University of Novi Sad, Serbia
Gia Sirbiladze	Iv. Javakhishvili Tbilisi State University, Georgia
Andrzej Skowron	University of Warsaw, Poland
Janusz Sobecki	Wrocław University of Technology, Poland
Kulwadee Somboonviwat	King Mongkut's Institute of Technology Ladkrabang, Thailand
Jerzy Stefanowski	Poznań Univeristy of Technology, Poland
Serge Stinckwich	UMI UMMISCO, France
Stanimir Stoyanov	University of Plovdiv Paisii Hilendarski, Bulgaria
Nidapan Sureerattanan	Thai-Nichi Institute of Technology, Thailand
Dejvuth Suwimonteerabuth	IBM Solutions Delivery, Thailand
Shinji Suzuki	University of Tokyo, Japan
Jerzy Świątek	Wrocław University of Technology, Poland
Edward Szczerbicki	University of Newcastle, Australia
Julian Szymański	Gdańsk University of Technology, Poland
Ryszard Tadeusiewicz	AGH University of Science and Technology, Poland
Yasufumi Takama	Tokyo Metropolitan University, Japan
Kay Chen Tan	National University of Singapore, Singapore

Name	Affiliation
Pham Dinh Tao	INSA-Rouen, France
Zbigniew Telec	Wrocław University of Technology, Poland
Thanaruk Theeramunkong	Thammasat University, Thailand
Krzysztof Tokarz	Silesian University of Technology, Poland
Behcet Ugur Toreyin	Cankaya University, Turkey
Bogdan Trawiński	Wrocław University of Technology, Poland
Krzysztof Trawiński	European Centre for Soft Computing, Spain
Konstantin Tretyakov	University of Tartu, Estonia
Iwan Tri Riyadi Yanto	Universiti Tun Hussein Onn Malaysia, Malaysia
Hong-Linh Truong	Vienna University of Technology, Austria
George A. Tsihrintzis	University of Piraeus, Greece
Alexey Tsymbal	Siemens AG, Germany
Rainer Unland	University of Duisburg-Essen, Germany
Olgierd Unold	Wrocław University of Technology, Poland
Arlette van Wissen	VU University Amsterdam, The Netherlands
Pandian Vasant	Universiti Teknologi Petronas, Malaysia
Emil Vassev	University of Limerick, Ireland
Jørgen Villadsen	Technical University of Denmark, Denmark
Maria Virvou	University of Piraeus, Greece
Gottfried Vossen	University of Münster, Germany
Wahidin Wahab	University of Indonesia, Indonesia
Kitsana Waiyamai	Kasetsart University, Thailand
Ali Wali	University of Kairouan, Tunisia
Botao Wang	Northeastern University, China
Yitong Wang	Fudan University, China
Yongkun Wang	Rakuten, Inc., Japan
Ukrit Watchareeruetai	King Mongkut's Institute of Technology Ladkrabang, Thailand
Izabela Wierzbowska	Gdynia Maritime University, Poland
Nuwee Wiwatwattana	Srinakharinwirot University, Thailand
Wayne Wobcke	University of New South Wales, Australia
Marek Wojciechowski	Poznań University of Technology, Poland
Dong-Min Woo	Myongji University, South Korea
Michał Woźniak	Wrocław University of Technology, Poland
Marian Wysocki	Rzeszow University of Technology, Poland
Guandong Xu	University of Technology, Sydney, Australia
Xin-She Yang	University of Cambridge, UK
Zhenglu Yang	University of Tokyo, Japan
Keem Siah Yap	Universiti Tenaga Nasional, Malaysia
Lean Yu	Chinese Academy of Sciences, AMSS, China
Sławomir Zadrożny	Systems Research Institute of Polish Academy of Science, Poland
Drago Žagar	University of Osijek, Croatia
Danuta Zakrzewska	Łódź University of Technology, Poland

Faisal Zaman	Kyushu Institute of Technology, Japan
Constantin-Bala Zamfirescu	Lucian Blaga University of Sibiu, Romania
Arkady Zaslavsky	CSIRO, Australia
Aleksander Zgrzywa	Wrocław University of Technology, Poland
Jianwei Zhang	Tsukuba University of Technology, Japan
Rui Zhang	Wuhan University of Technology, China
Zhongwei Zhang	University of Southern Queensland, Australia
Cui Zhihua	Complex System and Computational Intelligence Laboratory, China
Zhi-Hua Zhou	Nanjing University, China
Xingquan Zhu	University of Technology, Sydney, Australia

Program Committees of Special Sessions

Special Session on Multiple Model Approach to Machine Learning (MMAML 2014)

Jesús Alcalá-Fdez	University of Granada, Spain
Emili Balaguer-Ballester	Bournemouth University, UK
Abdelhamid Bouchachia	Bournemouth University, UK
Piotr Bródka	Wrocław University of Technology, Poland
Robert Burduk	Wrocław University of Technology, Poland
Oscar Castillo	Tijuana Institute of Technology, Mexico
Rung-Ching Chen	Chaoyang University of Technology, Taiwan
Suphamit Chittayasothorn	King Mongkut's Institute of Technology Ladkrabang, Thailand
José Alfredo F. Costa	Federal University (UFRN), Brazil
Bogusław Cyganek	AGH University of Science and Technology, Poland
Ireneusz Czarnowski	Gdynia Maritime University, Poland
Patrick Gallinari	Pierre et Marie Curie University, France
Fernando Gomide	State University of Campinas, Brazil
Francisco Herrera	University of Granada, Spain
Tzung-Pei Hong	National University of Kaohsiung, Taiwan
Tomasz Kajdanowicz	Wrocław University of Technology, Poland
Yong Seog Kim	Utah State University, USA
Mark Last	Ben-Gurion University of the Negev, Israel
Kun Chang Lee	Sungkyunkwan University, South Korea
Heitor S. Lopes	Federal University of Technology Paraná, Brazil
Edwin Lughofer	Johannes Kepler University Linz, Austria
Mustafa Mat Deris	Universiti Tun Hussein Onn Malaysia, Malaysia
Dragan Simić	University of Novi Sad, Serbia
Jerzy Stefanowski	Poznań University of Technology, Poland
Zbigniew Telec	Wrocław University of Technology, Poland

Bogdan Trawiński	Wrocław University of Technology,Poland
Olgierd Unold	Wrocław University of Technology, Poland
Pandian Vasant	University Technology Petronas, Malaysia
Michał Woźniak	Wrocław University of Technology, Poland
Faisal Zaman	Kyushu Institute of Technology, Japan
Zhongwei Zhang	University of Southern Queensland, Australia
Zhi-Hua Zhou	Nanjing University, China

Computational Intelligence (CI 2014)

Dariusz Barbucha	Gdynia Maritime University, Poland
Mariusz Boryczka	University of Silesia, Poland
Urszula Boryczka	University of Silesia, Poland
Longbing Cao	University of Technology Sydney, Australia
Bogusław Cyganek	AGH University of Science and Technology, Poland
Ireneusz Czarnowski	Gdynia Maritime University, Poland
Piotr Jędrzejowicz	Gdynia Maritime University, Poland
Tianrui Li	Southwest Jiaotong University, China
Alfonso Mateos Caballero	Universidad Politécnica de Madrid, Spain
Mikhail Moshkov	King Abdullah University of Science and Technology, Saudi Arabia
Agnieszka Nowak-Brzezińska	University of Silesia, Poland
Ewa Ratajczak-Ropel	Gdynia Maritime University, Poland
Rafał Różycki	Poznań University of Technology, Poland
Wiesław Sieńko	Gdynia Maritime University, Poland
Adam Słowik	Koszalin University of Technology, Poland
Rafał Skinderowicz	University of Silesia, Poland
Alicja Wakulicz-Deja	University of Silesia, Poland
Beata Zielosko	University of Silesia, Poland

Engineering Knowledge and Semantic Systems (IWEKSS 2014)

Gonzalo A. Aranda-Corral	Universidad de Sevilla, Spain
David Camacho	Autonomous University of Madrid, Spain
Fred Freitas	Universidade Federal de Pernambuco, Brazil
Daniela Godoy	Unicen University, Argentina
Tutut Herawan	University of Malaya, Malaysia
Adam Jatowt	Kyoto University, Japan
Jason J. Jung	Yeungnam University, Korea
Krzysztof Juszczyszyn	Wrocław University of Technology, Poland
Dariusz Król	Bournemouth University, UK
Monika Lanzenberger	Vienna University of Technology, Austria
Jinjiu Li	UTS, Australia

Innovations in Intelligent Computation and Applications (IICA 2014)

An-Zen Shih	Jinwen University of Science and Technology, Taiwan
Albert B. Jeng	Jinwen University of Science and Technology, Taiwan
Victor R. L. Shen	National Taipei University, New Taipei City, Taiwan
Jeng-Shyang Pan	National Kaohsiung University of Applications, Kaohsiung, Taiwan
Mong-Fong Horng	National Kaohsiung University of Applications, Kaohsiung, Taiwan
Huey-Ming Lee	Chinese Culture University, Taipei, Taiwan
Ying-Tung Hsiao	National Taipei University of Education, Taipei, Taiwan
Shou-Hsiung Cheng	Chienkuo Technology University, Changhua, Taiwan
Chun-Ming Tsai	Taipei Municipal University of Education, Taipei, Taiwan
Cheng-Yi Wang	National Taiwan University of Science and Technology, Taiwan
Shyi-Ming Chen	National Taiwan University of Science and Technology, Taiwan
Heng Li Yang	National Chenchi University, Taiwan
Jium-Ming Lin	Chung Hua University, Taiwan
Chun-Ming Tsai	University of Taipei, Taiwan
Yung-Fa Huang	Chaoyang University of Technology, Taiwan
Ho-Lung Hung	Chienkuo Technology University, Changhua, Taiwan
Chih-Hung Wu	National Taichung University of Education, Taiwan
Jyh-Horng Wen	Tunghai University, Taiwan
Jui-Chung Hung	University of Taipei, Taiwan

Modelling and Optimization Techniques in Information Systems, Database Systems and Industrial Systems (MOT-ACIIDS 2014)

Le Thi Hoai An	University of Lorraine, France
Pham Dinh Tao	INSA-Rouen, France
Pham Duc Truong	University of Cardiff, UK
Raymond Bisdorff	Université du Luxembourg, Luxembourg

Jin-Kao Hao — University of Angers, France
Joaquim Judice — University of Coimbra, Portugal
Yann Guermeur — LORIA, France
Boudjeloud Lydia — University of Lorraine, France
Conan-Guez Brieu — University of Lorraine, France
Gely Alain — University of Lorraine, France
Le Hoai Minh — University of Lorraine, France
Do Thanh Nghi — University of Can Tho, Vietnam
Alexandre Blansché — University of Lorraine, France
Nguyen Duc Manh — ENSTA Bretagne, France
Ta Anh Son — Hanoi University of Science and Technology, Vietnam
Tran Duc Quynh — Hanoi University of Agriculture, Vietnam

Innovation via Collective Intelligences and Globalization in Business Management (ICIGBM 2014)

Tzu-Fu Chiu — Aletheia University, Taiwan
Yi-Chih Lee — Chien Hsin University, Taiwan
Jian-Wei Lin — Chien Hsin University, Taiwan
Kuo-Sui Lin — Aletheia University, Taiwan
Tzu-En Lu — Chien Hsin University, Taiwan
Chia-Ling Hsu — TamKang University, Taiwan
Fang-Cheng Hsu — Aletheia University, Taiwan
Rahat Iqbal — Coventry University, UK
Irene Su — TamKang University, Taiwan
Ai-Ling Wang — TamKang University, Taiwan
Henry Wang — Institude of Software Chinese Academy of Sciences, China
Hung-Ming Wu — Aletheia University, Taiwan
Wei-Li Wu — Chien Hsin University, Taiwan
Feng-Sueng Yang — TamKang University, Taiwan

Intelligent Supply Chains (ISC 2014)

Areti Manataki — The University of Edinburgh, UK
Zbigniew Pasek — University of Windsor, Canada
Arkadiusz Kawa — Poznań University of Economics, Poland
Marcin Hajdul — Institute of Logistics and Warehousing, Poland
Paweł Pawlewski — Poznań University of Technology, Poland
Paulina Golińska — Poznań University of Technology, Poland

Human Motion: Acquisition, Processing, Analysis, Synthesis and Visualization for Massive Datasets (HMMD 2014)

Human Motion: Acquisition, Processing, Analysis, Synthesis and Visualization for Massive Datasets (HMMD 2014)

Aldona Drabik	Polish-Japanese Institute of Information Technology, Poland
André Gagalowicz	Inria, France
Ryszard Klempous	Wrocław University of Technology, Poland
Ryszard Kozera	Warsaw University of Life Science, Poland
Marek Kulbacki	Polish-Japanese Institute of Information Technology, Poland
Aleksander Nawrat	Silesian University of Technology, Poland
Lyle Noakes	The University of Western Australia, Australia
Jerzy Paweł Nowacki	Polish-Japanese Institute of Information Technology, Poland
Andrzej Polański	Polish-Japanese Institute of Information Technology, Poland
Andrzej Przybyszewski	University of Massachusetts, USA
Eric Petajan	LiveClips, USA
Jerzy Rozenblit	University of Arizona, Tucson, USA
Jakub Segen	Gest3D, USA
Aleksander Sieroń	Medical University of Silesia, Poland
Konrad Wojciechowski	Polish-Japanese Institute of Information Technology, Poland

Table of Contents – Part II

Machine Learning and Data Mining

Multiple Model Approach to Machine Learning (MMAML 2014)

Computational Intelligence (CI 2014)

Engineering Knowledge and Semantic Systems (IWEKSS 2014)

Innovations in Intelligent Computation and Applications (IICA 2014)

Modelling and Optimization Techniques in Information Systems, Database Systems and Industrial Systems (MOT 2014)

Innovation via Collective Intelligences and Globalization in Business Management (ICIGBM 2014)

Intelligent Supply Chains (ISC 2014)

Human Motion: Acquisition, Processing, Analysis, Synthesis and Visualization for Massive Datasets (HMMD 2014)

Table of Contents – Part I

Natural Language and Text Processing

Intelligent Information Retrieval

Semantic Web, Social Networks and Recommendation Systems

Intelligent Database Systems

Intelligent Information Systems

Decision Support Systems

Computer Vision Techniques

Intelligent Fuzzy Control with Multiple Constraints for a Model Car System with Multiplicative Disturbance

Wen-Jer Chang, Po-Hsun Chen, and Bo-Jyun Huang

Department of Marine Engineering
National Taiwan Ocean University
Keelung 202, Taiwan, R.O.C.
wjchang@mail.ntou.edu.tw

Abstract. A multiple constrained fuzzy controller design methodology is developed in this paper to achieve state variance constraint and passivity constraint for a model car system. The model car system considered in this paper is represented by a discrete-time Takagi-Sugeno fuzzy model with multiplicative disturbance. The proposed fuzzy controller is accomplished by using the technique of parallel distributed compensation. Based on the parallel distributed compensation technique, the sufficient conditions are derived to achieve variance constraint, passivity constraint and stability performance constraint, simultaneously. By solving these sufficient conditions with linear matrix inequality technique, the multiple constrained fuzzy controllers can be obtained for the model car system. At last, a numerical simulation example is provided to illustrate the feasibility and validity of the proposed fuzzy control method.

Keywords: Intelligent Fuzzy Control, Discrete Takagi-Sugeno Fuzzy Model, Model Car System, Multiplicative Disturbance.

1 Introduction

It is well known that fuzzy control has become a popular research in control engineering because it has made itself available not only in the laboratory work but also in industrial applications. In recent years, theoretical developments of fuzzy control have been proposed and the use of fuzzy controllers has been explored [1-6]. One important application of fuzzy control is in maritime, space and ground vehicles. In [1], a fuzzy control that uses rules on a skilled human operator's experience is applied to automatic train operations. For aircraft flight control, a fuzzy controller is proposed in [2] where the fuzzy rules are generated by interrogating an experienced pilot and asking him a number of highly structured questions. In [3], the authors have designed a fuzzy controller based on fuzzy modeling of a human operator's control actions to navigate and to park a car. By generating fuzzy rules using learning algorithms, the authors of [4] have developed fuzzy controllers for the truck backer upper to a loading dock problem from an arbitrary initial position by manipulating the steering. In [5], a fuzzy controller was proposed for an autonomous boat without initially having to develop a nonlinear dynamics model of a vehicle. In [6], the attenuation of the

N.T. Nguyen et al. (Eds.): ACIIDS 2014, Part II, LNAI 8398, pp. 1–12, 2014.

external disturbance energy and the fuzzy controller design problems for the ship steering systems was studied. This paper focuses on the design of a stabilizing fuzzy controller for the multiple constrained problems for a model car system. Such a dynamics model has been studied in [7].

The Takagi-Sugeno (T-S) fuzzy model is the most famous tool to approximate nonlinear systems in the fuzzy-model-based control. It is constructed by fuzzy blending linear subsystems via the IF-THEN fuzzy rules and corresponding membership functions. According to the T-S fuzzy models, the Parallel Distributed Compensation (PDC) concept [8-9] is applied to design the fuzzy controller for guaranteeing the system stability via the designed fuzzy controllers. The model car system considered in this paper is approximated by a T-S fuzzy model with multiplicative disturbance. Based on the T-S fuzzy model of the model car system, the outcome of the fuzzy control problem is parameterized in terms of a Linear Matrix Inequality (LMI) problem [10]. The LMI problem can be solved very efficiently by convex optimization techniques [10] to complete the fuzzy control design for the model car systems.

The performance constraints considered in this paper include stability, disturbance attenuation performance and individual state variance constraint. The stability performance of the closed-loop system is guaranteed by employing the Lyapunov theory. In addition, the passivity theory [6, 11-13] is employed in this paper to propose a general form for achieving disturbance attenuation performance. Passive systems theory can be traced back to the beginning of the 1970's, and its use in the feedback stabilization of nonlinear systems has recently gained renewed attention. In particular, the question of when a finite-dimensional nonlinear system can be rendered passive via state feedback was solved in [11]. In this paper, the concept of strictly input passive property is employed to analyze the stability of the T-S fuzzy systems for achieving attenuation performance. In addition to the passivity constraint, the individual state variance constraint is also considered in this paper. In the literature of system theory, covariance control theory [14-18] has been developed to provide control design methodologies for problems in which the objective is expressed as constraints on the variances of the states. In fact, this type of control that provides the ability to assign the second moment of the system state such as higher performance and improved robustness. Covariance control theory serves as a practical method for multiple objective control design as well as a foundation for linear system theory. However, it is difficult for the designers to assign a suitable state covariance matrix to achieve the Lyapunov equation condition.

The fuzzy control problem subject to passivity constraint and individual state variance constraint for the nonlinear T-S fuzzy models with multiplicative disturbance is not studied in literature. In order to discuss the proposed multiple performance constrained control problem, the sufficient conditions are derived via employing Lyapunov theory, passivity theory and covariance control scheme. Based on the Lyapunov stability theory, the common positive definite matrix is needed to be found for satisfying the conditions to guarantee the attenuation performance and individual state variance constraints. In order to solve these sufficient conditions, they should be converted into LMI problems. Applying the proposed fuzzy control design approach, the stability, disturbance attenuation performance and individual state variance constraint of nonlinear model car systems can be achieved, simultaneously.

2 System Description and Problem Statements

The fuzzy control for the autonomous mobile robot system was investigated in [19-20]. In this paper, a nonlinear model car system considered in [7] is employed to study the multiply constrained fuzzy control problem. The associated figure of model car is shown in Fig. 1 and the dynamic equations of this system are given as follows:

$$x_1(k+1)=x_1(k)+\frac{\upsilon T_s}{l}u(k)+0.1v_1(k)x_1(k)+0.05v_2(k)x_1(k) \quad (1a)$$

$$x_2(k+1)=x_2(k)+\upsilon T_s \sin(x_1(k))+0.05v_1(k)x_2(k)+0.1v_2(k)x_2(k)+w(k) \quad (1b)$$

$$y(k)=\alpha x_1(k)+\beta x_2(k) \quad (1c)$$

where $x_1(k)$ is the angle of the car, $x_2(k)$ is the vertical position of the rear end of the car, $u(k)$ is the steering angle, υ is the constant speed, T_s is the sampling time, l is the length of car, α and β are constant gains for output signals. Besides, the processes $v_1(k)$, $v_2(k)$ and $w(k)$ are all mutually independent zero-mean Gaussian white noise with intensity 1. In this example, it is assumed that $\upsilon=1$ (m/s); $T_s=1$ (s), $l=2.8$ (m); $\alpha=0.1708$ and $\beta=0.1708$. The purpose of this example is to control the car along a desired straight line of $x_2(k)=0$.

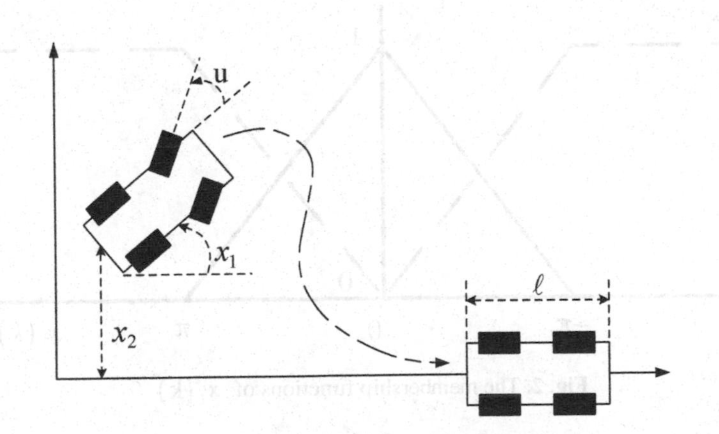

Fig. 1. A model car and its coordinate system

According to the fuzzy modeling technique [8], the nonlinear model car system can be transformed successfully into the following discrete-time two rules T-S fuzzy model and corresponding membership functions described in Fig. 2.

Rule1:

IF $x_1(k)$ is about 0,

$$\text{THEN} \quad x(k+1)=\left[\mathbf{A}_1+\sum_{e=1}^{2}\mathbf{N}_{e1}v_{e1}(k)\right]x(k)+\mathbf{B}_1u(k)+\mathbf{D}_1w(k)$$

$$y(k)=\mathbf{C}_1x(k)+\mathbf{H}_1w(k) \tag{2a}$$

Rule2:

IF $x_1(k)$ is about $\pm\pi$,

$$\text{THEN} \quad x(k+1)=\left[\mathbf{A}_2+\sum_{e=1}^{2}\mathbf{N}_{e2}v_{e2}(k)\right]x(k)+\mathbf{B}_2u(k)+\mathbf{D}_2w(k)$$

$$y(k)=\mathbf{C}_2x(k)+\mathbf{H}_2w(k) \tag{2b}$$

where $\mathbf{A}_1=\begin{bmatrix}1 & 0\\ 1 & 1\end{bmatrix}$, $\mathbf{A}_2=\begin{bmatrix}1 & 0\\ 0.003183 & 1\end{bmatrix}$, $\mathbf{N}_1=\begin{bmatrix}0.1 & 0\\ 0 & 0.05\end{bmatrix}$, $\mathbf{N}_2=\begin{bmatrix}0.05 & 0\\ 0 & 0.1\end{bmatrix}$, $\mathbf{B}_1=\mathbf{B}_2=\begin{bmatrix}0.357143\\ 0\end{bmatrix}$, $\mathbf{D}_1=\mathbf{D}_2=\begin{bmatrix}0\\ 0.05\end{bmatrix}$, $\mathbf{C}_1=\mathbf{C}_2=\begin{bmatrix}0.1708 & 0.0804\end{bmatrix}$ and $\mathbf{H}_1=\mathbf{H}_2=1$.

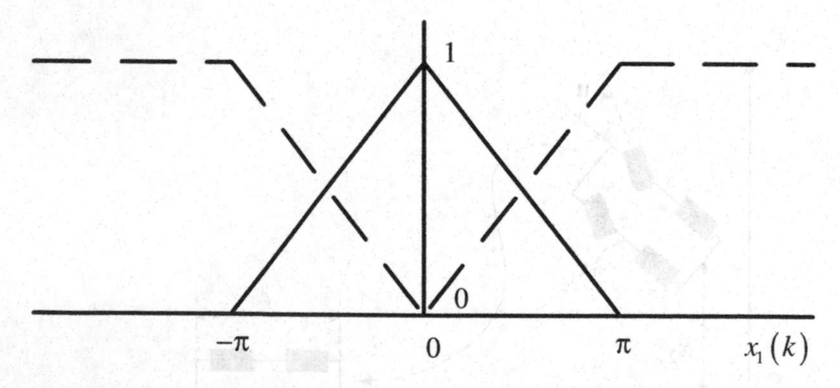

Fig. 2. The membership functions of $x_1(k)$

Applying a general T-S fuzzy model to represent the fuzzy system (2), one can obtain the following T-S fuzzy model with multiplicative disturbance:

$$x(k+1)=\sum_{i=1}^{r}h_i(z(k))\left\{\left[\mathbf{A}_i+\sum_{e=1}^{m}\mathbf{N}_{ei}v_{ei}(k)\right]x(k)+\mathbf{B}_i\,u(k)+\mathbf{D}_i\,w(k)\right\} \tag{3a}$$

$$y(k)=\sum_{i=1}^{r}h_i(z(k))\{\mathbf{C}_i x(k)+\mathbf{H}_i w(k)\} \tag{3b}$$

where $z_1(k)$, …, $z_q(k)$ are the premise variables, M_{iq} is the fuzzy set, q is the premise variable number, r is the number of fuzzy rules, $x(k)\in\Re^{n_x}$ is the state vector, $u(k)\in\Re^{n_u}$ is the input vector, $y(k)\in\Re^{n_y}$ is the output vector, the processes $w(k)\in\Re^{n_w}$ and $v_{ei}(k)\in\Re$ are mutually independent zero-mean Gaussian white noise with intensities $\mathbf{W}$ $(\mathbf{W}>0)$ and 1, respectively. The matrices $\mathbf{A}_i\in\Re^{n_x\times n_x}$, $\mathbf{B}_i\in\Re^{n_x\times n_u}$, $\mathbf{D}_i\in\Re^{n_x\times n_w}$, $\mathbf{N}_{ei}\in\Re^{n_x\times n_x}$, $\mathbf{C}_i\in\Re^{n_y\times n_x}$ and $\mathbf{H}_i\in\Re^{n_y\times n_w}$ are constant. Besides, $h_i(z(k))=\dfrac{\omega_i(z(k))}{\sum_{i=1}^{r}\omega_i(z(k))}$, $\omega_i(z(k))=\prod_{j=1}^{q}M_{ij}(z_j(k))$, $h_i(z(k))\geq 0$ and $\sum_{i=1}^{r}h_i(z(k))=1$.

Applying the concept of PDC, the fuzzy controller is designed to share the same IF part of the T-S fuzzy model (3). The proposed fuzzy controller can be represented as follows:

$$u(k)=\sum_{i=1}^{r}h_i(z(k))\{\mathbf{F}_i x(k)\} \tag{4}$$

Substituting (4) into (3), the closed-loop T-S fuzzy model can be obtained as follows:

$$x(k+1)=\sum_{i=1}^{r}\sum_{j=1}^{r}h_i(z(k))h_j(z(k))\left[\left(\mathbf{A}_i+\sum_{e=1}^{m}\mathbf{N}_{ei}v_{ei}(k)+\mathbf{B}_i\mathbf{F}_j\right)x(k)+\mathbf{D}_i w(k)\right] \tag{5}$$

For the system (5) with external disturbance $w(k)$ and output $y(k)$, the strict input passivity constraint is defined as follows:

$$E\left\{2\sum_{k=0}^{k_q}y^T(k)w(k)\right\}>E\left\{\sum_{k=0}^{k_q}\gamma w^T(k)w(k)\right\} \tag{6}$$

where $E\{\bullet\}$ denotes the expectation operator, $\gamma\geq 0$ and k_q is a positive integer that denotes terminal sampled time of control.

Considering each subsystem of the T-S fuzzy model (5), the steady state covariance matrix of the state vector has the following form:

$$\mathbf{X}_i=\mathbf{X}_i^T>0 \tag{7}$$

where $\mathbf{X}_i = \lim_{k\to\infty} E\left[x(k)x^T(k)\right]$ and $\mathbf{X}_i$ is the unique solution of the following Lyapunov equation for each rule [21].

$$(\mathbf{A}_i + \mathbf{B}_i\mathbf{F}_i)\mathbf{X}_i(\mathbf{A}_i + \mathbf{B}_i\mathbf{F}_i)^T - \mathbf{X}_i + \mathbf{D}_i\mathbf{W}\mathbf{D}_i^T + \sum_{e=1}^{m}\mathbf{N}_{ei}\mathbf{X}_i\mathbf{N}_{ei}^T = 0 \tag{8}$$

The state variance constraint considered in this paper is defined as follows:

$$[\mathbf{X}_i]_{kk} \le \sigma_k^{\ 2} \tag{9}$$

where $[\cdot]_{kk}$ denotes the kth diagonal element of matrix $[\cdot]$ and σ_k, $k = 1, 2, \cdots, n_x$, denote the root-mean-squared constraints for the variance of system states.

The purpose of this paper is to design the fuzzy controller (4) such that the passivity constraint (6) and the individual state variance constraint (9) are all achieved. In the next section, sufficient conditions for finding the above fuzzy controllers are provided.

3 Fuzzy Controller Design for Model Car System Based on T-S Fuzzy Models

In this section, the fuzzy controller design problem for achieving passivity constraint and individual state variance constraint is investigated by deriving sufficient conditions. By assigning a common upper bound matrix of the state covariance matrices for all fuzzy rules, the sufficient conditions are derived based on the Lyapunov theory and passivity theory. According to the closed-loop T-S fuzzy model (5), the sufficient conditions for achieving the stability, individual variance constraint and passivity constraint are derived in the following theorem.

3.1 Theorem 1

Given a dissipative rate γ. If there exists positive definite matrices $\bar{\mathbf{X}} > 0$, $\mathbf{P} > 0$ and feedback gains $\mathbf{F}_i$ satisfying the following sufficient conditions, then the closed-loop system (5) is asymptotically stable. In this case, strict input passivity constraint (6) and individual state variance constraint (9) are all satisfied.

$$\begin{bmatrix} \Phi_{ij} & \Pi_i^T \\ \Pi_i & \Lambda \end{bmatrix} < 0 \tag{10}$$

$$\bar{\mathbf{X}} - \mathrm{diag}\left(\sigma_1^{\ 2}, \ldots, \sigma_{n_x}^{\ 2}\right) < 0 \tag{11}$$

where

$$\Phi_{ij}=\begin{bmatrix} -\bar{\mathbf{X}}+\mathbf{D}_i\mathbf{W}\mathbf{D}_i^T+\sum_{e=1}^{m}\mathbf{N}_{ei}\bar{\mathbf{X}}\mathbf{N}_{ei}^T & \mathbf{G}_{ij} & 0 & 0 & 0 \\ \mathbf{G}_{ij}^T & -\mathbf{P} & -\mathbf{C}_i^T & 0 & \mathbf{G}_{ij}^T \\ 0 & -\mathbf{C}_i & \mathbf{D}_i^T\mathbf{P}\mathbf{D}_i+\gamma\mathbf{I}-\mathbf{H}_i-\mathbf{H}_i^T & \mathbf{D}_i^T\mathbf{P} & 0 \\ 0 & 0 & \mathbf{P}\mathbf{D}_i & -\mathbf{I} & 0 \\ 0 & \mathbf{G}_{ij} & 0 & 0 & -\mathbf{I} \end{bmatrix} \tag{12}$$

$$\Pi_i=\begin{bmatrix} 0 & \mathbf{N}_{1i} & 0 & 0 & 0 \\ \vdots & \vdots & \vdots & \vdots & \vdots \\ 0 & \mathbf{N}_{mi} & 0 & 0 & 0 \end{bmatrix} \tag{13}$$

$$\Lambda=\mathrm{diag}\left(\frac{\bar{\mathbf{X}}}{m},\ldots,\frac{\bar{\mathbf{X}}}{m}\right) \tag{14}$$

Besides, $\mathbf{G}_{ij}=\mathbf{A}_i+\mathbf{B}_i\mathbf{F}_j$, $\mathbf{P}=\bar{\mathbf{X}}^{-1}$ and $\mathrm{diag}(\sigma_1^{\ 2},\ldots,\sigma_{n_x}^{\ 2})$ denotes a diagonal matrix with the diagonal elements $\sigma_1^{\ 2}$, $\sigma_2^{\ 2}$,..., and $\sigma_{n_x}^{\ 2}$.

Proof:

Let us choose the Lyapunov function as $V(x(t))=x^T(t)\mathbf{P}x(t)$, where $\mathbf{P}=\mathbf{P}^T>0$, to represent the energy of system. The expected value of differential of the Lyapunov function $V(x(k))$ along the trajectories of (5) is given as follows:

$$\begin{aligned} &E\{\Delta V(x(k))\} \\ &=E\{x^T(k+1)\mathbf{P}x(k+1)-x^T(k)\mathbf{P}x(k)\} \\ &\le E\Bigg\{\sum_{i=1}^{r}\sum_{j=1}^{r}h_i(z(k))h_j(z(k)) \\ &\times\begin{bmatrix} x(k) \\ w(k)\end{bmatrix}^T\begin{bmatrix} (\mathbf{A}_i+\mathbf{B}_i\mathbf{F}_j)^T\mathbf{P}(\mathbf{A}_i+\mathbf{B}_i\mathbf{F}_j)-\mathbf{P}+(\mathbf{A}_i+\mathbf{B}_i\mathbf{F}_j)^T(\mathbf{A}_i+\mathbf{B}_i\mathbf{F}_j)+m\sum_{e=1}^{m}\mathbf{N}_{ei}^T\mathbf{P}\mathbf{N}_{ei} & 0 \\ 0 & \mathbf{D}_i^T\mathbf{P}\mathbf{P}\mathbf{D}_i+\mathbf{D}_i^T\mathbf{P}\mathbf{D}_i \end{bmatrix}\begin{bmatrix} x(k) \\ w(k)\end{bmatrix}\Bigg\}\end{aligned} \tag{15}$$

It is obvious that if condition (10) is satisfied, one can obtain the following inequality via Schur complement [10] and defining $\bar{\mathbf{X}} = \mathbf{P}^{-1}$.

$$(\mathbf{A}_i + \mathbf{B}_i\mathbf{F}_i)\bar{\mathbf{X}}(\mathbf{A}_i + \mathbf{B}_i\mathbf{F}_i)^T - \bar{\mathbf{X}} + \mathbf{D}_i\mathbf{W}\mathbf{D}_1^T + \sum_{e=1}^{m}\mathbf{N}_{ei}\bar{\mathbf{X}}\mathbf{N}_{ei}^T < 0 \tag{16}$$

Subtracting (8) from (16), one has

$$(\mathbf{A}_i - \mathbf{B}_i\mathbf{F}_i)(\bar{\mathbf{X}} - \mathbf{X}_i)(\mathbf{A}_i - \mathbf{B}_i\mathbf{F}_i)^T - (\bar{\mathbf{X}} - \mathbf{X}_i) + \sum_{e=1}^{m}\mathbf{N}_{ei}(\bar{\mathbf{X}} - \mathbf{X}_i)\mathbf{N}_{ei}^T < 0 \tag{17}$$

Due to the conditions (16) holds, one can obtain that the closed-loop system (5) is stable and the closed-loop system matrix $(\mathbf{A}_i - \mathbf{B}_i\mathbf{F}_i)$ is stable. In this case, it can be concluded that $\bar{\mathbf{X}} - \mathbf{X}_i \geq 0$ via the inequality (17). From the condition (11) and $\bar{\mathbf{X}} \geq \mathbf{X}_i$, one has $[\mathbf{X}_i]_{kk} \leq [\bar{\mathbf{X}}]_{kk} \leq \sigma_k^{\,2}$. Thus, the proof of individual state variance constraint (9) is completed.

For achieving the attenuating performance, the passivity theory provides a useful and effective tool to design the controller to achieve the energy constraints for the closed-loop systems. Considering the passivity constraint defined in (6), one can define a performance function such as

$$E\left\{\sum_{k=0}^{k_q}\left(\gamma w^T(k)w(k) - 2y^T(k)w(k)\right)\right\} \triangleq E\{K(x,w,k)\} \tag{18}$$

The proof of the strict input passivity constraint (6) is similar to the proof of Theorem 1 of Reference [6]. Due to the limitation of space, the detailed proof of strict input passivity constraint (6) is omitted here. At last, one can obtain the following result:

$$E\left\{2\sum_{k=0}^{k_q} y^T(k)w(k)\right\} > E\left\{\sum_{k=0}^{k_q}\gamma w^T(k)w(k)\right\} \tag{19}$$

It can be thus concluded that if the conditions (10) and (11) are satisfied, then the closed-loop system (5) is asymptotically stable, strictly input passive and the individual state variance constraint (9) is satisfied.

#

In Theorem 1, the conditions simultaneously includes variables $\mathbf{P}$ and $\mathbf{P}^{-1}$ such that the conditions are not a strictly LMI problem. For applying the LMI technique, let us introduce new variable $\bar{\mathbf{X}}$, such that $\mathbf{P}\bar{\mathbf{X}}=\mathbf{I}$. In order to solve the nonlinear minimization problem of (10), the cone complementarity technique [22] is applied to solve the following problem:

$$\text{Minimize trace}\left(\mathbf{P}\bar{\mathbf{X}}\right)\text{, subject to (10)-(11) and }\begin{bmatrix}\mathbf{P} & \mathbf{I}\\ \mathbf{I} & \bar{\mathbf{X}}\end{bmatrix}\geq 0 \tag{20}$$

Based on the nonlinear minimization problem stated in (20), the ILMI algorithm [23] can be employed to solve the sufficient conditions of Theorem 1. By solving the conditions of Theorem 1, the fuzzy controller can be obtained to achieve strict input passivity constraint (6) and individual state variance constraint (9) for the nonlinear model car systems, simultaneously.

4 Numerical Simulation for Control of a Model Car System

In order to indicate the efficiency of the present approach, a numerical example for the fuzzy control of a model car system (1) is illustrated in this section. According to the T-S fuzzy model (2) of the model car system, the fuzzy controller design method described in Section 3 can be used to find the fuzzy control gains. For starting analyzing and designing, we select the supply rate $\gamma=0.6$, $\mathbf{S}=\mathbf{I}$ with compatible dimension and assign the variance constraints as $\sigma_1^2=0.1$, $\sigma_2^2=0.2$. Solving the sufficient conditions of Theorem 1 via LMI technique [10], the matrix $\bar{\mathbf{X}}$ can be obtained as follows:

$$\bar{\mathbf{X}}=\begin{bmatrix}0.0047 & -0.0019\\ -0.0019 & 0.0891\end{bmatrix} \tag{21}$$

In addition, the fuzzy control gains $\mathbf{F}_1$ and $\mathbf{F}_2$ can be solved as follows:

$$\mathbf{F}_1=\begin{bmatrix}-2.8389 & -0.0647\end{bmatrix} \tag{22a}$$

$$\mathbf{F}_2=\begin{bmatrix}-2.7897 & -0.0626\end{bmatrix} \tag{22b}$$

Applying the fuzzy controller (4) with the above fuzzy control gains, the simulation responses of states are shown in Figs. 3-4 with initial condition $x(0)=\begin{bmatrix}-135^\circ & 2\end{bmatrix}^T$. Besides, the external disturbances $v_1(t)$, $v_2(t)$ and $w(t)$ are chosen as zero-mean white noises with variance one.

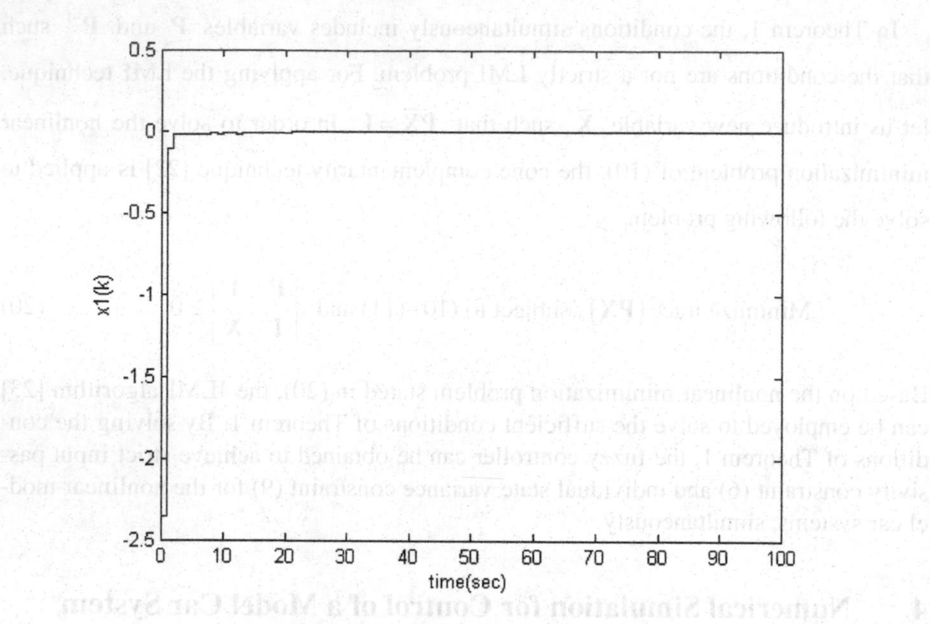

Fig. 3. The responses of state $x_1(k)$

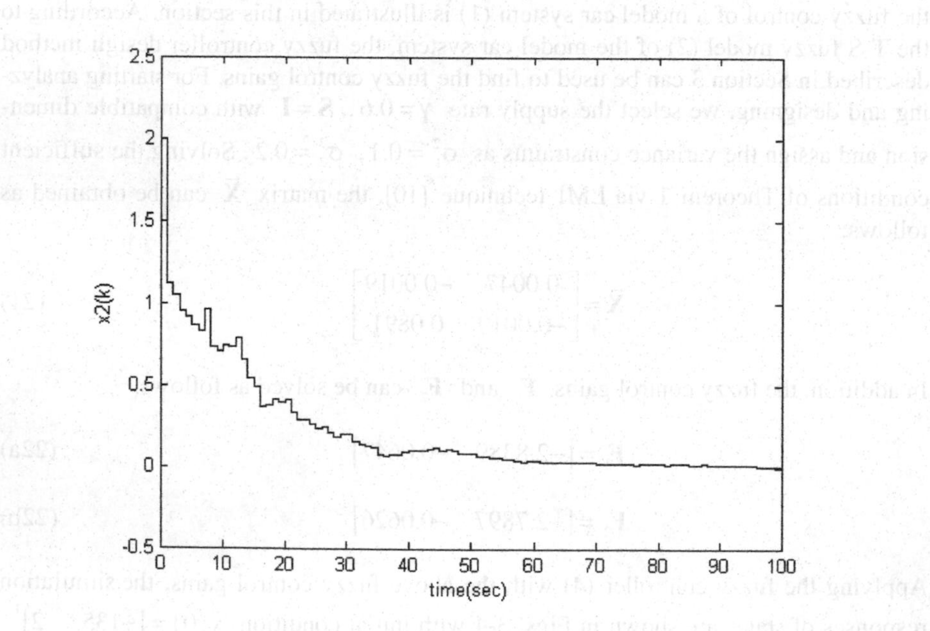

Fig. 4. The responses of state $x_2(k)$

From the simulation results, the effect of the external disturbance $w(t)$ on the proposed system can be criticized as follows:

$$\frac{E\left\{2\sum_{k=0}^{k_q} y^T(k)w(k)\right\}}{E\left\{\sum_{k=0}^{k_q} w^T(k)w(k)\right\}}=2.14 \tag{23}$$

The ratio value of (23) is bigger than determined dissipation rate $\gamma=0.6$, one can find that the passivity constraint (6) is satisfied. Besides, the variances of states $x_1(t)$ and $x_2(t)$ are 0.0549 and 0.1111, respectively. It is obvious that the individual state variance constraints are all satisfied. Thus, the considered model car system (1) with multiplicative disturbance can achieve desired multiple performance constraints by the designed fuzzy controller.

5 Conclusions

The intelligent fuzzy control problem with multiple constraints for a model car system with multiplicative disturbance has been investigated in this paper. The multiple constraints considered included stability, passivity and individual state variance constraints. Based on the T-S fuzzy models, the fuzzy controller design approach for nonlinear model car systems was developed. The sufficient conditions have been derived via the Lyapunov theory to achieve the above multiple constraints. By solving these sufficient conditions, a suitable fuzzy controller has been found for nonlinear model car system to achieve multiple constraints in the numerical example.

Acknowledgements. This work was supported by the National Science Council of the Republic of China under Contract NSC102-2221-E-019-051.

References

1. Yasunobu, S., Miyamoto, S., Ihara, H.: Fuzzy Control for Automatic Train Operation Systems. In: Proc. of the IFAC Control in Transportation Systems, pp. 33–39 (1983)
2. Larkin, L.I.: A Fuzzy Logic Controller for Aircraft Flight Control. In: Proc. of the Industrial Applications of Fuzzy Control, pp. 87–107. North-Holland, Amsterdam (1985)
3. Sugeno, M., Nishida, M.: Fuzzy Control of Model Car. Fuzzy Sets and Systems 16, 103–113 (1986)
4. Wang, L., Mendel, J.: Generating Fuzzy Rules by Learning from Examples. IEEE Trans. Systems, Man and Cybernetics 22, 1414–1427 (1992)
5. Waneck, T.W.: Fuzzy Guidance Controller for an Autonomous Boat. In: Proc. of the IEEE Control and Systems, vol. 17, pp. 43–51 (1997)

6. Chang, W.J., Chen, M.W., Ku, C.C.: Passive Fuzzy Controller Design for Discrete Ship Steering Systems via Takagi-Sugeno Fuzzy Model with Multiplicative Noises. Journal of Marine Science and Technology 21, 159–165 (2013)
7. Tanaka, K., Sano, M.: Trajectory Stabilization of a Model Car via Fuzzy Control. Fuzzy Sets and Systems 70, 155–170 (1995)
8. Tanaka, K., Wang, H.O.: Fuzzy Control Systems Design and Analysis: A Linear Matrix Inequality Approach. John Wiley & Sons, Inc., New York (2001)
9. Chang, W.J., Meng, Y.T., Tsai, K.H.: AQM Router Design for TCP Network via Input Constrained Fuzzy Control of Time-delay Affine Takagi-Sugeno Fuzzy Models. International Journal of Systems Science 43, 2297–2313 (2012)
10. Boyd, S., Ghaoui, L.E., Feron, E., Balakrishnan, V.: Linear Matrix Inequalities in System and Control Theory. SIAM, Philadelphia (1994)
11. Byrnes, C., Isidori, A., Willems, J.C.: Passivity, Feedback Equivalence, and the Global Stabilization of Minimum Phase Nonlinear Systems. IEEE Trans. Automatic Control 36, 1228–1240 (1991)
12. Lozano, R., Brogliato, B., Egeland, O., Maschke, B.: Dissipative Systems Analysis and Control Theory and Application. Springer, London (2000)
13. Chang, W.J., Liu, L.Z., Ku, C.C.: Passive Fuzzy Controller Design via Observer Feedback for Stochastic Takagi-Sugeno Fuzzy Models with Multiplicative Noises. International Journal of Control, Automation, and Systems 9, 550–557 (2011)
14. Hsieh, C., Skelton, R.E.: All Covariance Controllers for Linear Discrete-time Systems. IEEE Trans. Automatic Control 35, 908–915 (1990)
15. Skelton, R.E., Iwasaki, T.: Liapunov and Covariance Controllers. International Journal of Control 57, 519–536 (1993)
16. Chung, H.Y., Chang, W.J.: Extension of the Covariance Control Principle to Nonlinear Stochastic Systems. IEE Proceeding, Part D, Control Theory and Applications 141, 93–98 (1994)
17. Chang, W.J., Chung, H.Y.: A Covariance Controller Design Incorporating Optimal Estimation for Nonlinear Stochastic Systems. ASME, J. Dynamic Systems, Measurement and Control 118, 346–349 (1996)
18. Baromand, S., Khaloozadeh, H.: State Covariance Assignment Problem. IET Control Theory Applications 4, 391–402 (2010)
19. Castillo, O., Martinez-Marroquin, R., Melin, P., Valdez, F., Soria, J.: Comparative study of bio-inspired algorithms applied to the optimization of type-1 and type-2 fuzzy controllers for an autonomous mobile robot. Information Sciences 192, 19–38 (2012)
20. Melin, P., Astudillo, L., Castillo, O., Valdez, F., Garcia, M.: Optimal design of type-2 and type-1 fuzzy tracking controllers for autonomous mobile robots under perturbed torques using a new chemical optimization paradigm. Expert Systems with Applications 40, 3185–3195 (2013)
21. Chung, H.Y., Chang, W.J.: Input and State Covariance Control for Bilinear Stochastic Discrete Systems. Control-Theory and Advanced Technology 6, 655–667 (1990)
22. Ghaoui, L.E., Oustry, F., Rami, M.A.: A Cone Complementarity Linearization Algorithm for Static Output-Feedback and Related Problems. IEEE Trans. Automatic Control 42, 1171–1176 (1997)
23. Chang, W.J., Wu, W.Y., Ku, C.C.: H_∞ Constrained Fuzzy Control via State Observer Feedback for Discrete-time Takagi-Sugeno Fuzzy Systems with Multiplicative Noises. ISA Transactions 50, 37–43 (2011)

The Design of Knowledge-Based Medical Diagnosis System for Recognition and Classification of Dermatoglyphic Features

Hubert Wojtowicz[1], Jolanta Wojtowicz[1], and Wiesław Wajs[2]

[1] University of Rzeszow, Faculty of Mathematics and Nature, Chair of Computer Science, Rzeszow, Poland
[2] AGH University of Science and Technology, Faculty of Electrical Engineering, Institute of Automatics, Cracow, Poland

Abstract. Detecting Down's syndrome in newborns at an early stage helps in postnatal elimination or mitigation of negative effects of the resulting birth defects, which untreated may lead to infants' death or disability. This paper introduces an architecture of knowledge-based medical system for diagnosis of Down's syndrome in infants. The expert knowledge required for the design of the system was elicited from the scientific literature describing diagnostic scoring systems developed using statistical methods. The knowledge required for the calculation of diagnostic index score and determination of the genetic disorder presence is represented in the form of expert system rules. The diagnosis is carried out on the basis of results of pattern recognitions and measurements of dermatoglyphic features, which are passed to the expert system allowing determination of the value of diagnostic index score.

1 Introduction

Dermatoglyphics is a branch of anthropology, which finds use in contemporary medical diagnostics. The term dermatoglyphs, created by Cummins and Midlo in 1926 [1], was adopted by biologists around the world to describe skin ridges also called papillary lines. Arranging themselves into various patterns they create characteristic sculpture of fingerprints, palms and soles skin surfaces. The specific nature of dermatoglyphs is mainly due to their early stabilization in the fetal period and what follows their invariance after birth. Chromosomal aberrations affect the formation process of dermatoglyphic traits of particular fingerprints and morphological areas of palms and soles in the fetal period. Dermatoglyphs of infants with genetic disorder have a much higher incidence of specific combinations of patterns than healthy newborns. Combinations of dermatoglyphic patterns' types that deviate from normality were described in the scientific literature for most of the genetic disorders. On the basis of this knowledge diagnostic tests were designed using statistical analysis methods, which allow determination of the fact of particular genetic disorder incidence in the newborn.

N.T. Nguyen et al. (Eds.): ACIIDS 2014, Part II, LNAI 8398, pp. 13–22, 2014.

2 The Aim of the Work

The aim of the work is implementation of the medical decision support system inferring about the incidence of the genetic disorder on the basis of dermatoglyphic images data. Dermatoglyphs constitute an attractive diagnostic tool because their features can be analyzed and measured in prints acquired using traditional offline ink methods, without the recourse of employing invasive blood karyotyping methods, which are stressful for newborns. However due to the small size of the newborn ridges the patterns they create on all dermatoglyphic areas are difficult to analyze in the early postpartum period. The reduced thickness of the epidermis in this period is also leading to significant deformations of patterns upon slightest contact, which makes the traditional manual examination, usually involving magnifying scope, more difficult and demanding even for an experienced anthropologist. The digital equipment available nowadays enable acquisition of high-quality images, which can be then processed in order to give salience to their characteristic features and analyzed using pattern recognition and computational intelligence methods. These high-quality images collected moments after birth can be presented to the expert on the computer screen magnified and enhanced for convenient analysis. The image features measurement and pattern recognition components of the proposed system can, depending on the situation, support or even replace anthropologist in well-defined and usually mundane tasks such as calculation of the number of ridges between singular points in the chosen dermatoglyphic areas or determination of the dermatoglyphic patterns of prints of fingers or soles. The functioning of the medical decision support system proposed in the paper is governed by an expert system build on the basis of knowledge described in scientific literature and applied to knowledge-intensive tasks that require human anthropologist expertise to accomplish. The proposed decision support system relies on computational intelligence components for the determination of dermatoglyphic traits, which are used as values of premises of the rules comprising the expert system. The implementation of the expert system allows carrying out diagnosis on the basis of a set of digital images supporting the human expert in assessment of large amounts of medical data. Therefore the system improves the accessibility and quality of service available to the patients of the local hospital and also available to the other patients by enabling carrying out diagnosis through the web server containing the web page for uploading the necessary data and displaying the results of automatic analysis. The design of the support system in the form of an expert system allows the results of an inference process to be easily understood by humans making the system suitable for tasks of tutoring of medical students. One of the components of the system is designed to possess explanatory capabilities of the diagnosis in order to generate its description using natural language sentences and to explain the reasoning process the expert system is going through to achieve particular conclusions.

3 Architecture of Medical Decision Support System – The Proposed Approach

The architecture of the system has a modular design. The system is comprised of the following components: client application, www service, medical decision support server and database server (Fig. 1). These components work together in a client-server architecture and exchange information using TCP/IP protocol. A user gains access to the data through www application realized in Silverlight technology. This application is part of a three-tiered client-server type architecture. An end-user connects, using a web browser, with the www service located on the database server. WWW service acts as an intermediate layer between the database and the client application. Its main task is exchange of information between users (clients), the support decision system and the database management system. To the tasks of the www service belongs appropriate structuring of the data transferred to users and to decision support system. End application allows visualization of the data specified by the user. Form built into the client application responsible for displaying selected collections of images according to the user specifications allows viewing of examples of collections stored in the database belonging to the specific genetic disorders. It also allows viewing of examples of particular rare patterns on chosen dermatoglyphic areas. WWW service with consent of the person uploading data through the client application can keep them in order to create publicly accessible database containing sets of patterns typical for the particular chromosomal defects. These data could serve as a highly useful educational material for medical students. So far there exists no such digital database with dermatoglyphic images connected through the www service to the Internet and freely available for research and teaching purposes.

4 Design of the Medical Decision Support System

A medical diagnosis is generated by the medical decision support system for the detection of Down's syndrome. This system is comprised of three image pattern recognition modules and an expert system module [3]. The expert system module comprising of a set of rules is designed on the basis of dermatoglyphic nomogram [2]. Dermatoglyphic nomogram is a diagnostic index for the diagnosis of Down's syndrome. It has been developed utilizing discriminant function analysis to take the interrelationships of pattern areas into account. The dermatoglyphic nomogram first used calculation of the log-odds to weight the patterns on a scale from those characteristic of Down's syndrome to those more typical of controls (negative weight). Stepwise discriminant function analysis of 32 pattern areas was then used on the log-odds weights and resulted in a function using only four variables from which an index score and the graphical Dermatogram were derived. These variables correspond to the following four dermatoglyphic traits: pattern types of the right and left index fingers, pattern type of the hallucal area of the right sole and the ATD angle of the right hand. In the Dermatogram

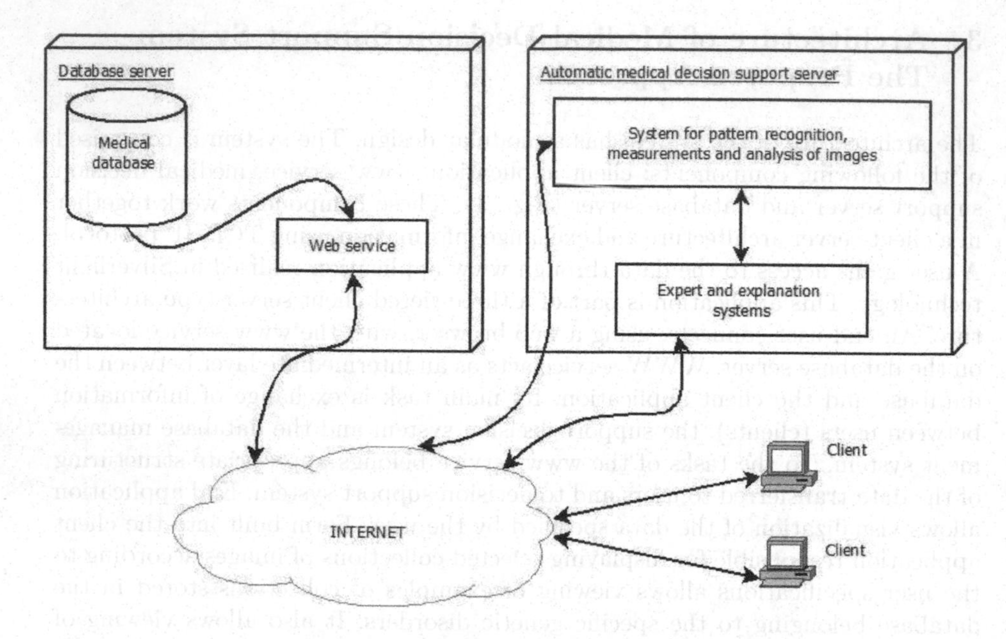

Fig. 1. Scheme of the system architecture

overlap boundaries are used to divide patients known to have Down's syndrome into three classes: correctly diagnosed, incorrectly diagnosed and those in the overlap area, which cannot be classified with certainty as belonging to either Down's syndrome or healthy infants groups.

Each rule in the set corresponds to the particular combination of the four dermatoglyphic pattern types. The set consists of 125 rules. Results of the pattern recognitions are passed as premises to the rules of the expert system. Conclusions returned by the system are translated to the natural language sentences describing particular test case in the form of an automatic medical diagnosis.

Pattern recognition and image measurement modules outputs yield values for the premises of the expert system rules. On the basis of the values passed from the pattern recognition modules the expert system calculates the diagnostic index score for the patient, which determines the probability of genetic disorder incidence. Three of the values required for the reasoning process are estimated by two pattern recognition modules, which have been implemented for carrying out classification tasks. Each pattern type in the classification problem has a corresponding unique value of weight assigned in the dermatogram. The first classification module supplies two of these values by determining the types of patterns present on the index fingers of left and right hand. The fingerprint patterns are classified according to the Henry scheme, which divides the set of fingerprint patterns into the following classes: left loop, right loop, whorl, arch and tented arch (Fig. 2).

The third value supplied to the dermatogram is determined by classification of the patterns of the hallucal area of the right sole. In this classification problem

Table 1. A partial set of expert system rules for the example combination of dermatoglyphic patterns

Combination	Right Index Finger	Left Index Finger	Right Hallucal Area	Right ATD Angle	Diagnostic Line Index
1	A	RL	LDL	⟨15;61⟩	Normal
	A	RL	LDL	⟨61;112⟩	NN
	A	RL	LDL	⟨112;120⟩	Down
2	A	RL	Other	⟨15;71⟩	Normal
	A	RL	Other	⟨71;120⟩	NN
	A	RL	Other	-	Down
3	A	RL	SDL	⟨15;41⟩	Normal
	A	RL	SDL	⟨41;92⟩	NN
	A	RL	SDL	⟨92;120⟩	Down
4	A	RL	TbA	-	Normal
	A	RL	TbA	⟨15;65⟩	NN
	A	RL	TbA	⟨65;120⟩	Down
5	A	RL	W or FL	⟨15;68⟩	Normal
	A	RL	W or FL	⟨68;118⟩	NN
	A	RL	W or FL	⟨118;120⟩	Down

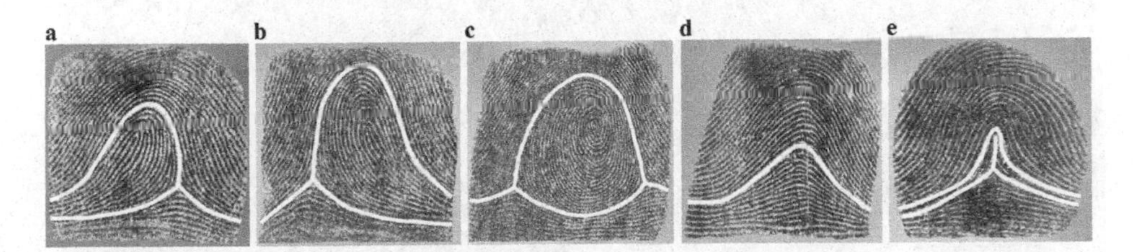

Fig. 2. Classification scheme of fingerprints: (a) left loop; (b) right loop; (c) whorl; (d) plain arch; (e) tented arch. (The white lines traced on the prints are the type lines, which define the unique skeletons of the patterns.)

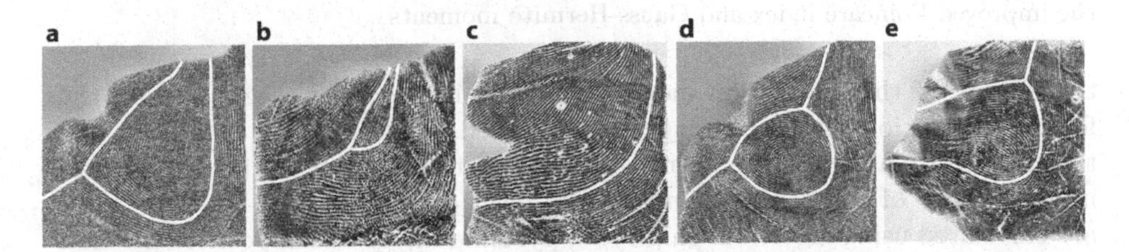

Fig. 3. Classification scheme of the hallucal area of the sole prints: (a) large distal loop, (b) small distal loop, (c) tibial arch, (d) whorl, (e) tibial loop

we take into account the following classes: large distal loop, small distal loop, tibial arch, whorl and tibial loop (Fig. 3). The raw images acquired from the offline impressions are preprocessed in order to remove the spurious background noise and enhance the structure of the dermatoglyphic pattern, which improves

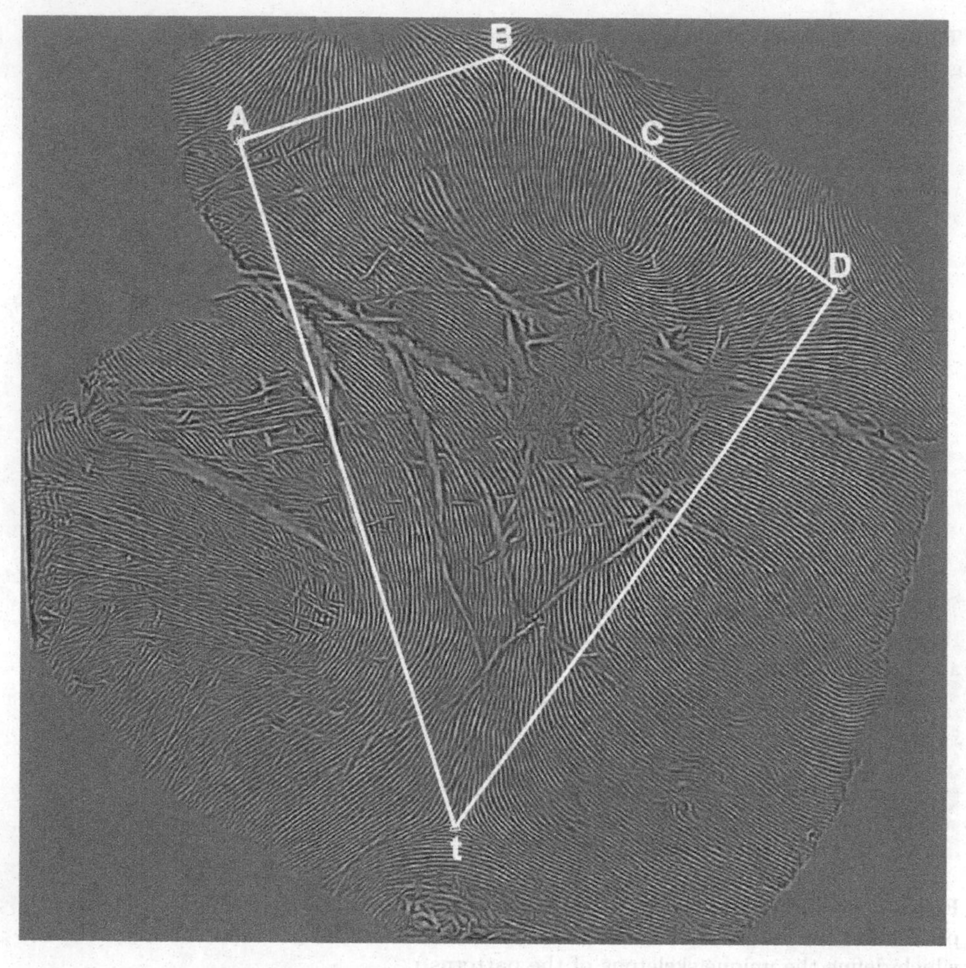

Fig. 4. Singular points of the palm print located with a two stage algorithm that uses the improved Poincare index and Gauss-Hermite moments

the quality of the orientation model generated in the later stages. Before the feature extraction phase an image is segmented, its contrast is enhanced and then the entire area of the pattern is submitted to the filtration procedure taking advantage of STFT algorithm. On the basis of an enhanced image features are calculated using algorithm employing pyramid image decomposition method and PCA for the estimation of local ridge orientations. Local ridge orientations represent the structures of patterns in the process of dermatoglyphic images classification. Maps of ridge orientations comprise the vectors classified by ensembles of SVM algorithms. Cross validation and grid search are used to select optimal values of parameters for the training of classifiers. Dermatoglyphic data sets have been tested on ensembles trained with RBF and triangular kernel types achieving accuracy ratio of 90% in the testing phase for both the problems of fingerprint and hallucal area of the sole patterns classification.

The ATD angle of the right palm print is constructed by drawing the lines from the digital triradius 'a' to the axial triradius 't' and from this to the digital triradius 'd' (Fig. 4). The width of the ATD angle is determined by the position of axial triradius along the distal direction of the palm. In comparison with healthy children the axial triradius of children with Down's syndrome is placed higher and the resulting ATD angle is in average twice as wide. Reliable determination of the ATD angle relies on detection of true triradii of the palm print. Therefore the procedure for the determination of the triradii locations relies on two separate criteria derived from different features of the image. The first of these criteria is a value of the improved formula of the Poincare index calculated on the basis of the map of local ridge orientations, which in turn are determined by the algorithm using pyramid image decomposition and PCA. The second criterion is a local value of a coherence map calculated from the texture of the analyzed image. The coherence map is created by convolution of two dimensional Gauss - Hermite moments with the local block of the image. The convolution operation is performed for each pixel in the image belonging to the area containing the structure of the pattern. After calculating the values of both of these criteria, it is checked if they are larger than the assumed thresholds that determine the characteristic point occurrence in the analyzed area of the pattern. Only if the values of both of these criteria are sufficiently large for the same point of the pattern area its location is marked as a true triradius.

5 Database Architecture

Medical database used in the project was developed using the PostgreSQL database system. The database contains data on patients with genetic disorders. The data describing a specific case consists of a collection of images containing dermatoglyphs of fingerprints, hallucal area of the sole prints, palm prints and descriptions of diagnoses made by anthropologists and descriptions of the diagnoses in natural language generated by the explanation module on the basis of results passed by the automatic system for diagnosis support. Images of dermatoglyphic areas are equipped with metadata collected in both automatic and manual manner. The data collected in a manual manner is entered by anthropologists in the form of natural language sentences. The data concern dermatoglyphic features such as: pattern type, number and width of lines, presence, direction and type of furrows, occurrence of abnormalities in the local structure of lines, and also values of dermatoglyphic indices determined on the basis of these observations. The description also contains information about the anatomical locations of the dermatoglyphic areas on the appropriate sides of the body. Data entered in a manual manner are stored in DIAGNOSIS table in the column named DESCRIPTION_DIAGNOSIS_DOCTOR (Fig. 5). Data collected in an automatic manner are generated by the diagnosis support system responsible for analysis of particular images of patients. The system provides detailed information about the types of patterns of the analyzed dermatoglyphic areas and the results of measurements of the angles determined by finding the locations of singular points of the palm prints. On the basis of this knowledge the system infers about the probabilities of genetic defects in the infant. Data supplied by the diagnosis support system are stored in the IMAGE_DATA table in the parametric form. Using this data explanation module generates final descriptive diagnostic formula in the form of natural language text. This formula contains detailed description of features and patterns of the particular dermatoglyphic areas and the text of diagnosis generated on the basis of this knowledge. Database in its current form is comprised of the following entities: PATIENT, DOCTOR, IMAGE_DATA, DIAGNOSIS, GENETIC_DISORDER, DERMATOGLYPHIC_AREA, BODY_PART, SIDE. PATIENT table contains patient's personal data such as: name, gender, date of birth and identifier of genetic disorder found in the patient (in particular case the identifier value defines a healthy person belonging to the so-called control group). Table DOCTOR contains information identifying a specialist using the system. Table IMAGE_DATA contains data of dermatoglyphic impressions in the form of images. Three versions of images are stored in this table: raw version, version described manually by anthropologist in the graphical form and the version processed by image enhancement algorithms.

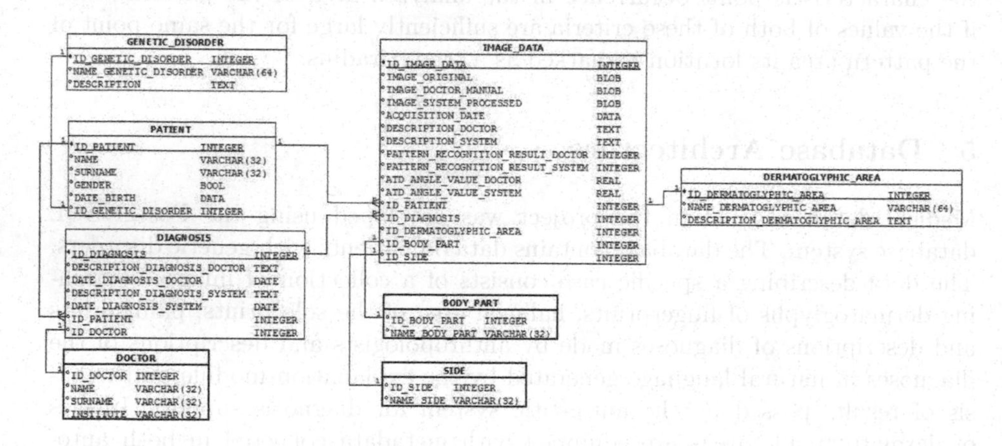

Fig. 5. Database scheme in the form of entity diagram

Additionally in this table metadata describing characteristics of a particular image are included, i.e. date of image acquisition, text description of an image made by an anthropologist (DESCRIPTION_DOCTOR), description of an image generated by explanation module, result of pattern recognition by anthropologist, result of pattern recognition generated by one of the classification modules of the decision support system, value of ATD angle determined manually by anthropologist (ATD_ANGLE_VALUE_DOCTOR) and determined automatically by the system (ATD_ANGLE_VALUE_SYSTEM). The table contains foreign keys, which determine: the patient for whom the image was taken (ID_PATIENT), identifier of diagnosis, which was carried out on the basis of collection, which the particular image is part of (ID_DIAGNOSIS), identifier of an image of a dermatoglyphic impression (ID_DERMATOGLYPHIC_AREA) and identifiers of the anatomical part and side of body from, which the image originates (ID_BODY_PART), (ID_SIDE). The patient can have multiple diagnoses made at different times by different specialists. Table DIAGNOSIS contains a single diagnosis of the patient in two forms: manual (DESCRIPTION_DIAGNOSIS_DOCTOR) and automatic (DESCRIPTION_DIAGNOSIS_SYSTEM), both stored as a text comprising of natural language sentences. Dates of diagnoses conducted in manual (DATE_DIAGNOSIS_DOCTOR) and automatic (DATE_DIAGNOSIS_SYSTEM) manner are also stored in the system. Table is also equipped with a foreign key identifying the patient, whom the diagnosis concerns (ID_PATIENT), and foreign key identifying a specialist conducting the diagnosis and entering its description to the database (ID_DOCTOR). The database is equipped with tables storing metadata. Table DERMATOGLYPHIC_AREA stores a list of dermatoglyphic areas along with descriptions of their names (NAME_DERMATOGLYPHIC_AREA) and topographical descriptions of particular areas (DESC_DERMATOGLYPHIC_AREA).

Table BODY_PART determines an anatomical belonging of relevant dermatoglyphic area - name of a body part (NAME_BODY_PART). Table SIDE determines anatomical side of patient's body. The granularity of the structure of meta-tables in the database was designed to increase flexibility of the database search process. It will help the specialist viewing data to retrieve collections of images, which meet specific criteria, i.e. images containing particular patters, images from particular anatomical parts and sides of body etc.

6 Results

The research conducted so far focused on two main areas. The first area of research concerned the design scheme of the algorithm for classification of patterns of fingerprints and the hallucal area of the soles patterns. Another task carried out in this research area was related to the design scheme of the algorithm for calculation of the ATD angle of the palm print. In the aforementioned research area implementation and testing of a number of image processing, feature extraction and computational intelligence algorithms for the classification of dermatoglyphic patterns was carried out using Matlab and C programming languages. The classifiers implemented in our system allow recognition of complex dermatoglyphic patterns with high 90% ratio accuracy for both classification problems [3] [4]. For the issue of the ATD angle calculation an algorithm for detection of characteristic points has been proposed based on two independent criteria in order to reliably determine the locations of only true triradii of the palm print. In the first area a research was also conducted on the construction of an expert system for diagnosis of Down syndrome in newborns. The expert system module has been implemented in Prolog language and functions as a higher level module in a hierarchical architecture of the decision support system. This design pattern concerning the role of the expert system module makes the system scalable to incorporate in the future new more complex decision modules working in parallel with the existing expert system module and allows easy interchanging of data between already implemented low level modules and these new decision modules.

7 Summary

In the paper principles of operation and architecture of an intelligent knowledge-based system for the recognition of dermatoglyphic features used in detection of Down's syndrome in infants were presented. Diagnosing a particular case requires processing a set of images, extraction of their features using computational intelligence or image features measurement techniques. Each of feature measurement or recognition tasks is realized by a separate module. Results of measurements and recognitions are passed to the hierarchically superior module carrying out, on the basis of those features' values, a diagnosis about the probability of genetic disorder incidence in infant. The high level module is designed in the form of an expert system based on the knowledge resulting from the statistical analysis of sets of dermatoglyphs of healthy children belonging to the control group and sets of dermatoglyphs of children with Down's syndrome. Modular design of the system allows easy expansion and in the future enables communication of already implemented modules with other more complex versions of the decision module taking advantage of greater number of possible dermatoglyphic features to generate more detailed diagnoses. All components of the system are wrapped in standard Web service interfaces and can be accessed by the users through the internet browser webpage. The future work includes implementation of new feature recognition modules and a more complex diagnostic module.

References

1. Cummins, H., Midlo, C.: Fingerprints, palms and soles - Introduction to Dermatoglyphics. Dover Publications Inc., New York (1961)
2. Reed, T.E., Borgaonkar, D.S., Conneally, P.M., Yu, P., Nance, W.E., Christian, J.C.: Dermatoglyphic nomogram for the diagnosis of Down's syndrome. The Journal of Pediatrics 77(6), 1024–1032 (1970)
3. Wojtowicz, H., Wajs, W.: Intelligent Information System for Interpretation of Dermatoglyphic Patterns of Down's Syndrome in Infants. In: Pan, J.-S., Chen, S.-M., Nguyen, N.T. (eds.) ACIIDS 2012, Part II. LNCS, vol. 7197, pp. 284–293. Springer, Heidelberg (2012)
4. Wojtowicz, H., Wajs, W.: Classification of Plantar Dermatoglyphic Patterns for the Diagnosis of Down's Syndrome. In: Selamat, A., Nguyen, N.T., Haron, H. (eds.) ACIIDS 2013, Part II. LNCS, vol. 7803, pp. 295–304. Springer, Heidelberg (2013)

Exact Discovery of Length-Range Motifs*

Yasser Mohammad[1] and Toyoaki Nishida[2]

[1] Assiut University, Egypt and Kyoto University, Japan
yasserm@aun.edu.eg, yasser@ii.ist.i.kyoto-u.ac.jp
[2] Kyoto University, Japan
nishida@i.kyoto-u.ac.jp

Abstract. Motif discovery is the problem of finding unknown patterns that appear frequently in real valued timeseries. Several approaches have been proposed to solve this problem with no a-priori knowledge of the timeseries or motif characteristics. MK algorithm is the de facto standard exact motif discovery algorithm but it can discover a single motif of a known length. In this paper, we argue that it is not trivial to extend this algorithm to handle multiple motifs of variable lengths when constraints of maximum overlap are to be satisfied which is the case in many real world applications. The paper proposes an extension of the MK algorithm called MK++ to handle these conditions. We compare this extensions with a recently proposed approximate solution and show that it is not only guaranteed to find the exact top pair-motifs but that it is also faster. The proposed algorithm is then applied to several real world time series.

1 Introduction

Discovering recurrent unknown patterns in time series data is known in data mining literature as motif discovery problem. Since the definition of the problem in [6], motif discovery became an active area of research in Data Mining. A time series motif is a pattern that consists of two or more similar subsequences based on some distance threshold. Several algorithms have been proposed for solving the motif discovery problem [9], [7], [6], [3].

These algorithms can be divided into two broad categories. The first category includes algorithms that discretize the input stream then apply a discrete version of motif discovery to the discretized data and finally localize the recovered motifs in the original time series. Many algorithms in this category are based on the PROJECTIONS algorithm that reduces the computational power required to calculate distances between candidate subsequences [1]. Members of this category include the algorithms proposed in [3] and [7]. The second category discovers the motifs in the real valued time-series directly. Example algorithms in this category can be found in [9] and [2]. For example, Minnen et al. [7] used a discretizing motif discovery (MD) algorithm based on PROJECTIONS for basic action discovery of exercises in activity logs and reported an accuracy of 78.2% with a precision

* The work reported in this paper is partially supported by JSPS Post-Doc Fellowship and Grant-In-Aid Programs of the Japanese Society for Promotion of Science.

N.T. Nguyen et al. (Eds.): ACIIDS 2014, Part II, LNAI 8398, pp. 23–32, 2014.

of 65% and recall of around 92%. The dataset used contained six exercises with roughly 144 examples of each exercises in a total of 20711 frames [7].

The second category tries to discover motifs in the original continuous timeseries directly. Example algorithms in this category can be found in [2] and [9]. For example, Mohammad and Nishida [8] proposed an algorithm called MC-Full based on comparison of short subsequences of candidates sampled from the distribution defined by a change point discovery algorithm. Most of these algorithms are *approximate* in the sense that discovered motif occurrences may not have the shortest possible distance of all subsequences of the same length in the timeseries.

Mueen et al. [12] proposed the MK algorithm for solving the *exact* motif discovery problem. This algorithms can efficiently find the pair-motif with smallest Euclidean distance (or any other metric) and can be extended to find top k motifs or motifs within a predefined distance *range*. Mohammad and Nishida [10] proposed an extension of the MK algorithm (called MK+) to discover top k pair-motifs efficiently. The proposed algorithm in this paper, builds upon the MK+ algorithm and extends it further to discover top k pair-motifs at different time lengths.

The main contribution of the paper is a proof that mean-normalized Euclidean distance satisfies a simple inequality for different motif lengths which allows us to use a simple extension of MK+ to achieve higher speeds without sacrificing the exactness of discovered motifs.

2 Problem Statement

A time series $x(t)$ is an ordered set of T real values. A subsequence $x_{i,j} = [x(i) : x(j)]$ is a continguous part of a time series x. In most cases, the distance between overlapping subsequences is considered to be infinitely high to avoid assuming that two sequences are matching just because they are shifted versions of each other (these are called trivial motifs [5]). There are many definitions in literature for motifs [13], [3] that are not always compatible. In this paper we utilize the following definitions:

Definition 1. *Motif:* Given a timeseries x of length T, a motif length L, a range R, and a distance function $D(.,.)$; a motif is a set M of n subsequences ($\{m_1, m_2, ..., m_n\}$) of length L where $D(m_i, m_{j_k}) < R$, for any pairs $m_i, m_k \in M$. Each $m_i \in M$ is called an *occurrence.*

Definition 2. *Pair-Motif:* A pair-motif is a motif with exactly 2 occurrences. We call the distance between these two occurrences, the *motif distance.*

Definition 3. *Exact Motif*: An Exact Motif is a pair-motif with lowest range for which a motif exists. That is a pair of subsequences $x_{i,i+l-1}, x_{j,j+l-1}$ of a time series x that are most similar. More formally, $\forall_{a,b,i,j}$ the pair $\{x_{i,i+l-1}, x_{j,j+l-1}\}$ is the exact motif iff $D(x_{i,i+l-1}, x_{j,j+l-1}) \leq D(x_{a,a+l-1}, x_{b,b+l-1})$, $|i-j| \geq w$ and $|a-b| \geq w$ for $w > 0$. This definition of an exact motif is the same as the one used in [12].

Using these definitions, the problem statement of this paper can be stated as: Given a time series x, minimum and maximum motif lengths (L_{min} and L_{max}), a maximum allowed within-motif overlap (wMO), and a maximum allowed between-motifs overlap (bMO), find the top k pair-motifs with smallest motif distance among all possible pairs of subsequences with the following constraints:

1. The overlap between the two occurrences of any pair-motif is less than or equal to wMO. Formally, $\forall M_k \forall m_1 \in M_k, m_2 \in M_k : overlap(m_1, m_2) \leq wMO$ where M_k is one of the top k motifs at length l, and $overlap(.,.)$ is a function that returns the number of points common to two subsequences divided by their length (l).
2. The overlap between any two pair-motifs is less than or equal to bMO. Formally, $\forall 1 \leq i, j \leq K \forall 1 \leq l, n \leq 2 : min(overlap(m_l^i, m_n^j)) \leq bMO$, where m_y^x is the occurrence number y in motif number x and $overlap(.,.)$ is defined as in the previous point.

This paper will provide an *exact* algorithm for solving this problem in the sense that *there are no possible pair-motifs that can be found in the time series with motif distance lower than the motif distance of the Kth motif at every length*. Given this solution, it is easy to find motifs that satisfy Definition 1 and in the same time find a data-driven meaningful range value by simply combining pair-motifs that share a common (or sufficiently overlapping) occurrences. Our solution is based on the MK algorithm of Mueen et al. in [12] and for this reason we start by introducing this algorithm and its previous extension in the following section.

3 MK and MK+ Algorithms

The MK algorithm finds the top pair-motif in a time series. The main idea behind MK algorithm [12] is to use the triangular inequality to prune large distances without the need for calculating them. For metrics $D(.,.)$ (including the Euclidean distance), the triangular inequality can be stated as:

$$D(A, B) - D(C, B) \leq D(A, C) \tag{1}$$

Assume that we have an upper limit on the distance between the two occurrences of the motif we are after (th) and we have the distance between two subsequences A and C and some reference point B. If subtracting the two distances leads a value greater than th, we know that A and C cannot be the motif we are after without ever calculating their distance. By careful selection of the order of distance calculations, MK algorithm can prune away most of the distance calculations required by a brute-force quadratic motif discovery algorithm. The availability of the upper limit on motif distance (th), is also used to stop the calculation of any Euclidean distance once it exceeds this limit. Combining these two factors, 60 folds speedup was reported in [12] compared with the brute-force approach.

The inputs to the algorithm are the time series x, its total length T, motif length L, and the number of reference points N_r.

The algorithm starts by selecting a random set of N_r reference points. The algorithm works in two phases:

The first phase (called hereafter referencing phase) is used to calculate both the upper limit on best motif distance and a lower limit on distances of all possible pairs. During this phase, distances between the subsequences of length L starting at the N_r reference points and all other $T - L + 1$ points in the time series are calculated resulting in a distance matrix of dimensions $N_r \times (T-L+1)$. The smallest distance encountered (D_{best}) and the corresponding subsequence locations are updated at every distance calculation.

The final phase of the algorithm (called scanning step hereafter) scans all pairs of subsequences in the order calculated in the referencing phase to ensure pruning most of the calculations. The scan progressed by comparing sequences that are k steps from each other in this ordered list and use the triangular inequality to calculate distances only if needed updating D_{best}. The value of k is increased from 1 to $T - L + 1$. Once a complete pass over the list is done with no update to d_{best}, it is safe to ignore all remaining pairs of subsequences and announce the pair corresponding to D_{best} to be the *exact* motif.

A naive way to extend the MK algorithm to discover top k pair-motifs rather than only the first is to apply the algorithm K times making sure in each step (i) to ignore the $(i-1)$ pair-motifs discovered in the previous calls. To resist trivial motif matches, a maximum allowable overlap between any two motifs can be provided and all candidates overlapping with an already discovered motif with more than this allowed maximum are also ignored. A major problem with this approach is that even though we require $K * (l_{max} - l_{min} + 1)$ pair-motifs, the number of times MK must be ran can be much higher because it is likely that the motif discovered in run $i + 1$ will be a trivial match of the motif discovered in run i. This is more likely to happen the longer and smoother motifs are. To extend the algorithm to discover motifs of multiple lengths, we simply apply the algorithm for every length sequentially. This algorithm will be called *NaiveMK* in the rest of this paper.

3.1 MK+

Mohammad and Nishida proposed an extension to the MK algorithm for discovering top k pair-motifs of the same length efficiently (MK+) [10] that keeps an ordered list of K candidate motifs (L_{bests}) and their corresponding motif distances (D_{bests}) rather than a single candidate motif during both the referencing and scanning phases. This allows MK+ to discover the top k motifs exactly in a single run. This algorithm was reported to provide an order of magnitude improvement in speed compared to rerunning MK [10].

MK+ though works with a single length. A possible extension of this algorithm to multiple lengths is by simply re-running it for each length. This is called *NMK+* in the rest of this paper. As will be shown in section 5, this approach is suboptimal in terms of speed and a faster approach is proposed in this paper.

4 Proposed Extension

The main idea behind the proposed extension is to use the fact that distance calculations at one length provide an upper bound on possible distances of lower lengths and a lower limit on distances of longer lengths. This fact can be utilized (in a similar way to the triangular inequality in the original MK algorithm) to speed up the calculations by removing the need to compare pairs that are *obviously* not candidates for being pair-motifs according to these bounds.

For this approach to work the distance function used to compare subsequences must be a metric (as is the case with the original MK algorithm). We also need to be sure that for any positive number s the following property is true:

$$D_l(i,j)D_{l+1}(i,j)\ \forall l > 0 \wedge 1 \leqslant i; j \leqslant T - l + 1 \tag{2}$$

where $x_{a:b}$ is the subsequence of the time series x starting at index a and ending at index b and all indices do not exceed the length of the time series and $D_l(i,j) = D\left(x_{i:i+l-1}, x_{j:j+l-1}\right)$. If the condition in Eq. 2 is not satisfied then distances at different lengths cannot be used to infer any bounds about distances of either larger or lower lengths and the best that could be done is *NMK+*. This puts a limitation on the types of distance functions that can be used. For example, Mueen et al. normalized all subsequences by subtracting the mean and dividing by the standard deviation prior to distance calculation [12]. This cannot be done in our case because this distance function does not respect the condition in Eq. 2. For this reason we normalize all subsequences only by subtracting the mean.

The effect of division by the standard deviation before distance calculation is to reduce the effect of variability in scale. In some applications; this is not needed or even desirable. For example, one of our primary goal domains is discovery of recurring motions in human motion and gesture datasets. A small change in scale between two occurrences of the same action can be handled with the Euclidean distance while a large scale difference corresponds in many cases to a quantitative difference in the action.

If subsequences are not normalized and the distance measure used is the Euclidean is used for $D(.,.)$, it is trivial to show that equation 2 holds. Subtracting the mean of the whole timeseries does not change this result. It is not immediately obvious that normalizing each subsequences by subtracting its own mean will result in a distance function that respects Equation 2.

Theorem 1. *Given that the distance function $D(.,.)$ in Definition 1 is defined for subsequences of length l as:*

$$D_l(i,j) = \sum_{k=0}^{l-1}\left[\left(x_{i+k} - \mu_i^l\right) - \left(x_{j+k} - \mu_j^l\right)\right]^2 \tag{3}$$

and $\mu_i^n = \frac{1}{n}\sum_{k=0}^{n-1} x_{i+k}$*; then*

$$D_l(i,j)D_{l+1}(i,j)\ \forall l > 0 \wedge 1 \leqslant i; j \leqslant T - l + 1$$

A sketch of the proof for Theorem 1 is given below: Let $\Delta_k \equiv \Delta_{j+l}^{i+k} \equiv x_{i+k} - x_{j+k}$ and $\mu_{ij}^l \equiv \mu_j^l - \mu_j^l$, then the definition of $D_l(i,j)$ can be written as:

$$D_l(i,j) = \sum_{k=0}^{l-1} \left(\Delta_k - \mu_{ij}^l\right)^2 = \sum_{k=0}^{l-1} (\Delta_k)^2 - 2\sum_{k=0}^{l-1} \left(\Delta_k \mu_{ij}^l\right) + \sum_{k=0}^{l-1} \left(\mu_{ij}^l\right)^2$$

$$D_l(i,j) = -l\left(\mu_{ij}^l\right)^2 + \sum_{k=0}^{l-1} (\Delta_k)^2$$

We used the definition of μ_{ij}^l in arriving at the last result. Using the same steps with $D_{l+1}(i,j)$, subtracting and with few manipulations we arrive at:

$$D_{l+1}(i,j) - D_l(i,j) = \Delta_l^2 + l\left(\mu_{ij}^l\right)^2 - (l+1)\left(\mu_{ij}^{l+1}\right)^2 \quad (4)$$

Using the definition of μ_{ij}^l, it is straight forward to show that:

$$(l+1)\,\mu_{ij}^{l+1} = l\mu_{ij}^l + \Delta_l \quad (5)$$

Substituting in Equation 4 and rearranging terms we get:

$$D_{l+1}(i,j) - D_l(i,j) = \Delta_l^2 + \left(\mu_{ij}^l\right)^2 - 2\mu_{ij}^l \Delta_l = \left(\Delta_l - \mu_{ij}^l\right)^2 \geqslant 0$$

This proves Theorem 1.

4.1 MK++

The following listing gives an overview of the proposed algorithm:

```
  Function MK++(x,{L},K,R,wMO,bMO)
1.Find motifs at shortest length
L_bests ← ∅, D_bests ← ∅
For(r ∈ Rand(R) ⋀ i = 1 : |x|)
      IF(|i − r| ≥ wMo × min(L))
                 UpdateBests (min(L), K, bMO, r, i)
      EndIf
EndFor
Zr(i) ← ordering of reference points by their distance variance
ref ← reference point with max. variance
Z(i) ← ordering of subsequences by distance to ref
Dist ← {D (ref, Z(i))}
ScanningPhase()
2.Find motifs at higher lengths
For(l ∈ {L})
       Update D_bests then re-sort
       IF(R)
                recalculate ref
```

```
        EndIf
        ScanningPhase()
EndFor
```

The MK++ algorithm starts by detecting 2-motifs at the shortest length (L_{min}) and progressively finds 2-motifs at higher lengths. The algorithm keeps three lists: D_{bests} representing a sorted list of K best distances encountered so far and L_{bests} representing the 2-occurrence motif corresponding to each member of D_{bests}, and μ_{bests} keeping track of the means of the subsequences in L_{bests}. The *best-so-far* variable of MK is always assigned to the maximum value in D_{bests}. During the referencing phase, the distance between the current reference subsequences and all other subsequences of length L_{min} that do not overlap it with more than $wMO \times L_{min}$ points are calculated. For each of these distances (d) we apply the following rules in order:

Rule1. If the new pair is overlapping the corresponding $L_{bests}(i)$ pair with more than $wMO \times L$ points, then this i is the index in D_{bests} to be considered

Rule2. If *Rule1* applies and $D < D_{bests}(i)$, then replace $L_{bests}(i)$ with P.

Rule3. If *Rule1* does not apply but $D < D_{bests}(i)$, then we search L_{bests} for all pairs $L_{bests}(i)$ for which *Rule1* applies and remove them from the list. After that the new pair P is inserted in the current location of L_{bests} and D in the corresponding location of D_{bests}

It is trivial to show that following these rules D_{bests} will always be a sorted list of *best-so-far* distances encountered that satisfy both between and within motif overlap constraints (wMO and bMO). The details of this calculation are given in the program listing below:

```
  Function updateBests(l,K,bMO,r,i)
done ← false, d ← D(x_{r:r+l-1}, x_{i:i+l-1})
IF (d > max(D_bests) ∧ |D_bests| < K)
        L_bests ← L_bests ∪ ⟨r,i⟩, D_bests ← D_bests ∪ d, done ← true
EndIf
For(j = 1 : K ∧ ¬done)
        ζ ← OverlapBoth(L_bests[j], ⟨r,i⟩)
        IF(ζ > bMO)
                IF(d < max(D_bests))
                        L_bests[j] ← ⟨r,i⟩, D_bests[j] ← d, done ← true
                EndIf
        Else
                IF(d < max(D_bests))
                        o ← w : w ∈ L_bests∧, OverlapBoth(w, ⟨r,i⟩) > bMO
                        IF(o = ∅) o ← K EndIf
                        remove o from L_bests, D_bests, L_bests ← L_bests∪⟨r,i⟩
                        D_bests ← D_bests ∪ d
                EndIf
        EndIf
EndFor
```

These steps are similar to MK+ [10] except that we carry along the means of subsequences in the list μ_{bests}. Once the top K 2-motifs of the first length are found, the algorithm progressively calculates the top K 2-motifs of longer lengths. Since our distance function defined in Equation 3 satisfies the condition of Equation 2 as Theorem 1 proves, we know that the distances in D_{best} from the previous length provide a lower bound on the distances at this length. This allows us to directly update the D_{best} array by appending the new distances after using Equation 5 to calculate the new mean and distance.

Updating the distances in D_{bests} may take them out of order. We sort L_{bests} (with corresponding sorting of D_{bests} and μ_{bests}) in ascending order and update the *best-so-far* accordingly. This gives us a –usually– tight upper bound on the possible distances for the top K 2-motifs at the current length. This is specially true if the increment in motif length is small and the timeseries is smooth. This is the source of the speedup achieved by MK++ over NMK+ which have to find these lists from scratch at every length.

The referencing step can either be recalculated for the new length or the older reference values can be used as lower bounds (again because of Theorem 1). Nevertheless, recalculating the reference distances can speedup the scanning phase because it provides a tighter bound. Because this depends on the data, we keep the choice for the algorithm user by providing a parameter R that controls whether this recalculation is carried out. It is important to notice that either choice will not affect the accuracy of the final results and in our experiments it had a negligible effect on the speed as well. The scanning step is the same as in MK and MK+ and for lack os space will not elaborated upon further. Please refer to [12] for its details.

5 Evaluation

The first evaluation considered comparing MK++ with NMK+. A set of 40 time series of lengths between 1000 and 10000 points each where generated using a random walk. This kind of random timeseries is the most challenging input for any *exact* motif discovery algorithm because distances between subsequences are similar which reduces the advantage of using lower and upper bounds. Both MK++ and NMK+ where used to find 10 pair-motifs of each length in the range 10 to 100. The average execution time for NMK+ was 3.8ms/point while the average execution of MK++ was 0.77ms/point achieving a speedup of more than 500%. Both algorithms returned the same motifs at all lengths. For completion we also applied NaiveMK to the same dataset which required an average execution time of 73.5ms/point (more than 60 times the proposed MK++ algorithm). Again both algorithms returned exactly the same motifs at all lengths.

5.1 Real World Data

To test MK++ on a real world dataset we used CMU Motion Capture Dataset available online from [4]. We used the time series corresponding to basketball

Fig. 1. Forward dribbling motion: discovered by the proposed algorithm

category. The occurrences of each recurring motion pattern in the time series of the fifteen available in this collection (21 different motions in total) were marked by hand from the videos using ANVIL [14] to get ground-truth motif locations.

In this paper, we report the results using AMC files only. The total number of frames in the fifteen time series was 8155. There were seven different motion patterns in the time series with an average of 8 occurrences each after removing motions that appeared only once or twice. Fig. 1 shows an example pattern.

The data from all sessions were concatenated to form a single timeseries and random data was added to it to make its length 10000 points. Before applying motif discovery algorithms, we reduced the dimensionality of the aggregated time series using Principal Component Analysis (PCA).

We applied MK++, shift-density based motif discovery (sdCMD) [15], two stem extension algorithms (GSteXS, GSteXB) [11], MCFull [9] and the recently proposed MOEN algorithm [13] with a motif length between 50 and 250 to the concatenated timeseries and calculated the fractions of discovered motifs that cover ground-truth patterns completely, and the fractions that cover partial patterns or multiple patterns. We also calculated the fraction of true patterns that were discovered by every algorithm (covered-fraction) and the extra parts appended to it (extra-motif). This data is summarized in Table 1.

Table 1. Evaluation of MD for CMU MoCap Dataset with the best and worst performance highlighted (worst in italics and best in bold font)

Algorithm	Correct	Partial	Multiple	Covered-fraction	Extra-motif	Accuracy
Proposed	0.3169	0.3761	**0.1089**	0.2543	3.7496	**0.6931**
sdCMD [15]	0.1594	*0.4783*	0.2899	0.1847	0.5264	0.6377
GSteXS [11]	0.1111	0.1296	0.6852	0.9796	8.8143	0.2407
GSteXB [11]	0.0323	0.2742	0.6129	0.8562	2.5713	0.3065
MCFull [9]	*0*	**0**	*1.0000*	**1.0000**	*115.8188*	*0.0*
MOEN [13]	**0.3333**	**0**	0.3333	*0.1299*	0.2001	0.3333

As Table 1 shows, the proposed algorithm provides a good balance between specificity (extra-motif=3.74) and sensitivity (covered-fraction=0.25) and provides the second best correct discovery rate (0.3169 compared with 0.3333 for MOEN) with best boundary separation between discovered motifs (multiple=0.1089). To get a sense of the accuracy of each algorithm we defined the total discovery rate as the summation of correct discovery rate and partial discoveries as it is always easy to use a motif-extension algorithm to extend partially discovered motifs. The proposed algorithm provides the highest total discovery rate of 69.31% compared with 63.77% for sdCMD, 33.33% for MOEN, 30.65%

for GSteXB, 24.07% for GSteXS and 0% for MCFull. Even though this result is based on a limited dataset, it provides a proof-of-applicability of the proposed algorithm for solving real world motif discovery problems. In the future, more extensive analysis with larger datasets will be conducted.

6 Conclusions

This paper presents an extension of the MK exact motif discovery algorithm called MK++ that can discover multiple motifs at multiple motif lengths simultaneously achieving a speed up of over 500% compared with repeated applications of the MK+ algorithm. The paper provides a proof that MK++ is still an exact algorithm for the mean-shifted Euclidean distance function. In the future, the proposed approach will be extended to zscore normalization.

References

1. Buhler, J., Tompa, M.: Finding motifs using random projections. In: 5th Internatinal Conference on Computational Biology, pp. 69–76 (2001)
2. Catalano, J., Armstrong, T., Oates, T.: Discovering patterns in real-valued time series. In: Fürnkranz, J., Scheffer, T., Spiliopoulou, M. (eds.) PKDD 2006. LNCS (LNAI), vol. 4213, pp. 462–469. Springer, Heidelberg (2006)
3. Chiu, B., Keogh, E., Lonardi, S.: Probabilistic discovery of time series motifs. In: ACM SIGKDD KDD, pp. 493–498 (2003)
4. CMU: Cmu motion capture dataset, `http://mocap.cs.cmu.edu`
5. Keogh, E., Lin, J., Fu, A.: Hot sax: efficiently finding the most unusual time series subsequence. In: Fifth IEEE ICDM, pp. 8–17 (2005)
6. Lin, J., Keogh, E., Lonardi, S., Patel, P.: Finding motifs in time series. In: The 2nd Workshop on Temporal Data Mining, at the 8th ACM SIGKDD International, pp. 53–68 (2002)
7. Minnen, D., Starner, T., Essa, I.A., Isbell Jr., C.L.: Improving activity discovery with automatic neighborhood estimation. In: IJCAI, vol. 7, pp. 2814–2819 (2007)
8. Mohammad, Y., Nishida, T.: Learning interaction protocols using augmented baysian networks applied to guided navigation. In: IEEE IROS, pp. 4119–4126 (2010)
9. Mohammad, Y., Nishida, T.: Constrained motif discovery in time series. New Generation Computing 27(4), 319–346 (2009)
10. Mohammad, Y., Nishida, T.: Unsupervised discovery of basic human actions from activity recording datasets. In: IEEE/SICE SII, pp. 402–409 (2012)
11. Mohammad, Y., Ohmoto, Y., Nishida, T.: G-SteX: Greedy stem extension for free-length constrained motif discovery. In: Jiang, H., Ding, W., Ali, M., Wu, X. (eds.) IEA/AIE 2012. LNCS, vol. 7345, pp. 417–426. Springer, Heidelberg (2012)
12. Mueen, A., Keogh, E.J., Zhu, Q., Cash, S., Westover, M.B.: Exact discovery of time series motifs. In: SDM, pp. 473–484 (2009)
13. Mueen, A.: Enumeration of Time Series Motifs of All Lengths. In: IEEE ICDM (2013)
14. Kipp, M.: Anvil – A Generic Annotation Tool for Multimodal Dialogue. In: Eurospeech, pp. 1367–1370 (2001)
15. Mohammad, Y., Nishida, T.: Approximately Recurring Motif Discovery Using Shift Density Estimation. In: Ali, M., Bosse, T., Hindriks, K.V., Hoogendoorn, M., Jonker, C.M., Treur, J. (eds.) IEA/AIE 2013. LNCS, vol. 7906, pp. 141–150. Springer, Heidelberg (2013)

Feature Reduction Using Standard Deviation with Different Subsets Selection in Sentiment Analysis

Alireza Yousefpour, Roliana Ibrahim*, Haza Nuzly Abdull Hamed, and Mohammad Sadegh Hajmohammadi

Faculty of Computing, Universiti Teknologi Malaysia (UTM), 81310 Skudai, Johor, Malaysia
{yalireza,shmohammad}@live.utm.my,
{roliana,haza}@utm.my

Abstract. The genesis of the internet and web has created huge information on the web, including users' digital or textual opinions and reviews. This leads to compiling many features in document-level. Consequently, we will have a high-dimensional feature space. In this paper, we propose an algorithm based on standard deviation method to solve the high-dimensional feature space. The algorithm constructs feature subsets based on dispersion of features. In other words, algorithm selects the features with higher value of standard deviation for construction of the subsets. To do this, the paper presents an experiment of performance estimation on sentiment analysis dataset using ensemble of classifiers when dimensionality reduction is performed on the input space using three different methods. Also different types of base classifiers and classifier combination rules were used.

Keywords: Sentiment analysis, Feature selection, Standard deviation, Classifier ensemble, Feature subsets.

1 Introduction

Sentiment analysis or opinion mining processes opinions, attitudes, sentiments, and emotions of people towards products, movies, entities, events, issues, topics, and their features. Sentiment analysis is basically analyzing the opinions to explore whether they are positive or negative. In fact, opinion mining is drawing out subjective information from a text corpus or reviews, while sentiment analysis is the evaluation of the extracted information. Studies that have been more recently conducted on sentiment analysis introduce different techniques to extract and analyze the individuals' sentiments [1, 4]. One of the key issues is identifying the semantic relationships between subject and sentiment-related expressions, which may be quietly different from each other. Sentiment expressions such as adjectives, adverbs, nouns, sentiment verbs and transfer verbs are used for sentiment analysis. In some papers, natural language processing, POS tagging, model of Markov-Model-Based tagger are used to demystify some multi-meaning expressions such as "like" that can denote a verb, adjective, or

* Corresponding author.

N.T. Nguyen et al. (Eds.): ACIIDS 2014, Part II, LNAI 8398, pp. 33–41, 2014.

preposition, and syntactic parsing is also used to identify relation among expression and subject term [4]. Several studies conducted in the last decade reported that accuracy of sentiment analysis can be improved by selecting the most confident features and their subsets, as well as using appropriate classifiers. In this research, the most confident features are selected through standard deviation method, and feature subsets are generated by heuristic method that uses a higher standard deviation.

2 Background of the Problem

Sentiment classification methods are divided into three methods: supervised, semi-supervised, and unsupervised learning. Pang et *al.* [1] introduced supervised approach and Turney [2] proposed the unsupervised approach. In supervised learning approach, a classifier first is trained by a large number of features that are labeled. Then, this classifier is used to identify and classify unlabeled data into two classes of positive and negative sentiments. Whereas in unsupervised learning, sentiment lexicon is used for identifying and classifying documents into polarity sentiment through calculating the word sentiment orientation using a dictionary or part-of-speech patterns [2, 7, 8, 13]. Though, the semi-supervised learning uses both labeled and unlabeled data for training classifiers instead of using only labeled data; this is because of the lack of sufficient labeling data. Some studies have proposed a graph-based semi-supervised learning algorithm for rating and labeling unlabeled document through calculating similarity rate between two documents [5, 6, 12].

Some researchers used several feature sets to improve the classification accuracy [1, 9, 10, 16,17]. Another work proposed a shallow parsing to select the appropriate feature set [14]. Prabowo and Thelwall [11] proposed a hybrid classification using combination rule-based classifiers, SVM algorithms, and statistic-based classifiers. Xia *et al.* [15] proposed an ensemble framework that included two different feature sets, namely part-of-speech and word-relation-based sets of features and used three algorithms in base-level, i.e., Naive Bayes, maximum entropy, and SVM for classifying each of the feature sets. Su *et al.* [3] introduced a classifier ensemble composed of five popular classifiers: Naive Bayes (NB), Centroid-Based classification (CB), K-nearest neighbor algorithms (KNN), Support Vector Machine (SVM), and Maximum Entropy model (ME) as base-level algorithms, and used stacking generalization in high-level or meta-level.

3 Feature Selection and Reduction

In the feature selection process, features are extracted, which leads to class separability using the univariate approach for rating and multivariate approach for optimizing a criterion function. Whereas in feature extraction process, high-dimensional space is reduced to lower feature space through a linear transformation or nonlinear transformation.

The aim of feature reduction is determining which features in a given set of training feature vectors is most useful for discriminating between the classes. In this paper, a comparative study is conducted on two popular methods: information gain (IG) and chi-square (CHI) in comparison with standard deviation (SD).

The phrase "most useful feature for discriminating" in above sentence has different meanings. In IG, the most useful features are those that have the most degree of difference in their presence and absence in positive and negative samples. It is noticeable that CHI selects the most useful features by a process similar to IG. However, SD chooses the most useful features based on higher standard deviation; this is because a higher standard deviation shows that features are extends out over a large range of values.

3.1 Feature Reduction Using Standard Deviation

Standard deviation is a statistical and probability method that calculates the amount of variation or dispersion of features from average (or mean). In fact, a high standard deviation shows that features are extended out over a large range of values and a low standard deviation indicates that feature points are located very close to the mean. As a result, using features with higher standard deviation provides a more accurate prediction in sentiment classification.

For example, assume that we have two features (x_1, x_2) and two classes (c_1, c_2). In Figure 1a, due to the fact that standard deviation value of feature x_2 is greater than feature x_1, we can predict only based on feature x_2. This is because samples are more depressed when they are corresponding to feature x_2 compared to when they are corresponding to feature x_1. Consequently, we can reduce the number of features from two to one. While in Figure 1b, this feature reduction is impossible because the value of both features is low and close to the mean. Therefore, for prediction, we should consider two features together.

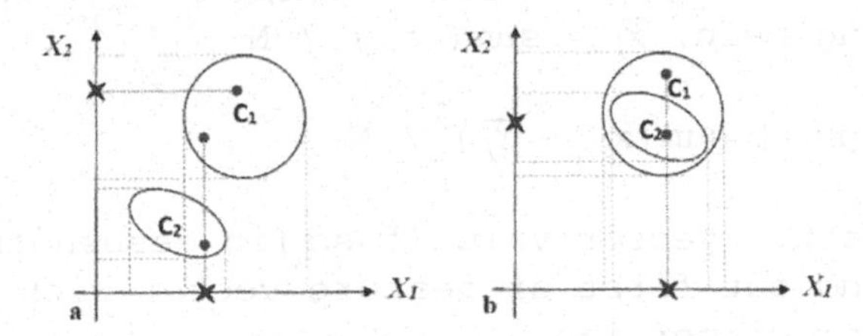

Fig. 1. A sample of standard deviation with two features and two classes

Assume that we have a weighted matrix of document-terms as follows:

Table 1. Weighted term-document matrix (TDM)

	$feature_1$	$feature_2$	$feature_3$	. . .	$feature_M$
Doc_1	x_{11}	x_{12}	x_{13}		x_{1N}
Doc_2	x_{21}	x_{22}	x_{23}		x_{2N}
Doc_3	x_{31}	x_{32}	x_{33}		x_{3N}
. . .					
Doc_N	x_{N1}	x_{N2}	x_{N3}		x_{NM}

Where xij is the feature weighting that exists in the set of documents, N signifies the number of documents and M denotes the number of features. There are two steps to be followed in order to calculate the standard deviation of the feature in the term - document matrix. It is calculated as follows:

$$\overline{x_j} = \frac{1}{N}\sum_{i=1}^{N} x_{ij} \;, j = 1,..,M \quad (1)$$

$$\sigma_j = \sqrt{\frac{1}{N}\sum_{i=1}^{N}\left(x_{ij} - \overline{x_j}\right)^2} \;, j = 1,..,M \quad (2)$$

Where xij signifies a weight calculated based on either Term Frequency (TF) scheme or a Term Frequency-Inverse Document Frequency (TF-IDF) scheme. In this research, we used TF-IDF for feature weighting. $\overline{x_j}$ represents average or means of jth feature, and σj is the standard deviation of jth feature.

We reduced feature space with selecting the P≤M where P denotes the desired value, e.g., 500, 1000, 1500, etc. For feature reduction, we used the following algorithm.

Algorithm: Feature reduction using a high standard deviation of features.

```
Input: feature vector, term-document matrix weight
Output: subsets of most confident of features

Calculating of standard deviation for each feature
(f_sd_j)
  For j = 1 to number of features(N)
  For i = 1 to number of document(M)
  calculating mean, x̄_j = sum( x_ij) / N
  next i
  f_sd_j = sqrt ( sum(x_ij - x̄_j )^2 / N
  next j
Sorting feature vector value(f_sd_j)in descending order
Select P and cut first of feature vector with the more
variation or dispersion
  {y_1, y_2, …, y_p}, P≤M
Generate subsets as follow:
  {y_1}, {y_1,y_2}, {y_1,y_2,y_3}, … , {y_1, y_2, …, y_p}
```

4 Sentiment Classification

Ensemble networks are used to find a solution for the same problem in a parallel independent. In this research, we have proposed classifier ensemble on feature subsets derived from the standard deviation method via base-level classifiers of SVM and NB and three types of ensemble methods, namely majority voting, sum rule, and the product rule in high-level at different time for comparison between them.

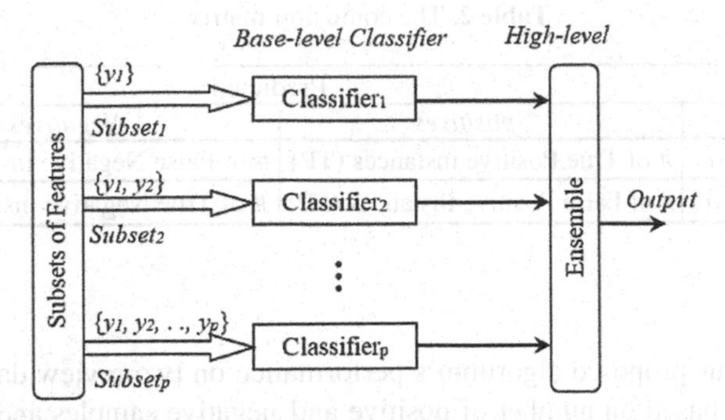

Fig. 2. Structure of classifier ensemble with different feature subsets

5 Experiment Results

To evaluate the performance of the proposed algorithm for feature reduction in sentiment analysis, we employed real datasets. All unigrams used as features and feature weight in the term-document matrix were set to the TF-IDF.

5.1 Datasets

In this research, two different data sets of book and health and personal care are used to investigate the performance of the proposed algorithm. Characteristics of these datasets are described in the following section.

- The dataset proposed by Blitzer *et al.* contains Amazon product reviews that cover different product types that belong to 25 different domains such as health and personal care, books, electronics, music reviews, etc. Each file contains a pseudo XML scheme for encoding the reviews. They are available in the link: http://www.cs.jhu.edu/~mdredze/datasets/sentiment/

5.2 The Performance Measure

Generally, the performance of sentiment classification is evaluated using index accuracy that can be computed through following equation:

$$Accuracy = \frac{TP+TN}{TP+TN+FP+FN} \quad (3)$$

$$F1 = \frac{2\times Precision\times Recall}{Precision+Recall} \quad (4)$$

$$Precision(pos) = \frac{TP}{TP+FP} \quad (5)$$

$$Precision(neg) = \frac{TN}{TN+FN} \quad (6)$$

$$Recall(neg) = \frac{TN}{TN+FP} \quad (7)$$

$$Recall(pos) = \frac{TP}{TP+FN} \quad (8)$$

The common way for computing this index is based on the confusion matrix shown in Table 2.

Table 2. The confusion matrix

		Predicted	
		positives	*Negatives*
Actual	*positives*	# of True Positive instances (TP)	# of False Negative instances (FN)
	negatives	# of False Positive instances (FP)	# of True Negative instances (TN)

5.3 Result

We examined our proposed algorithm's performance on two review datasets. Table 3 shows statistics based on number of positive and negative samples and number of all features after tokenizing and stop words removal. All of the features are presented in the unigram form.

Table 3. Reviews datasets statistics

	Number of positive samples	Number of negative samples	Number of all features	Number of positive features	Number of negative features
Book review	1000	1000	**23933**	15995	15607
Health and personal care review	1000	1000	**10260**	7184	6696

In this part, at first, we give the results of individual classifier to predict samples by two methods, namely Information Gain (IG) and Chi-square (CHI) on 5000 selected best features through 3-fold cross validation in three iterations. Finally, we give the results of individual classifier as base-level in classifier ensemble method by three types of ensemble techniques, i.e., majority voting, sum rule, and product rule through 3-fold cross validation in three iterations. In Figure 3, the results are presented and compared with the accuracy measure.

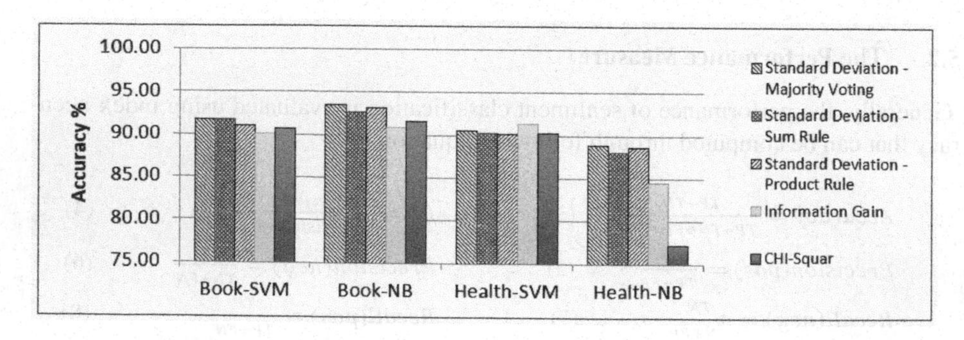

Fig. 3. A comparison between accuracy of three ensemble methods based on standard deviation with two feature reduction methods of Information Gain and Chi-Square on book and health and personal care reviews

Table 4,5,6, and 7 demonstrate the results of sentiment classification that show the measures of accuracy, precision, recall, and F1 values on positive and negative samples of book and health and personal care that were obtained by SVM and NB classifier using 3-fold cross validation in three iterations.

Table 4. SVM classifier as base-level in classifier ensemble on book reviews

			Positive			Negative		
Ensemble Method		Accuracy	Precision	Recall	F1	Precision	Recall	F1
Standard Deviation	Majority Voting	**91.70**	97.18	85.90	91.18	87.40	97.50	**92.17**
	Sum Rule	91.69	**97.22**	85.84	91.16	87.35	**97.53**	92.15
	Product Rule	91.09	97.11	84.70	90.47	86.47	97.47	91.63
Chi-Square		90.12	83.74	**99.60**	90.98	**99.51**	80.65	89.08
Information Gain		90.65	84.50	**99.60**	**91.43**	**99.51**	81.71	89.73

Table 5. NB classifier as base-level in classifier ensemble on book reviews

			Positive			Negative		
Ensemble Method		Accuracy	Precision	Recall	F1	Precision	Recall	F1
Standard Deviation	Majority Voting	**93.51**	**94.57**	**92.19**	**93.35**	**92.57**	**94.81**	**93.67**
	Sum Rule	92.71	94.09	91.16	92.59	91.46	94.27	92.83
	Product Rule	93.13	94.48	91.58	93.00	91.90	94.67	93.26
Chi-Square		90.94	93.72	87.77	90.64	88.52	94.10	91.22
Information Gain		91.62	92.93	90.11	91.49	90.41	93.14	91.75

Table 6. SVM classifier as base-level in classifier ensemble on health and personal care reviews

			Positive			Negative		
Ensemble Method		Accuracy	Precision	Recall	F1	Precision	Recall	F1
Standard Deviation	Majority Voting	90.60	98.95	82.02	89.68	84.74	99.13	**91.36**
	Sum Rule	90.34	98.93	81.57	89.40	84.35	99.11	91.13
	Product Rule	90.23	**99.00**	81.30	89.26	84.16	**99.17**	91.04
Chi-Square		**91.39**	85.73	99.35	**92.03**	99.23	83.44	90.64
Information Gain		90.37	84.22	**99.39**	91.17	**99.26**	81.35	89.41

Table 7. NB classifier as base-level in classifier ensemble on health and personal care reviews

Ensemble Method		Accuracy	Positive Precision	Positive Recall	Positive F1	Negative Precision	Negative Recall	Negative F1
Standard Deviation	Majority Voting	**88.90**	89.61	**87.71**	**88.64**	**88.27**	**90.06**	**89.15**
	Sum Rule	88.11	89.44	86.46	87.91	86.92	89.77	88.31
	Product Rule	88.61	**89.67**	87.24	88.43	87.65	89.77	88.78
Chi-Square		84.52	87.61	80.43	83.85	81.93	88.60	85.12
Information Gain		77.25	85.62	65.23	74.17	72.01	89.17	79.67

6 Conclusion

We proposed a statistical and probabilistic algorithm using standard deviation of features and different feature subset selection for high-dimensional problem on feature space in a sentiment analysis. We used two popular classifiers – Naive Bayes and Support Vector Machine – for sentiment classification. We made a comparison between the obtained results of our proposed algorithm with those of Information Gain and Chi-Square methods on book and health and personal care reviews. The results showed that classification by means of the standard deviation can be as accurate as and sometimes more accurate than Information Gain and Chi-Square methods in terms of sentiment classification.

References

1. Pang, B., Lee, L., Vaithyanathan, S.: Thumbs up? Sentiment classification using machine learning techniques. In: Proceedings of the ACL 2002 Conference on Empirical Methods in Natural Language Processing, vol. 10, pp. 79–86. Association for Computational Linguistics (2002)
2. Turney, P.D.: Thumbs up or thumbs down? Semantic orientation applied to unsupervised classification of reviews. In: Proceedings of the 40th Annual Meeting on Association for Computational Linguistics, pp. 417–424. Association for Computational Linguistics (2002)
3. Su, Y., Zhang, Y., Ji, D., Wang, Y., Wu, H.: Ensemble learning for sentiment classification. In: Ji, D., Xiao, G. (eds.) CLSW 2012. LNCS, vol. 7717, pp. 84–93. Springer, Heidelberg (2013)
4. Nasukawa, T., Yi, J.: Sentiment Analysis: Capturing favorability using natural language processing. In: Proceedings of the 2nd International Conference on Knowledge Capture, pp. 70–77. ACM (2003)
5. Godbole, N., Srinivasaiah, M., Skiena, S.: Large-Scale Sentiment Analysis for News and Blogs. In: ICWSM, vol. 7 (2007)
6. Goldberg, A.B., Zhu, X.: Seeing stars when there aren't many stars: graph-based semi-supervised learning for sentiment categorization. In: Proceedings of the First Workshop on Graph Based Methods for Natural Language Processing, pp. 45–52. Association for Computational Linguistics (2006)

7. Harb, A., Plantié, M., Dray, G., Roche, M., Trousset, F., Poncelet, P.: Web opinion mining: how to extract opinions from blogs? In: Proceedings of the 5th International Conference on Soft Computing as Transdisciplinary Science and Technology, pp. 211–217. ACM (2008)
8. Hu, X., Tang, J., Gao, H., Liu, H.: Unsupervised sentiment analysis with emotional signals. In: Proceedings of the 22nd International Conference on World Wide Web, pp. 607–618. International World Wide Web Conferences Steering Committee (2013)
9. Liu, B., Zhang, L.: A survey of opinion mining and sentiment analysis. In: Mining Text Data, pp. 415–463. Springer (2012)
10. Ortigosa-Hernández, J., Rodríguez, J.D., Alzate, L., Lucania, M., Inza, I., Lozano, J.A.: Approaching Sentiment Analysis by using semi-supervised learning of multi-dimensional classifiers. Neurocomputing 92, 98–115 (2012)
11. Prabowo, R., Thelwall, M.: Sentiment analysis: A combined approach. Journal of Informetrics 3, 143–157 (2009)
12. Sindhwani, V., Melville, P.: Document-word co-regularization for semi-supervised sentiment analysis. In: Eighth IEEE International Conference on Data Mining, ICDM 2008, pp. 1025–1030. IEEE (2008)
13. Taboada, M., Brooke, J., Tofiloski, M., Voll, K., Stede, M.: Lexicon-based methods for sentiment analysis. Computational Linguistics 37, 267–307 (2011)
14. Whitelaw, C., Garg, N., Argamon, S.: Using appraisal groups for sentiment analysis. In: Proceedings of the 14th ACM International Conference on Information and Knowledge Management, pp. 625–631. ACM (2005)
15. Xia, R., Zong, C., Li, S.: Ensemble of feature sets and classification algorithms for sentiment classification. Information Sciences 181, 1138–1152 (2011)
16. Zhou, S., Chen, Q., Wang, X.: Active deep learning method for semi-supervised sentiment classification. Neurocomputing (2013)
17. Zhu, S., Xu, B., Zheng, D., Zhao, T.: Chinese Microblog Sentiment Analysis Based on Semi-supervised Learning. In: Semantic Web and Web Science, pp. 325–331. Springer (2013)

A Support Vector Machine Approach to Detect Financial Statement Fraud in South Africa: A First Look

Stephen O. Moepya[1,2,*], Fulufhelo V. Nelwamondo[1,2], and Christiaan Van Der Walt[1]

[1] CSIR Modeling and Digital Science, Pretoria, South Africa
[2] Faculty of Engineering and the Built Environment, University of Johannesburg, South Africa
{smoepya,FNelwamondo,cvdwalt}@csir.co.za
http://www.csir.co.za

Abstract. Auditors face the difficult task of detecting companies that issue manipulated financial statements. In recent years, machine learning methods have provided a feasible solution to this task. This study develops support vector machine (SVM) models using published South African financial data. The input vectors are comprised of ratios derived from financial statements. The three SVM models are compared to the k-Nearest Neighbor (kNN) method and Logistic regression (LR). We compare the ability of two feature selection methods that provide an increase classification accuracy.

Keywords: financial statement fraud detection, machine learning, support vector machines.

1 Introduction

Financial statement fraud has the potential to destroy confidence in any economy. It involves deliberately falsifying or omitting material information in financial statements[1] with the intention to deceive creditors and investors. Enron Broadband and Worldcom are the two major scandals that have brought financial statement fraud to the foreground in recent years. The collapse of Enron alone caused a loss of approximately $70 billion in market capitalization which was devastating for a significant number of investors, employees and pensioners. The WorldCom collapse, caused by alleged financial statement fraud, is the biggest bankruptcy in United States' history [1]. The responsibility to detect fraudulent financial statements (FFS) has been left solely in the hands of auditors. This may cause an unnecessary burden in the case where the number

* Corresponding author.
[1] Financial statements are a company's basic documents that reflect its financial status [5].

N.T. Nguyen et al. (Eds.): ACIIDS 2014, Part II, LNAI 8398, pp. 42–51, 2014.

of companies increase substantially, especially in growing economies. Another facet to this problem is that fraud evolves with time since perpetrators who get caught find new and clever ways to circumvent procedures and controls that are implemented.

In recent years, data-driven approaches to solving the problem of financial statement fraud have been introduced. This entails using data mining[2] (Artificial Intelligence) techniques to predict/flag potential cases of FFS. Data mining has been utilized with great success in other fields such as money laundering, credit card fraud, insurance fraud and network intrusion detection [2]. The key to data-driven approaches is that they do not impose arbitrary assumptions on the input variables.

South Africa has the largest and most developed economy in Africa. It is the leading producer of platinum, gold, chromium and iron. The main objective of this paper is to test the predictive ability of support vector machine (SVM) models using financial statement data from the Johannesburg Stock Exchange (JSE). An additional aim is to observe the effect of two feature selection methods on classification accuracy. The remainder of this paper is organized as follows: Section 2 reviews prior research; Section 3 provides insight into the research methodology and sample data; Section 4 describes the development of models and gives analyses experimental results and Section 5 presents some concluding remarks.

2 Previous Work

Data mining techniques have been suggested as useful tools to detect fraudulent financial statements (FFS). Kirkos et al. [3], Gaganis [4] and Ravisankar et al. [5] provide excellent comparative studies on the effectiveness of data mining to predict the occurrence of FFS. The techniques include neural networks (NN), support vector machines (SVM), k-nearest neighbors (kNN), discriminant analysis (DA), and logistic regression (LR).

Recent studies have attempted to build predictive models using support vector machines (SVM). Deng [6] proposed a support vector machine-based classifier to detect FFS. The author used publicly listed Chinese company data to construct the models. The training data consisted of 88 companies (44 fraud) during the period 1999-2002. The validation sample comprised of 73 financial fraud and 99 non-fraud instances during the period 2003-2006. The kernel of choice for the SVM was the Radial Basis Function (RBF). The author showed that the SVM model achieved an overall accuracy rate of 82.19% on the holdout sample.

Öğüt et al. [7] compared the prediction ability of both SVM and probabilistic neural network (PNN) using Turkish financial data. The sample data comprised of 150 companies of which 75 had distorted their financial statements. For each fraudulent company, a corresponding same-sector non-fraudulent firm was chosen to be used in the sample. The eight variables chosen in their analysis were

[2] Data mining is an iterative process within which process is defined by discovery, either through automatic or manual methods [3].

suggested by Beneish [8]. The authors found SVM to be superior in overall classification accuracy with a score of 70% as well as with respect to specificity[3] and sensitivity[4].

A more recent study that introduced a novel SVM-based model was presented by Pai et al. [9]. The objective of this study was to assess the model's ability to minimise audit related risks by classifying FFS and presenting the auditor with comprehensible decision rules[5]. The features used in the classification were based on prior research [10,11]. Seventy-five[6] listed companies on the Taiwan stock market were used for this experiment. The authors showed that SVM outperformed multi-layer perceptron (MLP), decision trees (C4.5), and logistic regression (LR) achieving an average accuracy of 86.66%.

Doumpos et al. [11] examined SVM as a means to explain qualification in audit reports. The study used data involving 1754 companies in the U.K. during the period 1998-2003. The authors concluded that non-linear SVM models did not provide improved results compared to the linear model. It was also shown that the SVM model is robust since its performance did not deteriorate significantly when it was tested on future data[7].

From the literature surveyed, it can be noted that SVM play a crucial role in this research area. At present, there seems to be no literature on the detection of FFS using South African listed companies. Moreover, there is no study that has mentioned how to deal with missing values in the sample data.

3 Sample Data and Methodology

3.1 Support Vector Machines

We provide a brief description of support vector machine (SVM) models. SVM, proposed by Vapnik [12], is a binary classifier that is based on statistical learning theory and is used for classification, regression and density estimation. Its first application was Optical Character Recognition (OCR) but since then it has been widely used in diverse fields such as finance, bioinformatics and biometrics. SVM constructs a linear model to estimate a decision function using non-linear class boundaries based on support vectors[8]. The objective of SVM is to construct a decision function that distinguishes two classes in a data set. We will describe

[3] Also known as the true negative rate, this quantity measures the proportion of correctly classified non-fraud instances.

[4] Sensitivity is the true positive rate, in our case, the proportion of positively identified fraud cases.

[5] The model integrates SVM and a classification and regression tree (CART) to extract 'if-and-then' rules for the auditor to interpret.

[6] There was published indication that 25 of these companies were involved in issuing fraudulent financial statements.

[7] This analysis was performed using the walk-forward test.

[8] The support vectors are the training points closest to the optimal separating hyperplane.

the basic formalization of the SVM problem. Should the reader require a more in-depth explanation, we refer the reader to [12].

Suppose a training set and its corresponding labels is defined by (x_i, y_i), where $i = 1, \cdots, l$ and $y \in \{1, -1\}$. The SVM finds an optimal separating hyper-plane that distinguishes an instance in one class from another using the following optimization:

$$\min_{w,b,\xi} \frac{1}{2} w^T w + C \sum_{i=1}^{l} \xi_i \tag{1}$$

$$\text{subject to } y_i(w^T \phi(x_i) + b) \geq 1 - \xi_i,\ \xi_i \geq 0,$$

where the function $\phi(x_i)$ maps training data into high dimensional feature space, w is a weight vector, b is the bias term, ξ_i is a slack variable and C is the penalty for the error term. From a computational point of view, instead of solving the primal problem in equation (1), it is more convenient to consider its dual Lagrangian form:

$$\max_{\gamma} \sum_{i=1}^{l} \gamma_i - \frac{1}{2} \sum_{i,j=1}^{l} y_i y_j \gamma_i \gamma_j K(x_i, x_j) \tag{2}$$

$$\text{subject to } 0 \leq \gamma_i \leq C,\ i = 1, \ldots, l,$$

where the γ_i's are Lagrange multipliers and there exists a γ_i for each vector in the training set[9]. The mapping function $\phi(x_i)$ is defined through a symmetric positive kernel function $K(x_i, x_j)$. The kernel function is employed to transform the original input space into a high dimensional space. The decision function is expressed by the equation:

$$F(x) = sign\left[\left(\sum_{i=1}^{m} \alpha_i y_i K(x, x_i)\right) + b\right], \tag{3}$$

where α is a parameter and $K(x_i, x_j) = \phi(x_i)\phi(x_j)$ is the kernel function. The kernel function has the following alternatives:

$$K(x, x_i) = \exp\left\{-\|x - x_i\|^2 / 2\sigma^2\right\} \quad \text{(RBF kernel)} \tag{4}$$

$$K(x, x_i) = (1 + x \cdot x_i / c)^d \quad \text{(Polyn kernel of degree } d) \tag{5}$$

$$K(x, x_i) = \tanh(kx \cdot x_i + \theta) \quad \text{(MLP kernel)} \tag{6}$$

This concludes the brief discussion of SVM models. We will utilize the SVM models in experiments to show its effectiveness to predict financial statement fraud using a sample of financial data.

[9] The support vectors identify the surface of decision, and correspond to the subset of non-zero γ_i.

3.2 Dataset

The sample data was obtained from the financial database of BFA McGregor[10]. The sample selected contains financial ratios from 88 public companies who are/were listed on the Johannesburg Stock Exchange (JSE). Half of the companies selected were known to have received a qualified audit report. Each instance of financial statement fraud was matched with a corresponding company in the same sector which received an unqualified audit report. The justification for the matching is that a choice-based sample provides higher information content than a random sample [4].

The sample is divided into two separate sets. The training set consists of 48 companies which cover the period 2003-2005. The sample data for the period 2006-2007 is used as a test (holdout) set to fit the models on unseen samples. This is done to prevent over-fitting which can lead to a loss of generality.

Missing data values were a common feature in the given sample. Some financial ratios were stated as 'not available'. This is due to two possible reasons. The first being that the data provider did not compute that particular ratio for some unforeseen reason. A second explanation is that some of the values were not available in the published financial statements therefore a given ratio could not be computed[11]. The missing data problem can to be solved by means of imputation.

Imputation methods can generally be divided into two parts: statistical and machine learning. Statistical imputation include the mean[12], hot-deck[13] and multiple imputation[14] methods. Machine learning imputation methods, in general, create a predictive model to estimate values that will be substituted for missing items. Well known machine learning imputation algorithms are the multi-layer perceptron (MLP), self-organizing maps (SOM) and k-nearest neighbor (kNN) methods.

An interesting method that we chose to utilize (alternative to imputation) to fill in missing values was proposed by Candès and Recht [13]. Exact Matrix Completion (EMC) is a well known method in the field of compressed sensing. This method, through the use of convex optimization, completes a matrix that contains missing values, i.e. it perfectly recovers most low-rank matrices from what appears to be an incomplete set of entries [13]. We used EMC only on grouped instance of the same class.

---

[10] BFA McGregor is South Africa's leading provider of financial data feeds and analysis tools.

[11] An example of this is when one company issues dividends, for a given year, and another withholds dividends. Therefore any financial ratio (involving dividends) will not be available for the latter firm.

[12] Mean imputation imputes the mean values of each variable on the respective missing variables as an estimate of the missing value.

[13] In hot-deck imputation, a missing value is replaced by a similar (or other) candidate which contains data for that particular variable.

[14] The multiple imputation method replaces an unknown value with a set of plausible ones and uses an appropriate model that incorporates random variation.

4 Feature Selection

Feature selection is a key ingredient for the performance of data mining problems. It has been well studied in data samples that contain a large number of variables. The success of any machine learning algorithm is directly dependent on feature selection. If this step of the data mining process is performed poorly, it may lead to problems related to incomplete information, noisy or irrelevant features and poor classification accuracy [5].

The sample data consists of 48 features (financial ratios), which will be used as inputs in the predictive models to classify instances. These features cover all aspects of company performance such as: liquidity, leverage, profitability and efficiency. Most previous studies in this research domain, such as [7] [11] [15], do not completely agree on which variables are important in detecting financial statement fraud. Moreover, no study has been performed using South African market data. We take a first principle approach and allow a feature selection algorithm to decide the importance of each ratio. The financial ratios were derived from published financial statements and no other variables are considered.

In this particular study we consider two feature selection algorithms. ReliefF was used as a feature selection method by Kotsiantis et al. [10]. ReliefF assigns relevance to features based on their ability to disambiguate similar samples, where similarity is defined by proximity in the feature space[15]. For experimental purposes, we will extract the top eighteen and top ten features and use them to test classifier performance (such as in [5]).

The second feature selection method that we choose to utilize is the t-test statistic [14]. The t-test is one of the most commonly used feature selection techniques due to its simplicity. The t-statistic is computed for each feature (given a level of significance) and a p-value indicates whether a particular variable is statistically different from its counterpart in another class. The performance of the t-test and ReliefF are compared in the following section.

5 Results

In this section we will compare the predictive ability of three SVM models, kNN and Logistic Regression (LR). Also, a comparison between the t-test and ReliefF will be undertaken. We use MATLAB for the implementation of all the methods. Performance analysis is addressed by finding the average accuracy of each classifier when trained (data from 2003-2005) and tested using a holdout sample (period 2006-2007). Another aspect of performance analysis is assessing the sensitivity and specificity of each algorithm.

Table 1 lists the top ten (10) features ranked by ReliefF. These features along with eight others[16] are used as inputs to train the classifiers in the first

[15] The relevant features accumulate high positive weights, while irrelevant features retain near-zero weights.

[16] The eight other features are: Total Debt/ Cash flow, Leverage factor, Debt/Equity, Dividend/Share(c), Earnings Yield (%), Price/Share, Return on Average Assets and Inflation Adjusted Return on Average Total Assets.

experiment. In all the experiments, the parameter $C = 1$ and $\sigma = 1$ are used in the SVM models. With regard to the kNN method, $k = 3$ is selected for this study[17].

Table 1. Features selected by ReiefF

Variable	ReliefF Score
Debt/Assets	0.0211
Price/Cash	0.0169
Price/Cash flow	0.0167
Dir Rem % Pft BTax	0.0134
Book val/Share(c)	0.0127
Total Assets/Turnover	0.0122
N A V/ Share (c)	0.0120
Retention rate	0.0118
Cash flow/Share(c)	0.0099
Turnover/employee	0.0093

The results of our first experiment are presented in Table 2. UQ and Q represent companies who received unqualified and qualified audit reports respectively. During the training stage, the quadratic and RBF SVM models outperform the linear kernel each achieving an average accuracy of 100% . However, the performance of these models when fit on the test/holdout sample are significantly different. Logistic regression achieves the greatest average accuracy with a score

Table 2. Classification accuracy using training and testing samples using 18 best features selected by ReliefF

Method	Training			Testing		
	UQ	Q	Ave	UQ	Q	Ave
SVM (linear)	86.21	93.10	89.66	80.00	85.71	82.76
SVM (quadratic)	100.00	100.00	100.00	66.66	71.14	68.97
SVM (RBF)	100.00	100.00	100.00	60.00	78.57	68.97
kNN (3NN)	75.86	86.21	81.03	93.33	64.29	79.31
Logit (LR)	100.00	100.00	100.00	86.67	85.71	86.21

of 86.21%. Both the linear SVM and LR are found to be superior with regard to sensitivity as they were able to correctly classify 85.71% of qualified cases.

We repeat the same experiment using the top ten features ranked by ReliefF. A similar approach was taken by Ravisankar et al. [5]. Table 3 shows that both the RBF and quadratic kernels outperform all other models attaining the same average accuracy of 94.83% during training. Using the holdout sample, the

[17] All SVM parameters in this experiment are the default values given in the software package. The specified value $k = 3$ is adopted from the comparison in Kotsiantis et al. [10].

Table 3. Classification accuracy using training and testing samples using 10 best features selected by ReliefF

Method	Training			Testing		
	UQ	Q	Ave	UQ	Q	Ave
SVM (linear)	89.66	96.55	93.10	93.33	85.71	89.66
SVM (quadratic)	93.10	96.55	94.83	73.33	71.43	72.41
SVM (RBF)	89.66	100.00	94.83	73.33	85.71	79.31
*k*NN	75.86	79.31	77.59	86.67	64.29	75.86
Logit (LR)	89.66	93.10	91.38	86.67	85.71	86.21

results indicate that the linear SVM (with an average accuracy score of 89.66 %) outperforms the other models. The SVM RBF, SVM linear and LR models achieved the same result when comparing the ability to predict qualified cases.

Next we turn our attention to the classification performance where features were selected by the *t*-test. These features, together with their statistics, are given in Table 4. The *t*-test was performed using a 5% level of significance. An

Table 4. Features selected by *t*-test

Variable	Unqualified		Qualified		*p*-value
	mean	std dev	mean	std dev	
Book value/Share	1526.00	3181.70	184.25	468.39	0.0324
Debt/Assets	0.505	0.32	0.90	0.85	0.0239
Dir Rem % Pft BTax	25.26	69.91	-3.46	30.67	0.0497
Earnings/Share (c)	256.45	740.60	-49.12	201.27	0.0397
Earnings Yield	7.78	10.75	-85.57	242.49	0.0477
N A V/Share	1972.52	4192.03	189.24	479.66	0.0305
Price/ Cash flow	6.13	9.44	-0.56	14.11	0.0391
Retention rate	58.39	51.02	124.28	109.86	0.0056
Total Debt/Cash Flow	2.20	6.61	-4.90	15.44	0.0284
Price/Cash	37.65	80.75	195.51	347.06	0.0233

interesting point to note is that the *t*-test selected six features that are identical to those selected by ReliefF. These features are: Book value/Share, Debt/Assets, Direct Remuneration (%) Profit before Tax, Net Asset Value/ Share, Price/Cash flow, Retention rate and Price/Cash. We present the results given by these ratios in Table 5.

The training results in Table 5 show that the quadratic SVM model outperforms the other models where the average accuracy, sensitivity and specificity are concerned. The performance of the classifiers on the validation sample shows that the linear and RBF SVM models perform equally well with regard to average accuracy. Both models achieve 86.21% overall accuracy and the same accuracy with respect to sensitivity and specificity. During training, the performance of

Table 5. Classification accuracy using training and testing samples using features selected by the *t*-test

Method	Training			Testing		
	UQ	Q	Ave	UQ	Q	Ave
SVM linear	86.21	86.21	86.21	86.66	85.71	86.21
SVM quadratic	96.56	100.00	98.28	80.00	78.57	79.31
SVM RBF	93.10	86.21	89.66	86.66	85.71	86.21
*k*NN	89.66	89.66	89.66	80.00	71.14	75.86
Logit (LR)	86.21	96.55	91.38	86.67	78.57	82.76

*k*NN and the quadratic kernel are satisfactory. However their accuracy suffers greatly when fit on the holdout samples.

In summary, the linear SVM model shows the least amount of deterioration when fit onto the holdout sample, especially when using the features selected by the *t*-test. It outperformed all the other models (using average accuracy as a benchmark) using both feature selection techniques on the holdout samples. The linear model attains the greatest average accuracy when tested with inputs (10 features) selected by ReliefF. A reduction from eighteen to ten features using ReliefF shows an increase in classification performance using the test set. Features selected by the *t*-test produce a slightly better average accuracy when tested using the holdout sample. The overall performance of SVM as a tool to detect fraudulent financial statements is in line with previous studies. This shows the robustness of SVM in predicting fraudulent financial statements.

6 Conclusion

Financial fraud detection plays a crucial role in the stability of an economy. In recent years, machine learning has been introduced to predict companies that are likely to publish fraudulent financial statements with some degree of success. The current investigation shows the effectiveness of support vector machines as a tool to detect the manipulation of financial statements using publicly available data. The results indicate the robustness of the simple linear support vector machine as compared to other models using the holdout sample. The effect of feature selection on prediction accuracy is captured by an increase in model sensitivity and specificity.

Acknowledgment. The current work is being supported by the Department of Science and Technology (DST) and Council for Scientific and Industrial Research (CSIR). Special thanks to Prof. Fulufhelo V. Nelwamondo, Mr. Christiaan Van Der Walt and Mr. Cederick Jooste for their valuable suggestions regarding the paper's presentation.

References

1. Rezaee, Z.: Causes, consequences, and deterence of financial statement fraud. Critical Perspectives on Accounting 16(3), 227–298 (2005)
2. Phua, C., Lee, V., Smith, K., Gayler, R.: A comprehensive survey of data mining-based fraud detection research. arXiv preprint arXiv: 1009.6119 (2010)
3. Kirkos, E., Spathis, C., Manolopoulos, Y.: Data mining techniques for the detection of fraudulent financial statements. Expert Systems with Applications 32(4), 995–1003 (2007)
4. Gaganis, C.: Classification techniques for the identification of falsified financial statements: a comparative analysis. Intelligent Systems in Accounting, Finance and Management 16(3), 207–229 (2009)
5. Ravisankar, P., Ravi, V., Rao, G.R., Bose, I.: Detection of financial statement fraud and feature selection using data mining techniques. Decision Support Systems 50(2), 491–500 (2011)
6. Deng, Q.: Application of support vector machine in the detection of fraudulent financial statements. Computer Science & Education (2009)
7. Öğüt, H., Aktas, R., Alp, A., Doğanay, M.M.: Prediction of financial information manipulation by using support vector machine and probabilistic neural network. Expert Systems with Applications 36(3), 5419–5423 (2009)
8. Beneish, M.D.: The detection of earnings manipulations. Financial Analyst Journal, 24–36 (1999)
9. Pai, P.-F., Hsu, M.-F.: An enhanced support vector machines model for classification and rule generation. In: Koziel, S., Yang, X.-S. (eds.) COMA. SCI, vol. 356, pp. 241–258. Springer, Heidelberg (2011)
10. Kotsiantis, S., Koumanakos, E., Tzelepis, D., Tampakas, V.: Forecasting Fraudulent financial statements using data mining. International Journal of Computational Intelligence 3(2), 104–110 (2006)
11. Doumpos, M., Gaganis, C., Pasiouras, F.: Explaining qualifications in audit reports using a support vector machine methodology. Intelligent Systems in Accounting, Finance and Management 13(4), 197–215 (2005)
12. Cortes, C., Vapnik, V.: Support-vector networks. Machine Learning 20(3), 273–297 (1995)
13. Candès, E.J., Recht, B.: Exact matrix completion via convex optimization. Foundations of Computational Mathematics 9(6), 717–772 (2009)
14. Saeys, Y., Inza, I., Larranaga, P.: A review of feature selection techniques in bioinformatics. Bioinformatics 23(19), 2507–2517 (2007)
15. Checcini, M., Aytug, H., Koehler, G., Pathak, P.: Detecting management fraud in public companies. Management Science 56(7), 1146–1160 (2010)

Using Multi-agent Systems Simulations for Stock Market Predictions

Diana Dezsi[1], Emil Scarlat[1], Iulia Măries[1], and Ramona-Mihaela Păun[2]

[1] Dept. of Informatics and Economic Cybernetics, Bucharest University of Economics, Romania
[2] Webster University Thailand, Bangkok, Thailand
{dianadezsi,rmpaun26}@yahoo.com, emilscarlat@ymail.com, iulia.maries@hotmail.com

Abstract. The paper analyses the predictive power of heterogeneous agents interacting in an evolutionary agent-based model of the stock market when simulated through Altreva Adaptive Modeler multi-agent simulation software application. The paper tests the long-term prediction accuracy of an evolutionary agent-based model when simulating the S&P500 stock market index. The model incorporates 2,000 agents which trade amongst each other on an artificial stock market which uses a double auction trading mechanism. Within the evolutionary agent-based model, the population of agents is continuously adapting and evolving by using genetic programming in order to obtain new agents with better trading strategies generated from combining the trading strategies of the best performing agents and thus replacing the agents which have the worst performing trading strategies.

Keywords: heterogeneous agents, double auction, artificial stock market, price prediction.

1 Introduction

The aim of our research is to test the predictive power of an adaptive agent-based model of the stock market, when 2,000 heterogeneous agents are used to trade in an artificial stock market with a double auction trading mechanism. In order to achieve our research aim, we use the Adaptive Modeler [1] software to simulate the adaptive agent-based model for artificial stock market generation and price forecasting of real world S&P500 stock market index. Thus, heterogeneous agents trade a stock floated on the stock exchange market, placing orders depending on their budget constraints and trading rules, where the artificial market is simulated as a double auction market.

The population of agents continuously applies a breeding process by using evolutionary computing, namely the Strongly Typed Genetic Programming [2] so that the trading strategies of the agents in the population to adapt to the price change and to better predict future prices. This breeding process generates new agents with better trading strategies obtained from combining the trading strategies of the best performing agents and thus replacing the agents which have the worst performing trading

N.T. Nguyen et al. (Eds.): ACIIDS 2014, Part II, LNAI 8398, pp. 52–61, 2014.

strategies. This technique allows for adaptive, evolving and self-learning market modeling and forecasting solutions.

To explore and understand the complexity of the financial markets and trading behavior, the agent-based models have been successfully implemented, offering explanation for observed stylized facts and being able to reproduce many of them [3], [4]. Arthur et al. [5] from Santa Fe Institute, Ca., USA, developed an artificial stock market which allowed for testing of agent-based models with heterogeneous agents.

The double auction trading mechanism (also called double-sided auction, or bid-ask auction) represents an important development of the trading mechanism used in the agent-based models for simulating financial markets. Gode and Sunders [6] [7] have introduced the zero-intelligent trader concept, further developed by Cliff [8], Gjerstad and Dickhaut [9]. Rust [10] and Phelps et al. [11] have used in their experiments heterogeneous agents which change their strategies during the learning process, as follows: the unprofitable strategies are being replaced with the more profitable ones, thus developing adaptive models which use genetic algorithms to evolve. Walia [12] has studied the development of the agent-based models which use genetic programming, allowing for more flexibility and effectiveness in finding optimal solutions, programs being encoded as tree structures, thus crossover and mutation operators being applied easier. The later is similar with the learning process used in the hereto paper.

The wealth of an agent represents the total value of cash and shares that the agent holds. The literature review shows that many of the papers simulating agent-based models assume equal initial amount of wealth for the agents, which is an unrealistic assumption, as real markets involve heterogeneity in terms of wealth distribution of traders. According to empirical studies, the income distribution follows a power-law distribution, which we will also use in the hereto paper for the initial endowment of the population, as in [4].

The remainder of this paper is structured as follows. Section 2 presents the specifications of the adaptive agent-based model used in the simulations, Section 3 describes the datasets used in this study, while the results of the simulations that have been performed are presented in section 4, the paper ending with the conclusions and directions for future work.

2 The Specifications of the Evolutionary Agent-Based Model of a Multi-agent System

An agent-based model represents a computational model for simulating the actions and interactions among agents in a multi-agent system in order to analyze the effects on a complex system as a whole, being a powerful tool in the understanding of markets and trading behavior. An agent-based model of a stock market consists of a population of agents (representing investors) and a price discovery and clearing mechanism (representing an artificial stock market). The complex dynamics of these heterogeneous investors and the resulting price formation process require a simulation model of a multi-agent system and an artificial market.

The evolutionary agent-based model referred to in this paper is simulated in Adaptive Modeler software, which supports up to 2,000 agents and 20,000 simulation periods for each epoch of simulations. The agents are autonomous and heterogeneous entities representing the traders of the stock market, each having their own *wealth* and their own trading strategy called the *genome*.

For each epoch of simulation, the model is initialized as follows: the agents are endowed with an initial wealth according to a Pareto probability distribution, a well known power law distribution commonly used to describe wealth or income distributions, describing unequal distribution, where a large part of the total wealth is owned by a small percentage of individuals, which was first described by Pareto [13]. The agents are also endowed with a trading strategy which is called the genome, which is randomly created by taking in account the selected genes (which represent functions) using genetic programming. Broker fees are fixed at 10 points of value for each transaction. There is no market maker. All the parameters for each of the two models and their values are described in *Table 1*.

Each simulation period goes through a cycle, as follows:

- *Receive new quote bar.*
- *Agents evaluate trading rules and place orders:* Agents receive access to historical prices and evaluate the evolution of prices according to the technical analysis generated by their trading rules found in the genomes, resulting in a desired position as a percentage of wealth limited by the budget constraints, and a limit price. Agents are two-way traders during the simulations, meaning that they are allowed to both sell and buy during multiple simulation periods, and they are one-way traders during a single simulation period (in this case a day) corresponding to an auction, as they are able to submit only one order per auction, either buy or sell. The position is generated in a random manner, while the limit price is generated after a technical analysis has been performed, according to the genome structure which represents trading functions.
- *Artificial stock market clearing and forecast generation*: The artificial stock market determines the clearing price using a clearing house, which is a discrete time double-sided auction mechanism in which the artificial stock market collects all bids (buying orders) and asks (selling orders) submitted by the agents and then clears the market at a price where the supply quantity equals the demanded quantity, therefore the clearing price is the price for which the highest trading volume from limit orders can be matched, thus all agents establish their final positions and cash at the same time. In case the same highest trading volume can be matched at multiple prices, then the clearing price will be the average of the lowest and the highest of those prices. Market orders have no influence on the clearing price, only executed orders do. The artificial stock market also executes all executable orders and forecasts the price for the next simulation period, which is set equal to the clearing price. Also, a trading signal for the trading simulator of the model is generated based on the forecast.
- *Breeding*: During the breeding process, new agents are created from best performing agents in order to replace the worst performing agents, creating new genomes

by recombining the parent genomes through a crossover operation, and creating unique genomes by mutating a part of the genome. The breeding process repeats at each bar, with the condition that the agents must have a minimum breeding age of 80 simulation periods, in order to be able to assess the agents' performance.

Table 1. General settings of the models. Market and agents' parameters configuration in the simulations.

Parameter Type	Parameter Name	Parameter Value
Market Parameters	No. of simulation periods	15,800
	No. of agents	2,000
	Minimum price increment	0.01
	Average bid/ask spread	0.01%
	Fixed Broker fee	10
Agent Parameters	Wealth Distribution	Pareto distribution, Pareto index 2
	Position Distribution	Gaussian distribution
	Min. position unit	20%
	Max. genome size	1,000
	Max. genome depth	20
	Min. initial genome depth	2
	Max. initial genome depth	5
	Genes	CurPos, LevUnit, Rmarket, Vmarket, Long, Short, Cash, Bar, PndPos, IsMon, IsTue, IsWed, IsThu, IsFri, close, bid, ask, average, min, max, >, change, +, dir, isupbar, upbars, pos, lim, Advice, and, or, not, if
	Breeding Cycle Frequency	1 simulation period
	Minimum breeding age	80 simulation periods
	Initial selection: randomly select	100% of agents of minimum breeding age or older
	Parent selection	5% agents of initial selection will breed
	Mutation probability	10% per offspring

In order to obtain random seed, the Adaptive Modeler software uses the Mersenne Twister algorithm [14] to generate pseudo random number sequences for the initial creation of trading rules or genomes and for the crossover and mutation operators of the breeding process.

A double auction is a trading mechanism in which buyers and sellers can enter bid or ask limit orders and accept asks or bids entered by other traders [6]. This trading mechanism was chosen to be used for the artificial market simulation in the Adaptive Modeler models because most of the stock markets are organized as double auctions. In the double auction markets, agents introduce bid or ask orders, each order consisting of a price and quantity. The bids and asks orders received are put in the order book and an attempt is made to match them. The price of the trades arranged must lie in the bid-ask spread (interval between the selling price and buying price).

The genomes attached to the each agent uses a tree composed of genes which generates the trading strategies. The initial node in the genetic program tree combines the position desired in the security generated randomly, and the limit price value generated by a collection of functions working as a technical analysis on the historical prices, into a buy or a sell order advice. The *desired position value* ranges between -100% (short position, or selling position) and 100% (long position, or buying position) which is randomly generated from a uniform distribution. The *limit price value* is generated by a collection of functions which uses simple technical indicator initially generated in a random manner from the list of functions selected to be used in the model, which develop during the breeding process, in order to generate the limit price for the buy or sell order.

The buy or sell order is introduced in the market after comparing the desired position with the agent's current position and calculating the number of shares that need to be bought or sold, taking also in consideration the available cash. The trading rules of the model use historical price data as input from the artificial stock market, and return an advice consisting of a desired position, as a percentage of wealth, and an order limit price for buying or selling the security. Through evolution the trading rules are set to use the input data and functions (trading strategies) that have the most predictive value.

The agents' trading rules development is implemented in the software by using a special adaptive form of the *Strongly Typed Genetic Programming* (STGP) approach, and use the input data and functions that have the most predictive value in order for the agents with poor performance to be replaced by new agents whose trading rules are created by recombining and mutating the trading rules of the agents with good performance. In order to do this, a dynamic fitness function is used to evaluate the performance of the agents, and only the most recent simulation periods of the epoch are taken in consideration for computing the fitness function. As regards to the breeding process at each simulation period, the adaptive form of the STGP approach only takes in consideration a percentage of 5% of the total population of agents.

The STGP was introduced by Montana (2002) [2], with the scope of improving the genetic programming technique by introducing data types constraints for all the procedures, functions and variables, thus decreasing the search time and improving the generalization performance of the solution found. Therefore, the *genomes* (programs) represent the agents' trading rules and they contain *genes* (functions), thus agents trade the security on the artificial stock market based on their technical analysis of either the real market historical price data, either the historical clearing price data generated by the artificial stock market.

3 Data

The data used in this paper was retrieved from the Bloomberg application and contains daily data for the S&P500 stock index, including open, high, low and close values. The analysed period is January 3rd 1950 – March 12th 2013, meaning 15,800 trading days. Such a large period of time ensures the optimisation of the learning process and an increased level of adaptability and evolution of the agents' trading rules within the model, thus generating better simulation results.

The parameters used in the model are described in Section 2, Table 1, and remain constant during the simulations. The simulations will be processed by the Adaptive Modeler software application, using a double auction trading mechanism and heterogeneous agents which interact within the artificial stock market, while the population of agents adapts and evolves using genetic programming.

4 Simulation Results

The simulations in the hereto paper are conducted over 10 independent epochs, each covering 15,800 simulation periods, spanned over 63 years of daily quotations, using the same parameter configuration values but different random seeds for the initial wealth of the agents and for the trading strategies for each agent, allowing for a significant assessment of the robustness and accuracy of the simulation results. The market and agent parameters are explained in section 2 and summarized in Table 1.

In order to analyze the fit of the simulation results with the real data when forecasting the price for the next simulation period, the *Root Mean Squared Error (RMSE)* indicator was computed for each simulation epoch and illustrated in *Table 2*.

$$\text{RMSE} = \sqrt{\frac{1}{\text{n}}\sum_{i=0}^{n-1}(F_{t-i} - P_{t-i})^2} \tag{1}$$

where $\text{n} = \text{t} - \text{s} + 1$, when the calculation period is (s, t), F_i represents the forecast value generated by the artificial stock market for date *i*, P_i represents the real stock market value from the input data.

In order to analyze the percentage of simulation periods in an epoch, in which the forecasted price change was in the same direction as the real price change, the *Forecast Directional Accuracy (FDA)* indicator was computed for each simulation epoch and illustrated in *Table 2*.

$$\text{FDA} = \frac{\sum_{\text{i=0}}^{\text{n-1}} \text{FDA}_{\text{t-i}}}{\text{n}} \tag{2}$$

where $\text{n} = \text{t} - \text{s} + 1$, when the calculation period is (s, t) and

$$\mathrm{FDA_i} = \begin{cases} 1, if\ PCD_i = FD_i\ AND\ PCD_i \neq 0 \\ 0.5, if\ FD_i = 0\ OR\ PCD_i = 0 \\ 0, if\ PCD_i = -FD_i\ AND\ PCD_i \neq 0 \end{cases} \tag{3}$$

$$\mathrm{PCD_i} = \begin{cases} 1, if\ P_i > P_{i-1} \\ -1, if\ P_i < P_{i-1} \\ 0, if\ P_i = P_{i-1} \end{cases} \tag{4}$$

$$\mathrm{FD_i} = \begin{cases} 1, if\ F_i > P_{i-1} \\ -1, if\ F_i < P_{i-1} \\ 0, if\ F_i = P_{i-1} \end{cases} \tag{5}$$

where *PCD* represents the *Price Change Direction* and *FD* represents the *Forecast Direction*, F_i represents the forecast value generated by the artificial stock market for date *i*, P_i represents the real stock market value from the input data.

Table 2. The fit indicators of the simulation results with real data per each epoch generated by the trading simulator

Epoch No.	RMSE	FDA	Average annual return per epoch	Trading simulator's total return per epoch	Trading simulator's Excess return per epoch
1.	5.51%	52.4%	9.20%	15,516.00%	7,310.00%
2.	6.29%	55.0%	20.07%	3,393,771.00%	3,385,565.00%
3.	7.70%	54.5%	17.05%	984,471.60%	976,265.10%
4.	5.18%	54.6%	16.62%	872,291.00%	864,084.60%
5.	4.98%	54.5%	17.24%	1,548,997.00%	1,540,790.00%
6.	4.03%	54.8%	17.83%	1,385,545.00%	1,377,338.00%
7.	1.51%	54.8%	19.64%	3,705,214.00%	3,697,008.00%
8.	1.33%	55.1%	18.80%	2,742,407.00%	2,734,201.00%
9.	2.65%	54.6%	18.73%	2,691,233.00%	2,683,027.00%
10.	1.48%	53.3%	12.59%	113,141.00%	104,935.00%
S&P500	-	-	**8.50%**	**8,206.00%**	-

The trading simulator used for analyzing the long-term accuracy of prediction represents a trading strategy generated by the simulated model which offers buying and selling suggestions according to the agents' predictions. The annual return computed by the trading simulator is a hypothetical one, which would be achieved by following the trading simulator's suggestions.

The results show that the price forecast generated by the evolutionary agent-based model at each simulation period for the next simulation period (meaning at daily intervals) is highly accurate, given that the Root Mean Squared Error indicator is very low for the entire epoch, ranging between 1.48% and 7.70%. Also, the values of the Forecast Directional Accuracy indicator range between 52.4% and 55.1%, showing the percentage of simulation periods for which the forecasted price change was in the right direction, suggesting a higher rate of accurate prediction of the price change direction compared to the rate of wrong prediction of the price change direction.

In these conditions, we wonder if the daily forecast accuracy is high enough to over perform the S&P500 index over longer period of time. The answer is a positive one, as according to the trading simulator of the Adaptive Modeler simulation software, the trading strategy offered by the predictions made by the evolutionary agent-based model generates higher annual returns on average for each of the simulated epochs, over performing the S&P500 stock market index. Therefore, the average annual return of the simulated epoch ranges between 9.20% and 20.07% compared with an average annual return of 8.50% for the S&P500 stock market index, as illustrated in Table 2.

On the other hand, taking in consideration the long period of time in which we undergo the simulations, we should also observe the total return generated by the trading simulator per each epoch, shown in *Table 2*, and also the excess return of the trading simulator which is computed as the difference between the trading simulator's total return and the real stock market index return, which represents the benchmark. According to the results, the total return which could be obtained when following the trading simulator's suggestions is much higher compared to the benchmark, in all the simulated epochs.

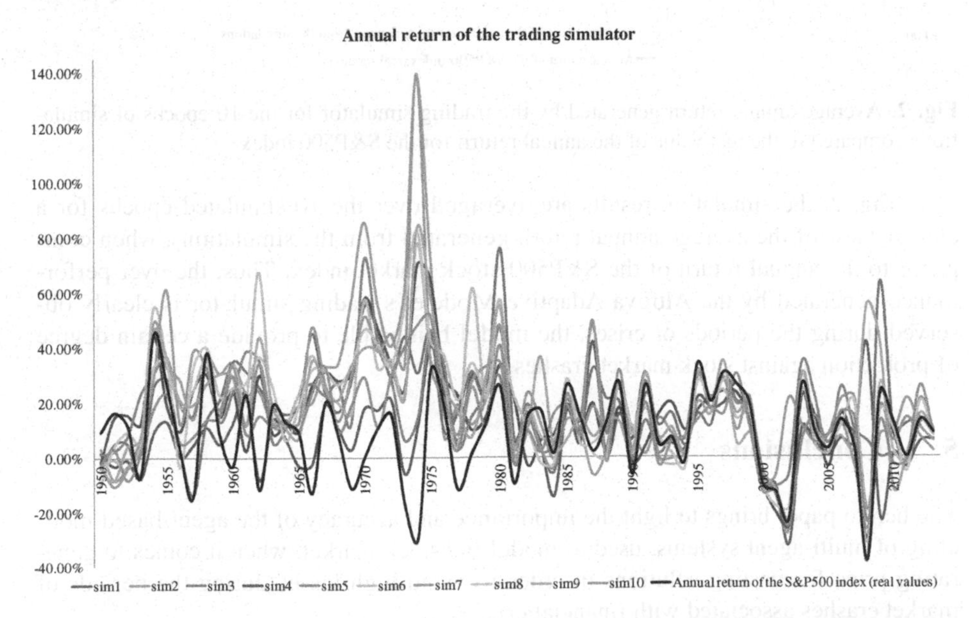

Fig. 1. Annual return generated by the trading simulator for the each of the 10 epochs of simulations compared to the real value of the annual return for the S&P500 index

In order to better visualize the results, Fig. 1 shows the annual returns for all the simulated epochs, for the simulated period starting from January 3rd, 1950, ending on March 12th, 2013. According to these results, the trading strategy suggested by the evolutionary agent-based model over performs the S&P500 stock market index in all the periods of crises, even though during the periods of boom in stock prices the returns generated by the trading simulator are not as high as the S&P500 annual returns. The long-term predictions generated by the model offer excess return for all the simulated epochs, for the entire period which is analyzed in this study, showing high predictive power on the long-term and also a promising degree of anticipation of the periods of crises.

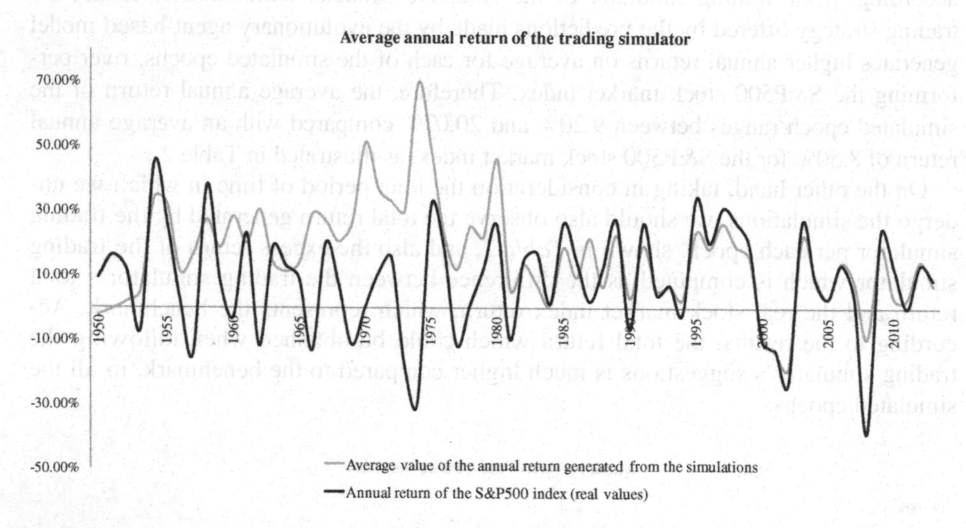

Fig. 2. Average annual return generated by the trading simulator for the 10 epochs of simulations compared to the real value of the annual return for the S&P500 index

In Fig. 2, the simulation results are averaged over the 10 simulated epochs for a clearer view of the average annual return generated from the simulations, when compared to the annual return of the S&P500 stock market index. Thus, the over performance generated by the Altreva Adaptive Modeler's trading simulator is clearly observed during the periods of crises, the model being able to provide a certain degree of protection against stock market crashes.

5 Conclusions

The hereto paper brings to light the importance and accuracy of the agent-based modeling of multi-agent systems, used to model the stock market, when it comes to generating price forecasting solutions in order to avoid high losses during the periods of market crashes associated with financial crises.

The results following the assessment of the long-term performance of the evolutionary agent-based model show promising excess return when following the advices of the Adaptive Modeler trading simulator. The conclusion comes from analyzing the fit of the models in terms of forecast abilities for daily quotes, as well as annual returns of the trading simulator.

Further research should focus on identifying common trading strategies which could be used for hedging purposes in case of future periods of market crashes.

References

1. Altreva, `http://www.altreva.com/technology.html`
2. Montana, D.: Strongly Typed Genetic Programming. Evolutionary Computation 3(2), 199–230 (2002)
3. Daniel, G.: Asynchronous simulations of a limit order book. University of Manchester. Ph.D. Thesis (2006)
4. Aloud, M., Fasli, M., Tsang, E., Dupuis, A., Olsen, R.: Modelling the High-Frequency FX Market: an Agent-Based Approach. University of Essex. Technical Reports, Colchester (2013)
5. Arthur, B., Holland, J., LeBaron, B., Palmer, R., Tayler, P.: Asset Pricing Under Endogenous Expectations in an Artificial Stock Market. In: Arthur, W.B., Durlauf, S.N., Lane, D. (eds.) The Economy as an Evolving, Complex System II, Redwood City, CA, pp. 15–44. Addison Wesley (1997)
6. Gode, D., Sunder, S.: Allocative Efficiency of Markets with Zero-Intelligence Traders: Market as a Partial Substitute for Individual Rationality. Journal of Political Economy 101, 119–137 (1993)
7. Gode, D., Sunder, S.: Lower Bounds for Efficiency of Surplus Extraction in Double Auctions. In The Double Auction Market: Institutions, Theories, and Evidence (1993)
8. Cliff, D.: Evolution of Market Mechanism Through a Continuous Space of Auction-Types. Technical report, Hewlett-Packard Research Laboratories, Bristol, England (2001)
9. Gjerstad, S., Dickhaut, J.: Price formation in double auctions. Games and Economic Behavior 22, 1–29 (1998)
10. Rust, J., Palmer, R., Miller, J.: A Double Auction Market for Computerized Traders. Santa Fe Institute, New Mexico (1989)
11. Phelps, S., Marcinkiewicz, M., Parsons, S., McBurney, P.: Using population-based search and evolutionary game theory to acquire better-response strategies for the double-auction market. In: Proceedings of IJCAI 2005 Workshop on Trading Agent Design and Analysis, TADA 2005 (2005)
12. Walia, V., Byde, A., Cliff, D.: Evolving Market Design in Zero-Intelligence Trader Markets. In: IEEE International Conference on E-Commerce (IEEE-CEC 2003), Newport Beach, CA., USA (2003)
13. Pareto, V.: Cours d'economie politique. Rouge, Lausanne (1897)
14. Matsumoto, M., Nishimura, T.: Mersenne Twister: A 623-dimensionally equidistributed uniform pseudorandom number generator. ACM Trans. on Modeling and Computer Simulation 8(1), 3–30 (1998)

Customer Lifetime Value and Defection Possibility Prediction Model Using Machine Learning: An Application to a Cloud-Based Software Company

Niken Prasasti[1,2], Masato Okada[2], Katsutoshi Kanamori[2], and Hayato Ohwada[2]

[1] School of Business and Management, Bandung Institute of Technology, Indonesia
[2] Department of Industrial Administration Department, Tokyo University of Science, Japan
niken.prasasti@sbm-itb.ac.id, okada@ohwada-lab.net,
{katsu,ohwada}@rs.tus.ac.jp

Abstract. This paper proposes an estimation of Customer Lifetime Value (CLV) for a cloud-based software company by using machine learning techniques. The purpose of this study is twofold. We classify the customers of one cloud-based software company by using two classifications methods: C4.5 and a support vector machine (SVM). We use machine learning primarily to estimate the frequency distribution of the customer defection possibility. The result shows that both the C4.5 and SVM classifications perform well, and by obtaining frequency distributions of the defection possibility, we can predict the number of customers defecting and the number of customers retained.

Keywords: Customer Lifetime Value, Machine Learning, Cloud-based Software, C4.5, SVM.

1 Introduction

Activities in a competitive market mostly depend on customers, so Customer Relationship Management (CRM) has been an essential business strategy. The main purpose of CRM is to accommodate the relationship with the customer so that it lasts and remains profitable. Over the past decade, a concept of CRM for measuring the achievement of CRM programs called Customer Lifetime Value (CLV) has emerged. CLV is commonly defined as the present value of all future profits earned from a customer through his or her relationship with a company.

Various models for calculating CLV have been generated by many researchers. Many companies and business sectors apply CLV calculation in order to succeed in their CRM programs, especially mobile telecommunication and internet service providers [1][2][3]. Some of the studies and calculations only consider past customer data, while others take their future behavior into account [4]. In this paper, we apply and estimate CLV calculations for a cloud-based software industry using machine-learning techniques.

Cloud-based software is distinct from other software or applications in that it "lives" on the Internet. By definition, the cloud is the delivery of computing services,

N.T. Nguyen et al. (Eds.): ACIIDS 2014, Part II, LNAI 8398, pp. 62–71, 2014.

such as storage or software, over the Internet as opposed to those services being hosted on a user's computer or provided within a local network [5]. It is commonly characterized by advanced features through which data is stored in a cloud infrastructure and can be accessed from a web browser or custom built applications installed on Internet-connected devices [6]. Some common examples include web-based file hosting services, social networking services, e-mail services, and cloud antivirus software.

The purpose of this study is twofold. First, machine-learning techniques are used to classify the customers of a cloud-based software company. We use two classification methods: C4.5 and a support vector machine (SVM). The performance results of both classifiers will be measured for accuracy, precision, recall, and F1 score.

Second, the frequency distribution of the customer defection possibility will be assessed. We examine the distribution of the customer defection rate based on the SVM result. From this distribution, we can determine whether the frequency distribution is appropriate in predicting the frequency of customers who will defect and be retained.

The remainder of this paper is organized as follows. Section 2 reviews the problem description and the customer lifetime value model and calculation. Section 3 presents the methodology used in doing this research. Section 4 defines the data description and data preparation methods for machine learning. Section 5 briefly presents the machine-learning procedures. The results of our experiments are provided in Section 6. Finally, a conclusion is provided in the last section.

2 Problem Description

We focused on the customer defection problem for a cloud-based software provider as our main distinction from previous research. For a mobile or telecommunications provider, a given customer may only be registered to one provider, while for a cloud-based software provider, it is possible for a given customer to have more than one product from more than one provider. For instance, one personal customer can have more than one web-based file hosting service account or can have more than one different cloud antivirus program available at one time.

The case study company offers three main products that vary by product price; these will be defined as LP (low-price), MP (middle-price), and HP (high-price). The company has an e-commerce site that sends a confirmation of auto-renewal e-mail to each customer at least twice between zero days and fifty days before their renewal time.

The option for the customer is to "opt-in" or "opt-out". If the customer chooses to opt-in, this indicates positively that they would like to be contacted with a particular form, in this case with a renewal form. In contrast, choosing opt-out indicates that they would prefer not to be in, or in other words it is a form of defection. There is a one-year contract between the customer and the company. Figure 1 illustrates the number of customers that chose the opt-out option by the time the e-commerce site sent the notification e-mail.

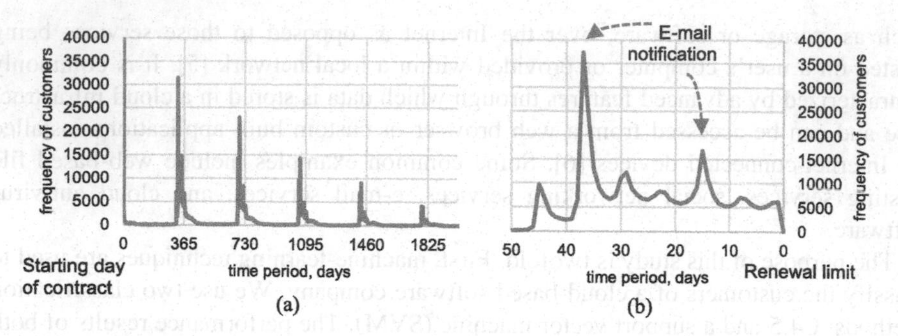

Fig. 1. Customer defection (a) in the long term (b) in the short term

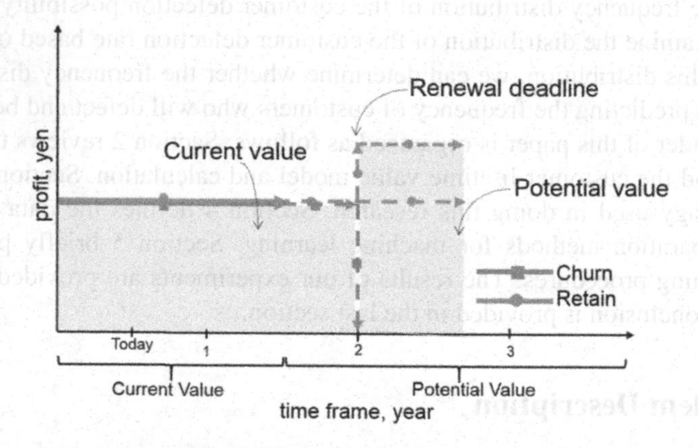

Fig. 2. Customer lifetime value

This paper calculates the lifetime value given in Eq. (1) by summing the customer lifetime value of each product with a different product ID (PID).

$$\text{CLV}=\sum_{PID}^{n} CLV(PID) \tag{1}$$

Figure 2 represents the customer lifetime value by its timeframe. We define the current value as the profit that the company gains from a customer over a certain period; in this case the current time is set to two years. Given the importance of considering up-selling probability in calculating the customer value, we define the potential customer value as the expected profits when that customer continues using the services during the current time period and starts using additional services from the company.

Customers may either churn or retain. The current value of a churning customer is the total profit for one year, while the potential value of these customers is zero. In contrast, a retained customer has a current value based on the total profit for two years, and a potential value based on the total profit summed with the up-selling profit for one year.

The sum of a customer's current value and potential value is defined as that customer's life time value (Eq. (2)).

$$\text{CLV} = \text{Current Value} + \text{Potential Value} \tag{2}$$

This paper calculates the risk of each customer using machine learning. We include risk as a variable in calculating the potential value, as defined in Eq. (3). The potential value can be defined by timing the profit to the possibility of retaining the customer. In particular, the risk is defined as customer defection.

$$Potential\ Value = (1 - \text{Risk}) x Profit \tag{3}$$

3 Methodology

One problem in the data is that some customers tend to opt-in for another product from the same company after they opt-out from the previous one (which should not be defined as defection), while the e-commerce site is only able to record the opt-out data. Therefore, data preparation is quite important in this research for classifying which data represent real defections.

This paper proceeds in the following steps. We collect data from the company's e-commerce site and then detect customer defection based on the data. We next construct a data preparation algorithm to determine which data represent real defections and prepare the training data for machine learning using the C4.5 and SVM algorithms. The misclassification rate and the ROC curve provide a comparative test of the data mining techniques used. Finally, we model an estimation method for defection possibility using frequency distribution. The flowchart of this research methodology is presented in Fig 3.

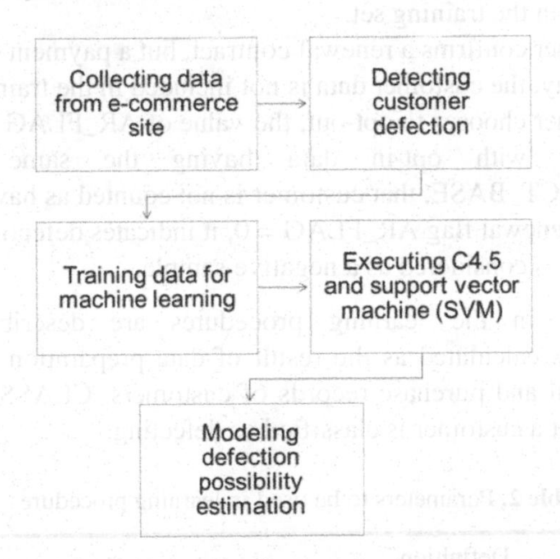

Fig. 3. Flowchart of research methodology

4 Data Preparation for Machine Learning

The study company is a cloud-based software provider. The dataset used in this paper is a six-year record of customer consent data (auto-renewal records) for 2007 through 2013. Before applying the prediction models, we prepared the data.

Table 1. Selected variables in data preparation

Variable	Definition
AR_KEY	Serial key of product
AR_FLAG	Latest renewal flag
OPTIN_DATE	Date of auto-renewal contract
AR_COUNT	Total count of renewal
CC_PRODUCT_ BASE	Product base
CC_PRODUCT_PRICE	Product price
CC_SUBS_DAY	Validity period of product
OPTIONAL_FLAG	Whether customer used optional service flag
ORG_FLAG	Type of customer, whether personal or company
MAIL_STATUS	Delivery status of e-mail

The recorded data are given in a trade log table named ConsentCapture. This table contains the pattern of cancellation of customers after they choose the opt-out option. The main purpose of this data preparation is to predict, and later to classify, which customer records represent actual defections.

The variables used in this data preparation method are listed in Table 1. There are five rules in data preparation:

Rule 1. If today occurs before the renewal deadline (see Fig. 2), the customer data is not included in the training set.

Rule 2. If the customer confirms a renewal contract, but a payment record is not identified by today, the customer data is not included in the training set.

Rule 3. If the customer chooses to opt-out, the value of AR_FLAG = 1, but if there is a customer with opt-in data having the same AR_KEY and CC_PRODUCT_BASE, that customer is not counted as having defected.

Rule 4. If the latest renewal flag AR_FLAG = 0, it indicates defection.

Rule 5. Otherwise, it is considered as a negative sample.

The variables used in the learning procedures are described in Table 2. UPDATE_COUNT is calculated as the result of data preparation and describes the total count of renewal and purchase records of customers. CLASS is a variable that defines whether or not a customer is classified as defecting.

Table 2. Parameters to be used in learning procedures

Variables	Definition
UPDATE_COUNT	Total count of renewals and purchases (first purchase is excluded)
CC_PRODUCT_PRICE	Recently purchased product price
OPTIONAL_FLAG	Whether customer used optional service flag
ORG_FLAG	Type of customer, whether personal or company
MAIL_STATUS	Delivery status of e-mail
CLASS	Type of customer (defecting or retained)

5 Machine-Learning Procedures

Machine-learning procedures in the form of classifiers are established with two machine-learning algorithms: a C4.5 decision tree and SVM. The C4.5 technique is based on the ID3 tree learning algorithm and is able to pre-define discrete classes from labeled examples [8]. The C4.5 algorithm generates a decision tree through learning from a training set. A decision tree's efficiency is heavily impacted by a large dataset.

SVM is essentially a binary classifier [7]. While applying SVM, we use the software package LIBSVM with –s svm and –t kernel type. We used epsilon-SVR in –s svm. Since the function expected a regression and classifier, the machine learned a value of 0 or 1, and the output of this process was continuous between 0 and 1. A radial basis function (RBF) is applied in the –t kernel type.

Parameters C (a penalty for misclassification) and gamma (a function of the deviation of the kernel) were determined by using different values of C and gamma in a 10-fold cross-validation of the training set. Other parameters are set to default values. We used a grid search to test various parameters and check their accuracy, recall, precision, and F1 score. As a result, we discovered better parameters that have higher F1 scores.

Table 3 lists the number of positive and negative examples used in each machine-learning procedure. The result from the classifier model is first measured by its accuracy, precision, recall, and F1 scores. Next, the SVM result is employed to produce an ROC graph and a frequency-distribution graph.

Table 3. Number of positive and negative samples used

	C4.5		SVM	
	Positive	Negative	Positive	Negative
LP	13709	5302	11778	7243
MP	8013	1764	6824	2953
HP	10961	2265	7699	5527

6 Results

6.1 Customer Classification

Table 4 compares the accuracy, recall, precision, and F1 scores of two classifiers based on the proposed methodology for three products. The table presents experiment results for all 10-fold cross validations. It can be safely concluded that no single model had the highest accuracy in all three products. Instead, the performance of every algorithm differed, depending on the characteristics and type of the data.

Table 4. Comparison of classification accuracies between C4.5 and SVM

Product	Classifier	Accuracy	Recall	Precision	F1 Score
LP	C.45	72.12%	83.91%	74.10%	78.71%
	SVM	82.61%	86.88%	88.05%	87.46%
MP	C.45	81.95%	85.80%	88.14%	86.96%
	SVM	72.68%	84.86%	74.54%	79.37%
HP	C.45	82.87%	76.39%	92.87%	84.15%
	SVM	82.94%	76.85%	92.58%	83.99%

6.2 ROC Curves

We constructed Receiver Operating Characteristics (ROC) curves from the results of both classifiers. The ROC curve is a useful technique for organizing classifiers and visualizing their performance [7]. ROC curves are plotted on two-dimensional graphs in which the True Positive (TP) rate is plotted on the Y axis and the False Positive (FP) rate is plotted on the X axis. The graphs in Fig. 4 plot two trend lines, the defecting rate and the random performance line. The random performance line (y=x) represents the strategy of randomly guessing a class. As we can see from Fig. 4, LP has the lowest true defecting rate with both classifiers. This is basically linked to the fact that, based on these three products, a customer that has subscribed for LP tends to move and choose HP in the end.

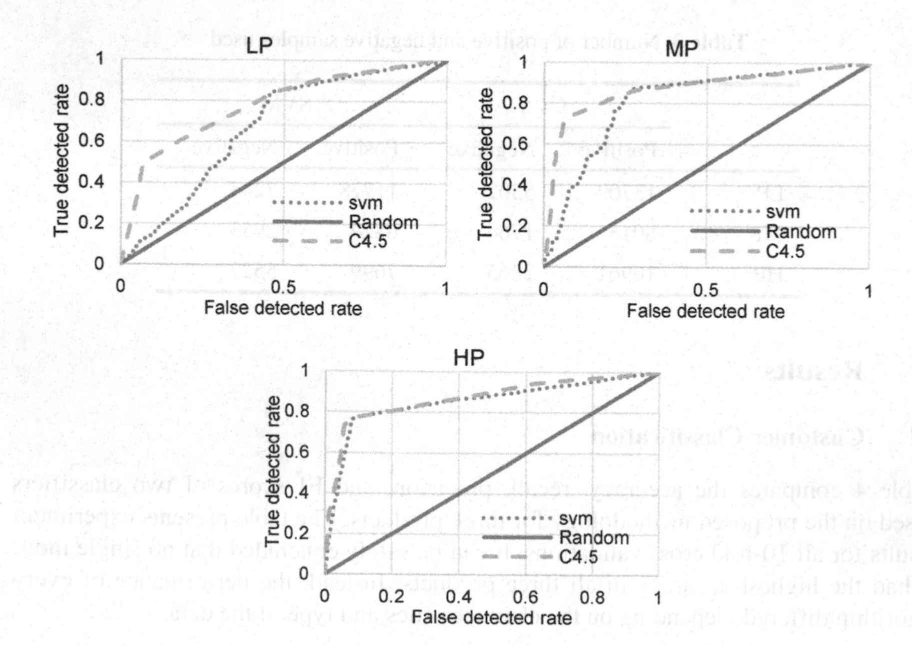

Fig. 4. ROC curves for each product

In all of the graphs, it is clear that the defecting rate trend line for each service appears in the upper left triangle, above the random trend line. The C4.5 and SVM classifiers thus both have good performance, as indicated by the defecting rate plot being skewed to the upper-left side.

6.3 Frequency Distribution

We obtained frequency distribution graphs that illustrate the distribution of the defection possibility. We set the threshold at 0.5: a customer who defects will have a defection possibility higher than 0.5; a customer who is retained will have a defection problem possibility lower than 0.5. Referring to Fig. 5, there is a clear distinction between the customer defection and customer retention distributions, which indicates that this frequency distribution has a positive impact on defection prediction, and that it is advantageous for predicting the frequency of customer defection and customer retention.

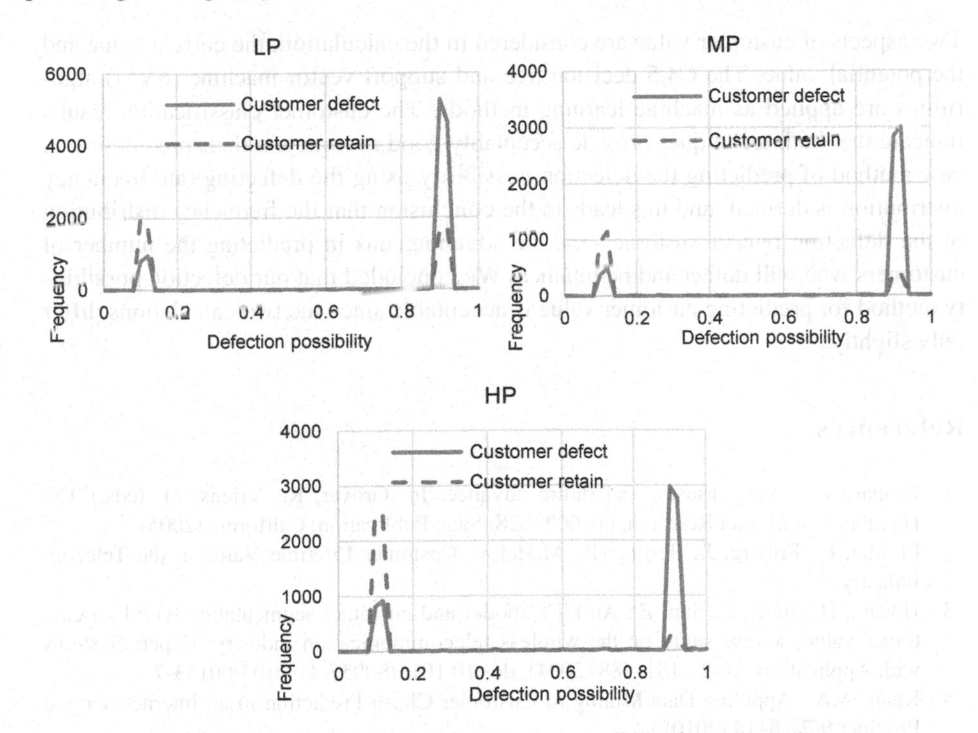

Fig. 5. Frequency distribution of defection possibility for each product

6.4 Customer Lifetime Value Estimation

The defection possibility is defined as risk in Eq. (3). We applied Eq. (3) and calculated the customer value based on the machine-learning result. In Fig. 6, each different height represents the percentage of the potential and current value relative to the total profit. We can see from the chart that LP gives the greatest potential value relative to total profit.

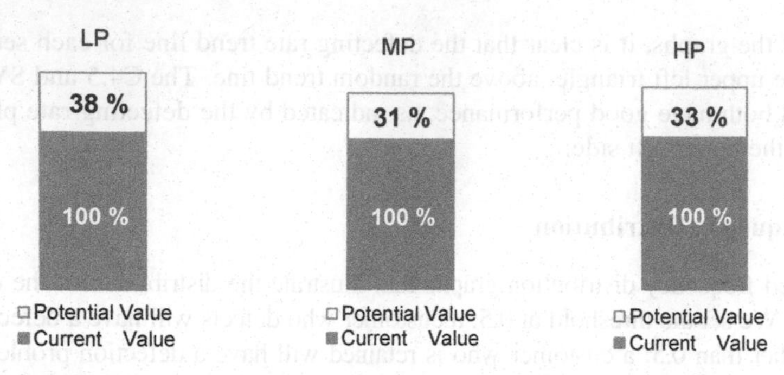

Fig. 6. Customer lifetime value for each product

7 Conclusion

Two aspects of customer value are considered in the calculation: the current value and the potential value. The C4.5 decision tree and support vector machine (SVM) algorithms are applied as machine learning methods. The customer classification results indicate that both techniques provide acceptably good accuracy in their predictions. A new method of predicting the defection possibility using the defecting rate frequency distribution is defined, and this leads to the conclusion that the frequency distribution of the defecting rate of customers can be advantageous in predicting the number of customers who will defect and be retained. We concluded that our defection possibility method for predicting customer value is acceptable since the two calculations differ only slightly.

References

1. Kumar, V.: Uses, misuses, and future advance. In: Grover, R., Vriens, M. (eds.) The Handbook of Market Research, pp. 602–628. Sage Publication, California (2005)
2. Flordal, P., Friberg, J., Berling, P.: Modeling Customer Lifetime Value in the Telecom Industry
3. Hwang, H., Jung, T., Suh, E.: An LTV model and customer segmentation based on customer value: a case study on the wireless telecommunication industry. Expert Systems with Applications 26(2), 181–188 (2004), doi:10.1016/S0957-4174(03)00133-7
4. Khan, A.A.: Applying Data Mining to Customer Churn Prediction in an Internet Service Provider 9(7), 8–14 (2010)
5. What are Cloud-Based Services? (2012), `http://www.colorado.edu/oit/node/7532` (retrieved)
6. Abubakr, T.: Cloud app vs. web app: Understanding the differences (2012), `http://www.techrepublic.com/blog/the-enterprise-cloud/cloud-app-vs-web-app-understanding-the-differences/5478/` (retrieved)

7. Fawcett, T.: ROC Graphs: Notes and Practical Considerations for Data Mining Researchers (2003)
8. Szarvas, G., Farkas, R., Kocsor, A.: A Multilingual Named Entity Recognition System Using Boosting and C4.5 Decision Tree Learning Algorithms. In: Todorovski, L., Lavrač, N., Jantke, K.P. (eds.) DS 2006. LNCS (LNAI), vol. 4265, pp. 267–278. Springer, Heidelberg (2006)
9. Huang, J., Lu, J., Ling, C.X.: Comparing naive Bayes, decision trees, and SVM with AUC and accuracy. In: Third IEEE International Conference on Data Mining, pp. 553–556 (2003), doi:10.1109/ICDM.2003.1250975
10. Kotsiantis, S.B.: Supervised Machine Learning: A Review of Classification Techniques 31, 249–268 (2007)

Mining Class Association Rules with the Difference of Obidsets

Loan T.T. Nguyen

VOV College, Ho Chi Minh City, Vietnam
nguyenthithuyloan@vov.org.vn

Abstract. In 2013, an efficient algorithm for mining class association rules, named CAR-Miner, has been proposed. It, however, still consumes much memory in storing Obidsets of itemsets and time in computing the intersection between two Obidsets. In this paper, we propose an improved algorithm for mining class association rules by using the difference between two Obidsets (d2O). Firstly, the d2O concept is developed. After that, a strategy for reducing the storage space and fast computing d2O is also derived. Experimental results show that the proposed algorithm is more efficient than CAR-Miner.

Keywords: classification, class association rules, data mining, difference of Obidsets.

1 Introduction

The rule-based classification is one of important methods in classification systems because it can give a high accuracy and interpretation. Some methods for mining rules used for the classification are decision tree [11-12], ILA [16-17], and associative classification (AC) [5-6]. Some studies showed that the classification based on AC is more accurate than decision tree and ILA [5, 18-20]. Therefore, a lot of algorithms for AC have been developed in recent years. They can be divided into two types. The first type considers the accuracy [1-6, 9-10, 13-15, 18-20, 22-25]. In these studies, the authors have developed some approaches to prune rules and predict the class of an unknown case based on the built classifiers. The second type is interested in the mining time of classification association rules (CARs) and the method for fast pruning redundant rules (the rule set after pruning is called pCARs) [7-8, 21].

This paper studies the solution for efficiently mining CARs in large datasets. Recently, CAR-Miner is one of efficient algorithms for mining CARs. CAR-Miner utilizes the MECR-tree and the technique divide-and-conquer in which the problem is transformed into sub-problems. However, the size of Obidsets (Object identifiers set) grows quickly in large datasets which means it needs much memory for Obidset storages and also takes much time to intersect two Obidsets. Consequently, reducing memory consumption and run time of CAR-Miner is an important issue needed to be solved. Therefore, this paper proposes an improved version for CAR-Miner. Instead of storing the whole intersection of two Obidsets, we store only the difference of

N.T. Nguyen et al. (Eds.): ACIIDS 2014, Part II, LNAI 8398, pp. 72–81, 2014.

Obidsets. A strategy for reducing the storage space of d2O (difference between two Obidsets) is also provided. The experimental results show that the proposed algorithm is superior to CAR-Miner.

2 Related Work

As mentioned above, the rule-based classifications are very efficient in terms of accuracy and interpretation. Thus, numerous rule-based classification methods have been developed recently. In 1986, Quinlan proposed the decision tree [11]. The algorithm ID3 which uses the information gain (IG) measure was also introduced. ID3 selects the attribute whose the highest IG to be the root node and partitions unclassified data into subsets of data. Each subset is computed recursively until none subset is generated. In 1992, C4.5 [12] was proposed with some improvements over ID3. C4.5 uses the Gain Ratio measure to handle data with missing attribute values. ILA, an inductive algorithm, was proposed in 1998 by Tolun et al. [16]. ILA often has higher accuracy than those of ID3 and C4.5.

C4.5 and ILA which are based on heuristics and greedy approaches generate rule sets that are either too general or too overfitting for a given dataset. They thus often yield high error ratios. Recently, a new method for classification, called the Classification Based on Associations (CBA) [5], has been proposed for mining class association rules (CARs). This method has more advantages than the heuristic and greedy methods in that the former could easily remove noise, and the accuracy is thus higher. It can additionally generate a rule set that is more complete than C4.5 and ILA. One of the weaknesses of mining CARs is that it consumes more time than C4.5 and ILA because it has to check the generated rules with the set of the other rules. At the result, lots of methods have been developed to reduce mining time and memory consumption of CBA. The first method for mining CARs was proposed by Liu et al. in 1998 [5]. In the paper, the authors proposed CBA-RG, an Apriori-based algorithm, for mining CARs. An algorithm for building the classification (CBA-CB) based on mined CARs was also proposed. Li et al. proposed a method, called CMAR, based on the FP-tree in 2001 [4]. CMAR uses an FP-tree to compress the dataset. It also uses the CR-tree to store rules. To predict unseen data, this method finds all rules that satisfy this data and uses a weighted $\chi 2$ measure to determine the class. Thabtah et al. proposed the method MMAC in 2004 [15]. MMAC uses multi-labels for each rule and multi-classes for prediction. Vo and Le [21] presented the algorithm ECR-CARM for quickly mining CARs in 2008. CAR-Miner, an enhanced version for ECR-CARM, has been developed in 2013 [8]. Some methods for pruning and sorting rules were also mentioned in [7, 24-25].

3 Definitions

Let D be the set of training data with n attributes $A_1, A_2, \ldots, A_n$ and $|D|$ objects (cases). Let $C = \{c_1, c_2, \ldots, c_k\}$ be a list of class labels. Specific values of attribute A_i and class C are denoted by lower case a and c, respectively.

Definition 1: An itemset is a set of some pairs of attributes and a specific value, denoted $\{(A_{i1}, a_{i1}), (A_{i2}, a_{i2}), \ldots, (A_{im}, a_{im})\}$.

Definition 2: A class association rule r has the form of $\{(A_{i1}, a_{i1}), \ldots, (A_{im}, a_{im})\} \rightarrow c$, where $\{(A_{i1}, a_{i1}), \ldots, (A_{im}, a_{im})\}$ is an itemset, and $c \in C$ is a class label.

Definition 3: The actual occurrence $ActOcc(r)$ of a rule r in D is the number of rows of D that match r's condition.

Definition 4: The support for r, denoted $Sup(r)$, is the number of rows of D that match r's condition and also belong to r's class.

Table 1. An example of training dataset

OID	*X*	*Y*	*Z*	*class*
1	*x1*	*y1*	*z1*	*1*
2	*x1*	*y2*	*z1*	*2*
3	*x2*	*y2*	*z1*	*2*
4	*x3*	*y3*	*z1*	*1*
5	*x3*	*y1*	*z2*	*2*
6	*x3*	*y3*	*z1*	*1*
7	*x1*	*y3*	*z2*	*1*
8	*x2*	*y2*	*z2*	*2*

For example, consider the rule $r = \{(X, x3)\} \rightarrow 1$ from the dataset in Table 1. We have $ActOcc(r) = 3$ and $Sup(r) = 2$ because there are three objects with $X = x3$, in that two objects have the same class 1.

4 Proposed Method

4.1 Tree Structure

In this section we describe the MECR-tree [8], a tree structure is used to efficiently mine CARs. Each node in the tree contains one itemset along with the following information:

a) *Obidset*: a set of object identifiers that contain the itemset.
b) count = $(\#c_1, \#c_2, \ldots, \#c_k)$ – where $\#c_i$ is the number of records in *Obidset* which belong to class c_i, and
c) pos – store the position of the class with the maximum count, i.e., pos = $\arg\max_{i \in [1,k]}\{c_i\}$

For example, consider the node containing itemset $R = \{(X,x3), (Y,y3)\}$. Because R is contained objects 4 and 6, where all of them belong to class 1, we have a node in the tree as $\underset{46(\underline{2},0)}{\{(X,x3),(Y,y3)\}}$ or more simply as $\underset{46(\underline{2},0)}{3 \times x3y3}$. The pos is 1 (underlined at position 1 of this node) because the count of class *1* is at a maximum (2 as compared to 0). For more details on the MECR-tree, readers can see in [8].

Theorem 1 [8]: Given two nodes $\frac{att_1 \times values_1}{Obidset_1(c_{11},...,c_{1k})}$ and $\frac{att_2 \times values_2}{Obidset_2(c_{21},...,c_{2k})}$, if $att_1 = att_2$ and $values_1 \neq values_2$, then $Obidset_1 \cap Obidset_2 = \varnothing$.

4.2 Difference between Two Obidsets

In [8], CAR-Miner computes the intersection between two Obidsets and uses this value to calculate the *count* information for each node in the tree. In this section, we present a different method for computing the *count* information by using the difference between two Obidsets.

Definition 5 (The difference between two Obidsets): Given two nodes X and Y containing k-itemsets have the same prefix (k-1)-itemset in MECR-tree in which *O1* is *Obidset* of *X* and *O2* is *Obidset* of *Y*. Assume that the generated node from these two nodes is Z. Let d2O(Z) be the difference between *O1* and *O2*, we have:

$$d2O(Z) = O1 \setminus O2$$

Example 1: Consider the dataset shown in Table 1. Let $X = \frac{1 \times x2}{38(0,\underline{2})}$ and $Y = \frac{2 \times y2}{238(0,\underline{3})}$, let *Z* be the generated node from *X* and $Y \Rightarrow$ d2O(Z) = $O1 \setminus O2 = \{3, 8\} \setminus \{2, 3, 8\} = \varnothing$.

Theorem 2: Given two nodes *XY* and XZ containing k-itemsets have the same prefix (k-1)-itemset in MECR-tree, we have:

$$d2O(XYZ) = d2O(XY) \setminus d2O(XZ) \quad (1)$$

Proof: We have d2O(*XYZ*) = *Obidset*(*XY*) \ *Obidset*(*XZ*). According to the definition of Obidset, we also have *Obidset*(*XY*) = *Obidset*(*X*) ∩ *Obidset*(*Y*) and *Obidset*(*XZ*) = *Obidset*(*X*) ∩ *Obidset*(*Z*). Therefore, d2O(*XYZ*) = [*Obidset*(*X*) ∩ *Obidset*(*Y*)] \ [*Obidset*(*X*) ∩ *Obidset*(*Z*)] = {*Obidset*(*X*) \ [*Obidset*(*X*) \ *Obidset*(*Y*)]} \ {*Obidset*(*X*) \ [*Obidset*(*X*) \ *Obidset*(*Z*)]} = [*Obidset*(*X*) \ *Obidset*(*X*)] ∪ {[*Obidset*(*X*) \ *Obidset*(*Z*)] \ [*Obidset*(*X*) \ *Obidset*(*Y*)]} = [*Obidset*(*X*) \ *Obidset*(*Z*)] \ [*Obidset*(*X*) \ *Obidset*(*Y*)] = d2O(*XZ*) \ d2O(*XY*).

Example 2: Consider the dataset shown in Table 1. Let $X = \frac{1 \times x2}{38(0,\underline{2})}$ and $Y = \frac{2 \times y2}{238(0,\underline{3})}$, let *Z* be the generated node from *X* and $Y \Rightarrow$ d2O(*3×x2y2*) = ∅. Similarly, we have d2O(*5×x2z1*) = {8}. At the result, d2O of *7×x2y2z1* can be calculated as follows:

$$d2O(Z) = d2O(Y) \setminus d2O(X) = \{8\} \setminus \{\varnothing\} = \{8\}.$$

To compute the *count* information of node *XYZ*, we use d2O(*XYZ*) as follows:

XYZ.count = *XY*.count – $\{\#c_i | \#c_i$ = the number of records in d2O(*XYZ*) which belong to class $c_i\}$ (2)

4.3 Algorithm

Based on above definitions, theorems, and two formulas (1)-(2), an efficient algorithm for mining CARs is proposed in this section. The main advantage of the proposed algorithm is that it uses d2O instead of Obidset to reduce storage space and time of computing the intersection between two Obidsets.

Figure 1 shows the algorithm for mining CARs using d2O. First, the root node contains single items (1-itemsets) which satisfy the *minSup* (Line 2). Then, the procedure **CAR-Miner-Diff-Recursive** is called with parameters L_r, *minSup*, and *minConf* to discover CARs from *D* which satisfy *minSup* and *minConf*.

```
Input: A dataset D, minSup, and minConf
Output: all CARs satisfy minSup, and minConf
Procedure:
CAR-Miner-Diff(Lr, minSup, minConf)
1.  CARs = ∅;
2.  Lr = {Single items whose supports satisfy minSup}
3.  Call CAR-Miner-Diff-Recursive(Lr, minSup, minConf)
4.  return CARs
CAR-Miner-Diff-Recursive(Lr, minSup, minConf)
5. for all li ∈ Lr.children do
6.    ENUMERATE-CAR(li, minConf )
7.    Pi = ∅;
8.    for all lj ∈ Lr.children, with j > i do
9.     if li.att ≠ lj.att then // by theorem 1
10.      O.att = li.att ∪ lj.att;
11.      O.itemset = li.values ∪ lj.values;
12.      if Lr = null then
13.        O.Obidset = li.Obidset \ lj.Obidset;
14.      else
15.        O.Obidset = li.Obidset \ lj.Obidset;
16.      O.count = li.count - {count(x ∈ O.Obidset |
   class(x) = ci, ∀i∈[1,k]};
17.      O.pos = arg max_{i∈[1,k]} {l.count_i} ;
18.      O.total = li.total - |O.Obidset|;
19.      if O.count[O.pos] ≥ minSup then
20.        Pi = Pi ∪ O ;
21.      CAR-Miner-Diff-Recursive(Pi, minSup, minConf)
ENUMERATE-CAR-Diff(l, minConf )
22. conf = l.count[l.pos] / l.total;
23. if conf ≥ minConf then
       CARs = CARs∪{l.itemset→c_pos(l.count[l.pos],conf)}
```

Fig. 1. CAR-Miner-Diff algorithm

The **CAR-Miner-Diff-Recursive** procedure (Figure 1) considers each node l_i with all other nodes l_j in L_r, with j > i (Lines 5 and 8) to generate a candidate child node *O*. With each pair (l_i, l_j), the algorithm checks whether l_i.att ≠ l_j.att or not (Line 9, using Theorem 1). If they are different, it computes the elements *att*, *values* for the new node *O* (Lines 10-11). Line 12 checks whether L_r is the root node. If this is true, then O.*Obidset* = l_i.*Obidset* \ l_j.*Obidset* regarding Definition 5 (Line 13); otherwise, O.*Obidset* = l_j.*Obidset* \ l_i.*Obidset* by Theorem 2 (Line 15). The algorithm computes the *O*.count by using l_i.count and *O*.*Obidset* (Line 16) and *O*.pos (Lines 17). After computing all of the information for node *O*, the algorithm adds it to P_i (P_i is initialized empty in Line 4) if *O*.count[*O*.pos] ≥ *minSup* (Lines 19-20). Finally, **CAR-Miner-Diff-Recursive** is recursively called with a new set P_i as its input parameter (Line 21).

The procedure **ENUMERATE-CAR-Diff** (*l*, *minConf*) generates a rule from node *l*. It firstly computes the confidence of the rule (Line 22), if the confidence of this rule satisfies *minConf* (Line 23), then it adds this rule into the set of CARs (Line 24). Note that when using d2O, |*l.Obidset*| is the cardinality of d2O instead of Obidset. Thus, we use a variable *total* to save the cardinality of Obidset. This variable is computed through the *total* of the parent node l_i (Line 18).

4.4 An Example

This section illustrates the process of **CAR-Miner-Diff** with *minSup* = 10% and *minConf* = 60%. Figure 2 shows the result of this process.

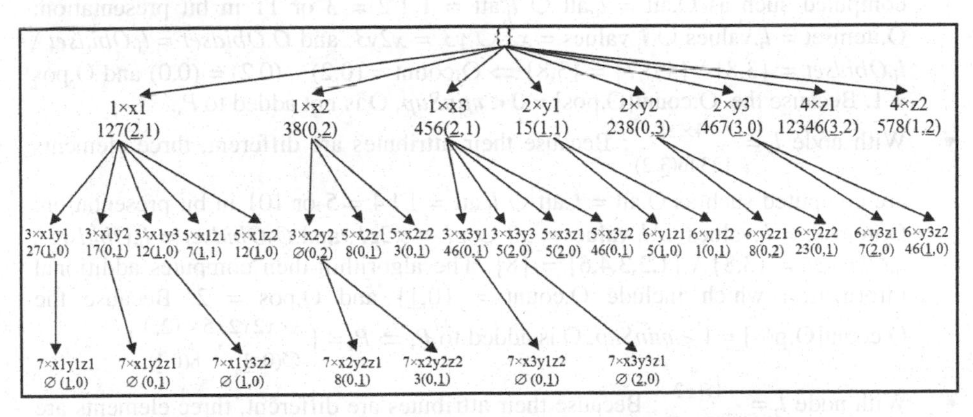

Fig. 2. MECR-tree for the dataset shown in Table 1 by using d2O

MECR-tree is built from the dataset shown in Table 1 as follows. Firstly, the root node L_r contains all frequent 1-itemsets such as $\begin{Bmatrix} 1\times x1 & 1\times x2 & 1\times x3 & 2\times y1 & 2\times y2 & 2\times y3 & 4\times z1 & 4\times z2 \\ 127(\underline{2},1) & 38(0,\underline{2}) & 456(\underline{2},1) & 15(\underline{1},1) & 238(0,\underline{3}) & 467(\underline{3},0) & 12346(\underline{3},2) & 578(1,\underline{2}) \end{Bmatrix}$. Secondly, the procedure **CAR-Miner-Diff** is called with the parameter (L_r, 1, 0.6).

Node $l_i = \frac{1\times x2}{38(0,\underline{2})}$ is used as an example for illustrating the **CAR-Miner-Diff** process. l_i joins with all other nodes following it in L_r:

- With node $l_j = \frac{1\times x3}{456(\underline{2},1)}$: They (l_i and l_j) have the same attribute but different values. According to Theorem 1, nothing is done from them.
- With node $l_j = \frac{2\times y1}{15(\underline{1},1)}$: Because their attributes are different, three elements are computed such as O.att = l_i.att ∪ l_j.att = 1 | 2 = 3 or 11 in bit presentation; O.itemset = l_i.values ∪ l_j.values = *x2* ∪ *y1* = *x2y1*, and *O.Obidset* = l_i.*Obidset* \ l_j.*Obidset* = {3,8} \ {1,5} = {3, 8} ⇒ O.count = (0,2) – (0,2) = (0,0) and O.pos = 1. Because O.count[O.pos] = 0 < *minSup*, O is not added to P_i.
- With node $l_j = \frac{2\times y2}{238(0,\underline{3})}$: Because their attributes are different, three elements are computed such as O.att = l_i.att ∪ l_j.att = 1 | 2 = 3 or 11 in bit presentation; O.itemset = l_i.values ∪ l_j.values = *x2* ∪ y2 = *x2y2*, and *O.Obidset* = l_i.*Obidset* \ l_j.*Obidset* = {3,8} \ {2,3,8} = {∅}. The algorithm computes additional information including O.count = (0,2) – (0,0) = (0,$\underline{2}$) and O.pos = 2. Because O.count[O.pos] = 2 > *minSup*, O is added to P_i ⇒ P_i = { $\frac{3\times x2y2}{\varnothing(0,\underline{2})}$ }.
- With node $l_j = \frac{2\times y3}{467(\underline{3},0)}$: Because their attributes are different, three elements are computed such as O.att = l_i.att ∪ l_j.att = 1 | 2 = 3 or 11 in bit presentation; O.itemset = l_i.values ∪ l_j.values = *x2* ∪ *y3* = *x2y3*, and *O.Obidset* = l_i.*Obidset* \ l_j.*Obidset* = {3,8} \ {4,6,7} = {3,8} ⇒ O.count = (0,2) – (0,2) = (0,0) and O.pos = 1. Because the O.count[O.pos] = 0 < *minSup*, O is not added to P_i.
- With node $l_j = \frac{4\times z1}{12346(\underline{3},2)}$: Because their attributes are different, three elements are computed such as O.att = l_i.att ∪ l_j.att = 1 | 4 = 5 or 101 in bit presentation; O.itemset = l_i.values ∪ l_j.values = *x2* ∪ *z1* = *x2z1*, and *O.Obidset* = l_i.*Obidset* \ l_j.*Obidset* = {3,8} \ {1,2,3,4,6} = {8}. The algorithm then computes additional information which include O.count = {0,$\underline{1}$} and O.pos = 2. Because the O.count[O.pos] = 1 ≥ *minSup*, O is added to P_i ⇒ P_i = { $\frac{3\times x2y2}{\varnothing(0,\underline{2})}$, $\frac{5\times x2z1}{8(0,\underline{1})}$ }.
- With node $l_j = \frac{4\times z2}{578(1,\underline{2})}$: Because their attributes are different, three elements are computed such as O.att = l_i.att ∪ l_j.att = 1 | 4 = 5 or 101 in bit presentation; O.itemset = l_i.values ∪ l_j.values = *x2* ∪ z2 = *x2z2*, and *O.Obidset* = l_i.*Obidset* \ l_j.*Obidset* = {3,8} \ {5,7,8} = {3}. The algorithm computes additional information including O.count = {0,$\underline{1}$} and O.pos = 2. Because the O.count[O.pos] = 1 ≥ *minSup*, O is added to P_i ⇒P_i ={ $\frac{3\times x2y2}{\varnothing(0,\underline{2})}$, $\frac{5\times x2z1}{8(0,\underline{1})}$, $\frac{5\times x2z2}{3(0,\underline{1})}$ }.
- After P_i is created, the procedure **CAR-Miner-Diff** is called recursively with parameters P_i, *minSup*, and *minConf* to create all children nodes of P_i

5 Experimental Results

The algorithms used in the experiments were coded on a personal computer with C#2008, Windows 7, Centrino 2×2.53 GHz, and 4 GB of RAM. The experiments were tested on the datasets obtained from the UCI Machine Learning Repository (http://mlearn.ics.uci.edu). Table 2 shows the characteristics of the experimental datasets.

Table 2. The characteristics of the experimental datasets

Dataset	*# of attrs*	*# of classes*	*# of distinct values*	# of Objects
German	21	2	1077	1000
Lymph	18	4	63	148
Connect	43	3	126	67557

The experimental datasets had different features. The German dataset had many attributes and distinctive (values) but had very few numbers of objects (or records), Lymph is the small dataset, Connect has many objects and attributes.

Experiments were then made to compare the execution time between CAR-Miner and CAR-Miner-Diff. The results are shown in Figures 3 to 5.

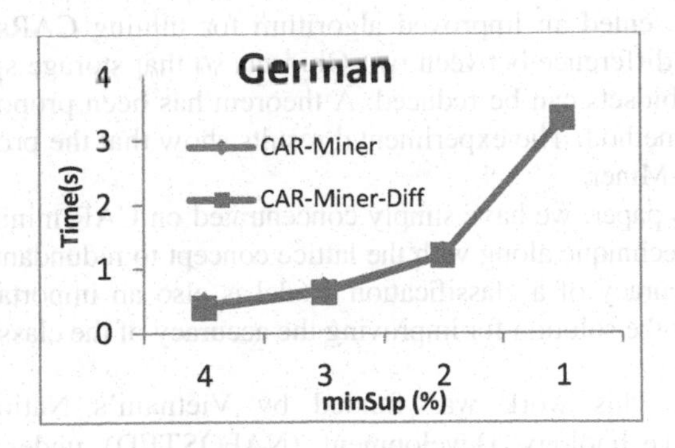

Fig. 3. The runtime of two algorithms in the German

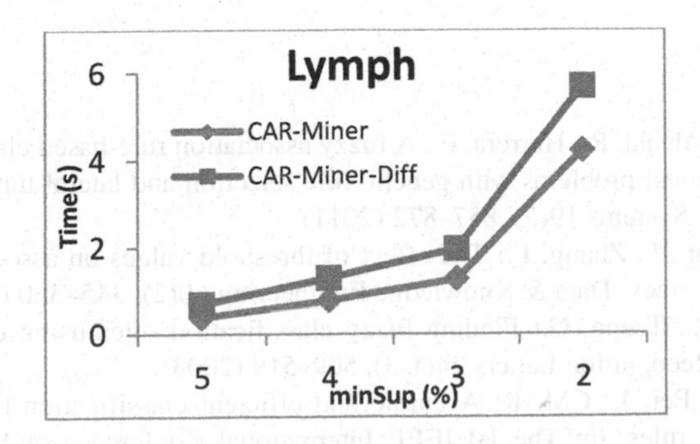

Fig. 4. The runtime of two algorithms in the Lymph

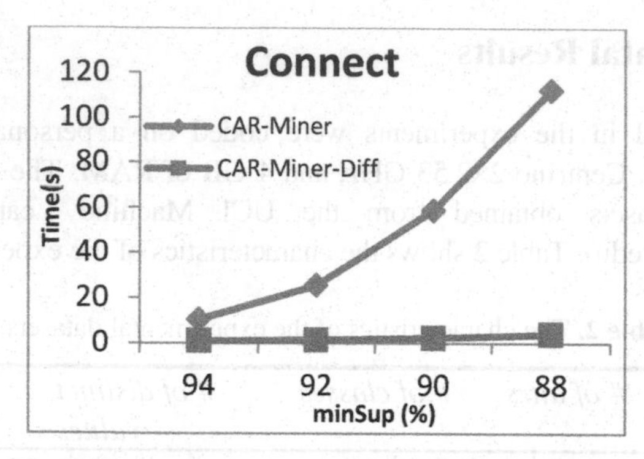

Fig. 5. The runtime of two algorithms in the Connect

Results from Figures 3 to 5 show that CAR-Miner-Diff is more efficient than CAR-Miner in large datasets. For example: Consider the Connect dataset with a *minSup* = 88%, the mining time for the CAR-Miner was 111.93(s), while that for the CAR-Miner-Diff was 3.852(s). The ratio was $\frac{3.852}{111.93} \times 100\% \approx 3.442\%$.

6 Conclusions and Future Work

This paper has presented an improved algorithm for mining CARs. The algorithm computes only the difference between two Obidsets so that storage space and time of intersecting two Obidsets can be reduced. A theorem has been proposed to prove the correctness of the method. The experimental results show that the proposed algorithm is better than CAR-Miner.

However, in this paper, we have simply concentrated on CAR mining. In the future, we will apply this technique along with the lattice concept to redundant rule pruning. In addition to, the accuracy of a classification model is also an important factor in AC. Thus, we also study the solution for improving the accuracy of the classifier.

Acknowledgment. This work was funded by Vietnam's National Foundation for Science and Technology Development (NAFOSTED) under Grant Number 102.01-2012.17.

References

1. Alcala-Fdez, J., Alcala, R., Herrera, F.: A fuzzy association rule-based classification model for high-dimensional problems with genetic rule selection and lateral tuning. IEEE Transactions on Fuzzy Systems 19(5), 857–872 (2011)
2. Coenen, F., Leng, P., Zhang, L.: The effect of threshold values on association rule based classification accuracy. Data & Knowledge Engineering 60(2), 345–360 (2007)
3. Hu, Y., Chen, R., Tzeng, G.: Finding fuzzy classification rules using data mining techniques. Pattern Recognition Letters 24(1-3), 509–519 (2003)
4. Li, W., Han, J., Pei, J.: CMAR: Accurate and efficient classification based on multiple class-association rules. In: The 1st IEEE International Conference on Data Mining, San Jose, California, USA, pp. 369–376 (2001)

5. Liu, B., Hsu, W., Ma, Y.: Integrating classification and association rule mining. In: The 4th International Conference on Knowledge Discovery and Data Mining, New York, USA, pp. 80–86 (1998)
6. Liu, B., Ma, Y., Wong, C.K.: Improving an association rule based classifier. In: Zighed, D.A., Komorowski, J., Żytkow, J.M. (eds.) PKDD 2000. LNCS (LNAI), vol. 1910, pp. 504–509. Springer, Heidelberg (2000)
7. Nguyen, T.T.L., Vo, B., Hong, T.P., Thanh, H.C.: Classification based on association rules: A lattice-based approach. Expert Systems with Applications 39(13), 11357–11366 (2012)
8. Nguyen, T.T.L., Vo, B., Hong, T.P., Thanh, H.C.: CAR-Miner: An efficient algorithm for mining class-association rules. Expert Systems with Applications 40(6), 2305–2311 (2013)
9. Pach, F., Gyenesei, A., Abonyi, J.: Compact fuzzy association rule-based classifier. Expert Systems with Applications 34(4), 2406–2416 (2008)
10. Priss, U.: A classification of associative and formal concepts. In: The Chicago Linguistic Society's 38th Annual Meeting, Chicago, USA, pp. 273–284 (2002)
11. Quinlan, J.R.: Introduction of decision tree. Machine Learning 1(1), 81–106 (1986)
12. Quinlan, J.R.: C4.5: program for machine learning. Morgan Kaufmann (1993)
13. Sun, Y., Wang, Y., Wong, A.K.C.: Boosting an associative classifier. IEEE Transactions on Knowledge and Data Engineering 18(7), 988–992 (2006)
14. Thabtah, F., Cowling, P., Hammoud, S.: Improving rule sorting, predictive accuracy and training time in associative classification. Expert Systems with Applications 31(2), 414–426 (2006)
15. Thabtah, F., Cowling, P., Peng, Y.: MMAC: A new multi-class, multi-label associative classification approach. In: The 4th IEEE International Conference on Data Mining, Brighton, UK, pp. 217–224 (2004)
16. Tolun, M.R., Abu-Soud, S.M.: ILA: An inductive learning algorithm for production rule discovery. Expert Systems with Applications 14(3), 361–370 (1998)
17. Tolun, M.R., Sever, H., Uludag, M., Abu-Soud, S.M.: ILA-2: An inductive learning algorithm for knowledge discovery. Cybernetics and Systems 30(7), 609–628 (1999)
18. Veloso, A., Meira Jr., W., Zaki, M.J.: Lazy associative classification. In: The 2006 IEEE International Conference on Data Mining (ICDM 2006), Hong Kong, China, pp. 645–654 (2006)
19. Veloso, A., Meira Jr., W., Gonçalves, M., Zaki, M.J.: Multi-label lazy associative classification. In: Kok, J.N., Koronacki, J., Lopez de Mantaras, R., Matwin, S., Mladenič, D., Skowron, A. (eds.) PKDD 2007. LNCS (LNAI), vol. 4702, pp. 605–612. Springer, Heidelberg (2007)
20. Veloso, A., Meira Jr., W., Goncalves, M., Almeida, H.M., Zaki, M.J.: Calibrated lazy associative classification. Information Sciences 181(13), 2656–2670 (2011)
21. Vo, B., Le, B.: A novel classification algorithm based on association rules mining. In: Richards, D., Kang, B.-H. (eds.) PKAW 2008. LNCS (LNAI), vol. 5465, pp. 61–75. Springer, Heidelberg (2009)
22. Yang, G., Mabu, S., Shimada, K., Hirasawa, K.: An evolutionary approach to rank class association rules with feedback mechanism. Expert Systems with Applications 38(12), 15040–15048 (2011)
23. Yin, X., Han, J.: CPAR: Classification based on predictive association rules. In: SIAM International Conference on Data Mining (SDM 2003), San Francisco, CA, USA, pp. 331–335 (2003)
24. Zhang, X., Chen, G., Wei, Q.: Building a highly-compact and accurate associative classifier. Applied Intelligence 34(1), 74–86 (2011)
25. Zhao, S., Tsang, E.C.C., Chen, D., Wang, X.Z.: Building a rule-based classifier - A fuzzy-rough set approach. IEEE Transactions on Knowledge and Data Engineering 22(5), 624–638 (2010)

Multi-Level Genetic-Fuzzy Mining with a Tuning Mechanism

Chun-Hao Chen[1], Yu Li[2], and Tzung-Pei Hong[2,3,*]

[1] Department of Computer Science and Information Engineering,
Tamkang University, Taipei, 251, Taiwan
chchen@mail.tku.edu.tw

[2] Department of Computer Science and Engineering,
National Sun Yat-sen University, Taiwan

[3] Department of Computer Science and Information Engineering,
National University of Kaohsiung, Taiwan

m023040059@student.nsysu.edu.tw, tphong@nuk.edu.tw

Abstract. In this paper, a two-stage multi-level genetic-fuzzy mining approach is proposed. In the first stage, the multi-level genetic-fuzzy mining (MLGFM) is utilized to derive membership functions of generalized items from the given taxonomy and transactions. In the second stage, the 2-tuples linguistic representation model is used to tune the derive membership functions. Experimental results on a simulated dataset show the effectiveness of the proposed approach.

Keywords: fuzzy set, genetic algorithm, genetic-fuzzy mining, 2-tuple linguistic representation, taxonomy.

1 Introduction

Association rule mining is most commonly used in attempt to derive relationship between items from given transaction data [1]. In real-world applications, items may have taxonomy. According to the given generalization hierarchies, many approaches have been proposed for mining multiple-concept-level fuzzy association rules [6, 8, 9, 10]. Lee first proposed a generalized fuzzy quantitative association rules mining algorithm [9], while Hong *et al.* then proposed a multiple-level fuzzy association rule mining approach [6]. Kaya *et al.* proposed a weighted fuzzy rule mining approach [8]. Lee et al. proposed a fuzzy mining algorithm for discovering generalized fuzzy association rules with multiple supports of items to extract implicit knowledge from quantitative transaction data [10].

Since given membership functions may have a critical influence for fuzzy data mining, various genetic-fuzzy mining (GFM) approaches have then proposed to derive appropriate membership functions and mining fuzzy association rules in recent years [2, 3, 5, 7, 11]. These approaches can be divided into two types, which are

* Corresponding author.

N.T. Nguyen et al. (Eds.): ACIIDS 2014, Part II, LNAI 8398, pp. 82–89, 2014.

single-concept-level GFM and multiple-concept-level GFM. In the single-concept-level GFM, Kaya *et al.* proposed a GA-based approach to derive a predefined number of membership functions for obtaining the maximum profit within a user specified interval of minimum supports [7]. Hong *et al.* also proposed a genetic-fuzzy data-mining algorithm for extracting both association rules and membership functions from quantitative transactions [5]. Alcalá-Fdez *et al.* then modified this approach to propose an enhanced approach based on the 2-tuples linguistic representation model [2]. Matthews et al. proposed a temporal fuzzy association rule mining with the 2-tuples linguistic representation [11]. As to multiple-concept-level GFM, only a few works have been proposed [3]. According to the given taxonomy, the multiple-level GFM (MLGFM) algorithm was proposed for mining membership functions of generalized items and fuzzy association rule with taxonomy [3].

In the mentioned approaches, various chromosome representations are designed for deriving useful knowledge. In some of them [3, 5], two parameters are used to encode the associated membership functions as in Parodi and Bonelli [12]. In [7], only centroid points of membership functions are encoded into chromosomes. Michigan encoding approach is utilized to represent a fuzzy rule and its membership functions . The 2-tuples linguistic representation model has also been used widely in GFM [2, 11], which uses two parameters to represent a membership function, namely linguistic term and lateral displacement. In other words, the linguistic term is known in advance and the lateral displacement needs to be adjusted. However, the problem is that whether the given linguistic terms are good enough

This study thus proposes a two-stage multi-level genetic-fuzzy mining approach with the 2-tuple linguistic representation model. In the first stage, the MLGFM is utilized to derive membership functions of generalized items (item classes) from the given taxonomy and transactions [3]. It first encodes the membership functions of each item class (category) into a chromosome. The fitness value of each individual is then evaluated by the summation of the large 1-itemsets of each item in different concept levels and the suitability of membership functions in each chromosome. When the GA process terminates, a more suitable set of membership functions are used as the initial membership functions of the next stage. In the second stage, the 2-tuples linguistic representation model is used to tune the derive membership functions. Two parameters are used to encode a membership function, which are the centroid point and lateral displacement of a membership function. The fitness function and the genetic operations are the same as those used in the first stage. After the GA tuning process, a better set of membership functions are used to induce multiple-level fuzzy association rules.

2 Preliminaries of MLGFM

Since the first stage of the proposed approach is MLGFM [3], the components of the MLGFM are stated in this section, including chromosome representation, fitness and selection, and genetic operators.

2.1 Chromosome Representation

Since each generalized item (item class) has its own membership functions, the genetic algorithm is utilized to optimize the different sets of membership functions of all item classes. Thus, R_{jk} denotes the membership function of the k-th linguistic term of *item class* IC_j, c_{jk} indicates the center abscissa of fuzzy region R_{jk}, and w_{jk} represents half the spread of the fuzzy region R_{jk}. As in Parodi and Bonelli [12], they represent each membership function as a pair (c, w), and all such pairs for a certain item class are concatenated to represent its membership functions. The set of membership functions MF_1 for the first item class IC_1 is represented as a substring of $c_{11}w_{11}\ldots c_{1|IC1|}w_{1|IC1|}$, where $|IC_1|$ is the number of linguistic terms of IC_1. The entire set of membership functions is then encoded by concatenating substrings of MF_1, MF_2,..., MF_j, ..., MF_m.

2.2 Fitness and Selection

In order to develop a good set of membership functions from an initial population, the genetic algorithm selects *parent* membership function sets with high fitness values to form next population. An evaluation function is used to qualify the derived membership function sets. The performance of the membership function sets is fed back to the genetic algorithm to control how the solution space is searched to promote the quality of the membership functions. In that paper, the fitness function is shown in equation (1).

$$f(C_q) = \frac{\sum_{k=1}^{levels} | L_{1q}^k |}{suitability(C_q)}, \quad (1)$$

where the $| L_{1q}^k |$ is the number of large 1-itemsets in level k of chromosome C_q, $1 \leq k \leq levels$, and the $suitability(C_q)$ is used in the fitness function can reduce the occurrence of the two bad kinds of membership functions, where the first one is too redundant, and the second one is too separate. The suitability of the membership functions in a chromosome C_q is defined as equation (2):

$$suitability(C_q) = \sum_{j=1}^{m} [\,overlap\_factor(C_{qj}) + coverage\_factor(C_{qj})], \quad (2)$$

where the overlap factor of the membership functions for an item class IC_j in the chromosome C_q is defined as equation (3):

$$overlap\_factor(C_{qj}) = \sum_{k \neq i} [\max((\frac{overlap(R_{jk}, R_{ji})}{\min(w_{jk}, w_{ji})}), 1) - 1], \quad (3)$$

where the overlap ratio $overlap(R_{jk}, R_{ji})$ of two membership functions R_{jk} and R_{ji} is defined as the overlap length divided by half the minimum span of the two functions.

If the overlap length is larger than half the span, then these two membership functions are thought of as a little redundant. Appropriate punishment must then be considered in this case. The coverage ratio of a set of membership functions for an item class IC_j is defined as the coverage range of the functions divided by the average value of maximum quantity in each level of that item class in the transactions. The larger the coverage ratio is, the better the derived membership functions are. Thus, the coverage factor of the membership functions for an item class IC_j in the chromosome C_q is defined as equation (4):

$$\text{coverage\_factor}(C_{qj}) = \frac{1}{\dfrac{range(R_{j1}, \dots, R_{jl})}{\max(IC_j)}}, \tag{4}$$

where $range(R_{j1}, R_{j2}, \dots, R_{jl})$ is the coverage range of the membership functions, l is the number of membership functions for IC_j, and avg$Max(IC_j)$ is the average value of maximum quantity in each level of that item class in the transactions. Note that the overlap factor is designed to avoiding the first bad case of being "too redundant", while the coverage factor is to avoid the second one of being "too separate".

2.3 Genetic Operations

Two genetic operators, the *max-min-arithmetical (MMA) crossover* proposed in Herrera *et al.* [4] and the *one-point mutation* are both used in the genetic fuzzy mining approach. The *max-min-arithmetical (MMA)* crossover operator will generate four candidate chromosomes from the given two chromosomes. The best two of the four candidates are then chosen as the offspring. The *one-point mutation* operator will create a new fuzzy membership function by adding a random value ε, for example $\pm w_{jk}$, to the center or the spread of an existing linguistic term, say R_{jk}. More details could be found in [3].

3 Chromosome Representation

In this section, the chromosome representation of the proposed approach in the second stage is stated. When the GA process of the first stage terminates, a suitable set of membership functions are derived. However, since those membership functions are for generalized items, they needs to be tuned. Thus, in this study, we take 2-tuple linguistic representation model into consideration to design the chromosome representation. Fig. 1. shows membership functions of an generalized item.

The membership functions show in Fig 1. with gray color are derived from first stage. Thus, the set of membership functions MF_1 for the first item class IC_j is represented as a substring of $c_{j1}w_{j1}\dots c_{j|ICj|}w_{j|ICj|}$, where $|IC_j|$ is the number of linguistic terms of IC_j. Since the aim of second stage is to tune the lateral displacement of membership function, the membership functions derived from first stage will be the initial membership functions of stage two. Thus, membership functions show in Fig 1.

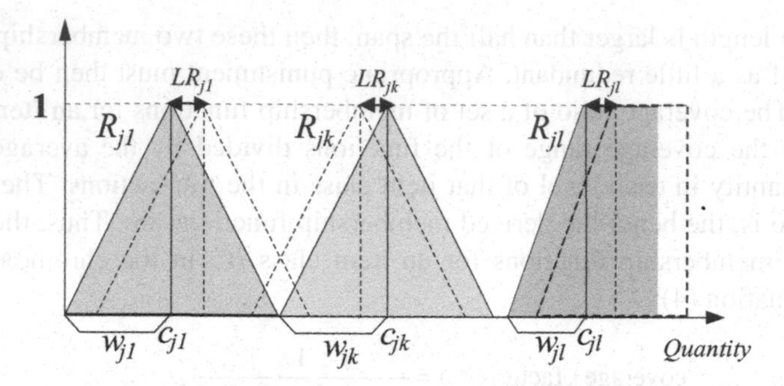

Fig. 1. Membership functions of a generalized item IC_j

with dotted lines are adjusted membership functions according to their lateral displacement LR_{jk}. Thus, the set of membership functions MF_1^2 for the first item class IC_j in second stage is represented as a substring of $c_{j1}LR_{j1}...c_{jk}LR_{jk}...c_{j|IC_j|}LR_{j|IC_j|}$. The entire set of membership functions is then encoded by concatenating substrings of $MF_1^2, MF_2^2, ..., MF_j^2$.

4 The Proposed Mining Algorithm

In this section, the proposed two-stage multi-level genetic-fuzzy mining algorithm by 2-tuple linguistic representation is described. Since the first stage of the proposed approach is the same with MLGFM, details of the first stage of the proposed approach could be found in [3], and the second stage of the proposed approach is stated as follows:

The proposed algorithm:

INPUT: A body of n quantitative transaction data, a set of m items, each with a number of linguistic terms, codes of item names, a predefined taxonomy, a support threshold α, and a confidence threshold λ, a set of derived membership functions from first stage $MF_1, MF_2, ..., MF_j,..., MF_m$.

OUTPUT: A set of multiple-level fuzzy association rules with its associated set of membership functions.

STEP 1: Encode the predefined taxonomy using a sequence of numbers and the symbol "*", with the t-th number represents the branch number of a certain item on level t.

STEP 2: Translate the item names in the transaction data according to the encoding scheme.

STEP 3: Find the level-k representation of the transaction data using following substeps:

SUBSTEP 3.1: Set $k = 1$, where k is used to store the level number being processed.

SUBSTEP 3.2: Group the items with the same first k digits in each transaction datum D_i, where $1 \le i \le n$. Denote the j-th group on level k in transaction D_i as v^k_{ij}.

SUBSTEP 3.3: If the level-k representation is not null, set $k = k + 1$ and go to SUBSTEP 3.1; otherwise, go to next step.

STEP 4: Generate a population of P individuals; each individual is a set of membership functions for generalized items in level-1 of the taxonomy. Each membership function represent by two parameters, namely c_{jk} and LR_{jk}. c_{jk} indicates the center abscissa of fuzzy region R_{jk} which is the center value of MF_j. LR_{jk} represents lateral displacement of the fuzzy region R_{jk}. The initial value of LR_{jk} is generated randomly.

STEP 5: Encode each set of membership functions into a string representation.

STEP 6: Calculate the fitness value of each chromosome by the following substeps:

SUBSTEP 6.1: Set $k = 1$, where k is used to store the level number being processed.

SUBSTEP 6.2: Transform the quantitative value v^k_{ij} of each transaction datum D_i (i =1 to n) for each encoded group name I_j^k into a fuzzy set f_{ij}^k represented as $\left(\frac{f_{ij1}^{(k)}}{R_{j1}^k} + \frac{f_{ij2}^{(k)}}{R_{j2}^k} + ... + \frac{f_{ijh}^{(k)}}{R_{jh}^k}\right)$, using the corresponding membership functions represented by the chromosome, where R^k_{jl} is the l-th fuzzy region of item I_j^k, $1 \le l \le h$, f_{ij}^k is v^k_{ij}'s fuzzy membership value in region R^k_{jl}, and l (= $|I_j^k|$) is the number of linguistic terms for I_j^k.

SUBSTEP 6.3: Calculate the scalar cardinality of each fuzzy region R^k_{jl} in the transactions as $count^k_{jl} = \sum_{i=1}^{n} f^k_{ijl}$.

SUBSTEP 6.4: Check whether the value of each fuzzy region R^k_{jl} is larger than or equal to the predefined minimum support value □. If the value of a fuzzy region R^k_{jl} is equal to or greater than the minimum support value, put it in the large 1-itemsets (L_1^k) at level k. That is: $L_1^k = \{ R_{jl}^k \mid count_{jl}^k \ge \alpha, 1 \le j \le m^k, 1 \le l \le h \}$.

SUBSTEP 6.5: If level-k representation is not null, set $k = k + 1$ and go to SUBSTEP 6.1; otherwise, go to next step.

STEP 7: Set the fitness value of the chromosome as the summation of the number of large 1-itemsets in L_1^k for all levels divided by its suitability *suitability*(C_q) as defined in equation (1).

STEP 8 to 10: Execute crossover and mutation operations on the population. Then, use the *selection* criteria to choose individuals for the next generation.

STEP 11: If the termination criterion is not satisfied, go to Step 6; otherwise, output the set of membership functions with the highest fitness value.

5 Experimental Results

In this section, experiments on a simulated dataset were made to show the performance of the proposed approach. The synthetic dataset had 64 purchased items (terminal nodes) in level 3, 16 generalized items in level 2, and four generalized items in level 1. Each non-terminal node had four branches, and only the terminal nodes could appear in transactions. The population size was set at 50, the crossover rate was set at 0.8, and the mutation rate was set at 0.001. The parameter d of the MMA crossover operator was set at 0.35 according to Herrera *et al.*'s paper [4]. Firstly, the comparison results of average fitness values (five runs) between MLGFM and the proposed approach are shown in Fig. 2.

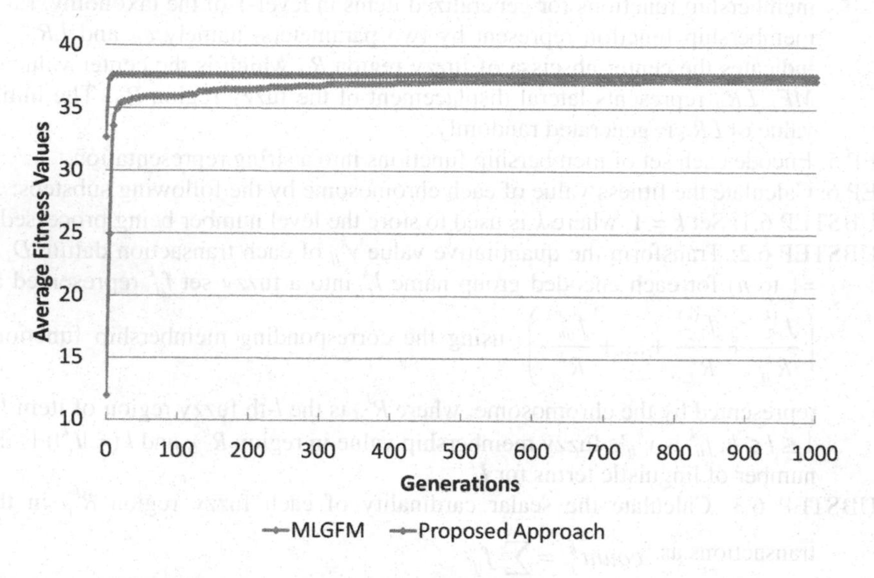

Fig. 2. The comparison between MLGFM and proposed approach

Fig. 2 shows that the curves of both the approaches gradually go upward, finally converging to a certain value. And, the proposed approach is better than MLGFM in terms of average fitness values. And, it also means that the membership functions tuned with the 2-tuple linguistic representation model are better than those without using the 2-tuple linguistic representation model. From the experimental results, it can be concluded that the proposed approach is effective.

6 Conclusion and Future Work

Since 2-tuple linguistic representation model is useful on representing membership functions, in this paper, we have proposed a multi-level genetic-fuzzy mining with a tuning mechanism. Firstly, the MLGFM is utilized to derive membership functions of generalized items from the given taxonomy and transactions. Then, the 2-tuples linguistic representation model is utilized to tune the derive membership functions. Two parameters are utilized to encode a membership function, namely the centroid point and lateral displacement of membership function. The fitness function and genetic operations are the same as those used in MLGFM. After the GA tuning process, a better set of membership functions are used to induce multi-level fuzzy association rules. Experimental results show that the proposed approach could not only reach better average fitness values but also derive more rules than that by MLGFM. In the future, we will enhance the proposed approach with more datasets and extend it to more complex problems.

Acknowledgment. This research was supported by the National Science Council of the Republic of China under grant NSC 100-2221-E-390-026-MY3, 102-2221-E-390-017 and 102-2221-E-032-056.

References

1. Agrawal, R., Srikant, R.: Fast algorithm for mining association rules. In: The International Conference on Very Large Databases, pp. 487–499 (1994)
2. Alcala-Fdez, J., Alcala, R., Gacto, M.J., Herrera, F.: Learning the membership function contexts for mining fuzzy association rules by using genetic algorithms. Fuzzy Sets and Systems 160(7), 905–921 (2009)
3. Chen, C.H., Hong, T.P., Lee, Y.C.: Genetic-Fuzzy Mining with Taxonomy. International Journal of Uncertainty, Fuzziness and Knowledge-Based Systems 20(2), 187–205 (2012)
4. Herrera, F., Lozano, M., Verdegay, J.L.: Fuzzy connectives based crossover operators to model genetic algorithms population diversity. Fuzzy Sets and Systems 92(1), 21–30 (1997)
5. Hong, T.P., Chen, C.H., Wu, Y.L., Lee, Y.C.: A GA-based fuzzy mining approach to achieve a trade-off between number of rules and suitability of membership functions. Soft Computing 10(11), 1091–1101 (2006)
6. Hong, T.P., Lin, K.Y., Chien, B.C.: Mining fuzzy multiple-level association rules from quantitative data. Applied Intelligence 18(1), 79–90 (2003)
7. Kaya, M., Alhajj, R.: Genetic algorithm based framework for mining fuzzy association rules. Fuzzy Sets and Systems 152(3), 587–601 (2005)
8. Kaya, M.A., Alhajj, R.: Effective mining of fuzzy multi-cross-level weighted association rules. In: Esposito, F., Raś, Z.W., Malerba, D., Semeraro, G. (eds.) ISMIS 2006. LNCS (LNAI), vol. 4203, pp. 399–408. Springer, Heidelberg (2006)
9. Lee, K.M.: Mining generalized fuzzy quantitative association rules with fuzzy generalization hierarchies. In: IFSA World Congress and 20th NAFIPS International Conference, vol. 5, pp. 2977–2982 (2001)
10. Lee, Y.C., Hong, T.P., Wang, T.C.: Multi-level fuzzy mining with multiple minimum supports. Expert Systems with Applications 34(1), 459–468 (2008)
11. Matthews, S.G., Gongora, M.A., Hopgood, A.A., Ahmadi, S.: Temporal fuzzy association rule mining with 2-tuple linguistic representation. In: The IEEE International Conference on Fuzzy Systems, pp. 1–8 (2012)
12. Parodi, A., Bonelli, P.: A new approach of fuzzy classifier systems. In: Proceedings of Fifth International Conference on Genetic Algorithms, pp. 223–230. Morgan Kaufmann, Los Altos (1993)

The Influence of a Classifiers' Diversity on the Quality of Weighted Aging Ensemble

Michał Woźniak[1], Piotr Cal[1], and Bogusław Cyganek[2]

[1] Department of Systems and Computer Networks
Wroclaw University of Technology
Wyb. Wyspianskiego 27, 50-370 Wroclaw, Poland
{michal.wozniak,piotr.cal}@pwr.wroc.pl

[2] AGH University of Science and Technology
Al. Mickiewicza 30, 30-059 Kraków, Poland
cyganek@agh.edu.pl

Abstract. The paper deals with the problem of data stream classification. In the previous works we proposed the WAE (*Weighted Aging Ensemble*) algorithm which may change the line-up of the classifier committee dynamically according to coming of new individual classifiers. The ensemble pruning method uses the diversity measure called the *Generalized Diversity* only. In this work we propose the modification of the WAE algorithm which applies the mentioned above pruning criterion by the linear combination of diversity measure and accuracy of the classifier ensemble. The proposed method was evaluated on the basis of computer experiments which were carried out on two benchmark databases. The main objective of the experiments was to answer the question if the chosen modified criterion based on the diversity measure and accuracy is an appropriate choice to prune the classifier ensemble dedicated to data stream classification task.

Keywords: classifier ensemble, data stream, incremental learning, ensemble pruning, forgetting.

1 Introduction

One of the main challenge of the modern computer classification systems is to propose efficient approach to analyze data stream. Such tool should take into consideration the following characteristics which separate the data stream classification task from the traditional canonical classification model:

- the statistical dependencies among the input features described given objects and their classifications may be changing,
- data can come flooding in the classifier what causes that it is impossible to label all incoming objects manually by human experts.

The first phenomena is called *concept drift* and according to [9] we can propose its taxonomy: gradual drift (smooth changes), sudden drift (abrupt changes), and

N.T. Nguyen et al. (Eds.): ACIIDS 2014, Part II, LNAI 8398, pp. 90–99, 2014.

reoccurring drift (changes are either periodical or unordered). When we face with the pattern classification task with the possibility of concept drift appearance then we can consider the two main approaches which strictly depend on the drift type:

- Detecting concept drift in new data and if these changes are significant, then retrain the classification model.
- Adopting a classification model to changes.

In this work we will focus on the last issue. The model is either updated (e.g., neural networks) or needs to be partially or completely rebuilt (as CVFDT algorithm [3]). Usually we analyze the data stream using so-called data chunks (successive portions of incoming data). The main question is how to adjust the data chunk size. On the one hand, a smaller chunk allows focusing on the emerging context, though data may not be representative for a longer lasting context. On the other hand, a bigger chunk may result in mixing the instances representing different contexts. One of the important group of algorithms dedicated to stream classification exploits strength of ensemble systems, which work pretty well in static environments [6]. An assumed strategy for generating the line-up of the classifier ensemble should guarantee its diversity improvement and consequently accuracy increasing.

The most popular ensemble approaches, as the *Streaming Ensemble Algorithm* (SEA) [13] or the *Accuracy Weighted Ensemble* (AWE)[14], keep a fixed-size set of classifiers. Incoming data are collected in data chunks, which are used to train new classifiers. If there is a free space in the ensemble, a new classifier joins the committee. Otherwise, all the classifiers are evaluated based on their accuracy and the worst one in the committee is replaced by a new one if the latter has higher accuracy. The SEA uses a majority voting strategy, whereas the AWE uses the more advanced weighted voting strategy. A similar formula for decision making is implemented in the *Dynamic Weighted Majority* (DWM) algorithm [5]. Nevertheless, unlike the former algorithms, the DWM modifies the weights and updates the ensemble in a more flexible manner. The weight of the classifier is reduced when the classifier makes an incorrect decision. Eventually the classifier is removed from the ensemble when its weight falls below a given threshold. Independently, a new classifier is added to the ensemble when the committee makes a wrong decision. Some evolving systems continuously adjust the model to incoming data, what is called implicit drift detection [7] as opposed to explicit drift detection methods that raise a signal to indicate change. In this work we propose the modification of the previously developed dynamic ensemble model called WAE (*Weighted Aging Ensemble*) which can modify the line-up of the classifier committee on the basis of the linear combination of diversity measure called *Generalized Diversity* and accuracy. Additionally the decision about object's label is made according to weighted voting, where weight of a given classifier depends on its accuracy and time spending in an ensemble. The detailed description of the algorithm is presented in the next section. In this work we would like to study how the method of individual classifier selection to a classifier committee could influence the compound classifier quality. Then we present preliminary results of computer experiments which were carried out on SEA and Hyper Plane Stream datasets. The last section concludes our research.

2 WAE - Classifier Ensemble for Data Stream Classification

Let's propose the idea of the WAE (*Weighted Aging Ensemble*), which was firstly presented in [15], then its modification will be presented. We assume that the data stream under consideration is given in a form of data chunks denotes as $\mathcal{DS}_k$, where k is the chunk index. The concept drift could appear in the incoming data chunks. We do not detect it, but we try to construct self-adapting classifier ensemble. Therefore on the basis of the each chunk some individuals are trained using different classifier models and we check if they could form valuable ensemble with the previously trained classification models. Because we assume the fixed size of the ensemble therefore we should select the most valuable classifier committee line-up on the basis of the exhaustive search (the number of the possible ensembles is not so high). In the previous versions our algorithm we proposed to use the *Generalized Diversity* (denoted as $\mathcal{GD}$) proposed by Partridge and Krzanowski [10] as the search criterion to assess all possible ensembles and to choose the best one. $\mathcal{GD}$ returns the maximum values in the case of failure of one classifier is accompanied by correct classification by the other one and minimum diversity occurs when failure of one classifier is accompanied by failure of the other.

$$\mathcal{GD}(\Pi) = 1 - \frac{\sum_{i=1}^{L} \frac{i(i-1)p_i}{L(L-1)}}{\sum_{i=1}^{L} \frac{ip_i}{L}} \tag{1}$$

where L is the cardinality of the classifier pool (number of individual classifiers) and p_i stands for the probability that i randomly chosen classifiers from Π will fail on randomly chosen example.

Lets $P_a(\Psi_i)$ denotes frequency of correct classification of classifier Ψ_i and $itter(\Psi_i)$ stands for number of iterations which Ψ_i has been spent in the ensemble. We propose to establish the classifier's weight $w(\Psi_i)$ according to the following formulae

$$w(\Psi_i) = \frac{P_a(\Psi_i)}{\sqrt{itter(\Psi_i)}} \tag{2}$$

and the final decision returned by the compound classifier Ψ is given by the following formulae

$$\Psi(x) = \arg\max_{j \in \mathcal{M}} \sum_{k=1}^{L} [\Psi_k(x) = j]\, w(\Psi_k), \tag{3}$$

where $\mathcal{M}$ denotes the set of possible labels, x is feature values, and [ ] stands for Inverson's bracket.

This proposition of classifier aging has its root in object weighting algorithms where an instance weight is usually inversely proportional to the time that has passed since the instance was read [4] and Accuracy Weighted Ensemble (AWE)[14], but the proposed method called Weighted Aging Ensemble (WAE) incudes two important modifications:

1. classifier weights depend on the individual classifier accuracies and time they have been spending in the ensemble,
2. individual classifier are chosen to the ensemble on the basis on the non-pairwise diversity measure.

In our work we propose tho replace $\mathcal{GD}$ (1) as the ensemble pruning criterion by the linear combination of the ensemble accuracy and the mentioned above measure

$$\mathcal{Q}(\Pi) = a\mathcal{GD}(\Pi) + (1-a)P_a(\Psi), \quad \text{where } a \in [0,1] \tag{4}$$

where Ψ is classifier ensemble using pool of individual classifiers Π, P_a denotes its accuracy, and a stands for arbitrary chosen factor.

The WAE pseudocode is presented in Alg.1 [15].

Algorithm 1. Weighted Aging Ensemble (WAE) based on heterogenous classifiers

---

Require: input data stream,
  data chunk size,
  k classifier training procedures,
  ensemble size L

1: $i := 1$
2: $\Pi = \emptyset$
3: **repeat**
4:   collect new data chunk DS_i
5:   **for** $j := 1$ **to** k **do**
6:     $\Psi_{i,j} \leftarrow$ classifier training procedure (DS_i,j)
7:     $\Pi := \Pi \cup \{\Psi_{i,j}\}$ to the classifier ensemble Π
8:   **end for**
9:   **if** $|\Pi| > L$ **then**
10:     choose the most valuable ensemble of L classifiers using (4)
11:   **end if**
12:   **for** $j = 1$ **to** L **do**
13:     calculate $w(\Psi_i)$ according to (2)
14:   **end for**
15: **until** end of the input data stream

---

3 Experimental Investigations

The aims of the experiment were:

– assessing if the proposed method of weighting and aging individual classifiers in the ensemble is valuable proposition compared with the methods which do not include aging or weighting techniques,
– establishing the dependency between the a factor value used in (4) and quality of the WAE algorithm

3.1 Set-Up

All experiments were carried out on two syntectic benchmark datasets:

– the **SEA** dataset [13] where each object belongs to the on of two classes and is described by 3 numeric attributes with value between 0 and 10, but only two of them are relevant. Object belongs to class 1 (TRUE) if $arg_1 + arg_2 < \phi$ and to class 2 (FALSE) otherwise. ϕ is a threshold between two classes, so different thresholds correspond to different concepts (models).Thus, all generated dataset is linearly separable, but we add 5% noise, which means that class label for some samples is changed, with expected value equal to 0. We simulated drift by instant random model change.
– **Hyper Plane Stream** [16] where each object belongs to one of the 5 classes and is described by 10 attributes. The dataset is a synthetic data stream containing gradually evolving (drifting) concepts. The drift is appeared each 800 observations.

For each of the experiments we decided to form heterogenous ensemble i.e., ensemble which consists of the classifier using the different models (to ensure its higher diversity) and we used the following models for individual classifiers:

– Naïve Bayes,
– decision tree trained by C4.5 [12],
– SVM with polynomial kernel trained by the sequential minimal optimization method (SMO) [11]
– nearest neighbour classifier,
– classifier using a multinominal logistic regression with a ridge classfier [8],
– OneR [2].

During each of the experiment we tried to evaluate dependency between data chunk sizes (which were fixed on 50, 100, 150, 200) and overall classifier quality (accuracy and standard deviation) and the diversity of the best ensemble for the following ensembles:

1. *simple* - an ensemble using majority voting without aging.
2. *weighted* - an ensemble using weighted voting without aging, where weight assigned to a given classifier is inversely proportional to its accuracy.

3. *aged* - an ensemble using weighted voting with aging, where weight assigned to a given classifier is calculated according to (2).

Method of ensemble pruning was the same for each ensembles and presented in (4). We run the experiments for different a values ($a \in \{0.0, 0.1, ..., 1.0\}$).

All experiments were carried out in the Java environment using Weka classifiers [1]. The new individual classifiers were trained on a given data chunk. The same chunk was used to prune the classifier committee, but the ensemble error was estimated on the basis on the next (unseen) portion of data.

3.2 Results

The results of experiment are presented in Fig.1-2 and in Tab. 1-2. The figures show the accuracies and diversity for different types of ensembles and different values of a factor and chunk size. Tab.1-2 present overall accuracy and standard deviation for the tested methods and how they depend on data chunk size. Unfortunately, because of the space limit we are not able to presents all extensive results, but they are available on demand.

Table 1. Classification accuracies and diversities for different sizes of data chunk for SEA dataset

chunk size	ensemble type	accuracy	sd	diversity	sd
50	*simple*	0.805	0.0059	0.476	0.0175
	weighted	0.893	0.0064	0.481	0.0165
	aged	0.895	0.0047	0.480	0.0136
100	*simple*	0.902	0.0048	0.466	0.0211
	weighted	0.904	0.0063	0.450	0.0170
	aged	0.906	0.0054	0.456	0.0196
150	*simple*	0.907	0.0075	0.437	0.0162
	weighted	0.910	0.0040	0.448	0.0306
	aged	0.908	0.0047	0.447	0.0297
200	*simple*	0.904	0.0046	0.459	0.0230
	weighted	0.899	0.0110	0.451	0.0355
	aged	0.904	0.0028	0.429	0.0268

3.3 Discussion of the Results

SEA dataset:

- The overall accuracies of the tested ensembles are stable according to the chunk sizes. We observed a slight accuracy improvement, but it is statistical significant for the chunk sizes 50 and 150 only (t-test). The standard deviation of the accuracies is unstable, but it is smallest for the chunk size 150. The observation is useful because the bigger size of data chunk means that effort dedicated to building new models is smaller because they are being built rarely.

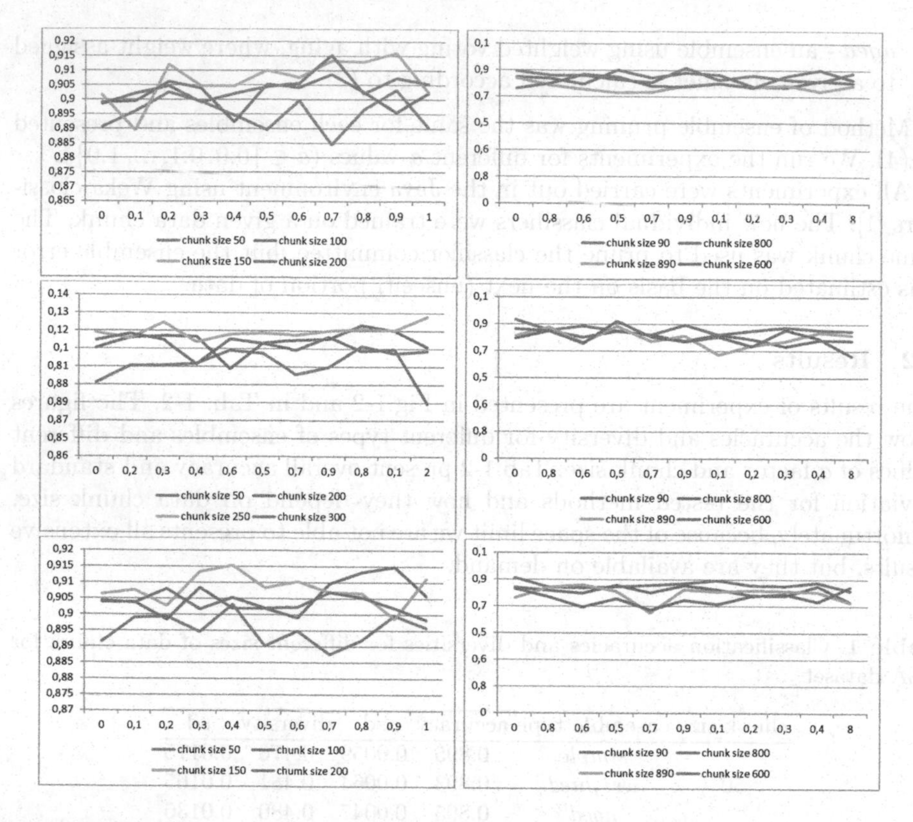

Fig. 1. The computer experimental results for the SEA dataset. Dependencies between *a* factor used in the pruning criterion (4) and ensembles' accuracies (left) diversities (right) for three type of classifier ensemble: *simple* (top), *weighted* (middle), and *aged* (bottom), and for 4 different sizes of data chunk.

- The overall diversities do not depend strongly on chunk size.
- Taking into consideration the mentioned above observations we may suggest that the best choice of chunk size is ca. 150, especially for *weighed* and *aged* ensemble.

Hyper Plane Stream dataset:

- The overall accuracies of the tested ensembles increase according to chunk size. The differences are statistically significant between the following pais of chunk sizes: 50 and 150, 50 and 200, 100 and 200 (t-test).
- The standard deviations of all ensemble accuracies increase according the chunk size.
- The ensemble diversity is decreasing according to the chunk sizes but the standard deviation is increasing.
- Taking into consideration the mentioned above observations we may suggest that the best choice of chunk size is also ca. 150.

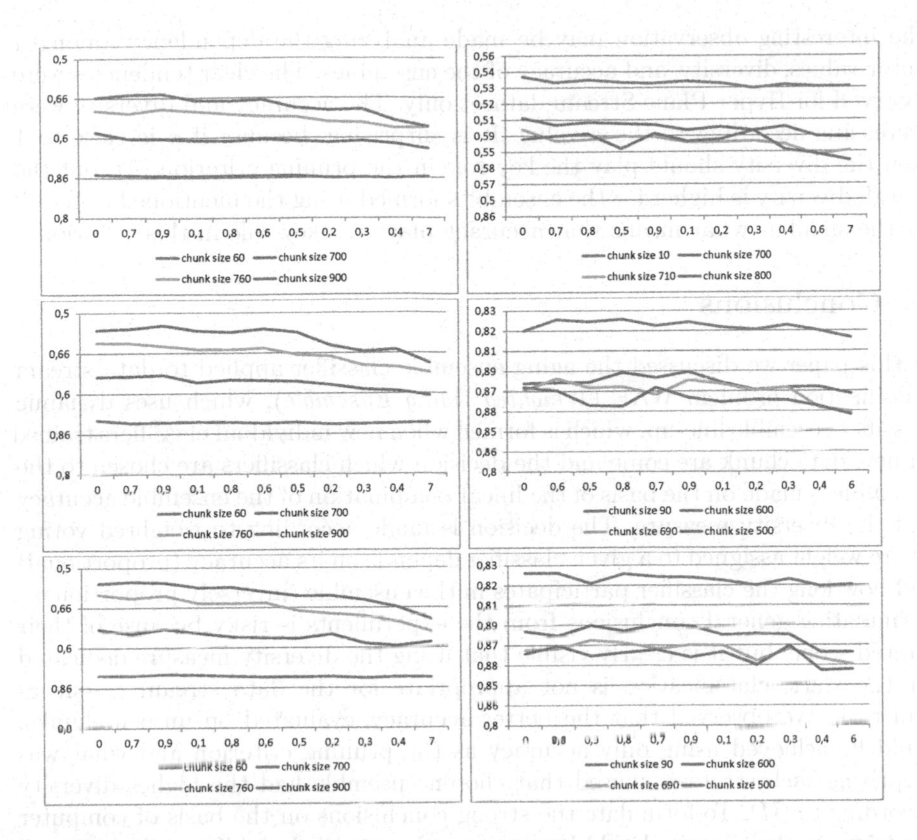

Fig. 2. The computer experimental results for the Hyper Plane Stream dataset. Dependencies between *a* factor used in the pruning criterion (4) and ensembles' accuracies (left) diversities (right) for three type of classifier ensemble: *simple* (top), *weighted* (middle), and *aged* (bottom), and for 4 different sizes of data chunk.

Table 2. Classification accuracies and diversities for different sizes of data chunk for Hyper Plane Stream dataset

chunk size	ensemble type	accuracy	sd	diversity	sd
	simple	0.452	0.0014	0.371	0.0021
50	*weighted*	0.463	0.0015	0.366	0.0028
	aged	0.463	0.0015	0.370	0.0025
	simple	0.486	0.0051	0.339	0.0037
100	*weighted*	0.497	0.0082	0.338	0.0034
	aged	0.507	0.0069	0.336	0.0048
	simple	0.513	0.0083	0.331	0.0043
150	*weighted*	0.526	0.0126	0.330	0.0052
	aged	0.520	0.0146	0.330	0.0039
	simple	0.514	0.0133	0.324	0.0060
200	*weighted*	0.537	0.0152	0.328	0.0043
	aged	0.535	0.0145	0.328	0.0043

The interesting observation may be made analyzing the dependency among a factor values, diversity, and accuracy of the ensembles. The clear tendencies were observed for Hyper Plane Stream dataset only. The accuracy and diversity were decreasing according to the a value. It is surprising, because if a is close to 1 then the diversity should play the key role in the pruning criterion (4), but the overall diversity is higher for the ensembles formed using the mentioned criterion for the small a (what means that accuracy plays the key role in this criterion).

4 Conclusions

In this paper we discussed the aging ensemble classifier applied to data stream classification problem WAE (*Weighted Aging Ensemble*), which uses dynamic classifier ensemble line-up, which is formed when new individual classifiers trained on new data chunk are come and the decision which classifiers are chosen to the ensemble is made on the basis of the linear combination of the ensemble accuracy and the diversity measure. The decision is made according to weighted voting where weight assigned to a given classifier depends on its accuracy (proportional) and how long the classifier participates in the ensemble (inversely proportional). Formulating general conclusions from the experiments is risky because of their limited scope, but it is clearly visible that using the diversity measure dedicated for the static classification is not appropriate for the data stream classification task. We observed that the better accuracy, evaluated on unseen chunks, could be achieved using only accuracy as the pruning criterion and what was surprising such strategy caused that chosen ensemble had the highes diversity according to $\mathcal{GD}$. To formulate the strong conclusions on the basis of computer experiments their scope should be significantly extended. Additionally, the used diversity measure does not seem to be appropriate for the data stream classification tasks, therefore we would like to extend the scope of experiments by using another non-pairwise diversity measures and maybe to propose a new one which can evaluate diversity taking into consideration the nature of the discussed pattern classification task.

It is worth noting that classifier ensemble is a promising research direction for aforementioned problem, but its combination with a drift detection algorithm could have a higher impact to the classification performance.

Acknowledgment. The work was supported by EC under FP7, Coordination and Support Action, Grant Agreement Number 316097, ENGINE European Research Centre of Network Intelligence for Innovation Enhancement (http://engine.pwr.wroc.pl/).

References

1. Hall, M., Frank, E., Holmes, G., Pfahringer, B., Reutemann, P., Witten, I.H.: The weka data mining software: an update. SIGKDD Explor. Newsl. 11(1), 10–18 (2009)
2. Holte, R.C.: Very simple classification rules perform well on most commonly used datasets. Mach. Learn. 11(1), 63–90 (1993)

3. Hulten, G., Spencer, L., Domingos, P.: Mining time-changing data streams. In: Proc. of the 7th ACM SIGKDD Int. Conf. on Knowledge Discovery and Data Mining, pp. 97–106 (2001)
4. Klinkenberg, R., Renz, I.: Adaptive information filtering: Learning in the presence of concept drifts, pp. 33–40 (1998)
5. Kolter, J.Z., Maloof, M.A.: Dynamic weighted majority: a new ensemble method for tracking concept drift. In: Third IEEE International Conference on Data Mining, pp. 123–130 (November 2003)
6. Kuncheva, L.I.: Combining Pattern Classifiers: Methods and Algorithms. Wiley-Interscience (2004)
7. Kuncheva, L.I.: Classifier ensembles for detecting concept change in streaming data: Overview and perspectives. In: 2nd Workshop SUEMA 2008 (ECAI 2008), pp. 5–10 (2008)
8. Le Cessie, S., Van Houwelingen, J.C.: Ridge estimators in logistic regression. Applied Statistics, 191–201 (1992)
9. Narasimhamurthy, A., Kuncheva, L.I.: A framework for generating data to simulate changing environments. In: Proceedings of the 25th IASTED International Multi-Conference: Artificial Intelligence and Applications, AIAP 2007, Anaheim, CA, USA, pp. 384–389. ACTA Press (2007)
10. Partridge, D., Krzanowski, W.: Software diversity: practical statistics for its measurement and exploitation. Information and Software Technology 39(10), 707–717 (1997)
11. Platt, J.C.: Fast training of support vector machines using sequential minimal optimization. In: Advances in Kernel Methods, pp. 185–208. MIT Press, Cambridge (1999)
12. Quinlan, J.R.: C4.5: Programs for Machine Learning. Morgan Kaufmann Publishers (1993)
13. Street, W.N., Kim, Y.: A streaming ensemble algorithm (sea) for large-scale classification. In: Proceedings of the Seventh ACM SIGKDD International Conference on Knowledge Discovery and Data Mining, KDD 2001, pp. 377–382. ACM, New York (2001)
14. Wang, H., Fan, W., Yu, P.S., Han, J.: Mining concept-drifting data streams using ensemble classifiers. In: Proceedings of the Ninth ACM SIGKDD International Conference on Knowledge Discovery and Data Mining, KDD 2003, pp. 226–235. ACM, New York (2003)
15. Woźniak, M., Kasprzak, A., Cal, P.: Weighted aging classifier ensemble for the incremental drifted data streams. In: Larsen, H.L., Martin-Bautista, M.J., Vila, M.A., Andreasen, T., Christiansen, H. (eds.) FQAS 2013. LNCS, vol. 8132, pp. 579–588. Springer, Heidelberg (2013)
16. Xu, X.: Stream data mining repository (2010), `http://www.cse.fau.edu/~xqzhu/stream.html`

Comparison of Ensemble Approaches: Mixture of Experts and AdaBoost for a Regression Problem

Tadeusz Lasota[1], Bartosz Londzin[2], Zbigniew Telec[2], and Bogdan Trawiński[2]

[1] Wroclaw University of Environmental and Life Sciences, Dept. of Spatial Management
ul. Norwida 25/27, 50-375 Wroclaw, Poland
[2] Wrocław University of Technology, Institute of Informatics,
Wybrzeże Wyspiańskiego 27, 50-370 Wrocław, Poland
tadeusz.lasota@wp.pl, bartek.londzin@gmail.com,
{zbigniew.telec,bogdan.trawinski}@pwr.wroc.pl

Abstract. Two machine learning approaches: mixture of experts and AdaBoost.R2 were adjusted to the real-world regression problem of predicting the prices of residential premises based on historical data of sales/purchase transactions. The computationally intensive experiments were conducted aimed to compare empirically the prediction accuracy of ensemble models generated by the methods. The analysis of the results was performed using statistical methodology including nonparametric tests followed by post-hoc procedures designed especially for multiple $n \times n$ comparisons. No statistically significant differences were observed among the best ensembles: two generated by mixture of experts and two by AdaBoost.R2 employing multilayer perceptrons and general linear models as base learning algorithms.

Keywords: mixture of experts, AdaBoost.R2, mlp, glm, svr, real estate appraisal, Matlab.

1 Introduction

Ensemble systems have been gaining a large attention of machine learning community for the last two decades. They combine the output of machine learning algorithms, called “weak learners”, in order to get smaller prediction errors (in regression) or lower error rates (in classification). There are several reasons for using ensemble systems [1], [2]. They may decrease the risk of unlucky selecting a poorly performing learner. They may allow for efficient analysing large volumes of data by applying single learners to smaller partitions of data and combining their outputs. On the other hand, using resampling techniques may be very effective in the case of lacking adequate quantity of training data. Ensembles allow also for handling the problems which are too complex for a single learner to solve. Individual classifiers may learn over simpler partitions and the final solution may be achieved by by an appropriate combination of their output. Finally, data coming from different sources and characterizing by heterogeneous features can be used to train different classifiers. Applying suitable fusion mechanisms may provide successful results.

N.T. Nguyen et al. (Eds.): ACIIDS 2014, Part II, LNAI 8398, pp. 100–109, 2014.

For a few years we have been exploring methods for developing an intelligent system to assist with of real estate appraisal designed for a broad spectrum of users interested in the premises management. The outline of the system to be exploited on a cloud computing platform is presented in Fig. 1. Public registers and cadastral systems create a complex data source for the intelligent system of real estate market. The core of the system are valuation models constructed according to the professional standards as well as models generated using machine learning algorithms.

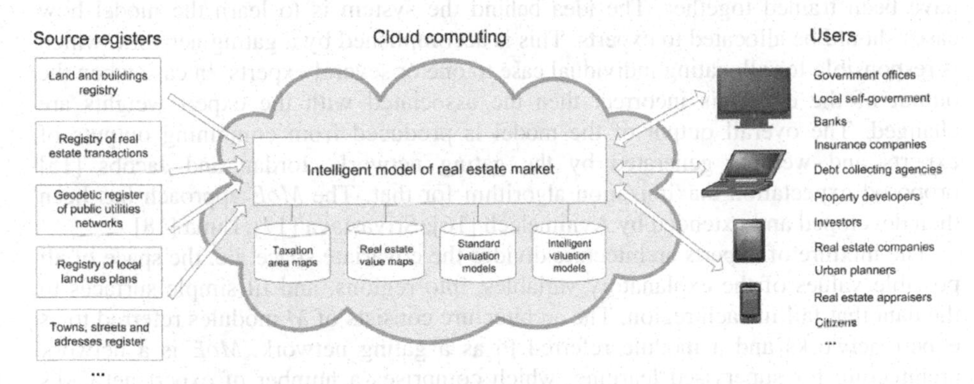

Fig. 1. Outline of the intelligent system of real estate market

So far, we have studied several ensemble approaches to construct regression models to assist with real estate appraisal using as base learning algorithms various fuzzy systems, neural networks, support vector machines, regression trees, and statistical regression [3], [4], [5], [6]. We have also developed an approach to predict from a data stream of real estate sales transactions using ensembles of genetic fuzzy systems and neural networks [7], [8], [9], [10].

The goals of the study presented in this paper are twofold. First, we compared empirically two ensemble machine learning methods less frequently applied to solve regression problems, namely mixture of experts (*MoE*) and AdaBoost.R2 (*AR2*). Second, we investigated the usefulness of both approaches to real world application such as an internet system to assist in property valuation. *MoE* and *AR2* are much less popular in regression than various resampling techniques including bagging, random subspace, and random forests [11], [12], [13]. We implemented our own versions of the methods. In the case of *MoE* the algorithms multilayer perceptron (*mlp*) general linear model (*glm*), and support vector regression (*svr*) were employed as experts and *mlp* and *glm* as the gating network. In turn, *mlp*, *glm*, and regression tree were used as base learners for *AR2*. For *mlp* and *glm* we worked out an algorithm of replicating the instances of a training sets according to the prediction accuracy they provided. This algorithm was applied instead of reweighting of the instances because both *mlp* and *glm* cannot handle the weighted examples.

All experiments were performed using 27 real-world datasets composed of data taken from a cadastral system and GIS data derived from a cadastral map. The analysis of the results was performed using statistical methodology including nonparametric tests followed by post-hoc procedures designed especially for multiple $n \times n$ comparisons.

2 Mixture of Expert Approach

The main idea of mixture of experts is based on "divide and conquer" principle it consists in partitioning a hard to solve problem into a number of simpler subproblems, whose solutions can be easily combined to obtain resolution of the original problem.

The study in this area has been initiated by Jacob et al. [14]. They devised the mixture of experts model, in which the set of expert networks and a gating network have been trained together. The idea behind the system is to learn the model how cases should be allocated to experts. This is accomplished by a gating network, which is responsible for allocating individual case to one or several experts. In case when the output of the expert is incorrect then the associated with the expert weights are changed. The overall output of the model is produced from combining outputs of experts and weights generated by the gating network. Jordan and Jacobs [15] proposed expectation-maximization algorithm for that. The *MoE* approach has been then developed and extended by Avnimelech [16], Srivastava [17], Lima [18].

The mixture of experts architecture divides the covariate space, i.e. the space of all possible values of the explanatory variables, into regions, and fit simple surfaces to the data that fall in each region. The architecture consists of M modules referred to as expert networks and a module referred to as a gating network. *MoE* is a network architecture for supervised learning, which comprises a number of expert networks and a gating network (see Fig. 2). Expert networks approximate the data within each region of the input space: expert network i maps its input, the input vector $\boldsymbol{x}$, to an output vector y_i. It is assumed that different expert networks are appropriate in different regions of the input space. The architecture contains a module, referred to as a gating network, that identifies for any input $\boldsymbol{x}$, the expert or mixture of experts whose output is most likely to approximate the corresponding target output y. The task of a gating network is to combine the various experts by assigning weights to individual networks, which are not constant but are functions of the input instances. Both expert and the gating networks are fed with the input vector $\boldsymbol{x}$ and the gating network produces a set of scalar coefficients w_i that weight the contributions of the various experts. For each input $\boldsymbol{x}$, these coefficients are constrained to be nonnegative and to sum to one. The total output of the architecture, given by

$$y(\boldsymbol{x}) = \sum_{j=1}^{M} w_j(\boldsymbol{x}) y_j(\boldsymbol{x})$$

is a convex combination of the expert outputs for each $\boldsymbol{x}$. The output of *MoE* is the weighed sum of the expert outputs. The expectation-maximization (*EM*) algorithm is usually applied to learn the parameters of the *MoE* architecture.

So far the authors have investigated the architectures of *MoE* with *mlp, glm*, and *svr* playing the role of expert networks and *mlp*, *glm*, and gaussian mixture model (*gmm*) serving as gating networks [19], [20]. In the current contribution we compare the *MoE* approach with boosting one in terms of prediction accuracy over a new real-world cadastral dataset. We applied *mlp, glm*, and *svr* as the experts, whereas *mlp* and *glm* as the gating network. *Svr* algorithm cannot be used as the network gating since it returns only one value at any given time, which means that it does not take into account the dependencies in the output. The architecture of mixture of experts used in the experiments reported in the paper is given in Fig. 2.

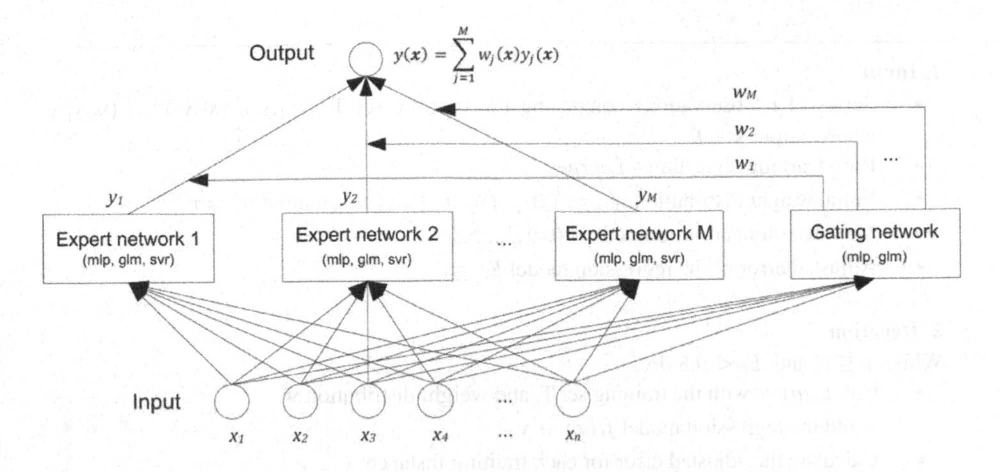

Fig. 2. The architecture of mixture of experts used in the experiments

3 AdaBoost Agorithms for Regression

AdaBoost was devised by Freund and Shapire [21] originally to solve binary classification problems. It is a powerful classification algorithm that has enjoyed remarkable attention of many researchers and practical success in many various areas, e.g. [22], [23]. AdaBoost constructs a strong composite learner by iteratively adding weak learners. The core of the algorithm is reweighting the training instances at each iteration in such a way that the wrongly predicted examples get higher weights and correctly predicted examples get lower weights. Thus, the greater focus is given on examples that were misclassified at previous iterations. Hence, AdaBoost belongs to adaptive approaches where individual learners are trained sequentially. The version of AdaBoost devoted to a multiclass case was extended Freund and Shapire [21] to boosting regression problems. The algorithm called AdaBoost.R reduced regression problems into multiclass classification ones. In turn, Drucker [24] proposed the algorithm AdaBoost.R2 in which the degree to which a given instance was reweighted in a given iteration depended on how large the error on this instance was relative to the error on the worst instance. The AdaBoost.RT algorithm developed by Shrestha and Solomatine [25], on the other hand, labelled each output as correct or incorrect using a relative error threshold. Different boosting methods for regression proposed also Zemel and Pitassi [26], Duffy and Hembold [27], Song and Zhang [28], Pardoe and Stone [29].

In our implementation of AdaBoost.R2 as base learners *mlp*, *glm*, and a regression tree were used. For *mlp* and *glm* we developed an algorithm replicating the instances of a training sets according to the prediction accuracy they provided. The procedure was adjusted to the data used in the experiments and based on the fact that the prices of premises were mapped into the 0 to 1 range as the result of min-max normalization procedure (see Fig. 3). In turn, for the regression tree the original version of *AR2* reweighting the training instances was employed.

---

1. Input

- Series of n observations composing the training set T_1: (x_1,y_1), (x_2,y_2),…, (x_n,y_n), where output $y \in R$
- Base learning algorithm – *Learner*
- Initial weight distribution $w_{1i} = 1/n_1 \ \ for\ 1 \leq i \leq n_1\ , where\ n_1 = n$
- Maximal number of iterations N (t=1,2,..,N)
- Adjusted error of the regression model $E_t = 0$

2. Iteration

While $t \leq N$ and $E_t < 0.5$ do:

- Call *Learner* with the training set T_t and weight distribution w_t
- Build the regression model $f_t(x) \to y$
- Calculate the adjusted error for each training instance:
  - Let $D_t = \max(e_{ti})\ for\ 1 \leq i \leq n_t$
  - then $e_{ti} = (f_t(x_i) - y_i)^2 / D_t^2$, square loss function was selected
- Calculate the adjusted error for of f_t: $E_t = \sum_{i=1}^{n_t} e_{ti}\, w_{ti}$
- Let $\beta_t = E_t/(1 - E_t)$
- Update the weight wector $w_{(t+1)i} = w_{ti}\beta_t^{1-e_{ti}}/Z_t$, where Z_t is a normalizing constant
- Update the training vector T_t by removing or replicating observations according to updated weight values to obtain T_{t+1}
- For i=1 to n_t do
  - $wInt_{(t+1)i}=round(w_{(t+1)i} *1000)$
  - If $wInt_{(t+1)i}=0$ then remove the observation from the training set
  - If $wInt_{(t+1)i}=1$ then leave the observation unchanged
  - If $wInt_{(t+1)i}=c$ then replicate the observation *c* times in the training set with the accompanying weight $w_{(t+1)i}$
- Calculate n_{t+1} the number of observations in T_{t+1}
- Set $t = t + 1$

3. Output

The final regression model (final hypothesis) is

$f_{Fin}(x)$ = weighted median of $f_t(x)\ for\ 1 \leq t \leq N$

using $\ln(1/\beta_t)$ as the weight for hypothesis f_t

Fig. 3. Pseudo code of AdaBoost.R2 used in the experiments

4 Experimental Setup and Results

The investigation was conducted with our experimental system developed in Matlab. We extended the system to include the AdaBoost.R2 method and base algorithms we implemented using Matlab functions: *mlp, glm,* and *classregtree*. For processing mixture of experts we employed the MIXLAB library written by Moerland [30] and and our own implementation of the algorithm *svr* as an expert network in *MoE*.

Real-world dataset used in experiments was taken from a cadastral system and referred to residential premises transactions accomplished in one Polish big city with the population of 640 000 within 14 years from 1998 to 2011. The final dataset after cleansing and outlier removing counted the 9795 instances. Four following attributes were pointed out as main price drivers by professional appraisers: usable area of a flat (*Area*), age of a building construction (*Age*), number of storeys in the building (*Storeys*), the distance of the building from the city centre (*Centre*), in turn, price of premises (*Price*) was the output variable.

Due to the fact that the prices of premises change substantially in the course of time, the whole 14-year dataset cannot be used to create data-driven models using machine learning. Therefore it was split into subsets covering individual half years, and we might assume that within a half year the prices of premises with similar attributes were roughly comparable. Starting from the beginning of 1998 the prices were updated for the last day of subsequent half years using the trends modelled by polynomials of degree four. We might assume that half year datasets differed from each other and might constitute different observation points to compare the accuracy of ensemble models in our study and carry out statistical tests. The data attributes with their descriptive statistics are presented in Table 1 and the sizes of 27 half year datasets are given in Table 2.

All data were normalized using min-max technique. The mixture of expert models and single algorithms were run in Matlab individually for each dataset using 10-fold cross validation and the prediction accuracy was measured with the mean square error (*MSE*).

During the preliminary phase of the experiments we examined many different parameter settings of ensemble architectures and base algorithms to determine the appropriate set of parameter values providing the lowest prediction error. As the result we selected three architectures of *MoE* composed of *mlp, glm*, and *svr* as expert

Table 1. Features of residential premises (flats) supplemented with GIS data

Name	Max	Min	Avg	Std	Med	Description
Area	199.9	12.9	51.2	20.2	48.3	usable area of premises (in m^2)
Age	159	0	47.3	38.3	43	age of building construction
Storeys	13	1	5.7	2.6	5	no. of storeys in a building
Centre	13,206	94	2,918	1,816	2,344	distance from the centre of a city
Price	1,089,000	10,000	189,784	122,097	158,000	price of premises (in PLN)
Price 1sqm	8,300	500	3,752	1,926	3,103	square metre price of premises

Table 2. Data subsets comprising sales transactions ordered by date used in experiments

Dataset	# Inst	Dataset	# Inst	Dataset	# Inst	Dataset	# Inst
1998-01	185	2001-02	366	2005-01	331	2008-02	516
1998-02	261	2002-01	346	2005-02	409	2009-01	334
1999-01	257	2002-02	282	2006-01	395	2009-02	487
1999-02	389	2003-01	326	2006-02	381	2010-01	365
2000-01	246	2003-02	464	2007-01	237	2010-02	931
2000-02	308	2004-01	363	2007-02	205	2011-01	577
2001-01	280	2004-02	411	2008-01	218		

networks and *mlp, glm*, and again *glm* as corresponding gating networks. In turn, in AdaBoost.R2 *mlp, glm*, and *tre* (*classregtree* function) were employed and the 250 iterations produced satisfactory output. Each base algorithm contained 4 input variables and one output. Due to lacking space we omit detailed presentation of selected parameter values for individual algorithms. In the main run the ensemble models were confronted with their single (base) counterparts. Then, we compared all six ensembles in terms of their prediction accuracy.

Statistical analysis of the results of experiments was performed using a software available on the web page of Research Group "Soft Computing and Intelligent Information Systems" at the University of Granada (http://sci2s.ugr.es/sicidm). This JAVA program calculates nonparametric Friedman, Iman-Davenport, Nemenyi, Holm, Shaffer, and Bergmann-Hommel tests according to the multiple comparison methodology described in [31], [32], [33], [34]. In turn the paired Wilcoxon test was performed using the Statistica package.

For each pair of single and mixture of experts models for *mlp, glm*, and *svr* Wilcoxon's matched pairs test proved that ensemble models outperform single ones and the differences in prediction accuracy are statistically significant. The same results were obtained when Wilcoxon test was applied to the pairs of single and Adaboost.R2 models. Ensemble models surpassed significantly the single ones in prediction accuracy in any case.

Statistical tests adequate to multiple comparisons were performed for three *MoE* models and three *AR2* models altogether. For Friedman and Iman-Davenport tests, the calculated values of χ^2 and F statistics were 71.97 and 29.69, respectively, whereas the critical values at α=0.05 are $\chi^2(5)$=12.83 and F(5,130)=2.28, so the null-hypothesis were rejected. It means that there are significant differences between some models. Average ranks of individual models are shown in Table 3, where the lower rank value the better model. The denotations *MoE-mlp, MoE-glm*, and *MoE-svr* indicate the type of algorithms used to constitute the expert networks. It could be seen there are minor differences among *AR2-mlp, MoE-mlp, AR2-glm*, and *MoE-glm* algorithms and these take clearly lower positions than *MoE-svr* and *AR2-tre* algorithms.

Table 3. Average rank positions of ensembles determined during Friedman test

1st	1st	3rd	4th	5th	6th
AR2-mlp	MoE-mlp	AR2-glm	MoE-glm	MoE-svr	AR2-tre
(2.56)	(2.56)	(2.67)	(2.81)	(4.67)	(5.74)

Thus, we were justified in proceeding to post-hoc procedures. In Table 4 p-values for paired Wilcoxon test and adjusted p-values for Nemenyi, Holm, Shaffer, and Bergmann-Hommel tests are placed for $n \times n$ comparisons for all possible 15 pairs of ensemble methods. The p-values below 0.05 indicate that respective algorithms differ significantly in prediction errors; they were marked with the italic font. Following main observations can be done: there are not significant differences among *AR2-mlp, MoE-mlp, AR2-glm*, and *MoE-glm* models. In turn, *MoE-svr* and *AR2-tre* models reveal significantly worse performance than any of first four ensembles in the ranking. In this research it was also shown that the paired Wilcoxon test allowed for the rejection of a greater number of null hypotheses than post-hoc procedures, thus it could lead to over-optimistic decisions.

Table 4. Adjusted p-values for $n \times n$ comparisons of ensembles over 27 half-year datasets

Alg vs Alg	pWilcox	pNeme	pHolm	pShaf	pBerg
AR2-mlp vs AR2-tre	*0.000008*	*5.94E-09*	*5.94E-09*	*5.94E-09*	*5.94E-09*
MoE-mlp vs AR2-tre	*0.000006*	*5.94E-09*	*5.94E-09*	*5.94E-09*	*5.94E-09*
AR2-glm vs AR2-tre	*0.000009*	*2.35E-08*	*2.04E-08*	*1.57E-08*	*1.10E-08*
MoE-glm vs AR2-tre	*0.000006*	*1.37E-07*	*1.09E-07*	*9.12E-08*	*6.38E-08*
MoE-mlp vs MoE-svr	*0.000286*	*0.000507*	*0.000372*	*0.000338*	*0.000338*
AR2-mlp vs MoE-svr	*0.000377*	*0.000507*	*0.000372*	*0.000338*	*0.000338*
AR2-glm vs MoE-svr	*0.000081*	*0.001285*	*0.000771*	*0.000600*	*0.000343*
MoE-glm vs MoE-svr	*0.000006*	*0.004138*	*0.002207*	*0.001931*	*0.001103*
MoE-svr vs AR2-tre	*0.000147*	0.523600	0.244347	0.244347	0.244347
MoE-mlp vs MoE-glm	0.107470	1.000000	1.000000	1.000000	1.000000
MoE-mlp vs AR2-glm	0.211556	1.000000	1.000000	1.000000	1.000000
AR2-mlp vs AR2-glm	0.361270	1.000000	1.000000	1.000000	1.000000
AR2-mlp vs MoE-glm	0.442013	1.000000	1.000000	1.000000	1.000000
AR2-mlp vs MoE-mlp	0.532201	1.000000	1.000000	1.000000	1.000000
AR2-glm vs MoE-glm	0.736607	1.000000	1.000000	1.000000	1.000000

5 Conclusions and Future Work

Several experiments were conducted in order to compare empirically in terms of prediction accuracy two ensemble machine learning methods less frequently applied to solve regression problems, namely mixture of experts (*MoE*) and AdaBoost.R2 (*AR2*). 27 real-world datasets composed of data taken from a cadastral system, registry of property sales/purchase transaction and GIS data derived from a cadastral map were employed in the study. The analysis of the results was performed using recently proposed statistical methodology including nonparametric tests followed by post-hoc procedures designed especially for multiple $n \times n$ comparisons.

Following general conclusions could be drawn on the basis of the experiments. All variants of *MoE* and *AR2* revealed better performance than their single (base) algorithms including multilayer perceptron, general linear model, support vector regression, and decision tree for regression. No significant differences were observed among between *MoE* and *AR2* ensembles built using multilayer perceptron and general linear model. These four variants outperformed significantly *MoE* with support vector regression and *AR2* with decision tree.

The study proved also the usefulness of *MoE* and *AR2* ensemble approaches to an online internet system assisting with real estate appraisal. We strive to incorporate in this system a wide variety of intelligent valuation models to enable the users to make final decisions based on the results provided by different methods. Further research is planning to explore various architectures of mixture of experts and AdaBoost the context of regression problems and to compare them with other ensemble techniques structured in a parallel topology such as bagging, random subspace and random forest.

Acknowledgments. This paper was partially supported by the "Młoda Kadra" funds of the Wrocław University of Technology, Poland and the Polish National Science Centre under grant no. N N516 483840.

References

1. Woźniak, M., Graña, M., Corchado, E.: A survey of multiple classifier systems as hybrid systems. Information Fusion 16, 3–17 (2014)
2. Polikar, R.: Ensemble Based Systems in Decision making. IEEE Circuits and Systems Magazine 6(3), 21–45 (2006)
3. Lasota, T., Telec, Z., Trawiński, B., Trawiński, K.: Exploration of Bagging Ensembles Comprising Genetic Fuzzy Models to Assist with Real Estate Appraisals. In: Corchado, E., Yin, H. (eds.) IDEAL 2009. LNCS, vol. 5788, pp. 554–561. Springer, Heidelberg (2009)
4. Lasota, T., Telec, Z., Trawiński, B., Trawiński, K.: A Multi-agent System to Assist with Real Estate Appraisals Using Bagging Ensembles. In: Nguyen, N.T., Kowalczyk, R., Chen, S.-M. (eds.) ICCCI 2009. LNCS (LNAI), vol. 5796, pp. 813–824. Springer, Heidelberg (2009)
5. Graczyk, M., Lasota, T., Trawiński, B., Trawiński, K.: Comparison of Bagging, Boosting and Stacking Ensembles Applied to Real Estate Appraisal. In: Nguyen, N.T., Le, M.T., Świątek, J. (eds.) ACIIDS 2010, Part II. LNCS (LNAI), vol. 5991, pp. 340–350. Springer, Heidelberg (2010)
6. Krzystanek, M., Lasota, T., Telec, Z., Trawiński, B.: Analysis of Bagging Ensembles of Fuzzy Models for Premises Valuation. In: Nguyen, N.T., Le, M.T., Świątek, J. (eds.) ACIIDS 2010, Part II. LNCS (LNAI), vol. 5991, pp. 330–339. Springer, Heidelberg (2010)
7. Trawiński, B., Lasota, T., Smętek, M., Trawiński, G.: An Attempt to Employ Genetic Fuzzy Systems to Predict from a Data Stream of Premises Transactions. In: Hüllermeier, E., Link, S., Fober, T., Seeger, B. (eds.) SUM 2012. LNCS (LNAI), vol. 7520, pp. 127–140. Springer, Heidelberg (2012)
8. Trawiński, B.: Evolutionary fuzzy system ensemble approach to model real estate market based on data stream exploration. J. Univers. Comput. Sci. 19(4), 539–562 (2013)
9. Telec, Z., Lasota, T., Trawiński, B., Trawiński, G.: An Analysis of Change Trends by Predicting from a Data Stream Using Neural Networks. In: Larsen, H.L., Martin-Bautista, M.J., Vila, M.A., Andreasen, T., Christiansen, H. (eds.) FQAS 2013. LNCS (LNAI), vol. 8132, pp. 589–600. Springer, Heidelberg (2013)
10. Trawiński, B., Lasota, T., Smętek, M., Trawiński, G.: Weighting Component Models by Predicting from Data Streams Using Ensembles of Genetic Fuzzy Systems. In: Larsen, H.L., Martin-Bautista, M.J., Vila, M.A., Andreasen, T., Christiansen, H. (eds.) FQAS 2013. LNCS (LNAI), vol. 8132, pp. 567–578. Springer, Heidelberg (2013)
11. Lasota, T., Łuczak, T., Trawiński, B.: Investigation of Random Subspace and Random Forest Methods Applied to Property Valuation Data. In: Jędrzejowicz, P., Nguyen, N.T., Hoang, K. (eds.) ICCCI 2011, Part I. LNCS, vol. 6922, pp. 142–151. Springer, Heidelberg (2011)
12. Lasota, T., Telec, Z., Trawiński, G., Trawiński, B.: Empirical Comparison of Resampling Methods Using Genetic Fuzzy Systems for a Regression Problem. In: Yin, H., Wang, W., Rayward-Smith, V. (eds.) IDEAL 2011. LNCS, vol. 6936, pp. 17–24. Springer, Heidelberg (2011)
13. Lasota, T., Telec, Z., Trawiński, B., Trawiński, G.: Investigation of Random Subspace and Random Forest Regression Models Using Data with Injected Noise. In: Graña, M., Toro, C., Howlett, R.J., Jain, L.C. (eds.) KES 2012. LNCS (LNAI), vol. 7828, pp. 1–10. Springer, Heidelberg (2013)
14. Jacobs, R.A., Jordan, M.I., Nowlan, S.J., Hinton, G.E.: Adaptive mixtures of local experts. Neural Computation 3, 79–87 (1991)
15. Jordan, M.I., Jacobs, R.A.: Hierarchical mixtures of experts and the EM algorithm. Neural Computation 6, 181–214 (1994)

16. Avnimelech, R., Intrator, N.: Boosted mixture of experts: An ensemble learning scheme. Neural Computation 11(2), 483–497 (1999)
17. Srivastava, A.N., Su, R., Weigend, A.S.: Data mining for features using scale-sensitive gated experts. IEEE Transactions on Pattern Analysis and Machine Intelligence 21, 1268–1279 (1999)
18. Lima, C.A.M., Coelho, A.L.V., Von Zuben, F.J.: Hybridizing mixtures of experts with support vector machines: Investigation into nonlinear dynamic systems identification. Information Sciences 177(10), 2049–2074 (2007)
19. Graczyk, M., Lasota, T., Telec, Z., Trawiński, B.: Application of mixture of experts to construct real estate appraisal models. In: Graña Romay, M., Corchado, E., Garcia Sebastian, M.T. (eds.) HAIS 2010, Part I. LNCS (LNAI), vol. 6076, pp. 581–589. Springer, Heidelberg (2010)
20. Lasota, T., Londzin, B., Trawiński, B., Telec, Z.: Investigation of Mixture of Experts Applied to Residential Premises Valuation. In: Selamat, A., Nguyen, N.T., Haron, H. (eds.) ACIIDS 2013, Part II. LNCS (LNAI), vol. 7803, pp. 225–235. Springer, Heidelberg (2013)
21. Freund, Y., Schapire, R.E.: Decision-theoretic generalization of on-line learning and an application to boosting. Journal of Computer and System Sciences 55(1), 119–139 (1997)
22. Burduk, R.: New AdaBoost Algorithm Based on Interval-Valued Fuzzy Sets. In: Yin, H., Costa, J.A.F., Barreto, G. (eds.) IDEAL 2012. LNCS, vol. 7435, pp. 794–801. Springer, Heidelberg (2012)
23. Kajdanowicz, T., Kazienko, P.: Boosting-based Multi-label Classification. Journal of Universal Computer Science 19(4), 502–520 (2013)
24. Drucker, H.: Improving Regressors using Boosting Techniques. In: Fisher Jr., D.H. (ed.) Proceedings of the Fourteenth International Conference on Machine Learning, pp. 107–115. Morgan Kaufmann (1997)
25. Shrestha, D.L., Solomatine, D.P.: Experiments with AdaBoost.RT, an improved boosting scheme for regression. Neural Computing 18(7), 1678–1710 (2006)
26. Zemel, R.S., Pitassi, T.: A gradient-based boosting algorithm for regression problems. In: Advances in Neural Information Processing Systems 13. MIT Press (2001)
27. Duffy, N., Helmbold, D.: Boosting methods for regression. Machine Learning 47, 153–200 (2002)
28. Song, Y., Zhang, C.: New Boosting Methods of Gaussian Processes for Regression. In: Proceedings of International Joint Conference on Neural Networks, Montreal, Canada (2005)
29. Pardoe, D., Stone, P.: Boosting for Regression Transfer. In: Proceedings of the 27th International Conference on Machine Learning, Haifa, Israel (2010)
30. Moerland, P.: Some methods for training mixtures of experts. Technical Report IDIAP-Com 97-05, IDIAP Research Institute (1997)
31. Demšar, J.: Statistical comparisons of classifiers over multiple data sets. Journal of Machine Learning Research 7, 1–30 (2006)
32. García, S., Herrera, F.: An Extension on "Statistical Comparisons of Classifiers over Multiple Data Sets" for all Pairwise Comparisons. Journal of Machine Learning Research 9, 2677–2694 (2008)
33. Graczyk, M., Lasota, T., Telec, Z., Trawiński, B.: Nonparametric Statistical Analysis of Machine Learning Algorithms for Regression Problems. In: Setchi, R., Jordanov, I., Howlett, R.J., Jain, L.C. (eds.) KES 2010, Part I. LNCS (LNAI), vol. 6276, pp. 111–120. Springer, Heidelberg (2010)
34. Trawiński, B., Smętek, M., Telec, Z., Lasota, T.: Nonparametric Statistical Analysis for Multiple Comparison of Machine Learning Regression Algorithms. International Journal of Applied Mathematics and Computer Science 22(4), 867–881 (2012)

The AdaBoost Algorithm with the Imprecision Determine the Weights of the Observations

Robert Burduk

Department of Systems and Computer Networks,
Wroclaw University of Technology,
Wybrzeze Wyspianskiego 27, 50-370 Wroclaw, Poland
robert.burduk@pwr.wroc.pl

Abstract. This paper presents the AdaBoost algorithm that provides for the imprecision in the calculation of weights. In our approach the obtained values of weights are changed within a certain range of values. This range represents the uncertainty of the calculation of the weight of each element of the learning set. In our study we use the boosting by the reweighting method where each weak classifier is based on the recursive partitioning method. A number of experiments have been carried out on eight data sets available in the UCI repository and on two randomly generated data sets. The obtained results are compared with the original AdaBoost algorithm using appropriate statistical tests.

Keywords: AdaBoost algorithm, weight of the observation, machine learning.

1 Introduction

Boosting is a machine learning effective method of producing a very accurate classification rule by combining a weak classifiers [1]. The weak classifier is defined to be a classifier which is only slightly correlated with the true classification i.e. it can classify the object better than a random classifier. In boosting, the weak classifier is learns on various training examples sampled from the original learning set. The sampling procedure is based on the weight of each example. In each iteration, the weights of examples are changing. The final decision of the boosting algorithm is determined on the ensemble of classifiers derived from each iteration of the algorithm. One of the fundamental problems of the development of different boosting algorithms is choosing the weights and defining rules for an ensemble of classifiers. In recent years, many authors presented various concepts based on the boosting idea [2], [3], [4], [5]. There are also many studies showing the application of this method in the medical diagnosis problem [6] or in the multi-label classification problem [7].

In this article we present a new extension of the AdaBoost [8] algorithm. This extension is for the weights used in samples of the training sets. The original weights are the real number from the interval $[0, 1]$. We propose two approaches to this problem. In one of them in the early iterations weights are larger than in

N.T. Nguyen et al. (Eds.): ACIIDS 2014, Part II, LNAI 8398, pp. 110–116, 2014.

the original algorithm. In the second, in the early iterations weights are smaller than in the original algorithm.

This paper is organized as follows: In section 2 the AdaBoost algorithm is presented. In section 3 the our modification of the AdaBoost algorithm are presented. Section 4 presents the experiment results comparing AdaBoost with our modification. Finally, some conclusions are presented.

2 AdaBoost Algorithm

The first algorithm utilising the idea of boosting was proposed by Schapire in 1990 [9]. It concerned the binary classification problem, for which a set of three classifiers was proposed, and the final response of that set of classifiers was determined at the basis of simple majority of votes. The first of the component classifiers was a weak classifier, for the second one a half of the learning sample was constituted by misclassified observations by the first classifier. The third one used as the learning set those observations from the sample, which were placed in various groups by the two earlier classifiers.

Later, modifications of the original boosting algorithm were proposed. The first one concerned simultaneous combining of many weak classifiers [10]. In 1997, the AdaBoost algorithm was presented [8], which solved several practical difficulties noticed earlier. Its name is an acronym derived from Adaptive Boosting concept. In this case, adaptation concerns readjustment to errors of its component classifiers which result from their activity. Now, we are going to discuss the AdaBoost algorithm action for a case of two classes with Ψ_b classifier assuming values from the set $\{-1, 1\}$. For those assumptions steps of the algorithm look as follows [13] (See Tab. 1):

Table 1. The AdaBoost algorithm

1.	Let $w_{1,1} = ... = w_{1,n} = 1/n$
2.	For $t = 1, 2, ...T$ do:
a.	Fit f_t using weights $w_{t,1}, ..., w_{t,n}$, and compute the error e_t
b.	Compute $c_t = \ln((1 - e_t)/e_t)$.
c.	Update the observations weights: $w_{t+1,i} = w_{t,i} \exp(c_t, I_{t,i}) / \sum_{j=1}^{n}(w_{t,i} \exp(c_t, I_{t,i})), \quad i = 1, ..n.$
3.	Output the final classifier: $\hat{y}_i = F(x_i) = sign(\sum_{t=1}^{T} c_t f_t(x_i))$.

Action of the AdaBoost algorithm begins with assigning all objects from a learning set the corresponding weights reflecting the difficulty degree in correct classifying of a given case. At the beginning, weights are equal and amount to $1/n$, where n is the number of elements in a learning set. In the main loop of the algorithm - point 2 - so many component classifiers is created how many boosting iterations were foreseen. In the 2c step the level of error for Ψ_b qualifier is estimated, which takes into account weights of individual elements from a

Table 2. Notation of the AdaBoost algorithm

i	Observation number, $i = 1, ..., n$.
t	Stage number, $t = 1, ..., T$.
x_i	A p-dimensional vector containing the quantitative variables of the ith observation.
y_i	A scalar quantity representing the class membership of the ith observation, $y_i = -1, 1$.
f_t	The weak classifier at the tth stage.
$f_t(x_i)$	The class estimate of the ith observation at the tth stage.
$w_{t,i}$	The weight of the ith observation at the tth stage, $\sum_i w_{t,i} = 1$.
$I_{t,i}$	The indicator function, $I(f_t(x_i) \neq y_i)$.
e_t	The classification error at the tth stage, $\sum_i w_{t,i} I_{t,i}$.
c_t	The weight of f_t.
$\text{sign}(x)$	$= 1$ if $x \geq 0$ and $= -1$ otherwise.

learning set. Thus, it is a weighted sum and not a fraction of misqualified observations. Further, the c_b factor is being determined, used for weights updating. New values of weights are normalised to a unit. The c_b coefficient is selected so that the observation weights misclassified by Ψ_b are increased, and instead, the correctly classified are decreased. Due to this, in subsequent algorithm iterations increases the probability with which an object misclassified in b iteration will be drawn to bootstrap sample LS_n^{b+1}. In subsequent iterations a component classifier is focused on more difficult samples. It result form the fact, that subsequent bootstrap sample is drawn from a distribution depending on weights of individual samples - point 2a.

The final decision of combined classifier is also dependant on the c_b coefficient. It can be said that each classifier receives its weight which is equal to that coefficient, and a classifier with higher prediction correctness has greater share in final decision of the combined classifier.

It should be also noted, that AdaBoost algorithm, in opposite to the bagging algorithm, can not be implemented at many machines simultaneously. This is caused by the fact that each subsequent component classifier is depending on results of its predecessor.

The AdaBoost algorithm presented above may be used for classifiers returning their response in a form of class label. In the work [11] a general form of the algorithm was proposed for the binary classification problem called the *real* version of AdaBoost. In this case a response of the component classifiers are estimators of a posteriori probability $\hat{p}(1|x), \hat{p}(-1|x) = 1 - \hat{p}(1|x)$.

The boosting algorithms presented above are based at resampling procedure [12]. In each of the B iterations n observations is being drawn with replacement with probability proportional to their current weights (step 2a). As earlier mentioned, weights are updated so, as to increase a share of misclassified samples in the learning set.

In case the component classifiers are able to benefit directly from weights of individual observations, than we talk of a boosting variation by reweighting [12].

In this approach, each of the component classifiers receives weighted information on each element of a learning set. Thus, there are no various learning sets, in understanding of the appearance of individual observations, for subsequent iterations of the algorithm. Each learning set LS_n^b contains the same observations, and instead, their weights are changing. An algorithm utilising the reweighted version is fully deterministic, as sampling is not present in this case.

3 AdaBoost Algorithm with the Imprecision Determine the Weights of the Observations

As we have previously described one of the main problems of the development of different boosting algorithms is the choice of weights. They concern the weights of the observation $w_{t,i}$ and are needed to determine the weighted error e_t of each learned classifier. Now we present two cases of changes in the obtained weights (step 2a in algorithm 1). In one of them in the early iterations weights are larger than in the original algorithm, but in the final iterations smaller than in the original algorithm. In order to change the received weights in the original AdaBoost algorithm the λ parameter is introduced. It defines uncertainty as it received the original weights. The algorithm in this case labeled as lsw-AdaBoost and it is as follows:

Table 3. The lsw-AdaBoost algorithm

1.	Determine the value of λ
2.	Let $w_{1,1} = ... = w_{1,n} = 1/n$
3.	For $t = 1, 2, ...T$ do:
a.	Fit f_t using weights $w_{t,1}, ..., w_{t,n}$, and compute the error e_t
b.	Fit $e_t = e_t * ((1 + (T/2 - t)) * \lambda)$
c.	Compute $c_t = \ln((1 - e_t)/e_t)$.
d.	Update the observations weights:
	$w_{t+1,i} = w_{t,i} \exp(c_t, I_{t,i}) / \sum_{j=1}^{n}(w_{t,i} \exp(c_t, I_{t,i})), \quad i = 1, ..n.$
3.	Output the final classifier:
	$\hat{y}_i = F(x_i) = sign(\sum_{t=1}^{T} c_t f_t(x_i))$.

In the second case in the finaly iterations weights are larger than in the original algorithm, but in the early iterations smaller than in the original algorithm. In this case, we change the step 3b of the algorithm 3. It is labeled now as slw-AdaBoost and it is as follows: In both of these cases, only in the middle iteration weights are the same as in the original AdaBoost algorithm.

4 Experiments

In the experiential research ten data sets were tested. Eight data sets come from the UCI repository [14] and two are generated randomly. One of them is called the banana distribution and has objects generated according to the procedure [15], the second one, instead, has random objects drawn in accordance with

Table 4. The slw-AdaBoost algorithm

1.	Determine the value of λ
2.	Let $w_{1,1} = ... = w_{1,n} = 1/n$
3.	For $t = 1, 2, ...T$ do:
a.	Fit f_t using weights $w_{t,1}, ..., w_{t,n}$, and compute the error e_t
b.	Fit $e_t = e_t * ((1 + (t - T/2)) * \lambda)$
c.	Compute $c_t = \ln((1 - e_t)/e_t)$.
d.	Update the observations weights:
	$w_{t+1,i} = w_{t,i} \exp(c_t, I_{t,i}) / \sum_{j=1}^{n} (w_{t,i} \exp(c_t, I_{t,i})), \quad i = 1, ..n.$
3.	Output the final classifier:
	$\hat{y}_i = F(x_i) = sign(\sum_{t=1}^{T} c_t f_t(x_i))$.

the procedure [16] – Higleyman distribution. In both cases the a priori probability for two classes equals 0.5, and for each class 200 elements were generated. The numbers of attributes, classes and available examples of all the data sets are presented in Tab. 5. The results are obtained via 10-fold-cross-validation method. In the experiments, the value of the parameter λ was set at 0.004. The value of this parameter determines how to change the weight of the observation. In this case, they are increased by the value of 0.004. The same value of weight, as in the original AdaBoost algorithm , is in the half of the assumed iterations, it is on $T/2$. That is, in the initial iterations of the weight are smaller than in the original AdaBoost algorithm, and after half iteration larger than in the original.

Table 5. Description of data sets selected for the experiments

Data set	example	attribute	class
Banana	400	2	2
Breast Cancer Wis.(Original)	699	10(8)	2
Haberman's Survival	306	3	2
Highleyman	400	2	2
ILPD (Indian Liver Patient)	583	10	2
Mammographic Mass	961	6	2
Parkinsons	197	22(23)	2
Pima Indians Diabetes	768	8	2
Sonar (Mines vs. Rocks)	208	60	2
Statlog (Heart)	270	13	2

Tab. 6 shows the results of classification for the AdaBoost algorithm and its modifications proposed in the work. The results are for 30, 40 and 50 iterations of the algorithms. Tab. 6 shows additionally the average ranks which were obtained in accordance with the Friedman test [17], [18]. The resulting average

Table 6. Classification error for different number iterations of AdaBoost algorithms and average rank produced by Friedman test

	Algorithm								
	AB	lsw-AB	slw-AB	AB	lsw-AB	slw-AB	AB	lsw-AB	slw-AB
	After 50 iter.			After 40 iter.			After 30 iter.		
Banan	0.044	0.045	0.042	0.045	0.047	0.043	0.047	0.050	0.044
Cancer	0.061	0.060	0.057	0.063	0.058	0.057	0.063	0.061	0.056
Haber.	0.288	0.270	0.286	0.288	0.274	0.287	0.288	0.279	0.287
Hig.	0.082	0.071	0.083	0.079	0.073	0.083	0.077	0.068	0.082
Liver	0.325	0.301	0.309	0.321	0.304	0.308	0.326	0.304	0.311
Mam.	0.186	0.191	0.199	0.186	0.190	0.199	0.186	0.208	0.199
Park.	0.145	0.140	0.124	0.145	0.136	0.121	0.145	0.150	0.117
Pima	0.240	0.239	0.250	0.237	0.240	0.243	0.239	0.250	0.237
Sonar	0.197	0.189	0.211	0.203	0.197	0.206	0.195	0.194	0.208
Statlog	0.356	0.356	0.337	0.356	0.388	0.338	0.356	0.388	0.338
Av. rank	2.35	1.65	2.00	2.20	1.80	2.00	2.20	2.10	1.70

rank shows an improvement in classification obtained by the proposed in the paper modification of the AdaBoost algorithm.

In order to determine whether the proposed modification method differs from the orginal AdaBoost algorithm the post-hoc Bonferroni-Dunn test was performed. The critical difference for the described experiments equals 0.87 at $p = 0.1$. So, we can conclude that no statistically significant differences in classification error were observed between the proposed modifications of the AdaBoost algorithms and the standard AdaBoost algorithm. However, the received mean ranks of 50 iterations are close to the critical difference.

5 Conclusions

In this paper we presented modification of the AdaBoost algorithm. We consider the situation where the weights are changed within a certain range of values. The paper proposes two approaches. In one of them in the early iterations weights are larger than in the original algorithm. In the second, in the early iteration weights are smaller than in the original algorithm. In our study we use boosting by the reweighting method where each weak classifier is based on the recursive partitioning method. The received results for ten data sets show improvement in classification obtained by the proposed in the paper modification of the AdaBoost algorithm.

Acknowledgments. The work was supported in part by the statutory funds of the Department of Systems and Computer Networks, Wroclaw University of Technology and by the by The Polish National Science Centre under the grant N N519 650440 which is being realized in years 2011–2014.

References

1. Kearns, M., Valiant, L.: Cryptographic limitations on learning boolean formulae and finite automata. J. Assoc. Comput. Mach. 41(1), 67–95 (1994)
2. Chunhua, S., Hanxi, L.: On the Dual Formulation of Boosting Algorithms. IEEE Transactions on Pattern Analysis and Machine Intelligence 32(12), 2216–2231 (2010)
3. Oza, N.C.: Boosting with Averaged Weight Vectors. In: Windeatt, T., Roli, F. (eds.) MCS 2003. LNCS, vol. 2709, pp. 15–24. Springer, Heidelberg (2003)
4. Freund, Y., Schapire, R.: Experiments with a new boosting algorithm. In: Proceedings of the Thirteenth International Conference on Machine Learning, Bari, Italy, pp. 148–156 (1996)
5. Wozniak, M.: Proposition of Boosting Algorithm for Probabilistic Decision Support System. In: Bubak, M., van Albada, G.D., Sloot, P.M.A., Dongarra, J. (eds.) ICCS 2004. LNCS, vol. 3036, pp. 675–678. Springer, Heidelberg (2004)
6. Wozniak, M.: Boosted decision trees for diagnosis type of hypertension. In: Oliveira, J.L., Maojo, V., Martín-Sánchez, F., Pereira, A.S. (eds.) ISBMDA 2005. LNCS (LNBI), vol. 3745, pp. 223–230. Springer, Heidelberg (2005)
7. Kajdanowicz, T., Kazienko, P.: Boosting-based Multi-label Classification. Journal of Universal Computer Science 19(4), 502–520 (2013)
8. Freund, Y., Schapire, R.: A decision-theoretic generalization of on-line learning and an application to boostin. Journal of Computer and System Scienses 55(1), 119–139 (1997)
9. Schapire, R.E.: The Strenght of Weak Learnability. Machine Learning 5, 197–227 (1990)
10. Freund, Y.: Boosting a Weak Learning Algorithm by Majority. Information and Computation 121, 256–285 (1995)
11. Friedman, J., Hastie, T., Tibshirani, R.: Additive Logistic Regression: A Statistical View of Boosting. The Annals of Statistics 38, 337–374 (2000)
12. Seiffert, C., Khoshgoftaar, T.M., Hulse, J.V., Napolitano, A.: Resampling or Reweighting: A Comparison of Boosting Implementations. In: 2008 20th IEEE International Conference on Tools with Artificial Intelligence, pp. 445–451 (2008)
13. Dmitrienko, A., Chuang-Stein, C.: Pharmaceutical Statistics Using SAS: A Practical Guide. SAS Press (2007)
14. Murphy, P.M., Aha, D.W.: UCI repository for machine learning databases. Technical Report, Department of Information and Computer Science, University of California, Irvine (1994), http://www.ics.uci.edu/~mlearn/databases/
15. Duin, R.P.W., Juszczak, P., Paclik, P., Pekalska, E., de Ridder, D., Tax, D., Verzakov, S.: PR-Tools4.1, A Matlab Toolbox for Pattern Recognition. Delft University of Technology (2007)
16. Highleyman, W.H.: The design and analysis of pattern recognition experiments. Bell System Technical Journal 41, 723–744 (1962)
17. Derrac, J., Garcia, S., Molina, D., Herrera, F.: A practical tutorial on the use of nonparametric statistical tests as a methodology for comparing evolutionary and swarm intelligence algorithms. Swarm and Evolutionary Computation 1(1), 3–18 (2011)
18. Trawinski, B., Smetek, M., Telec, Z., Lasota, T.: Nonparametric statistical analysis for multiple comparison of machine learning regression algorithms. International Journal of Applied Mathematics and Computer Science 22(4), 867–881 (2012)

Vehicle Logo Recognition with an Ensemble of Classifiers

Bogusław Cyganek[1] and Michał Woźniak[2]

[1] AGH University of Science and Technology, Al. Mickiewicza 30, 30-059 Kraków, Poland
cyganek@agh.edu.pl
[2] Wrocław University of Technology, Wybrzeże Wyspiańskiego 27, 50-370 Wrocław, Poland
Michal.Wozniak@pwr.wroc.pl

Abstract. The paper presents a system for vehicle logo recognition from real digital images. The process starts with license plates localization, followed by vehicle logo detection. For this purpose the structural tensor is employed which allows fast and reliable detections even in low quality images. Detected logo areas are classified to the car brands with help of the classifier operating in the multi-dimensional tensor spaces. These are obtained after the Higher-Order Singular Value Decomposition of the prototype logo tensors. The proposed method shows high accuracy and fast operation, as verified by the experiments.

Keywords: Vehicle logo recognition, logo classification, tensor classifiers.

1 Introduction

Recognition of a vehicle brand and model has found interest in recent years [12][13]. Many of the reported systems concentrate on the appearance based approaches which require high computational load. However, recognition of vehicle logos is an alternative way of determining type of a car. This is especially welcome in the intelligent surveillance systems which frequently operate with low quality images and require fast and reliable response. However, the existing solutions lack efficacy or are not suitable for reliable operations with poor quality images. For example in the system proposed by Petrovic and Cootes [12] edge gradient with a refined matching algorithm are proposed. However, the results are sensitive to viewpoint change as well as plane rotation. On the other hand, the statistical moments, as proposed in the paper by Dai *et al.* [5], are very sensitive to noise and geometrical distortions. Other methods rely on object recognition with sparse features. In this respect the SIFT is one of the most frequently used detectors, such as in the system by Psyllos *et al.* [13]. However, computation burden in this case is high, whereas the method is not free from illumination and viewpoint variations, as well as depends on many parameters.

In this paper a system for frontal and rear-view vehicle logo recognition is proposed. The process starts with license plates localization, followed by car logo detection. This is achieved first computing the structural tensor and then looking for highly structured areas which usually correspond to the license plate and logo areas. Detected areas are then fed to the classifier operating in the multi-dimensional pattern tensor subspaces. These are obtained after the Higher-Order Singular Value Decomposition of the

N.T. Nguyen et al (Eds.): ACIIDS 2014, Part II, LNAI 8398, pp. 117–126, 2014.

prototype logo tensors. The proposed method shows high accuracy and fast operation, as verified by the experiments. Apart from the logo localization, also the license plates areas are returned, as will be discussed in the next sections of this paper.

2 System Architecture

Fig. 1 shows the general architecture of the presented system for vehicle logo localization and recognition.

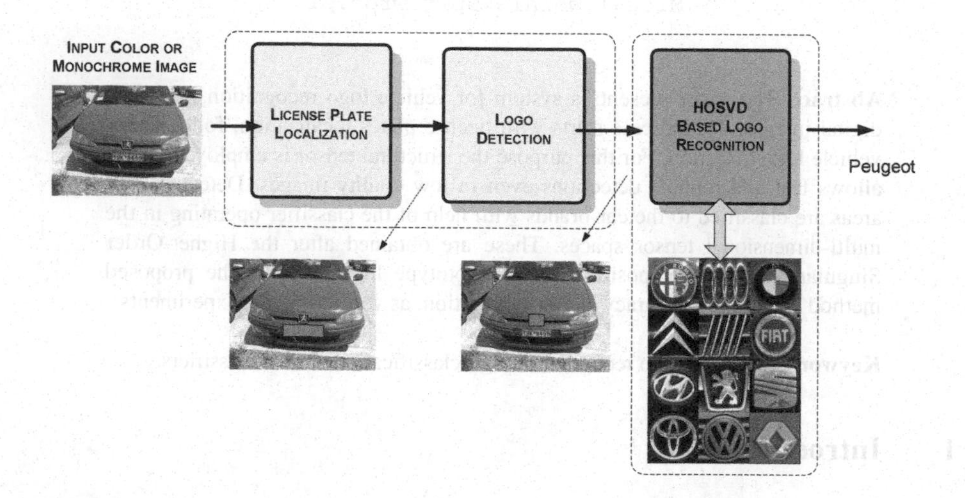

Fig. 1. Architecture of the system for vehicle logo recognition. The processing path includes license plate detection, vehicle logo areas detection, and finally logo classification.

Basically there are two processing stages: vehicle logo detection and classification. However, the former is split into car license plates detection which guides logo area localization. The two detections, i.e. license plates followed by logo areas, are performed with the same algorithm which greatly facilitates the operation. On the other hand, the recognition relies on provided database of the vehicle logos, as depicted in Fig. 1. All three stages are discussed in subsequent sections of this paper.

3 Logo Detection Method

In the presented system, the logo detection follows license plate detection, as shown in the diagram in Fig. 2. However, the two detections are done by the same algorithm which greatly facilitates implementation, as well as improves computation time, as will be discussed. The detection method works the same for the color, as well as monochrome images, since color information is not used. Therefore, if the input image is color, then it is converted to the monochrome version by the weighted averaging of the pixel values.

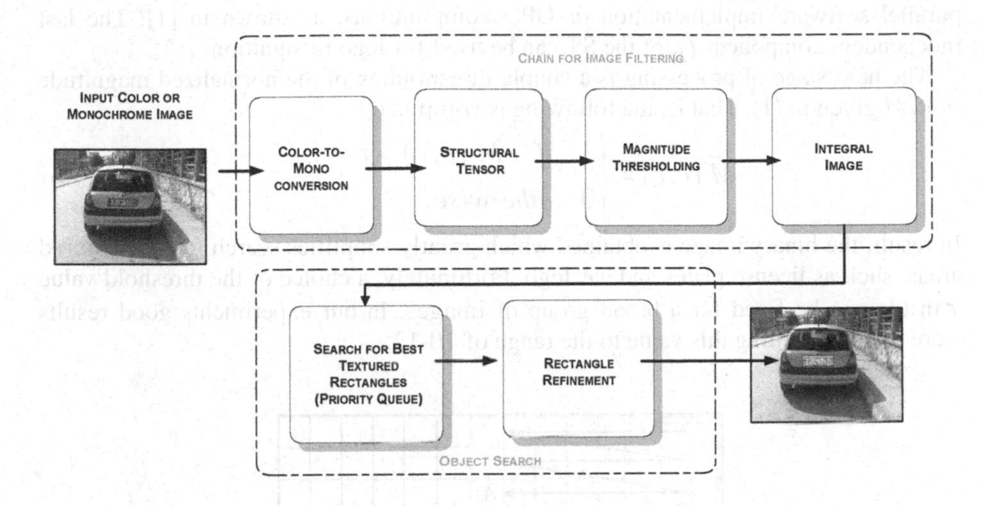

Fig. 2. Architecture of the license plate and logo detector. All information is computed in the monochrome version of the input image thanks to computation of the structural tensor. Places of license plates and car logo are detected searching for rectangular areas fulfilling specific texture conditions.

In the monochrome version of the input image the structural tensor (ST) is computed. Details of this method, as well as the rationale behind using and interpreting components of ST can be found in literature [7][3]. Below, presented are only the formulas which are directly used in the proposed system.

The components of the structural tensor are used to compute the gradient magnitude field *M* given at each pixel position (*c*,*r*), which is given as follows

$$M(c,r) = T_{xx}(c,r) + T_{yy}(c,r), \tag{1}$$

where

$$T_{xx} = \int_{-\infty}^{+\infty} F(x,y)\frac{\partial I}{\partial x}\frac{\partial I}{\partial x}dxdy \text{, and } T_{yy} = \int_{-\infty}^{+\infty} F(x,y)\frac{\partial I}{\partial y}\frac{\partial I}{\partial y}dxdy, \tag{2}$$

denote two of three independent structural tensor components, *h* denotes a smoothing operator and *I* stands for image intensity. Certainly, for digital images the above need to be converted into the domain of discrete signals, as follows

$$\hat{T}_{xx} = \hat{F}\left(\hat{G}_x(I)\cdot\hat{G}_x(I)\right) \text{, and } \hat{T}_{yy} = \hat{F}\left(\hat{G}_y(I)\cdot\hat{G}_y(I)\right). \tag{3}$$

In other words, computation of the two structural tensor components requires subsequent computations of the x and y gradient fields, respectively, followed by their multiplications. For the gradient filters $\hat{G}$ the Simoncelli filters were chosen, whereas $\hat{F}$ is a simple binomial filtering [3]. Since the aforementioned filters are separable, the above operations can be significantly speeded up either with help of the multi-core parallel software implementation or GPU computations, as shown in [1]. The last independent component T_{xy} of the ST can be used for logo recognition.

The next stage of processing is a simple thresholding of the normalized magnitude field M given in (1). That is, the following is computed

$$\bar{M}(c,r)=\begin{cases}1 & if \quad M(c,r)>\tau \\ 0 & otherwise.\end{cases} \tag{4}$$

In result, the binary image is obtained which greatly simplifies search for the textured areas, such as license plates and car logo. Fortunately, a choice of the threshold value τ in (4) can be fixed for a broad group of images. In our experiments good results were obtained setting this value to the range of 10-12.

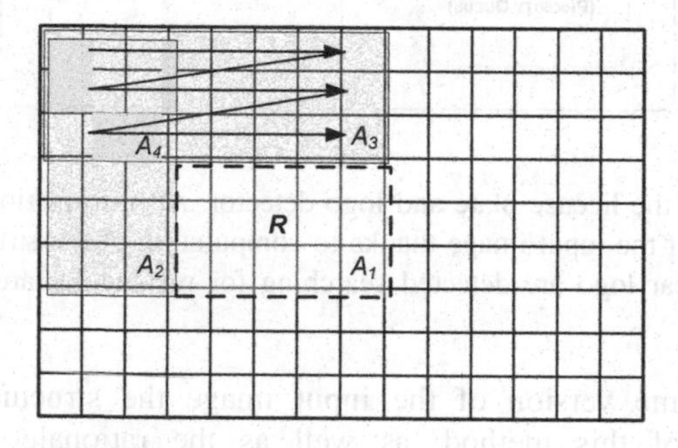

Fig. 3. Computation of the sum of pixel values in the region R requires only three additions in the space of the integral image: $R=(A_1+A_4)-(A_2+A_3)$

Search for license plates and car logo location is performed in the binarized field $\bar{M}$. For this purpose, in each location a sum of internal pixels is computed. However, to speed-up computations, before traversing the whole space, the integral image data structure is computed. Thanks to this, for any rectangular regions R, a sum of all its pixels is computed simply with only two additions and one subtraction, as follows

$$\sum_{(c,r)\in R} \bar{M}(c,r)=(A_1+A_4)-(A_2+A_3), \tag{5}$$

where $A_i(c,r)$ stands for a sum of all pixels above and to the left of the point with coordinates (c,r), as shown in Fig. 3.

For all rectangles of a predefined size the sum of all their pixels is computed, as described. The rectangles are successively inserted to the priority queue depending on their sum value. In other words, due to the properties of the structural tensor, the higher this sum, the more regularly structured region is detected, as expected for the car license plates and logos. In our experiments size of the checked rectangles follows expected size of the license plate which was set to *plate_h* = {34, 44, 64} and *plate_w* = 4.7 * *plate_h*, as denoted in Fig. 4. Dimensions of the license plate are set depending on an average size of cars expected in the images. Dimensions of the logo search area are computed based on the license plate dimensions. In our experiments the latter were set *logo_w* = *plate_w*, whereas *logo_h* = 4 * *plate_h*.

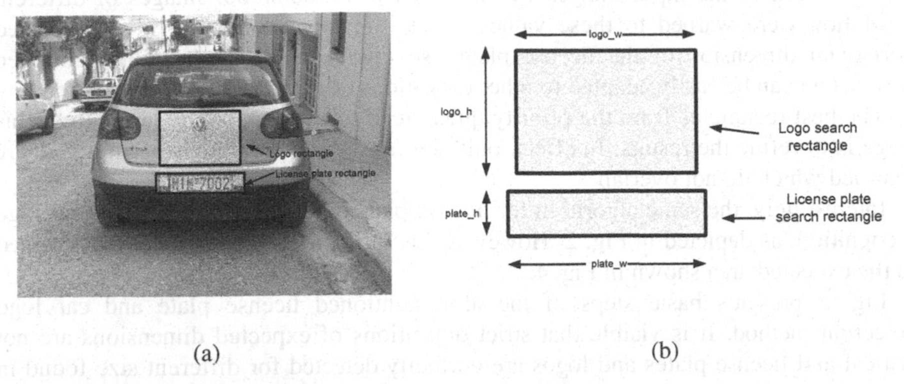

Fig. 4. A model for license plate and logo search areas (a). Dimensions of the license plate are set depending on an average size of cars expected in the images. Dimensions of the logo search area are computed based on the license plate dimensions (b).

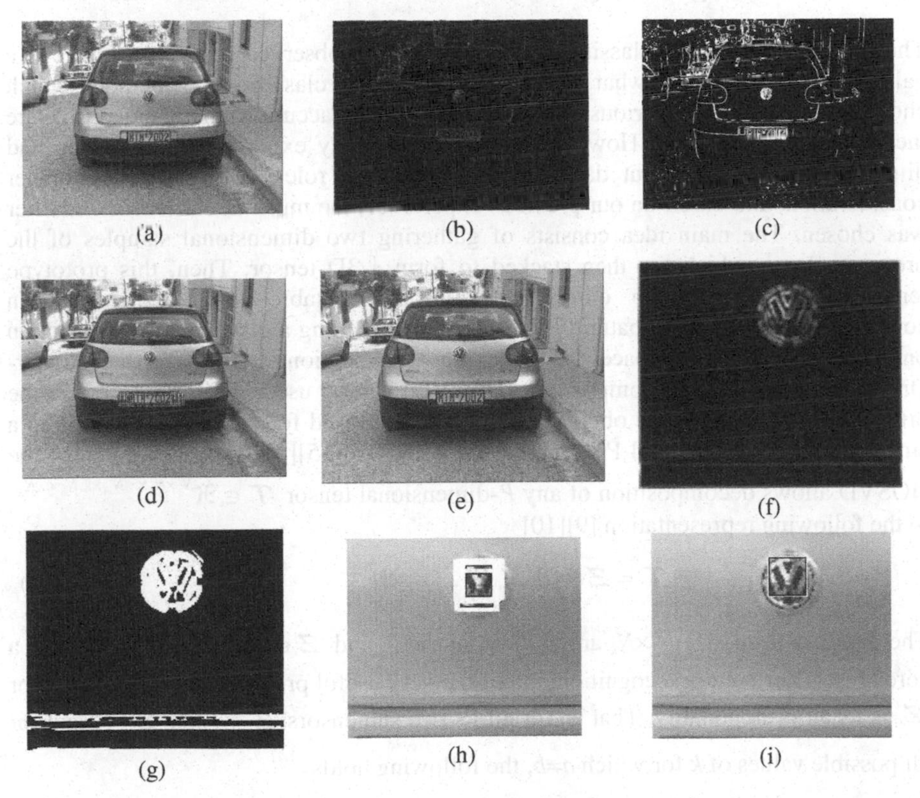

Fig. 5. Basic steps of license plate and car logo detection. The input image (a). The tensor magnitude field M (b). Binarized field M (c). License plate best candidate rectangles (d). The selected rectangle after refinement (e). The car logo search rectangle (f). Binarized field of the car logo (g). Candidate rectangles (h). Final logo localization marked with a fixed window size (i).

Resolution of the input images was assumed to be 600x800. Images of different resolution were warped to these values. Thus, the input parameter is an expected horizontal dimension of the license plate. Nevertheless, all of the aforementioned parameters can be easily adapted to other conditions of image acquisition front-end.

The best rectangles from the priority queue tend to overlap, so the last step of this stage is to refine the results. In effect, only the most structured areas of an image are returned which do not overlap.

Interestingly, the same algorithm for license plate detection is used for the car logo recognition, as depicted in Fig. 2. However, this time the search region is constrained to the expected area shown in Fig. 4.

Fig. 5 presents basic steps of the aforementioned license plate and car logo detection method. It is visible that strict definitions of expected dimensions are not critical and license plates and logos are correctly detected for different size found in an image.

4 Vehicle Logo Recognition with the Tensor Based Classifiers

The task of vehicle logo classification is to tell if an observed part of an image is a valid logo and, if so, of what class. There are many classification methods which choice is determined by various factors. From these the accuracy and respond time are the most important once. However, as shown in many experiments, also type and dimensionality of the input data plays an important role. Taking all these under consideration, and based on our previous experience, the multi-linear tensor classifier was chosen. The main idea consists of gathering two dimensional samples of the prototype logos which are then stacked to form a 3D tensor. Then, this prototype tensor is decomposed to obtain the subspace suitable for classification. In consequence, an unknown pattern is classified by checking a distance of its projection onto tensor spanned subspace. To build the aforementioned subspace, the Higher-Order Singular Value Decomposition (HOSVD) can be used [1][9][8]. In effect the orthogonal tensor bases are obtained which are then used for pattern recognition in a similar way to the standard PCA based classifiers [6][15][16]. Let us recall that the HOSVD allows decomposition of any P-dimensional tensor $\mathcal{T} \in \Re^{N_1 \times N_2 \times \ldots N_m \times \ldots N_n \times \ldots N_P}$ to the following representation [9][10]

$$\mathcal{T} = \mathcal{Z} \times_1 \mathbf{S}_1 \times_2 \mathbf{S}_2 \ldots \times_P \mathbf{S}_P . \tag{6}$$

The matrices $\mathbf{S}_k$ of size $N_k \times N_k$ are unitary matrices, and $\mathcal{Z} \in \Re^{N_1 \times N_2 \times \ldots N_m \times \ldots N_n \times \ldots N_P}$ is a core tensor. For pattern recognition one of the very useful properties of the core tensor $\mathcal{Z}$ is its all-orthogonality. That is, for all its two subtensors $\mathcal{Z}_{n_k=a}$ and $\mathcal{Z}_{n_k=b}$, and for all possible values of k for which $a \neq b$, the following holds

$$\mathcal{Z}_{n_k=a} \cdot \mathcal{Z}_{n_k=b} = 0 . \tag{7}$$

An algorithm for computation of the HOSVD is based on successive computations of the SVD decomposition of the matrices actually being the flattened versions of the tensor $\mathcal{T}$. The algorithm requires a number of the SVD computations which is equal to the valence of that tensor.

Further, thanks to the commutative properties of the k-mode tensor multiplication [14], the following sum can be constructed for each mode matrix $\mathbf{S}_i$ in (6)

$$\mathcal{T} = \sum_{h=1}^{N_P} \mathcal{T}_h \times_P \mathbf{s}_P^h , \tag{8}$$

where the tensors

$$\mathcal{T}_h = \mathcal{Z} \times_1 \mathbf{S}_1 \times_2 \mathbf{S}_2 \ldots \times_{P-1} \mathbf{S}_{P-1} \tag{9}$$

create the orthogonal tensors basis. The vectors $\mathbf{s}^h{}_P$ are just columns of the unitary matrix $\mathbf{S}_P$. Because each $\mathcal{T}_h$ is of dimension P-1 then $\times_P$ in (8) represents an outer product, which is a product of two tensors of dimensions P-1 and 1. Moreover, due to the mentioned all-orthogonality property (7) of the core tensor $\mathcal{Z}$ in (9), $\mathcal{T}_h$ are also orthogonal. Thanks to this, they span and orthogonal subspace which is used for pattern classification, as described.

In the tensor subspace, pattern recognition consists of computing a distance of a test pattern $\mathbf{P}_x$ to its projections onto each of the spaces spanned by the set of the bases $\mathcal{T}_h$ in (9). As shown by Savas and Eldén, this can be written as the following minimization problem [14]

$$\min_{i,c_h^i} \underbrace{\left\| \mathbf{P}_x - \sum_{h=1}^{H} c_h^i \mathcal{T}_h^i \right\|^2}_{Q_i} . \tag{10}$$

In the above, the scalars $c^i{}_h$ are unknown coordinates of $\mathbf{P}_x$ in the subspace spanned by $\mathcal{T}_h^i$. The value of $H \le N_P$ controls a number of chosen dominating components.

It can be shown that to minimize (10) the following value needs to be maximized

$$\hat{\rho}_i = \sum_{h=1}^{H} \left\langle \hat{\mathcal{T}}_h^i, \hat{\mathbf{P}}_x \right\rangle^2 , \tag{11}$$

Thanks to the above, the HOSVD classifier returns a class i for which its ρ_i from (11) is the largest. The C++ implementation of the above classification algorithm is discussed in [2].

5 Experimental Results

The presented method was implemented in C++ using the *HIL* and *DeRecLib* software libraries [2][3]. Experiments were run on the PC with 8 GB RAM and Pentium Q 820 microprocessor. All of the presented databases come from the Medialab site [11].

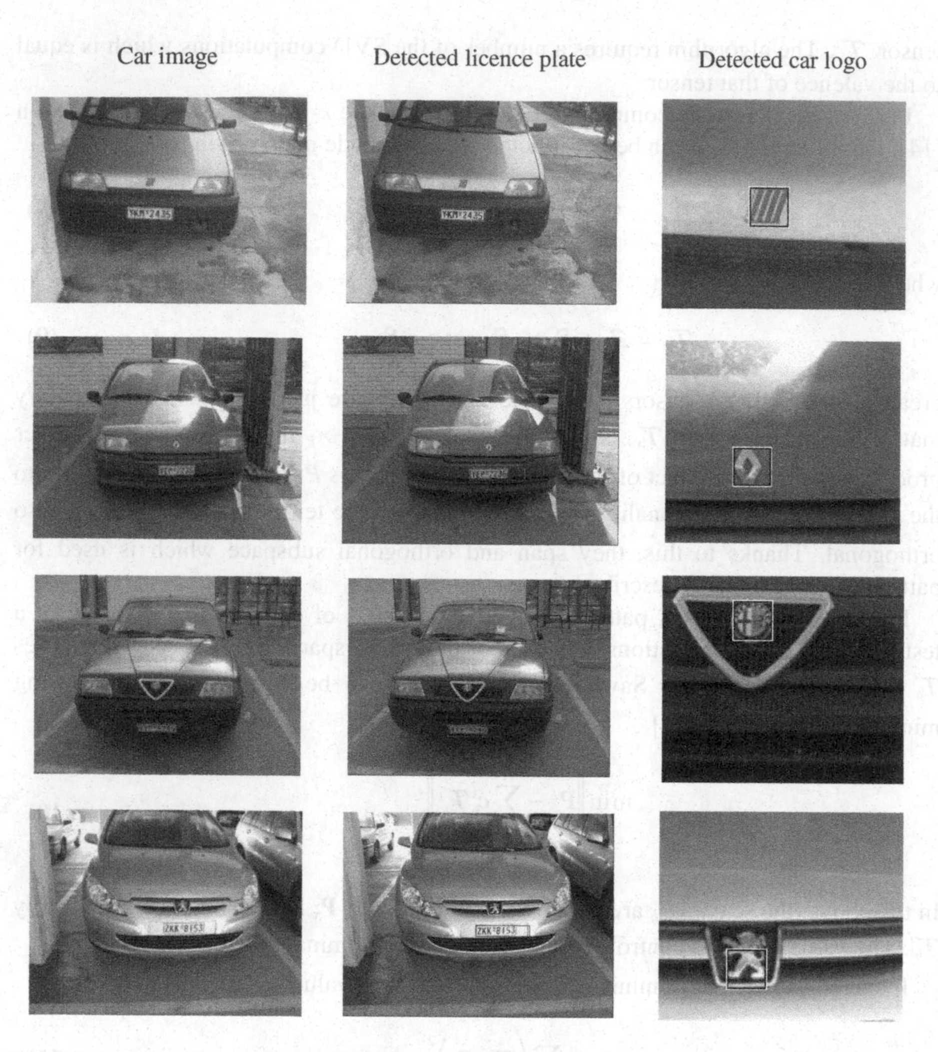

Fig. 6. Examples of correctly detected initial locations of the license plates and car logos from the car examples in the Medialab database [11]. These are then refined with the adaptive window growing method to tightly outline the detected logo.

In the experiments we separately address the accuracy of logo detection, as well as its subsequent classification. It is obvious that false detection always lead to incorrect classification. Fig. 6 shows results of the license plate and car logo detection for some exemplary patterns from the Medialab database [11]. The assumed logo size is 34x34. Therefore, after a successful detection this region is grown by the adaptive window growing method, as described in [3]. However, detections not always are correct, as shown in Fig. 7. Usually such situations arise in the cases of poorly visible license plates, low quality images or other texture areas which erroneously are taken to be license places. However, this happens relatively rarely and in the future research we plan to make the method resistant to such situations.

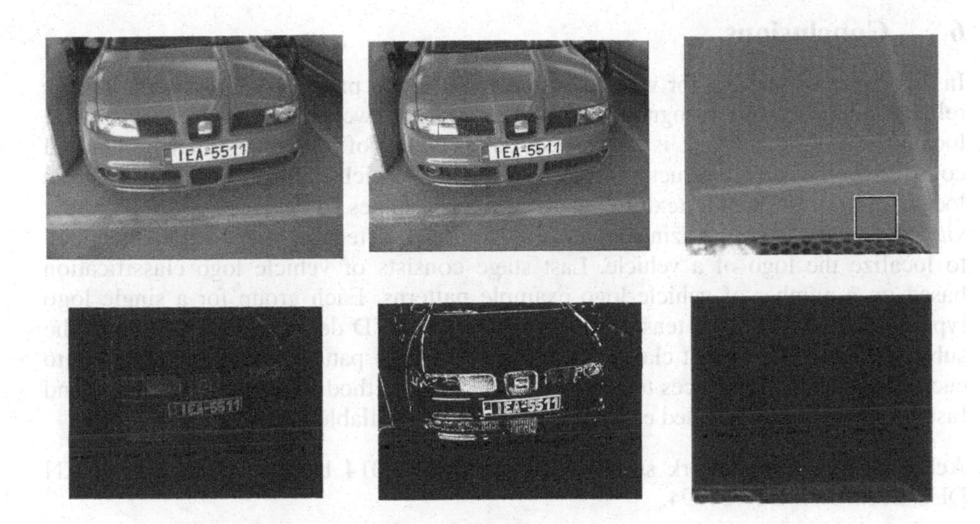

Fig. 7. An example of erroneously detected license places which forbids logo recognition

Last processing stage consists of car logo recognition with the HOSVD multi-class classifier, trained with the car logo prototypes presented in Fig. 8 as described in the previous section.

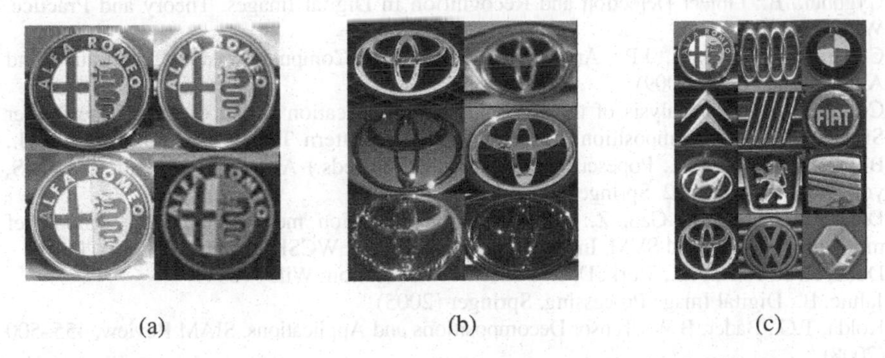

(a) (b) (c)

Fig. 8. Databases of the car logos used to train the tensor based multi-classification system. The Alfa Romeo brand (a), Toyota (b), all types of logos used in the presented experiments (c).

The detection and recognition accuracies measured independently on the Medialab database are as follows:

- Logo detection: 94%
- Logo recognition: 92%

Thus, the overall recognition accuracy of the system can be assessed as 87%, although sometimes false recognitions happen not only due to false detection but also for properly detected logos. The benefit of the presented system is the execution time which allows processing of 8-10 frames per second in pure software implementation, while 30 frames per second with the GPU acceleration of the ST [1]. As an output the system returns both, the recognized logo of a car, as well as its license plate.

6 Conclusions

In this paper the system for vehicle logo recognition is presented. The system allows reliable and very fast recognition of a logo type, as well as provides license plate localization. Thc system is organized as a chain of classifiers. The front-end constitutes the novel structural places detector which allows fast and reliable localization of the highly textured areas of license plates in the front, as well as rear views of a car. After localizing a plate area, the same detection method is then applied to localize the logo of a vehicle. Last stage consists of vehicle logo classification based on a number of vehicle logo example patterns. Each group for a single logo type is stacked to form a tensor. This after the HOSVD decomposition provides the subspace particular to that class of a logo. A test logo pattern is then projected onto each of the tensor subspaces to provide its class. A method allows high accuracy and fast computations as verified experimentally on the available databases.

Acknowledgements. Work supported in the year 2014 by the Polish grant NCN DEC-2011/01/B/ST6/01994.

References

1. Bugaj, M., Cyganek, B.: GPU Based Computation of the Structural Tensor for Real-Time Figure Detection. In: Proceedings of the 20th International Conference on Computer Graphics, Visualization and Computer Vision (WSCG 2012), Czech Republic (2012)
2. Cyganek, B.: Object Detection and Recognition in Digital Images. Theory and Practice. Wiley (2013)
3. Cyganek, B., Siebert, J.P.: An Introduction to 3D Computer Vision Techniques and Algorithms. Wiley (2009)
4. Cyganek, B.: An Analysis of the Road Signs Classification Based on the Higher-Order Singular Value Decomposition of the Deformable Pattern Tensors. In: Blanc-Talon, J., Bone, D., Philips, W., Popescu, D., Scheunders, P. (eds.) ACIVS 2010, Part II. LNCS, vol. 6475, pp. 191–202. Springer, Heidelberg (2010)
5. Dai, S., Huang, H., Gao, Z.: Vehicle-logo recognition method based on Tchebichef moment invariants and SVM. In: Software Engineering, WCSE 2009, pp. 18–21 (2009)
6. Duda, R.O., Hart, P.E., Stork, D.G.: Pattern Classification. Wiley (2001)
7. Jahne, B.: Digital Image Processing. Springer (2005)
8. Kolda, T.G., Bader, B.W.: Tensor Decompositions and Applications. SIAM Review, 455–500 (2008)
9. de Lathauwer, L.: Signal Processing Based on Multilinear Algebra. PhD dissertation, Katholieke Universiteit Leuven (1997)
10. de Lathauwer, L., Moor de, B., Vandewalle, J.: A Multilinear Singular Value Decomposition. SIAM Journal of Matrix Analysis and Applications 21(4), 1253–1278 (2000)
11. Medialabdatabase (2013), `http://www.medialab.ntua.gr/research/LPRdatabase.html`
12. Petrovic, V.S., Cootes, T.F.: Analysis of features for rigid structure vehicle type recognition. In: Proc. BMVC (2004)
13. Psyllos, A.P., Anagnostopoulos, C.-N.E., Kayafas, E.: Vehicle logo recognition using a sift-based enhanced matching scheme. IEEE ITS 11(2), 322–328 (2010)
14. Savas, B., Eldén, L.: Handwritten digit classification using higher order singular value decomposition. Pattern Recognition 40, 993–1003 (2007)
15. Theodoridis, S., Koutroumbas, K.: Pattern Recognition, 4th edn. Academic Press (2009)
16. Turk, M., Pentland, A.: Eigenfaces for recognition. Journal of Cognitive Neuroscience 3(1), 71–86 (1991)

Optimization Algorithms for One-Class Classification Ensemble Pruning

Bartosz Krawczyk and Michał Woźniak

Department of Systems and Computer Networks,
Wrocław University of Technology, Wrocław, Poland
{bartosz.krawczyk,michal.wozniak}@pwr.wroc.pl

Abstract. One-class classification is considered as one of the most challenging topics in the contemporary machine learning. Creating Multiple Classifier Systems for this task has proven itself as a promising research direction. Here arises a problem on how to select valuable members to the committee - so far a largely unexplored area in one-class classification. Recently, a novel scheme utilizing a multi-objective ensemble pruning was proposed. It combines selecting best individual classifiers with maintaining the diversity of the committee pool. As it relies strongly on the search algorithm applied, we investigate here the performance of different methods. Five algorithms are examined - genetic algorithm, simulated annealing, tabu search and hybrid methods, combining the mentioned approaches in the form of memetic algorithms. Using compound optimization methods leads to a significant improvement over standard search methods. Experimental results carried on a number of benchmark datasets proves that careful examination of the search algorithms for one-class ensemble pruning may greatly contribute to the quality of the committee being formed.

Keywords: machine learning, one-class classification, classifier ensemble, ensemble pruning, classifier selection, diversity.

1 Introduction

One-class classification (OCC) is a specific subfield of machine learning. During the classifier training step there are at disposal only objects from a single class, called the target concept. It is assumed that at the exploitation phase of such a classifier there may appear new, unseen objects, called outliers. Therefore OCC aims at establishing a boundary that separates the target objects from possible outliers [21]. The term single-class classification was introduced in [10], but also outlier detection or novelty detection [3] are used to name this field of study.

OCC is a difficult task and there are many open problems related to it. One of the most prominent is how the target class boundary should be tuned - in case of being too general unwanted outliers would be accepted, in case of being too matched to the training set a strong overfitting may occur. From this one may see that it is risky to rely only on a single given model. In recent years there

N.T. Nguyen et al. (Eds.): ACIIDS 2014, Part II, LNAI 8398, pp. 127–136, 2014.

have been several successful attempts on how to improve the quality of one-class recognition systems. One of them is the ensemble approach utilizing outputs of more than one model [16,22].

Multiple classifier systems (MCSs) despite the well established status they are the subject of ongoing intense research. Among many factors entangled in the design process of MCS one of the uttermost importance is how to select classifiers to ensure the high quality of the ensemble. Combining similar classifiers do not contribute to the system being constructed, apart from increasing the computational complexity. In some cases (e.g. voting fusion methods) it may even decrease the quality of the committee. Therefore selected members should display characteristics unique to each of them in order to improve the quality of the collective decision. There are two main research trends in this area - how to assure the diversity among the individual classifiers in the pool and how to measure it efficiently [2].

One should bore in mind that diversity itself is not the perfect criterion for classifier selection. It is easy to imagine a situation in which two classifiers have a high diversity in comparison to each other but at the same time one of them (or even both) is of low quality. When using diversity for the ensemble pruning such models would be selected but at the same time the quality of the MCS will drop. In an ideal situation the committee should consist of models that are competent and mutually complementary i.e., each of the classifiers should display a high individual accuracy and a high diversity when compared to other members.

In our previous works [13,14], we proposed a new way of the classifier selection designed for the specific nature of OCC. We simultaneously utilized two criteria - one responsible for the predictive quality of classifiers and one for the diversity of the ensemble. This way we prevent our committee from consisting of models too weak or too similar to each other.

This was achieved by solving the optimization problem, stated as finding the best subset of classifiers from the pool. Our findings showed, that using proposed compound ensemble pruning algorithm lead to a significant improvement of the overall quality of the one-class classifier committee. However, we noted that the optimization algorithm being used may have a crucial impact on the selection process.

In this paper, we present a follow-up to this research direction. Based on the proposed multi-criteria classifier selection model, several different popular optimization algorithms are examined. We check the behavior of genetic algorithm, simulated annealing, tabu search and hybrid methods, combining the mentioned approaches in the form of memetic algorithms.

The extensive experimental results, backed up with the test of a statistical significance prove that a careful tuning of the search algorithm may lead to an improved ensemble quality.

2 Combining One-Class Classifiers

The problem of building MCSs on the basis of one-class still awaits for proper attention. There are some papers dealing with the proposals on how to combine one-class classifiers [22], but most of them are oriented on the practical application [7], not on theoretical advances.

One-class boundary methods base their output on computing the distance the object x and the decision boundary that encloses the target class ω_T. Therefore, it is impossible to directly apply a fusion method which requires the values of support functions. Hence, we need some way to map the distance into probability.

We propose to use the following heuristic solution:

$$\widehat{F}(x, \omega_T) = \frac{1}{c_1} exp(-d(x|\omega_T)/c_2), \tag{1}$$

which models a Gaussian distribution around the classifier, where $d(x|\omega_T)$ is an Euclidean distance metric between the considered object and a decision boundary, c_1 is the normalization constant and c_2 is the scale parameter. Parameters c_1 and c_2 should be fitted to the target class distribution.

After such a mapping it is possible to combine OCC models based on their support functions [20]. Let us assume that there are L OCC classifiers in the pool. In this paper we use the mean of the estimated probabilities which is expressed by:

$$y_{mp}(x) = \frac{1}{L} \sum_k (P_k(x|\omega_T). \tag{2}$$

This fusion method assumes that the outliers objects distribution is independent of x and thus uniform in the area around the target concept.

3 Model Selection for One-Class Classification

Ensemble pruning (also referred to as classifier selection) is a process of high importance for forming ensembles. During it, one should take two main criteria under consideration: Classifier selection (known also as ensemble pruning) plays an important role in the process of the MCS design.

- accuracy - adding weak classifiers with a low competence will decrease the overall quality of the MCS;
- diversity - adding similar classifiers will contribute nothing to the ensemble apart from increasing the computational cost.

Each of these criteria has its drawbacks - highly accurate classifiers may not be diverse, while diverse classifiers may not be accurate in their own. Both of these criteria are popular in ensemble pruning for multi-class cases. Yet for the specific problem of OCC there is still little work so far on how to efficiently evaluate the given pool of classifiers. In [4] it is suggested to prune the ensemble according to the individual performance of classifiers, while in our previous works we have proposed a novel measures for describing diversity for OCC [15].

In this paper, two separate criteria are used - consistency measure to rank classifiers according to their quality (as we have only the objects coming from the target class, we cannot use standard accuracy measure) and energy measure which we introduce as a novel diversity measure tuned to the nature of OCC.

3.1 Consistency Measure

The consistency measure indicates how consistent a pool of classifiers is in rejecting fraction f of the target data [19]. One may compute it by comparing the rejected fraction f with an estimate of the error on the target class $\widehat{\varepsilon^t}$:

$$Cons(\Pi^l) = |\widehat{\varepsilon^t} - f|, \tag{3}$$

where Π^l is the tested pool of classifiers. This is an unsupervised measure well suitable for OCC problems as we need only the estimation of error on the target class - no information about outliers is required.

3.2 One-Class Energy Measure

Energy approach is an effective measure of fuzziness, successfully implemented in many practical applications such as ECG analysis [5]. Let's assume that there are L classifiers in the pool, out of which S classifiers can correctly classify a given training object $x_j \in X$ to ω_T. Additionally a threshold $\lambda \in [0, 1]$ is introduced. It's role is to filter insignificant degrees of membership, that may otherwise contribute to decreasing the stability of the proposed measure. The energy measure is described as follows:

$$EN_{oc}(\Pi^l) = \int_X \sum_{i=1}^{L} f_\lambda(x)dx, \tag{4}$$

where

$$f_\lambda(x) = f(x) \Leftrightarrow \frac{\sum_{m=1}^{M} \sum_{i=1}^{i_m} \delta(\Psi_{i_M}^M(x), \Psi^*(x))}{L} > \lambda, \tag{5}$$

and $\Psi^*(x)$ denotes a classifier correctly classifying the object x and $f(x) : [0, 1] \rightarrow R_+$ is an increasing function in interval [0,1] for $f(0) = 0$.

4 Proposed Method

In this paper, we propose to select OCC classifiers to the committee according to the combination of both of these criteria, hoping that this will allow to combine their strong points while becoming more robust to flaws exhibited by each of them.

Let us formulate the multi-objective optimization criterion as:

$$maximize\ g(\Pi^l) = Cons(\Pi^l) + EN_{oc}(\Pi^l),, \tag{6}$$

where Π^l is the given pool of classifiers that will undergo an ensemble pruning procedure, $Cons(\Pi^l)$ stands for the overall consistency of the given ensemble and $EN_{oc}(\Pi^l)$ is the diversity of the considered ensemble expressed by the mentioned One-Class Energy Measure.

The selected pool can be encoded for the search algorithms as a binary mask:

$$C = [C_1, C_2, ..., C_L], \tag{7}$$

with 1s indicating the chosen individual classifiers (i.e., if we have 10 classifiers then 0010110010 would indicate that classifiers 3, 5, 6, and 9 are chosen for the ensemble).

To achieve this we propose to use a multi-objective optimization, conducted with the usage of one of the following search algorithms. Apart from individual setting, each of the examined algorithms uses the following parameters:

- N_c - the upper limit of algorithm cycles,
- V - the upper limit of algorithm iterations without quality improvement.

We have examined the following five optimization procedures:

- Genetic Algorithm (GA) - a popular nature-inspired search heuristic. It is based on a population of candidates, which is evolved toward better solutions. Each candidate solution has a set of properties which can be mutated and altered. Previous works showed that GA are effective for OCC classifier selection [12]. The control parameters of the GA algorithm are as follows:
  - N_p - the population quantity,
  - β - the mutation probability,
  - γ - the crossover probability,
  - Δ_m - the mutation range factor,
- Simulated Annealing (SA) - a probabilistic metaheuristic for the global optimization. At each step it considers neighbouring state of the current state, and probabilistically decides between moving the system from state to state or remaining in the present location. These probabilities ultimately lead the system to move to states of lower energy. The control parameters of the SA algorithm are as follows:
  - T - the temperature
  - α - the cooling factor
- Tabu Search (TS) - uses a local or neighborhood search procedure to iteratively move towards an improved solution. It uses memory structures form, known as the tabu list. It is a set of rules and banned solutions used to filter which solutions will be admitted to the neighborhood to be explored by the search. The control parameters of the TS algorithm are as follows:
  - S - the tabu list size
- Memetic Algorithm (MA) - be seen as a hybrid solution that fuses together different metaheuristics in hope to use gain advantage from combining their strong points [8]. The idea of MAs is based on the individual improvement plus population cooperation. Unlike traditional Evolutionary Algorithms (EA), MAs are biased towards exploiting all the knowledge about the problem under study. By this they may be seen as less random and more directed search method. In this work we examine two types of memetic algorithms:
  - using GA for exploration and SA for exploitation,
  - using GA for exploration and TS for exploitation.

5 Experimental Results

The aims of the experiment was to check the level of importance of selecting a proper search algorithm to prune the one-class classifier ensembles - to how much the choice of the optimization approach determines the quality of the ensemble. Additionally, we waned to see if using a hybrid optimization (memtic algorithms) leads to a significantly better performance in comparison with standard solutions.

We have chosen 10 binary datasets in total - 9 coming from UCI Repository and an additional one, originating from chemoinformatics domain and describing the process of discovering pharmaceutically useful isoforms of CYP 2C19 molecule. The dataset is available for download at [17].

The objects from the minor class were used as the target concept, while objects from the major class as outliers.

Details of the chosen datasets are given in Table 1.

Table 1. Details of datasets used in the experimental investigation. Numbers in parentheses indicates the number of objects in the minor class in case of binary problems

No.	Name	Objects	Features	Classes
1	Breast-cancer	286 (85)	9	2
2	Breast-Wisconsin	699 (241)	9	2
3	Colic	368 (191)	22	2
4	Diabetes	768 (268)	8	2
5	Heart-statlog	270 (120)	13	2
6	Hepatitis	155 (32)	19	2
7	Ionosphere	351(124)	34	2
8	Sonar	208 (97)	60	2
9	Voting records	435 (168)	16	2
10	CYP2C19 isoform	837 (181)	242	2

For the experiment a Support Vector Data Description [18] with a polynomial kernel is used as a base classifier. The pool of classifiers were homogeneous, i.e. consisted of classifiers of the same type.

The pool of classifiers consisted in total of 30 models build on the basis of a Random Subspace [9] approach with each subspace consisting of 40 % of original features.

The thershold parameter λ for One-class Energy Measure was set to 0.1 and a hyperbolic tangent was selected as the $f(x)$ function.

The combined 5x2 cv F test [1], tuned to OCC problems according to a scheme presented in [11], was carried out to asses the statistical significance of obtained results. Additionally the Friedman ranking test [6] was done for comparison over multiple segmentations datasets.

Table 2. Details of optimization algorithm parameters, used in the experiments. Values presented are initial ranges / best values.

Parameter	GA	SA	TS	MA (GA+SA)	MA (GA + TS)
N_c	[100:500] / 400	[500:2000] / 1500	[100:500] / 500	[100:500] / 300	[100:500] / 300
V	[10:50] / 30	[10:50] / 40	[10:50] / 20	[10:50] / 35	[10:50] / 25
N_p	[20:100] / 50	-	-	[20:100] / 40	[20:100] / 40
β	[0.2:0.8] / 0.7	-	-	[0.2:0.8] / 0.7	[0.2:0.8] / 0.7
γ	[0.2:0.8] / 0.3	-	-	[0.2:0.8] / 0.3	[0.2:0.8] / 0.3
Δ_m	[0.1:0.5] / 0.2	-	-	[0.1:0.5] / 0.2	[0.1:0.5] / 0.2
T	-	[1000:10000] / 5500	-	[1000:10000] / 3500	-
α	-	[0.5:0.9] / 0.9	-	[0.5:0.9] / 0.9	-
S	-	-	[3:10] / 9	-	[3:10] / 7

The parameters for the optimization procedures were selected with the usage of the grid-search procedure. The detailed ranges of tested values and best selected parameters are given in Table 2.The initial range values were selected on the basis of our previous experienced with optimization-based ensembles [23].

The results are presented in Tab. 3.

Table 3. Results of the experimental results with the respect to the accuracy [%] and statistical significance

No.	GA[1]	SA[2]	TS[3]	MA (GA+SA)[4]	MA (GA+TS)[5]
1.	54.21 3	54.03 3	53.41 –	58.54 1,2,3	60.03 *ALL*
2.	86.34 –	87.87 1,3	86.79 –	89.43 1,2,3	89.91 1,2,3
3.	73.23 –	73.46 –	73.20 –	75.00 1,2,3	76.02 *ALL*
4.	58.54 3	59.12 3	57.43 –	63.45 1,2,3	63.85 1,2,3
5.	86.12 –	86.33 –	85.96 –	87.85 1,2,3	89.57 *ALL*
6.	64.32 2	62.08 –	63.21 2	66.32 1,2,3	68.00 *ALL*
7.	75.87 –	75.73 –	75.87 –	76.16 –	76.29 –
8.	87.34 1,2	85.32 –	86.14 –	93.54 *ALL*	92.45 1,2,3
9.	86.81 –	86.43 –	87.94 1,2	88.21 1,2	90.13 *ALL*
10.	75.54 3	76.12 3	73.84 –	80.06 1,2,3	82.06 *ALL*
Avg. rank	3.29	3.62	4.33	2.10	1.66

5.1 Results Discussion

The experimental investigations clearly prove that the selected search method has a crucial influence on the quality of the formed ensemble. Only for a single case all the optimization method returned similar results.

Out of all tested methods, the weakest performance was returned by the Tabu Search. This may be explained by its inability to efficiently explore the search space and tendency towards exploiting local neighborhood.

Genetic algorithm and simulated annealing returned similar results, as they both put an emphasis on efficient exploration of the set of states (which is achieved in GA with cross-over and mutation, and in SA with rapid decrease of energy).

Interestingly, the simpler approaches were in most cases outperformed by hybrid memetic algorithm. This results prove, that population implementation allows for an efficient exploration, while local improvement is more stable than in traditional GA.

Better results were returned by the combination of GA and TS, which is quite interesting considering the unsatisfactory performance of single TS. This can be explained by the fact, that GA deals with finding local minimums and requires an efficient local search to fully exploit them. In such situations TS can more easily find good solutions, as the search space is reduced and it may effectively use its tabu list.

6 Conclusions

The paper discussed the idea of pruning one-class ensembles. In this work we have introduced a novel approach for selecting OCC models to the committee using the fusion of two separate criteria - consistency and diversity. This allowed to select classifiers to the OCC ensemble in such a way that they at the same time display a high recognition quality and are not similar to each other. The performance of several multi-objective optimization algorithms was tested. We wanted to see to what extend the used search method influences the performance of the ensemble.

Experimental investigations, executed on benchmark datasets, proved the quality of our approach. We conclude, that using hybrid memetic algorithms can lead to a statistically significant boost in the ensemble quality, while making it robust to weak models in the pool.

Acknowledgment. Bartosz Krawczyk was supported by the statutory funds of the Department of Systems and Computer Networks, Faculty of Electronics, Wrocław University of Technology dedicated for Young Scientists.

Michał Woźniak was supported by EC under FP7, Coordination and Support Action, Grant Agreement Number 316097, ENGINE European Research Centre of Network Intelligence for Innovation Enhancement (http:// engine.pwr.wroc.pl/).

References

1. Alpaydin, E.: Combined 5 x 2 cv f test for comparing supervised classification learning algorithms. Neural Computation 11(8), 1885–1892 (1999)
2. Bi, Y.: The impact of diversity on the accuracy of evidential classifier ensembles. International Journal of Approximate Reasoning 53(4), 584–607 (2012)
3. Bishop, C.M.: Novelty detection and neural network validation. IEE Proceedings: Vision, Image and Signal Processing 141(4), 217–222 (1994)
4. Cheplygina, V., Tax, D.M.J.: Pruned random subspace method for one-class classifiers. In: Sansone, C., Kittler, J., Roli, F. (eds.) MCS 2011. LNCS, vol. 6713, pp. 96–105. Springer, Heidelberg (2011)
5. Czogala, E., Leski, J.: Application of entropy and energy measures of fuzziness to processing of ecg signal. Fuzzy Sets and Systems 97(1), 9–18 (1998)
6. Demšar, J.: Statistical comparisons of classifiers over multiple data sets. J. Mach. Learn. Res. 7, 1–30 (2006)
7. Giacinto, G., Perdisci, R., Del Rio, M., Roli, F.: Intrusion detection in computer networks by a modular ensemble of one-class classifiers. Inf. Fusion 9, 69–82 (2008)
8. Harman, M., McMinn, P.: A theoretical and empirical study of search-based testing: Local, global, and hybrid search. IEEE Transactions on Software Engineering 36(2), 226–247 (2010)
9. Ho, T.K.: The random subspace method for constructing decision forests. IEEE Trans. Pattern Anal. Mach. Intell. 20, 832–844 (1998)
10. Koch, M.W., Moya, M.M., Hostetler, L.D., Fogler, R.J.: Cueing, feature discovery, and one-class learning for synthetic aperture radar automatic target recognition. Neural Networks 8(7-8), 1081–1102 (1995)
11. Krawczyk, B.: Diversity in ensembles for one-class classification. In: Pechenizkiy, M., Wojciechowski, M. (eds.) New Trends in Databases & Inform. Sys. AISC, vol. 185, pp. 119–129. Springer, Heidelberg (2012)
12. Krawczyk, B., Woźniak, M.: Combining diverse one-class classifiers. In: Corchado, E., Snášel, V., Abraham, A., Woźniak, M., Graña, M., Cho, S.-B. (eds.) HAIS 2012, Part II. LNCS, vol. 7209, pp. 590–601. Springer, Heidelberg (2012)
13. Krawczyk, B., Woźniak, M.: Accuracy and diversity in classifier selection for one-class classification ensembles. In: 2013 IEEE Symposium on Computational Intelligence and Ensemble Learning (CIEL), pp. 46–51 (2013)
14. Krawczyk, B., Woźniak, M.: Pruning one-class classifier ensembles by combining sphere intersection and consistency measures. In: Rutkowski, L., Korytkowski, M., Scherer, R., Tadeusiewicz, R., Zadeh, L.A., Zurada, J.M. (eds.) ICAISC 2013, Part I. LNCS, vol. 7894, pp. 426–436. Springer, Heidelberg (2013)
15. Krawczyk, B., Woźniak, M.: Diversity measures for one-class classifier ensembles. Neurocomputing 126, 36–44 (2014)
16. Krawczyk, B., Woźniak, M., Cyganek, B.: Clustering-based ensembles for one-class classification. Information Sciences (2014)
17. SIAM. Proceedings of the Eleventh SIAM International Conference on Data Mining, SDM 2011, April 28-30, Mesa, Arizona, USA. SIAM Omnipress (2011)
18. Tax, D.M.J., Duin, R.P.W.: Support vector data description. Machine Learning 54(1), 45–66 (2004)

19. Tax, D.M.J., Müller, K.: A consistency-based model selection for one-class classification. In: Proceedings - International Conference on Pattern Recognition, vol. 3, pp. 363–366 (2004)
20. Tax, D.M.J., Duin, R.P.W.: Combining one-class classifiers. In: Kittler, J., Roli, F. (eds.) MCS 2001. LNCS, vol. 2096, pp. 299–308. Springer, Heidelberg (2001)
21. Tax, D.M.J., Duin, R.P.W.: Characterizing one-class datasets. In: Proceedings of the Sixteenth Annual Symposium of the Pattern Recognition Association of South Africa, pp. 21–26 (2005)
22. Wilk, T., Woźniak, M.: Soft computing methods applied to combination of one-class classifiers. Neurocomput. 75, 185–193 (2012)
23. Woźniak, M., Krawczyk, B.: Combined classifier based on feature space partitioning. Journal of Applied Mathematics and Computer Science 22(4), 855–866 (2012)

Evaluation of Fuzzy System Ensemble Approach to Predict from a Data Stream

Bogdan Trawiński[1], Magdalena Smętek[1], Tadeusz Lasota[2], and Grzegorz Trawiński[3]

[1] Wrocław University of Technology, Institute of Informatics, Wybrzeże Wyspiańskiego 27, 50-370 Wrocław, Poland
[2] Wrocław University of Environmental and Life Sciences, Dept. of Spatial Management ul. Norwida 25/27, 50-375 Wrocław, Poland
[3] Wrocław University of Technology, Faculty of Electronics, Wybrzeże S. Wyspiańskiego 27, 50-370 Wrocław, Poland
{bogdan.trawinski,magdalena.smetek}@pwr.wroc.pl, tadeusz.lasota@up.wroc.pl, grzegorz.trawinsky@gmail.com

Abstract. In the paper we present extensive experiments to evaluate our recently proposed method applying the ensembles of genetic fuzzy systems to build reliable predictive models from a data stream of real estate transactions. The method relies on building models over the chunks of a data stream determined by a sliding time window and incrementally expanding an ensemble by systematically generated models in the course of time. The aged models are utilized to compose ensembles and their output is updated with trend functions reflecting the changes of prices in the market. The experiments aimed at examining the impact of the number of aged models used to compose an ensemble on the accuracy and the influence of degree of polynomial trend functions applied to modify the results on the performance of single models and ensembles. The analysis of experimental results was made employing statistical approach including nonparametric tests followed by post-hoc procedures devised for multiple *N*×*N* comparisons.

Keywords: genetic fuzzy systems, data stream, sliding windows, ensembles, trend functions, property valuation.

1 Introduction

Processing data streams poses a considerable challenge because it requires taking into account memory limitations, short processing times, and single scans of arriving data. Many strategies and techniques for mining data streams have been devised. Gaber in his recent overview paper categorizes them into four main groups: two-phase techniques, Hoeffding bound-based, symbolic approximation-based, and granularity-based ones [1]. Much effort is devoted to the issue of concept drift which occurs when data distributions and definitions of target classes change over time [2], [3], [4]. Comprehensive reviews of ensemble based methods for handling concept drift in data streams can be found in [5], [6].

N.T. Nguyen et al. (Eds.): ACIIDS 2014, Part II, LNAI 8398, pp. 137–146, 2014.

For several years we have been investigating methods for generating regression models to assist with real estate appraisal based on fuzzy approach: i.e. genetic fuzzy systems as both single models [7] and ensembles built using various resampling techniques [8], [9]. An especially good performance revealed evolving fuzzy models applied to cadastral data [10], [11]. Evolving fuzzy systems are appropriate for modelling the dynamics of real estate market because they can be systematically updated on demand based on new incoming samples and the data of property sales ordered by the transaction date can be treated as a data stream.

In this paper we present the results of our further study on the method to predict from a data stream of real estate sales transactions based on ensembles of regression models [12], [13], [14], [15]. Having prepared a new real-world dataset we investigated the impact of the number of aged models used to compose an ensemble on the accuracy and the influence of degree of polynomial trend functions applied to modify the results on the performance of single models and ensembles. The scope of extensive experiments was enough to conduct advanced statistical analysis of results obtained including nonparametric tests followed by post-hoc procedures devised for multiple $N \times N$ comparisons.

2 Motivation and GFS Ensemble Approach

The approach based on fuzzy logic is especially suitable for property valuation because professional appraisers are forced to use many, very often inconsistent and imprecise sources of information. Their familiarity with a real estate market and the land where properties are located is frequently incomplete. Moreover, they have to consider various price drivers and complex interrelation among them. The appraisers should make on-site inspection to estimate qualitative attributes of a given property as well as its neighbourhood. They have also to assess such subjective factors as location attractiveness and current trend and vogue. So, their estimations are to a great extent subjective and are based on uncertain knowledge, experience, and intuition rather than on objective data. In the paper an evolutionary fuzzy approach to explore data streams to model dynamic real estate market is presented. The problem is not trivial because on the one hand a genetic fuzzy system needs a number of samples to be trained and on the other hand the time window to determine a chunk of training data should be as small as possible to retain the model accuracy at an acceptable level. The processing time in this case is not a decisive issue because property valuation models need not to be updated and/or generated from scratch in an on-line mode.

The outline of the our ensemble approach to predict from a data stream is illustrated in Fig. 1. The data stream is partitioned into data chunks of a constant length t_c. The sliding window, which length is a multiple of a data chunk, delineates training sets. We consider a point of time t_0 at which the current model was built over data that came in between time $t_0 - 2t_c$ and t_0. The models created earlier that have aged gradually are utilized to compose an ensemble so that the current test set is applied to each component model. However, in order to compensate ageing, their output produced for the current test set is updated using trend functions determined over all data since the beginning of the stream. As the functions to model the trends of price changes the polynomials of the degree from one to five were employed: $Ti(t)$, where i

denotes the degree. The method of updating the prices of premises with the trends is based on the difference between a price and a trend value in a given time point. More detailed description of the approach presented in the paper can be found in [15].

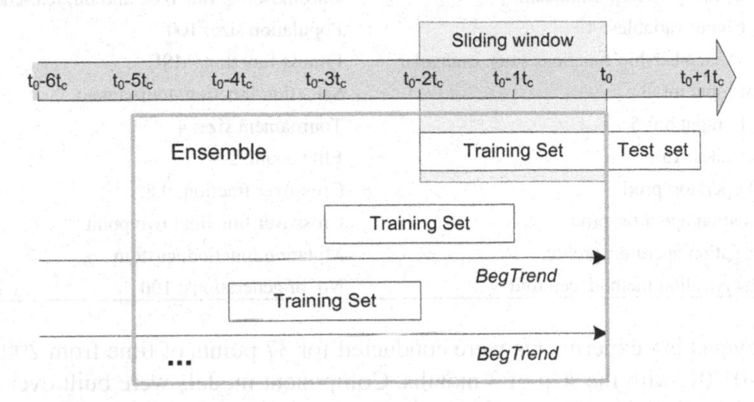

Fig. 1. Outline of ensemble approach to predict from a data stream

3 Experimental Setup

The investigation was conducted with our experimental system implemented in Matlab devoted to carry out research into ensembles of regression models built over data streams. The data driven models, considered in the paper, were generated using real-world data on sales transactions taken from a cadastral system and a public registry of real estate transactions. Real-world dataset used in experiments was drawn from an unrefined dataset containing above 100 000 records referring to residential premises transactions accomplished in one Polish big city with the population of 640 000 within 14 years from 1998 to 2011. In this period the majority of transactions were made with non-market prices when the council was selling flats to their current tenants on preferential terms. First of all, transactional records referring to residential premises sold at market prices were selected. Then, the dataset was confined to sales transaction data of residential premises (apartments) where the land was leased on terms of perpetual usufruct. The other transactions of premises with the ownership of the land were omitted due to the conviction of professional appraisers stating that the land ownership and lease affect substantially the prices of apartments and therefore they should be used separately for sales comparison valuation methods. The final dataset counted 9795 samples. Due to the fact we possessed the exact date of each transaction we were able to order all instances in the dataset by time, so that it can be regarded as a data stream. Four following attributes were pointed out as main price drivers by professional appraisers: usable area of a flat (*Area*), age of a building construction (*Age*), number of storeys in the building (*Storeys*), the distance of the building from the city centre (*Centre*), in turn, price of premises (*Price*) was the output variable.

The parameters of the architecture of fuzzy systems as well as genetic algorithms are listed in Table 1. Similar designs are described in [7], [16].

Table 1. Parameters of GFS used in experiments

Fuzzy system	Genetic Algorithm
Type of fuzzy system: Mamdani	Chromosome: rule base and mf, real-coded
No. of input variables: 4	Population size: 100
Type of membership functions (mf): triangular	Fitness function: MSE
No. of input mf: 3	Selection function: tournament
No. of output mf: 5	Tournament size: 4
No. of rules: 15	Elite count: 2
AND operator: prod	Crossover fraction: 0.8
Implication operator: prod	Crossover function: two point
Aggregation operator: probor	Mutation function: custom
Defuzzyfication method: centroid	No. of generations: 100

The evaluating experiments were conducted for 37 points of time from 2001-01-01 to 2011-01-01, with the step of 3 months. Component models were built over training data delineated by the sliding windows of constant length of 12 months .The sliding window was shifted by one month along the data stream. The test datasets, current for a given time point, determined by the interval of 3 months were applied to each ensemble. As the accuracy measure the root mean squared error (*RMSE*) was employed. The resulting output of the ensemble for a given time point was computed as the arithmetic mean of the results produced by the component models and corrected by corresponding trend functions.

The analysis of the results was performed using statistical methodology including nonparametric tests followed by post-hoc procedures designed especially for multiple $N \times N$ comparisons [17], [18], [19], [20]. The routine starts with the nonparametric Friedman test, which detect the presence of differences among all algorithms compared. After the null-hypotheses have been rejected the post-hoc procedures should be applied in order to point out the particular pairs of algorithms which produce significant differences. For $N \times N$ comparisons nonparametric Nemenyi's, Holm's, Shaffer's, and Bergamnn-Hommel's procedures are recommended. Due to space limitation the Freidman tests followed by one of the most powerful post-hoc procedures, namely Shaffer's one is consistently applied in the paper.

4 Analysis of Experimental Results

4.1 Comparison of GFS Ensembles of Different Size

The performance of models of five selected sizes, i.e. single ones and ensembles comprehending 6, 12, 18 and 24 models for no trend correction (*noT*) and *T4* trend functions is illustrated in Figures 2 and 3, respectively. The values of *RMSE* are given in thousand PLN. However, the differences among the models are not visually apparent, therefore one should refer to statistical tests of significance.

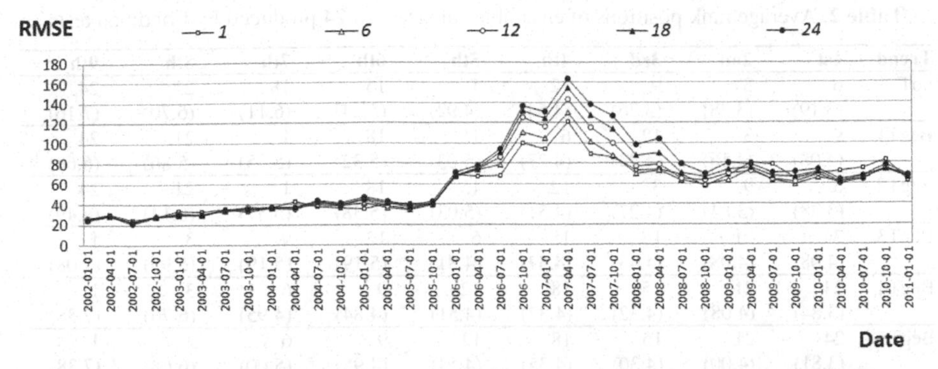

Fig. 2. Performance of GFS ensembles of different size for no trend correction (*noT*)

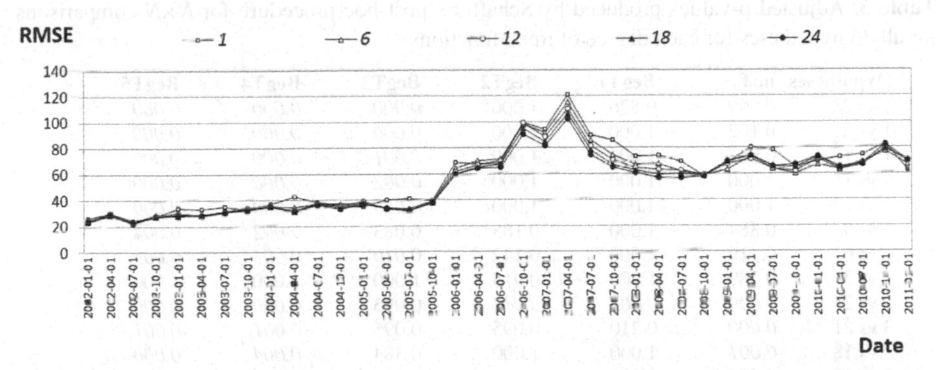

Fig. 3. Performance of GFS ensembles of different size for correction with *T4* trend functions

The Friedman test performed in respect of *RMSE* error measure showed that there were significant differences among ensembles in each case. Average ranks of individual models for polynomial trend functions of degrees from *T1* to *T5* produced by the test are shown in Table 2, where the lower rank value the better model. Adjusted *p-values* for the Schaffer's post-hoc procedure are shown in Table 3. The *p-values* indicating statistically significant differences between given pairs of models are marked with italics. The significance level considered for the null hypothesis rejection was assumed 0.05. Following main observations could be done based on the results of both Friedman tests and Shaffer's post-hoc procedures: the greater number of models with corrected outputs with *T4* and *T5* in an ensemble the better performance. The ensembles composed of 15 to 24 models without output correction (*noT*) exhibit the inverse behaviour. The number of null hypotheses rejected by Shaffer's post-hoc procedure out of 36 is equal to 4, 5, 6, 12, 12, and 13 for *T1*, *T2*, *T3*, *T4*, *T5*, and *noT*, respectively. The following analysis applies only to models with corrected outputs. The models with no output correction reveal statistically worse performance than the ones corrected. The ensembles embracing from 12 to 24 models outperform significantly the single models for *T3*, *T4*, and *T5* trend functions. Moreover, for *T4* and *T5* trend functions, the ensembles of size from 12 to 24 models surpass single models. In turn, no significant difference is observed among ensembles composed of 6 to 24 models for *T3*, *T4*, and *T5* trend functions.

Table 2. Average rank positions of ensembles of size 1 to 24 produced by Friedman tests

Trend	1st	2nd	3rd	4th	5th	6th	7th	8th	9th
noT	6	3	9	12	1	15	18	21	24
	(3.19)	(3.38)	(3.76)	(4.54)	(4.92)	(5.24)	(6.11)	(6.70)	(7.16)
BegT1	9	3	12	6	15	18	1	21	24
	(4.08)	(4.16)	(4.24)	(4.27)	(5.03)	(5.32)	(5.35)	(5.86)	(6.68)
BegT2	6	9	3	12	15	18	1	21	24
	(3.78)	(3.84)	(4.27)	(4.51)	(5.03)	(5.38)	(5.57)	(6.14)	(6.49)
BegT3	24	21	18	15	6	12	9	3	1
	(4.08)	(4.08)	(4.43)	(4.57)	(4.81)	(5.19)	(5.19)	(5.95)	(7.08)
BegT4	24	21	15	18	12	9	6	3	1
	(3.84)	(4.08)	(4.32)	(4.32)	(4.51)	(4.84)	(4.95)	(6.76)	(7.38)
BegT5	24	21	15	18	12	9	6	3	1
	(3.81)	(4.00)	(4.30)	(4.35)	(4.54)	(4.95)	(5.00)	(6.68)	(7.38)

Table 3. Adjusted p-values produced by Schaffer's post-hoc procedure for $N \times N$ comparisons for all 36 hypotheses for each degree of trend functions

Hypotheses	noT	BegT1	BegT2	BegT3	BegT4	BegT5
1 vs 24	*0.009*	0.826	1.000	*0.000*	*0.000*	*0.000*
1 vs 21	0.107	1.000	1.000	*0.000*	*0.000*	*0.000*
1 vs 18	0.803	1.000	1.000	*0.001*	*0.000*	*0.000*
1 vs 15	1.000	1.000	1.000	*0.002*	*0.000*	*0.000*
1 vs 12	1.000	1.000	1.000	*0.010*	*0.000*	*0.000*
1 vs 9	0.884	1.000	0.185	0.083	*0.002*	*0.004*
1 vs 6	0.119	1.000	0.142	*0.010*	*0.004*	*0.005*
1 vs 3	0.280	1.000	0.915	1.000	1.000	1.000
3 vs 24	*0.000*	*0.002*	*0.014*	0.095	*0.000*	*0.000*
3 vs 21	*0.000*	0.210	0.095	0.095	*0.001*	*0.001*
3 vs 18	*0.001*	1.000	1.000	0.384	*0.004*	*0.006*
3 vs 15	0.075	1.000	1.000	0.669	*0.004*	*0.005*
3 vs 12	0.884	1.000	1.000	1.000	*0.009*	*0.018*
3 vs 9	1.000	1.000	1.000	1.000	0.057	0.145
3 vs 6	1.000	1.000	1.000	1.000	0.098	0.187
6 vs 24	*0.000*	*0.004*	*0.001*	1.000	1.000	1.000
6 vs 21	*0.000*	0.343	*0.006*	1.000	1.000	1.000
6 vs 18	*0.000*	1.000	0.270	1.000	1.000	1.000
6 vs 15	*0.028*	1.000	1.000	1.000	1.000	1.000
6 vs 12	0.507	1.000	1.000	1.000	1.000	1.000
6 vs 9	1.000	1.000	1.000	1.000	1.000	1.000
9 vs 24	*0.000*	*0.002*	*0.001*	1.000	1.000	1.000
9 vs 21	*0.000*	0.142	*0.009*	1.000	1.000	1.000
9 vs 18	*0.005*	1.000	0.342	1.000	1.000	1.000
9 vs 15	0.313	1.000	1.000	1.000	1.000	1.000
9 vs 12	1.000	1.000	1.000	1.000	1.000	1.000
12 vs 24	*0.001*	*0.004*	0.054	1.000	1.000	1.000
12 vs 21	*0.015*	0.304	0.239	1.000	1.000	1.000
12 vs 18	0.249	1.000	1.000	1.000	1.000	1.000
12 vs 15	1.000	1.000	1.000	1.000	1.000	1.000
15 vs 24	0.057	0.269	0.482	1.000	1.000	1.000
15 vs 21	0.350	1.000	1.000	1.000	1.000	1.000
15 vs 18	1.000	1.000	1.000	1.000	1.000	1.000
18 vs 24	1.000	0.744	1.000	1.000	1.000	1.000
18 vs 21	1.000	1.000	1.000	1.000	1.000	1.000
21 vs 24	1.000	1.000	1.000	1.000	1.000	1.000
Rejected	**13**	**4**	**5**	**6**	**12**	**12**

4.2 Comparison of GFS Ensembles Using Trend Functions of Different Degrees

The performance of ensembles comprising models with corrected output using *T1*, *T2*, *T3*, *T4*, and *T5* trend functions and without output correction (*noT*) of sizes from 1 to 24 is illustrated in Figures 4 and 5, respectively. The values of *RMSE* are given in thousand PLN. However, the differences among the models are not visually apparent, therefore one should refer to statistical tests of significance.

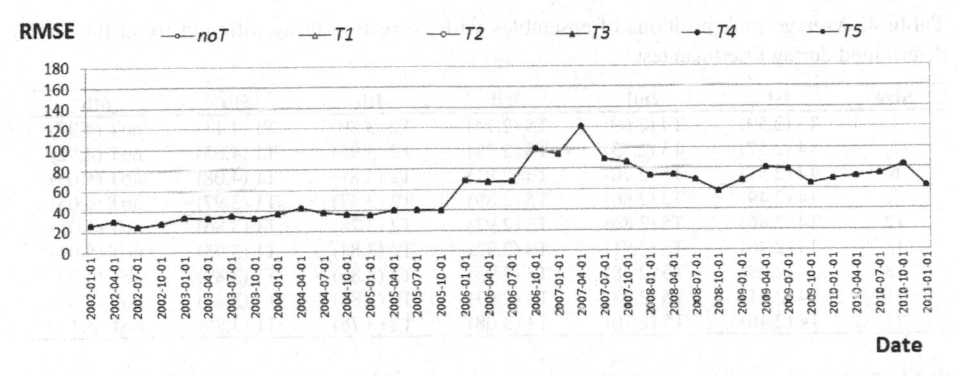

Fig. 4. Performance of GFS ensembles with correction using different trend functions for single models

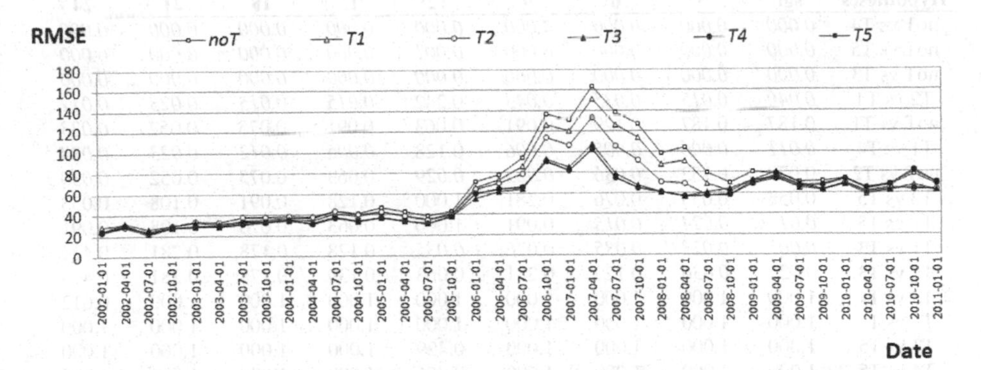

Fig. 5. Performance of GFS ensembles with correction using different trend functions for Size=24

The Friedman test performed in respect of *RMSE* values provided by the ensembles showed that there were significant differences among models because *p-value* was much lower than 0.05 in each case. Average ranks of individual models for individual ensemble sizes produced by the test are shown in Table 4, where the lower rank value the better model. Adjusted p-values for one of the Schaffer's post-hoc procedure are shown in Table 5. The *p-values* indicating the statistically significant differences between given pairs of models are marked with italics. Following main observations could be done based on the results of both Friedman tests and Shaffer's post-hoc procedures: the ensembles comprising 3 and more models which output was corrected with *T4* trend functions are in the first position. Next, the ensembles comprising 3, 9 and more models which output was corrected with *T5* trend functions are in the second position, and then the smaller degree of polynomial functions the worse position. However, no significant differences among *T3*, *T4*, and *T5* trend functions can be observed. The models with no output correction reveal statistically worse performance than the ones corrected with *T3*, *T4*, and *T5* trend functions for all ensemble sizes. Moreover, there are no significant differences among *T1*, *T2* trend functions, and *noT*, except for the ensembles composed of 24 models.

Table 4. Average rank positions of ensembles with correction using different trend functions determined during Friedman test

Size	1st	2nd	3rd	4th	5th	6th
1	T3 (2.59)	T4 (2.69)	T5 (2.74)	T2 (3.89)	T1 (4.11)	noT (4.97)
3	T4 (2.57)	T3 (2.73)	T5 (2.78)	T2 (3.95)	T1 (4.05)	noT (4.92)
6	T4 (2.57)	T5 (2.70)	T3 (2.81)	T2 (3.81)	T1 (4.08)	noT (5.03)
9	T4 (2.49)	T3 (2.86)	T5 (2.89)	T2 (3.37)	T1 (3.97)	noT (5.05)
12	T4 (2.46)	T5 (2.86)	T3 (2.97)	T2 (3.78)	T1 (3.86)	noT (5.05)
15	T4 (2.46)	T5 (2.81)	T3 (2.92)	T2 (3.84)	T1 (3.95)	noT (5.03)
18	T4 (2.49)	T5 (2.78)	T3 (2.95)	T2 (3.86)	T1 (3.89)	noT (5.03)
21	T4 (2.51)	T5 (2.78)	T3 (2.97)	T2 (3.84)	T1 (3.84)	noT (5.05)
24	T4 (2.46)	T5 (2.70)	T3 (3.08)	T2 (3.78)	T1 (3.86)	noT (5.11)

Table 5. Adjusted p-values produced by Schaffer's post-hoc procedure for *N*×*N* comparisons for all 15 hypotheses for each ensemble size

Hypotheses	sgl	3	6	9	12	15	18	21	24
noT vs T4	*0.000*	*0.000*	*0.000*	*0.000*	*0.000*	*0.000*	*0.000*	*0.000*	*0.000*
noT vs T5	*0.000*	*0.000*	*0.000*	*0.000*	*0.001*	*0.000*	*0.000*	*0.000*	*0.000*
noT vs T3	*0.000*	*0.000*	*0.000*	*0.000*	*0.000*	*0.000*	*0.000*	*0.000*	*0.000*
T2 vs T4	*0.040*	*0.015*	*0.035*	*0.043*	0.242	*0.015*	*0.015*	*0.023*	*0.012*
noT vs T1	0.187	0.187	0.129	0.091	0.063	0.091	0.075	0.052	*0.023*
T1 vs T4	*0.011*	*0.006*	*0.005*	*0.006*	0.128	*0.006*	*0.012*	*0.023*	*0.023*
noT vs T2	0.078	0.152	*0.036*	*0.023*	*0.029*	0.063	0.075	0.052	*0.030*
T2 vs T5	0.058	0.053	0.076	0.281	1.000	0.128	0.091	0.108	0.053
T1 vs T5	*0.017*	*0.024*	*0.015*	0.091	1.000	0.063	0.076	0.108	0.091
T1 vs T3	*0.005*	*0.023*	*0.035*	0.076	*0.035*	0.128	0.178	0.281	0.429
T2 vs T3	*0.020*	*0.036*	0.129	0.281	0.063	0.139	0.178	0.281	0.429
T3 vs T4	1.000	1.000	1.000	1.000	1.000	1.000	1.000	1.000	0.612
T1 vs T2	1.000	1.000	1.000	1.000	1.000	1.000	1.000	1.000	1.000
T3 vs T5	1.000	1.000	1.000	1.000	0.429	1.000	1.000	1.000	1.000
T4 vs T5	1.000	1.000	1.000	1.000	0.856	1.000	1.000	1.000	1.000
Rejected	**8**	**8**	**8**	**7**	**5**	**5**	**5**	**5**	**7**

5 Conclusions and Future Work

Our further investigation into the method to predict from a data stream of real estate sales transactions based on ensembles of regression models is reported in the paper. The core of our approach is incremental expanding an ensemble by models built from scratch over successive chunks of a data stream determined by a sliding window. In order to compensate ageing the output produced by individual component models for the current test dataset is updated using trend functions which reflect the changes of the market. In our research we employed genetic fuzzy systems of the Mamdani type as the base machine learning algorithms and the trends were modelled over data that came in from the beginning of a stream.

The experiments aimed at examining the impact of the number of aged models used to compose an ensemble on the accuracy and the influence of degree of polynomial trend functions applied to modify the results on the accuracy of single models and ensembles. The data driven models, considered in the paper, were generated using real-world data of sales transactions taken from a cadastral system and a public registry of real estate transactions. The whole dataset, which after cleansing counted 9,795 samples, was ordered by transaction date forming a sort of a data stream. The comparative experiments consisted in generating ensembles of GFS models for 37 points of time within the period of 10 years using the sliding window one year long which delineated training sets. The predictive accuracy of GFS ensembles for different variants of ensemble sizes and polynomial trend functions, was compared using nonparametric tests of statistical significance adequate for multiple comparisons. The ensembles consisted of 3, 6, 9, 12, 15, 18, 21, and 24 component GFS models; for comparison single models were also utilized. As the functions to model the trends of price changes, the polynomials of degree from one to five were employed.

The results proved the usefulness of ensemble approach incorporating the correction of individual component model output. For the majority of cases the bigger ensembles encompassing from 12 to 24 GFS models produced more accurate predictions than the smaller ensembles. Moreover, they outperformed significantly the single models. As for correcting the output of component models, the need to apply trend functions to update the results provided by ageing models is indisputable. However, the selection the most suitable trend function in terms of the polynomial degree has not been definitely resolved. As the matter of fact in majority of cases the trend functions of higher degree, i.e. four and five led to better accuracy. However, the differences were not statistically significant. Therefore, further study is needed for example into the selection of correcting functions dynamically depending on the nature of price changes.

We plan to conduct experiments employing other algorithms capable of learning from concept drifts such as: decision trees, recurrent neural networks, support vector regression, etc. Moreover, we intend to compare the results provided by data-driven regression models with the predictions made by professional appraisers using standard procedures.

Acknowledgments. This paper was partially supported by the Polish National Science Centre under grant no. N N516 483840.

References

1. Gaber, M.M.: Advances in data stream mining. Wiley Interdisciplinary Reviews: Data Mining and Knowledge Discovery 2(1), 79–85 (2012)
2. Brzeziński, D., Stefanowski, J.: Accuracy Updated Ensemble for Data Streams with Concept Drift. In: Corchado, E., Kurzyński, M., Woźniak, M. (eds.) HAIS 2011, Part II. LNCS (LNAI), vol. 6679, pp. 155–163. Springer, Heidelberg (2011)
3. Sobolewski, P., Woźniak, M.: Concept Drift Detection and Model Selection with Simulated Recurrence and Ensembles of Statistical Detectors. Journal for Universal Computer Science 19(4), 462–483 (2013)
4. Tsymbal, A.: The problem of concept drift: Definitions and related work. Technical Report. Department of Computer Science, Trinity College, Dublin (2004)

5. Kuncheva, L.I.: Classifier ensembles for changing environments. In: Roli, F., Kittler, J., Windeatt, T. (eds.) MCS 2004. LNCS, vol. 3077, pp. 1–15. Springer, Heidelberg (2004)
6. Minku, L.L., White, A.P., Yao, X.: The Impact of Diversity on Online Ensemble Learning in the Presence of Concept Drift. IEEE Transactions on Knowledge and Data Engineering 22(5), 730–742 (2010)
7. Król, D., Lasota, T., Trawiński, B., Trawiński, K.: Investigation of Evolutionary Optimization Methods of TSK Fuzzy Model for Real Estate Appraisal. International Journal of Hybrid Intelligent Systems 5(3), 111–128 (2008)
8. Kempa, O., Lasota, T., Telec, Z., Trawiński, B.: Investigation of bagging ensembles of genetic neural networks and fuzzy systems for real estate appraisal. In: Nguyen, N.T., Kim, C.-G., Janiak, A. (eds.) ACIIDS 2011, Part II. LNCS (LNAI), vol. 6592, pp. 323–332. Springer, Heidelberg (2011)
9. Lasota, T., Telec, Z., Trawiński, G., Trawiński, B.: Empirical Comparison of Resampling Methods Using Genetic Fuzzy Systems for a Regression Problem. In: Yin, H., Wang, W., Rayward-Smith, V. (eds.) IDEAL 2011. LNCS, vol. 6936, pp. 17–24. Springer, Heidelberg (2011)
10. Lasota, T., Telec, Z., Trawiński, B., Trawiński, K.: Investigation of the eTS Evolving Fuzzy Systems Applied to Real Estate Appraisal. Journal of Multiple-Valued Logic and Soft Computing 17(2-3), 229–253 (2011)
11. Lughofer, E., Trawiński, B., Trawiński, K., Kempa, O., Lasota, T.: On Employing Fuzzy Modeling Algorithms for the Valuation of Residential Premises. Information Sciences 181, 5123–5142 (2011)
12. Trawiński, B., Lasota, T., Smętek, M., Trawiński, G.: An Attempt to Employ Genetic Fuzzy Systems to Predict from a Data Stream of Premises Transactions. In: Hüllermeier, E., Link, S., Fober, T., Seeger, B. (eds.) SUM 2012. LNCS (LNAI), vol. 7520, pp. 127–140. Springer, Heidelberg (2012)
13. Trawiński, B., Lasota, T., Smętek, M., Trawiński, G.: An Analysis of Change Trends by Predicting from a Data Stream Using Genetic Fuzzy Systems. In: Nguyen, N.-T., Hoang, K., Jędrzejowicz, P. (eds.) ICCCI 2012, Part I. LNCS (LNAI), vol. 7653, pp. 220–229. Springer, Heidelberg (2012)
14. Trawiński, B., Lasota, T., Smętek, M., Trawiński, G.: Weighting Component Models by Predicting from Data Streams Using Ensembles of Genetic Fuzzy Systems. In: Larsen, H.L., Martin-Bautista, M.J., Vila, M.A., Andreasen, T., Christiansen, H. (eds.) FQAS 2013. LNCS (LNAI), vol. 8132, pp. 567–578. Springer, Heidelberg (2013)
15. Trawiński, B.: Evolutionary Fuzzy System Ensemble Approach to Model Real Estate Market based on Data Stream Exploration. Journal of Universal Computer Science 19(4), 539–562 (2013)
16. Cordón, O., Herrera, F.: A Two-Stage Evolutionary Process for Designing TSK Fuzzy Rule-Based Systems. IEEE Tr. on Sys., Man and Cyber., Part B 29(6), 703–715 (1999)
17. Demšar, J.: Statistical comparisons of classifiers over multiple data sets. Journal of Machine Learning Research 7, 1–30 (2006)
18. García, S., Herrera, F.: An Extension on "Statistical Comparisons of Classifiers over Multiple Data Sets" for all Pairwise Comparisons. Journal of Machine Learning Research 9, 2677–2694 (2008)
19. Graczyk, M., Lasota, T., Telec, Z., Trawiński, B.: Nonparametric Statistical Analysis of Machine Learning Algorithms for Regression Problems. In: Setchi, R., Jordanov, I., Howlett, R.J., Jain, L.C. (eds.) KES 2010, Part I. LNCS (LNAI), vol. 6276, pp. 111–120. Springer, Heidelberg (2010)
20. Trawiński, B., Smętek, M., Telec, Z., Lasota, T.: Nonparametric Statistical Analysis for Multiple Comparison of Machine Learning Regression Algorithms. International Journal of Applied Mathematics and Computer Science 22(4), 867–881 (2012)

Genetic Transformation Techniques in Cryptanalysis

Urszula Boryczka and Kamil Dworak

University of Silesia, Institute of Computer Science,
Bedzinska. 39, 41200 Sosnowiec, Poland
urszula.boryczka@us.edu.pl,
kamil.dworak@us.edu.pl
http://ii.us.edu.pl/en

Abstract. This paper shows how evolutionary algorithms (*EAs*) can be used to speed up the process of cryptanalysis. The purpose of this study is to demonstrate that evolutionary algorithms can effectively be used for cryptanalysis attacks in order to obtain major speed and memory optimization. A classical transposition cipher was applied in all of the research tests in order to prove this hypothesis. All ciphertexts will be subject to cryptanalysis attacks extended by EAs.

Keywords: cryptanalysis, evolutionary algorithms, optimization techniques, cryptology, genetic operators.

1 Introduction

Security based on authentication and access control is already insufficient. We have heard of many cases when the message was intercepted, overheard or substituted. Due to the danger coming either from hackers, database administrators, or developers, it has become necessary to use cryptography. Instead of hiding or smuggling, the message can be encrypted to such an extent that the contents will be known only to the sender and the destined recipient. Cryptanalysis was born with the rise of cryptography. It is a science of analyzing and breaking the ciphers of cryptographic systems. It is a field that is closely related with anycryptology attacks, whose main goal is to obtain the key that was used to decrypt the encrypted message or to locate weaknesses in the cryptographic system. Cryptanalysis processes are very time-consuming. They mostly come down to testing all available solutions. It would be useful to use some optimizations in order to accelerate this process. One of the best ways of determining the approximate solution, in a fairly acceptable period of time, is to use Evolutionary Computation (*EC*).

Techniques taken from the field of Artificial Intelligence, such as EC, are steadily becoming more present in the area of computer security. In recent years, many approaches that use applications based on EC, such as Particle Swarm Optimization or Differential Evolution, have been proposed, for example, in the design and analysis of a number of new cryptographic primitives ranging from

N.T. Nguyen et al. (Eds.): ACIIDS 2014, Part II, LNAI 8398, pp. 147–156, 2014.

pseudo-random number generators to block ciphers in the cryptanalysis of state-of-the-art cryptosystems and in the detection of network attaching, to name but a few. There is growing interest in the computer security community towards EC techniques as a result of recent successes, but there is still a number of open problems in this field that should be addressed.

The rest of the paper is organized as follows. The next section presents basic information about EC and a survey about previous work on the use of certain evolutionary techniques in cryptanalysis. The third section presents cryptography and how it works. Also, there is a description of the classical transposition cipher which will be used for cryptanalysis research. The next section shows a sample proposed evolutionary attack which will be used to break a ciphertext encrypted by a transposition cipher. The fifth section presents the observations, results and a comparison of the tests on the proposed attack and an attack prepared from the work of Toemeh and Arumugam [14]. The last section contains a summary and conclusions of this work.

2 Popularity of Evolutionary Algorithms

In 1993 Robet Spillman, with other scientists, tried to introduce EAs into cryptanalysis processes for a simple substitution cipher [11].Keys (24 signs of length), as individuals of the population, are subject to genetic operations which are specially modified for that kind of problem. After 100 iterations a key was successfully found by searching only 1000 keys. Jun Song, in his work, used the simulated annealing algorithm to break the transposition's cipher [12].A thesis written by Bethany Delman showed many samples of using EAs in cryptanalysis [4]. In another work, Mishra and Bali presented an algorithm which was then used in cryptography to generate public and private keys [9]. Andrew John Clark successfully broke a classical cipher by using EAs, Tabu Search and Simulated Annealing attacks [3].

EAs are optimization techniques which are mainly based on metaphors of biological processes, i.e. they are based on the principles of natural evolution, heredity, and the replication of the phenomena and principles of the laws of nature [8]. Chromosomes are subject to a process of heredity, which allows descendant chromosomes to inherit a portion of the genetic material from the parent chromosomes [1]. An algorithm contains a set of potential solutions to the problem through the population. Some of them will be used to create a new population by using special genetic operators, such as crossover or mutation. The following sets are generated until the stop criterion is reached [5].

3 Introduction to Cryptography

Cryptography is used for the encryption and decryption of messages and documents [7]. Not without reason was it named the best method of data hiding. It is based, in particular, on mathematical transitions and swaps. Cryptography obscures the message content, not information about its existence [10]. Cryptography allows secret information, even though it may be in the hands of an intruder, to be so secure that it is impossible to read.

Encryption involves converting the text in such a way that it is not readable. In most cases cryptography is carried out by using two keys, the encryption key K_1 and the K_2 decryption key. By using a suitable decryption key the person receiving the message is able to recover the original, legible information.

In this paper, a classic transposition cipher is used as an example. Each plaintext sign is exactly one sign in the cryptogram. Marks remain the same, and only the order is modified. Each key represents a single permutation of length m as well, so:

$$E_\pi(P_1, ..., P_m) = (P_{\pi(1)}, ..., P_{\pi(m)}), \tag{1}$$

$$D_\pi(C_1, ..., C_m) = (C_{\pi^{-1}(1)}, ..., C_{\pi^{-1}(m)}), \tag{2}$$

where:
π - a permutation of all the letters of the alphabet,
E - an encryption function,
D - a decryption function,
P - a plaintext,
C - a ciphertext,
$\imath$ - an index of the sign in the plaintext or ciphertext [4]

The message is written into the table with a width equal to the length of π (Table 1). The table must be fully completed. If the message does not completely fill it in, additional random signs can be added.

Table 1. Transposition cipher - encryption process

5	2	3	7	1	6	4
t	o		b	e		o
r		n	o	t		t
o		b	e		t	h
a	t		i	s		t
h	e		q	u	e	s
t	i	o	n	e	a	d

Then we create the ciphertext. If the signs in the table are read according to the number of columns from top to bottom, a ciphertext is produced:

ET SUEO TEI NB OOTHTSDTROAHT T EABOEIQN

The drawn signs have been underlined. The process of decrypting the ciphertext is not complicated. It is based on an inverse operation. By knowing the key we can complement the content of the table by filling in the part of the ciphertext in the respective columns. We are then able to read the message by following the table from left to right.

4 Breaking the Ciphers - Cryptanalysis

Cryptanalysis is the science of deciphering a plaintext without knowing the key [10]. In most cases, the main goal of the attacker is to obtain the decryption key [13]. The simplest attack is a brute force attack. This approach involves checking all of the available keys [6]. No detailed analysis is required. The percentage rebound depends on the power of the computer by means of which more solutions are generated and on the complexity of the key [2].

One of the basic concepts of cryptanalysis is frequency analysis. This involves counting the frequency of all signs in the selected ciphertext and then creating a ranking of the most common symbols [2]. A comparison is made between the distribution of the ciphertext symbols and the comparable text. This way the attacker tries to infer the cryptogram equivalents of the signs. Frequency analysis does not mean only checking the individual signs. The notation n-gram is often used. It is based on the fact that we search for the prevalence of two-character (digrams/bigrams) or three-character (trigrams) pieces of text and then compare the frequency of their occurrence in a natural language, such as English [6].

When using EC in the field of cryptanalysis, the main issue is how to define a fitness used to find the best keys. In 1998 Andrew John Clark [3] presented an interesting method based on n-grams which was used to compute a fitness function's value. An attacker receives the encrypted message and the key's length by which it is encoded. The initial population is made by generating a random permutation of the key length. The fitness function value is calculated based on the following:

$$F_f = \beta \cdot \Sigma_{i,j \in A} |K^b_{(i,j)} - D^b_{(i,j)}| + \gamma \cdot \Sigma_{i,j,k \in A} |K^t_{(i,j,k)} - D^t_{(i,j,k)}| \qquad (3)$$

where:
A - the language alphabet,
$i,\ j,\ k$ - next indexes of the message,
F_f - the fitness function,
K - known language statistics,
D - the frequency of occurrence in the decrypted text,
$b,\ t$ - the indices to denote bigram/trigram statistics,
β, γ - weight parameters, determining the priority of the statistics (the condition $\beta + \gamma = 1$ must be fulfilled).

The unigrams statistics are completely omitted here. A transposition cipher has the same number of signs before and after the encryption process. Comparing theise statistics does not make sense, thus it can be left out [3]. The value of the fitness function is based on the incidence of n-grams in the decrypted text. The total sum of bigrams and trigrams is substracted from the known language statistics. In addition, there are two parameters, β and γ, allowing to assign the weight which determines the usefulness of this statistic, i.e. of n-gram. In some cases, calculating the bigrams may be sufficient and more effective. Trigrams increase the algorithm's complexity to $O(N^3)$ [3]. The smaller value of the function suggests that the differences between the statistics and the decrypted text are insignificant. It can thus be concluded that the correct decryption key has the smallest possible value of the function. Due to the additional random signs generated during the encryption process, obtaining the value 0 becomes practically impossible.

The classical cryptosystems, nowadays entirely broken, had a general property where the real key producing the encryptions/decryptions was close to original ciphertext/plaintext. This significantly aided in defining the fitness functions. Modern ciphers do not exhibit this property. This is a major problem and the reason behind the relative lack of applications of these heuristic techniques into modern cryptosystems, where while testing a key that has 255 right bits out of 256 (is 99.6% correct). The resulting plaintext would then appear completely random due to a property that has been named the Avalanche Effect.

In 2007 Toemeh and Arumugam presented a modification [14], named the "original attack" in this work, of Clark's genetic algorithm applied to break a transposition cipher. The authors used the same fitness function calculation as was proposed before by Clark [3].

The original crossover's operator is presented on Fig. 1.

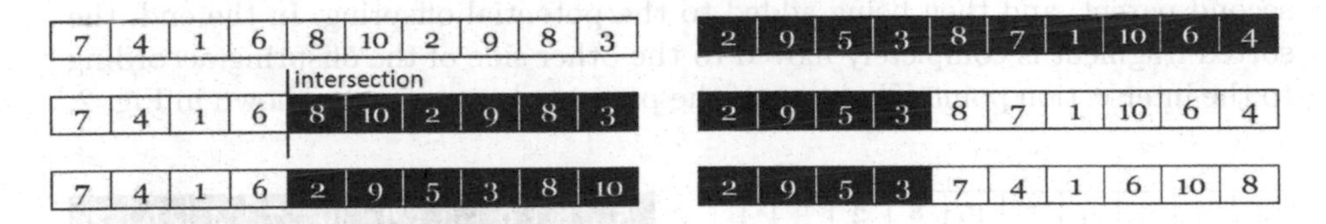

Fig. 1. Original crossover operator proposed in Clark's scientific work [3]

The intersection point is selected at random. In the case of the first child, the cut off part is sorted in order of the second parent, for the second child in order of the first parent. The mutation operator selects two random elements from the descendant and swaps them.

After longer observations of the algorithm performance it was noted that the diversity of solutions in the population is not a major change, i.e. only one part of the key is modified. In some cases (especially when the point of intersection is near the edge), a descendant looks the same as the parent, or differs from the parent only slightly. The other part of the key almost always remains unchanged. The only chance for a modification is a mutation. The probability of its occurrence in the EA is negligible. In addition, the mutation in the best of cases will be reduced to a change of one or two key's signs, depending on which letters were drawn.

Algorithm 1: Original attack's crossover

```
intersectionPoint := random(0, parentLength - 1)
for i := 0 To intersectionPoint do
  child1[i] := parent1[i]
  child2[i] := parent2[i]

idx1 := idx2 := intersectionPoint
for i := 0 To parentLength - 1 do
  if child1 not contain parent2[i] then
    child1[idx1] := parent2[i]
    idx1 := idx1 + 1
  if child2 not contain parent1[i] then
    child2[idx2] := parent1[i]
    idx2 := idx2 + 1
```

It became necessary to use a crossover capable of producing greater diversity in the population than in [14]. Several modifications proposed in the literature [5,8] of crossover operators were tested, such as one-point, two-point, heuristic and staple. The best results were obtained by a one-point crossover, therefore, that kind of crossover is used in the attack presented in this work (named "proposed attack" in this work). The intersection point is chosen at random. In the case of the first parent the right part is cut off, in the second parent the left part is cut off. In the next step the separated fragment is sorted in the order set out by the second parent, and then being added to the potential offspring. In the end, the sorted fragment is completely moved to the other side of the offspring according to the intersection point. The steps of the presented operator are shown in Fig. 2.

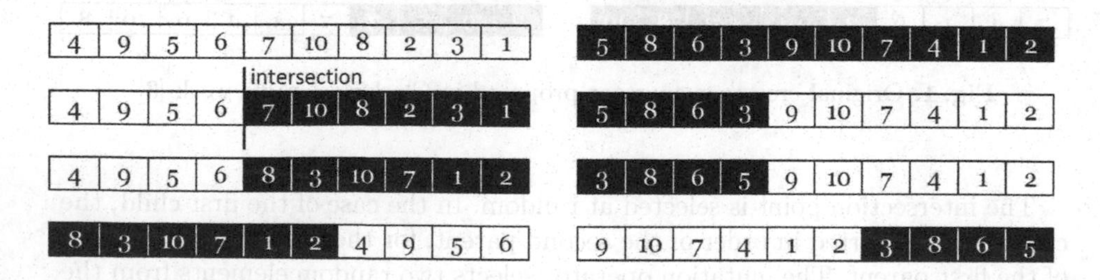

Fig. 2. Modification of the original attack's crossover

While working on the algorithms as described above and in [14] it was noted that the correct solution can be found quite quickly, although it is shifted; for example, for a valid key "12345678" the solutions which were found were "56781234" or "23456781". The mutation operator, instead of replacing two random signs, transfers the whole string one position to the left.

Algorithm 2: Proposed attack's crossover

```
intersectionPoint := random(0, parentLength - 1)
for i := 0 To parentLength - 1 do
   if i < intersectionPoint then
      idx1 := parentLength - intersectionPoint
      child1[idx1] := parent1[i]
   else
      idx2 := i - intersectionPoint
      child2[idx2] := parent2[i]

idx1 := 0
idx2 := parentLength - intersectionPoint
for i := 0 To parentLength - 1 do
   if child1 not contain parent2[i] then
      child1[idx1] := parent2[i]
      idx1 := idx1 + 1
   if child2 not contain parent1[i] then
      child2[idx2] := parent1[i]
      idx2 := idx2 + 1
```

5 Experiments and Results

This section presents a comparison of the original attack with the proposed one. Both algorithms were implemented and tested on the same machine. Due to the continuous change in the length of the key, only a given number of the population's keys are going to be changed. All parameters are shown in Table 2.

Table 2. Parameters of the evolutionary algorithms

Number of Iterations	Unlimited (time limit to 5 minutes)
Crossover's Probability	0.7
Mutation's Probability	0.25
Selection's Type	roulette wheel method
The Weight Parameter Beta	0.5
The Weight Parameter Alpha	0.5

Algorithms were implemented in C# programming language and tested on a PC with a processor of core frequency 3.4GHz and 4GB RAM memory. The tested texts contain 3850 signs. A total of 900 tests were performed (30 different ciphertexts with several keys and 30 tests for each ciphertext) for both algorithms (the ciphertexts were also tested on the classic cryptanalysis attack; no cryptogram was decrypted in under 5 minutes). It was decided to select the five most interesting tests for each algorithm. The results obtained by the original

Table 3. Results obtained by the original attack [14]

ID	N	K_L	$F_f(MAX)$	$F_f(MIN)$	$F_f(Median)$	Iteration	Time
5	300	11	4620	3057	3848	10	00:22
10	450	13	4422	3048	3715	10	00:36
17	1000	15	4329	3055	3675	14	01:47
23	2000	17	4321	3049	3570	15	03:16
30	3000	19	4538	3292	3700	14	05:00

attack [14] are presented in Table 3, the results from the attack described in this work are presented in Table 4.

where:
ID - identity of a single test,
N - number of solutions in the population,
K_L - key's length,
$F_f(MIN)/F_f(MAX)$ - the best/worst value in the population,
$F_f(Median)$ - median value adjustment of individuals in the population,
Iteration - number of the iteration which found the best solution,
Time - duration of the algorithm (for finding the correct key).

Table 4. Results obtained in the proposed evolutionary attack

ID	N	K_L	$F_f(MAX)$	$F_f(MIN)$	$F_f(Median)$	Iteration	Time
5	100	11	4462	3057	3654	18	00:18
10	200	13	4650	3048	3681	15	00:32
17	350	15	4688	3055	3459	24	01:17
23	500	17	4667	3049	3458	29	02:20
30	1000	19	4706	3060	3471	24	03:27

The roulette wheel method gave the best result for both algorithms. While testing the transposition's cipher cryptanalysis, the minimum value ($F_f(MIN)$) always coincided with the best individual in the population. This function counts the number of inconsistencies that arise with respect to the occurrence of digrams and trigrams between English and the decrypted message. It can be concluded that the smaller the value of the function for a particular individual, the more reliable the solution is.

When analyzing the convergence graphs in Fig. 3, results presented in Table 3 and Table 4 it can easily be seen that the method described in [14] requires a much larger number of keys in the population in contrast to the attack proposed in this work. The calculated total keys usage for both algorithms is presented in Table 5.

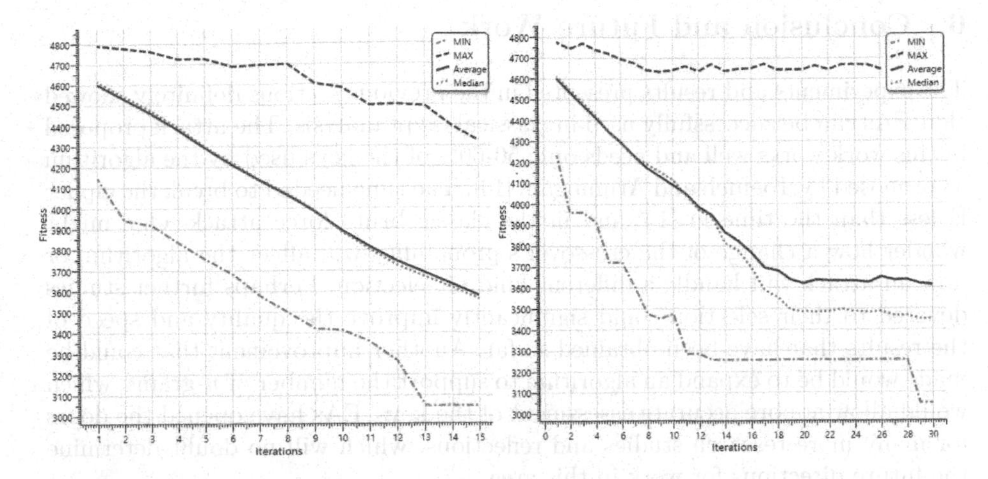

Fig. 3. Convergence's chart for test 10 for both algorithms - original attack on left, proposed on right

Table 5. Total keys usage for both algorithms

ID	$\sum_1$	$\sum_2$	Keys usage
5	3000	1800	60%
10	4500	3000	66%
17	14000	8400	60%
23	30000	14500	48%
30	42000	24000	57%

where:
ID - an identity of the test,
$\sum_1$ - a total cost of algorithm [14],
$\sum_2$ - a total cost of the algorithm proposed in this work,
Keys usage - percentage of keys from [14] in the proposed attack.

A total of 30 000 keys were used in the original attack and 14 500 keys were used in the proposed attack in order to find the correct key. The algorithm proposed in this work needs 50% of the keys needed to find a solution using Toemeh and Arumugam's algorithm.

The attack proposed in this document works slightly faster for keys from the 11 - 15 range, but later (which is visible from the tests of ids 23 and 30) it gets more performance. Also, a mutation sliding the key by one position is very helpful. On the 28th iteration the algorithm improves its solutions (leaves the convergence), and in the next generation it gets the correct key. The proposed attack has to use more iterations of the algorithm but does not need that many keys in the population as the original attack. It also works faster.

6 Conclusion and Future Work

The experiments and results presented in the previous sections definitely showed that EAs can be successfully used in classical cryptanalysis. The attack proposed in this work works well and needs only 50-60% of the keys used by the algorithm as proposed by Toemeh and Arumugam [14]. The time needed to break the cipher is less than the time in [14] and in the classic brute force attack. One might wonder how a change of the crossover's probability will affect the algorithm or how an attack will handle a different kind of selection. Perhaps further studies devoted to their selection could significantly improve the quality and speed of the results that have been obtained so far. Another improvement that could be made would be to expand an algorithm to support the number of n-grams, which would allow a more accurate assessment of the keys. EAs have opened the doors for many more research studies and reflections, which will no doubt determine the future directions for work in this area.

References

1. Alba, E., Cotta, C.: Evolutionary Algorithms. Chapman & Hall/CRC, Boca Raton (2006)
2. Bauer, F.L.: Decrypted Secrets. Methods and Maxims of Cryptology. Springer, Heidelberg (2002)
3. Clark, A.J.: Optimisation Heuristics for Cryptology. PhD thesis, Information Research Centre Faculty of Information Technology Queensland (1998)
4. Delman, B.: Genetic Algorithms in Cryptography. Rochester Institute of Technology, Rochester (2004)
5. Goldberg, D.E.: Genetic Algorithms in Search Optimization, and Machine Learning. Addison-Wesley, Boston (1989)
6. Kahn, D.: The Code-breakers. Scribner, New York (1996)
7. Kenan, K.: Cryptography in the databases. The Last Line of Defense. Addison Wesley Publishing Company (2005)
8. Michalewicz, Z.: Genetic Algorithms + Data Structures = Evolution Programs. Springer, London (1996)
9. Mishra, S., Bali, S.: Public Key Cryptography Using Genetic Algorithm. International Journal of Recent Technology and Engineering 2(2) (2013)
10. Schneier, B.: Applied Cryptography: Protocols, Algorithms, and Source Code in C, 2nd edn. Wiley (1996)
11. Spillman, R., Janssen, M., Nelson, B., Kepner, M.: Use of a Genetic Algorithm in the Cryptanalysis of Simple Substitution Cipher. Cryptologia (January 1993)
12. Song, J., Yang, F., Wang, M., Zhang, H.: Cryptanalysis of Transposition Cipher Using Simulated Annealing Genetic Algorithm. China University of Geosciences, China
13. Stinson, D.S.: Cryptography. Theory and Practice. Chapman & Hall/CRC, Taylor & Francis Group, Boca Raton (2006)
14. Toemeh, R., Arumugam, S.: Breaking Transposition Cipher with Genetic Algorithm. Electronics and Electrical Engineering (7(79)) (2007) ISSN 1392-1215

On-the-Go Adaptability in the New Ant Colony Decision Forest Approach

Urszula Boryczka and Jan Kozak

Institute of Computer Science, University of Silesia, Będzińska 39,
41–200 Sosnowiec, Poland
{urszula.boryczka,jan.kozak}@us.edu.pl

Abstract. In this article we present a new and effective adaptive ant colony algorithm that we employ to construct a decision forest (aACDF). The aim of this proposition is to create an adaptive meta-ensemble based on data sets created during the algorithm runtime. This on-the-go approach allows to construct classifiers in which wrongly classified objects obtain a greater probability of being chosen for the pseudo-samples in subsequent iterations. Every pseudo-sample is created on the basis of training data. Our results confirm the standpoint that this new adaptive ACDF slightly reduces the accuracy of classification as well as builds considerably smaller decision trees.

Keywords: Ant Colony Decision Forest, Boosting, Ant Colony Optimization, Decision Forest, ACDT.

1 Introduction

Our goal was to create a self-adaptive ensemble method with incorporation of the ant colony metaphor. In our proposition, pseudo-samples (training subsets) are created for each population of virtual ants. It is worth mentioning that this creation is based on results that have been obtained until now, during decision tree construction. Our starting point is to improve the classification quality of the decision trees. This value depends on the classification accuracy and decision tree growth observed in the generated forest.

Ant Colony Optimization (ACO) is a metaheuristic approach to solving many different optimization problems by using the principles of communicative behaviour observed in real ant colonies. Ants can communicate with one another about the paths they have traversed by using reinforcement mechanisms of pheromone trails laid on the appropriate edges. The pheromone trails can lead other ants to the food sources. ACO was introduced in [7]. It is a population-based approach in which several generations of virtual ants search for good solutions. Subsequent ants of the next generation are attracted by the pheromone changes so that they can search in the solution space near attractive solutions, i.e. in specific subspaces.

Data mining and machine learning have been subjects of increasing attention over the past 30 years. Ensemble methods, which are popular in machine learning

N.T. Nguyen et al. (Eds.): ACIIDS 2014, Part II, LNAI 8398, pp. 157–166, 2014.

and pattern recognition, are learning algorithms that construct a set of many individual classifiers, called base learners, and combine them to classify new data points or samples by taking a weighted or unweighted vote of their predictions. It is now well known that ensembles are often much more accurate than the individual classifiers that make them up. The success of ensemble approaches on many benchmark data sets has raised considerable interest in understanding why such methods succeed and in identifying the circumstances in which they can be expected to produce good results.

This article is organised as follows. Section 1 comprises an introduction to the subject of this article. In section 2, Decision Trees and Ant Colony Decision Trees are presented. Section 3 and 4 describe decision forests and, in particular, the Random Forest and the on-the-go adaptive approach. Section 5 describes the Ant Colony Decision Forest approach, especially the version where a meaningful similarity to RF is presented. Section 6 focuses on the on-the-go adaptive ACDF approach. Section 7 presents the experimental study that was conducted to evaluate the performance of the on-the-go adaptive ACDF by taking into consideration twelve data sets. Finally, we conclude with general remarks on this work, and a few directions for future research are pointed out.

2 Ant Colony Decision Trees

There seem to be some reasons for improvement of the classical approach by incorporating the nondeterministic and stochastic algorithm called the Ant Colony Optimization Technique. A decision tree is used to determine the optimum course of action in situations having several possible alternatives with uncertain outcomes. The resulting diagram displays the structure of a particular decision and the interrelationships and interplay between different alternatives, decisions, and possible outcomes. Decision trees are commonly used in operational research, specifically in decision analysis, for identifying the optimal strategy of reaching a goal. The evaluation function for decision trees will be calculated according to the following formula:

$$Q(T) = \phi \cdot w(T) + \psi \cdot a(T, P) \tag{1}$$

where:
$w(T)$ – the size (numer of nodes) of the decision tree T,
$a(T, P)$ – the accuracy of the classification object from a test set P by the tree T,
ϕ and ψ – constants determining the relative importance of $w(T)$ and $a(T, P)$.

The Ant Colony Decision Tree (ACDT) algorithm [1,2] employs ant colony optimization techniques [6,8] for constructing decision trees and decision forests. Ant Colony Optimization is a branch of a newly developed form of artificial intelligence called swarm intelligence. Swarm intelligence is a form of emergent collective intelligence of groups of simple individuals, e.g. ants, termites or bees, in which the form of indirect communication via pheromones was observed. Pheromone values encourage ants to follow a path in order to build good solutions of the analyzed problem, and the learning process occurring in this situation is called positive feedback or auto-catalysis.

In each ACDT step an ant chooses an attribute and its value to split the samples in the current node of the constructed decision tree. The choice is made according to a heuristic function and the pheromone values. The heuristic function is based on the Twoing criterion (eq. (4)), which helps ants select an attribute-value pair which adequately divides the samples into two disjoint sets, i.e. with the intention that samples belonging to the same decision class should be in the same subset. The best splitting is observed when a similar number of samples exists in the left subtree and in the right subtree, and samples belonging to the same decision class are in the same subtree. The pheromone values indicate the best way (connection) from the superior to the subordinate nodes – all possible combinations are taken into account.

As it was mentioned before, the value of the heuristic function is determined according to the splitting rule employed in the CART approach (see formula (4)). The probability of choosing the appropriate split in the node is calculated according to a classical probability used in ACO [9]:

$$p_{i,j} = \frac{\tau_{m,m_{L(i,j)}}(t)^{\alpha} \cdot \eta_{i,j}^{\beta}}{\sum_i^a \sum_j^{b_i} \tau_{m,m_{L(i,j)}}(t)^{\alpha} \cdot \eta_{i,j}^{\beta}}, \tag{2}$$

where:

$\eta_{i,j}$ – a heuristic value for the split using the attribute i and value j,

$\tau_{m,m_{L(i,j)}}$ – an amount of pheromone currently available at time t on the connection between nodes m and m_L, (it concerns the attribute i and value j),

α, β – the relative importance with experimentally determined values 1 and 3, respectively.

The pheromone trail is updated (3) by increasing the pheromone levels on the edges connecting each tree node with its parent node:

$$\tau_{m,m_L}(t+1) = (1-\gamma) \cdot \tau_{m,m_L}(t) + Q(T), \tag{3}$$

where $Q(T)$ is a quality of the decision tree (see formula (1)), and γ is a parameter representing the evaporation rate, equal to 0.1.

The Twoing criterion will search for two classes that will together make up more than 50% of the data. The Twoing splitting rule maximises the following change-of-impurity measure, which implies the following maximisation problem for the nodes m_l, m_r:

$$\underset{a_j \leq a_j^R, j=1,\ldots,M}{\arg\max} \left(\frac{P_l P_r}{4} \left[\sum_{k=1}^{K} |p(k|m_l) - p(k|m_r)| \right]^2 \right), \tag{4}$$

where:
$p(k|m_l)$ – the conditional probability of the class k provided in node m_l,
P_l – the probability of transition objects into the left node m_l,
P_r – the probability of transition objects into the right node m_r,
K – number of decision classes,
a_j – j–th variable,
a_j^R– the best splitting value of variable a_j.

3 Decision Forests and Random Forests

A decision forest is a collection of decision trees [4,5,13]. We defined the decision forest by the following formula:

$$DF = \{d_j : X \to \{1, 2, ..., g\}\}_{j=1,2,...,J}, \quad (5)$$

where J is a number of decision trees j ($J \geqslant 2$); g is a number of decision class and X is the set of the samples.

In decision forests, predictions of decision trees are combined to make the overall prediction for the forest. Classification is done by a simple voting. Each decision tree votes on the decision for the sample and the decision with the highest number of votes is chosen. The classifier created by decision forest DF, denoted as $dDF : X \to 1, 2, ..., g$, uses the following voting rule:

$$dDF(x) := \arg\max_k N_k(x), \quad (6)$$

where k is a decision class, such that $k \in \{1, 2, \ldots, g\}$; $N_k(x)$ is the number of votes for the sample $x \in X$ classification in to class k, such that $N_k(x) := \#\{j : d_j(x) = k\}$.

Some ensemble methods such as Random Forests are particularly useful for high-dimensional data sets due to increased classification accuracy. This can be achieved by generating multiple prediction models, each with a different subset of learning data consisted of attribute subsets [4].

Breiman provides a general framework for tree ensembles called "random forests" [4]. Each tree depends on the values of randomly chosen attributes, independently for each node or tree and with the same distribution for all trees. Thus, a random forest is a classifier (ensemble) that consists of many decision trees. Each splitting rule is performed independently for different subsets of attributes. As a result, m attributes could be chosen from p descriptions of the learning samples. Assuming that $m \ll p$, and according to the performed experiments, good results should be obtained when $m = \sqrt{p}$. Let us assume that $\frac{1}{3}$ of the samples cannot be chosen for the training sample (in accordance with the probability equal to $(1-n)^n \approx e^{-n}$), so only $\frac{1}{3}$ of the trees in the analyzed forest are constructed without this sample. In this situation Breiman proposed that it would be well grounded to apply the unencumbered estimator of misclassification probability obtained by the decision tree [4].

4 On-the-Go Adaptive Approach

In this article we focus on an approach in which we concentrate on the weak or difficult samples to discriminate or classify. For this reason we introduce the boosting algorithm, which reduces sensitivity to training samples and tries to force the analyzed approach to change the performance by constructing a new distribution over samples based on previously generated results.

This approach is firstly proposed by Schapire in 1990 [15], which was inspired by Kearns [12]. In particular, this approach is connected with the creation of a good learning set based on weak learning subsets. The newest boosting version, AdaBoost, was created in 1995 and presented by Freund and Schapire in publications from 1996 and 1997 [10,11]. This algorithm is still under further development, which was shown in [14].

A weight is assigned (initially equal to $\frac{1}{n}$) for each element of the learning set. This value determines the element's probability of being chosen for the pseudo-samples. In the next step the classifier is created. Then, for each element that is wrongly classified to the decision class the weight coefficient is increased, so that during the next pseudo-sample creation this weak element (object) will more likely be chosen.

In the AdaBoost approach the weights of elements belonging to the learning set are modified depending on the classification error coming from the individual classifier. This factor consists of the sum of the weak element's weights:

$$\epsilon(j) = \sum_{x_i} we_i[k_i \neq k_i^j], \tag{7}$$

where we_i represents the weight of element x_i and k_i^j is a decision class of the analyzed object.

The modification is performed when the classification error is smaller or equal to 0.5. The weights are multiplied by the coefficient (8) and then normalized.

$$\kappa(j) = \frac{1 - \epsilon(j)}{\epsilon(j)} \tag{8}$$

5 Ant Colony Decision Forest Algorithm

The ACDF algorithm is based on two approaches: Random Forest and ACDT. The ACDF algorithm can be applied for difficult data set analyzes by adding randomness to the process of choosing which set of features or attributes will be distinguished during construction of the decision trees [3].

In case of the ACDF, agent-ants create a collection of hypotheses in a random manner by complying to the threshold or rule to split on. The challenge is to introduce a new random subspace method for growing collections of decision trees – this means that the agent-ants can create a collection of hypotheses from the hypothesis space by using random-proportional rules. At each node of the tree the agent-ant can choose from the random subset (random pseudo-samples) of attributes and then constrain the tree-growing hypothesis to choose its splitting rule from among this subset. Because of the re-labelled randomness proposed in our approach we have resigned from having different subsets of attributes chosen for each agent-ant or colony in favour of greater stability of the undertaken hypotheses. This is a consequence of the proposition that was firstly used in Random Forest.

The ACDT approach that suffers from the problem of excessive size is said to have good diversity in the (random pseudo-samples) training and testing samples as well as balance in decision making. ACDF is described as an algorithm with high diversity because the agent-ants make a cascade of choices consisting of attributes and values chosen at each internal node in the decision tree (to create a special hypothesis). Consequently, ensembles of decision tree classifiers perform better than individuasubset of saml decision trees. This is due to independently performed exploration/exploitation of the subspace of hypotheses.

In the presented experiments the ACDF algorithm was employed (described in [3] as the 6th version). The best decision trees from each population (local best trees) constitute a decision forest; and for each colony of ants the pseudo-samples are the same and each decision tree depends on the values of randomly chosen attributes, independently for each node (similar to the Random Forest).

6 An Adaptive ACDF Algorithm

The ACDF algorithm was a prototype for the proposed adaptive ACDF algorithm (aACDF), described in section 5, where the similarities to Random Forests are emphasised. We propose a new method of generating pseudo-samples for each population of virtual ants. The on-the-go, dynamic pseudo-samples are chosen according to a previously obtained classification quality. The adaptability is focused on weak samples. The choice of objects is done by sampling with a replacement from the n-objects' set, and always consists of n objects.

The original probability of choice is equal to $\frac{1}{n}$. In the following population of virtual ants the value of this probability depends on the weight of this object. In case of an incorrect classification this coefficient is increased according to the formula:

$$we_i = \begin{cases} 1, & \text{if the object is well classified} \\ 1 + \lambda \cdot n, & \text{otherwise,} \end{cases} \tag{9}$$

Meanwhile, the probability of choosing the object is calculated according to the formula:

$$pp(x_i) = \frac{we_i}{\sum_{j=1}^{n} we_j}. \tag{10}$$

7 Experiments

A variety of experiments was conducted to test the performance and behaviour of the proposed algorithm. In this section we will consider an experimental study (see Table 1, Figs. 1 and 2 – the best results are presented in bold) performed for the following adjustments. We performed 30 experiments for each data set. Each experiment included 1250 generations with the population size of the ant colony equal to 50. In each case the decision forest consisted of 25 trees. A comparative study of the ACDT algorithm (described in [2]) and the ACDF algorithm (described in [3] – version 6) with an adaptive ACDF algorithm (aACDF) for five different values of the λ parameter described in section 7.2 was performed to examine this new approach.

Table 1. Comparative study – accuracy rate

Data set	ACDT – forest		ACDF_6		aACDF	
					$\lambda = 1.0$	
	acc	#n	acc	#n	acc	#n
heart	0.7628	253.4	**0.8391**	194.8	0.8372	**101.39**
breast-cancer	0.7284	82.1	0.7414	96.0	**0.7434**	**53.80**
balance-scale	0.8003	1198.8	0.8559	884.3	**0.8613**	**365.70**
dermatology	0.9314	**113.7**	**0.9690**	297.7	0.9667	244.79
hepatitis	0.8005	43.4	**0.8210**	58.4	0.8125	**41.46**
breast-tissue	0.4611	205.1	0.4919	146.0	0.4682	**109.37**
cleveland	0.5456	338.7	0.5737	202.9	0.5732	**137.71**
b-c-w	0.9306	**103.2**	0.9280	275.1	**0.9319**	221.03
lymphography	0.7821	**161.6**	0.8265	318.5	0.8193	208.29
shuttle	**0.9975**	1508	0.9972	2458	0.9945	**1253**
mushroom	0.6313	1004	0.6475	1007	0.6469	950
optdigits	0.8751	4273	**0.9442**	3942	0.8985	**1203**

Data set	aACDF							
	$\lambda = 0.5$		$\lambda = 0.1$		$\lambda = 0.05$		$\lambda = 0.005$	
	acc	#n	acc	#n	acc	#n	acc	#n
heart	0.8348	109.65	0.8344	145.99	0.8385	169.10	0.8323	193.96
breast-cancer	0.7374	58.05	0.7393	74.12	0.7411	78.04	0.7389	95.77
balance-scale	0.8597	412.33	0.8601	617.18	0.8565	722.26	0.8565	870.88
dermatology	0.9661	264.29	0.9661	291.10	**0.9690**	302.16	0.9681	299.41
hepatitis	0.8130	43.97	0.8134	52.12	0.8203	55.42	0.8188	56.26
breast-tissue	0.4696	115.22	0.4922	135.29	**0.4954**	140.85	0.4922	147.77
cleveland	0.5687	138.89	0.5733	155.34	0.5710	167.29	**0.5751**	197.87
b-c-w	0.9309	244.23	0.9306	265.52	0.9310	282.60	0.9262	265.40
lymphography	0.8231	238.93	**0.8282**	291.70	0.8211	309.53	0.8197	319.77
shuttle	0.9951	1585	0.9960	2104	0.9962	2162	0.9970	2500
mushroom	**0.6492**	944	0.6485	**940**	0.6488	947	0.6458	1043
optdigits	0.9024	1331	0.9193	2171	0.9290	2707	0.9422	3828

Abbrev.: acc – accuracy rate; #n – number of nodes.

7.1 Data Sets

An evaluation of the performance behavior of aACDF was performed using 12 public-domain data sets from the UCI (University of California at Irvine) data set repository. The experiments were carried out on an Intel Core i5 2.27 GHz Computer with 2.9 GB RAM.

Data sets larger than 1000 samples are divided into two groups in a random way: training and testing sets, appropriately. Data sets with fewer than 1000 samples are estimated by 10-fold cross-validation. In both cases we also used an additional data set – a clean set. The results are tested on a clean set that has not been used to build the classifier.

7.2 Comparative Study – An Examination of λ Parameter Values

The proposed version of the adaptive ACDF demands, above all, comparison with its predecessors: ACDF-forest and ACDF (based on Random Forest) [1,2,3] in the context of different λ parameter values. These values of parameter λ were established arbitrarily. We examined the following values of the λ parameter: $1.0, 0.5, 0.1, 0.05$ and 0.005.

7.3 Experiments and Results

The experimental results confirm that adaptively applied subsets of samples allow to obtain significantly better results (Tab. 1). For a majority of the data sets for most cases we managed to construct decision forests consisting of smaller decision tress (with a decreased number of nodes), as compared to the ACDT-forest and ACDF algorithms, which can be observed in Fig. 2. A decrease in the decision trees' growth depended on the λ parameter value. As the λ parameter value rose, smaller decision trees and forests were created. We also observed the influence of the λ parameter on the quality of the classification (see Fig. 1), therefore, using λ parameter values greater than 1.0 causes a deterioration of the classification quality.

The best results were obtained for $\lambda = 1.0$ and $\lambda = 0.5$. In Fig. 3 we can notice the relationship between decision tree growth and classification accuracy depending on the applied algorithms.

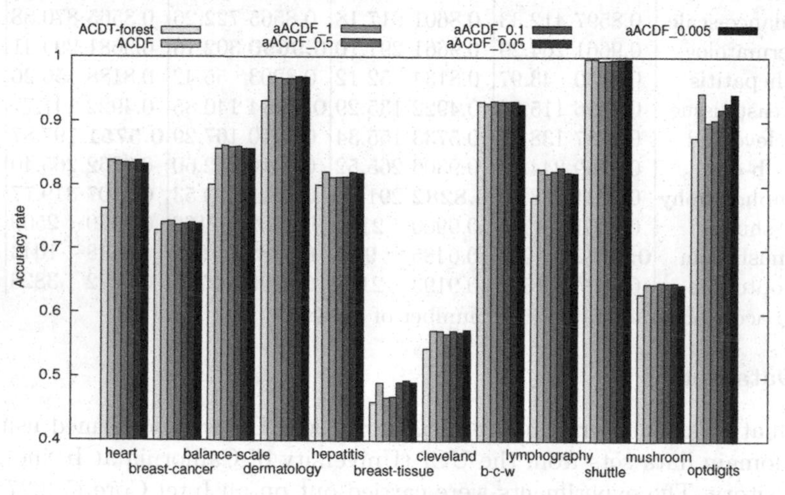

Fig. 1. Accuracy rate

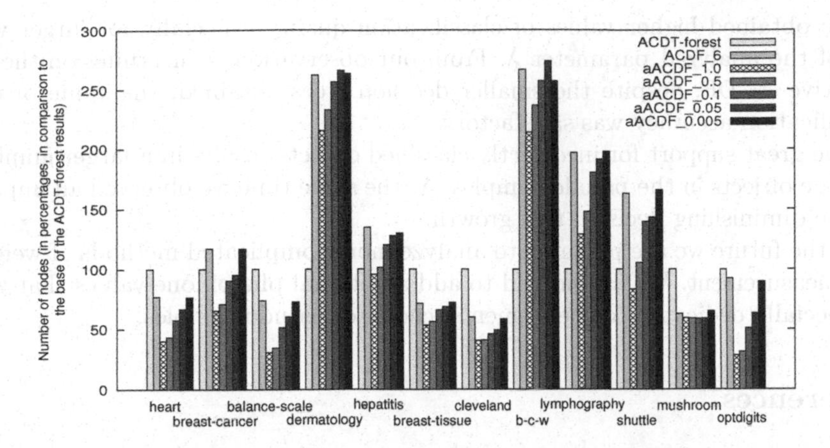

Fig. 2. Number of nodes (in percentages, in comparison to the base of the ACDT-forest results)

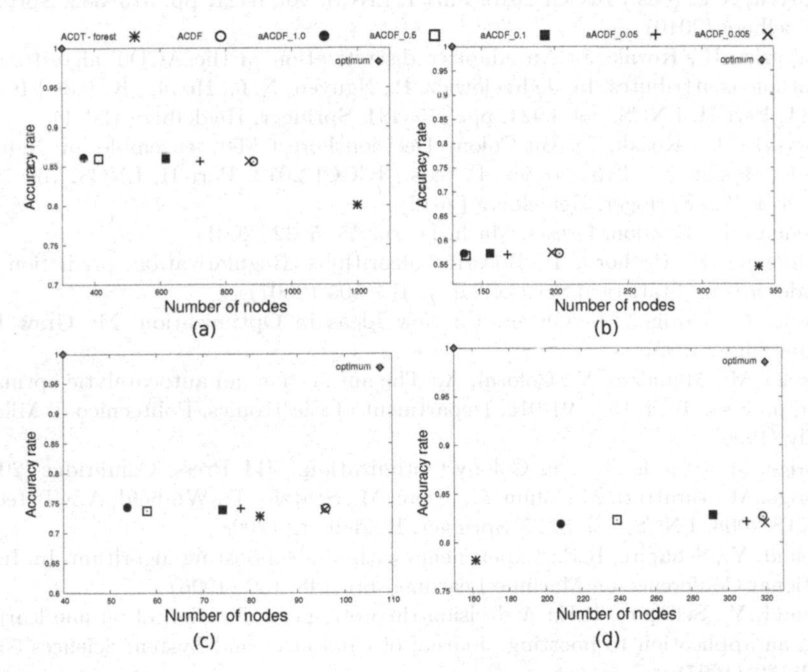

Fig. 3. Relationship between decision tree growth and classification accuracy depending on the applied algorithms for a data set: (a) balance-scale (b) cleveland (c) breast-cancer (d) lymphography. Figures (a) and (b) show the characteristics of the most frequently chosen samples for the analyzed data sets

8 Conclusions

Adaptive ACDF algorithms allowed the situation to get better results obtained in our experiments in comparison to all previously presented approaches.

We obtained higher values of classification quality especially for larger values of the analyzed parameter λ. From our observations concerning on-the-go adaptive ACDF, despite the smaller decision trees obtained, the value of the classification accuracy was satisfactory.

The great support for incorrectly classified objects results in a larger number of these objects in the pseudo-samples. At the same time we observed an impact on the diminishing decision tree growth.

In the future we are planning to analyze more complicated methods of weight we_i measurement. We also intend to add additional pheromone values that will be specially dedicated to the elements' choice of pseudo-samples.

References

1. Boryczka, U., Kozak, J.: Ant colony decision trees – A new method for constructing decision trees based on ant colony optimization. In: Pan, J.-S., Chen, S.-M., Nguyen, N.T. (eds.) ICCCI 2010, Part I. LNCS, vol. 6421, pp. 373–382. Springer, Heidelberg (2010)
2. Boryczka, U., Kozak, J.: An adaptive discretization in the ACDT algorithm for continuous attributes. In: Jędrzejowicz, P., Nguyen, N.T., Hoang, K. (eds.) ICCCI 2011, Part II. LNCS, vol. 6923, pp. 475–484. Springer, Heidelberg (2011)
3. Boryczka, U., Kozak, J.: Ant Colony Decision Forest Meta-ensemble. In: Nguyen, N.-T., Hoang, K., Jędrzejowicz, P. (eds.) ICCCI 2012, Part II. LNCS, vol. 7654, pp. 473–482. Springer, Heidelberg (2012)
4. Breiman, L.: Random forests. Mach. Learn. 45, 5–32 (2001)
5. Bühlmann, P., Hothorn, T.: Boosting algorithms: Regularization, prediction and model fitting. Statistical Science 22(4), 477–505 (2007)
6. Corne, D., Dorigo, M., Glover, F.: New Ideas in Optimization. Mc Graw–Hill, Cambridge (1999)
7. Dorigo, M., Maniezzo, V., Colorni, A.: The ant system: an autocatalytic optimization process. Tech. Rep. 91-016, Department of Electronics, Politecnico di Milano, Italy (1996)
8. Dorigo, M., Stützle, T.: Ant Colony Optimization. MIT Press, Cambridge (2004)
9. Dorigo, M., Birattari, M., Blum, C., Clerc, M., Stützle, T., Winfield, A.F.T. (eds.): ANTS 2008. LNCS, vol. 5217. Springer, Heidelberg (2008)
10. Freund, Y., Schapire, R.E.: Experiments with a new boosting algorithm. In: International Conference on Machine Learning, pp. 148–156 (1996)
11. Freund, Y., Schapire, R.E.: A decision-theoretic generalization of on-line learning and an application to boosting. Journal of Computer and System Sciences 55(1), 119–139 (1997)
12. Kearns, M.: Thoughts on hypothesis boosting, project for Ron Rivest's machine learning course at MIT (1988)
13. Rokach, L., Maimon, O.: Data Mining With Decision Trees: Theory and Applications. World Scientific Publishing (2008)
14. Rudin, C., Schapire, R.E.: Margin-based ranking and an equivalence between AdaBoost and RankBoost. J. Mach. Learn. Res. 10, 2193–2232 (2009)
15. Schapire, R.E.: The strength of weak learnability. Machine Learning 5, 197–227 (1990)

Bootstrapping and Rule-Based Model for Recognizing Vietnamese Named Entity

Hieu Le Trung[1,*], Vu Le Anh[2], and Kien Le Trung[3]

[1] Duy Tan University, Da Nang, Vietnam
[2] Nguyen Tat Thanh University, Ho Chi Minh, Vietnam
[3] Hue University of Sciences, Hue, Vietnam
letrunghieu@dtu.edu.vn

Abstract This paper intends to address and solve the problem Vietnamese Named Entity recognition and classification (VNER) by using the bootstrapping algorithm and rule-based model. The rule-based model relies on contextual rules to provide contextual evidence that a VNE belongs to a category. These rules exploit linguistic constraints of category are constructed by using the bootstrapping algorithm. Bootstrapping algorithm starts with a handful of seed VNEs of a given category and accumulate all contextual rules found around these seeds in a large corpus. These rules are ranked and used to find new VNEs.

Our experimented corpus is generated from about 250.034 online news articles and over 9.000 literatures. Our VNER system consists 27 categories and more 300.000 VNEs which are recognized and categorized. The accuracy of the recognizing and classifying algorithm is about 95%.

1 Introduction

Named Entity Recognition (NER) problem has become a major task of Natural Language Processing (NLP). Named Entities (NE) represent important parts of the meaning of human-written sentences, such as persons, places and objects [1]. In the sixth and the seventh editions of the Conference on Computational Natural Language Learning (CoNLL 2002 and CoNLL 2003) the NER task was defined as *to determine the proper names existing within an open domain text and classify them as one of the following four classes: Person; Location; Organization; and Miscellaneous.*

To correctly identify all of the named entities is a very difficult task for any language, since the level of difficulty depends on the diversity of language settings. As to Vietnamese, there are many difficult problems in NER. Firstly, Vietnamese Named Entity (VNE) doesn't have some given special syllables for names. Secondly, VNE is an open class and the number of its component is very large such that it is very hard to enumerate all of them.

Our work intends to address and solve the problem Vietnamese named entity recognition and classification(VNER).

* Corresponding author.

N.T. Nguyen et al. (Eds.): ACIIDS 2014, Part II, LNAI 8398, pp. 167–176, 2014.

Our approach for two problems is based on *the bootstrapping algorithm*[4, 5, 6] and *rule-based model*. The rule-based model relies on contextual rules to provide contextual evidence that a VNE belongs to a category. These rules exploit linguistic constraints and contextual information in identifying VNEs, is constructed based on *models of word recognition*[2] and *theory of word order patterns*[3] in Vietnamese. In which, each word and phrase surrounding VNE in sentence is recognized with the high accuracy, and frequency sequence of words are ranked.

The collected VNEs and contextual rules are ranked based on *confidence functions* and *similarity functions*. In which, confidence function expresses a numerical confidence (reliably) that the contextual rule will extract members of the category, and similarity function expresses similarity between a VNE and a category.

The bootstrapping algorithm involves a small set of seed VNEs for each category, for starting the learning process. A set of seed VNEs is generated by sorting the VNEs in the corpus and manually identifying. First, the system searches for sentences or phrases that contain these VNEs and tries to identify some contextual rules. Then, the system tries to find other instances of VNEs appearing in similar contexts. The learning process is then reapplied to the newly found VNEs, so as to discover new relevant contexts.

Finally, the VNE recognition and classification algorithm is performed based on *model matching* of context surrounding of VNE in sentence with linguistic constraints of each category.

Our contribution is a new approach for solving the VNER problem. In which, the model of the problem VNER with the bootstrapping algorithm for recognizing a category, and the automatic algorithm identify VNEs in sentence are described and discussed. The system of categories of VNEs are constructed with high accuracy and its linguistic constraints are important data for solving different tasks in Vietnamese natural language processing.

The rest of this paper is structured as follows. In section 2, we introduce and describe the problem recognition and classification Vietnamese named entities(VNER). In section 3, we define a contextual rule and it's construction. Based on a set of contextual rules, the C-VNE recognition algorithm is proposed to recognize a C-VNE syllable, a Vietnamese Named Entity which has a category C, and the VNE recognition algorithm is proposed to recognize a VNE-syllable, a Vietnamese Named Entity. Section 4 presents the results of our experiments that support our approach. Section 5 concludes the paper.

2 VNE Recognition Problem

The purpose of the *Vietnamese Named Entities Recognition* (VNER) problem[7, 8] is to recognize a *Vietnamese Named Entity* (VNE) in documents, and determine whether the VNE is a member of predefined categories of interested as: *person names*, *organization names*, *location names*, etc. Simply, the VNER problem is constructed as follows:

1. Syllable, sentence and corpus are given by following descriptions:
   - *Syllable* s is an original syllable (such as *"của"*, *"đã"*) or a linking syllable (such as *"công_ việc"*, *"Hồ_ Chí_ Minh"*, *"Hà_ Nội"*). We call *"của"* the first component of the syllable *"của"*, *"công"* and *"việc"* the first and second components of the syllable *"công_ việc"*.
   - *Sentence* S is an ordered sequence of syllables, $S = s_1 s_2 \ldots s_n$. $n = |S|$ is called the *length* of the sentence S.
   - *Corpus* $\mathcal{C} = \{(S, s, k)\}$ is a finite set of triple objects: S is a sentence, s is a syllable in S, and k is a position of s in S, $1 \leq k \leq |S|$. S is called *the context* of the triple. For example, assuming that there is a sentence

$$S = \text{“Chúng\_ tôi yêu Việt\_ Nam”}$$

   in the corpus, so $\mathcal{C}$ has three elements $(S, \textit{Chúng\_ tôi}, 1)$, $(S, \textit{yêu}, 2)$, and $(S, \textit{Việt\_ Nam}, 3)$.

   A structure of the corpus $\mathcal{C}$ is described carefully in [2].
2. Let $\mathcal{V} \subset \mathcal{C}$ is a set of all *Vietnamese Named Entities* in the corpus $\mathcal{C}$. The set $\mathcal{V}$ is determined by a map $f_{rec} : \mathcal{C} \rightarrow \{0; 1\}$ for which

$$f_{rec}(S, s, k) = \begin{cases} 1 & \text{if } (S, s, k) \in \mathcal{V} \\ 0 & \text{otherwise.} \end{cases} \tag{1}$$

   The map f_{rec} is called the ***recognition function***.
3. Let $\mathcal{E} = \{C_1, C_2, \ldots, C_m\}$ is a *category* of VNE, and a map $f_{cla} : \mathcal{V} \rightarrow \mathcal{E}$ which is called a ***classification function***. For each category $C \in \mathcal{E}$, a VNE of which category is C is called shortly a *C-VNE*.

The VNER aims to solve two problems:

Problem 1. *Determining the recognition function f_{rec} so that for all $(S, s, k) \in \mathcal{C}$, we recognize the syllable s is VNE or not.*

Problem 2. *Determining the classification function f_{cla} so that for all VNE $(S, s, k) \in \mathcal{V}$, we identify a category of the syllable s, $f_{cla}(S, s, k)$.*

For each category $C \in \mathcal{E}$, let us denote $\mathcal{V}(C) = \{(S, s, k) \in \mathcal{V} \mid f_{cla}(S, s, k) = C\}$, the set of all C-VNEs. The key of our works is how we gain the *grammar-rules* of the syllables in $\mathcal{V}(C)$ to recognize a syllable is in $\mathcal{V}(C)$ or not. Precisely, we consider a problem:

Problem 3. *Let $C \in \mathcal{E}$ be a category, and $S(C) \subset \mathcal{V}(C)$ be a given set of some C-VNEs. Based on $S(C)$, constructing some grammar-rules $R(C)$ to determine a map $f_C : \mathcal{C} \rightarrow \{0; 1\}$ for which*

$$f_C(S, s, k) = \begin{cases} 1 & \text{if } (S, s, k) \in \mathcal{V}(C) \\ 0 & \text{otherwise.} \end{cases} \tag{2}$$

f_C is called a C-VNE recognition function.

First of all, we introduce a simple grammar-rule for which we can easily recognize a lot of non-VNE and non-C-VNE syllables. This grammar-rule based on a following concept.

Definition 1. *A syllable s is called a candidate if the first letters of all its components are capitalization.*

An example of a candidate is a syllable *"Hồ_ Chí_ Minh"*, and an example of a non-candidate is a syllable *"Chủ_ tịch"*. In this work, we mention that a VNE should be a candidate. Thus,

Grammar-Rule 1 *Given* (S, s, k) *in the corpus. If s is not a candidate,*(S, s, k) *is not a VNE,* $f_{rec}(S, s, k) = 0$*, and of course for all category* $C \in \mathcal{E}$*,* (S, s, k) *is not a* C*-VNE,* $f_C(S, s, k) = 0$.

In next sections, we would like to study how we utilize a given set of some C-VNEs to construct other grammar-rules for determining the C-VNE recognition function f_C.

3 Contextual Rules and Recognition

Let $C \in \mathcal{E}$ be a category, and $S(C)$ be a given set of some C-VNEs. The C-VNE recognition function f_C is determined based on the set of *contextual rules*, which represent linguistic constraints for deciding another syllable $(\bar{S}, \bar{s}, \bar{k}) \notin S(C)$ is a C-VNE or not. Thus, the aims of this section is study how to construct the set of contextual rules of the category C from the given set $S(C)$ and how to determine the function f_C from these contextual rules. The first aim is done based on a *confidence function*, a *similar function* and the idea of *bootstrapping algorithm.* The second aim is based on a probability model which measures a similar structure between a target syllable and some given C-VNE syllables. These measuring are depended on the roles of contextual rules.

3.1 Contextual Rule

Let us consider an example: Given two sentences

S_1 = *"Ông Nam và ông Nhân đã thực_ hiện thành_ công dự_ án này."*
S_2 = *"Tôi cùng ông Nghĩa đã thực_ hiện một nhiệm_ vụ quan trọng."*

Assuming that two candidates $(S_1$, *"Nhân"*$, 5)$ and $(S_2$, *"Nghĩa"*$, 4)$ belong to $\mathcal{V}$(C), where C is the "*names of people*". We have an idea: For any candidate $(S, s, k) \in \mathcal{C}$, if

"ông s đã thực_ hiện" is a subsequence of S

can we decide the candidate $(S, s, k) \in \mathcal{V}$(C)? If we can decide, the subsequence *"ông s đã thực_ hiện"* where s is a candidate is called a *contextual rule* of the *'names of people'*-VNE. Formally, let $C \in \mathcal{E}$ be a category, a contextual rule of C is defined by:

Definition 2. *A contextual rule p of C is a pair of subsequences $(\mathbf{s}_l, \mathbf{s}_r)$, where $\mathbf{s}_l$ is called a left context and $\mathbf{s}_r$ is called a right context.*

Considering the above example, $p =$ (*"ông", "đã thực_ hiện"*) is a contextual rule of the category *names of people*, and $\mathbf{s}_l =$ *"ông"* is the left context, $\mathbf{s}_r =$ *"đã thực_ hiện"* is the right context of p.

For any candidate (S, s, k) and a contextual rule $p = (\mathbf{s}_l, \mathbf{s}_r) \in \mathcal{R}$, we define

Definition 3. *(S, s, k) satisfies the contextual rule p if $\mathbf{s}_l\ s\ \mathbf{s}_r \subset S$. $p(S, s, k) = 1$ is denoted for (S, s, k) satisfies p, and $p(S, s, k) = 0$ is denoted for another case.*

3.2 Constructing Contextual Rules

Assuming that $\mathcal{R}$ is a set of some contextual rules of the category C. Let us denote $\sigma(\mathcal{V}_\mathrm{C})$ a set of all subsets of $\mathcal{V}_\mathrm{C}$, $\mathcal{V}_\mathrm{C}(\mathrm{p}) = \{(S, s, k) \in \mathcal{V}_\mathrm{C} \mid p(S, s, k) = 1\}$ $\mathcal{V}_\mathrm{C}(\mathrm{p}, \mathrm{W}) = \{(S, s, k) \in W | p(S, s, k) = 1\}$, where $p \in \mathcal{R}$, $W \subseteq \mathcal{V}_\mathrm{C}$, and $\mathcal{R}(\mathrm{S}, \mathrm{s}, \mathrm{k})$ $= \{p \in \mathcal{R} \mid p(S, s, k) = 1\}$.

Definition 4. *F_{con} is called a confidence function corresponding with $\mathcal{R}$ if $F_{con} : \mathcal{R} \times \sigma(\mathcal{V}_\mathrm{C}) \rightarrow \mathrm{R}$ for which $\forall p \in \mathcal{R}$, $\mathrm{W} \in \sigma(\mathcal{V}_\mathrm{C})$,*

$$F_{con}(p, W) = \frac{|\mathcal{V}_\mathrm{C}(\mathrm{p}, \mathrm{W})|}{|\mathcal{V}_\mathrm{C}(\mathrm{p})|} \log |\mathcal{V}_\mathrm{C}(\mathrm{p}, \mathrm{W})|, \tag{3}$$

where $|\cdot|$ denote the size of a set.

The confidence function receives a high value if a high percentage of the contextual rule's extractions are members of W, or if a moderate percentage of the rule's extractions are members of W and it extracts a lot of them.

Definition 5. *F_{sim} is called a similarity function corresponding with $\mathcal{C}$ if $F_{sim} : \mathcal{C} \times \sigma(\mathcal{V}_\mathrm{C}) \rightarrow \mathrm{R}$ for which $\forall (S, s, k) \in \mathcal{C}$, $\mathrm{W} \in \sigma(\mathcal{V}_\mathrm{C})$,*

$$F_{sim}(S, s, k, W) = \frac{\sum_{p \in \mathcal{R}(\mathrm{S,s,k})} |\mathcal{V}_\mathrm{C}(\mathrm{p}, \mathrm{W})|}{|\mathcal{R}(\mathrm{S}, \mathrm{s}, \mathrm{k})|} \tag{4}$$

where $|\cdot|$ denote the size of a set.

The similarity function measures a similarity between a candidate (S, s, k) and other candidates in W corresponding with $\mathcal{R}$. It receives a high value if (S, s, k) satisfies contextual rules in $\mathcal{R}$ that also have a tendency to extract the members of W.

The significance of the confidence function and the similarity function is explained as follows: If we have a set W of some C-VNEs, the confidence function shows that from W it can find some contextual rules for recognizing C-VNEs. Next, when the confidence function gives us a set $\mathcal{R}$ of some contextual rules for recognizing C-VNEs, the similarity function will recognize other C-VNEs outside W. This is an idea for constructing the set of contextual rules and also recognizing the set $\mathcal{V}(C)$ of all C-VNEs.

Concretely, let $S(C) \subset \mathcal{V}(C)$ be a small set of given C-VNEs. The constructing contextual rules and recognizing C-VNEs algorithm consists two steps. First, the algorithm searches for sentences that contain C-VNEs in $S(C)$ and tries to identify some contextual rules to the best examples. Next, it tries to find new C-VNEs appearing in similar contexts. The learning process is then reapplied to the newly found examples. By repeating this process, a large number of C-VNEs and a large number of contextual rules will eventually be gathered. This algorithm is called a *C-VNEs recognition algorithm* and given as follows

---

The C-VNEs Recognition Algorithm

Input: $A = S(C)$
Output: $\mathcal{V}(C)$ - the set of all *C-VNEs*

$R(C) = \emptyset$
1. **for** $i = 1$ **to** n **do**
2. Setting the parameters to (θ_i^c, θ_i^s)
3. **repeat**
4. **for each** contextual rule p **in** $\mathcal{R}$
5. Calculate $F_{con}(p, A)$
6. **if** $F_{con}(p, A) \geq \theta_i^c$ **then**
7. Add p to $R(C)$
8. **for each** (S, s, k) **in** $\mathcal{C} \setminus A$
9. Evaluated $F_{sim}(S, s, k, A)$
10. **if** $F_{sim}(S, s, k, A) \geq \theta_i^s$ **then**
11. Add (S, s, k) to A
12. **until** no contextual rules add to $R(C)$

The bootstrapping process loops n-times with n parameters, which converge to desired one. The way we choose these parameters guarantees the quality of learning process. At each iteration, the process adds the new contextual rules, which have confidence value higher than θ_i^c to $R(C)$. All of its extractions, which have similarity value higher than θ_i^s, are inferred to be category members and added to A. Then the next best contextual rule is identified, based on both the original seed VNEs and the new VNEs that were just added to the category, and the process repeats. Adding the best contextual rules help to more accurately identify value of similarity function of new VNEs with the set of C-VNEs A.

The results of learning process are the set of all C-VNEs A, and the set of all contextual rules $R(C)$ for category C.

3.3 Recognition of VNEs

Let $\mathcal{E}$ be a category set. For each category $C \in \mathcal{E}$, let $R(C)$ be a set of all contextual rules of C. We define a *C-VNEs recognition score*, a function on the space of candidates and categories $\mathcal{C} \times \mathcal{E}$, as follows:

$$Score(S, s, k, C) = \sum_{p \in R(C)} F_{con}(p, \mathcal{V}(C)) * p(S, s, k) * F_{occ}(S, s, k, \mathcal{V}(C)) \quad (5)$$

where, $F_{con}(p, \mathcal{V}(C))$ is a *confidence function* of the contextual rule p with $\mathcal{V}(C)$, and $F_{occ}(S, s, k, \mathcal{V}(C))$ is an *occurrence function* of candidate (S, s, k) in $\mathcal{V}(C)$ is determined by following formula:

$$F_{occ}(S, s, k, \mathcal{V}(C)) = \begin{cases} 1 - \epsilon & \text{if } (S, s, k) \in \mathcal{V}(C), \\ \epsilon & \text{otherwise} \end{cases} \quad (6)$$

where, $0 \leq \epsilon \leq 1$. By definition, the candidate (S, s, k) has high value of score C-VNEs recognition if it satisfies some high confidence value contextual rules in

$R(C)$ and is a C-VNEs. However, if (S, s, k) is not in $\mathcal{V}(C)$, we are still calculating its C-VNEs recognition score with the coefficient ϵ. In this case, the candidate (S, s, k) satisfies a lot of contextual rules of which their confidence values are very high, then we recognize a new C-VNE.

Given a parameter $\theta \geq 0$, the recognition function and the classification function are determined based on the C-VNE recognition score. These formulae are given: for all $(S, s, k) \in \mathcal{C}$,

$$f_{rec}(S, s, k) = \begin{cases} 1 & \text{if } \exists C \in \mathcal{E},\ Score(S, s, k, C) \geq \theta, \\ 0 & \text{otherwise} \end{cases} \tag{7}$$

and

$$f_{cla}(S, s, k) = \begin{cases} \oslash & \text{if } f_{rec}(S, s, k) = 0, \\ argmax_{C \in \mathcal{E}} Score(S, s, k, C) & \text{otherwise} \end{cases} \tag{8}$$

The *VNE recognition algorithm* is given as follows:

The VNE Recognition Algorithm

Input: Candidate (S, s, k), $\theta > 0$
Output: $f_{rec}(S, s, k)$ and $f_{cla}(S, s, k)$

$max = 0$
1. **for each** category C **in** $\mathcal{E}$ **do**
2. Calculate $Score(S, s, k, C)$
3. **if** $Score(S, s, k, C) > max$ **then**
4. $max = Score(S, s, k, C)$
5. $C_{max} = C$
6. **if** $max < \theta$ **then**
7. **return** $f_{rec}(S, s, k) = 0$ **and** $f_{cla}(S, s, k) = \oslash$
8. **return** $f_{rec}(S, s, k) = 1$ **and** $f_{cla}(S, s, k) = C_{max}$

4 Evaluation

Corpus and Contextual Rules. Our data of sentences is collected from 250.034 articles in the TuoiTre online newspaper (http://www.tuoitre.com.vn) and more than 9.000 novels in VNThuquan website (http://www.vnthuquan.net). We preprocessed the data first by applying the data normalization: fixing the code fonts and repairing spelling mistakes of syllables and segmenting sentences[2]. The initial corpus has 31.565.364 sentences whose total length is 411.623.127 syllables.

Categories and The Seed VNEs. Each category of VNEs is specified by a small number of the seed VNEs. The seed VNEs of a given category must be satisfy 2 conditions: (i) must be frequent in the corpus for the bootstrapping algorithm to work well; and (ii) satisfy the condition of many contextual rules in linguistic constraints of given category, do not satisfy the condition of ambiguous rules, which has high frequency and extracts many VNEs of many distinct categories. The seed VNEs was generated by sorting candidates in the corpus and manually

Table 1. Seed Named Entities Lists

Category	Seed NEs
NEs for Nations:	Mỹ, Pháp, Anh, Đức, Nga, Nhật, Úc, Trung_Quốc, Ý, Việt_Nam.
NEs for Cities:	Sài_Gòn, Huế, Đà_Nẵng, Cần_Thơ, Đà_Lạt, Hồ_Chí_Minh, Bình_Dương, Quảng_Nam, Nha_Trang, Đồng_Nai.
NEs for Rivers:	Hương, Hàn, Gianh, Cửu_Long, Nhà_Bè, Cầu, Đà, Kim_Ngưu.
NEs for Streets:	Lê_Duẩn, Chi_Lăng, Đống_Đa, Cách_Mạng_Tháng_Tám, Phố_Huế, Hai_Bà_Trưng, Trần_Hưng_Đạo, Bến_Nghé, Bạch_Đằng.

Table 2. Top 15 extraction patterns for semantic categories

NEs for Nations	NEs for Cities	NEs for Rivers	NEs for Humans
trận gặp [NE]	và tỉnh [NE]	dọc sông [NE]	là anh [NE]
hàng phòng_ngự [NE]	địa_bàn tỉnh [NE]	vượt sông [NE]	[B] ông [NE]
vào thị_trường [NE]	địa_phận tỉnh [NE]	tả_ngạn sông [NE]	của ông [NE]
với đội_tuyển [NE]	lãnh_đạo tỉnh [NE]	của sông [NE]	là ông [NE]
[B] quốc_tịch [NE]	cho tỉnh [NE]	thượng_nguồn sông [NE]	hỏi thằng [NE]
của tuyển [NE]	từ tỉnh [NE]	hạ_lưu sông [NE]	gọi chị [NE]
sang thị_trường [NE]	chính_quyền tỉnh [NE]	trên dòng_sông [NE]	tìm anh [NE]
nền kinh_tế [NE]	của tỉnh [NE]	dọc_theo sông [NE]	lúc anh [NE]
tại thị_trường [NE]	địa_bàn thành_phố [NE]	dọc bờ_sông [NE]	gặp_lại chị [NE]
nhập quốc_tịch [NE]	tại tỉnh [NE]	bên dòng_sông [NE]	và cậu [NE]
nhập_khẩu từ [NE]	toàn tỉnh [NE]	bên bờ_sông [NE]	mời chú [NE]
[B] du_khách [NE]	trên địa_bàn [NE]	và sông [NE]	nhìn ông [NE]
tiền_đạo người [NE]	ở tỉnh [NE]	ven sông [NE]	để bác [NE]
[B] người [NE]	chi_nhánh tại [NE]	nguồn sông [NE]	nhìn bà [NE]
xuất_khẩu sang [NE]	thuộc tỉnh [NE]	đoạn sông [NE]	của mợ [NE]

identifying. The seed VNEs lists for some categories that we used are shown in Table 1.

Contextual rules for category. We ran the bootstrapping algorithm for 50 iterations. Contextual rules produced by the last iteration were the output of the system. Through the process of checking list of the best contextual rules of 20 categories, some results and ideas are collected:

– Collected contextual rules reflect correctly linguistic constraints of each category. Ambiguous rules such as: *"của [NE]"*, *"là [NE]"* do not appear in list. The experimental results reflect correctly the accuracy and efficiency of the our measurements and algorithms.
– Process recognition and classification categories depends only on the seed VNEs, help increase the flexibility and efficiency of the system. It would indeed be appropriate to classify all VNEs, which are separated by different type.

Table 2 shows the top 20 contextual rules for some categories produced by bootstrapping after 50 iterations. In which, [B] is label for beginning of phrase, and [NE] is label for VNE. Most of these contextual rules are clearly useful linguistic constraints for extracting VNEs for each catogory.

Table 3. Accuracy of the Semantic Lexicons

	Iter 1	Iter 10	Iter 20	Iter 30	Iter 40	Iter 50
NEs for Nations	5/5(1)	43/50(.86)	82/100(.82)	134/150(.89)	178/200(.89)	219/250(.87)
NEs for Cities	5/5(1)	44/50(.88)	91/100(.91)	139/150(.93)	183/200(.92)	221/250(.88)
NEs for Rivers	5/5(1)	49/50(.98)	98/100(.98)	146/150(.97)	185/200(.97)	239/250(.96)
NEs for Streets	5/5(1)	50/50(1)	99/100(.99)	149/150(.99)	197/200(.98)	245/250(.98)
NEs for Persons	5/5(1)	49/50(.98)	99/100(.99)	149/150(.99)	199/200(.99)	248/250(.99)

The Bootstrapping Algorithm. Table 3 shows the accuracy of some categories after the 1st iteration of bootstrapping and after each 10th iteration. Each cell shows the number of true category members among the entries generated thus far. For example, 50 VNEs were recognized as named entities of cities after tenth iteration and 44 of those (88%) were true VNEs of cities. Table 3 shows that bootstrapping identified 219 VNEs of nations, 221 VNEs of cities, 239 VNEs of rivers, 245 VNEs of streets and 248 VNEs of persons. The accuracy of processing recognize VNEs is very high.

Vietnamese Named Person Recognition. Experiment for Vietnamese named person recognition ran bootstrapping algorithm for 200 iterations. The results identified 31.427 VNEs of person with the accuracy about 98%, cả short name as *"Vũ, Kiên, Hiếu"*, and full name as *"Lê_ Trung_ Hiếu"*, *"Nguyễn_ Thu_ Phương"*. Besides, the selection of appropriate seed VNEs help identify good named persons for man, woman, or named persons for singer, actor, politicians,... These results are an important data for problems such as Extraction Information, Abstract Document, Classification Documents,...

5 Conclusion

We have presented rule-based model for recognizing and classifying Vietnamese named entities. The model is constructed by using the bootstrapping algorithm. The rule-based model relies on contextual rules to provide contextual evidence that a VNE belongs to a category. The model works well on a large corpus and can extract many categories, such as, name of person, nations, cities, streets, rivers, places, oranizations,... with the high accuracy.

Vietnamese language is not explained and described well by grammar rules. However, according to research results have demonstrated the ability to apply models Vietnamese word recognition and theory word order patterns for solving some tasks in Vietnamese language processing. One of our reseach direction is

"Finding the most common formulas of Vietnamese sentence". We believe that with the huge corpus, we can solve many problems of Vietnamese language processing based on statistic.

References

1. Chen, C., Lee, H.J.: A Three-Phase System for Chinese Named Entity Recognition. In: Proceedings of ROCLING XVI, pp. 39–48 (2004)
2. Le Trung, H., Le Anh, V., Le Trung, K.: An Unsupervised Learning and Statistical Approach for Vietnamese Word Recognition and Segmentation. In: Nguyen, N.T., Le, M.T., Świątek, J. (eds.) ACIIDS 2010, Part II. LNCS (LNAI), vol. 5991, pp. 195–204. Springer, Heidelberg (2010)
3. Le Trung, H., Le Anh, V., Dang, V.-H., Hoang, H.V.: Recognizing and Tagging Vietnamese Words Based on Statistics and Word Order Patterns. In: Nguyen, N.T., Trawiński, B., Katarzyniak, R., Jo, G.-S. (eds.) Adv. Methods for Comput. Collective Intelligence. SCI, vol. 457, pp. 3–12. Springer, Heidelberg (2013)
4. Lin, W., Yangarber, R., Grishman, R.: Bootstrapped learning of semantic classes from positive and negative examples. In: Proceedings of ICMLK 2003 Workshop on the Continuum from Labeled to Unlabeled Data (2003)
5. Micheal, T., Riloff, E.: A Bootstrapping Method for Learning Semantic Lexicon using Extraction Pattern Contexts. In: Proceedings of the ACL 2002 conference on Empirical Methods in Natural Language Processing, pp. 214–221 (2002)
6. Riloff, E., Jones, R.: Learning Dictionaries for Information Extraction by Multi-level Bootstrapping. In: Proceedings of the Sixteenth National Conference on the Artificial Intelligence and the Eleventh Innovative Applications of Artificial Intelligence Conference, pp. 474–479 (1999)
7. Tran, Q.T., Pham, T.X.T., Ngo, Q.H., Dinh, D., Collier, N.: Named Entity Recognition in Vietnamese documents. Progress in Informatics Journal, 5–13 (2007)
8. Pham, T.X.T., Kawazoe, A., Dinh, D., Collier, N., Tran, Q.T.: Construction of a Vietnamese Corpora for Named Entity Recognition. In: RIAO 2007, 8th International Conference, pp. 719–724. Carnegie Mellon University, Pittsburgh (2007)

A Family of the Online Distance-Based Classifiers

Joanna Jędrzejowicz[1] and Piotr Jędrzejowicz[2]

[1] Institute of Informatics, Gdańsk University,
Wita Stwosza 57, 80-952 Gdańsk, Poland
jj@inf.ug.edu.pl

[2] Department of Information Systems, Gdynia Maritime University,
Morska 83, 81-225 Gdynia, Poland
pj@am.gdynia.pl

Abstract. In this paper a family of algorithms for the online learning and classification is considered. These algorithms work in rounds, where at each round a new instance is given and the algorithm makes a prediction. After the true class of the instance is revealed, the learning algorithm updates its internal hypothesis. The proposed algorithms are based on fuzzy C-means clustering followed by calculation of distances between cluster centroids and the incoming instance for which the class label is to be predicted. Simple distance-based classifiers thus obtained serve as basic classifiers for the implemented rotation forest kernel. The proposed approach is validated experimentally. Experiment results show that proposed classifiers perform well against competitive approaches.

Keywords: online learning, fuzzy C-means clustering, Rotation Forest.

1 Introduction

One of the data mining basic tasks is classification that is identification of some unknown object or phenomenon as a member of a known class of objects or phenomena. In machine learning and statistics, classification is usually understood as the problem of identifying to which of a set of categories (subpopulations) a new observation belongs, on the basis of a training set of data containing instances (observations) whose category membership is known. The idea in machine learning is to produce a so called classifier, which can be viewed as the function induced by a classification algorithm that maps input data to a category.

In machine learning there are two basic approaches to inducing classifiers: static and dynamic one. Static approach is based on two assumptions. First one assumes that a training set of the adequate size required to induce classifier and consisting of instances with known class labels, is available in advance, that is before a classifier is constructed. Second assumption requires that data instances arriving in the future have a stationary distribution identical with the distribution of data in the training set.

N.T. Nguyen et al. (Eds.): ACIIDS 2014, Part II, LNAI 8398, pp. 177–186, 2014.

In case of the dynamic approach assumptions are weaker. Usually, some limited number of training examples with known class labels is required at the outset, however future incoming instances, after their true class label is revealed, can be added to the available training set extending it, or perhaps replacing some of its earlier arriving members. Second assumption does not have to hold, hence data distribution does not have to be a stationary one.

Dynamic approach is a natural way of dealing with important online learning problems. Online learning is considered to be of increasing importance to deal with never ending and usually massive stream of received data such as sensor data, traffic information, economic indexes, video streams, etc. [18]. Online approach is, as a rule, required when the amount of data collected over time is increasing rapidly. This is especially true in the data stream model where the data arrive at high speed so that the algorithms used for mining the data streams must process them in very strict constraints of space and time [15]. A data stream can roughly be thought of as an ordered sequence of data items, where the input arrives more or less continuously as time progresses (see for example [7]). Reviews of algorithms and approaches to data stream mining can be found in [5], [6] and [15]. Online classifiers are induced from the initially available dataset as in case of the static approach. However, in addition, there is also some adaptation mechanism providing for a classifier evolution after the classification task has been initiated and started. In each round a class label of the incoming instance is predicted and afterwards information as to whether the prediction was correct or not, becomes available. Based on this information adaptation mechanism may decide to leave a classifier unchanged, or modify it, or induce a new one.

Usual approach to deal with the online classification problems is to design and implement an online classifier incorporating some incremental learning algorithm [12], [16]. According to [14] an algorithm is incremental if it results in a sequence of classifiers with different hypothesis for a sequence of training requirements. Among requirements an incremental learning algorithm should meet are ability to detect concept drifts, ability to recover its accuracy, ability to adjust itself to the current concept and use past experience when needed [19]. Examples of some state of the art classifiers belonging to the discussed class include approaches proposed in [18] and [2].

In this paper a family of algorithms for the online learning and classification is considered. These algorithms work in rounds, where at each round a new instance is given and the algorithm makes a prediction. After the true class of the instance is revealed, the learning algorithm updates its internal hypothesis. The proposed algorithms are based on fuzzy C-means clustering followed by calculation of distances between cluster centroids and the incoming instance for which the class label is to be predicted. Simple distance-based classifiers thus obtained serve as basic classifiers for the implemented Rotation Forest kernel.

The paper is organized as follows. Section 1 contains introduction. Section 2 contains algorithms description and analysis of their complexity. Section 3 presents validation experiment settings and experiment results. Section 4 contains conclusions and suggestions for future research.

2 Classification Algorithms

The general data classification problem is formulated as follows. Let C be the set of categorical classes which are denoted $1, \ldots, |C|$. The learning algorithm is provided with the learning instances $LD = \{< d, c > \mid d \in D, c \in C\} \subset D \times C$, where D is the space of attribute vectors $d = (w_1^d, \ldots, w_N^d)$ with w_i^d being a numeric value, N - the number of attributes. The algorithm is used to find the best possible approximation $\hat{f}$ of the unknown function f such that $f(d) = c$. Then $\hat{f}$ can be used to find the class $c = \hat{f}(d)$ for any d such that $(d, c) \notin LD$, that is the algorithm will allow to classify instances not seen in the process of learning.

2.1 Online Algorithms

The first step of the considered algorithms is fuzzy C-means clustering (see [4]), that is an iterative method which allows one row of data to belong to two or more clusters. The method is based on minimization of the objective function

$$J_m = \sum_{i=1}^{M} \sum_{j=1}^{noCl} u_{ij}^m \cdot d(r_i, c_j)$$

where m is a fixed number greater than 1 (in the experiments the value was fixed and equal 2), M is the number of data rows, $noCl$ is the number of clusters, c_j is the center of the j-th cluster, u_{ij} is the degree of membership of the i-th data row x_i in cluster j and d is a fixed metric to calculate the distance from the data row r_i to cluster centroid c_j. Fuzzy C-means clustering is an iterative process. In each iteration step the membership factors u_{ij} and cluster centers c_j are updated.

Given the partition of the training data via C-means clustering and a data row r to be classified, to assign the class label to r, firstly the distances from r to all cluster elements are calculated. The distances are sorted in a non-decreasing order and the coefficient measuring the sum of distances from x nearest neighbours (x is a parameter) is calculated. The class for which the coefficient is minimal, is assigned to r. The details of the partition as well as class assignment are given in [11].

Concept of the training dataset used in the paper is similar to the sliding window idea used in various incremental algorithms, see for example [5]. For the proposed algorithms the size of the training set is kept constant. Online algorithms work in rounds. Starting with the initial training set IN, in each round a new instance (not seen before) from LD is considered and the algorithm makes a prediction. After the true class of the instance is revealed, the learning algorithm updates the training set, according to (1).

Algorithm 1. Online algorithms OL-rand, OL-fifo, OL-max

Require: training data LD, containing k data rows, Y - correct class labels of the training dataset, x - number of neighbours, initial training set IN,
Ensure: qc - quality of classification
1: initialize training set $TD \leftarrow IN$, $testSize \leftarrow 0$, $correctClsf \leftarrow 0$
2: **while** $testSize < k$ **do**
3: $testSize \leftarrow testSize + 1$
4: let (r, c) stand for the next row from LD (not considered before)
5: perform C-means clustering on TD, calculate classification coefficient using x neighbours of r and find class $\hat{c}$ for row r
6: **if** $c = \hat{c}$ **then**
7: $correctClsf \leftarrow correctClfs + 1$
8: **end if**
9: update TD according to (1)
10: **end while**
11: $qc \leftarrow \frac{correctClfs}{testSize}$

In case of algorithm OL-rand, the new row replaces the randomly chosen one, in algorithm OL-fifo it replaces one on the basis of first-in-first-out queue. Finally, for the algorithm OL-max the new instance (r, c) replaces the row from the cluster of r which is in maxiumum distance from the respective centroid.

$$update(TD) = \begin{cases} \text{replace random row by (r,c)} & \text{alg. OL-rand} \\ \text{replace row from fifo by (r,c)} & \text{alg. OL-fifo} \\ \text{replace row maximally distant by (r,c)} & \text{alg. OL-max} \end{cases} \quad (1)$$

For each of the algorithms the quality of classification is calculated. The details of all three algorithms are given in **Algorithm 1**.

2.2 Principal Component Analysis and Rotation Forest Ensemble Method

Rotation Forest method creates a family of base classifiers, each based on a different axis rotation of attributes. To create the training dataset, the set of attributes is randomly split into a given number of subsets and Principal Component Analysis (PCA) is applied to each subset. The algorithm for PCA is given in **Algorithm 2**. Let M stand for the number of data rows, each with N attributes.

The matrix with eigenvectors used in Algorithm 3, is defined in (2).

$$E_i^{cl} = \begin{bmatrix} \mathbf{a_1^1}\ \mathbf{a_1^2} \cdots \mathbf{a_1^{M_1}} & [0] & \cdots & [0] \\ [0] & \mathbf{a_2^1}\ \mathbf{a_2^2} \cdots \mathbf{a_2^{M_2}} & \cdots & [0] \\ \vdots & \vdots & \ddots & \vdots \\ [0] & [0] & \cdots & \mathbf{a_K^1}\ \mathbf{a_K^2} \cdots \mathbf{a_K^{M_K}} \end{bmatrix} \quad (2)$$

Algorithm 2. Algorithm PCA

Require: dataset - a matrix X consisting of M rows and N columns
Ensure: the set of M eigenvectors
1: transform the data matrix X into a dataset whose mean is 0 - calculate the mean value for each column and subtract the mean from each row in the given column
2: find the covariance matrix $Cov = \frac{1}{N-1} X \cdot X^T$
3: compute M eigenvectors of the covariance matrix Cov

Algorithm 3. Algorithm Rotation Forest

Require: TD training dataset with M data rows, each described by N attributes (TD is a matrix of size $M \times N$), Y - correct class labels of the training dataset TD (Y is a matrix of size $N \times 1$ of elements from C), K - the number of subsets.
Ensure: rotation matrix R
1: split the attribute set F into K subsets $\{F_j\}_{j \leq K}$ each with M_j attributes
2: **for** $j = 1$ to K **do**
3: let X_j be the dataset TD for the attributes in F_j
4: transform X_j by deleting 25 % rows, randomly
5: apply algorithm PCA to the transformed X_j to obtain eigenvectors $\mathbf{a}_j^1, \ldots, \mathbf{a}_j^{M_j}$
6: **end for**
7: arrange the eigenvectors in matrix E as shown in (2)
8: construct the rotation matrix R from E by rearranging the columns to match the order in F

Algorithm 4 applies Rotation Forest method to online classifiers. The proposed approach preserves main features of the original Rotation Forest method. The main heuristic is to apply feature extraction and to subsequently reconstruct a full feature set for each classifier in the family. This is done through dividing randomly the feature set into a number of subsets, applying principal component analysis to each subset and constructing new set of features by combining all principal components.

2.3 Computational Complexity of the Algorithms

As shown in [11] the computational complexity of fuzzy C-means clustering is $O(t \cdot M \cdot noCl)$, where t is the number of iterations of the C-means algorithm, M is the number of data rows and $noCl$ is the number of clusters. Classification based on calculating the classification coefficient demands sorting of M values which sums up to complexity $O(M^2)$. **Algorithm 1** performs C-means clustering and classification for each data row which gives $O(t \cdot M^3)$, since $noCl$ is much smaller than M.

To estimate the complexity of the algorithm based on Rotation Forest consider first **Algorithm 2**. It needs the computation of M eigenvectors of a symmetric matrix *cov*. In the computational experiments, for this step the Java library

Algorithm 4. Algorithm Rotation Forest with online kernels,OL-RF

Require: LD training dataset with M data rows, each described by N attributes (LD is a matrix of size $M \times N$), Y - correct class labels of the training dataset LD (Y is a matrix of size $N \times 1$ of elements from C), initial training set IN, L - the number of classifiers in the ensemble, K - the number of subsets.
Ensure: qc - quality of classification
1: initialize training set $TD \leftarrow IN$, $testSize \leftarrow 0$, $correctClsf \leftarrow 0$
2: **while** $testSize < M$ **do**
3: $testSize \leftarrow testSize + 1$
4: let (r, c) stand for the next row from LD (not considered before)
5: **for** $i = 1$ to L **do**
6: apply Algorithm 3 to training dataset TD
7: let R_i stand for the obtained rotation matrix
8: perform C-means clustering on $TD \cdot R_i$
9: calculate classification coefficient using x neighbours
10: find class $\hat{c}_i$ for row r
11: **end for**
12: use majority vote on $\hat{c}_1, \ldots, \hat{c}_L$ to define class $\hat{c}$ for r
13: **if** $c = \hat{c}$ **then**
14: $correctClsf \leftarrow correctClfs + 1$
15: **end if**
16: update TD according to (1) for case OL_rand
17: **end while**
18: $qc \leftarrow \frac{correctClfs}{testSize}$

Apache Commons Mathematics Library was used. It is assumed that a standard method based on iterative approach is used and it requires $O(p \cdot M^2)$ steps, where p is the number of iterations. **Algorithm 4** for each data row generates L classifiers which first applies PCA to the data matrix and then proceeds as in previously considered online algorithm. Thus the complexity of OL-RF is $O(T \cdot L \cdot M^3)$, where T is the maximum of t and p.

3 Computational Experiment Results

To test performance of the proposed online algorithms we run them using a set of publicly available benchmark datasets including data often used to test incremental learning algorithms. Datasets used in the experiment are shown in Table 1.

In Table 2 classification accuracies averaged over 20 runs obtained by the proposed algorithms are shown. In case of the Rotation Forest, basic classifiers have been produced through using the proposed distance-based online classifier with random replacement strategy. Number of such classifiers has been set to $L = 5$. Number of attribute partitions in the rotation forest kernel has been set to 3 for datasets where the number of attributes is smaller than 10 and to 5 otherwise. Window refers to the training dataset size and x denotes number of

Table 1. Benchmark datasets used in the experiment

Dataset	Instances	Attributes	Dataset	Instances	Attributes
Banana [8]	5300	16	Bank Marketing [1]	4522	17
Breast [1]	263	10	Chess [21]	503	9
Diabetes [1]	768	9	WBC [1]	630	11
Heart [1]	303	14	Hepatitis [1]	155	20
Luxembourg [21]	1901	32	Spam [1]	4601	58
Twonorm [8]	7400	21	Ionosphere [1]	351	35
SEA [17]	50000	4	Magic [1]	19020	11
Electricity [13]	44976	6	Image [1]	2086	19

the considered neighbours as explained in Section 2. In all cases distances have been calculated using the Manhattan metrics.

In Table 3 some of the results from obtained by the proposed distance-based online algorithms are compared with results obtained by incremental (I) and batch (B) learning algorithms published recently in the literature.

Classification rate measured in terms of number of instances per second depends on several factors as it was shown in Section 3. Apart from the number of attributes the critical factor is the sliding window size, that is the number of instances in the training set. For example, for the Banana dataset the average classification rate with sliding window size of 100 was 12 inst./ sec and with the size of 200 it droped down to 7 inst./ sec in case of OL-rnd, OL-fifo and OL-max. For the Rotation Forest variant the classification rate was down to 2 inst./ sec. For the Spam dataset the respective rates were 90, 30 and 11. All computations have been performed on PC computer with Intel Core i7 processor.

4 Conclusions

The paper contributes through extending a family of the online classifiers proposed originally in [11]. The extension involves proposing and implementing three different strategies for updating a current training set of examples as well as adding the Rotation Forest kernel with a view to improve accuracy of classification. Computational complexity analysis has shown that the approach assures classification in polynomial time. Proposed classifiers have been validated experimentally. The reported computational experiment proves that OL family of algorithms performs well in terms of the classification accuracy. It can be also concluded that OL algorithms supported by the Rotation Forest kernel outperform some state-of-the-art incremental classifiers reported in the literature. The experiment does not allow to draw any conclusions as to the comparison of the OL classifiers with the best batch classifiers. However, it is worth observing that in a few instances our algorithms performed as well as a very good batch classifiers with accuracy measured over 10-cross-validation scheme.

Table 2. Average accuracies (% correct) of the proposed distance-based online classifiers

Dataset	Window	x	OL-rnd	OL-fifo	OL-max	OL-RF
Banana	200	5	87.1	87.2	87.1	**88.1**
Banana	400	5	88.0	88.1	87.9	**88.5**
Bank M.	100	5	88.1	88.3	88.0	**89.7**
Bank M.	400	5	91.1	91.2	91.1	**91.4**
Breast	50	10	77.0	**77.1**	76.8	76.9
Breast	70	5	77.3	**79.2**	77.1	77.4
Chess	200	5	75.7	76.1	75.9	**76.8**
Diabetes	200	10	74.3	**75.2**	75.1	75.1
WBC	100	5	97.8	98.0	97.7	**98.1**
Heart	50	5	80.7	80.9	80.5	**81.3**
Heart	100	5	83.3	84.3	81.6	**85.2**
Hepatitis	50	5	83.6	83.4	82.8	**87.5**
Hepatitis	70	5	83.5	83.9	83.1	**88.1**
Luxembourg	200	5	74.1	81.1	80.9	**81.9**
Spam	200	5	70.5	72.1	70.3	**72.9**
Twonorm	200	5	95.7	96.1	94.5	**96.9**
Twonorm	400	10	96.1	96.4	96.1	**97.6**
Ionosphere	100	5	86.9	87.3	87.1	**88.6**
Ionosphere	200	5	87.7	89.0	88.3	**89.2**
SEA	50	5	71.1	78.3	78.2	**78.9**
SEA	200	5	80.4	**81.1**	80.1	**81.1**
Magic	200	5	70.9	73.1	71.8	**74.3**
Electricity	50	5	88,9	89.1	88.7	**89.2**
Electricity	100	5	89.1	89.8	88.7	**90.7**
Image	50	5	77.6	78.2	77.5	**79.3**
Image	100	5	77.8	79.4	77.5	**80.3**

Table 3. Comparison of accuracies (% correct)

Dataset	Best OL	Literature reported
Banana	88.5	**89.3** - IncSVM, [18] (I)
Bank M.	**91.4**	86.9 LibSVM; 76.5 J48 [20](B)
Breast	**77.4**	72.2 - IncSVM, [18](I)
Chess	**76.8**	71.8 EDDM [21] (I)
Diabetes	75.2	**77.5** IncSVM [18] (I)
Heart	**85.2**	83.8 IncSVM [18] (I)
Hepatitis	**88.1**	82.1 RF GEP [10] (B)
Twonorm	**97.6**	**97.6** FPA [18] (I)
Ionosphere	89.2	**96.9** GEPC-ad [9] (B)
SEA	81.1	**96.6** KAOGINC [2] (I)
Electricity	**90.7**	88,5 Inc.L. [3] (I)

It is worth noting that the main factor affecting classifier performance it is the sliding windoe size. Increasing this size usually results in accuracy of classification improvement at a cost of increased computation time.

Future research will focus on refining adaptation mechanism used to update and control current training set used by the incremental classifier.

References

1. Asuncion, A., Newman, D.J.: UCI Machine Learning Repository. University of California, School of Information and Computer Science (2007), http://www.ics.uci.edu/~mlearn/MLRepository.html
2. Bertini, J.R., Zhao, L., Lopes, A.: An incremental learning algorithm based on the K-associated graph for non-stationary data classification. Information Sciences 246, 52–68 (2013)
3. Ditzler, G., Polikar, R.: Incremental learning of concept drift from streaming imbalanced data. IEEE Transactions on Knowledge and Data Engineering 25(10), 2283–2301 (2013)
4. Dunn, J.C.: A fuzzy relative of the ISODATA process and its use in detecting compact well-separated clusters. Journal of Cybernetics 3, 32–57 (1973)
5. Gaber, M.M., Zaslavsky, A., Krishnaswamy, S.: Mining data streams: a review. ACM SIGMOD Record 34(1), 18–26 (2005)
6. Gaber, M.M., Zaslavsky, A., Krishnaswamy, S.: Data Stream Mining. In: Maimon, O., Rokach, L. (eds.) Data Mining and Knowledge Discovery Handbook, Part 6, pp. 759–787 (2010)
7. Gama, J., Gaber, M.M.: Learning from Data Streams. Springer, Berlin (2007)
8. IDA Benchmark Repository, http://mldata.org/repository/tags/data/IDA_Benchmark_Repository/ (January 12, 2013)
9. Jędrzejowicz, J., Jędrzejowicz, P.: Cellular GEP-Induced Classifiers. In: Pan, J.-S., Chen, S.-M., Nguyen, N.T. (eds.) ICCCI 2010, Part I. LNCS (LNAI), vol. 6421, pp. 343–352. Springer, Heidelberg (2010)
10. Jędrzejowicz, J., Jędrzejowicz, P.: Rotation Forest with GEP-Induced Expression Trees. In: O'Shea, J., Nguyen, N.T., Crockett, K., Howlett, R.J., Jain, L.C. (eds.) KES-AMSTA 2011. LNCS (LNAI), vol. 6682, pp. 495–503. Springer, Heidelberg (2011)
11. Jędrzejowicz, J., Jędrzejowicz, P.: Online classifiers based on fuzzy C-means Clustering. In: Bădică, C., Nguyen, N.T., Brezovan, M. (eds.) ICCCI 2013. LNCS (LNAI), vol. 8083, pp. 427–436. Springer, Heidelberg (2013)
12. Last, M.: Online classification of nonstationary data streams. Intelligent Data Analysis 6, 129–147 (2002)
13. http://moa.cms.waikato.ac.nz/datasets/ (September 03, 2013)
14. Murata, N., Kawanabe, N., Ziehe, A., Muller, K.R., Amari, S.: On-line learning in changing environments with application in supervised and unsupervised learning. Neural Networks 15, 743–760 (2002)
15. Pramod, S., Vyas, O.P.: Data Stream Mining: A Review on Windowing Approach. Global Journal of Computer Science and Technology Software & Data Engineering 12(11), 26–30 (2012)
16. Sung, J., Kim, D.: Adaptive acting appearance model with incremental learning. Pattern Recognition Letters 30, 359–367 (2009)

17. Street, W., Kim, Y.: A streaming ensemble algorithm (SEA) for large-scale classification. In: 7th International Conference on Knowledge Discovery and Data Mining, KDD 2001, San Francisco, CA, pp. 377–382 (August 2001)
18. Wang, L., Ji, H.-B., Jin, Y.: Fuzzy Passive-Aggressive classification: A robust and efficient algorithm for online classification problems. Information Sciences 220, 46–63 (2013)
19. Widmar, G., Kubat, M.: Learning in the presence of concept drift and hidden contexts. Machine Learning 23, 69–101 (1996)
20. Wisaeng, K.: A Comparison of Different Classification Techniques for Bank Direct Marketing. International Journal of Soft Computing and Engineering 3(4), 116–119 (2013)
21. Žliobaitė, I.E.: Controlled Permutations for Testing Adaptive Classifiers. In: Elomaa, T., Hollmén, J., Mannila, H. (eds.) DS 2011. LNCS, vol. 6926, pp. 365–379. Springer, Heidelberg (2011)

Online Learning Based on Prototypes

Ireneusz Czarnowski and Piotr Jędrzejowicz

Department of Information Systems, Gdynia Maritime University
Morska 83, 81-225 Gdynia, Poland
{irek,pj}@am.gdynia.pl

Abstract. The problem addressed in this paper concerns learning form data streams with concept drift. The goal of the paper is to propose a framework for the online learning. It is assumed that classifiers are induced from incoming blocks of prototypes, called data chunks. Each data chunk consists of prototypes including also information as to whether the class prediction of these instances was correct or not. When a new data chunk is formed, classifier ensembles formed at an earlier stage are updated. Three online learning algorithms for performing machine learning on data streams based on three different prototype selection approaches to forming data chunks are considered. The proposed approach is validated experimentally and the computational experiment results are discussed.

Keywords: online learning, incremental learning, data streams.

1 Introduction

The key objective of the machine learning is to design algorithms that are able to improve performance at some task through experience [23]. Such algorithms are called learners. The learner uses examples of a particular task to learn. Learning from examples is one of most popular paradigms of the machine learning. It deals with the problem of identifying regularities between a number of independent variables (attributes) and a target or dependent categorical variable observing and analyzing some given dataset [30].

Learning from examples is understood as the process of finding a model (or function) that describes and distinguishes data classes. The model produced under the machine learning process is called a *classifier*. It should be also noted that in case a dataset with examples is categorical, then learning from examples is based on the existence of certain real-world *concepts* which might or might not be stable during the process of learning [10], [23].

The most common approach to learning classifier from data is based on the assumption that the data are static and the learning process remains unchanged during the time of the learner operation. However, in many real-world situations, the environment in which a learner works is dynamic, i.e. the target concept and its statistical properties change over time and these changes cannot be predicted in advance. Such a property is typical in case of the, so called, *data streams*, where the

N.T. Nguyen et al. (Eds.): ACIIDS 2014, Part II, LNAI 8398, pp. 187–196, 2014.

class distribution of the streaming data is imbalanced. Changes of data properties occurring in time are usually referred to as a data drift or a dynamic character of the data source [7], [25]. Such changes are also known as a *concept drift* [34] or dataset shift [31], and learning in such case is referred to as *learning drift concept* [18] or *learning classifiers from the streaming data* [37].

Learning classifiers from data streams is a one of the recent challenges in data mining. The main goal of the paper is to propose an approach to the prototype-based online learning. It is assumed that a classifier is induced from coming blocks of prototypes, called data chunks. Each data chunk consists of prototypes. The prototypes are selected from available instances including information as to whether the prediction of incoming instances has been correct or not. When a new instance arrives, a new data chunk is formed. When a new data chunk is formed it enforces updating of the classifier. In this paper we attempt to answer the question how the prototype selection method influences the online classification performance. To answer the above a comparative experiment is carried-out with the three versions of different prototype selection methods.

The paper is organized as follows. The next section discusses several features of the online learning from data streams problem. It also includes a short review of the related works on the online learning for data streams. Section 3 gives details of the proposed framework. Section 4 provides details on the computational experiment results. In the final section conclusions are presented.

2 Data Streams and Online Learning

In this paper the problem of online learning is seen through the prism of three components. The first one is dealing with data streams, from which the online learning model is built, the second one with the classification and the third one with the online learning. In this section a selected techniques for data streams, so-called data summarization techniques, are briefly reviewed. The second part of the section deals with the problems of the online learning and the classification.

2.1 Data Streams

A data stream is understood as a sequence of instances that arrive to the system at various points of time. Processing streams of continuously incoming data implies a new computational requirement concerning a limited amount of memory and a short processing time, especially when data streams are large [6].

In [6] several constrains applicable to the data stream model and processing methods have been formulated. One of them states that it is not possible to store all the data from streams and only a small part of the data can be stored and used for computations within a limited time span. Essentially, the arrival speed of the incoming instances from data streams enforce their processing in the real time. Finally changes over time in the distribution of data require incorporating some adaptation mechanisms into learning algorithms and classifiers that operate on data streams [6].

To deal with the above described constraints, a typical and standard approach is to apply to the data stream analysis some summarization techniques. Example of such a simple summarization technique is to reduce data stream size. Sampling is a most common technique used to decrease data size. Because, the data stream model is not stationary and in times even unbounded, the sampling approach must be modified online. Such modification could base on the analysis of the data in each pass and removal of some instances from the training set with a given probability instead of periodically selecting them [33]. In [8] the idea has been extended to the case of weighted sampling. Among other sampling-based approaches are clustering of data streams and sampling within the sliding window model [29]. The last one bases on the assumption that analysis of the data stream is limited to the most recent instances, thus only limited number of instances is used to the learner training. In a simple approach sliding windows are of the fixed size and include most recent instances. With each new instance the oldest instance that does not fit in the window is removed. When the window has a small size the classifier may react quickly to changes. Otherwise the classifier fails to adapt as rapidly as required to changes of data properties. Of course decreasing the sliding window size may lead to a loss of accuracy.

Several alternative approaches based on windowing technique have been proposed. Weighted windows is one of them. The idea of weighted windows is to assign a weight to each instance, however older instances receive smaller weights [11]. For VFDT (Very Fast Decision Tree) the size of the window is determined based on distribution-free Hoeffding's bound [4].

In case of FISH family of algorithms the size of window is established based on distances between instances. Consecutive versions of the algorithms in this family allow to dynamically establish the size of the training window [38].

The sliding window algorithm, called ADWIN has been proposed in [5]. This algorithm updates the size of the window analyzing statistical hypotheses with respect to subwindows. Other algorithm, called OLIN, dynamically adjusts the size of the window and the number of new instances that should be considered due to change in the concept [22]. Features of the adaptive sliding window approaches have been discussed in [18], [34].

Examples of the window-based algorithms include also the FLORA family of algorithms [34], FRANN [19], and Time-Windowed Forgetting (TWF) [26]. As it has been indicated, some algorithms use windows of the fixed size, while others use heuristics to adjust the window size.

An adaptive window size has been discussed in [18]. An interesting approach to online learning using info-fuzzy network as a base classifier has been also proposed in [22]. The approach repeatedly applies info-fuzzy network algorithm to a sliding window training data and dynamically adapts the size of the training window.

Another technique to data stream analysis are sketching techniques. These base on a statistical summary of the data stream [22].

Data summarization techniques can be also merged with the drift detection. The aim of the drift detection techniques is to detect changes in the concept and inform the system that a learner should be updated or rebuild. A simple approach to drift detection is based on a statistical verification, for example, on evaluating the class distribution. Examples of approaches based on a statistical tests are Drift Detection Method (DDM), and its modification called EDDM [3].

It can be also observed, that typical drift detector approaches operate on blocks of data. Thus the recognition of concept drift in data streams can be merged with sliding-window approaches.

Another approach, which is dealing with a concept drift is the forgetting mechanism. The forgetting mechanism allows in a simple way to react to concept changes. The main idea is to select adequate data to remember. The approach assumes forgetting training instances at a constant rate and using only a window of the latest instances to train the classifier [21]. Alternatively, selection of instances according to their class distribution can be used as a forgetting mechanism.

2.2 Online Learning and Classification

There are two basic groups of approaches to solve classification tasks through applying machine learning techniques. The first group includes the so-called batch approaches for which training dataset is known and used by a machine learning algorithm to induce the classifier before taking any classification decision. The second group involves the online learning, also termed incremental learning. The incremental learning approaches are designed to sequentially learn a prediction model based on the feedback from answers to previous questions and possible additional side information [27]. In other words, during the classification, a class label of the incoming instance is predicted in each round, and afterwards information as to whether the prediction was correct or not is available. Based on this information the classifier can be updated to accommodate a new training instance [16]. Thus, idea of the online learning means that upon receiving a new instances the existing model is updated. It is much less expensive to update the existing model than to build a new one. The decision on updating the classifier is depending on the implemented adaptation mechanism. The pseudo-code of the basic online learning model is shown as Algorithm 1.

Algorithm 1. The basic online learning schema
Input: X - input stream of instances $x_1, x_2, \ldots, x_i, \ldots$
1. For each instance from X apply the classifier and predict a label y'_i
2. Determine the true label y_i for x_i
3. If $y'_i \neq y_i$ then update the prediction rules of the classifier based on x_i, y_i and y'_i

Feature of the incremental learning is that the data are read in blocks at a time, not being available at the beginning. It is typical that there is only one pass through the data. It is also true that the online learning algorithms must process the data in a very strict constraints of space and time when the data arrive at high speed [21].

The incremental learning algorithm should be able to deal with changes of the concept drift over time. Several methods have been proposed for learning in the presence of drift. Trigger-based methods use a change detector. If the change is detected then the classifier is modified and updated. Evolving methods attempt to update a classifier without the drift detection. Ensemble methods are examples of evolving approaches. In general, the ensemble methods build a new classifier when a new block of data arrives. Such a new classifier replaces the worst component in the ensemble. A family of ensemble algorithms for evolving data streams includes:

Accuracy Weighted Ensemble, Adaptive Classifier Ensemble, Batch Weighted Ensemble, Streaming Ensemble Algorithm, Accuracy Diversified Ensemble and Weighted Majority Algorithm [13], [20], [21], [28].

Ensemble-based online learning schema using AdaBoost as a means to re-weight classifiers has been proposed in [14]. The implementation is based on learning a new classifier from the weighted new data, which means that when the new data appear, the weight updating rule is used to both "re-weight" previous classifiers and generate weighted training set. The disadvantage of such method is that all classifiers previously learned must be retained in the memory.

In [32] it has been proposed to differentiate incrementally build ensemble members using the so-called data chunks. In this approach the learner processes the incoming streams in data chunks. It is also observed that the size of these chunks is an important parameter and have influence on the learning quality.

A promising approach to online learning is to use adaptive algorithms [9]. An example of such adaptive algorithm is the RBF network. In traditional RBF network, the structure of the network is usually fixed at some compromise setting, and only the output weights are updated over time. In the online case, the size of the RBF model is may grow or be pruned based on the incoming data. The RBF network approaches are also suitable to handle the problem of online learning in the distributed environment. This feature results from ability of the RBF models to control dynamical changes in each site in parallel.

3 A Framework for Data Streams

The paper deals with the problem of online learning from data streams. The main goal is to find the optimal online learning model using instance selection, data chunks and evolving approaches.

The proposed approach uses three components: classification, learning and data summarization. The proposed approach is outlined in Fig. 1. The role of the classification component is to predict the class of instances whose class label are unknown. The data summarization component is responsible for extracting data chunks from the data stream instances. Afterwards data chunks are used to induce a classifier. This task is carried-out by the learning component.

Data chunks are formed from available instances and from sequence of incoming instances for which predictions were incorrect. It is assumed that the size of those chunks is not greater than the acceptable threshold. When a data chunk size is smaller than the threshold size incoming instances are being added to data chunk. When the size of a data chunk reaches the threshold, the chunk is updated.

Updating a data chunk involves two steps. At the first one instances from a chunk which do not pass the criterion of selection are removed. Instances that are to be retained are selected using the prototype selection technique. It is obvious that selection can be carried-out through implementation of different prototype selection algorithms. In such manner prototype selection is a tool allowing formation of data chunks. At the second step an incoming new instance is added to data chunk.

When a new data chunk is available, a new classifier is induced from it. In this paper it is assumed that the learning component may consist of several classifiers, each produced from independent sequentially arriving data chunks. Such approach is classified as an evolving method with block-based ensembles. Next, each new induced classier is compared with other classifiers in the ensemble. The worst classifier from the ensemble is replaced by the new classifier. The proposed approach allows also using different strategies for updating block ensembles. Also a very simple approach based on removing from the ensemble oldest classifiers is acceptable.

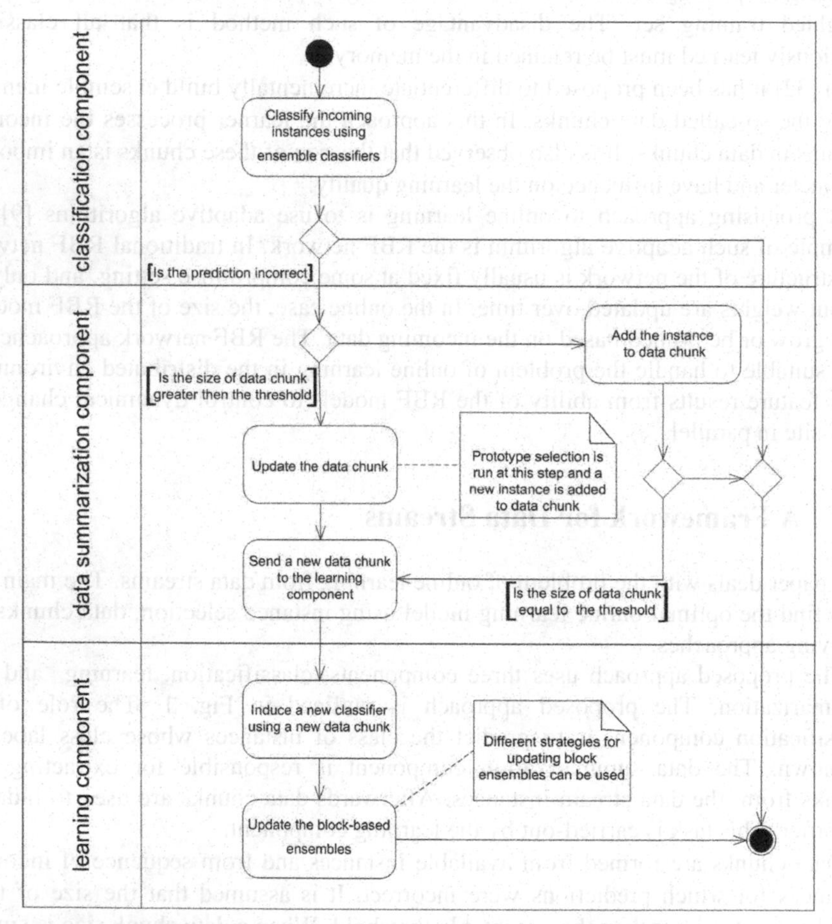

Fig. 1. Activity diagram of the proposed online learning approach

4 Computational Experiment

This section contains the results of several computational experiments carried out with a view to evaluate the performance of the proposed approach measured in terms of the classification accuracy.

In particular, the reported experiment aimed at answering the question whether the proposed approach can be useful tool to solve the online classification problem. The experiment allowed also to study and evaluate how the choice of the prototype selection method for forming data chunks may influence the quality of the online classification system.

In this paper the proposed approach has been denoted as OLP (Online Learning based on Prototypes). In the reported research the following variants of the proposed approaches have been compared:

- OLP based on the Condensed Nearest Neighbor algorithm [15] used for forming data chunks – OLP_{CNN},
- OLP based on the Edited Nearest Neighbor algorithm [35] used for forming data chunks – OLP_{ENN},
- OLP based on the Instance Based learning algorithm 2 [1] used for forming data chunks – OLP_{IB2}.

For all of the above listed variants of algorithm it has been decided to use a simple ensemble model. Block of ensembles is updated by removing the oldest classifier. The final output decision is produced using a simple majority voting to combine member decisions.

The proposed approach has been also compared with other selected approaches to online classification i.e. Accuracy Weighted Ensemble (AWE), Hoeffding Option Tree (HOT) and iOVFDT (Incrementally Optimized Very Fast Decision Tree), which are implemented as extensions of the Massive Online Analysis package within WEKA environment [36].

Generalization accuracy has been used as the performance criterion. The learning tool was the C4.5 algorithm [24]. The C4.5 algorithm has been also applied to induce all of the base models for all ensemble classifiers. The value of the parameter denoting the number of base classifiers has been set arbitrarily and was equal to 5. The ENN algorithm has been run with the number of neighbors equal to 3 (set arbitrarily). All implemented prototype selection algorithms (ENN, CNN and IB2) have been applied using the Euclidean metric. Thresholds for the size of data blocks (chunks) have been also set up arbitrarily. Values of the threshold are shown in Table 1.

In all cases algorithms have been applied to solve the respective problems using several benchmark datasets obtained from the UCI Machine Learning Repository [2]. Basic characteristics of these datasets are shown in Table 1. Table 1 consists also some results reported in literature obtained using batch classifiers.

An experiment plan has involved 30 repetitions of the proposed schema. The instances for the initial training set have been selected randomly from each considered dataset providing their number is not greater than a threshold. In each round, when a new instance arrives the algorithm predicts its class label. When the prediction is wrong an arrived instance is added to a data chunk or replace other one within the current data chunk according to the proposed approach.

Table 2 shows mean values of the classification accuracy of the classifiers obtained using the OLP approach (i.e. using the set of prototypes found by selecting instances). In Table 2 performances of the proposed approaches are also compared with performance of other online learning algorithms and two batch classifiers.

When the versions of the OLP algorithm are compared as shown in Table 2, it can be observed that the best results have been obtained by the OLP_{IB2} algorithm. It should be also noted that the experiment supported hypotheses that the prototype selection method influences the quality of the online classification system.

Table 1. Datasets used in the experiment

Dataset	Number of instances	Number of attributes	Number of classes	Best reported results classification accuracy	Threshold (in % with respect to the original data set)
Heart	303	13	2	90.0% [8]	10%
Diabetes	768	8	2	77.34%[17]	5%
WBC	699	9	2	97.5% [2]	5%
Australian credit (ACredit)	690	15	2	86.9% [2]	9%
German credit (GCredit)	1000	20	2	77.47%[17]	10%
Sonar	208	60	2	97.1% [2]	10%
Satellite	6435	36	6	-	10%

Table 2. Accuracy of the classification results (%)

Algorithm	Heart	Diabetes	WBC	ACredit	GCredit	Sonar	Satellite
OLP_{CNN}	78.14	73.2	70.1	81.52	70.06	76.81	80.45
OLP_{ENN}	80.4	72.82	71.21	82.4	71.84	75.44	78.25
OLP_{IB2}	81.5	71.22	72.4	84.06	71.3	77.51	76.14
AWE	78.01	72.5	72.81	84.5	73.5	77.02	82.4
HOT	81.4	80.42	72.67	82.41	72.87	76.05	83.4
iOVFDT	81.7	77.4	71.04	84.5	75.21	75.38	81.54
C 4.5 [12]	77.8	73.0	94.7	84.5	70.5	76.09	-
SVM [12]	81.5	77.0	97.2	84.81	72.5	90.41	85.00

5 Conclusions

The paper introduces a framework for online learning form data streams with concept drift. The framework is based on block-based ensembles, where classifiers are induced from data chunks. Each data chunk is formed from incoming instances and using data selection methods. In the paper an initial results of application of the proposed approach are shown. Three version of the OLP (Online Learning based on Prototypes) proposed algorithm differ in their approach to forming of the data chunks have been evaluated and compared with some other approaches.

From the reported experiment it can be concluded that the proposed approach is a useful tool allowing to obtain quite good results, which can be comparable with other results obtained by known approaches to online classification. The proposed approach can be also considered as a useful extension of a range available online classifiers.

The experiment, due to its limited scope, allows only for a preliminary validation of the approach. Further research should aim at investigating different techniques for producing data chunks and different ensemble models as well as carrying-out more extensive experiments.

References

1. Aha, D.W., Kibler, D., Albert, M.K.: Instance-based Learning Algorithms. Machine Learning 6, 37–66 (1991)
2. Asuncion, A., Newman, D.J.: UCI Machine Learning Repository. University of California, School of Information and Computer Science, Irvine, CA (2007), http://www.ics.uci.edu/~mlearn/MLRepository.html
3. Baena-Garcia, M., del Campo-Avila, J., Fidalgo, R., Bifet, A., Gavaldà, R., Morales-Bueno, R.: Early drift detection method. In: Fourth International Workshop on Knowledge Discovery from Data Streams (2006)
4. Bifet, A., Gavaldà, R.: Learning from time-changing data with adaptive windowing. In: SDM. SIAM (2007)
5. Bifet, A., Gavaldà, R.: Kalman filters and adaptive windows for learning in data streams. In: Todorovski, L., Lavrač, N., Jantke, K.P. (eds.) DS 2006. LNCS (LNAI), vol. 4265, pp. 29–40. Springer, Heidelberg (2006)
6. Bifet, A.: Adaptive learning and mining for data streams and frequent patterns. PhD thesis, Universitat Politecnica de Catalunya (2009)
7. Caragea, D., Silvescu, A., Honavar, V.: Agents That Learn from Distributed Dynamic Data Sources. In: ECML 2000/Agents 2000 Workshop on Learning Agents, Barcelona, Spain (2000)
8. Chaudhuri, S., Motwani, R., Narasayya, V.R.: On random sampling over joins. In: Delis, A., Faloutsos, C., Ghandeharizadeh, S. (eds.) SIGMOD Conference, pp. 263–274. ACM Press (1999)
9. Chen, H., Gong, Y., Hong, X., Chen, S.: A Fast Adaptive Tunable RBF Network For Non-stationary Systems. Transactions on Systems, Man, and Cybernetics, Part B (2012) (to appear)
10. Cichosz, P.: Systemy uczące się. Wydawnictwo Naukowo-Techniczne, Warszawa (2000) (in Polish)
11. Cohen, E., Strauss, M.J.: Maintaining time-decaying stream aggregates. Journal of Algorithms 59(1), 19–36 (2006)
12. Datasets used for classification: comparison of results. Directory of Data Sets, http://www.is.umk.pl/projects/datasets.html (accessed September 1, 2009)
13. Deckert, M., Stefanowski, J.: Comparing Block Ensembles for Data Streams with Concept Drift. In: Pechenizkiy, M., Wojciechowski, M., (eds.) New Trends in Databases & Inform. Syst. AISC, vol. 185, pp. 69–78. Springer, Heidelberg (2012)
14. Fan, W., Stolfo, S.J., Zhang, J.: The application of AdaBoost for distributed, scalable and on-line learning. In: KDD 1999, Proceedings of the Fifth ACM SIGKDD International Conference on Knowledge Discovery and Data Mining, pp. 362–366. ACM, New York (1999)
15. Hart, P.E.: The Condensed Nearest Neighbour Rule. IEEE Transactions on Information Theory 14, 515–516 (1968)
16. Jędrzejowicz, J., Jędrzejowicz, P.: Online Classifiers Based on Fuzzy C-means Clustering. In: Bădică, C., Nguyen, N.T., Brezovan, M. (eds.) ICCCI 2013. LNCS (LNAI), vol. 8083, pp. 427–436. Springer, Heidelberg (2013)
17. Jędrzejowicz, J., Jędrzejowicz, P.: Cellular GEP-Induced Classifiers. In: Pan, J.-S., Chen, S.-M., Nguyen, N.T. (eds.) ICCCI 2010, Part I. LNCS, vol. 6421, pp. 343–352. Springer, Heidelberg (2010)

18. Klinkenberg, R.: Learning Drifting Concepts: Example Selection vs. Example Weighting. Intelligent Data Analysis. Incremental Learning Systems Capable of Dealing with Concept Drift 8(3), 281–300 (2004)
19. Kubat, M., Widmer, G.: Adapting to drift in continuous domains. Tech. Report ÖFAI-TR-94-27, Austrian Research Institute for Artificial Intelligence, Vienna (1994)
20. Kuncheva, L., Whitaker: Measures of diversity in classifier ensembles. Machine Learning 51, 181–207 (2003)
21. Kuncheva, L.I.: Classifier ensembles for changing environments. In: Roli, F., Kittler, J., Windeatt, T. (eds.) MCS 2004. LNCS, vol. 3077, pp. 1–15. Springer, Heidelberg (2004)
22. Last, M.: Online Classification of Nonstationary Data Streams. Intelligent Data Analysis 6(2), 129–147 (2002)
23. Mitchell, T.: Machine Learning. McGraw-Hill, New York (1997)
24. Quinlan, J.R.: C4.5: Programs for Machine Learning. Morgan Kaufmann, SanMateo (1993)
25. Sahel, Z., Bouchachia, A., Gabrys, B., Rogers, P.: Adaptive Mechanisms for Classification Problems with Drifting Data. In: Apolloni, B., Howlett, R.J., Jain, L. (eds.) KES 2007, Part II. LNCS (LNAI), vol. 4693, pp. 419–426. Springer, Heidelberg (2007)
26. Salganicoff, M.: Tolerating concept and sampling shift in lazy learning using prediction error context switching. AI Review, Special Issue on Lazy Learning 11(1-5), 133–155 (1997)
27. Shalev-Shwartz, S.: Online learning: Theory, Algorithms, and Applications. PhD thesis (2007)
28. Stefanowski, J.: Multiple and Hybrid Classifiers. In: Polkowski, L. (ed.) Formal Methods and Intelligent Techniques in Control, Decision Making. Multimedia and Robotics, Warszawa, pp. 174–188 (2001)
29. Guha, S., Mishra, N., Motwani, R., O'Callaghan, L.: Clustering data streams. In: FOCS, pp. 359–366 (2000)
30. Tsoumakas, G., Angelis, L., Vlahavas, I.: Clustering Classifiers for Knowledge Discovery from Physically Distributed Databases. Data & Knowledge Engineering 49, 223–242 (2004)
31. Tsymbal, A.: The Problem of Concept Drift: Definitions and Related work. Tech. Rep. TCD-CS-2004-15, Department of Computer Science, Trinity College Dublin, Dublin, Ireland (2004)
32. Venkatesh, G., Gehrke, J., Ramakrishnan, R.: Mining Data Streams under Block Evolution. SIGKDD Explorations 3(2), 1–10 (2002)
33. Vitter, J.S.: Random sampling with a reservoir. ACM Trans. Math. Software 11(1), 37–57 (1985)
34. Widmer, G., Kubat, M.: Learning in the Presence of Concept Drift and Hidden Contexts. Machine Learning 23(1), 69–101 (1996)
35. Wilson, D.R., Martinez, T.R.: Reduction Techniques for Instance-based Learning Algorithm. Machine Learning 33(3), 257–286 (2000)
36. Witten, I.H., Frank, E.: Data Mining: Practical Machine Learning Tools and Techniques, 2nd edn. Morgan Kaufman, San Francisco (2005)
37. Zhu, X., Zhang, P., Lin, X., Shi, Y.: Active Learning from Data Streams. In: Proceedings of the Seventh IEEE International Conference on Data Mining, pp. 757–762 (2007)
38. Zliobaite, I.: Adaptive training set formation. PhD thesis, Vilnius University, Vilnius (2010)

Reinforcement Learning Strategy for Solving the Resource-Constrained Project Scheduling Problem by a Team of A-Teams

Piotr Jędrzejowicz and Ewa Ratajczak-Ropel

Department of Information Systems, Gdynia Maritime University
Morska 83, 81-225 Gdynia, Poland
{p.jedrzejowicz,e.ratajczak-ropel}@wpit.am.gdynia.pl

Abstract. In this paper the Team of A-Teams for solving the resource-constrained project scheduling problem (RCPSP) using the reinforcement learning interactive strategy is proposed. RCPSP belongs to the NP-hard problem class. To solve this problem a parallel cooperating A-Teams consisting of the asynchronous agents implemented using JABAT middleware have been proposed. Within each of the A-Team the interaction strategy using reinforcement learning is used. To evaluate the proposed approach computational experiment has been carried out.

Keywords: resource-constrained project scheduling, optimization, agent system, team of A-Teams, reinforcement learning.

1 Introduction

Resource Constrained Project Scheduling Problem (RCPSP) have attracted a lot of attention and many exact and heuristic algorithms have been proposed for solving it ([11,1,13,21]). The current approaches to solve this problem produce either approximate solutions or can be only applied for solving instances of the limited size. Hence, searching for a more effective algorithms and solutions to the RCPSP is still a lively field of research. One of the promising directions of such research is to take advantage of the parallel and distributed computation solutions, which are the common feature of the contemporary multiple-agent systems.

The multiple-agent systems are an important and intensively expanding area of research and development. There is a number of multiple-agent approaches proposed to solve different types of optimization problems. One of them is the concept of an asynchronous team (A-Team), originally introduced by [23]. The idea of A-Team was used to develop the JADE-based environment for solving a variety of computationally hard optimization problems called JABAT ([12,2]). JABAT is a middleware supporting the construction of the dedicated A-Team architectures allowing to implement the population-based algorithms. The mobile agents used in JABAT allow for decentralization of computations and use of multiple hardware platforms in parallel, resulting eventually in more effective use of the available resources and reduction of the computation time.

N.T. Nguyen et al. (Eds.): ACIIDS 2014, Part II, LNAI 8398, pp. 197–206, 2014.

In [3] an extended version of JABAT has been proposed. The idea of that approach is extending the JABAT environment through integrating the team of asynchronous agent paradigm with the island-based genetic algorithm concept first introduced in [10]. In the resulting Team of A-Teams (TA-Teams) architecture two levels of agent cooperation are introduced. Cooperation at the lower level takes place within a single A-Team. Cooperation at the upper level involves communication, that is information exchange, between cooperating A-Teams belonging to the TA-Teams.

Reinforcement Learning (RL) [5,22,17] belongs to the category of unsupervised machine learning algorithms. It is concerned with how software agents ought to take actions so as to achieve one or more goals. The learning process takes place through interaction with an environment. RL is often used for solving combinatorial optimization problems [25].

The other field where RL is commonly used is the Multi-Agent Reinforcement Learning (MARL) where multiple reinforcement learning agents act together in a common environment [9,24]. In this paper the RL is used to support strategy of searching for the optimal solution by a single team of agents.

In [15] the TA-Teams approach for solving RCPSP problem was proposed and experimentally validated. Within each A-Team a static strategy for cooperation was used. In [16] the RL strategy based on utility values was proposed and used in one A-Team. In this paper the dynamic RL strategy for cooperation within the A-Team and static strategy for cooperation between A-Teams is used for solving instances of the RCPSP problem. It is expected that two-levels of cooperation between A-Teams supported by the reinforcement learning will result in obtaining high quality solutions in an efficient manner.

Optimization agents used to produce solutions to the RCPSP instances represent heuristic algorithms such as the tabu search or path relinking algorithm. A behavior of the single A-Team is defined by the, so called, interaction strategy and cooperation between A-Teams by the migration strategy. The approach extends the earlier research results described in [13,3,15,16].

The paper is constructed as follows: Section 2 of the paper contains the RCPSP problem formulation. Section 3 gives some information on extended JABAT environment. Section 4 provides details of the TA-Teams using RL interaction strategy designed for solving the RCPSP instances. Section 5 describes settings of the computational experiment carried-out with a view to validate the proposed approach and a discussion of the computational experiment results. Finally, Section 6 contains conclusions and suggestions for future research.

2 Problem Formulation

A single-mode resource-constrained project scheduling problem consists of a set of n activities, where each activity has to be processed without interruption to complete the project. The dummy activities 1 and n represent the beginning and the end of the project. The duration of an activity j, $j = 1, \ldots, n$ is denoted by d_j where $d_1 = d_n = 0$. There are r renewable resource types. The availability of each resource type k in each time period is r_k units, $k = 1, \ldots, r$. Each activity j requires r_{jk} units of resource k during each period of its duration, where $r_{1k} = r_{nk} = 0$, $k = 1, ..., r$. All parameters are non-negative integers. There are precedence relations of the finish-start type with a zero parameter value (i.e. $FS = 0$) defined between the activities. In other words activity i precedes activity j if j cannot start until i has been completed. The structure of a project can be represented by an activity-on-node network $G = (SV, SA)$, where SV is the set of activities and SA is the set of precedence relationships. SS_j (SP_j) is the set of successors (predecessors) of activity j, $j = 1, \ldots, n$. It is further assumed that $1 \in SP_j$, $j = 2, \ldots, n$, and $n \in SS_j$, $j = 1, \ldots, n-1$.

The objective is to find a schedule S of activities starting times $[s_1, \ldots, s_n]$, where $s_1 = 0$ and resource constraints are satisfied, such that the schedule duration $T(S) = s_n$ is minimized.

The above formulated problem as a generalization of the classical job shop scheduling problem belongs to the class of NP-hard optimization problems [7]. It is noted as $PS|prec|C_{max}$ [8].

3 The Extended JABAT Environment

JABAT is a middleware allowing to design and implement A-Team architectures for solving various combinatorial optimization problems. The problem-solving paradigm which the proposed system uses can be best defined as the population-based approach.

JABAT produces solutions to combinatorial optimization problems using a set of optimization agents, each representing an improvement algorithm. Each improvement (optimization) algorithm when supplied with a potential solution to the problem at hand, tries to improve this solution. The initial population of solutions (individuals) is generated or constructed. Individuals forming the initial population are, at the following computation stages, improved by independently acting optimization agents. Main functionality of the environment includes organizing and conducting the process of search for the best solution.

In the extended version of JABAT a Team of A-Teams can be implemented and used similarly to a single A-Team. Each A-Team in the TA-Teams uses one population of individuals and a fixed number of optimization agents. Individuals can migrate between populations in accordance with a user-defined migration strategy. The earlier experiments using the TA-Teams architecture were described in [3,15].

To implement the proposed architecture, the most important are the following classes of agents: SolutionManager, MigrationManager and OptiAgent. SolutionManager represents and manages one A-Team. MigrationManager manages the communication between A-Teams represented by SolutionMangers. OptiAgent represents an improvement algorithm.

Other important classes include: Task representing an instance or a set of instances of the problem and Solution representing the solution. To initialize the agents and maintain the system the TaskManager and PlatformManager classes are used. Objects of the above classes also act as agents.

In the extended JABAT the MigrationManger supervises the process of communication between solution managers with their common memories where populations of solutions are stored. The migration is asynchronous. With a given frequency the MigrationManager sends messages to SolutionManagers pointing out to which SolutionManager messages with the best solution or solutions should be send to. Then each thus informed SolutionManager resends the best current solution or solutions to the respective common memory. A single SolutionManager controls the process of solving a single problem instance (task) in accordance with the interaction strategy. Interaction strategy is a set of rules applicable to managing and maintaining a population of current solutions in the common memory.

JABAT and extended JABAT environment have been designed and implemented using JADE (Java Agent Development Framework), which is a software framework proposed by TILAB ([6]) supporting the implementation of the multiagent systems. More detailed information about JABAT environment and its implementations can be found in [12], [2] and [3].

4 TA-Teams Using RL Strategy for Solving the RCPSP

JABAT environment was successfully used by the authors for solving the RCPSP, MRCPSP and RCPSP/max ([13,14,3]). The TA-Teams architecture was used for solving RCPSP with static interaction strategy for each A-Team as described in [15]. The dynamic RL interaction strategy for one A-Team was proposed in [16]. In the proposed approach the dynamic RL interaction strategy is used for each A-Team in TA-Teams.

Classes describing the problem are responsible for reading and preprocessing the data and generating random instances of the problem. They include: RCPSPTask inheriting from the Task class and representing the instance of the problem, RCPSPSolution inheriting from the Solution class and representing the solution of the problem instance, Activity representing the activity of the problem, Resource representing the renewable resource.

A special set includes classes describing the optimization agents. All of them are inheriting from OptiAgent class. In the proposed TA-Teams this set includes the following classes:

- OptiLSA denoting the Local Search Algorithm (LSA),
- OptiTSA denoting the Tabu Search Algorithm (TSA),
- OptiCA denoting Crossover Algorithm (CA),
- OptiPRA denoting Path Relinking Algorithm (PRA),

All optimization agents (OptiAgents) co-operate together using their A-Team common memory managed by the SolutionManager. An individual is represented as schedule of activities S. The final solution is obtained from the schedule by forward or backward Serial Generation Scheme (serial SGS) procedure [18]. The basic RL interaction strategy is based on the Blocking2 strategy used in [16]. The basic RL interaction strategy details are as follows:

- All individuals in the initial population of solutions are generated randomly and stored in the common memory.
- Individuals for improvement are selected from the common memory randomly and blocked which means that once selected individual (or individuals) cannot be selected again until the OptiAgent to which they have been sent returns the solution.
- Returning individual which represent the feasible solution replaces its version before the attempted improvement. It means that the solutions are blocked for the particular OptiAgent and the returning solution replaces the blocked one or the worst from the blocked once. If none is worse a random one is replaced. All solutions blocked for the considered OptiAgent are released and returned to the common memory.
- The new solution is generated randomly with probability 0.2 to replace the worse one or random one from the population.
- For each level of learning the environment state is remembered. This state includes: the best individual and the population average diversity. The state is calculated every fixed number of iteration $itNS = \left\lfloor \frac{PopulationSize}{AgentsNumber} \right\rfloor$. Diversity of two solutions for RCPSP is calculated as the relative difference between the number of activities n and the number of activities with the same starting time in both solutions. To reduce the computation time, average diversity of the population is calculated by comparison with the best solution only.
- The A-Team stops computations when the average diversity in it's population is smaller then 0.01%.

The values of the parameters are chosen experimentally.

The three RL rules are formulated and adopted to the basic RL strategy:

RL1 - in which, RL controls the replacement of one individual from the population by another randomly generated one.

RL2 - in which, the method of choosing the individual for replacement is controlled by RL.

RL3 - in which, individuals in the population are grouped according to certain features, and next the procedure of choosing individuals to be forwarded to optimization agents from respective groups is controlled by RL.

As a result the RL123 in which the three rules are used is considered and tested experimentally. The proposed RL rules are based on utility values reinforcement proposed in [19,4].

All A-Teams (managed by the SolutionMangers) exchange the best solutions according to the migration strategy carried-out by the MigrationManager. The migration strategy is based on the randomized topology in which the one (source) SolutionManager asks for a new solution. The source SolutionManager sends appropriate message to the MigrationManager. It chooses randomly one other (target) SolutionManager and asks it to send its best solution to the source SolutionManager. The migration strategy is defined as follows:

- SolutionManager asks for a new solution when the current best solution in its common memory has not been changed for a fixed time.
- In each case one individual is sent from the one (source) SolutionManager to the other (target) SolutionManager.
- The best solution taken from the source SolutionManager replaces the worst solution in the common memory of the target SolutionManager.
- All A-Teams stop computation when one of them stops due to its interaction strategy.

5 Computational Experiment

5.1 Settings

To evaluate the effectiveness of the proposed approach and compare the results, depending on the number of SolutionManagers used, the computational experiment has been carried out using benchmark instances of RCPSP from PSPLIB[1] - test sets: sm30 (single mode, 30 activities), sm60, sm90, sm120. Each of the first three sets includes 480 problem instances while set sm120 includes 600. The experiment involved computation with the fixed number of optimization agents, fixed population size, and the limited time.

In the experiment 4 sets of parameters have been used as presented in Table 1. The values are set experimentally. In each set the total number of individuals in all populations is 80 and the total number of optimization agents working for all SolutionManagers is 32. The fixed time after which SolutionManager asks for a new solution is 1 minute.

The no improvement iteration gap has been set to 2 minutes, and the fixed part of no improvement time gap after which the SolutionManager asks for new solution is half of that time. The number of reviews in the interaction strategy is 5.

Table 1. Parameters setting

#SolutionManager-s (islands)	#OptiAgents for one SolutionManager	Population size for one SolutionManager
1	8x4	80
2	4x4	40
4	2x4	20
8	1x4	10

The proposed TA-Teams includes 4 kinds of optimization agents representing the LSA, TSA, CA and PRA algorithms mentioned in Section 4.

The experiment has been carried out using nodes of the cluster Holk of the Tricity Academic Computer Network built of 256 Intel Itanium 2 Dual Core 1.4 GHz with 12 MB L3 cache processors. During the computation one node per eight optimization agents was used.

[1] See PSPLIB at `http://www.om-db.wi.tum.de/psplib/`

5.2 Results

During the experiment the following characteristics of the computational results have been calculated and recorded: mean relative error (MRE) calculated as the deviation from the optimal solution for sm30 set or from the best known solution and critical path lower bound (CPLB) for sm60, sm90 and sm120 sets, mean computation time required to find the best solution (Mean CT) and mean total computation time (Mean total CT). Each instance has been solved five times and the results have been averaged over these solutions. In each case the 100% of feasible solutions has been obtained. The computation results are presented in Tables 2-5.

The experiment results show that using A-Teams with RL interaction strategy within the proposed TA-Teams architecture is beneficial. The results obtained using more solution managers are in most cases better than the results obtained using only one SolutionManager with a similar parameter settings.

Table 2. Results for benchmark test set sm30 (RE from optimal solution)

#SolutionManagers	MRE from optimal solution	Mean CT [s]	Mean total CT [s]
1	0.019%	2.89	35.02
2	0.016%	3.12	33.48
4	0.012%	5.31	34.04
8	0.014%	8.28	34.16

Table 3. Results for benchmark test set sm60

#SolutionManagers	MRE from best known solution	MRE from CPLB	Mean CT [s]	Mean total CT [s]
1	0.62%	11.16 %	14.72	63.41
2	0.58%	10.98 %	15.36	65.24
4	0.50%	10.83 %	16.83	66.19
8	0.51%	10.86 %	16.91	67.48

Table 4. Results for benchmark test set sm90

#SolutionManagers	MRE from best known solution	MRE from CPLB	Mean CT [s]	Mean total CT [s]
1	0.98%	10.90%	26.07	76.72
2	0.96%	10.85%	26.12	75.45
4	0.95%	10.79%	25.48	77.08
8	0.95%	10.81%	26.67	76.26

Table 5. Results for benchmark test set sm120

#SolutionManagers	MRE from best known solution	MRE from CPLB	Mean CT [s]	Mean total CT [s]
1	2.90%	33.25 %	88.05	197.10
2	2.89%	33.23 %	90.58	186.25
4	2.86%	33.21 %	89.33	191.56
8	2.87%	33.22 %	90.09	199.15

Table 6. Literature reported results [11], [1]

Set	Algorithm	Authors	MRE[a]	Mean CT [s]	Computer
sm30	Decompos. & local opt	Palpant et al.	0.00	10.26	2.3 GHz
	VNS–activity list	Fleszar, Hindi	0.01	5.9	1.0 GHz
	Local search–critical	Valls et al.	0.06	1.61	400 MHz
sm60	PSO	Tchomte et al.	9.01	–	–
	Decompos. & local opt	Palpant et al.	10.81	38.8	2.3 GHz
	Population–based	Valls et al.	10.89	3.7	400 MHz
	Local search–critical	Valls et al.	11.45	2.8	400 MHz
sm90	Filter and fan	Ranjbar	10.11	–	–
	Decomposition based GA	Debels, Vanhoucke	10.35	–	–
	GA–hybrid, FBI	Valls at al.	10.46	–	–
sm120	Filter and fan	Ranjbar	31.42	–	–
	Population-based	Valls et al.	31.58	59.4	400 MHz
	Decompos. & local opt.	Palpant et al	32.41	207.9	2.3 GHz
	Local search–critical	Valls et al.	34.53	17.0	400 MHz

[a] For the set sm30 the MRE from optimal solution and for the remaining sets the MRE from CPLB are reported.

The presented results are compared with the results reported in [11,1] as presented in Table 6. The comparison includes the results with known computation times and processor clocks mainly. In the proposed agent-based approach computation times as well as number of schedules differ between nodes and optimization agent algorithms work in parallel. The results obtained by a single agent may or may not influence the results obtained by the other agents. Additionally the computation time includes the time used by agents to prepare, send and receive messages.

6 Conclusions

Experiment results show that the proposed implementation of TA-Teams with RL interaction strategy is an effective and competitive tool for solving instances of the RCPSP problem. Presented results are comparable with solutions known from the literature and in some cases outperform them. It can be also noted that they have been obtained in a comparable time. Time comparisons in this case might be misleading since the proposed TA-Teams have been run using different numbers and kinds of processors. In case of the agent-based environments the significant part of the time is used for agent communication which has an influence on both - computation time and quality of the results.

Future research will concentrate on using reinforcement learning at the upper level in communication between cooperating A-Teams e.g. to manage the migration strategy within TA-Teams.

Since the JABAT gives a possibility to run more than one copy of each agent it is interesting which agents should or should not be replicated to improve the results. Additionally, testing and adding to JABAT more different optimization agents and improving the existing ones will be considered.

References

1. Agarwal, A., Colak, S., Erenguc, S.: A Neurogenetic Approach for the Resource-Constrained Project Scheduling Problem. Computers & Operations Research 38, 44–50 (2011)
2. Barbucha, D., Czarnowski, I., Jedrzejowicz, P., Ratajczak-Ropel, E., Wierzbowska, I.: E-JABAT – An Implementation of the Web-Based A-Team. In: Nguyen, N.T., Jain, L.C. (eds.) Intelligent Agents in the Evolution of Web and Applications. SCI, vol. 167, pp. 57–86. Springer, Heidelberg (2009)
3. Barbucha, D., Czarnowski, I., Jędrzejowicz, P., Ratajczak-Ropel, E., Wierzbowska, I.: Parallel Cooperating A-Teams. In: Jędrzejowicz, P., Nguyen, N.T., Hoang, K. (eds.) ICCCI 2011, Part II. LNCS (LNAI), vol. 6923, pp. 322–331. Springer, Heidelberg (2011)
4. Barbucha, D.: Search Modes for the Cooperative Multi-agent System Solving the Vehicle Routing Problem, Intelligent and Autonomous Systems. Neurocomputing 88, 13–23 (2012)
5. Barto, A.G., Sutton, R.S., Anderson, C.W.: Neuronlike adaptive elements that can solve difficult learning control problems. IEEE Transactions on Systems, Man, and Cybernetics, SMC-13, 835–846 (1983)
6. Bellifemine, F., Caire, G., Poggi, A., Rimassa, G.: JADE. A White Paper, Exp. 3(3), 6–20 (2003)
7. Blazewicz, J., Lenstra, J., Rinnooy, A.: Scheduling subject to resource constraints: Classification and complexity. Discrete Applied Mathematics 5, 11–24 (1983)
8. Brucker, P., Drexl, A., Mohring, R., Neumann, K., Pesch, E.: Resource-Constrained Project Scheduling: Notation, Classification, Models, and Methods. European Journal of Operational Research 112, 3–41 (1999)
9. Busoniu, L., Babuska, R., De Schutter, B.: A Comprehensive Survey of Multiagent Reinforcement Learning. IEEE Transactions on Systems, Man, and Cybernetics, Part C: Applications and Reviews 38(2), 156–172 (2008)
10. Cohoon, J.P., Hegde, S.U., Martin, W.N., Richards, D.: Punctuated Equilibria: a Parallel Genetic Algorithm. In: Proceedings of the Second International Conference on Genetic Algorithms, pp. 148–154. Lawrence Erlbaum Associates, Hillsdale (1987)
11. Hartmann, S., Kolisch, R.: Experimental Investigation of Heuristics for Resource-Constrained Project Scheduling: An Update. European Journal of Operational Research 174, 23–37 (2006)

12. Jędrzejowicz, P., Wierzbowska, I.: JADE-Based A-Team Environment. In: Alexandrov, V.N., van Albada, G.D., Sloot, P.M.A., Dongarra, J. (eds.) ICCS 2006. LNCS, vol. 3993, pp. 719–726. Springer, Heidelberg (2006)
13. Jedrzejowicz, P., Ratajczak-Ropel, E.: New Generation A-Team for Solving the Resource Constrained Project Scheduling. In: Proc. the Eleventh International Workshop on Project Management and Scheduling, Istanbul, pp. 156–159 (2008)
14. Jedrzejowicz, P., Ratajczak-Ropel, E.: Solving the RCPSP/max Problem by the Team of Agents. In: Håkansson, A., Nguyen, N.T., Hartung, R.L., Howlett, R.J., Jain, L.C. (eds.) KES-AMSTA 2009. LNCS (LNAI), vol. 5559, pp. 734–743. Springer, Heidelberg (2009)
15. Jedrzejowicz, P., Ratajczak-Ropel, E.: Team of A-Teams for Solving the Resource-Constrained Project Scheduling Problem. In: Advances in Knowledge-Based and Intelligent Information and Engineering Systems, Frontiers in Artificial Intelligence and Applications, vol. 243, pp. 1201–1210. IOS Press Ebooks (2012)
16. Jędrzejowicz, P., Ratajczak-Ropel, E.: Reinforcement Learning Strategy for A-Team Solving the Resource-Constrained Project Scheduling Problem. In: Bădică, C., Nguyen, N.T., Brezovan, M. (eds.) ICCCI 2013. LNCS, vol. 8083, pp. 457–466. Springer, Heidelberg (2013)
17. Kaelbling, L.P., Littman, M.L., Moore, A.W.: Reinforcement learning: A survey. Journal of Artificial Intelligence Research 4, 237–285 (1996)
18. Kolisch, R.: Serial and parallel Resource-Constrained Project Scheduling Methods Revisited: Theory and Computation. European Journal of Operational Research 43, 23–40 (1996)
19. Nareyek, A.: Choosing search heuristics by non-stationary reinforcement learning. In: Metaheuristics: Computer Decision-making, pp. 523–544. Kluwer Academic Publishers (2001)
20. PSPLIB, `http://www.om-db.wi.tum.de/psplib/`
21. Schutt, A., Feydy, T., Stuckey, P.J.: Explaining Time-Table-Edge-Finding Propagation for the Cumulative Resource Constraint. In: Gomes, C., Sellmann, M. (eds.) CPAIOR 2013. LNCS, vol. 7874, pp. 234–250. Springer, Heidelberg (2013)
22. Sutton, R.S., Barto, A.G.: Reinforcement Learning: An Introduction. MIT Press, Cambridge (1998)
23. Talukdar, S., Baerentzen, L., Gove, A., de Souza, P.: Asynchronous Teams: Cooperation Schemes for Autonomous, Computer-Based Agents. Technical Report EDRC 18-59-96. Carnegie Mellon University, Pittsburgh (1996)
24. Tuyls, K., Weiss, G.: Multiagent learning: Basics, challenges, prospects. AI Magazine 33(3), 41–53 (2012)
25. Wauters, T.: Reinforcement learning enhanced heuristic search for combinatorial optimization. Doctoral thesis, Department of Computer Science, KU Leuven (2012)

A Hybrid Cooperative and Reactive Search Approach for Vehicle Routing Problem with Time Windows

Dariusz Barbucha

Department of Information Systems
Gdynia Maritime University
Morska 83, 81-225 Gdynia, Poland
d.barbucha@wpit.am.gdynia.pl

Abstract. The paper proposes a new hybrid approach for Vehicle Routing Problem with Time Windows (VRPTW), which combines a cooperative problem solving paradigm with the reactive search. Process of searching for the best solution is performed by a set of cooperating heuristics working on a population of individuals. Additionally, at each iteration, a history of search of each heuristic and its performance are recorded, and next used for dynamic management of the set of heuristics and adjustment their parameters. Computational experiment which has been carried out confirmed the effectiveness of the proposed approach.

Keywords: cooperative search, reactive search, multi-agent systems, vehicle routing problem with time windows.

1 Introduction

Main force which can drive metaheuristic applications to a higher performance are mechanisms of effective and efficient exploration of the search space involved. Such a search carried-out by the metaheuristic algorithms should focus enough on both, intensive exploration of the areas of the promising search space with high quality solutions (*intensification*), and on movement to the unexplored areas of the search space when necessary (*diversification*) [4]. In order to achieve a good quality of the metaheuristic-based solutions to a particular optimization problem, it is necessary to provide and dynamically control a balance between intensification and diversification strategies. It often requires modification (or tuning) of some parameters in response to the local properties of the search space and/or algorithm's behavior (its history) during the execution. The key feature of the *reactive search* proposed by Battiti and Brunato [2] is the online adaptation of the algorithm parameters to various external and internal factors observed during searching for the best solution.

An interesting and effective way of controlling the balance between intensification and diversification is to use a set of search programs (instead of a single one), running in parallel, and combined into an effective problem-solving system.

N.T. Nguyen et al. (Eds.): ACIIDS 2014, Part II, LNAI 8398, pp. 207–216, 2014.

Such an approach to solving optimization problems is known as the *cooperative search* [12]. Since different heuristics have different strengths and weaknesses, it is expected that combining them, the strengths of one heuristic can compensate the weaknesses of another. Moreover, joined execution of several heuristics can also possibly create a synergy effect, in which the combined effect of cooperation between programs (heuristics) is greater than the sum of their separate effects.

The paper proposes a new agent-based hybrid cooperative and reactive search approach to the Vehicle Routing Problem with Time Windows (VRPTW). It uses a list of heuristics (represented as software agents) which collectively solve instances of the problem by operating on individuals stored in the common, sharable memory. This research is especially focused on intelligent management of the set of heuristics during the whole process of searching for the best solution with a view to make sych search more effective and efficient. In particular, at each iteration one can decide what heuristics should be used in order to either intensify or diversify the exploration or exploitation of the search region. One of the idea for managing the available hauristics is to used the proposed tabu list of heuristics. Such a list includes heuristics which are excluded from the computation for a given period of time. Which heuristic is added to the tabu list and how long it stays on this list is determined according to the trajectory of search, performance of each heuristic and its contribution to the improvement of the best solution.

The paper is organized as follows. Section 2 defines the Vehicle Routing Problem with Time Windows. Details of the proposed approach are presented in Section 3. Sections 4 and 5 present the results of the computational experiment and main conclusions, respectively.

2 Vehicle Routing Problem with Time Windows

The Vehicle Routing Problem with Time Windows (VRPTW) can be formulated as the problem of determining optimal routes through a given set of locations (customers) and defined on an undirected graph $G = (V, E)$, where $V = \{0, 1, \ldots, N\}$ is the set of nodes and $E = \{(i, j)|i, j \in V\}$ is a set of edges. Node 0 is a central depot with NV identical vehicles of capacity W. Each other node $i \in V \setminus \{0\}$ denotes customer characterized by a non-negative demand d_i, and a service time of the customer s_i. Moreover, with each customer $i \in V$, a time window $[e_i, l_i]$ is associated, wherein the customer has to be supplied. Here e_i is the earliest possible departure (ready time), and l_i - the latest time a service to the customer has to be started. The time window at the depot ($[e_0, l_0]$) is called the scheduling horizon. Each link $(i, j) \in E$ denotes the shortest path from customer i to j and is described by the cost c_{ij} of travel from i to j by shortest path ($i, j \in V$). It is assumed that $c_{ij} = c_{ji}$ ($i, j \in V$). It is also often assumed that c_{ij} is equal to travel time t_{ij}.

The goal is to minimize the number of vehicles and the total distance needed to supply all customers (minimization of the fleet size is considered to be the primary objective of the VRPTW), such that each route starts and ends at the depot, each customer $i \in V \setminus \{0\}$ is serviced exactly once by a single vehicle, the total load on any vehicle associated with a given route does not exceed vehicle capacity, each customer $i \in V \setminus \{0\}$ has to be supplied within a time window associated with him (a vehicle arriving before the lower limit of the time window causes additional waiting time on the route), and each route must start and end within the time window associated with the depot.

Because of the NP-hardness of this problem, a majority of aproaches proposed for solving it has heuristic nature. The wide spectrum of the methods belonging to this class, ranging from simple local search methods to more complex meta-heuristics, has confirmed their effectiveness and practical usefulness in solving this problem [8,9].

Both, cooperative and reactive search, have been proposed independently, for solving VRPTW by different authors (for example, cooperative search: Bouthillier and Crainic [5] and Bouthillier et al. [6], reactive search: Chiang and Russell [10], and Bräysy [7]). However, as far as the author is aware, the hybridization of both techniques have not yet been proposed in the literature for VRPTW. On the other hand, to the best author's knowledge, only Masegosa et al. [11] focused their work on hybridization of cooperative and reactive search for solving different optimization problem. Their approach, using fuzzy and reactive rules incorporated into cooperative strategy, has been tested on the Uncapacitated Single Allocation p-Hub Median Problem, producing promising results.

3 A Cooperative and Reactive Search Approach to the VRPTW

3.1 Model

The proposed approach belongs to an agent-based cooperative population-based methods group. It assumes the existence of a *set of heuristics* $H = \{H_1, H_2, \ldots, H_M\}$, where each H_j $(j = 1, 2, \ldots, M)$ is represented by optimizing agent - `OptiAgent` (`OA(`H_j`)`). Each `OptiAgent` is an implementation of the approximation method dedicated for solving the VRPTW. During the search, `OptiAgents` operate on a population of individuals $P = \{p_1, p_2, \ldots, p_{|P|}\}$, stored in the common memory, where each p_i $(i = 1, 2, \ldots, |P|)$ represents a solution of the problem. As `OAs` try to improve individuals forming the population, it evolves over time and the best individual from the population is taken as a final result at the end of the problem solving process. The process of searching for the best solution is supervised by a dedicated agent, called `SolutionManager` (`SMa`), which role is to maintain and manage the population of individuals P, and to coordinate the work of the set of `OptiAgents`.

The whole process of solving the problem includes two phases: *initialization* and *improvement*. Whereas the former one focuses on initialization of all parameters and creation of the initial population of solutions, the second one includes steps performed towards improvement of the individuals - members of the population. In the proposed approach, the following steps constitue the algorithm:

Step 1. At first the initial population of solutions is created using a modified version of the Solomon's *I1* constructive heuristic [13]

Step 2. Repeat the following activities in the loop until a stopping criterion (predefined time of computation) is met [in parallel]:

a. `OAs` which are ready to act, informs `SMa` about it.
b. `SMa` randomly selects a number of individuals (solutions) from the common memory and sends it to selected `OA`, which has just announced its readiness.
c. After receiving the individual, `OA` tries to improve it using built-in dedicated method, and next it sends it back to the `SMa`.
d. If the received solution has been improved by `OA` then `SMa` merges it with the population by replacing the individual previously sent to the `OA`.

Step 3. Finally, the best solution from the population is taken as a result.

In general, the process of communication between `SMa` and `OAs` is organized in an asynchronous way. It means that each `OA`, after finishing the process of improvement of the current solution and anouncing its readiness to work to `SMa`, receives immediately a new solution taken from the population by `SMa` to be improved, regardless of behavior of other `OAs`. It implies that the iddle time of each `OA` is minimized and, in case of sufficient amount of available resources, its utilization by agents is higher. On the other hand, according to the author's previous observations [1], the effectiveness of heuristics decreases, as the process of searching progresses. Although heuristics spend much time on calculations, the time between subsequent improvements grows and no improvements are observed more often.

3.2 Tabu List of Agents

In order to use `OptiAgents` in more efficient way, the proposed approach uses also the tabu list of them - T_{OA} of a fixed length $M-1$, where M is the number of `OAs`. Each element belonging to the T_{OA} includes: the *id* of a given `OA`, time t_{add} in which this `OA` has been added to the T_{OA} and a tabu duration t_{dur}, meant as a period of time in which this `OA` spends on T_{OA}, and therefore, is not available to the `SolutionManager`. It is also assumed that at any moment of time, the tabu list contains at most a single instance of each `OA`, and at least one `OA` must be tabu inactive (in case where the list is full, and some `OA` has to be added to the list, an `OA` with the shorter remaining time to be tabu inactive, is removed from the list).

When using tabu list, the fundamental decisions to be made are if and when a given `OA` should be added to T_{OA} as well as what tabu duration value works best for a given problem instance. In the proposed approach, the above decisions are made dynamically by `SMa` basing on the performance of each `OA`.

3.3 Performance Measures and Reacting Rules

Let T be the total time in which all OAs spend on improving solutions, and let nt be a number of time slots $t_1, t_2, \ldots t_{|nt|}$, where each of them has length T/nt. Each slot begins at t_k^b, and finishes at t_k^f $(k = 1, \ldots, nt)$.

During the whole process of searching, the SMa monitors the behaviour of each OA by storing information about its performance. In particular, each agent a has assigned two online measures of its effectiveness on local and global level. Whereas the first one, *Local Effectiveness* (LE) concentrates on comparison of the value of goal function (and time) after and before improvement by this agent at iteration t, *Global Effectiveness*(GE) compares the value of goal function (and time) after improvement by the agent a at iteration t with the best one known before improvement:

$$LE(a,t) = \frac{|f(sR(a,t)) - f(sS(a,t))|}{tR(a,t) - tS(a,t)} \tag{1}$$

$$GE(a,t) = \frac{\max(f(sB(t) - f(sR(a,t)), 0)}{tR(a,t) - tS(a,t)} \tag{2}$$

$f(sS(a,t))$ $(f(sR(a,t)))$ is the value of the goal function of the solution sent to (received from) the agent a at iteration t, $f(sB(t))$ is the value of the goal function of the best solution known at iteration t, and $tS(a,t)$ $(tR(a,t))$ - time in which a solution $sS(a,t)$ $(sR(a,t))$ has been sent to (received from) the agent a.

It is easy to see, that both measures (LE and GE) concentrate on the performance of each agent at current iteration without any considering its performance in the past. Hence by averaging both measure over all iterations within the slot t_k, one can obtain *Average Local Effectiveness* ($AvgLE(a,t_k)$) and *Average Global Effectiveness* ($AvgGE(a,t_k)$) of the agent a at time slot t_k, respectively. Both measures take into account a history of search of each agent:

$$AvgLE(a,t_k) = \frac{\sum_{t:t_k^b \le t \le t_k^f} LE(a,t)}{n(a,t_k)} \tag{3}$$

$$AvgGE(a,t_k) = \frac{\sum_{t:t_k^b \le t \le t_k^f} GE(a,t)}{n(a,t_k)} \tag{4}$$

where $n(a,t_k)$ is a number of iterations (number of calling) of the agent a within the slot t_k.

Additionaly, the following two measures of the number of local (LI) and global (GI) improvements of each agent a within each slot t_k are recorded:

$$LI(a,t_k) = \frac{n_{LI}(a,t_k)}{n(a,t_k)} \tag{5}$$

$$GI(a,t_k) = \frac{n_{GI}(a,t_k)}{n(a,t_k)} \tag{6}$$

where $n(a, t_k)$ has the same meaning as previously, and $n_{LI}(a, t_k)$ ($n_{GI}(a, t_k)$) tells how many times the agent a has locally (globally) improved solutions within the time slot t_k.

Adding the OptiAgent to Tabu List. In order to decide what heuristc (if any) should be placed on tabu list, at the end of each time slot t_k ($k = 1, \ldots, nt$) the following rule is considered:

Rule (Add to Tabu List):

IF $(GI(a, t_k) = 0)$ OR
$((AvgLE(a, t_k) < AvgLE(a, t_{k-1})) - p_{LE})$ OR
$((AvgGE(a, t_k) < AvgGE(a, t_{k-1})) - p_{GE})$
THEN $T_{OA} \leftarrow T_{OA} \cup \{a\}$

It says that if the agent a did not improve the best solution in slot t_k or average local (or global) effectiveness of the agent a has significantly decreased (controlled by thresholds p_{LE}, and p_{GE}, respectively) within the time slot t_k then the agent a is added to T_{OA}. If all heuristics satisfy the above rule, the agent with the best global performance is choosen and still remains non-tabu. Time duration of each heuristic added to the T_{OA} is equal to length of a single slot.

Removing the OptiAgent from Tabu List. At the end of each slot t_k ($k = 1, \ldots, nt$) the following rule is checked for each agent belonging to T_{OA}:

Rule (Remove from Tabu List):

IF $(t_{add}(a) + t_{dur}(a) > t_{now})$
THEN $T_{OA} \leftarrow T_{OA} \setminus \{a\}$

If for given agent a included in T_{OA}, the condition $(t_{add}(a) + t_{dur}(a) > t_{now})$ holds (t_{now} is the current time), it becomes tabu inactive, and is removed from T_{OA}, otherwise it still remains tabu active. OA which has been just removed from the T_{OA} remains tabu inactive by at least one time slot.

4 Computational Experiment

To validate the proposed approach computational experiment has been carried out. The main goal was to evaluate to what extent the proposed intelligent management of the set of heuristics within the cooperative approach for VRPTW influences the computational results. The quality of the results obtained by the proposed approach has been evaluated using two measures: the number of vehicles needed to serve all requests, and the total distance needed to be covered by vehicles in order to supply all customers. It was also assumed that a solution with a fewer routes is preferred over one with more routes and that in case of the tie in the number of routes, a solution with the shortest distance is chosen.

The approach has been implemented in Java programming language and using JADE (Java Agent Development Framework) [3]. The experiment involved two datasets of instances of Solomon (R1, R2) [13] (available at [14]) with 100 customers uniformly distributed over the plane. Each instance was repeatedly solved 10 times. All computations have been carried out on PC computer Intel Core i5-2540M CPU 2.60 GHz with 8 GB RAM running under MS Windows 7 operating system.

The following `OptiAgents` representing simple improvemement heuristics, operating on a single selected solution, have been used:

- Four local search heuristics: `Relocate1R`, `Swap1R`, `Relocate2R`, `Swap2R`, which operate on a single (`Relocate1R`, `Swap1R`) or two (`Relocate2R`, `Swap2R`) randomly selected route(-s). They relocate customer(-s) from selected position to another one, within the same route (`Relocate1R`) or between routes (`Relocate2R`), or swap two selected customers belonging to the same (`Swap1R`) or different route(-s) (`Swap2R`).
- Four simulated annealing implementations: `SA(Relocate1R)`, `SA(Swap1R)`, `SA(Relocate2R)`, `SA(Swap2R)`, based on the four moves defined above, respectively.
- Two implementations of one (and two-) point crossover evolutionary operators: `Cross1` and `Cross2`, operating on two randomly selected routes.
- Implementation of simple path relinking procedure (`PathRel`).

It was also assumed that the process of searching stops after $T = 5$ minutes of computation, and the total time has been divided on 15 time slots, which implies that the length of each time slot is equal to 20 seconds. After the preliminary experiment, p_{LE} and p_{GE} have been set to 0.2.

The results of the computational experiment are presented in Tables 1 and 4. Table 1 includes statistics of each agent for exemplary instance (R201 in the experiment), which tell how many times each agent did try to improve the received solutions (Received Solutions), how many times it improved solution locally (Improved Local) and globally (Improved Global).

Table 4 presents the results obtained by the proposed approach for two tested verions of the cooperative algorithm (without and with implemented reacting rules) for entire R1 and R2 datasets. Besides the name of the instance, the next columns include the best known results for each instance (avaialble at [15], together with their sources), the results obtained by the proposed cooperative approach without (CS) and with (CS+RS) implemented reacting rules, and a mark '*', which indicates whether CS+RS outperforms CS for particular instance.

Results presented in Table 1 show that implemented agents have different complexity and different amount of time is required by each of them for computation in a single iteration. By focusing observation on CS one can see that each agent differently contributes in improvements of received (or/and the best

Table 1. Statistics of opimizing agents (instance R201)

	CS			CS+RS		
Agents	Received Solutions	Improved Local	Improved Global	Received Solutions	Improved Local	Improved Global
Relocate1R	949	171	2	664	110	3
Relocate2R	808	167	2	464	142	6
Swap1R	799	138	3	618	103	0
Swap2R	757	148	5	527	168	7
SA(Relocate1R)	253	109	4	146	53	1
SA(Relocate2R)	286	60	3	158	33	3
SA(Swap1R)	28	12	0	16	11	0
SA(Swap2R)	18	5	0	9	3	1
Cross1	930	121	1	627	144	4
Cross2	865	0	0	201	0	0
PathRel	852	0	0	309	0	0
Total Improvements		931	20		767	25

one) individuals. Some agents (SA(Swap1R), SA(Swap2R), Cross1) very rarely or never improve the best solutions, however their contribution to cooperative search may be significant, especially at early stages of computation. On the other hand, there is a group of agents for which improvements (local or/and global) are observed more often (Swap1R, Swap2R, SA(Relocate1R), SA(Relocate2R)).

Results for the second version, with implemented RS rules (CS+RS), show that, due to the decrease in the effectiveness of some agents, they are occasionally eliminated from the search in some time slots, hence the number of their calls decreases. Only agents, which significanlty positively influence the computational results are active most of the computation time. An interesting observation is that, although the number of callings all agents is less than in case of CS, the total number of global improvements increased. A few agents without any improvements observed in the first case (CS), now (CS+RS) have been able to achieve some improvements. Also, the number of global improvements for several agents have increased.

Analysis of the results presented in Table 4 allows one to observe that the proposed approach (in both version: CS and CS+RS) produces quite good results close to the best known ones. Although the overall best known results have not been reached for any instance, the difference between the number of vehicles of best solution and obtained one does not exceed one for all instances. Also, the deviation between the best distance and the distance produced by the proposed approach does not exceed a few percent for all tested instances. By comparing the results for both versions (CS and CS+RS) one can conlude that average results over all instances are slightly better for CS version with RS (15 out of 23 instances, which is rqual to 65%). However, at this stage, it is hard to conclude that this difference is significant.

Table 2. Results (min. number of vehicles/distance) obtained by the proposed cooperative approach for tested configurations with and without reacting rules (groups R1, R2)

Instance	Best known		CS		CS+RS		RS?
	NV	Distance	NV	Distance	NV	Distance	
R101	19	1650.80	19	1719.57	19	1715.09	*
R102	17	1486.12	17	1549.25	18	1544.00	
R103	13	1292.68	14	1278.14	14	1265.16	*
R104	9	1007.31	10	1006.49	10	1028.32	
R105	14	1377.11	15	1421.55	14	1406.95	*
R106	12	1252.03	13	1283.96	13	1287.87	
R107	10	1104.66	11	1110.63	11	1106.52	*
R108	9	960.88	10	968.09	10	962.88	*
R109	11	1194.73	12	1195.71	12	1179.53	*
R110	10	1118.84	12	1133.69	11	1152.24	*
R111	10	1096.72	12	1112.04	11	1123.80	*
R112	9	982.14	10	979.47	10	974.95	*
Average	11.92	1210.34	12.92	1229.88	12.75	1228.94	
R201	4	1252.37	4	1257.37	4	1259.32	
R202	3	1191.70	4	1094.95	4	1091.30	*
R203	3	939.50	3	957.95	3	942.65	*
R204	2	825.52	3	767.05	3	772.78	
R205	3	994.42	3	1048.76	3	1028.68	*
R206	3	906.14	3	924.52	3	924.83	
R207	2	890.61	3	817.30	3	819.61	
R208	2	726.82	3	721.55	2	751.68	*
R209	3	909.16	3	926.48	3	915.60	*
R210	3	939.37	3	962.59	3	976.58	
R211	2	885.71	3	791.64	3	782.86	*
Average	2.73	951.03	3.18	933.65	3.09	933.26	

5 Conclusions

A new hybrid approach for Vehicle Routing Problem with Time Windows has been proposed in the paper. It combines the cooperative problem solving paradigm with the reactive search. Process of searching for the best solution is performed by the set of cooperating agents working on the population of individuals. Additionally, at each iteration the history of search of each heuristic and its performance are recorded, which next are used for dynamic management of the set of heuristics engaged in process of searching and adjustment of their parameters.

Computational experiment confirmed that intelligent management of the set of heuristics using the tabu list gives an opportunity to more effective use of them. Moreover, the results obtained by the cooperative approach with management are promissing, and the benefit of using RS is observed, even if not always necesarilly significant. The future experiments will aim at testing the proposed approach on different groups of instances (for example with customers clustered on the plane), which will allow one to draw more general conclussions. Another direction of future work is to propose and implement the set of new reacting rules.

Acknowledgments. The research has been supported by the Polish National Science Centre grant no. 2011/01/B/ST6/06986 (2011-2013).

References

1. Barbucha, D.: Search modes for the cooperative multi-agent system solving the vehicle routing problem. Neurocomputing 88, 13–23 (2012)
2. Battiti, R., Brunato, M.: Reactive Search: Machine Learning for Memory-Based Heuristics. In: Gonzalez, T.F. (ed.) Handbook of Approximation Algorithms and Metaheuristics. Computer and Information Science Series, pp. 21.1–21.17. Chapman & Hall, CRC (2007)
3. Bellifemine, F., Caire, G., Greenwood, D.: Developing Multi-Agent Systems with Jade. John Wiley & Sons, Chichester (2007)
4. Blum, C., Roli, A.: Metaheuristics in combinatorial optimization: Overview and conceptual comparison. ACM Computing Surveys 35(3), 268–308 (2003)
5. Le Bouthillier, A., Crainic, T.G.: A cooperative parallel meta-heuristic for the vehicle routing problem with time windows. Computers & Operations Research 32(7), 1685–1708 (2005)
6. Le Bouthillier, A., Crainic, T.G., Kropf, P.: A Guided Cooperative Search for the Vehicle Routing Problem with Time Windows. IEEE Intelligent Systems 20(4), 36–42 (2005)
7. Bräysy, O.: A reactive variable neighborhood search for the vehicle-routing problem with time windows. INFORMS Journal on Computing 15(4), 347–368 (2003)
8. Bräysy, O., Gendreau, M.: Vehicle Routing Problem with Time Windows, Part I: Route Construction and Local Search Algorithms. Transportation Science 39, 104–118 (2005)
9. Bräysy, O., Gendreau, M.: Vehicle Routing Problem with Time Windows, Part II: Metaheuristics. Transportation Science 39, 119–139 (2005)
10. Chiang, W., Russell, R.: A reactive tabu search metaheuristic for the vehicle routing problem with time windows. INFORMS Journal on Computing 9, 417–430 (1997)
11. Masegosa, A.D., Mascia, F., Pelta, D., Brunato, M.: Cooperative Strategies and Reactive Search: A Hybrid Model Proposal. In: Stützle, T. (ed.) LION 3. LNCS, vol. 5851, pp. 206–220. Springer, Heidelberg (2009)
12. Toulouse, M., Thulasiraman, K., Glover, F.: Multi-level Cooperative Search: A New Paradigm for Combinatorial Optimization and an Application to Graph Partitioning. In: Amestoy, P.R., Berger, P., Daydé, M., Duff, I.S., Frayssé, V., Giraud, L., Ruiz, D., et al. (eds.) Euro-Par 1999. LNCS, vol. 1685, pp. 533–542. Springer, Heidelberg (1999)
13. Solomon, M.: Algorithms for the Vehicle Routing and Scheduling Problems with Time Window Constraints. Operations Research 35, 254–265 (1987)
14. Solomon, M.: VRPTW Benchmark problems, `http://w.cba.neu.edu/~msolomon/problems.htm`
15. Transportation Optimization Portal, VRPTW instances, `http://www.sintef.no/Projectweb/TOP/VRPTW/Solomon-benchmark/100-customers/`

Comparative Study on Bio-inspired Global Optimization Algorithms in Minimal Phase Digital Filters Design

Adam Słowik

Department of Electronics and Computer Science
Koszalin University of Technology
ul. Sniadeckich 2, 75-453 Koszalin, Poland
aslowik@ie.tu.koszalin.pl

Abstract. In this paper, a comparative study is presented on various bio-inspired global optimization algorithms in the problem of digital filters design. The designed digital filters are minimal phase infinite impulse response digital filters with non-standard amplitude characteristics. Due to the non-standard amplitude characteristics, typical filter approximations cannot be used to solve this design problem. In our comparative study, we took into consideration the four most popular bio-inspired global optimization techniques. We examined bio-inspired algorithms such as: an ant colony optimization algorithm for a continuous domain, a particle swarm optimization algorithm, a genetic algorithm and a differential evolution algorithm. After experiments, we observed that the differential evolution algorithm is the most effective one for the problem of the digital filters design.

1 Introduction

Digital signal processing is a very important issue in modern digital technologies. Among digital signal processing techniques, the most important part is signal filtering. Signal filtering is currently used in many applications, like for example: in equalizers, in hearing aid systems, in audiophile systems etc. In digital signal filtering, we possess two most popular architectures of digital filters: FIR (Finite Impulse Response) and IIR (Infinite Impulse Response). The main advantage of FIR digital filters is that the phase characteristics of these filters is linear, but FIR filters are slower than IIR digital filters. The main advantage of IIR digital filters is that they are faster than FIR filters, and therefore we can perform signal filtering in real-time [1, 2].

If we want to design a digital filter that possesses typical amplitude characteristics, we may use one of the existing approximations as for example: the Butterworth approximation, the Chebyshev I and Chebyshev II approximations, or the Cauer approximation. Nevertheless, the problem of filter design is complex: if we wish to obtain a digital filter that possesses non-standard amplitude characteristics, typical approximation cannot be used. Therefore, in the recent

N.T. Nguyen et al. (Eds.): ACIIDS 2014, Part II, LNAI 8398, pp. 217–226, 2014.

years, some bio-inspired techniques have been used to solve the digital filters design problem. Moreover, the problem of digital filter design is represented by a multi-modal function [3]. Therefore, methods based on a function gradient can easily stick in the local extremum. In last years, many bio-inspired global optimization techniques have been developed. We may mention for example such techniques as: ant colony optimization algorithms for a continuous domain [11], particle swarm optimization algorithms [6, 7, 8], genetic algorithms [4, 5] and differential evolution algorithms [9, 10]. Many of them have been used in digital filters design with non-standard characteristics [13, 14, 20-24] or to solve other problems connected with digital circuits design [12, 17, 18]. Sometimes, when we are not quite certain whether the set of admissible solutions in the search space is non-empty, as in an examination of the problem solvability, we can use constraint programming methods [15, 16] before the application of an adequate optimization algorithm. In this paper, we would like try to find an answer to the question: which is the most efficient bio-inspired global optimization algorithm for the digital filter design? Therefore, we have implemented four most popular bio-inspired global optimization techniques, and next these techniques were used in the design of IIR minimal phase digital filters with non-standard amplitude characteristics. We compared the following algorithms: the ant colony optimization algorithm for a continuous domain, the particle swarm optimization algorithm, the genetic algorithm, and the differential evolution algorithm. When using these algorithms, we designed four IIR minimal phase digital filters with amplitude characteristics: linearly falling, linearly growing, non-linearly falling, and non-linearly growing. Also, it is worth to noticed that we can design any IIR digital filter with any type of non-standard amplitude characteristics using the bio-inspired methods presented in this paper. No comparison is included in this paper between bio-inspired optimization techniques and other optimization methods that are not bio-inspired.

2 IIR Minimal Phase Digital Filters

The IIR minimal phase digital filter is described using the transfer function in the z domain as follows:

$$H(z) = \frac{Y(z)}{X(z)} = \frac{b_0 + b_1 \cdot z^{-1} + b_2 \cdot z^{-2} + ... + b_{n-1} \cdot z^{-(n-1)} + b_n \cdot z^{-n}}{1 - (a_1 \cdot z^{-1} + a_2 \cdot z^{-2} + ... + a_{n-1} \cdot z^{-(n-1)} + a_n \cdot z^{-n})} \quad (1)$$

The symbols are as follows: a_i, b_i - filter coefficients, z^{-1}- delay element, *X(z)* - the vector of input samples in z domain, *Y(z)* - the vector of output samples in z domain, n - filter order.

The main goal of the any optimization algorithm in the digital filter design problem is to find a such a set of digital filter coefficients a_i, and b_i in order to fulfill all the design assumptions. Furthermore, the designed filter should be a stable filter, therefore all the poles of the transfer function should be located in a unitary circle in z domain. Also, the designed filter should be a minimal phase digital filter, therefore all the zeros of the transfer function should be located in the unitary circle in the z plane.

3 Bio-inspired Global Optimization Algorithms

In this section, we presented a short brief concerning four bio-inspired global optimization algorithms which were used in our comparative study. The particular algorithm descriptions are adapted to the problem of the IIR minimal phase digital filters design.

3.1 Ant Colony Optimization Algorithm for Continuous Domain

(1) Table T that consists of T_{max} digital filters is created. For each digital filter, the values of all filter coefficients are randomly selected from range $[-Coeff_{max}, +Coeff_{max}]$.
(2) All the filters from set T are evaluated using the objective function (see section 4).
(3) All the digital filters from set T are sorted. The best digital filter (one that has the lowest value of the objective function) possesses index 1. The worst digital filter (one that has the highest value of the objective function) possesses index T_{max}.
(4) The value of P_i is computed according to the formula:

$$P_i = \frac{1}{q \cdot T_{max} \cdot \sqrt{2 \cdot \pi}} \cdot exp\left(-\frac{(i-1)^2}{2 \cdot q^2 \cdot T_{max}^2}\right) \tag{2}$$

where: P_i is the value of the probability of the selection of the $i-th$ digital filter from set T, q is a coefficient to level out of the worst digital filters, T_{max} is the number of digital filters in set T.

Furthermore, the value of P_i essentially defines the weight to be a value of the Gaussian function with argument i, mean 1.0, and standard deviation $q \cdot T_{max}$. When q is small, the best-ranked solutions are strongly preferred, and when it is large, the probability becomes more uniform. The influence of this parameter on this algorithm (ACO_R [11]) is similar to adjusting the balance between the iteration-best and the best-so-far pheromone updates used in ACO [11].
(5) The artificial ant population that consists of M ants is created. Each ant, based on the value of the probability P_i selects a given digital filter from set T using a roulette selection method. Next, for each filter coefficient from the selected digital filter, the value of the standard deviation is computed according to the formula:

$$\sigma_j = \epsilon \cdot \sum_{e=1}^{T_{max}} \frac{|t_{e,j} - t_{h,j}|}{T_{max} - 1} \tag{3}$$

where: σ_j is the value of standard deviation which is computed for $j-th$ filter coefficient in $h-th$ digital filter, ϵ is the positive real number which increases the sensitivity of the pheromone, t is a digital filter from the set T.

Each artificial ant creates one new digital filter using formula:

$$S_j = \mu_j \cdot randGauss + \sigma_j \tag{4}$$

where: S_j is the value of the $j-th$ filter coefficient for the newly created digital filter S, μ_j is the actual value of the $j-th$ filter coefficient for the selected digital filter, $randGauss$ is the random value from range [0, 1] which is selected using the normal distribution.
(6) Newly created digital filters M are evaluated using the objective function.
(7) The newly created digital filters are added at the end of set T.
(8) After addition, the digital filters in set T are sorted (identically as in the third step). Next, M worst digital filters are removed from set T.
(9) The termination criteria is checked. If the termination criteria is fulfilled, the first digital filter from set T is returned as an algorithm result, and the algorithm is stopped. In other cases, the algorithm jumps to the (4).

3.2 Particle Swarm Optimization Algorithm

(1) T_{max} digital filters are created and stored in set T. The value of each filter coefficient is randomly selected from range $[-Coeff_{max}, +Coeff_{max}]$ for each digital filter stored in set T. Next, for each filter, the coefficient is a randomly selected value of velocity from range $[-V_{max}, +V_{max}]$. The values of all the velocities for all the filters in set T are stored in set V.
(2) The digital filters stored in set T are evaluated using the objective function (see section 4).
(3) For each $i-th$ digital filter, its best solution $Pbest_i$ is selected which has been found so far. Also, the best global solution which has been found so far is remembered in variable *Gbest*.
(4) The new values of velocity are computed according to the formula:

$$newV_i = w \cdot V_i + c1 \cdot r1 \cdot (Gbest - T_i) + c2 \cdot r2 \cdot (Pbest_i - T_i) \tag{5}$$

where: $newV_i$ is a new vector of velocity computed for the $i-th$ digital filter from set T, w is an inertia weight coefficient: $w = 0.5 * (c1 + c2) - 1$, V_i is the actual velocity vector for the $i-th$ digital filter from set T, $c1, c2$ are the values of global and social coefficients, $r1, r2$ are the randomly selected values from range [0, 1]. Equation (5) was developed by Shi and Eberhart [19] to better control the exploration and exploitation properties of the particle swarm optimization algorithm.
(5) Each value of velocity vector V_i for the $i-th$ digital filter is replaced by velocity vector $newV_i$. Next, the new digital filter $newT_i$ is created using the following formula:

$$newT_i = T_i + V_i \tag{6}$$

The newly created digital filter $newT_i$ replaces the old digital filter T_i in set T.
(6) The newly created digital filters are evaluated using the objective function.
(7) The termination criteria is verified. If the termination criteria is fulfilled, the first digital filter stored in variable *Gbest* is returned as an algorithm result, and the algorithm is stopped. In other cases, the algorithm jumps to the (3).

3.3 Genetic Algorithm

(1) T_{max} digital filters are created and stored in set T. The value of each filter coefficient is randomly selected from range $[-Coeff_{max}, +Coeff_{max}]$ for each digital filter stored in set T.
(2) All the digital filters from set T are evaluated using the objective function.
(3) The selection of individuals for the new population is performed. As a selection operator, the roulette selection method has been chosen.
(4) The cross-over of the individuals is performed with the cross-over probability being equal to pc. As a cross-over operator, the single point cross-over operator has been chosen.
(5) The mutation of the individuals is performed with the mutation probability being equal to pm. The simple mutation operator has been used as a mutation operator.
(6) All the digital filters from set T are evaluated using the objective function.
(7) The termination criteria is checked. If termination criteria is fulfilled then the best digital filter (with the lowest value of the objective function) stored in set T is returned as an algorithm result, and next the algorithm is stopped. In other cases the algorithm jumps to the (3).

3.4 Differential Evolution Algorithm

(1) The T_{max} digital filters are randomly created and stored in set T. For each filter coefficient in each digital filter, the initial value is randomly taken from range $[-Coeff_{max}, +Coeff_{max}]$.
(2) All the digital filters stored in set T are evaluated using the objective function.
(3) The best digital filter which has been found so far is remembered in variable T_{r1}, and next vector V_i is generated for each $i-th$ digital filter from set T using the following formula:

$$V_i = T_{r1} + F \cdot (T_{r2} - T_{r3}) \tag{7}$$

where: F is a randomly taken value from range [0, 2], T_{r1} is the best digital filter from set T, T_{r2} and T_{r3} are randomly selected digital filters from set T, V_i is a vector created for digital filter T_i from set T. Also, it is assumed that:

$$T_i \neq T_{r1} \neq T_{r2} \neq T_{r3} \tag{8}$$

(4) The cross-over of digital filter T_i with vector V_i corresponding to it, is performed. As a cross-over operation result, vector U_i is created using the following scheme:

- randomly generate one real number $randj \in [0, 1)$ for each $j-th$ digital filter coefficient from T_i
- if the value of $rand_j$ is lower than the value of CR factor, then $U_{i,j} := V_{i,j}$
- if the value of $rand_j$ is higher or equal to the value of CR factor, then $U_{i,j} := T_{i,j}$
where: CR factor is an algorithm parameter ($CR \in [0, 1)$).

(5) The newly created digital filters U_i are evaluated using the objective function, and next the selection of the newly created digital filters is performed using the following scheme:
- if the objective function value for digital filter U_i is lower than the objective function value for digital filter T_i, replace digital filter T_i with digital filter U_i in set T
- in other cases, leave digital filter T_i in set T.

(6) The termination criteria is verified. If the termination criteria is fulfilled, the digital filter stored in variable T_{r1} is returned as an algorithm result, and the algorithm is stopped. In other cases, the algorithm jumps to the (3).

4 Objective Function

Objective function FC is computed as follows (in equations (9-15), index i represents the i-th digital filter in set T):

$$FC_i = (AmplitudeError_i + 1) \cdot (StabError_i + MinPhaseError_i + 1) - 1 \tag{9}$$

$$AmplitudeError_i = \sum_{k=1}^{R} AmpErr_{i,k} \tag{10}$$

$$AmpErr_{i,k} = \begin{cases} H(f_k)_i - Upper_{i,k}, & when H(f_k)_i > Upper_{i,k} \\ Lower_{i,k} - H(f_k)_i, & when H(f_k)_i < Lower_{i,k} \\ 0, & otherwise \end{cases} \tag{11}$$

$$StabError_i = \sum_{j=1}^{J} StabErr_{i,j} \tag{12}$$

$$StabErr_{i,j} = \begin{cases} |p_{i,j}| - 1, & when\, |p_{i,j}| \geq 1 \\ 0, & otherwise \end{cases} \tag{13}$$

$$MinPhaseError_i = \sum_{q=1}^{Q} PhaseErr_{i,q} \tag{14}$$

$$PhaseErr_{i,q} = \begin{cases} |z_{i,q}| - 1, & when\, |z_{i,q}| \geq 1 \\ 0, & otherwise \end{cases} \tag{15}$$

where: $AmpErr_{i,k}$ is a partial value of the amplitude characteristics error for the k-th value of the normalized frequency, $H(f_k)_i$ is the value of the amplitude characteristics for the k-th value of the normalized frequency f, $Lower_{i,k}$ is the value of the lower constraint for the amplitude characteristics value for the k-th value of the normalized frequency, $Upper_{i,k}$ is the value of the upper constraint for the amplitude characteristics value for the k-th value of the normalized frequency, $StabErr_{i,j}$ is a partial filter stability error for the j-th pole of the transfer function, J is the number of the poles of the transfer function, $|p_{i,j}|$ is the value of the module for the j-th pole of the transfer function, $PhaseErr_{i,q}$ is a partial filter minimal phase error for the q-th zero of the transfer function, Q is the number of zeros of transfer function, $|z_{i,q}|$ is the value of module for q-th zero of the transfer function.

In order to obtain the value of objective function FC for the i-th digital filter in set T, first the amplitude characteristics $H(f)_i$ whose coefficients are stored in the i-th individual is computed. The amplitude characteristics is computed using R values of normalized frequency $f \in [0;1]$ (where 1 represents the Nyquist frequency; in the method proposed, the normalized frequency is divided into R points). Also, the poles and zeros of the transfer function (see Equation 1) are computed for each digital filter in set T. When we have the amplitude characteristics and the values of poles and zeros of the transfer function for the i-th digital filter, we can compute the objective function FC. Equation (9) has been determined experimentally. Of course, at the start of each algorithm, we do not have a guarantee that the digital filter obtained is a stable digital filter. The digital filter is stable and it fulfills all the design assumption when the value of FC is equal to 0.

5 Description of Experiments and Obtained Results

As a test, the four minimal phase digital filters with non-standard amplitude characteristics were designed using each of the algorithms which were described in Section 3. We have assumed the following amplitude characteristics: linearly falling (Figure 1a), linearly growing (Figure 1b), quadratic non-linearly falling (Figure 1c), and quadratic non-linearly growing (Figure 1d). The parameters of these characteristics are presented in Figure 1. Axis x represents the normalized frequency in range [0, 1], where value 1 represents the Nyquist frequency. Axis y represents the attenuation of amplitude characteristics in dB.
The parameters of the algorithms are as follows:

- ant colony optimization algorithm for a continuous domain
$T_{max} = 300$, $Coeff_{max} = 2000$, $\epsilon = 0.085$, $q = 0.1$, $M = 10$
- particle swarm optimization algorithm
$T_{max} = 300$, $Coeff_{max} = 2000$, $V_{max} = 7000$, $c1 = 0.3$, $c2 = 0.2$
- genetic algorithm
$T_{max} = 300$, $Coeff_{max} = 2000$, $pc = 0.5$, $pm = 0.1$
- differential evolution algorithm
$T_{max} = 300$, $Coeff_{max} = 2000$, F is a random value from the range [0, 2], $CR = 0.2$

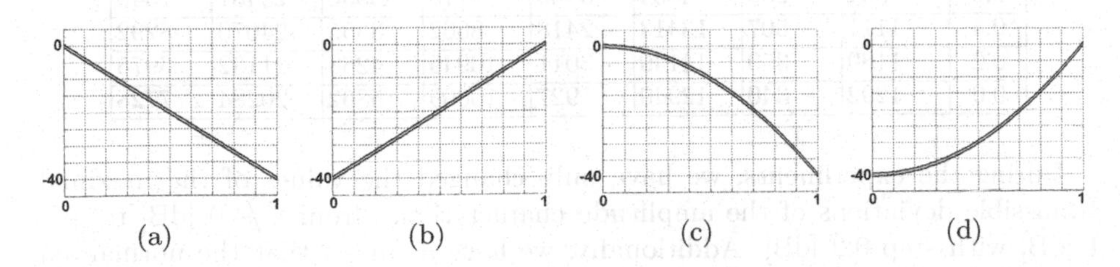

Fig. 1. Shape of amplitude characteristics of designed digital filters: linearly falling (a), linearly growing (b), non-linearly falling (c), and non-linearly growing (d)

Table 1. The results obtained for two linear amplitude characteristics

The results obtained for linearly falling amplitude characteristics								
ADev	DE		GA		PSO		ACO_R	
[dB]	Average	StdDev	Average	StdDev	Average	StdDev	Average	StdDev
1.0	366	357	9536	882	6173	1299	12685	5049
0.8	722	316	9697	719	6367	1337	14826	8330
0.6	750	320	11029	2220	6430	1280	13541	4077
0.4	832	206	10838	2069	6112	1060	14292	5029
0.2	710	222	10061	2133	7123	2912	9902	4218
0.0	1136	337	10965	2547	6719	2556	14291	5621
The results obtained for linearly growing amplitude characteristics								
ADev	DE		GA		PSO		ACO_R	
[dB]	Average	StdDev	Average	StdDev	Average	StdDev	Average	StdDev
1.0	460	236	9629	641	5994	3733	15681	4743
0.8	494	310	10073	1502	6823	3385	15651	3949
0.6	647	286	10953	2420	5362	1903	14784	6843
0.4	659	176	10759	2010	7462	2801	16496	5735
0.2	748	220	9744	622	7769	3261	18813	4299
0.0	866	311	10238	1221	6621	2627	16137	5734

Table 2. The results obtained for two non-linear amplitude characteristics

The results obtained for non-linearly falling amplitude characteristics								
ADev	DE		GA		PSO		ACO_R	
[dB]	Average	StdDev	Average	StdDev	Average	StdDev	Average	StdDev
1.0	705	323	7177	636	4332	1351	8489	3578
0.8	1036	318	7471	918	3451	1637	12102	3259
0.6	1073	320	7251	1056	4222	2550	10913	5008
0.4	1092	320	7754	1716	4270	1275	7482	3283
0.2	888	302	7410	1823	3954	905	10917	3860
0.0	849	137	7830	1163	4075	1735	9327	1964
The results obtained for non-linearly growing amplitude characteristics								
ADev	DE		GA		PSO		ACO_R	
[dB]	Average	StdDev	Average	StdDev	Average	StdDev	Average	StdDev
1.0	753	310	13728	3180	7651	3029	17921	7209
0.8	907	257	13084	2820	9698	2349	17864	5031
0.6	954	260	14334	3355	9710	3936	23487	7549
0.4	1111	507	13413	2418	8662	3655	24070	8292
0.2	1189	350	14140	3011	10949	4268	21132	6975
0.0	1292	340	12940	927	10069	5862	20284	7728

During the experiments, we have only changed the values of the maximal admissible deviations of the amplitude characteristics from +/- 0 [dB] to +/- 1 [dB] with step 0.2 [dB]. Additionally, we have assumed that the normalized frequency was divided into 512 points (R=512), and that 8-th order IIR digital filters are designed (n=8). The execution time for each algorithm was equal to 3 minutes.

In tables 1-2, the average values of the objective function (from the 10-fold repetition) together with standard deviations were presented for each algorithm, and for a given kind of the shape of amplitude characteristics.

The symbols in Tables 1-2 are as follows: *ADev* - acceptable deviations of amplitude characteristics, *DE* - differential evolution algorithm, *GA* - genetic algorithm, *PSO* - particle swarm optimization algorithm, ACO_R - ant colony optimization algorithm for a real domain, *Average* - the average value of the objective function from the 10-fold repetition of a given algorithm, *StdDev* - standard deviation.

6 Conclusions

In this paper, a comparative study on bio-inspired global optimization algorithms in minimal phase digital filters design has been presented. Based on the results which were shown in Tables 1-2, we can see that for all the cases, the results obtained using differential evolution algorithm (DE) are best. The worst results were obtained for the ant colony optimization algorithm for a real domain (ACO_R). We think that this is related with the fact that the ACO_R algorithm is the most complex algorithm as compared with the four algorithms presented. Therefore, during the same computational time, the lowest number of iterations is computed. In our future research, we want to study different variations of the differential evolution algorithm in the application to the problem of the minimal phase digital filters design.

References

1. Shenoi, B.A.: Introduction to Digital Signal Processing and Filter Design. John Wiley & Sons, New Jersey (2006)
2. Lyons, R.G.: Understanding Digital Signal Processing. Prentice Hall (2004)
3. Chen, S., Istepanian, R.H., Luk, B.L.: Digital IIR filter design using adaptive simulated annealing. Digital Signal Processing 11(3), 241–251 (2001)
4. Michalewicz, Z.: Genetic Algorithms + Data Structures = Evolution Programs. Springer, Heidelberg (1992)
5. Goldberg, D.E.: Genetic Algorithms in Search, Optimization, and Machine Learning. Addison-Wesley Publishing Company Inc. (1989)
6. Eberhart, R.C., Kennedy, J.: A new optimizer using particle swarm theory. In: Proceedings of the Sixth International Symposium on Micromachine and Human Science, Nagoya, Japan, pp. 39–43 (1995)
7. Kennedy, J., Eberhart, R.C.: Particle swarm optimization. In: Proceedings of IEEE International Conference on Neural Networks, Piscataway, NJ, pp. 1942–1948 (1995)
8. Kennedy, J., Eberhart, R.C., Shi, Y.: Swarm intelligence. Morgan Kaufmann Publishers, San Francisco (2001)
9. Rainer, S., Price, K.: Differential Evolution - A Simple and Efficient Heuristic for Global Optimization over Continuous Spaces. Journal of Global Optimization 11, 341–359 (1997)

10. Price, K.: An Introduction to Differential Evolution. In: Corne, D., Dorigo, M., Glover, F. (eds.) New Ideas in Optimization, pp. 79–108. McGraw-Hill, London (1999)
11. Socha, K., Doringo, M.: Ant colony optimization for continous domains. European Journal of Operational Research 185(3), 1155–1173 (2008)
12. Słowik, A., Białko, M.: Partitioning of VLSI Circuits on Subcircuits with Minimal Number of Connections Using Evolutionary Algorithm. In: Rutkowski, L., Tadeusiewicz, R., Zadeh, L.A., Żurada, J.M. (eds.) ICAISC 2006. LNCS (LNAI), vol. 4029, pp. 470–478. Springer, Heidelberg (2006)
13. Slowik, A.: Application of Evolutionary Algorithm to Design of Minimal Phase Digital Filters with Non-Standard Amplitude Characteristics and Finite Bits Word Length. Bulletin of The Polish Academy of Science - Technical Science 59(2), 125–135 (2011), doi:10.2478/v10175-011-0016-z
14. Slowik, A., Bialko, M.: Design and Optimization of IIR Digital Filters with Nonstandard Characteristics Using Continuous Ant Colony Optimization Algorithm. In: Darzentas, J., Vouros, G.A., Vosinakis, S., Arnellos, A. (eds.) SETN 2008. LNCS (LNAI), vol. 5138, pp. 395–400. Springer, Heidelberg (2008)
15. Bocewicz, G., Wójcik, R., Banaszak, Z.: AGVs distributed control subject to imprecise operation times. In: Nguyen, N.T., Jo, G.-S., Howlett, R.J., Jain, L.C. (eds.) KES-AMSTA 2008. LNCS (LNAI), vol. 4953, pp. 421–430. Springer, Heidelberg (2008)
16. Bocewicz, G., Banaszak, Z.: Declarative modeling of multimodal cyclic processes. In: Golinska, P., Fertsch, M., Marx-Gomez, J. (eds.) Information Technologies in Environmental Engineering. Environmental Science and Engineering - Environmental Engineering, vol. 3, pp. 551–568. Springer, Heidelberg (2011)
17. Ratuszniak, P.: Processor Array Design with the Use of Genetic Algorithm. In: Lirkov, I., Margenov, S., Waśniewski, J. (eds.) LSSC 2011. LNCS, vol. 7116, pp. 238–246. Springer, Heidelberg (2012)
18. Słowik, A., Białko, M.: Design and Optimization of Combinational Digital Circuits Using Modified Evolutionary Algorithm. In: Rutkowski, L., Siekmann, J.H., Tadeusiewicz, R., Zadeh, L.A. (eds.) ICAISC 2004. LNCS (LNAI), vol. 3070, pp. 468–473. Springer, Heidelberg (2004)
19. Shi, Y., Eberhart, R.C.: A Modified Particle Swarm Optimizer. In: Proceedings of the IEEE Congress on Evolutionary Computation, pp. 69–73 (May 1998)
20. Karaboga, N.: Digital IIR filter design using differential evolution algorithm. EURASIP Journal on Applied Signal Processing 8, 1269–1276 (2005)
21. Karaboga, N., Cetinkaya, B.: Performance comparison of genetic and differential Evolution algorithms for digital FIR filter design. In: Yakhno, T. (ed.) ADVIS 2004. LNCS, vol. 3261, pp. 482–488. Springer, Heidelberg (2004)
22. Karaboga, N., Kalinli, A., Karaboga, D.: Designing digital IIR filters using ant colony optimisation algorithm. Engineering Applications of Artificial Intelligence 17(3), 301–309 (2004)
23. Dai, C., Chen, W., Zhu, Y.: Seeker optimization algorithm for digital iir filter design. IEEE Transactions on Industrial Electronics 51, 1710–1718 (2010)
24. Karaboga, N.: A new design method based on artificial bee colony algorithm for digital iir filters. Journal of the Franklin Institute 346, 328–348 (2009)

On Modelling Social Propagation Phenomenon

Dariusz Król[1,2]

[1] Smart Technology Research Centre, Faculty of Science and Technology, Bournemouth University, United Kingdom
[2] Institute of Informatics, Wrocław University of Technology, Poland
Dariusz.Krol@pwr.wroc.pl

Abstract. Propagation phenomenon is an important problem that has been studied within varied research fields and application domains, leading to the development of propagation based models and techniques in social informatics. These models are briefly surveyed in this paper. This paper discusses common features and two selected scenarios of propagation mechanisms that frequently occur in social networks. In summary, a list of the most recent open issues on social propagation is presented.

Keywords: cascading behaviour, information diffusion, spreading model.

1 Introduction

Propagation phenomona have become a pervasive and significant feature of contemporary complex networks. By studying these phenomena, we can better understand the behaviour cascading in a system. Social networks based on e-mail contacts like *Hotmail*, *Yahoo* and social networking services like *Facebook*, *Twitter*, *YouTube*, for example, are generally modelled as complex structures whose nodes represent individuals, and whose links represent interaction, collaboration, or influence between them. They are designed for information sharing and above all for information spreading. As an obvious result, we can observe various propagation effects. Here we ask, what are the practical consequences of this day-to-day existence and interaction.

Propagation mechanisms could be very useful, depending on the different types of communication networks and their purpose. For instance in online networks, there are several popular methods frequently built on 'positive' spreading mechanism such as: capturing structures [1], detecting local communities [2], predicting future network structure and user's behaviours [3], discovering rumors [4], tracking collective blogging and argumentations [5], monitoring opinions [6], recommending trust and measuring influence of authorities [7], and finally maximizing immunization of the network [8]. On the other hand, propagation in the form of gossiping [9], misinformation [10], fraudulent reputation building or targeted cyber-attacks [11] could have extremely serious 'negative' effects. What is more, the in-depth understanding of propagation phenomenon could enable early identification of social crises, and provide methods to mitigate potential catastrophes.

N.T. Nguyen et al. (Eds.): ACIIDS 2014, Part II, LNAI 8398, pp. 227–236, 2014.

From a scientific point of view a lot of paramount questions w.r.t. social propagation have been raised and analysed. A few survey works on propagation mainly in online social networks have already been done [12, 13]. However, important questions about how to best fit a propagation model and efficiency indicators are still relevant. Due to lack of other relevant analyses our work provides some steps toward this direction. Though we do not present a full survey, the aim of this paper is an attempt to respond to realistic features of propagation issues that the social networks confront. The results aim to aid researchers and engineers involved in the development of new technologies to better understand propagation phenomena principles, its power and limitations.

From a business perspective, that might be useful for many services to aptly adopt the propagation mechanism to make the information (action) spreading faster and wider. Accurate pushing-pulling selected information to cooperators and business partners can speed-up communication and significantly lower the load costs. On the other hand, good knowledge of propagation phenomenon could be helpful to stop diffusing the undesirable things which can threaten individuals or organizations.

There are three major parts concerning modelling of social propagation: the major principles, an appropriate model, and finally a generic algorithm and its variants. In this paper, our goal is to find answer to the following problems: what properties of social propagation are essential for efficiency and robustness, and how a propagation process can directly correlate with time and underlaying structure. To the best of our knowledge, these questions have not been addressed by the research community at large.

2 Principles of Social Propagation Mechanisms

What are the major principles associated with propagation phenomena in social informatics?

The general idea behind social propagation is that individuals interact repeatedly under the reciprocal influence of other individuals, modelled as a *social influence score*, which may often generate a flood of action across the connections. Elementary social examples include rumor dissemination, forwarding memes, joining events or groups, hashtags re-tweeting, and purchasing products. We have investigated these situations in detail, and our main observations are summarized in Table 1.

2.1 The Classical Approach to Propagation

Many propagation models, also called diffusion models [22], have been studied to date. To describe social and biological propagation, for instance, for spread of infectious diseases and computer viruses, the diffusion of innovations, political upheavals, and the dissemination of religious doctrine, the generalized contagion model was developed [23]. This non-graph based model exhibits behaviour that falls into one of two basic classes called respectively: epidemic threshold, and

Table 1. Principles of social propagation mechanisms

Topology-dependent. Propagation utilizes to the utmost the underlying structure of a network. Most social networks are characterized, among other qualities, with the *'small-world' phenomenon*, *triadic closure*, *assortative mixing* and *broad degree distributions*. Recently, more fascinating properties have been discovered, like over time *shrinking diameters*, *homophily tendency*, *aging effect*, *temporal dynamics* and the so-called *densification power law* [14], including advanced graph theoretic measures [15]. In the vast majority these features can significantly speed up the diffusion process.
Complexity. Social data are often interconnected, have overlapping communities and are coupled across time through processes. This evolving idea has been successfully implemented by *Google+ circles* and *Facebook smart lists*. When nodes in one network depend on nodes in another, a small disturbance in one network can cascade through the entire system often growing to sufficient size to collapse it [16].
Fast influence by ties. Propagation extends Granovetter's hypothesis that the majority of influence is collectively generated by weak ties, although strong ties are individually more influential [17]. Very often action spreads only partially overlapping the network, but extremely fast (in sublogarithmic time [18]).
Push-pull activity. For social propagation mechanisms, not only the topology of links, but also the communication activity is highly relevant [18]. The standard protocol is based on symmetric push-pull activity. It pushes the information in case it has, and pulls the information in case the neighbor has. A common form of positive and negative propagation is spreading of breaking news, rumors, fads, beliefs, sentiments and norms. While positive users' opinions promote an action, negative opinions suppress its adoption.
Survive or perish. Social informatics are ubiquitous and this applies also to propagation. Individuals tend to adopt the behavior of their peers, so first propagation happens locally in their neighborhood. Then, this behaviour might become global and survive or decay and finally perish as it crosses the network [19].
Epidemic-type and cascading-type. Propagation can be conceptualized either as an *epidemic-type dynamic*, where a node in an infected state may infect a neighbor independently of the status of the other nodes [20], or as a *cascading-type dynamic*, where the change may depend on the reinforcement caused by changes in other nodes [21].

critical mass. The first class is based on the idea of a threshold: an adoption depends on the fraction of neighbors which exceed a specific critical value. In the second class the population can be infected if the earliest outbreak size makes up a 'critical mass'. This contagion model can be identified with two seminal models for the spread of disease: the so-called $susceptible \rightarrow infective \rightarrow recovered$ (SIR) model and simplified $susceptible \rightarrow infective \rightarrow susceptible$ (SIS) model [24]. As a rule, all nodes in SIR model are in one of three states: susceptible (able to be infected), infected, or recovered (no longer able to infect or be infected). These three classes in specific rumor model correspond to ignorant, spreader, and stifler nodes. At each time step, nodes infected in the last time step can infect any of its neighbors who are in a susceptible state with a probability p. In this model, two transition rates $S \rightarrow I$ (the contact rate) and $I \rightarrow R$ (the rate of recovery) determine the cumulative number of infected nodes.

Many others model variations have been studied for social informatics. However, the two most widely employed models are: the Independent Cascade Model (ICM) and the Linear Threshold Model (LTM) [25]. In these graph-based models, each node is either active or inactive. An active node never becomes inactive again. The propagation process repeats until no more activations are possible. The spread represented in the ICM and LTM are very similar and proceeds in discrete steps as follows.

In the ICM, a process starts with an initial set of active nodes A_0, called later seed nodes. It is assumed that nodes can switch their states only from inactive to active, but not in the opposite direction. When a node i first becomes active in step t, it has a single chance of influencing each inactive neighbor j with probability $p(i,j)$. If node i succeeds, node j becomes active in step $t+1$. If node j has incoming links from a few newly activated nodes, every propagation effort is randomly sequenced. When all the influence probabilities are equal to one, the ICM becomes equivalent to the deterministic model, in which every active node unconditionally activates all its neighbors.

In the LTM, each node i is described by a threshold θ_i from the range $[0,1]$. This threshold represents the fraction of i's neighbors, denoted by $deg^{-}(i)$, that must be active in order for node i to become active. Here 'active' means that the node adopts the action and 'inactive' means otherwise. Given the thresholds and seed nodes A_0, the process unfolds deterministically in discrete steps. In step $t+1$, all nodes that were active in step t remain active, and we activate any node j for which the total influence power of his active neighbors is at least $\theta_j : \sum p(i,j) \geqslant \theta_j$. Thus, the threshold θ_j represents the trend of a node j to adopt the action when his neighbors do.

Both models have parameter attached to each directional link, i.e., propagation probability in the ICM and weight in the LTM. In the ICM-based virus-spreading model, the probability of being infected is proportional to the number of neighbors infected. Although both models appeared to be comparative, there are important differences. Intuitively, only a small number of nodes overlap for these models. Furthermore, it is more difficult for the LTM to transmit action to hub nodes than the ICM does.

2.2 Our Generic Approach

With respect to existing propagation models and aforementioned six principles, this directs us to the following generic model definition. A propagation $\wp(\mathfrak{S}, G)$ with action $\mathfrak{S}$ over the network is described by a directed connected graph $G = (N, L)$ where N is the node set and L is the set of directed links containing connected pairs of nodes (i,j) unfolds in discrete time-steps $t \geqslant 0$ and is defined as an ordered sequence of triplets (i,j,t). Each triplet corresponds to a single interaction event at time-step t between a pair (i,j), where $i,j = 1..N$. We assume that no individual adopts the same action more than once. In Twitter, for instance, user j adopts tweet posted by i at time t and re-tweets it, then the tweet becomes available to his followers. The total number of triplets is given by $|\wp(\mathfrak{S}, G)|$. At the empirical level, propagation with action over the network is a set of interaction events recorded every δ (the sampling resolution) during an interval Δt. This leads us to the notion of the propagation graph. The corresponding static propagation graph $G_{\wp}^{\Delta t}$ is obtained by aggregating all the same interactions events within Δt into a graph structure with links connecting in the direction of propagation. This propagation graph is now a directed acyclic graph (DAG) which may have disconnected components.

To facilitate the propagation dynamics, particularly its transitivity, in order to merge propagations along each path, to select one out of multiple propagations, or to combine propagations from all paths, a set of primitives based on path algebra [26] that effectively operate on a given graph should be formalized. For instance, for propagating trust presented as a triple $(belief, distrust, uncertainty)$ three associative operators: *concatenation*, *selection*, and *aggregation* are analysed. The description of these operators goes beyond the scope of this paper, however, the interested reader is referred to [27] for details.

Note that our basic model fulfils the requirements of both models: ICM and LTM. Contrary to the classic models, instead of considering sets of nodes, links and probabilities separately, the proposed model focuses on time-stamped propagation triplets, leading to a reduction of the complexity in propagation scenarios. Our proposal seems to be adjusted to any action-propagation model, including variations on compartmental models in epidemiology.

3 The Propagation Algorithm

How does the propagation process normally proceed?
Propagation $\wp(\mathcal{G}, G)$ in the form of triplets shows pathways for the transmission of infection, rumor, trust, or other quantities. Common algorithms for crawling or sampling social networks include, first of all, variants of breadth-first search and depth-first search. Using only social ties and forward paths we do not necessarily crawl an entire network. We only explore the connected component reachable from the set of seed nodes. The random walk algorithm and snowball sampling [19] for directed graph are the simplest implementations for propagation process by passing action from one node to its neighbor. There are numerous variations like following a randomly selected triadic node (friend, casual friend or grandparent). Another possible implementation is the preferential walk to follow someone from whom or through whom they have adopted actions using a back-propagation mechanism. It has been found in [22] that a combined strategy with triadic closure and coincided traffic-based shortcuts yield the best accuracy.

A general propagation algorithm has initialization including seed node(s) selection, propagation loop with threshold condition, output update and termination condition(s). By applying different social-based loop and influence conditions, we can consider various strategies. The choice of the strategy is generally dependent on the communication flow schema like (1) push and (2) pull. The first one accommodates the propagation along a sender-centered approach such as ICM, the second, receiver-centered approach such as LTM, which combines the propagation from different nodes.

Recall that a threshold value is associated with each node (i.e., social entity or person). This value represents the positive or negative influence of that individual. To get a better understanding we illustrate the key parts of propagation advancement with the aid of Algorithm 1. In the algorithm, the power to influence neighbors is modelled as a propagation probability denoted by $p(i, j)$. Note that this probability p could be time-varying t and algorithm α dependent as follows $p(i, j) = \alpha^t(i, j)$. In every step t, each node i belongs to one of the three sets: *waiting*, *active* and *inactive*. For readability, we omit the *inactive* set from our algorithm. The algorithm terminates, if the current time-step t reaches the time limit T, or there are no more *waiting* nodes, which mean that no more activations are possible.

Input: graph G, threshold θ_i for node, seed nodes A_0, time limit T
Output: *triplets*

$active, triplets \leftarrow \varnothing$;
$waiting \leftarrow A_0$;
$t \leftarrow 0$;
while *termination condition not satisfied* **do**
  foreach *node* $i \in waiting$ **do**
    $t \leftarrow t + 1, waiting \leftarrow waiting \backslash \{i\}, active \leftarrow active \cup \{i\}$;
    `propagation loop` with `threshold condition`
  end
end

Algorithm 1. Generic propagation algorithm

In the algorithm, we use `propagation loop` with `threshold condition`, that determines how to select next node to activate it in the process to maximize the spread. In the push variant of propagation loop and condition (Algorithm 2), propagator one-to-many using deg^+ operator is required, e.g., sending photo to all my group members.

foreach *node* $j \in deg^+(i)$ **do**
  if $\alpha^t(i, j) \geqslant \theta_j \wedge j \notin active, waiting$ **then**
    $waiting \leftarrow waiting \cup \{j\}$;
    $triplets \leftarrow triplets \cup \{i, j, t\}$;
  end
end

Algorithm 2. Push variant of propagation loop with threshold condition

In the pull variant, see Algorithm 3, each node's tendency to become active increases monotonically as more of its neighbors become active. This time the propagator requires a many-to-one operator using deg^- which corresponds to a decision taken by an expert committee.

foreach *node* $j \in deg^-(i)$ **do**
  if $\sum_{k \in deg^-(j) \wedge k \in active} \alpha^t(k,j) \geqslant \theta_j \wedge j \notin active, waiting$ **then**
    $waiting \leftarrow waiting \cup \{j\}$;
    $triplets \leftarrow triplets \cup \{i,j,t\}$;
  end
end

Algorithm 3. Pull variant of propagation loop with threshold condition

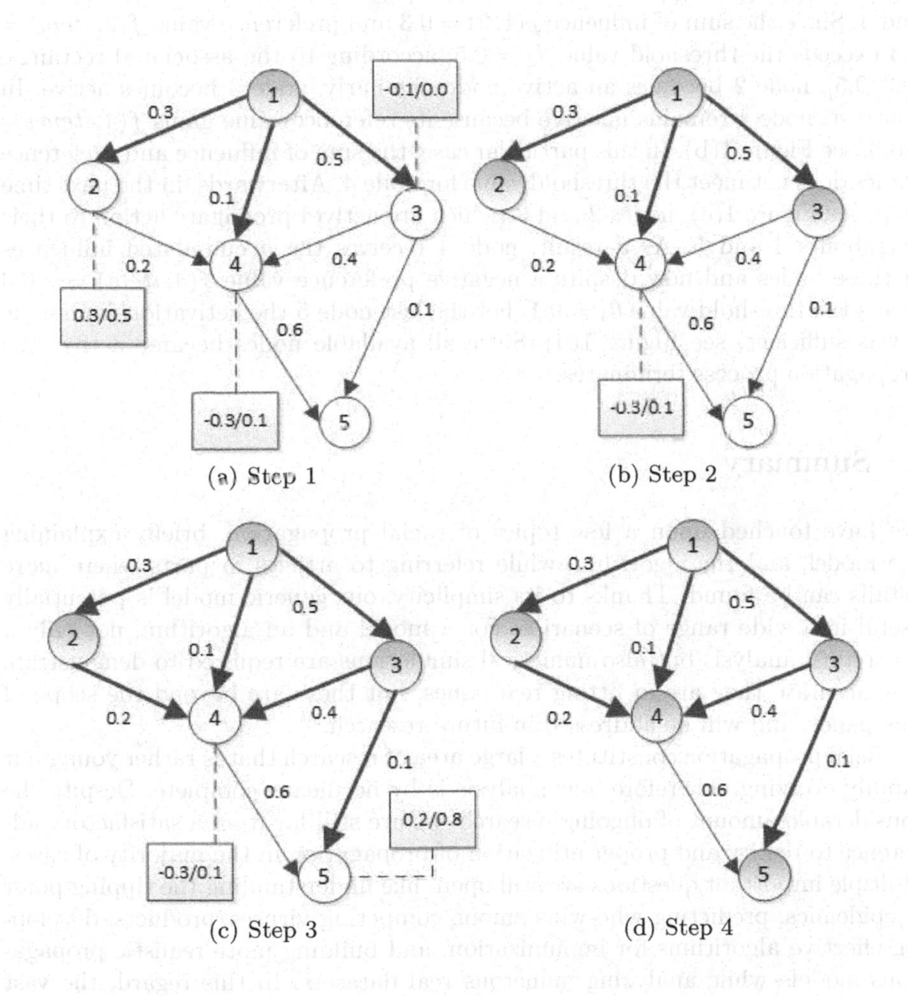

Fig. 1. Preference-pull variant of propagation algorithm

In order to show the diversity of threshold conditions we employ one more example. The preference variant of condition could incorporate the factor positive/negative influence $f(i,m)$ of node i corresponding to propagated item m, where $f(i,m) \in [-1,1]$. This idea of using preferences is motivating by bundling multiple items in viral marketing. Then, in the pull case, a node i is activated when the sum of the preference scores on the propagated items $\sum_{m=1}^{k} f(j,m)$ and the influence scores from a non-empty set of neighbour active nodes, is greater or equal to the threshold θ_i. In addition, the influence factor f can be changed by using a temporal decay function as in [19]. Additionally, we can immunize selected nodes by adding them to *inactive* set.

In order to have a better understanding of the propagation algorithm, we capture the dynamics of it's pull variant with preference in a step-by-step fashion. As shown in Figure 1(a), the seed node 1 propagates action to all neighbours 2, 3 and 4. Since the sum of influence $p(1,2) = 0.3$ and preference value $f(2, item) = 0.3$ exceeds the threshold value $\theta_2 = 0.5$, according to the associated rectangle [0.3/0.5], node 2 becomes an active node. Similarly, node 3 becomes active. In contrast, node 4 remains inactive because its reference value gains $f(4, item) = -0.3$, see Figure 1(b). In this particular case, the sum of influence and preference values does not meet the threshold value for node 4. Afterwards, in the next time step, see Figure 1(c), nodes 2 and 3 (which are active) propagate action to their neighbours 4 and 5. As a result, node 4 receives the accumulated influences of three nodes and now despite a negative preference value $f(4, item) = -0.3$ meets the threshold value $\theta_4 = 0.1$. For the last node 5 the activation from node 3 was sufficient, see Figure 1(d). Since all available nodes became active, the propagation process terminates.

4 Summary

We have touched upon a few topics of social propagation, briefly explaining the model, and the algorithm, while referring to articles in parts where more details can be found. Thanks to its simplicity, our generic model is potentially useful in a wide range of scenarios. For a model and an algorithm, not only a theoretical analysis but also numerical simulations are required to demonstrate how accurate they are in fitting real issues, but these are beyond the scope of this paper, and will be addressed in future research.

Social propagation constitutes a large area of research that is rather young but rapidly evolving. Therefore, our analysis is by no means complete. Despite the considerable amount of ongoing research, we are still far from a satisfactory adherence to reality and proper utilization of propagation in the majority of cases. Multiple important questions are still open, like understanding the tipping point of epidemics, predicting who-wins among competing viruses/products, developing effective algorithms for immunization, and building more realistic propagations models while analyzing numerous real datasets. In this regard, the vast number of propagation techniques, which still remain unexplored, need to be thoroughly investigated in order to improve propagation capabilities and enhance system's efficiency.

One interesting but to date unsolved problem is how to learn on the fly the time-varying elements of propagation by mining the present and archived log of past propagations. The simple and insufficient algorithm for capturing the influence probabilities among the nodes and the prediction time by which an influenced node will perform an action after its neighbors have performed the action is presented in [28].

Another open issue is the low efficiency of greedy algorithms to compute the influence maximization problem to large networks. Even with recent optimizations, it still takes several hours. Improving the greedy algorithm is difficult, so this leads to a second possibility - the quest for appropriate heuristic.

Social propagation occurs in various forms. Some of them, like social influence mining, community detection, locating and repairing faults, finding effectors and maximizing influence, constitute the key components that enable useful insights into network behaviour and developing future services.

We hope that this paper provides several advanced points for engineers and scientists to use the propagation methods in more effective manner. However, it is possible to exploit appropriately this potential only after making further investigations on real data what we plan soon.

Acknowledgments. This research was supported by a Marie Curie Intra European Fellowship within the 7th European Community Framework Programme under grant FP7-PEOPLE-2010-IEF-274375-EPP.

References

1. Fu, X., Wang, C., Wang, Z., Ming, Z.: Threshold random walkers for community structure detection in complex networks. Journal of Software 8(2) (2013)
2. Barbieri, N., Bonchi, F., Manco, G.: Cascade-based community detection. In: Proceedings of the 6th International Conference on Web Search and Data Mining, WSDM 2013, pp. 33–42. ACM, New York (2013)
3. Kim, H., Tang, J., Anderson, R., Mascolo, C.: Centrality prediction in dynamic human contact networks. Comput. Netw. 56(3), 983–996 (2012)
4. Doerr, B., Fouz, M., Friedrich, T.: Why rumors spread so quickly in social networks. Commun. ACM 55(6), 70–75 (2012)
5. Zhao, L., Guan, X., Yuan, R.: Modeling collective blogging dynamics of popular incidental topics. Knowledge and Information Systems 31(2), 371–387 (2012)
6. Garg, P., King, I., Lyu, M.R.: Information propagation in social rating networks. In: Proceedings of the 21st ACM International Conference on Information and Knowledge Management, CIKM 2012, pp. 2279–2282. ACM, New York (2012)
7. Chen, Y.C., Peng, W.C., Lee, S.Y.: Efficient algorithms for influence maximization in social networks. Knowledge and Information Systems 33(3), 577–601 (2012)
8. Gao, C., Liu, J., Zhong, N.: Network immunization and virus propagation in email networks: experimental evaluation and analysis. Knowledge and Information Systems 27(2), 253–279 (2011)
9. Liu, D., Chen, X.: Rumor propagation in online social networks like twitter – a simulation study. In: 2011 Third International Conference on Multimedia Information Networking and Security (MINES), pp. 278–282 (2011)
10. Nguyen, N.P., Yan, G., Thai, M.T., Eidenbenz, S.: Containment of misinformation spread in online social networks. In: Proceedings of the 3rd Annual ACM Web Science Conference, WebSci 2012, pp. 213–222. ACM, New York (2012)

11. Piraveenan, M., Uddin, S., Chung, K.: Measuring topological robustness of networks under sustained targeted attacks. In: IEEE/ACM International Conference on Advances in Social Networks Analysis and Mining (ASONAM), pp. 38–45 (2012)
12. Bonchi, F.: Influence propagation in social networks: A data mining perspective. IEEE Intelligent Informatics Bulletin 12(1), 8–16 (2011)
13. Sun, J., Tang, J.: A survey of models and algorithms for social influence analysis. In: Aggarwal, C.C. (ed.) Social Network Data Analytics, pp. 177–214. Springer US (2011)
14. Leskovec, J., Kleinberg, J., Faloutsos, C.: Graphs over time: densification laws, shrinking diameters and possible explanations. In: Proceedings of the 11th Int. Conf. on Knowledge Discovery and Data Mining, KDD 2005, pp. 177–187. ACM (2005)
15. Fay, D., Haddadi, H., Thomason, A., Moore, A.W., Mortier, R., Jamakovic, A., Uhlig, S., Rio, M.: Weighted spectral distribution for internet topology analysis: theory and applications. IEEE/ACM Trans. Netw. 18(1), 164–176 (2010)
16. Buldyrev, S.V., Parshani, R., Paul, G., Stanley, H.E., Havlin, S.: Catastrophic cascade of failures in interdependent networks. Nature 464(7291), 1025–1028 (2010)
17. Granovetter, M.: Threshold models of collective behavior. American Journal of Sociology 83(6), 1420–1443 (1978)
18. Doerr, B., Fouz, M., Friedrich, T.: Social networks spread rumors in sublogarithmic time. In: Proceedings of the 43rd Annual ACM Symposium on Theory of Computing, STOC 2011, pp. 21–30. ACM, New York (2011)
19. Castellano, C., Fortunato, S., Loreto, V.: Statistical physics of social dynamics. Rev. Mod. Phys. 81, 591–646 (2009)
20. Shah, D., Zaman, T.: Detecting sources of computer viruses in networks: theory and experiment. SIGMETRICS Perform. Eval. Rev. 38(1), 203–214 (2010)
21. Borge-Holthoefer, J., Banos, R.A., Gonzalez-Bailon, S., Moreno, Y.: Cascading behaviour in complex socio-technical networks. Journal of Complex Networks 1(1), 3–24 (2013)
22. Weng, L., Ratkiewicz, J., Perra, N., Goncalves, B., Castillo, C., Bonchi, F., Schifanella, R., Menczer, F., Flammini, A.: The role of information diffusion in the evolution of social networks. In: Proceedings of the 19th International Conference on Knowledge Discovery and Data Mining, KDD 2013. ACM (2013)
23. Dodds, P.S., Watts, D.J.: A generalized model of social and biological contagion. Journal of Theoretical Biology 232(4), 587–604 (2005)
24. Jacquez, J.A., O'Neill, P.: Reproduction numbers and thresholds in stochastic epidemic models. Math. Biosciences 107(2), 161–186 (1991)
25. Kempe, D., Kleinberg, J., Tardos, E.: Maximizing the spread of influence through a social network. In: Proceedings of the 9th International Conference on Knowledge Discovery and Data Mining, KDD 2003, pp. 137–146. ACM (2003)
26. Rodriguez, M.A., Neubauer, P.: A path algebra for multi-relational graphs. In: Proceedings of the 27th International Conference on Data Engineering Workshops, ICDEW 2011, pp. 128–131. IEEE Computer Society, Washington, DC (2011)
27. Hang, C.W., Wang, Y., Singh, M.P.: Operators for propagating trust and their evaluation in social networks. In: Proceedings of the 8th International Conference on Autonomous Agents and Multiagent Systems, AAMAS 2009, pp. 1025–1032 (2009)
28. Goyal, A., Bonchi, F., Lakshmanan, L.V.: Learning influence probabilities in social networks. In: Proceedings of the 3rd ACM International Conference on Web Search and Data Mining, WSDM 2010, pp. 241–250. ACM, New York (2010)

Design of a Performance Analyzer for Electric Vehicle Taxi Systems*

Junghoon Lee[1], Chan Jung Park[2], and Gyung-Leen Park[1,**]

[1] Dept. of Computer Science and Statistics,
[2] Dept. of Computer Education,
Jeju National University, Republic of Korea
{jhlee,cjpark,glpark}@jejunu.ac.kr

Abstract. This paper designs a performance analysis framework for electric vehicle taxis, aiming at promoting their wide deployment. Consisting of an event tracker, a stream handler, object interfaces, and strategy integrator, the analysis procedure can measure the performance of a dispatch and relocation strategy in terms of dispatch latency, customer waiting time, and the number of daily fast charging operations. Each pick-up and drop-off record from the actual call taxi system is associated with the corresponding taxi and charger object. It can host a new dispatch strategy to test and revise, while a specific road network and a future demand prediction model can be incorporated for better accuracy. This framework finds out that most battery charging can be done using slow chargers through the out-of-service intervals under the control of an intelligent coordinator for the fleet of member taxis, avoiding the significant increase in power load brought by fast charging operations.

Keywords: Electric vehicle, taxi business, performance analysis framework, dispatch-relocation strategy.

1 Introduction

Smart transportation, one of the major areas in smart grid, is pursuing energy efficiency in the transport system. Electric vehicles, or EVs in short, are its key element and get energy from battery-stored electricity. As they can avoid burning fossil fuels, it is possible to significantly reduce air pollution and greenhouse gas emissions. However, their driving range is less than 120 km in practice and it takes about 6 $\sim$ 7 hours to fully charge an EV battery with slow chargers and about 20 minutes to charge up to 80 % with fast chargers [1]. Here, due to high voltage injection, fast chargers can shorten the battery life and it is recommended that fast charging is done at most once a day. Just like other smart grid entities

* This research was financially supported by the Ministry of Trade, Industry and Energy (MOTIE), Korea Institute for Advancement of Technology (KIAT) through the Inter-ER Cooperation Projects.
** Corresponding author.

N.T. Nguyen et al. (Eds.): ACIIDS 2014, Part II, LNAI 8398, pp. 237–244, 2014.

[2], EVs can take advantage of intelligent computer algorithms to alleviate such problems, mainly by means of sophisticated charging plans [3].

Thanks to the energy efficiency and eco-friendliness of EVs, many cities are accelerating the deployment of EVs. Not just in personal ownership, EVs will be exploited in taxis, rent-a-cars, shared vehicles, logistics, and other transport businesses [4]. Among these, EV taxis are very attractive, as the total moving distance of taxis will be largest in most urban cities, compared with other types of vehicles [5]. However, for an EV taxi, the driving distance during its working hours is almost always longer than the driving range of a fully charged EV, and EV taxis must be charged more often than other EVs. Actually, taxis do not always carry passengers and charging can be done hopefully during this period. Moreover, to further reduce battery consumption, it is desirable for a dispatch center to assign a taxi to a pick-up request, not allowing taxis to roam randomly looking for a customer.

In EV taxi systems, the dispatch strategy is very important to customers' satisfaction and system efficiency, which can be measured by waiting time, dispatch overhead, and service ratio [6]. They will essentially depend on the pick-up request pattern from customers, but the demand dynamics are not known, as the EV taxi is still immature business. We think that the demand patterns will not be different for both EV and regular taxis. Here, using the location records accumulated in most call taxi systems and tracing the field indicating whether a taxi is carrying passengers, we can know where a taxi picks up and drops off customers. By this record, we can develop and test a dispatch strategy capable of intelligently guiding drivers when and where to charge for each EV. In this regard, this paper designs a performance analysis framework for EV taxis with location records, incorporating the road network and charging station distribution, specifically on Jeju city, Republic of Korea.

2 Related Works

[7] presents a multiagent taxi dispatch scheme capable of improving group level customer satisfaction by concurrently dispatching multiple taxis to multiple customers in the same region. Not processing requests one by one in the FIFO order, this scheme simultaneously assigns taxis to all customer bookings made within the time window. Each taxi is associated with an agent. A taxi agent announces the availability of its taxi in a new operation area to the dispatch center, negotiates assignments of booking requests with other taxi agents, and finally informs the dispatch center of the driver's decision to accept or reject the assignment. The collaborative reasoning process begins with arbitrarily initialized permutation and exchanges assigned taxis each round. Impressively, this system is targeting at the Singapore taxi system currently in operation.

[8] focuses on how to select the best taxi stand after a passenger drop-off, considering such factors as how many taxis are already waiting and the future demand in each stand. Their prediction model is built upon time-varying Poisson process and ARIMA (AutoRegressive Integrative Moving Average) models.

First, the Poisson process is incorporated with a weighted average model to give more weight to the correlation with the demand pattern in the last week. The ARIMA model, being able to versatilely represent different types of time series, makes it possible to take the future value of a variable as a linear function of several observations and random errors. This model was applied to the city of Porto, Portugal, where a city-wide taxi network is currently running with 63 taxi stands and 441 vehicles. Matching between a taxi and a taxi stand provides a good reference for matching between an EV and a charging stand in EV taxi systems.

As an example of dispatch schemes for EV taxis, [5] combines charging plans to legacy dispatching strategies. Taxi assignment is carried out taking into account the remaining power of EVs, the availability of the battery charging or swap stations, and future taxi demand. For the charging plans, the dispatching center suggests the taxi driver whether to replace depleted batteries immediately or charge batteries, according to the future demand prediction. The dispatch center, continuously tracking the location of all EVs and receiving a request from a customer, first finds a set of taxis that pass the reachability test. Their main idea is that if the future demand is high in the destination area, a taxi having enough remaining power (or can be charged quickly) is dispatched. Here, a reservation mechanism is assumed in charging stations. The authors simulate the EV taxi fleet management for Taipei city, Taiwan.

3 Analysis Framework Design

Figure 1 shows our analysis framework architecture, which consists of an event-driven tracker, a stream handler, dispatch/relocation strategy modules, and object interfaces. The even-driven tracker is a discrete event simulator which associates a time-ordered event to the corresponding objects such as EVs and chargers mainly based on the object ids. The stream handler takes pick-up and drop-off records, which can be created either randomly or from exiting taxi operation data. The dispatch/relocation strategy module implements a specific EV taxi management technology and interacts with other analysis framework parts. Next, the taxi object pool maintains all information for dispatch and relocation strategies. In addition to the current location and SoC (State of Charge) of each taxi, charging history is included to avoid fast charging as possible as we can. The charger object pool includes the location, current queue length, and type of each charger, to help allocate a charger to an empty taxi.

In Jeju city, location records had been massively collected from 50 taxis in Jeju Taxi Telematics System during its launching test period [9]. Each record basically contains time and location stamps as well as a flag indicating whether taxi is carrying a customer. By tracing the change in this flag, we found 81,813 trips out of the spatio-temporal stream lasting for the 1.5 month. Each trip consists of a pick-up record and a drop-off record, having its own time and location stamps. A pick-up record is considered as a pick-up request in our analysis framework and invokes a dispatch module. For the dispatch center to assign an EV taxi to the pick-up request, the current distribution of taxis and their SoC information must be available. To this end, the event handler tracks the move of each EV via the series of location records and estimates the battery consumption based on moving distance. Here, the tracker ignores the move not carrying customers, as it is not allowed in EV taxis.

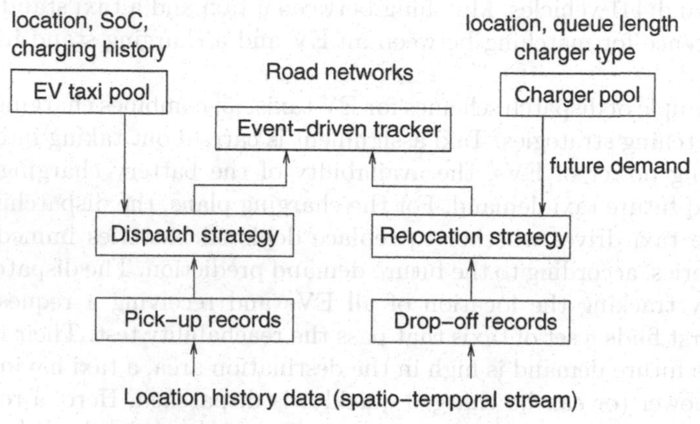

Fig. 1. Analysis framework architecture

A drop-off record makes the taxi empty. An empty taxi is forced to go to a charging stand until assigned to a pick-up request again. It is true that a taxi can pick up a customer during its move to a charging stand, but this situation is ignored to focus on the relocation strategy. The dispatch center assigns a charging stand according to how far it is from the current EV location, whether it is already occupied by another EV, and most importantly, how future demand will be. In Jeju city, the charging station tracking system is now in operation to monitor the real-time status of each charger over the city, as shown in Figure 2. By this geographic information, realistic distance between an EV and a charging stand can be calculated. Whether a charging stand is being occupied can be hardly traced solely by our analysis framework, so we define an occupation probability for system behavior tracing.

The event-driven tracker maintains the current status of each member taxi and charger allocation. Every event is ordered by the time stamp and includes the taxi id field. On each event trigger, through the difference from the previous event, the tracker updates SoC of all taxis. For taxis currently carrying passengers, SoC decreases according to the distance from the previous location. The distance between two location stamps is estimated by the well-known A* algorithm on the road network. Here, it is necessary convert a location point to the map coordinate consisting of latitude and longitude. As contrast, for taxis plugged-in to a charging stand, SoC will increase according to the interval between the previous and current time stamps. Such integration can host any dispatch and relocation strategy into our analysis framework.

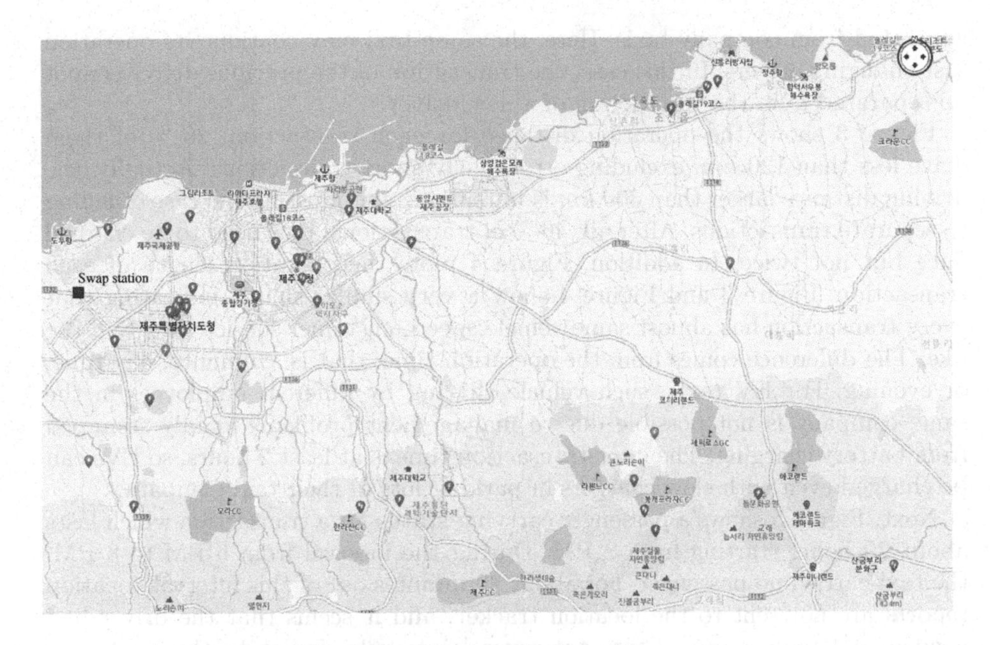

Fig. 2. Charging facility distribution

Next, the relocation strategy allocates taxis to chargers and makes them prepare for the next service in the area where the future pick-up demand will be high. An efficient allocation can reduce both spatial and temporal dispatch overhead, much improving the customer-side convenience. To this end, it is necessary to either divide the whole area into several continuous subareas or filter a set of subareas having hot demand. Each subarea is required to have sufficient number of chargers and has its own pick-up request dynamics. It is now possible to develop demand fitting and prediction models for each subarea, but those models need a huge amount of pick-up records. With the expansion of EV taxi businesses, such big data will be available and we can expect the development of a more accurate prediction model [10].

4 Analysis Results

To begin with, the basic question is whether regular taxis can be replaced by EVs and how much intelligent control is necessary. Our experiment selects the vehicle having the largest number of location records and traces its operation behavior. Here, it must be mentioned that some records are missed as drivers forget to turn on the in-vehicle telematics device. In addition, we observe that even during the normal operation period, some records do not appear due to communication errors. Anyway, the analysis procedure divides the records if the time stamp of a location record is more than 2 hours away from its previous record. It is regarded as the start of a new transaction. In addition, these location records are generated from company taxis. Here, the same taxi may continue its operation just changing drivers. In this case, the transaction of the previous driver cannot be separated from the transaction of a new driver.

Figure 3 shows the operation distance for each transaction. 26 % of them drive less than 120 *km*, excluding erroneously short transactions. Actually, the driving distance larger than 300 *km* is unrealistic and must be the case of failing to separate transactions. After all, 46 % of transactions, EVs need to be charged once but not twice. In addition, Figure 4 plots the operation length of each transaction. Figure 3 and Figure 4 show a very similar shape, indicating that every transaction has almost same vehicle speed, customer service rate, and the like. The difference comes from the operation time, that is, commute, day time, or evening. For EV taxis, such vehicle sharing by different employees in the same company is not possible due to management problems mainly stemmed from battery charging. The inter-transaction time is at least 2 hours, so EVs can be charged even with slow chargers in parking lots of their taxi company.

Next, Figure 5 shows a passenger carrying status for a transaction which lasts about 20 hours starting from 2 PM. During the interval from 6 PM to 8 PM, the taxi carried no passenger. For about 80 minutes out of this interval, location reports are not sent to the location tracker, and it seems that the driver had a meal and took a rest. There are several intervals similar to this situation,

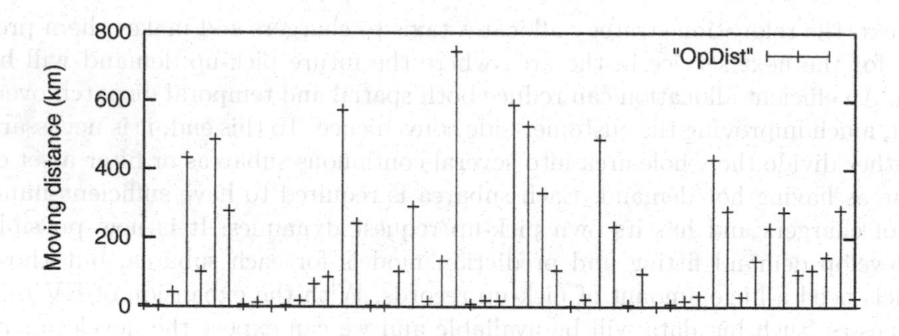

Fig. 3. Moving distance in transactions

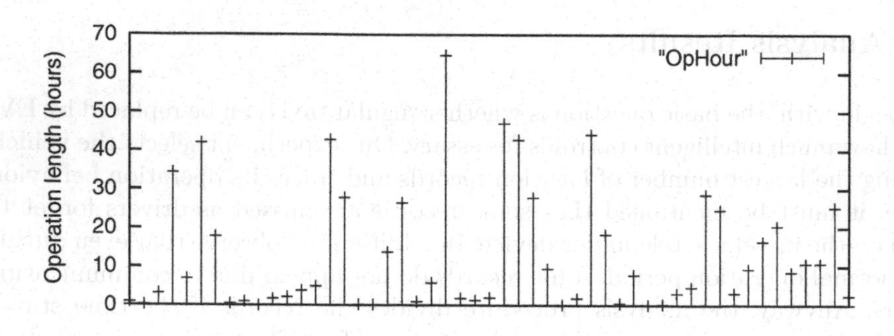

Fig. 4. Operation length behavior

indicating that even for a transaction having long time, the EV can be charged by slow chargers with an efficient charging schedule without being unable to provide service due to battery shortage. Actually, one of the most important goals of the dispatch strategy is to charge with slow chargers as many times as possible. Multiple taxis are simultaneously in operation for most intervals, and the charging coordinator can make drivers take a rest one by one considering the SoC of respective EV taxis.

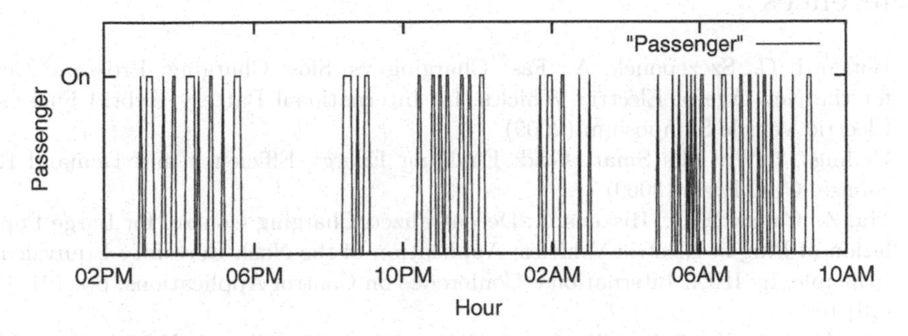

Fig. 5. Passenger carrying status

5 Concluding Remarks

EVs can achieve energy efficiency in modern transport systems and it is necessary to promote their deployment. Replacing taxi with EVs will have a great impact on energy efficiency and environment, but their daily driving distance is usually longer than the driving range, which is much smaller than legacy gasoline-powered vehicles. To cope with this problem taking advantage of computational intelligence just like other smart grid entities, an intelligent charging and operation planning scheme is essential. In this paper, we have designed a performance analysis framework for EV taxis based on the location history data obtained from a legacy call taxi system. Consisting of an event tracker, a stream handler, object interfaces, and strategy integrator, the analysis procedure can measure the performance of a dispatch/relocation strategy in terms of dispatch latency, customer waiting time, and the number of daily fast charging operations.

By this framework, we have found that an EV taxi needs to be charged about once a day on the condition that it is fully charged before it starts its operation. In addition, out of the operation hours, drivers have meals and take a rest. During this interval, it is possible to charge EVs by slow chargers. Here, the portion of fast charging must be kept as small as possible, as it can impose a significant burden on the power system, especially in day time. To this end, an intelligent controller must coordinate the charging and operation plan for the fleet of EV taxis, assigning them to appropriate charging stands. Our framework helps to develop this planner without expensive filed tests.

As future work, we are planning to develop a dispatch scheme for EV taxi systems, particularly targeting at Jeju city. A heuristic-based approach is now being considered in selecting an EV taxi for each call, according to whether the call is an in-town trip, a long-distance trip, and others. In addition, as this area has high potential for wind energy, the integration of such renewable energies is another important issue [11]. Finally, our research team is currently developing a battery consumption model for the road network in this area, and we can integrate this model to elaborate our dispatch strategy.

References

1. Botsford, C., Szczepanek, A.: Fast Charging vs. Slow Charging: Pros and Cons for the New Age of Electric Vehicles. In: International Battery Hybrid Fuel Cell Electric Vehicle Symposium (2009)
2. Gellings, C.W.: The Smart Grid: Enabling Energy Efficiency and Demand Response. CRC Press (2009)
3. Ma, Z., Callaway, D., Hiskens, I.: Decentralized Charging Control for Large Population of Plug-in Electric Vehicles: Application of the Nash Certainty Equivalence Principle. In: IEEE International Conference on Control Applications, pp. 191–195 (2010)
4. Lee, J., Park, G.: A Tour Recommendation Service for Electric Vehicles Based on a Hybrid Orienteering Model. In: ACM Symposium on Applied Computing, pp. 1652–1654 (2013)
5. Lu, J., Yeh, M., Hsu, Y., Yang, S., Gan, C., Chen, M.: Operating Electric Taxi Fleets: A New Dispatching Strategy with Charging Plans. In: IEEE International Electric Vehicle Conference (2012)
6. Chen, W., Chang, J., Chen, Y., Chou, W.: Cloud Computing Based Taxi Dispatching Service. In: International Conference on ITS Telecommunications, pp. 433–436 (2012)
7. Seow, K., Dang, N., Lee, D.: A Collaborative Multiagent Taxi-Dispatch System. IEEE Transactions on Automatic Science and Engineering 7(3), 607–616 (2010)
8. Moreira-Matias, L., Gama, J., Ferreira, M., Damas, L.: A Predictive model for the Passenger Demand on a Taxi Network. In: International IEEE Conference on Intelligent Transport Systems, pp. 1014–1019 (2012)
9. Lee, J., Park, G.-L., Kim, H., Yang, Y.-K., Kim, P., Kim, S.-W.: A Telematics Service System Based on the Linux Cluster. In: Shi, Y., van Albada, G.D., Dongarra, J., Sloot, P.M.A. (eds.) ICCS 2007, Part IV. LNCS, vol. 4490, pp. 660–667. Springer, Heidelberg (2007)
10. Fang, X., Yang, D., Xue, G.: Evolving Smart Grid Information Management Cloudward: A Cloud Optimization Perspective. IEEE Transactions on Smart Grid 4(1), 111–119 (2013)
11. Goebel, C., Callaway, D.: Using ICT-Controlled Plug-in Electric Vehicles to Supply Grid Regulation in California at Different Renewable Integration Levels. IEEE Transactions on Smart Grid 4(2), 729–740 (2013)

Item-Based Collaborative Filtering with Attribute Correlation: A Case Study on Movie Recommendation

Parivash Pirasteh, Jason J. Jung*, and Dosam Hwang

Department of Computer Engineering,
Yeungnam University, Korea
{parivash63,j2jung,dosamhwang}@gmail.com

Abstract. User-based collaborative filtering (CF) is a widely used technique to generate recommendations. Lacking sufficient ratings will prevent CF from modeling user preference effectively and finding trustworthy similar users. To alleviate this problems, item-based CF was introduced. However, when number of co-rated items is not enough or new item is added to the system, item-based CF result is not reliable, too. This paper presents a new method based on movies similarity that focuses on improving recommendation performance when dataset is sparse. In this way, we express a new method to measure the similarity between items by utilizing the genre and director of movies. Experiments show the superiority of the measure in cold start condition.

Keywords: Recommender systems, Item-based collaborative filtering, Attribute correlation.

1 Introduction

Recommender systems are based on one of two strategies. The content filtering approach creates a profile for each user or product to characterize its nature. For example, a movie profile could include attributes regarding its genre, the participating actors, and so forth. User profiles might include demographic information or answers provided on a suitable questionnaire. The profiles allow programs to associate users with matching products [1]. Of course, content-based strategies require gathering external information that might not be available or easy to collect.

An alternative to content filtering relies only on past user behavior, for example, previous transactions or product ratings without requiring the creation of explicit profiles. This approach is known as collaborative filtering [2]. CF analyzes relationships between users and inter-dependencies among products to identify new user-item associations.

Regardless of CF success in many application settings, it encounters three serious limitations, namely sparsity, cold start and scalability [3]. Data sparsity refers to the problem of insufficient data, or sparseness, in the user-item matrix. The introduction of new users or new items can cause the cold start problem, as there will be insufficient data on these new entries for the collaborative filtering to work accurately. As the numbers of users and items grow, traditional CF algorithms will suffer serious scalability problems.

* Corresponding author.

N.T. Nguyen et al. (Eds.): ACIIDS 2014, Part II, LNAI 8398, pp. 245–252, 2014.

For example, Amazon has more than many millions of products and users. Hence, the matrix size is very large; the calculation over the matrix will be very expensive.

In order to overcome scalability issue, Sarwar et al. [4] introduced item based that is a method to recommend similar products from previously purchase by the same customer but focusing on item aspect rather than users. item-based techniques first analyze the ⟨user,item⟩ matrix to identify relationships between different items, and then use these relationships to indirectly compute recommendations for users.Although the item-based CF technique has been successfully applied in many applications, its major limitations including sparsity and cold start item problems have restricted its widespread use.

In this paper, we try to overcome the sparsity and cold start problem. In addition to using ⟨user,item⟩ matrix to calculate similarity between items, we propose new similarity measure between items, based on genre and director of movies. This information is more reliable than the user rating, since genre information provided by movie expert. Moreover, users tend to watch movies of certain directors could benefit from additional similarity measure. The main advantage of this work is that newly added item always has genre and director information, allowing avoidance of cold start problem.

The reminder of paper is organized as follow. In the Sect. 2 we summarize the related work. After the proposed method are described in detail in Sect. 3 and Sect. 4, we present experimental evaluation on the dataset in Sect. 5. Finally, we conclude our paper and propose future work in Sect. 6.

2 Related Work

Methods specifically designed to alleviate the sparsity and cold start problem usually consist of a hybrid method between a collaborative filter and a content based recommender [5–7].

Our method differs in that we do not use content based method to extract user preference. Melville et al. [8] employed a content-based predictor to enhance existing user data, and then provided personalized suggestions through collaborative filtering. Basilico and Hofmann [9] developed a framework that incorporates all available information by using a suitable kernel or similarity function between user-item pairs. Pazzani [10] proposed a hybrid method that recommends items based on vote of four different algorithms: user-user collaborative filtering, content-based, demographic-based, and collaboration via content. This approach can provide new user recommendation by assembling several independent models. Papagelis [3] focuses on developing a computational model that permits the exploration of transitive user similarities based on trust inferences for addressing the sparsity problem.

Using genre similarity to recommend movie to user has been developed by [11]. Their goal was to find similar genre to user preferred genre and calculate recommendation point. Although their system needs that user provide his/her own preferred genres while our system does not require any extra information.

3 Movie Similarity Methods

In this section different methods for measuring the similarity between movies have been introduced, first method is based on genre information. Another item used for measuring the similarity between movies is director. Uluyagmur et al. [12] showed that director feature gives the best recommendation performance among different features (actor, genre and director) in content based recommendation.

3.1 Genre-Based Movie Similarity

This section presents different movie similarity based on genre. It consists of:

- Movie similarity based on Jaccard coefficient between genres
- Movie similarity based on genre correlation.

Jaccard Similarity Coefficient. The Jaccard similarity is a common index for binary variables. It is defined as the quotient between the intersection and the union of the pair-wise compared variables among two objects. The Jaccard similarity coefficient between genres is defined as follow:

$$J(X,Y) = \frac{G_x \bigcap G_y}{G_x \bigcup G_y} \tag{1}$$

where G_x represents genres of movie x and G_y represents genres of movie y.

Genre Correlation. Genre correlation can be calculated based on genre co-occurrence of movies in database. One movie can have the characteristics of various genres at once. For example, *Inception* has the characteristic of five genres of action, adventure, mystery, Sci-Fi, and thriller. Let us consider vector M_i containing all genres belonged to movie j.

$$M_i = \{G_j | j \in [1, 22]\} \tag{2}$$

If movie i contains Genre j then Genre j is equal to 1 otherwise 0. By using M_i, it obtains similarity value among genres with Pearson correlation coefficient.

Once genre correlation is ready, similarity between movies can be calculated as follows:

$$Sim(M_i, M_j) = \frac{MGC(GM_i, GM_j)}{MGN(GM_i, GM_j)} \tag{3}$$

where GM_i expresses genres of movie M_i, MGC is maximum genre correlation between M_i and M_j, MGN is the maximum genre number of Movies M_i and M_j. Therefore, if M_i has genres like drama and romance and M_j has genres like romance, family and war the similarity between them would be 0.39, which is calculated by adding 1 (because of one common genre-romance) to correlation between drama and war (0.1957) which is higher than correlation between drama and family, divided by 3 (maximum genre number between two movies).

3.2 Director-Based Movie Similarity

Some users like to watch movies of certain directors, rather than of certain actors/actress or certain type. Director similarity factor can help them to find unseen movies by their favorite directors.

Director Correlation. If we determine director-genre matrix D, (consisting of N director and m genres), in which $d_{i,g}$ indicates the extent of potency that director i has in genre g, we can define $d_{i,g}$ as below:

$$d_{i,g} = \frac{\sum_{g=1}^{m} r_{d,g}}{M_d} \tag{4}$$

where $r_{d,g}$ shows the rate of movies directed by d in genre g. And M_d is the number of director d's movies. We can define the general director to director similarity between director i and j using cosine similarity as shown below:

$$Sim(i,j) = \frac{\sum_{g=1}^{m} d_{i,g} d_{j,g}}{\sqrt{\sum_{g=1}^{m} d_{i,g}^2}\sqrt{\sum_{g=1}^{m} d_{j,g}^2}} \tag{5}$$

4 New Approach to Calculating Similarity Measures

In this section our approach for movie recommendation based on movie similarity has been discussed. We illustrate how making use of genres can help to assess movie similarity. Our method recommends movies which are similar to movies that user already rated using genre information. Movie similarity function can be formulated by integrating genre similarity term and director similarity term.

$$Sim_{attribute}(i,j) = Sim_g(i,j) + \beta Sim_d(i,j) \tag{6}$$

β is parameter for weighting the contribution of director similarity. The traditional item similarity measure can be formulated as below:

$$Sim_{item}(i,j) = \frac{\sum_{u \in U}(R_{u,i} - \overline{R}_i)(R_{u,j} - \overline{R}_j)}{\sqrt{\sum_{u \in U}(R_{u,i} - \overline{R}_i)^2}\sqrt{\sum_{u \in U}(R_{u,j} - \overline{R}_j)^2}} \tag{7}$$

Let the set of users who both rated i and j are denoted by U. Here $R_(u,i)$ denotes the rating of user u on item, $\overline{R}_i$ is the average rating of the i-th item. It is common for item to have high correlated neighbors that are based on very few co-rated item. These neighbors based on small number of overlapping tend to be bad predictor. Along with this, there are other situations where Correlation-based item similarity cannot compute accurately the closeness:

- New items which have not rated by users.
- If all ratings of an item are equal, e.g. <1,1,1,1,1>, correlation cannot be calculated, since the numerator of formula becomes zero.

In order to address the above issues, we incorporated the traditional correlation similarity measure with our proposed similarity function. A unified item similarity model can be formulated by taking into account the item similarity based on the genre and director as an additional term.

$$Sim(i,j) = Sim_{item}(i,j) + \alpha \times Sim_{attribute}(i,j) \quad (8)$$

The larger α is, the greater the emphasis on obtaining the similarity based on movie information.

4.1 Prediction Computation

The most important step in collaborative filtering system is to generate the output interface in terms of prediction. Once we collect set of most similar items based on movie similarity, the next step is to look into target user ratings and use a technique to obtain prediction.

Weighted Sum. As the name implies, this method compute the prediction on an item i for user u by computing the sum of ratings given by the user on items similar to i. Each rating is weighted by corresponding similarity $S(i,j)$ which is obtained from Equ. 8 between items i and j. We can denote the prediction $P(u,i)$ as:

$$P_{u,i} = \frac{\sum_{allsimilaritems,N}(S_{i,N} * R_{u,N})}{\sum_{allsimilaritems,N}(|S_{i,N}|)} \quad (9)$$

Basically, this approach tries to capture how the active user rates the similar items. The weighted sum is scaled by the sum of the similarity terms to make sure the prediction is within the predefined range [4].

5 Experiments

5.1 Dataset

We experiment with integrating metadata of movies from two different data sources: IMDB and MovieLens . the Information like director and genre of movies is accessible in IMDB. The MovieLens dataset contains user ratings for movies.

5.2 Metrics

We use Root Mean Square Error (RMSE) and Mean Absolute Error (MAE) to measure the prediction quality of our proposed approach in the comparison with collaborative filtering.

$$RMSE = \sqrt{\frac{1}{T}\sum_{i,j}(R_{i,j} - \overline{R}_{i,j})^2} \quad (10)$$

The metric MAE is defined as:

$$MAE = \frac{1}{T}\sum_{i,j}|R_{ij} - \overline{R}_{ij}| \tag{11}$$

Where R_{ij} denotes the rating that user i gave to item j, $\overline{R}_{i,j}$ denotes the rating user i gave to item j as predicted by method, and T denotes the number of tested ratings.

5.3 Performance Comparison

In this section, we compare the performance of our proposed method versus individual item similarity methods.

- CIS: Correlation-based item similarity (Equ. 7)
- GJ: Genre similarity between items based on the Jaccard index.
- GD: Similarity measure based on the genre and director movies (Equ. 6)
- CIS-GD: Combination of our proposed approach and correlation-based item similarity

Table 1. Comparison of recommendation performance between the proposed algorithm and other approaches

METHODS	RMSE	MAE
CIS	0.95	0.86
GJ	1.6	1.3
GD	1.2	0.97
GIS-GD	**0.86**	**0.8**

From the result we can notice that our method(GIS-GD) outperforms other approached. It is observed that movie similarity based on genre correlation show better results than movie similarity based on Jaccard coefficient which does not consider genre correlation between movies.

5.4 Impact of Parameters α and β

In our method proposed in this paper, the parameter α play very important role. It controls how much item similarity based on movie information should incorporate in similarity measure. If we use a very small value of α, we only mine similarity based on Pearson correlation. In contrast, for large α value the impact of attribute similarity is more substantial than Pearson correlation similarity . The optimal α is nearly 0.4 so, as we increase the value of α up to 0.4 we can see reduction in RMSE. the parameter β in the proposed algorithm influences the relative contribution from director similarity measurement. Fig. 2 illustrates the impact of β on the evaluation metric. we can see that the optimal β is nearly 0.2.

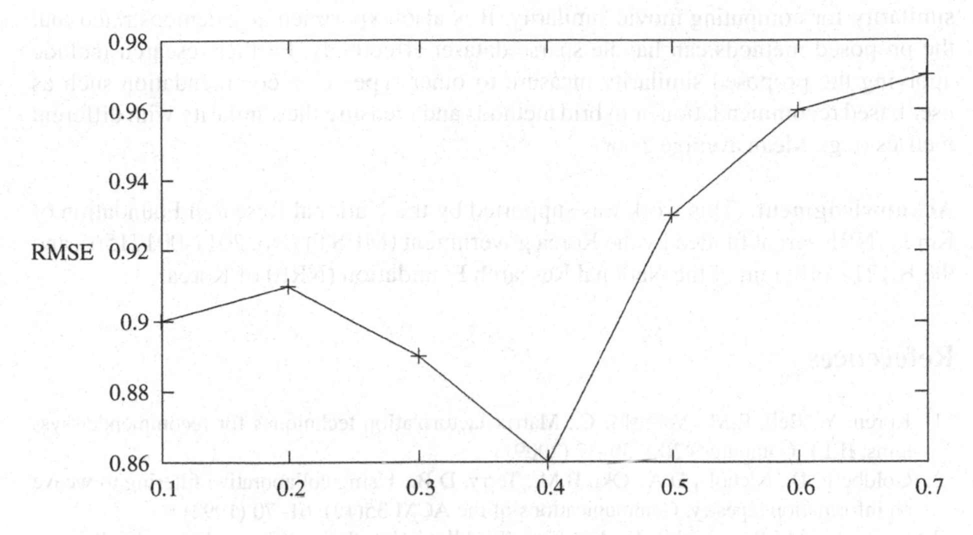

Fig. 1. Impact of parameter α

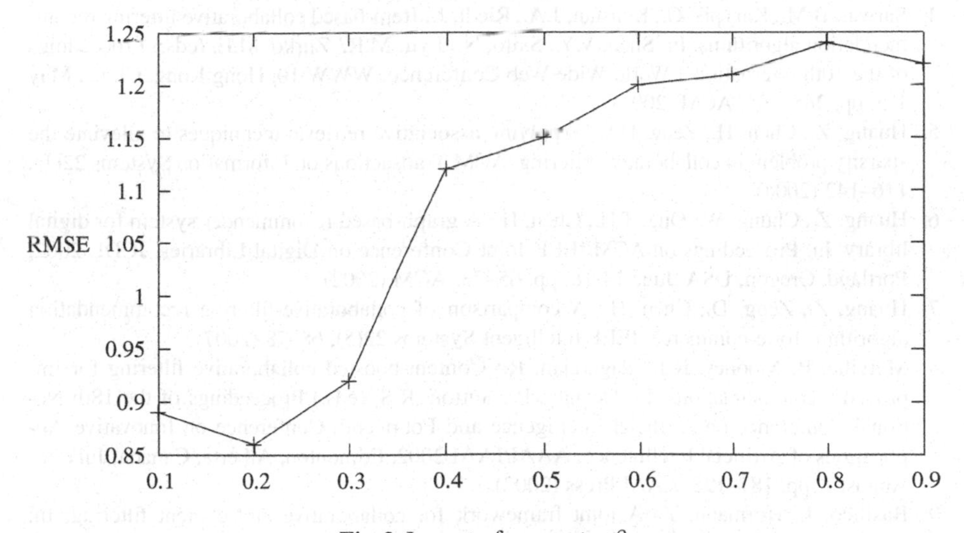

Fig. 2. Impact of parameter β

6 Conclusion

The movie similarity function in this paper plays as an important rule in directing users to choose similar movies to movies they like. Small numbers of co-rated items lead to many misleadingly in high correlations. To solve this issue, we proposed a similarity measure which is independent of user's ratings. Furthermore, since we compute the similarity measure based on genres and directors, we will not confront with cold start problem. From Fig. 2 we concluded that genre similarity is more reliable than director similarity for computing movie similarity. It is also experimentally demonstrated that the proposed methods can handle sparse dataset effectively. Further research include applying the proposed similarity measure to other types of recommendation such as user based recommendation or hybrid methods and measure the similarity with different metrics (e.g., Mean average error).

Acknowledgment. This work was supported by the National Rcsearch Foundation of Korea (NRF) grant funded by the Korea government (MEST) (No. 2011-0017156), and the BK21+ program of the National Research Foundation (NRF) of Korea.

References

1. Koren, Y., Bell, R.M., Volinsky, C.: Matrix factorization techniques for recommender systems. IEEE Computer 42(8), 30–37 (2009)
2. Goldberg, D., Nichols, D.A., Oki, B.M., Terry, D.B.: Using collaborative filtering to weave an information tapestry. Communications of the ACM 35(12), 61–70 (1992)
3. Papagelis, M., Plexousakis, D., Kutsuras, T.: Alleviating the sparsity problem of collaborative filtering using trust inferences. In: Herrmann, P., Issarny, V., Shiu, S. (eds.) iTrust 2005. LNCS, vol. 3477, pp. 224–239. Springer, Heidelberg (2005)
4. Sarwar, B.M., Karypis, G., Konstan, J.A., Riedl, J.: Item-based collaborative filtering recommendation algorithms. In: Shen, V.Y., Saito, N., Lyu, M.R., Zurko, M.E. (eds.) Proceedings of the 10th International World Wide Web Conference, WWW 10, Hong Kong, China, May 1-5, pp. 285–295. ACM (2001)
5. Huang, Z., Chen, H., Zeng, D.D.: Applying associative retrieval techniques to alleviate the sparsity problem in collaborative filtering. ACM Transactions on Information Systems 22(1), 116–142 (2004)
6. Huang, Z., Chung, W., Ong, T.H., Chen, H.: A graph-based recommender system for digital library. In: Proceedings on ACM/IEEE Joint Conference on Digital Libraries, JCDL 2002, Portland, Oregon, USA, June 14-18, pp. 65–73. ACM (2002)
7. Huang, Z., Zeng, D., Chen, H.: A comparison of collaborative-filtering recommendation algorithms for e-commerce. IEEE Intelligent Systems 22(5), 68–78 (2007)
8. Melville, P., Mooney, R.J., Nagarajan, R.: Content-boosted collaborative filtering for improved recommendations. In: Dechter, R., Sutton, R.S. (eds.) Proceedings of the 18th National Conference on Artificial Intelligence and Fourteenth Conference on Innovative Applications of Artificial Intelligence, AAAI/IAAI 2002, Edmonton, Alberta, Canada, July 28-August 1, pp. 187–192. AAAI Press (2002)
9. Basilico, J., Hofmann, T.: A joint framework for collaborative and content filtering. In: Sanderson, M., Järvelin, K., Allan, J., Bruza, P. (eds.) Proceedings of the 27th Annual International ACM SIGIR Conference on Research and Development in Information Retrieval, Sheffield, UK, July 25-29, pp. 550–551. ACM (2004)
10. Pazzani, M.J.: A framework for collaborative, content-based and demographic filtering. Artificial Intelligence Review 13(5-6), 393–408 (1999)
11. Choi, S.M., Ko, S.K., Han, Y.S.: A movie recommendation algorithm based on genre correlations. Expert Systems with Applications 39(9), 8079–8085 (2012)
12. Uluyagmur, M., Cataltepe, Z., Tayfur, E.: Content-based movie recommendation using different feature sets. In: Proceedings of the World Congress on Engineering and Computer Science, WCECS 2012, San Francisco, USA, October 24-26, vol. 1 (2012)

Linked Data Mashups: A Review on Technologies, Applications and Challenges

Tuan Nhat Tran, Duy Khanh Truong, Hanh Huu Hoang, and Thanh Manh Le

Hue University
3 Le Loi Street, Hue City, Vietnam
hhhanh@hueuni.edu.vn

Abstract. To remedy the data integration issues of the traditional Web mashups, the Semantic Web technology uses the Linked Data based on RDF data model as the unified data model for combining, aggregating and transforms data from heterogeneous data resources to build Linked Data Mashups. There have been tremendous amount of efforts of Semantic Web community to enable Linked Data Mashups but there still lack of a systematic survey on concepts, technologies, applications and challenges. Therefore, this paper gives an over view of Linked Data Mashups and conducts a state-of-the-art survey about technologies, applications and challenges on Linked Data Mashups.

Keywords: Linked Data, Mashups, RDF, Semantic Web.

1 Introduction

The development of generic Web applications is well understood and supported by many traditional computer science domains, such as classical database applications. In current Web application development data integration and access are typically dealt with by fairly sophisticated abstractions and tools which support rapid application development and the generation of reusable and maintainable software components. The task of programming such applications has become the task of combining existing components from well-established component libraries, i.e., customizing and extending them for application-specific tasks. Typically, such applications are built relying on a set of standard architectural styles which shall lower the number of bugs and ensure code that is easy to understand and maintain.

Historically, the process of writing new queries and creating new graphic interfaces has been something that has been left to the experts. A small team of experts with a limit skill-sets would create applications, and all users would have to use what was available, even if it did not quite fit their needs [16]. A *mashup* is an (web) application that offers new functionality by combining, aggregating and transforming resources and services available on the web [16]. Therefore, mashups are an attempt to move control over data closer to the user and closer to the point of use. Although mashups are technically similar to the data integration techniques that preceded them, they are philosophically quite different. While data integration has historically been about

N.T. Nguyen et al. (Eds.): ACIIDS 2014, Part II, LNAI 8398, pp. 253–262, 2014.

allowing the expert owners of data to connect their data together in well-planned, well-structured ways, mashups are about allowing arbitrary parties to create applications by repurposing a number of existing data sources, without the creators of that data having to be involved [5]. Therefore, mashup enabling technologies not only reduce the effort of building a new application by reusing available data sources and systems but also allow the developers to create novel applications beyond imagination of the data creators. However, the traditional web mashups still suffers the heterogeneity of data coming from different sources having different formats and data schema. To remedy the data integration issues of the traditional web mashups, the Semantic Web technology uses the Linked Data based on RDF data model as the unified data model for combining, aggregating and transforms data from heterogeneous data resources to build Linked Data Mashups. Powered by tools and technologies having been developed by the Semantic Web community, there are various applications domains building applications with Linked Data Mashups.

There has not been any work that give a comprehensive survey about technologies and applications of Linked Data Mashups as well the challenges for building Linked Data Mashups. This shortcoming comes from several following reasons. Typical Linked mashups are data-intensive and require the combination and integration of RDF data from distributed data sources. In contrast to that, data-intensive applications using RDF are currently mostly custom-built with limited support for reuse and standard functionalities are frequently re-implemented from scratch. While the use of powerful tools such as SPARQL processors, takes the edge off of some of the problems, a lot of classical software development problems remain. Also such applications are not yet built according to agreed architectural styles which are mainly a problem of use rather than existence of such styles. This problem though is well addressed in classical Web applications. For example, before the introduction of the standard 3-tier model for database-oriented Web applications and its support by application development frameworks, the situation was similar to a lot the situation that we see now with RDF-based applications [1].

This paper aims at giving a systematic view about Linked Data Mashups. It will give an overview of Linked Data Mashups in Section 2. Then, in Section 3, the paper gives a survey about enabling technologies for Linked Data Mashups such as data integration, mashup execution engines, interactive programing and visualization. Section 3 will introduce a series of applications domains for Linked Data Mashups. The challenges of building Linked Data Mashups are discussed in Section 4.

2 Linked Data Mashups

2.1 Linked Data

The term *Linked Data* refers to a set of best practices for publishing and linking structured data on the Web. These best practices were introduced by Tim Berners-Lee in his Web architecture note namely *Linked Data* [17] and have become known as the *Linked Data principles*. These principles are descripted as: the basic idea of Linked Data is to apply the general architecture of the World Wide Web to the task of sharing structured data on global scale [7].

Linked Data principles firstly advocates using URI references to identify, not just Web documents and digital content, but also real world objects and abstract concepts. These may include tangible things such as people, places and cars, or those that are more abstract, such as the relationship type of *knowing somebody*. Linked Data use the HTTP protocol for Web resources access mechanism with the use of HTTP URIs to identify objects and abstract concepts, enabling these URIs to be *dereferenced* (i.e., looked up) over the HTTP protocol into a description of the identified object or concept. Linked Data principle also advocates use of a single data model for publishing structured data on the Web – the Resource Description Framework (RDF), a simple graph-based data model that has been designed for use in the context of the Web [7]. Lastly, Linked Data uses of hyperlinks to connect not only Web documents, but any type of thing. For example, a hyperlink may be set between a person and a place, or between a place and a company. Hyperlinks that connect things in a Linked Data context have types which describe the relationship between the things. For example, a hyperlink of the type *friend of* may be set between two people, or a hyperlink of the type *based near* may be set between a person and a place. Hyperlinks in the Linked Data context are called *RDF links* in order to distinguish them from hyperlinks between classic Web documents.

The RDF data model represents information as node-and-arc-labelled directed graphs. The data model is designed for the integrated representation of information that originates from multiple sources, is heterogeneously structured, and is represented using different schemata [7, 12]. Data is represented in RDF as RDF *triples*. The RDF data model is described in detail as part of the W3C RDF Primer[1].

2.2 Linked Data Mashups

Linked Data Mashups are created in the similar fashion as web mashups whilst they use a unified data model, RDF model for combining, aggregating and transforming data from heterogeneous data resources. Using a single data model for data manipulation operations enables a simpler abstraction of application logics for mashup developers. The RDF data model is driven by vocabularies or ontologies which play the role as the common understanding among machines, developers, domain experts and users.

A Linked Data Mashup is composed from different piece of technologies. The first type of technologies is data integration which covers data transformation, storage, accessing and application APIs based on RDF data model. The second type of technologies is mashup execution engines which provide the execution eviroments for computing the mashup processing workflow. The third type of technologies is interactive programing and visualization which provide a composing and exploring environments for mashup developer to build data processing workflow for a mashup (Section 3).

One simple example of a Linked Data Mashup is an aggregated Sales application that integrates customer relationship management (CRM) and financial data with functionality from the Web and corporate backend data. This example mashup would

[1] W3C RDF Primer, `http://www.w3.org/TR/rdf-primer/`

employ real-time information, streaming content, and Web services to form a coordinated application using all of these data sources. Integrated sales information for the traveling sales person could be available from their smart phone or laptop. The data integration tools are responsible for transforming streams real-time Web information of financial and CRM data and Background information and Request for Information (RFI) documents to Linked Data. Internally, Internal, proprietary customer data about installed products, contracts, and upsell possibilities can be exposed as Linked Data via RDFisers [7]. When all the data are accessible as Linked Data and can be queried via SPARQL, there is a series of front-end application can be built. The facet browsers for Linked Data [3, 14] enable combining financial, CRM and other data with online maps to visually identify, locate and categorise customers for each geographical location. Using Google Maps or Mapquest[2] APIs, each customer site appears on the map and allows the sales person to drill down using the map paradigm to identify customer sites to expose new sales or possible upsell opportunities. Background information and RFI documents could be generated partly using semantically rich content from DBpedia (http://dbpedia.org/), the semantically structured content from Wikipedia. Integrated and updated glossary definitions of domain vernacular, references to partners and competitors could come together as competitive analysis documents. Prospective customers could read marketing evaluations combined with general reference content, and links to trusted independent blogger opinions, all from a single document. Customer data can be integrated with the maps, reference information, and sales database to provide personalised content for customers.

3 Technologies Enabling Linked Data Mashups

3.1 Data Integration

Data integration technologies for Linked Data Mashups involve all solutions and tools to enable data from heterogeneous sources accessible as Linked Data. The representative architecture for data integration of Linked Data Mashups is depicted in Fig. 3.

In this architecture, the publishing layer provides all tools to expose traditional data sources in RDF data formats. They include wrappers for the databases, RDFizers for transforming data from other format like XML, JSON, HTML into RDF. Then when all data is accessible as Linked Data it might be stored in storages or accessed via Web APIs such as SPARQL Endpoints, called Web Linked Data. These data might be manipulated and integrated to access in a refined form via a SPARQL query interface by application code in the application layer.

3.2 Mashup Execution Engines

A mashup is usually constructed in a formal language to representing the computing process that generates the output for the mashup. Then the mashup represented in a

[2] http://www.mapquest.com/

such language is executed in an execution engine. In this section, we introduce two popular execution engines, MashMaker [5] and DERI Pipes [1]. MashMaker uses functional programing language whilst DERI Pipes uses Domain Specific Language (DSL) in XML

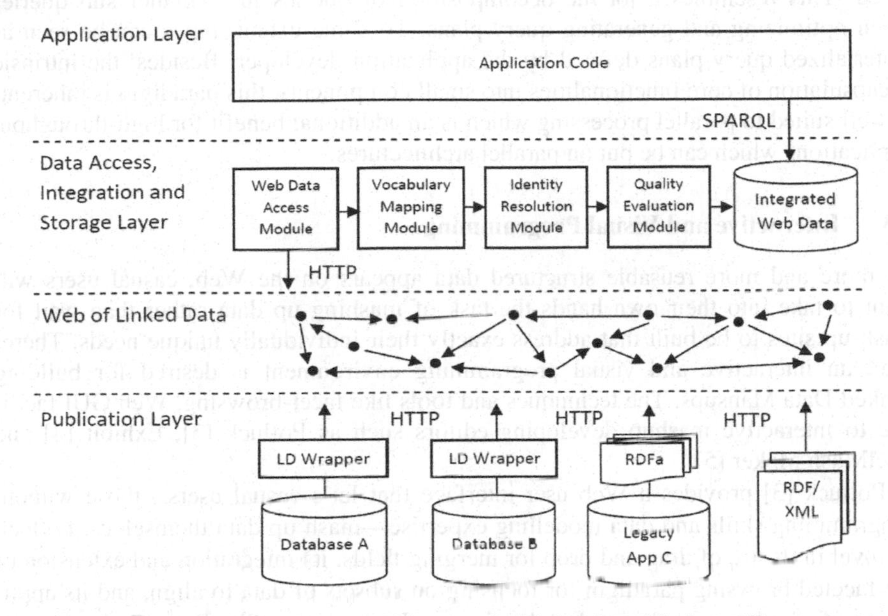

Fig. 1. Data integration architecture for Linked Data Mashups [7]

MashMaker provides a modern functional programming language with non-side effecting expressions, higher order functions, and lazy evaluation. MashMaker programs can be manipulated either textually, or through an interactive tree representation, in which a program is presented together with the values it produces. MashMaker expressions are evaluated lazily. The current consensus in the programming language community seems to now be that lazy evaluation is the wrong evaluation model for conventional programming languages. This is because the bookkeeping overhead of lazy evaluation makes programs run slowly, the complex evaluation behavior makes performance hard to predict, and programmers often have to battle with space leaks due to long chains of lazy thunks. In the case of web mashups, the bookkeeping cost of remembering how to evaluate something is tiny compared to the massive cost of fetching and scraping a web site, thus it is only necessary for a very small number of expressions to be unneeded for the bookkeeping cost to be more than paid back. Even if fetching a web site was cheap, it is important for us to minimize the number of queries we make to a remote server, to avoid overwhelming a server. Typical mashup programs work with relatively small amounts of data that are not directly presented to the user, and so space leaks are far less of a problem.

DERI Pipes [1] proposes a flexible architectural style for the fast development of reliable data intensive applications using RDF data. Architectural styles have been around for several decades and have been the subject of intensive research in other domains such as software engineering and databases. Le-Phuoc et al. [1] bases their work on the classical pipe abstraction and extends it to meet the requirements of (semantic) Web applications using RDF. The pipe concept lends itself naturally to the data-intensive tasks at hand by its intrinsic concept of decomposing an overall data-integration and processing task into a set of smaller steps which can be freely combined. This resembles a lot the decomposition of queries into smaller sub queries when optimizing and generating query plans. To some extent, pipes can be seen as materialized query plans defined by the application developer. Besides, the intrinsic encapsulation of core functionalities into small components, this paradigm is inherently well suited to parallel processing which is an additional benefit for high-throughput applications which can be put on parallel architectures.

3.3 Interactive and Visual Programming

As more and more reusable structured data appears on the Web, casual users will want to take into their own hands the task of mashing up data rather than wait for mash-up sites to be built that address exactly their individually unique needs. Therefore, an interactive and visual programming environment is desired for building Linked Data Mahsups. The techniques and tools like facet-browsing, Web GUI facilitate to interactive mashup developing editors such as Potluck [3], Exhibit [4] and IntelMash Maker [5].

Potluck [3] provides a Web user interface that let's casual users—those without programming skills and data modelling expertise—mash up data themselves. Potluck is novel in its use of drag and drop for merging fields, its integration and extension of the faceted browsing paradigm for focusing on subsets of data to align, and its application of simultaneous editing for cleaning up data syntactically. Potluck also lets the user construct rich visualizations of data in-place as the user aligns and cleans up the data. This iterative process of integrating the data while constructing useful visualizations is desirable when the user is unfamiliar with the data at the beginning—a common case—and wishes to get immediate value out of the data without having to spend the overhead of completely and perfectly integrating the data first.

Exhibit [14] is a lightweight framework for publishing structured data on standard web servers that requires no installation, database administration, or programming. Exhibit lets authors with relatively limited skills-those same enthusiasts who could write HTML pages for the early Web-publish richly interactive pages that exploit the structure of their data for better browsing and visualization. Such structured publishing in turn makes that data more useful to all of its consumers: individual readers get more powerful interfaces, mashup creators can more easily repurpose the data, and Semantic Web enthusiasts can feed the data to the nascent Semantic Web.

IntelMash Maker [5] does this by making mashup creation part of the normal browsing process. Rather than having a reasonably skilled user create a mashup in advance as a mashup site that other users browse to, MashMaker instead creates personalized mashups for the user inside their web browser. Rather than requiring that a user tell a mashup tool what they want to create, MashMaker instead watches what information the user looks at, correlates the user's behaviour with that of other users, and guesses a mashup application that the user would find useful, without the user even having to realize they wanted for a mashup.

4 Application Domains

4.1 DBpedia Mashups

If you see Wikipedia as a main place where the knowledge of mankind is concentrated, then DBpedia [12]—which is extracted from Wikipedia—is the best place to find the machine representation of that knowledge. DBpedia constitutes a major part of the semantic data on the web. Its sheer size and wide coverage enables you to use it in many kind of mashups: it contains biographical, geographical, bibliographical data; as well as discographies, movie metadata, technical specifications, and links to social media profiles and much more. Just like Wikipedia, DBpedia is a truly cross-language effort, e.g., it provides descriptions and other information in various languages. DBpedia is an unavoidable resource for applications dealing with commonly known entities like notable persons, places; and for others looking for a rich hub connecting other semantic resources.

4.2 Mashups for Internet of Things

Internet of Things (IoT) has been creating vast amount of distributed stream data which can be modelled using RDF data model called Linked Stream Data. Linked Stream Data is becoming new valuable data sources for Linked Data Mashups. Therefore, the Web of Things (WoT) together with mashup-like applications is gaining popularity with the development of the Internet towards a network of interconnected objects, ranging from cars and transportation cargos to electrical appliances.

A long the same line, cities are alive: they rise, grow, and evolve like living beings, WoT allows a wide rage of Smart City applications. In essence, the state of a city changes continuously, influenced by a lot of factors, both human (people moving in the city or extending it) and natural ones (rain or climate changes). Cities are potentially huge sources of data of any kind and for the last years a lot of effort has been put in order to create and extract those sources. This scenario offers a lot of opportunities for mashup developers: by combining and processing the huge amount of data (both public and private) is possible to create new services for urban stakeholders—citizens, tourists, etc. called urban mashups [9].

Another application domain for IoT is emergency management [10]. Emergency management applications support a command staff in disruptive disaster situations, such as earthquakes, large-scale flooding or fires. One crucial requirement to emergency management systems is to provide decision makers with the relevant information to support their decisions. Mashups can help here by providing flexible and easily understandable views on up-to-date information.

4.3 Tourism Mashups

Web 2.0 has revolutionized the way users interact with information, by adding a vast amount of services, where end users explicitly and implicitly, and as a side effect of their use, generate content that feeds back into optimization of these services.

The resulting (integrated) platforms support users in and across different facets of life, including discovery and exploration, travel and tourism. Linked Data Mashup enables the creation and use of Travel Mashups, defined based on the varied travel information needs of different end users, spanning temporal, social and spatial dimensions [8]. The RDF-based travel mashups are created for bridging these dimensions, through the definition and use of composite, web- and mobile-based services. Their applications elicit the information need of an end user exploring an unfamiliar location, and demonstrates how the Topica Travel Mashup leverages social streams to provide a topical profile of Points of Interest that satisfies these user's requirements.

4.4 Biological and Life Science Domains

Semantic Web technologies provide a valid framework for building mashups in the life sciences. Ontology-driven integration represents a flexible, sustainable and extensible solution to the integration of large volumes of information. Additional resources, which enable the creation of mappings between information sources, are required to compensate for heterogeneity across namespaces. For instance, [6] uses an ontology-driven approach to integrate two gene resources (Entrez Gene and HomoloGene) and three pathway resources (KEGG, Reactome and BioCyc), for five organisms, including humans. Satya et al. [6] created the Entrez Knowledge Model (EKoM), an information model in OWL for the gene resources, and integrated it with the extant BioPAX ontology designed for pathway resources. The integrated schema is populated with data from the pathway resources, publicly available in BioPAX-compatible format, and gene resources for which a population procedure was created. The SPARQL query language is used to formulate queries over the integrated knowledge base to answer the three biological queries. Simple SPARQL queries could easily identify hub genes, i.e., those genes whose gene products participate in many pathways or interact with many other gene products. The identification of the genes expressed in the brain turned out to be more difficult, due to the lack of a common identification scheme for proteins.

5 Open Challenges

Even there has been a plenty of technology and research achievements of Linked Data community to enable Linked Data Mashups, there are a number of challenges to address when building mashups from different sources. The challenges can be classified into four groups: Entity extraction from text, object identification and consolidation, abstraction level mismatch, data quality.

Transforming Text Data to Symbolic Data for Linked Data Entities. A large portion of data is described in text. Human language is often ambiguous - the same company might be referred to in several variations (e.g. IBM, International Business Machines, and Big Blue). The ambiguity makes cross-linking with structured data difficult. In addition, data expressed in human language is difficult to process via software programs. Hence overcoming the mismatch between documents and data to extract RDF-based entities is still emerging challenges.

Object Identification and Consolidation. Structured data are available in a plethora of formats. Lifting the data to a common data format is thus the first step. But even if all data is available in a common format, in practice sources differ in how they state what is essentially the same fact. The differences exist both on the level of individual objects and the schema level. As an example for a mismatch on the object level, consider the following: the SEC uses a so-called Central Index Key (CIK) to identify people (CEOs, CFOs), companies, and financial instruments while other sources, such as DBpedia, use URIs to identify entities. In addition, each source typically uses its own schema and idiosyncrasies for stating what is essentially the same fact. Thus, methods have to be in place for reconciling different representations of objects and schemata.

Abstraction Levels. Data sources provide data at incompatible levels of abstraction or classify their data according to taxonomies pertinent to a certain sector. Since data is being published at different levels of abstraction (e.g. person, company, country, or sector), data aggregated for the individual viewpoint may not match data e.g. from statistical offices. Also, there are differences in geographic aggregation (e.g. region data from one source and country-level data from another). A related issue is the use of local currencies (USD vs. EUR) which have to be reconciled in order to make data from disparate sources comparable and amenable for analysis.

Data Quality. Data quality is a general challenge when automatically integrating data from autonomous sources. In an open environment the data aggregator has little to no influence on the data publisher. Data is often erroneous, and combining data often aggravates the problem. Especially when performing reasoning (automatically inferring new data from existing data), erroneous data has potentially devastating impact on the overall quality of the resulting dataset. Hence, a challenge is how data publishers can coordinate in order to fix problems in the data or blacklist sites which do not provide reliable data. Methods and techniques are needed to; check integrity, accuracy, highlight, identify and sanity check, corroborating evidence; assess the probability that a given statement is true, equate weight differences between market sectors or companies; act as clearing houses for raising and settling disputes between competing (and possibly conflicting) data providers and interact with messy erroneous Web data of potentially dubious provenance and quality. In summary, errors in signage, amounts, labeling, and classification can seriously impede the utility of systems operating over such data.

6 Conclusion

In this paper we have investigated state-of-the-art approaches in Linked Data Mashups in terms of technologies and application domain. From analytical reviews on approaches, we have drawn up open challenges for Linked Data Mashups. This review is a first steps of our research aiming at pointing out research trends in building up real applications. They are based on the open linked data in order to make a new leash of intelligent applications that utilise and facilitate the advantages of RDF data model and Linked Data for new generation of Linked Data Mashups application line.

References

[1] Le-Phuoc, D., Polleres, A., Hauswirth, M., Tummarello, G., Morbidoni, C.: Rapid prototyping of semantic mash-ups through semantic web pipes. In: Proceedings of the 18th International Confcrence on World Wide Web (WWW 2009) (2009)
[2] Huynh, D.F., Karger, D.R., Miller, R.C.: Exhibit: lightwcight structured data publishing. In: Proceedings of the 16th International Conference on World Wide Web (WWW 2007), pp. 737–746. ACM, New York (2007)
[3] Huynh, D.F., Miller, R.C., Karger, D.R.: Potluck: Data Mash-Up Tool for Casual Users. In: Aberer, K., et al. (eds.) ISWC/ASWC 2007. LNCS, vol. 4825, pp. 239–252. Springer, Heidelberg (2007)
[4] Liu, D., Li, N., Pedrinaci, C., Kopecký, J., Maleshkova, M., Domingue, J.: An approach to construct dynamic service mashups using lightweight semantics. In: Harth, A., Koch, N. (eds.) ICWE 2011. LNCS, vol. 7059, pp. 13–24. Springer, Heidelberg (2012)
[5] Ennals, R., Brewer, E., Garofalakis, M., Shadle, M., Gandhi, P.: Intel Mash Maker: join the web. SIGMOD Rec. 36(4), 27–33 (2007)
[6] Sahoo, S.S., Bodenreider, O., Rutter, J.L., Skinner, K.J., Sheth, A.P.: An ontology-driven semantic mashup of gene and biological pathway information: Application to the domain of nicotine dependence. J. of Biomedical Informatics 41(5), 752–765 (2008)
[7] Heath, T., Bizer, C.: Linked Data: Evolving the Web into a Global Data Space, 1st edn. Synthesis Lectures on the Semantic Web: Theory and Technology, vol. 1(1), pp. 1–136. Morgan & Claypool (2011)
[8] Cano, A.E., Dadzie, A.-S., Ciravegna, F.: Travel Mashups. In: Semantic Mashups (2013)
[9] Dell'Aglio, D., Celino, I., Della Valle, E.: Urban Mashups. In: Semantic Mashups (2013)
[10] Sosins, A., Zviedris, M.: Mashups for the Emergency Management Domain. In: Semantic Mashups (2013)
[11] Kenda, K., Fortuna, C., Moraru, A., Mladenić, D., Fortuna, B., Grobelnik, M.: Mashups for the Web of Things. In: Semantic Mashups (2013)
[12] Héder, M., Solt, I.: DBpedia Mashups. In: Semantic Mashups (2013)
[13] Papadakis, I., Apostolatos, I.: Mashups for Web Search Engines. In: Semantic Mashups (2013)
[14] Huynh, D.F., Karger, D.R., Miller, R.C.: Exhibit: lightweight structured data publishing. In: Proceedings of the 16th International Conference on World Wide Web (WWW 2007), pp. 737–746. ACM, New York (2007)
[15] Quan, D., Huynh, D., Karger, D.R.: Haystack: A Platform for Authoring End User Semantic Web Applications. In: Fensel, D., Sycara, K., Mylopoulos, J. (eds.) ISWC 2003. LNCS, vol. 2870, pp. 738–753. Springer, Heidelberg (2003)
[16] Endres-Niggemeyer, B.: Semantic mashups. Springer (2013) ISBN 978-3-642-36403-7
[17] Berners-Lee, T.: Linked Data - Design Issues (2006), http://www.w3.org/DesignIssues/LinkedData.html

Belief Propagation Method for Word Sentiment in WordNet 3.0

Andrzej Misiaszek, Przemysław Kazienko, Marcin Kulisiewicz, Łukasz Augustyniak, Włodzimierz Tuligłowicz, Adrian Popiel, and Tomasz Kajdanowicz

Faculty of Computer Science and Management, Institute of Informatics, Wroclaw University of Technology, Wroclaw, Poland
{andrzej.misiaszek,kazienko,marcin.kulisiewicz,lukasz.augustyniak, wlodzimierz.tuliglowicz,adrian.popiel,tomasz.kajdanowicz}@pwr.wroc.pl

Abstract. The main goal of the paper is to present that word's sentiment can be discovered from propagation through well-defined word networks such as Word–Net. Therefore a new method for propagation of sentiment from a given word seed - Micro-WNOp corpus over the word network (WordNet 3.0) has been proposed and evaluated. The experimental studies proved that WordNet has a great potential in sentiment propagation, even if types of links (e.g. hyponymy, heteronymy etc.) and semantic meaning of words are not taken into consideration.

Keywords: sentiment analysis, belief propagation, relational propagation, wordnet, sentiwordnet, complex networks, linguistic network.

1 Introduction

Sentiment analysis has become a very popular research filed in recent years. Overall, sentiment analysis is typically used to classify opinions about various things ranging from movie reviews to political opinions about elections. Basically, it is a method to tag an opinionated text. In its simplest representation we can add some values like positive, negative or objective to a given text. This gives us understanding if an opinion is more in favour of the subject described in the text or not. Sentiment assessment might be considered with various granularity, starting from single word and ending on whole documents. The problem of sentiment assignment for a single word is considered in this paper. For this purpose it is proposed a new method for sentiment propagation, which can assign words' sentiment regardless of the WordNet's language. Recently introduced approaches were specific and limited only to English language. In the paper we recall basic related work (Section 2), introduce the sentiment problem (Section 3) and propose a method for sentiment propagation (Section 4), which is followed by the results of its evaluation.

2 Related Work

Sentiment analysis has a great value in text analysis and direct applications driven by modern business, especially placed in the Internet. The Internet seems

N.T. Nguyen et al. (Eds.): ACIIDS 2014, Part II, LNAI 8398, pp. 263–272, 2014.

to be an ideal place for sentiment mining, not only because of its accessibility, but mainly due to its nature as a place for massive social communication. Simultaneously, the Internet is a near infinite resource if it comes to texts that contain opinions. Among others, Ohana and Tierney presented sentiment analysis in film reviews [2] using SentiWordNet. SentiWordNet is a network of synsets that were taken from WordNet and evaluated by experts in terms of sentiment. Authors took terms from movie reviews in order to determine their sentiment. In the paper it was proved that SentiWordNet can be used as resource for sentiment analysis. Typical representation of synsets in a graph is shown in Figure 1.

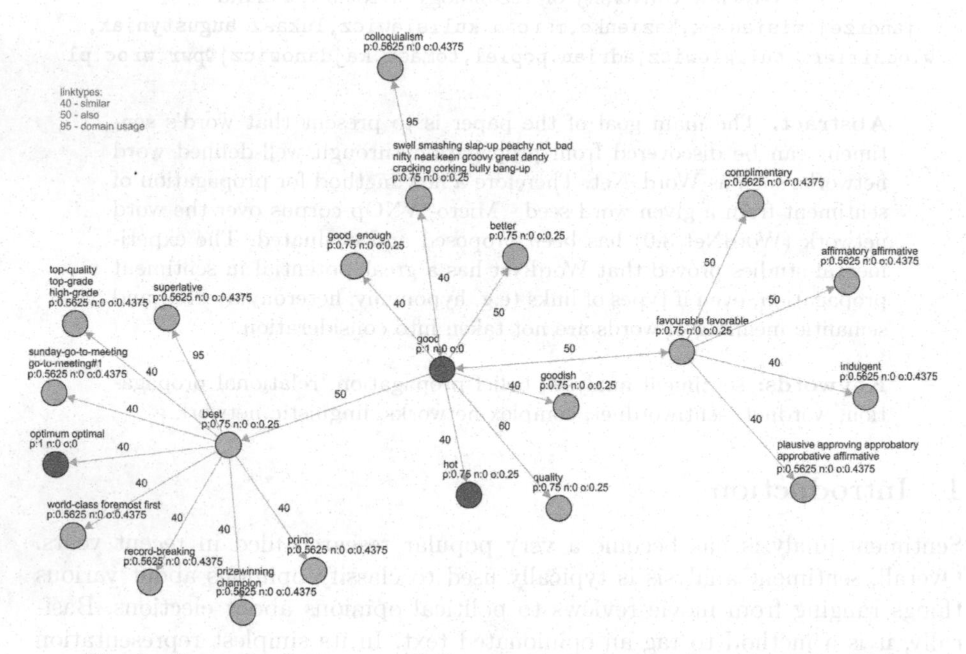

Fig. 1. Example of synsets with its relations extracted from SentiWordNet

Pang showed in his work [1] that simple classification methods can outperform human produced baseline in distinct sentiment analysis tasks, but are to simple in order to reflect whole phenomena. Another approach was proposed by Prabowo and Thelwall who combined rule-based classification and supervised learning into new method [6]. This approach used on movie reviews and social network comments showed that hybrid classification can improve performance in sentiment analysis. That work reaffirmed the classification approach in the field.

WordNet as a network of synsets was used to measure a distance or similarity between words. This measures focused mostly on taxonomic relations. Kamps used synonymy relation to derive semantic orientation of adjectives [4].

Sentiment and emotional propagation was mentioned in [3]. Fan et al. observed that some emotions like anger are more influential in network than another.

However problem of word sentiment propagation without partition on part of speech or without usage of some kind of hybrid supervised learning was rarely studied.

3 Problem Description

Sentiment analysis, as mentioned earlier, is a process of identifying subjective information from mostly text data. Before the process can start, it needs to be identified whether the text is either subjective or objective. This process is called opinion mining. Only the opinionated texts can be used for sentiment analysis since only author's subjective statements about a given topic are valuable. The result of sentiment analysis is classification of a textual units-chunks (a document, comment, paragraph or just a word) to one of three classes: positive, negative or objective) [8].

The scope of sentiment analysis and related activities is very diverse and consists of many component tasks described in Figure 2.

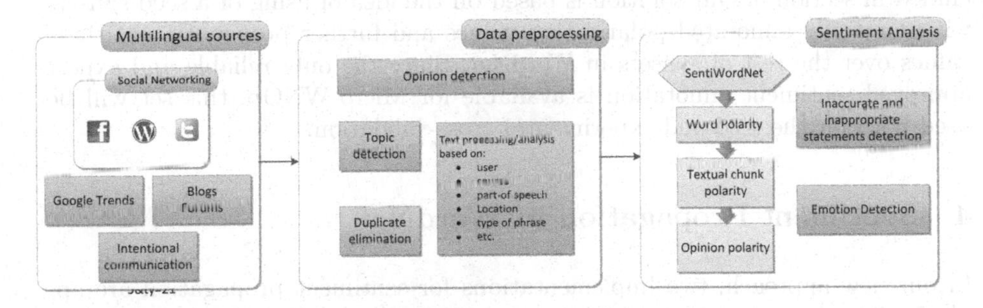

Fig. 2. Sentiment analysis model

The main source of polarity knowledge used in sentiment analysis could be SentiWordNet. For English, it can be based on Princeton WordNet. WordNet is a large lexical database, in which nouns, verbs, adjectives and adverbs are grouped into sets of cognitive synonyms (synsets), each expressing a distinct concept. Synsets are interlinked by means of conceptual semantic and lexical relations, see section 5.1. Every synset may also be tagged with a gloss that describes it.

The WordNet can be extended with sentiment information assigned to every synset. This assignment for all synsets may be performed in many ways. Esuli [7] used a small seed that was taken from research of Turney and Littman [8]. This seed contained seven words that are positive and seven negative. Next, they used gloss for sentiment propagation in WordNet. If a new synsets gloss had any word that were already tagged with sentiment then a new word would also be tagged with a sentiment value.

The general problem with this approach lies in accessibility to a reliable resource in languages other than English that can be used for WordNet sentiment annotation. For English, there exists SentiWordNet but there are no automatic methods or manually worked out SentiWordNets for other than English languages. The solution for that is the development of generic methods for tagging of any WordNet with sentiment. Since non-English WordNets are not coherent with Englis h one, the reliable polarity transfer from English SentiWordNet to other languages is hardy possible. Key differences and limitations in SentiWordNet creation are as follows:

- language specific lexicons,
- sentiment score should be updated in time,
- high cost of resources needed for manual annotation,
- copying sentiment values is forbidden,
- different connections between synsets,
- source lexicon usually is incompatible.

Authors of SentiWordNet used a manually annotated resource (Micro-WNOp) for synset sentiment evaluation. An additional description of this corpus is included in section 5. Our solution is based on the idea of using of a seed synsets with manually annotated polarization values and further propagation of these values over the rest of synsets in WordNet. Since the only reliable and expert approved sentiment annotation is available for Micro-WNOp, this set will be used both for the seed and experimental cross-validation.

4 Sentiment Propagation in WordNet

In our new approach, two implementations for sentiment propagation are applied. The first one is Belief Propagation (BP). Sen et all. [5] classified BP as algorithm based on global formulations in Markov Random Field (MRF). MRF is a graph where every node represents a Random Variable, and edges represent relations between nodes. In MRF, the input X denotes the observed variables (random variables network nodes with a known value) and output Y are unobserved variables (unknown nodes). The algorithm needs to assign values to these unknown variables based on the observed variables and edges between all nodes in network.

The algorithm of Belief Propagation is a simple message-passing algorithm. In each iteration, each random variable Y_i in the MRF sends a message m_{ij} to a neighbouring random variable Y_j . Every message consists of belief which value should have a node that receives message. The algorithm stops when message passing stops showing any change. In our case, the role of MRF plays the WordNet network. As X observed variables synsets from Micro-WNOp were taken. As Y unobserved variables remaining synsets from Wordnet were used.

The idea of the proposed algorithm is depicted in Figure 3. For research R language was used with RStudio. As the seed data from Micro-WNOp corpus was taken and randomized into 10 equal sets. For propagation basic rules were applied:

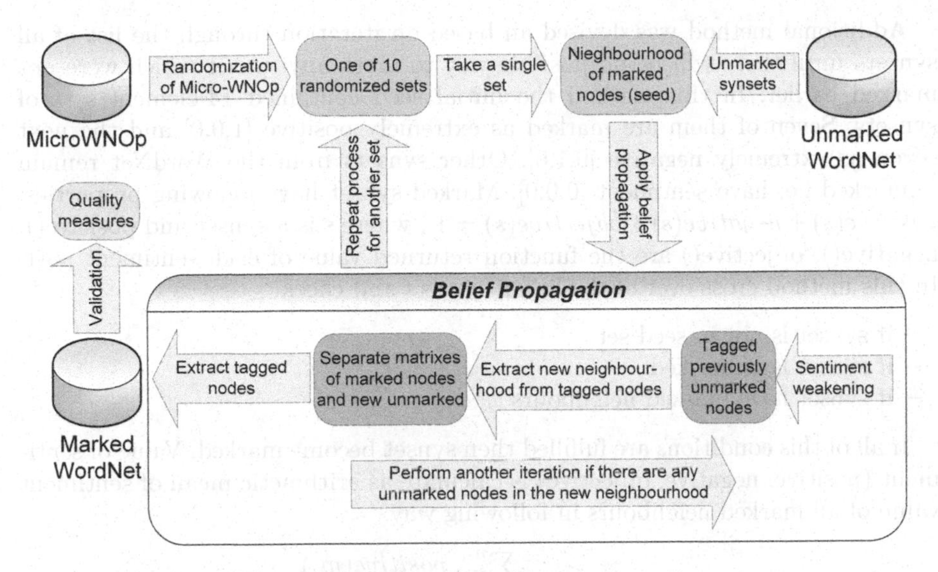

Fig. 3. Algorithm of sentiment propagation

- from every seed set (roughly 106 synsets in each) the sentiment value has been propagated to all other nodes in the WordNet,
- any node that has already been tagged with sentiment was not annotated again,
- during the runtime, a matrix was storing information about passed network paths which prevented the algorithm from reverse loops,
- the distance between an unmarked node and the seed node matters. If the distance is greater, bigger sentiment weakening was implemented,
- the algorithm stops when all possible nodes in the graph have been considered and every possible node has been annotated with sentiment.

Weaking of sentiment value can be represented by:

$$Sentiment = NeighbourSentiment * \lambda \qquad (1)$$

In experiments 0.75 was assigned as λ.

Note that the propagation is performed over the entire WordNet network (by means of its links) even though the quality of validation is limited to only reference nodes: the Micro-WNOp remaining nodes, i.e. the nodes that do not belong to the seed and are initially treatred as unknown.

Only one seed set in every iteration was used. The first step was to take the neighborhood of the seed and propagated (with weakening) the sentiment value to its neighbors. After that the sentiment value was propagated further to a new set of neighbors. At the end, ten sets of synsets were received with sentiment values added. Every node from these sets would have a positive and negative value ranging from 0 to 1. If a node had values 0 as positive and 0 as negative it would be classified as objective.

Additional method was devised an based on iteration through the list of all synsets until there will be no possible way to mark any synset which were not marked earlier. In this method the initial seed contained 14-element sets of synsets. Seven of them are marked as extremely positive [1,0,0] and the next seven as extremely negative [0,1,0]. Other synsets from the WordNet remain unmarked i.e. have sentiment [0,0,0]. Marked synset have following properties: $positive(s) + negative(s) + objective(s) = 1$, where s is a synset and positive(), negative(), objective() are the function returned value of each sentiment part. In this method cross over of the list of synsets and check:

– if synset is not in seed set
– if synset is unmarked
– if synset have marked neighbours

If all of this conditions are fulfilled then synset become marked. Value of sentiment (positive, negative, objective) is calculate as arithmetic mean of sentiment value of all marked neighbours in following way:

$$positive(s) = \frac{\sum_{i=1}^{n} positive(m_i)}{n}, \tag{2}$$

$$negative(s) = \frac{\sum_{i=1}^{n} negative(m_i)}{n}, \tag{3}$$

$$objective(s) = \frac{\sum_{i=1}^{n} objective(m_i)}{n}, \tag{4}$$

where m_i is element of n-element set M which is set of all marked neighbors of synset s. In case where none of the above conditions is met this synset is bypassed. If in the first iterations unmarked synsets are present the process begins anew. This method was named as Neighbourhood Sentiment Propagation (NSP) and results for this method are presented in tables 8 and 9.

5 Validation

5.1 Datasets

Two datasets, WordNet 3.0 and Micro-WNOp were used in experiments. The sources are available at http://wordnet.princeton.edu/wordnet for WordNet and http://www-3.unipv.it/wnop for Micro-WNOp.

In this paper, connections between various synsets are very important factor. Hence, below some statistics for both of datasets are presented.

WordNet 3.0. Synsets in WordNet 3.0 are connected by both semantic and lexical types of relations. There are 117,659 various synsets that are consisted of 147,306 words in this database. Synsets are connected by 22 different types of semantic relations and 12 types of lexical ones. For the purpose of this study, were used only semantic relations. There are 285,348 semantic links between all synsets:

- 89,089 both hypernym and hyponym,
- 21,386 similarity,
- 12,293 both holonym and meronym,
- 61,198 others connections.

Micro-WNOp. The Micro-WNOp corpus is composed of 1,105 WordNet synsets (but it was designed for earlier WordNet versions). It was created using synsets from General Inquirer lexicon. However, after mapping Micro-WNOp with Word- Net 3.0 49 synsets were omitted. Hence, this is a dataset of 1,056 synsets from WordNet 3.0 which were evaluated by human experts as positive, negative or objective. The set consists of 3 groups of synsets. The first of them (covering approximately 10% of synsets) was evaluated by all 5 experts. The second group of synsets (ca. 45% of synsets) was evaluated by 3 experts (each synset consists of 3 pairs of positive and negative sentiment values). The third group (about 45% of synsets) was evaluated by the remaining 2 experts. Its purpose was to evaluate results of sentiment propagation in SentiWordNet 1.0, 1.1, 2.0 and 3.0. Due to the human expert engagement, Micro-WNOp appers to be a reliable reference set for validation. Note that some synsets may be annotated as both positive and negative simultaneously.

Experimental Dataset. As a result of running the algorithm 10 datasets were obtained, which contained Micro-WNOp seed propagation on the whole WordNet. It was not possible to validate our findings on SentiWordNet since it does not have every node marked correctly and reliably. For the purpose of evaluation a cross-validation method was devised on the Micro- WNOp dataset.

From each output dataset synsets were extracted that were present in Micro-WNOp. Ten datasets were received containing around 950 each. After that all datasets were combined into one. This dataset contained a mean value for every synsets from the propagation. Next step was to cross-validate the results from propagation with same synsets in Micro-WNOp. Our final assessment would be based on matching polarity from both datasets. If same synsets have identical values it would be classified as a good result.

5.2 Evaluation Measures

Classification. In order to evaluate the sentiment propagation two classifications were used. Separate classifications for positive and negative polarity were performed. Each classification were divided into 5 classes. Table 1 presents assignment to the classes.

Comparison. Both the study and the Micro-WNOp database were embraced by a classification. It allowed to evaluate obtained during the experiment results. During classification of Micro-WNOp average sentiment polarity for each synset was used. Initially Micro-WNOp contained several values (sometimes different values) of sentiment, see more in Section 5.1. Similar results averaging for

Table 1. Sentiment polarity classification

Class	Sentiment range
0	0
1	(0, 0.25]
2	(0.25, 0.5]
3	(0.5, 0.75]
4	(0.75, 1]

sub- sequent cross-validation folds was performed. In addition, several methods were used for evaluation:

Method 1 - the average values for synsets of all 10 folds from cross-validation.
Method 2 - the values obtain by Method 1 were normalized.
Method 3 - the values which has not been marked by algorithm were skipped.

Finally, results as percentage of correct sentiment polarity assignment were calculated. Outcome is presented in table 10.

5.3 Results

The obtained results are presented in tables from 2 to 7. The „MNW" means sentiment class assignment for Micro-WNOp and „Method" implies sentiment classification of results for chosen methods in table description. For example, number 65 from table 2 means: 65 synsets assigned as class 1 in Micro-WNOp, were assigned to class 2 by Belief Propagation algorithm used in this experiment. Hence, the best assignments are at diagonal in tables. In table 10 final results are presented.

Accuracy of sentiment propagation using presented BP implementation gives outcomes at percentage ranging from 22 to 30. The highest results (positive correctness = 30.05% and negative correctness = 26.35%) were obtained by Method 2. However, this method has the broadest range of obtained classes. This is obvious result of using normalization. The best outcomes should present high concentration near the diagonal line (in tables from 2 to 7). Hence, as

Table 2. Method 1 - positive

MWN	Method				
	0	1	2	3	4
0	**166**	24	5	3	7
1	490	**112**	65	70	78
2	1	0	**0**	3	29
3	0	0	0	**0**	2
4	0	0	0	0	**0**

Table 3. Method 1 - negative

MWN	Method				
	0	1	2	3	4
0	**143**	11	7	4	8
1	545	**94**	59	63	114
2	0	0	**0**	1	6
3	0	0	0	**0**	0
4	0	0	0	0	**0**

Table 4. Method 2 - positive

MWN	Method 0	1	2	3	4
0	**165**	8	0	5	24
1	219	**44**	10	0	0
2	216	66	**36**	19	5
3	50	15	19	**34**	49
4	7	3	5	18	**38**

Table 5. Method 2 - negative

MWN	Method 0	1	2	3	4
0	**146**	7	2	3	30
1	231	**31**	9	1	0
2	244	55	**33**	24	4
3	62	12	20	**22**	48
4	5	0	2	18	**46**

Table 6. Method 3 - positive

MWN	Method 0	1	2	3	4
0	**164**	8	0	0	0
1	435	**110**	46	19	5
2	57	18	**24**	52	87
3	1	0	0	**5**	17
4	0	0	0	0	**7**

Table 7. Method 3 - negative

MWN	Method 0	1	2	3	4
0	**143**	7	2	0	0
1	475	**86**	42	25	4
2	67	12	**22**	40	94
3	3	0	0	**3**	18
4	0	0	0	0	**12**

Table 8. NSP - positive

MWN	Method 0	1	2	3	4
0	**119**	27	11	18	43
1	14	**3**	0	0	2
2	87	17	**17**	16	17
3	94	30	12	**14**	20
4	343	59	30	28	**34**

Table 9. NSP - negative

MWN	Method 0	1	2	3	4
0	**190**	23	11	12	19
1	237	**33**	27	23	36
2	131	33	**14**	13	32
3	41	6	9	**10**	16
4	89	10	5	10	**25**

a best result could be chosen Method 3 (with positive correctness = 29.38% and negative correctness = 25.21%). Method 3 tables do not contain ranges as wide as Method 2 of obtained classes, i.e. for Method 2 negative cells from 0 to 4 contain 30 synsets and corresponding cell for Method 3 negative contain 0 synsets.

Based on the obtained results, basic BP can provide only low acvuracy for sentiment propagation. Better accuracy could be achieved by using semantic links in WordNet. Some of them have can have significant impact on synsets sentiment. However, analysis on which semantic links have more influence for sentiment than others has not yet been examined.

Table 10. Correct sentiment assignment

Synset sentiment	Method 1	Method 2	Method 3	Simple NSP
Positive	26.35%	30.05%	29.38%	17.73%
Negative	22.46%	26.35%	25.21%	25.78%

6 Conclusions and Future Works

The general idea of the paper was to test if sentiment can propagate through well-defined word networks such as WordNet. For that purpose a new method for sentiment propagation through WordNet by semantic connections and use BP algorithm was proposed. The experiemental studies revealed that sentiment propagation is possible by means of knowledge hidden in the WordNet links. However, this approach requires further research. Sentiment analysis is a topic of active research, but still a lot can be done in this area. More advanced algorithms will be used for better sentiment propagation in WordNet.

Appendix

This work is partially funded by the European Commission under the 7th Framework Programme, Coordination and Support Action, Grant Agreement Number 316097, European research centre of Network intelliGence for INnovation Enhancement (ENGINE).

References

[1] Bo, P., Lillian, L., Shivakumar, V.: Thumbs up?: sentiment classification using machine learning techniques. In: Proceedings of the ACL 2002 Conference on Empirical Methods in Natural Language Processing, vol. 10, pp. 79–86. Association for Computational Linguistics (2002)
[2] Bruno, O., Brendan, T.: Sentiment classification of reviews using sentiwordnet. In: 9th IT & T Conference (2009)
[3] Fan, R., Zhao, J., Chen, Y., Xu, K.: Anger is more influential than joy: Sentiment correlation in weibo (2013)
[4] Kamps, J., Marx, M., Mokken, R.J., de Rijke, M.: Using WordNet to measure semantic orientation of adjectives. In: LREC (2004)
[5] Prithviraj, S., Galileo, N., Mustafa, B., Lise, G.: Collective classification. In: Encyclopedia of Machine Learning, pp. 189–193. Springer (2010)
[6] Rudy, P., Mike, T.: Sentiment analysis: A combined approach. Journal of Informetrics 3(2), 143–157 (2009)
[7] Stefano, B., Andrea, E., Fabrizio, S.: Sentiwordnet 3.0: An enhanced lexical resource for sentiment analysis and opinion mining. In: LREC 2010, vol. 10, pp. 2200–2204 (2010)
[8] Turney, P.D., Littman, M.L.: Measuring praise and criticism: Inference of semantic orientation from association. ACM Transactions on Information Systems 21(4), 315–346 (2003)

A New Method for Autocratic Decision Making Using Group Recommendations Based on Intervals of Linguistic Terms and Likelihood-Based Comparison Relations

Shyi-Ming Chen and Bing-Han Tsai

Department of Computer Science and Information Engineering,
National Taiwan University of Science and Technology,
Taipei, Taiwan

Abstract. This paper presents a new method for autocratic decision making using group recommendations based on intervals of linguistic terms and likelihood-based comparison relations. It can overcome the drawbacks of the existing methods for autocratic decision making using group recommendations. The proposed method provides us with a useful way for autocratic decision making using group recommendations based on intervals of linguistic terms and likelihood-based comparison relations.

Keywords: Consensus measures, group decision making, intervals of linguistic terms, likelihood-based comparison relations.

1 Introduction

Some methods have been presented for dealing with group decision making problems [1]-[15], [17]. In [2], Ben-Arieh and Chen presented the fuzzy linguistic order weighted average (FLOWA) operator and applied it for autocratic decision making using group recommendations. In [5], Chen and Lee presented a method for autocratic decision making using group recommendations based on the interval linguistic labels ordered weighted average (ILLOWA) operator and likelihood-based comparison relations. However, Ben-Arieh and Chen's method [2] has the drawbacks that it cannot get a correct preference order of alternatives in some situations and it cannot deal with interval linguistic terms. Moreover, Chen and Lee's method [5] has the drawback that it cannot deal with the preferring ordering of the alternatives in some situations. Therefore, we must develop a new method for autocratic decision making using group recommendations based on intervals of linguistic terms and likelihood-based comparison relations to overcome the drawbacks of Ben-Arieh and Chen's method [2] and Chen and Lee's method [5].

In this paper, we present a new method for autocratic decision making using group recommendations based on intervals of linguistic terms and likelihood-based

N.T. Nguyen et al. (Eds.): ACIIDS 2014, Part II, LNAI 8398, pp. 273–281, 2014.

comparison relations. It can overcome the drawbacks of Ben-Arieh and Chen's method [2] and Chen and Lee's method [5] for autocratic decision making using group recommendations.

2 Likelihood-Based Comparison Relations

In [16], Xu and Da presented the concept of likelihood-based comparison relations and applied it to deal with interval linguistic terms for group decision making. Assume that there is a set $S = \{s_0, s_1, \cdots, s_t\}$ of linguistic terms and assume that there are two interval linguistic terms $\tilde{f}_1 = [s_{x_1}, s_{x_2}]$ and $\tilde{f}_2 = [s_{y_1}, s_{y_2}]$, where $0 \le x_1 \le x_2 \le t$ and $0 \le y_1 \le y_2 \le t$, then the likelihood-based comparison relation $p(\tilde{f}_1 \ge \tilde{f}_2)$ between the interval linguistic terms $\tilde{f}_1 = [s_{x_1}, s_{x_2}]$ and $\tilde{f}_2 = [s_{y_1}, s_{y_2}]$ is defined as follows:

$$p(\tilde{f}_1 \ge \tilde{f}_2) = max\left(1 - max\left(\frac{y_2 - x_1}{L(\tilde{f}_1) + L(\tilde{f}_2)}, 0\right), 0\right), \tag{1}$$

where $L(\tilde{f}_1) = x_2 - x_1$ and $L(\tilde{f}_2) = y_2 - y_1$.

3 A New Method for Autocratic Decision Making Using Group Recommendations

In this section, we present a new method for autocratic decision making using group recommendations based on intervals of linguistic terms and likelihood-based comparison relations. Assume that there is a linguistic terms set $S = \{s_0, s_1, \cdots, s_t\}$, assume that there are n alternatives $A_1, A_2, \cdots, A_n$, and assume that there are m experts $E_1, E_2, \cdots, E_m$. Assume that the target $T = [s_0, s_t]$. Let $|s_p|$ denote the subscript of the linguistic term s_p, where $|s_p| = p$ and $1 \le p \le t$. In the following, we presented the addition operation and the division operation between intervals of linguistic terms. Let X and Y be two intervals of linguistic terms, where $X = [s_a, s_b]$ and $Y = [s_c, s_d]$, and let m be a constant. The addition operation between the intervals of linguistic terms X and Y is defined as follows:

$$X + Y = [s_a, s_b] + [s_c, s_d] = [s_{a+c}, s_{b+d}].$$

The division operation between the interval of linguistic terms X and the constant m is defined as follows:

$$X/m = [s_a, s_b]/m = [s_{\lfloor \frac{a}{m} \rfloor}, s_{\lceil \frac{b}{m} \rceil}],$$

where "$\left\lfloor \frac{a}{m} \right\rfloor$" denotes the round down value of $\frac{a}{m}$ and "$\left\lceil \frac{b}{m} \right\rceil$" denotes the round up value of $\frac{b}{m}$. Assume that the interval linguistic preference matrix P_k for expert E_k is shown as follows:

$$P_k = (\tilde{f}_{ij}^k)_{n\times n} = \begin{array}{c} \\ A_1 \\ A_2 \\ \vdots \\ A_n \end{array} \begin{array}{c} \begin{array}{cccc} A_1 & A_2 & \cdots & A_n \end{array} \\ \begin{bmatrix} - & \tilde{f}_{12}^k & \cdots & \tilde{f}_{1n}^k \\ \tilde{f}_{21}^k & - & \cdots & \tilde{f}_{2n}^k \\ \vdots & \vdots & \ddots & \vdots \\ \tilde{f}_{n1}^k & \tilde{f}_{n2}^k & \cdots & - \end{bmatrix} \end{array}$$

where $\tilde{f}_{ij}^k = [s_\alpha^k, s_\beta^k]$, $\tilde{f}_{ji}^k = [s_{t-\beta}^k, s_{t-\alpha}^k]$, $\tilde{f}_{ij}^{k(L)}$ denotes the left linguistic term of $\tilde{f}_{ij}^k$, $\tilde{f}_{ij}^{k(L)} = s_\alpha^k$, $\tilde{f}_{ij}^{k(R)}$ denotes the right linguistic term of $\tilde{f}_{ij}^k$, $\tilde{f}_{ij}^{k(R)} = s_\beta^k, i \neq j, 0 \leq \alpha \leq \beta \leq t$, and $1 \leq k \leq m$. The proposed method is now presented as follows:

Step 1: Initially, let $r = 1$. Aggregate the interval linguistic preference matrices P_1, P_2, …, and P_m into the collective interval linguistic preference matrix $G = [g_{ij}]_{n\times n}$, where

$$g_{ij} = \left[g_{ij}^{(L)}, g_{ij}^{(R)}\right] = \left[s_{\left\lfloor \frac{\sum_{k=1}^m \left|\tilde{f}_{ij}^{k(L)}\right|}{m} \right\rfloor}, s_{\left\lceil \frac{\sum_{k=1}^m \left|\tilde{f}_{ij}^{k(R)}\right|}{m} \right\rceil}\right] \tag{2}$$

$g_{ij}^{(L)} = s_{\left\lfloor \frac{\sum_{k=1}^m \left|\tilde{f}_{ij}^{k(L)}\right|}{m} \right\rfloor}$, $g_{ij}^{(R)} = s_{\left\lceil \frac{\sum_{k=1}^m \left|\tilde{f}_{ij}^{k(R)}\right|}{m} \right\rceil}$, $\tilde{f}_{ij}^{k(L)}$ denotes the left linguistic term of $\tilde{f}_{ij}^k$, $\tilde{f}_{ij}^{k(R)}$ denotes the right linguistic term of $\tilde{f}_{ij}^k$, $\left|\tilde{f}_{ij}^{k(L)}\right|$ denotes the subscript of the linguistic term $\tilde{f}_{ij}^{k(L)}$, $\left|\tilde{f}_{ij}^{k(R)}\right|$ denotes the subscript of the linguistic term $\tilde{f}_{ij}^{k(R)}$, "$\left\lfloor \frac{\sum_{k=1}^m \left|\tilde{f}_{ij}^{k(L)}\right|}{m} \right\rfloor$" denotes the round down value of $\frac{\sum_{k=1}^m \left|\tilde{f}_{ij}^{k(L)}\right|}{m}$, "$\left\lceil \frac{\sum_{k=1}^m \left|\tilde{f}_{ij}^{k(R)}\right|}{m} \right\rceil$" denotes the round up value of $\frac{\sum_{k=1}^m \left|\tilde{f}_{ij}^{k(R)}\right|}{m}$, $1 \leq i \leq n, 1 \leq j \leq n,\ i \neq j$, and $1 \leq k \leq m$.

Step 2: Construct the preference matrix $R_k = \left[r_{ij}^k\right]_{n\times n}$ for each expert E_k and construct the collective preference matrix $RG = \left[rg_{ij}\right]_{n\times n}$ for all experts, where

$$r_{ij}^k = p(\tilde{f}_{ij}^k \geq T) = \max\left(1 - \max\left(\frac{t - \left|\tilde{f}_{ij}^{k(L)}\right|}{\left(\left|\tilde{f}_{ij}^{k(R)}\right| - \left|\tilde{f}_{ij}^{k(L)}\right|\right) + t}, 0\right), 0\right), \tag{3}$$

$$rg_{ij} = p(g_{ij} \geq T) = \max\left(1 - \max\left(\frac{t - \left|g_{ij}^{(L)}\right|}{\left(\left|g_{ij}^{(R)}\right| - \left|g_{ij}^{(L)}\right|\right) + t}, 0\right), 0\right), \tag{4}$$

where t denotes the subscript of the largest linguistic term in the linguistic terms set S, $\tilde{f}_{ij}^{k(L)}$ denotes the left linguistic term of $\tilde{f}_{ij}^{k}$, $\tilde{f}_{ij}^{k(R)}$ denotes the right linguistic term of $\tilde{f}_{ij}^{k}$, $\left|\tilde{f}_{ij}^{k(L)}\right|$ denotes the subscript of the linguistic term $\tilde{f}_{ij}^{k(L)}$, $\left|\tilde{f}_{ij}^{k(R)}\right|$ denotes the subscript of the linguistic term $\tilde{f}_{ij}^{k(R)}$, $g_{ij}^{(L)}$ denotes the left linguistic term of g_{ij} in the collective interval linguistic preference matrix G, $g_{ij}^{(R)}$ denotes the right linguistic term of g_{ij} in the collective interval linguistic preference matrix G, $\left|g_{ij}^{(L)}\right|$ denotes the subscript of the linguistic term $g_{ij}^{(L)}$ in the collective interval linguistic preference matrix G, $\left|g_{ij}^{(R)}\right|$ denotes the subscript of the linguistic term $g_{ij}^{(R)}$ in the collective interval linguistic preference matrix G, $1 \leq i \leq n$, $1 \leq j \leq n$, $i \neq j$, and $1 \leq k \leq m$.

Step 3: Calculate the score V_i of alternative A_i, where the score V_i is equal to the summation of the values at the ith row of the constructed collective preference matrix RG, where $1 \leq i \leq n$.

Step 4: Construct the consensus matrix $C_k = [c_{ij}^k]_{n \times n}$ for each expert E_k, where

$$c_{ij}^k = 1 - \left| rg_{ij} - r_{ij}^k \right|, \tag{5}$$

$1 \leq i \leq n$, $1 \leq j \leq n$, $i \neq j$, and $1 \leq k \leq m$.

Step 5: Calculate the consensus degree CD_k for each expert E_k, shown as follows:

$$CD_k = \sum_{j=1}^{n}\sum_{i=1}^{n} \frac{c_{ij}^k}{n(n-1)}, \tag{6}$$

where $1 \leq i \leq n$, $1 \leq j \leq n$, $i \neq j$, and $1 \leq k \leq m$.

Step 6: Calculate the group consensus degree C_G for all experts, shown as follows:

$$C_G = \sum_{k=1}^{m} \frac{CD_k}{m}. \tag{7}$$

If the group consensus degree C_G is larger than or equal to a predefined group consensus threshold value δ, where $\delta \in [0, 1]$, then the larger the score V_i of alternative A_i, the better the preference order of the alternative A_i, where $1 \leq i \leq n$; **Stop**. Otherwise, go to **Step 7**.

Step 7: Construct the proximity matrix $PM_k = \left[pm_{ij}^k\right]_{n\times n}$ for each expert E_k, where

$$pm_{ij}^k = \left[pm_{ij}^{k(L)}, pm_{ij}^{k(R)}\right] = \left[\left|\left|\tilde{f}_{ij}^{k(L)}\right| - \left|g_{ij}^{(L)}\right|\right|, \left|\left|\tilde{f}_{ij}^{k(R)}\right| - \left|g_{ij}^{(R)}\right|\right|\right], \tag{8}$$

$pm_{ij}^{k(L)} = \left|\tilde{f}_{ij}^{k(L)}\right| - \left|g_{ij}^{(L)}\right|$, $pm_{ij}^{k(R)} = \left|\tilde{f}_{ij}^{k(R)}\right| - \left|g_{ij}^{(R)}\right|$, $\tilde{f}_{ij}^{k(L)}$ denotes the left linguistic term of $\tilde{f}_{ij}^{k}$, $\tilde{f}_{ij}^{k(R)}$ denotes the right linguistic term of $\tilde{f}_{ij}^{k}$, $\left|\tilde{f}_{ij}^{k(L)}\right|$ denotes the subscript of the linguistic term $\tilde{f}_{ij}^{k(L)}$, $\left|\tilde{f}_{ij}^{k(R)}\right|$ denotes the subscript of the linguistic term $\tilde{f}_{ij}^{k(R)}$, $g_{ij}^{(L)}$ denotes the left linguistic term of g_{ij} in the collective interval linguistic preference matrix G, $g_{ij}^{(R)}$ denotes the right linguistic term of g_{ij} in the collective interval linguistic preference matrix G, $\left|g_{ij}^{(L)}\right|$ denotes the subscript of the linguistic term $g_{ij}^{(L)}$ in the collective interval linguistic preference matrix G, $\left|g_{ij}^{(R)}\right|$ denotes the subscript of the linguistic term $g_{ij}^{(R)}$ in the collective interval linguistic preference matrix G, $1 \le i \le n, 1 \le j \le n, i \ne j$, and $1 \le k \le m$.

Step 8: Get the set Q_k of the pairs (i, j) of alternatives A_i and A_j for expert E_k, shown as follows:

$$Q_k = \{(i, j) \mid CD_k < C_G \text{ and } c_{ij}^k < C_G\}, \tag{9}$$

where CD_k denotes the consensus degree for expert E_k, c_{ij}^k is the element at the ith row and the jth column of the consensus matrix C_k, C_G is the group consensus degree for all experts, $1 \le i \le n, 1 \le j \le n, i \ne j$, and $1 \le k \le m$.

Step 9: Based on the set Q_k of the pairs (i, j) of alternatives A_i and A_j, adjust the interval of linguistic term $\tilde{f}_{ij}^k = [\tilde{f}_{ij}^{k(L)}, \tilde{f}_{ij}^{k(R)}]$ in the interval linguistic preference matrix P_k for expert E_k, shown as follows:

$$\tilde{f}_{ij}^{k(L)} = \begin{cases} s_{\left|\tilde{f}_{ij}^{k(L)}\right|+1}, & \text{if } \left|\tilde{f}_{ij}^{k(L)}\right| - \left|g_{ij}^{(L)}\right| = pm_{ij}^{k(L)} < 0 \\ s_{\left|\tilde{f}_{ij}^{k(L)}\right|-1}, & \text{if } \left|\tilde{f}_{ij}^{k(L)}\right| - \left|g_{ij}^{(L)}\right| = pm_{ij}^{k(L)} > 0 \\ s_{\left|\tilde{f}_{ij}^{k(L)}\right|}, & \text{if } \left|\tilde{f}_{ij}^{k(L)}\right| - \left|g_{ij}^{(L)}\right| = pm_{ij}^{k(L)} = 0 \end{cases} \tag{10}$$

$$\tilde{f}_{ij}^{k(R)} = \begin{cases} s_{\left|\tilde{f}_{ij}^{k(R)}\right|+1}, \text{if } \left|\tilde{f}_{ij}^{k(R)}\right| - \left|g_{ij}^{(R)}\right| = pm_{ij}^{k(R)} < 0 \\ s_{\left|\tilde{f}_{ij}^{k(R)}\right|-1}, \text{ if } \left|\tilde{f}_{ij}^{k(R)}\right| - \left|g_{ij}^{(R)}\right| = pm_{ij}^{k(R)} > 0 \\ s_{\left|\tilde{f}_{ij}^{k(R)}\right|}, \quad \text{if } \left|\tilde{f}_{ij}^{k(R)}\right| - \left|g_{ij}^{k(R)}\right| = pm_{ij}^{k(R)} = 0 \end{cases} \tag{11}$$

where $\tilde{f}_{ij}^{k(L)}$ denotes the left linguistic term of $\tilde{f}_{ij}^{k}$, $\tilde{f}_{ij}^{k(R)}$ denotes the right linguistic term of $\tilde{f}_{ij}^{k}$, $\left|\tilde{f}_{ij}^{k(L)}\right|$ denotes the subscript of the linguistic term $\tilde{f}_{ij}^{k(L)}$, $\left|\tilde{f}_{ij}^{k(R)}\right|$ denotes the subscript of the linguistic term $\tilde{f}_{ij}^{k(R)}$, $g_{ij}^{(L)}$ denotes the left linguistic term of g_{ij} in the collective interval linguistic preference matrix G, $g_{ij}^{(R)}$ denotes the right linguistic term of g_{ij} in the collective interval linguistic preference matrix G, $\left|g_{ij}^{(L)}\right|$ denotes the subscript of the linguistic term $g_{ij}^{(L)}$ in the collective interval linguistic preference matrix G, $\left|g_{ij}^{(R)}\right|$ denotes the subscript of the linguistic term $g_{ij}^{(R)}$ in the collective interval linguistic preference matrix G, $1 \leq i \leq n$, $1 \leq j \leq n$, $i \neq j$, and $1 \leq k \leq m$. Let $r = r + 1$.

Step 10: Based on Eqs. (2)-(4), update the collective interval linguistic preference matrix G, update the preference matrix R_k for expert E_k, and update the collective preference matrix RG for all experts, respectively, where $1 \leq k \leq m$.

Step 11: Calculate the score V_i of alternative A_i, where the score V_i is equal to the summation of the values at the ith row of the collective preference matrix RG, where $1 \leq i \leq n$.

Step 12: Based on Eq. (5), update the consensus matrix C_k for expert E_k, where $1 \leq k \leq m$.

Step 13: Based on Eqs. (6) and (7), calculate the consensus degree CD_k for each expert E_k, where $1 \leq k \leq m$, and calculate the group consensus degree C_G for all experts, respectively. If the group consensus degree C_G is larger than or equal to a predefined group consensus threshold value δ, where $\delta \in [0, 1]$, then the larger the score V_i of alternative A_i, the better the preference order of the alternative A_i, where $1 \leq i \leq n$; **Stop**. Otherwise, go to **Step 14**.

Step 14: Based on Eq. (8), update the proximity matrix PM_k for each expert E_k, where $1 \leq k \leq m$. Go to **Step 8**.

In the following, we use an example to illustrate the group decision making process of the proposed method.

Example 3.1 [5]: Assume that there is a set S of nine linguistic terms, $S = \{s_0, s_1, s_2, s_3, s_4, s_5, s_6, s_7, s_8\}$, where s_0 = "Incomparable", s_1 = "Significantly Worse", s_2 = "Worse", s_3 = "Somewhat Inferior", s_4 = "Equivalent", s_5 = "Somewhat Better", s_6 = "Superior", s_7 = "Significantly Superior" and s_8 = "Certainly Superior". Assume that there are four alternatives A_1, A_2, A_3, A_4 and assume that there are four experts E_1, E_2, E_3, E_4, assume that the target $T = [s_0, s_8]$, and assume that the predefined threshold value $\delta = 0.998$. Assume that the interval linguistic preference matrices P_1, P_2, P_3 and P_4 given by the experts E_1, E_2, E_3 and E_4, respectively, are shown as follows:

$$P_1 = \begin{array}{c} \\ A_1 \\ A_2 \\ A_3 \\ A_4 \end{array} \begin{array}{c} \begin{array}{cccc} A_1 & A_2 & A_3 & A_4 \end{array} \\ \begin{bmatrix} - & [s_3, s_4] & [s_5, s_7] & [s_1, s_1] \\ [s_5, s_5] & - & [s_7, s_8] & [s_3, s_3] \\ [s_3, s_4] & [s_1, s_2] & - & [s_0, s_2] \\ [s_7, s_7] & [s_5, s_6] & [s_8, s_8] & - \end{bmatrix} \end{array} \quad P_2 = \begin{array}{c} \\ A_1 \\ A_2 \\ A_3 \\ A_4 \end{array} \begin{array}{c} \begin{array}{cccc} A_1 & A_2 & A_3 & A_4 \end{array} \\ \begin{bmatrix} - & [s_1, s_3] & [s_3, s_4] & [s_3, s_4] \\ [s_7, s_7] & - & [s_6, s_7] & [s_5, s_5] \\ [s_5, s_6] & [s_2, s_3] & - & [s_3, s_3] \\ [s_5, s_7] & [s_3, s_3] & [s_5, s_6] & - \end{bmatrix} \end{array},$$

$$P_3 = \begin{array}{c} \\ A_1 \\ A_2 \\ A_3 \\ A_4 \end{array} \begin{array}{c} \begin{array}{cccc} A_1 & A_2 & A_3 & A_4 \end{array} \\ \begin{bmatrix} - & [s_5, s_5] & [s_7, s_8] & [s_3, s_4] \\ [s_3, s_4] & - & [s_6, s_8] & [s_2, s_3] \\ [s_1, s_1] & [s_2, s_3] & - & [s_0, s_1] \\ [s_5, s_6] & [s_6, s_6] & [s_8, s_8] & - \end{bmatrix} \end{array}, \quad P_4 = \begin{array}{c} \\ A_1 \\ A_2 \\ A_3 \\ A_4 \end{array} \begin{array}{c} \begin{array}{cccc} A_1 & A_2 & A_3 & A_4 \end{array} \\ \begin{bmatrix} - & [s_6, s_7] & [s_7, s_7] & [s_5, s_7] \\ [s_2, s_3] & - & [s_4, s_5] & [s_3, s_3] \\ [s_1, s_2] & [s_4, s_4] & - & [s_0, s_2] \\ [s_3, s_3] & [s_5, s_7] & [s_8, s_8] & - \end{bmatrix} \end{array}.$$

Table 1 shows the preference order of the alternatives for each round by the proposed method. Table 2 shows a comparison of the preference orders of the alternatives for different methods. From Table 2, we can see that the proposed method and Chen and Lee's method [5] get the same preference order of the alternatives, i.e., $A_4 > A_2 > A_1 > A_3$, whereas Ben-Arieh and Chen's Method [2] cannot obtain the preference order of the alternatives in this situation.

Table 1. Preference order of the alternatives for each round by the proposed method

Round Number r	Group Consensus Degree C_G	Scores	Preference Order
1	0.8660	$V_1 = 1.6444$, $V_2 = 1.7000$, $V_3 = 0.9333$, $V_4 = 2.1556$.	$A_4 > A_2 > A_1 > A_3$
2	0.9073	$V_1 = 1.6000$, $V_2 = 1.8000$, $V_3 = 0.9333$, $V_4 = 2.2222$.	$A_4 > A_2 > A_1 > A_3$
3	0.9345	$V_1 = 1.6000$, $V_2 = 1.8444$, $V_3 = 0.9333$, $V_4 = 2.2556$.	$A_4 > A_2 > A_1 > A_3$
4	0.9605	$V_1 = 1.6000$, $V_2 = 1.8444$, $V_3 = 0.9333$, $V_4 = 2.2556$.	$A_4 > A_2 > A_1 > A_3$
5	0.9809	$V_1 = 1.6000$, $V_2 = 1.8444$, $V_3 = 0.9333$, $V_4 = 2.2556$.	$A_4 > A_2 > A_1 > A_3$
6	0.9928	$V_1 = 1.6000$, $V_2 = 1.8444$, $V_3 = 0.9333$, $V_4 = 2.2889$.	$A_4 > A_2 > A_1 > A_3$
7	0.9977	$V_1 = 1.6000$, $V_2 = 1.8444$, $V_3 = 0.9333$, $V_4 = 2.2889$.	$A_4 > A_2 > A_1 > A_3$
8	1.0000	$V_1 = 1.6000$, $V_2 = 1.8444$, $V_3 = 0.9333$, $V_4 = 2.2889$.	$A_4 > A_2 > A_1 > A_3$

Table 2. A comparison of the preference orders of the alternatives for different methods

Methods	Preference Order
Ben-Arieh and Chen's Method [2]	N/A
Chen and Lee's Method [5]	$A_4 > A_2 > A_1 > A_3$
The Proposed Method	$A_4 > A_2 > A_1 > A_3$

Note: "N/A" denotes "cannot be obtained".

4 Conclusions

We have present a new method for autocratic decision making using group recommendations based on intervals of linguistic terms and likelihood-based comparison relations. It can overcome the drawbacks of Ben-Arieh and Chen's method [2] and Chen and Lee's method [5] for autocratic decision making using group recommendations based on intervals of linguistic terms and likelihood-based comparison relations.

Acknowledgments. This work was supported in part by the National Science Council, Republic of China, under Grant NSC 101-2221-E-011-171-MY2.

References

1. Alonso, S., Herrera-Viedma, E., Chiclana, F., Herrera, F.: A web based consensus support system for group decision making problems and incomplete preferences. Information Sciences 180(23), 4477–4495 (2010)
2. Ben-Arieh, D., Chen, Z.: Linguistic-labels aggregation and consensus measure for autocratic decision making using group recommendations. IEEE Transactions on Systems, Man, and Cybernetics-Part A: Systems and Humans 36(3), 558–568 (2006)
3. Cabrerizo, F.J., Alonso, S., Herrera-Viedma, E.: A consensus model for group decision making problems with unbalanced fuzzy linguistic information. International Journal Information Technology Decision Making 8(1), 109–131 (2009)
4. Cabrerizo, F.J., Perez, I.J., Herrera-Viedma, E.: Managing the consensus in group decision making in an unbalanced fuzzy linguistic context with incomplete information. Knowledge-Based Systems 23(2), 169–181 (2010)
5. Chen, S.M., Lee, L.W.: Autocratic decision making using group recommendations based on the ILLOWA operators and likelihood-based comparison relations. IEEE Transactions on Systems, Man, and Cybernetics-Part A: Systems and Humans 42(1), 115–129 (2012)
6. Chen, S.M., Lee, L.W., Yang, S.W., Sheu, T.W.: Adaptive consensus support model for group decision making systems. Expert Systems with Applications 39(16), 12580–12588 (2012)
7. Herrera-Viedma, E., Alonso, S., Chiclana, F., Herrera, F.: A consensus model for group decision making with incomplete fuzzy preference relations. IEEE Transactions on Fuzzy Systems 15(5), 863–877 (2007)

8. Herrera-Viedma, E., Herrera, F., Chiclana, F.: A consensus model for multiperson decision making with different preference structures. IEEE Transactions on Systems, Man, and Cybernetics-Part A: Systems and Humans 32(3), 394–402 (2002)
9. Herrera, F., Herrera-Viedma, E., Verdegay, J.L.: A sequential selection process in group decision making with a linguistic assessment approach. Information Sciences 85(4), 223–239 (1995)
10. Herrera, F., Herrera-Viedma, E., Verdegay, J.L.: A model of consensus in group decision making under linguistic assessments. Fuzzy Sets and Systems 78(1), 73–87 (1996)
11. Herrera-Viedma, E., Martinez, L., Mata, F., Chiclana, F.: A consensus support system model for group decision-making problems with multi-granular linguistic preference relations. IEEE Transactions on Fuzzy Systems 13(5), 644–658 (2005)
12. Kacprzyk, J., Fedrizzi, M., Nurmi, H.: Group decision making and consensus under fuzzy preferences and fuzzy majority. Fuzzy Sets and Systems 49(1), 21–31 (1992)
13. Mata, F., Martinez, L., Herrera-Viedma, E.: An adaptive consensus support model for group decision-making problems in a multigranular fuzzy linguistic context. IEEE Transactions on Fuzzy Systems 17(2), 279–290 (2009)
14. Perez, I.J., Cabrerizo, F.J., Herrera-Viedma, E.: A mobile decision support system for dynamic group decision-making problems. IEEE Transactions on Systems, Man, and Cybernetics-Part A: Systems and Humans 40(6), 1244–1256 (2010)
15. Tapia-Garcia, J.M., del Moral, M.J., Martinez, M.A., Herrera-Viedma, E.: A consensus model for group decision making problems with linguistic interval fuzzy preference relations. Expert Systems with Applications 39(11), 10022–10030 (2012)
16. Xu, Z.S., Da, Q.L.: A likelihood-based method for priorities of interval judgment matrices. Chinese Journal of Management Science 11(1), 63–65 (2003)
17. Xu, G., Liu, F.: An approach to group decision making based on interval multiplicative and fuzzy preference relations by using projection. Applied Mathematical Modelling 37(6), 3929–3943 (2013)

Applying Fuzzy AHP to Understand the Factors of Cloud Storage Adoption

Shiang-Lin Lin[1,*], Chen-Shu Wang[2], and Heng-Li Yang[1]

[1] Department of Management Information Systems,
National Chengchi University, Taipei, Taiwan
{102356505,yanh}@nccu.edu.tw

[2] Graduate Institute of Information and Logistics Management,
National Taipei University of Technology, Taipei, Taiwan
wangcs@ntut.edu.tw

Abstract. Cloud storage service (CSS) is one of the widely applications in cloud computing. However, most users tend to doubt the mechanism of the data security and privacy protection, which may reduce the expected benefits of CSS. Therefore, it`s an important issue to understand the factors that affect the intention of user to adopt CSS. This study summarize various risk reliever strategies that could be taken by general users when they face a newly emerged technological service. Therefore, it total 4 evaluation dimensions and 10 factors were finally chosen. Then, the Fuzzy Analytic Hierarchy Process method was used to analyze. From the results, it is found the user's acquaintance with the cloud service provider, the positive online word-of-mouth about the CSS, and the CSS provider's statement and warranty of data security and privacy protection are three key factors for the users when they consider whether or not to use this service.

Keywords: Cloud storage service, Risk reliever strategy, Fuzzy theory, Analytic Hierarchy Process.

1 Introduction

Cloud computing is an emerging application in recent years. It provides user with a new perspective of operational model. The general definition of cloud computing is the user in an Internet-accessible environment can quickly share or access network resources (e.g., remote servers, storage spaces and network service applications) through some easy operating interfaces and management modes (Ghormley, 2012). IT companies provide effective and efficient software and hardware deployment, with which users can make use of various cloud services directly via networks without the need of purchasing expensive software or hardware (Kim, 2009). Through cloud technology, it is able to more effectively utilize various resources and reduce IT cost expenditure (Marston et al., 2011).

N.T. Nguyen et al. (Eds.): ACIIDS 2014, Part II, LNAI 8398, pp. 282–291, 2014.

In 2009, the investigation by ABI Research indicated that cloud services are deemed an important trend in future network applications[1]. Gartner Group also identifies cloud computing as the first of the top 10 IT industries in the future[2]. According to the forecast provided by International Data Corporation (IDC) in 2013, the growing expenditure for cloud services mainly included spending for five major product/service types, i.e. application software(38%), apps development and deployment(13%), infrastructure of software (20%), storage service(14%), and server(15%)[3]. Among others, the storage service is most closely related to the network users' need.

Cloud storage service (CSS) is an emerging application that provides user with a new perspective of operational model. In the past, all types of files must be stored in users' personal computers or storage media. Nevertheless, there are many frequent data change, when editing these data on different computer devices, inconsistency of data might occur. Now, the problem of data inconsistency edited or processed on different computers or storage media can be solved by using CSS. By storing files in the web hard disk, the user may use any Internet-connectable device to access data at any time and any place, which largely increases the user's flexibility in data access (Bowers et al., 2009).

In July 2010, the Foreseeing Innovative New Digiservices (FIND) website of the Institute for Information Industry (III) conducted an investigation to realize the requirements of the users for CSS[4]. The research results indicated that regarding the users' intention of using cloud storage, more than 60% users claimed that the intention of using cloud storage in near future. The research results also implied that cloud storage would gradually replace the traditional storage by using hard disk or USB flash drive. Besides, the convenience in data access, the speed and stability of data upload/download, as well as the data integrity are also important factors that are taken into consideration by users before they decide to use the CSS. On the other hand, a large part of the other users who had no intention of using CSS had worries about the potential risks of privacy and leakage of important information. This means that data security and privacy protection are important factors to users if they intend to use the CSS.

Presently, most users tend to doubt the data security and privacy protection in the cloud environment, which prevents the users from storing their important or confidential data in the cloud storage space and accordingly, largely limits the benefits of cloud service that were expected to achieve. Further, many consumers are uncertain of the performance that can be expected with the newly developed products or services, and even worry about some unpredictable loss possibly caused by the use of such new products or services (Dowling & Staelin, 1994), which leads to customer's perceived risk of using such products or services (Bauer, 1960; Cox & Rich, 1964; Roselius, 1971). Moreover, being a type of network application service, CSS carries more perceived risk (e.g., security and privacy risk of data) than other network services (Biswas & Biswas, 2004). Therefore, these user's considerations would have deep influence on the usage of cloud-based information technology.

---

[1] ABI Research, http://www.abiresearch.com/research/1003385, 2013.

[2] Gartner Group, http://www.gartner.com/it/page.jsp?id=777212, 2013.

[3] IDC, http://blogs.idc.com/ie/?p=543, 2013.

[4] FIND, http://books.find.org.tw/newbook_disp.asp?book_id=142, 2013.

Recently, most of researches emphasized on the theoretical discussion of cloud computing (Marston et al., 2011), while few of them are directed to the factors that are considered by users before they decide to use the CSS. In addition, most of the relevant studies having been conducted in recent years are aimed mainly at the analysis of the advantages/disadvantages and the benefits of business-level cloud computing technology (Ghormley, 2012), while few of the related literature are directed to the analysis of personal users' consideration of using the cloud services (Obeidat & Turgay, 2013).

In facing a high-risk web service like CSS, this study intends to explore the factors that are important to general users when they consider whether to use CSS or not, and the risk reliever strategies that would be taken as key factors by general users to lower the perceived risk in using CSS. Based on the chosen key factors, a hierarchical structure of evaluation for this study is constructed. Then, Fuzzy Analytic Hierarchy Process (FAHP) is employed in this study to find out the key factors that are most frequently considered by the users before they make any decision on the use of the CSS.

2 Literature Review

2.1 Cloud Storage Service

The concept of cloud computing was originated in the early 1980s from the motto "The Network is the Computer" of Sun Microsystems. Today, the era of cloud computing has come, and there is a variety of online cloud services available for network users. By cloud storage, it means cloud service providers offer web hard disk storage and users pay an amount of money for buying or leasing required storage capacity from the cloud service providers. In this manner, the users can be connected to the cloud storage space via Internet from anywhere at any time to flexibly access and back up their data. In addition, cloud storage allows the users to synchronize files across different devices in real time and share files faster.

Currently, there are quite many cloud service providers who offer the cloud storage service (CSS) to personal users, such as Google, Microsoft, Apple, and so on. All these cloud service providers provide free cloud storage space for trial use by users. Users who require a storage capacity larger than the free cloud storage space may lease cloud hard disk from the providers according to their requirements.

2.2 Fuzzy Analytical Hierarchy Process

The assessment of the information system is a multi-criteria decision-making problem (Alonso et al., 2000). In 1977, Saaty presented that the Analytical Hierarchy Process (AHP) is the most widely used and practical methods to solve such problems. It can structure the decision-making problems, and provide guidelines, goal weight, and alternatives, to give determination of consistency for various criteria (Mahdavi et al., 2008).

Generally speaking, people have not clear boundaries when they make any decisions. Therefore, when the experts decide the importance of elements, the results may be biased. For this reason, in 1985, Buckley presented the Fuzzy Analytic Hierarchy Process (FAHP). This method combines the concept of Fuzzy Theory and AHP, which complete the disadvantage of ambiguity on the two assessment elements for the importance by experts, and more truly reflect the actual situation.

2.3 Risk Reliever Strategy

Perceived risk is the potential risk a consumer perceived when making a decision on using a new product/service. Consumer's intention of adopting the new product/service is largely affected by higher perceived risk, when consumer is not skilled in using or knows little about the new product/service. According to Lakhani et al. (2013), perceived risk is the potential loss in online shopping or using online services subjectively perceived by a user. The research of Mitchell & Greatorex (1993) indicated that services are perceived as possessing more inherent risk than physical goods.

Consumers will take various risk reliever strategies, when they are worrying about what if the new product/service does not perform as well as they expected. Thus, risk reliever strategies are strategies taken by consumers in making a decision on using a new product or service for ensuring higher product/service use satisfaction and lower risk of incurring any loss (Roselius, 1971). According to Cox & Rich (1964), risk reliever strategies are aimed to reduce consumers' perceived uncertainty and potential loss. When the customers are facing different types of products/services, they will take different strategies to reduce their perceived risks (Nelson, 1970; Kim, 2010). These risk reliever strategies reflect the factors that have influences on the consumers' decisions on using the products or services. For general users, cloud storage is a service having higher risks. Therefore, it is absolutely necessary for the cloud service providers to analyze the risk reliever strategies taken by users in response to the high perceived risk in using cloud hard disk.

3 Research Design and Research Model

3.1 Research Design

In this study, key factors having an influence on users' adoption of the CSS were carefully chosen through literature review and market research as shown in section 3.2. Then, a hierarchical structure of evaluation was constructed based on the chosen key factors, and a corresponding AHP expert questionnaire was designed. Futher, an one-month deep interviews were implemented via the designed expert questionnaire. Finally, total 40 questionnaires were received. The returned questionnairs passing the consistency test can be regarded as effective questionaires, which were analyzed using FAHP to find the relative fuzzy weights for these key factors. Further, by way of series of hierarchical, the priority of these key factors was decided. Finally, the results of this study were analyzed and discussed. The research model are shown as Fig. 1.

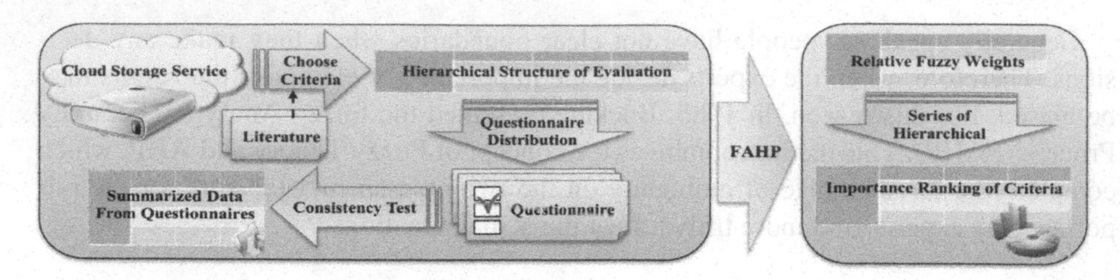

Fig. 1. Research model

3.2 Development of Evaluation Factors

In this study, we refer the strategy classification by Tan (1999), who conducted a research with regard to online shopping. In his research, consumers' preferred risk reliever strategies were divided into "reference group", "brand image" and "warranty". Our research adopts the above strategy classification by Tan, and incorporates other additional factors form the market research on existing CSSs. Further, we grouped four measurement dimensions, i.e. "group opinion", "service and brand image", "service warranty" and "functionality". These dimensions are defined as follows:

1. "Group opinion" refers to the opinions and evaluation on the CSSs from some groups of persons around the user.
2. "Service and brand image" refers to the user's past experiences and the impression in using cloud storage and existing cloud service providers.
3. "Service warranty" refers to the warranty statement that is given by the CSS providers to consumers.
4. "Functionality" refers to the functions that can be provided by the CSS and the benefits the user can have from this service.

By summarizing the reviewed literature, ten evaluation factors for each of the measurement dimensions are further developed. Definition and description of these evaluation factors are shown in Table 1 below.

Table 1. Definition & description of evaluation factors

Dimension	Evaluation factor	Factor Definition & Description
D1. Group Opinion	C1. Endorsement by experts or celebrities	Recommendation or introduction of cloud hard disk by experts in information technology or professionals good at cloud hard disk.
	C2. Online word-of-mouth	Positive/negative online word-of-mouth and web users' recommendation of cloud hard disk in network communities or web forums.
	C3. Recommendation from families & friends	Recommendation of cloud hard disk from the user's families, relatives or friends.

Table 1. (*continued*)

Dimension	Evaluation factor	Factor Definition & Description
D2. Service and Brand Image	C4. Acquaintance with the service and brand	The user's knowledge of and acquaintance with cloud hard disk or the impression the user had on existing cloud service brands and providers.
	C5. Renowned supplier	The cloud hard disk is provided by a renowned film or sold under a famous brand.
D3. Service Warranty	C6. Warranties of free trial use	The free trial use provided by the cloud hard disk providers to new users.
	C7. Warranties of data security and privacy	The cloud hard disk providers' guarantee on the absolute privacy protection of users' information uploaded to the cloud space against access or illegal use by other persons.
	C8. Warranty of data access speed	The cloud hard disk providers' guarantee on the stable and efficient access speed during data upload and download by users without long waiting time.
D4. Functionality	C9. Convenience	The cloud storage can be used to replace the existing storage media (e.g. flash drive), and user can use device at hand, such as notebook or smartphone, allowing the user immediately access required data from the cloud space in an emergency or if the data is important but not remembered by the user at any web-accessible place without spatial limitation
	C10. Integration	Data having been accessed, edited and modified by a user on different devices, such as notebook, tablet PC and smartphone, can be integrated via the cloud hard disk with ensured data consistency and integrity.

This study developed four measurement dimensions and ten evaluation factors; based on them, a hierarchical structure of evaluation as shown in Fig. 2. Further, the hierarchical structure of evaluation is analyzed using FAHP, in order to understand the factors that are deemed important by users when they are considering whether or not to use CSS. The weights and the ranking of these factors are also analyzed in this study.

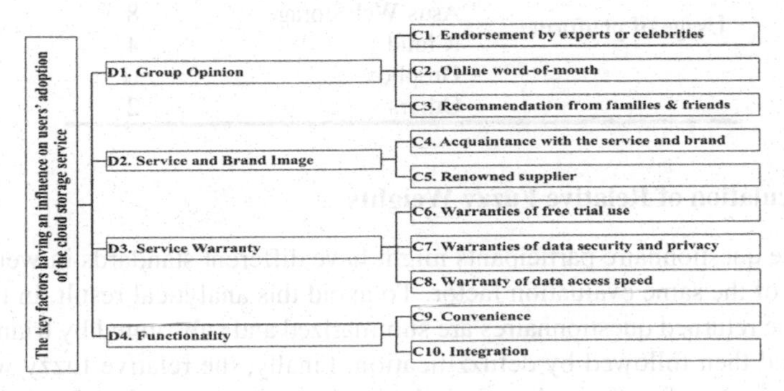

Fig. 2. Hierarchical structure of evaluation

4 Results Analysis

4.1 Consistency Verification

The AHP questionnaire for this study is designed based on the hierarchical structure of evaluation shown in Figure 1. The questionnaire participants must be users who are currently using the CSS. In making a pair-wise comparison between the evaluation factors, according to Saaty's suggestion in 1977, the consistency index (C.I.) is preferably smaller than or equal to 0.1. The order of the pair-wise comparison matrix would increase with the complexity of the problem being studied, which in turn results in difficulty in determining the consistency of the matrix. Therefore, Saaty further proposed the concept of random consistency index (R.I.) and the comparison between the consistency index and the random consistency index for deriving a consistency ratio (C.R.). The value of C.R. can be used to adjust the change in different C.I. values for the pair-wise comparison matrix of different levels. The value of C.R. should also be smaller than 0.1 to represent the questionnaire passes the consistency test.

In this study, the questionnaire is carried out in both the business and the academic field, and an in-depth interview is conducted for each of the questionnaire participants, total 40 participants replied the questionnaire. All the returned questionnaires are subjected to a questionnaire consistency test for verifying the consistency of answers to all questions in the questionnaire, so as to ensure the effectiveness of each replied questionnaire. Sample data of questionnaire participants is shown in Table 2 below.

Table 2. Demographical sample data of questionnaire participants

Catalog	Sub-cat.	Count
Gender of expert	Male	28
	Female	12
Usage of device	Smartphone	36
	Tablet	17
	Notebook	24
	Desktop	34
Usage of platform	Google Drive	29
	SkyDrive	14
	Asus WebStorage	8
	iCloud	4
	Dropbox	31
	Others	2

4.2 Calculation of Relative Fuzzy Weights

It is possible questionnaire participants might have different standards in weighing the importance of the same evaluation factor. To avoid this analytical result, in this study, data from the returned questionnaires are summarized and substituted by triangle fuzzy numbers, and then followed by defuzzification. Finally, the relative fuzzy weights of different evaluation factors in the whole hierarchical structure of evaluation are generated in Table 3.

As can be seen in Table 3, in measuring the evaluation factors influencing the users' adoption of the CSS, the "Functionality" in the dimension hierarchy has the highest weight (0.433). Since the CSS can integrate users' different IT devices and operating systems to bring a lot of conveniences to users in managing their data, the functionality of CSS is deemed by users the most important measurement dimension in making their decision on using the CSS.

Table 3. Relative fuzzy weights of evaluation factors

Dimensions and Factors	Triangle fuzzy numbers	Defuzzication	Relative weights
D1. Group Opinion	(0.068, 0.129, 0.291)	**0.163**	**0.174**
C1. Endorsement by experts or celebrities	(0.081, 0.141, 0.311)	0.178	0.189
C2. Online word-of-mouth	(0.185, 0.372, 0.786)	0.447	0.476
C3. Recommendation from families & friends	(0.130, 0.267, 0.549)	0.315	0.335
D2. Service and Brand Image	(0.050, 0.098, 0.220)	**0.123**	**0.131**
C4. Acquaintance with the service and brand	(0.311, 0.546, 1.048)	0.635	0.642
C5. Renowned supplier	(0.173, 0.308, 0.582)	0.354	0.358
D3. Service Warranty	(0.083, 0.165, 0.368)	**0.205**	**0.219**
C6. Warranties of free trial use	(0.088, 0.145, 0.285)	0.173	0.206
C7. Warranties of data security and privacy	(0.159, 0.292, 0.571)	0.341	0.406
C8. Warranty of data access speed	(0.150, 0.279, 0.545)	0.325	0.387
D4. Functionality	(0.182, 0.391, 0.766)	**0.446**	**0.476**
C9. Convenience	(0.253, 0.398, 0.721)	0.457	0.518
C10. Integration	(0.230, 0.392, 0.654)	0.425	0.482

4.3 Priority of Evaluation Factors

In this study, a relative fuzzy weight for each of the evaluation factors is obtained by means of fuzzy weight calculation. Further, for the purpose of obtaining the importance ranking of all evaluation factors, we multiply the relative fuzzy weight of each layer so as to calculate the absolute fuzzy weight for each factor in the whole hierarchical structure of evaluation. Then the evaluation factors are ranked in importance according to their absolute fuzzy weights, as shown in Table 4.

From the priority of evaluation factors, it is found that "Acquaintance with the service and brand" (0.135), "Online word-of-mouth" (0.130) and "Warranties of data security and privacy" (0.122) are the first three most important key factors of users' adoption of the CSS.

Table 4. The priority of evaluation factors

Evaluation factor	absolute fuzzy weight	Priority
C1. Endorsement by experts or celebrities	0.051	10
C2. Online word-of-mouth	0.130	2
C3. Recommendation from families & friends	0.092	7
C4. Acquaintance with the service and brand	0.135	1
C5. Renowned supplier	0.075	8
C6. Warranties of free trial use	0.061	9
C7. Warranties of data security and privacy	0.122	3
C8. Warranty of data access speed	0.117	4
C9. Convenience	0.112	5
C10. Integration	0.104	6

5 Conclusions and Future Research

Cloud computing is a big change in the field of network application in recent years. Among various cloud services, cloud storage is a service most closely related to web users' need, because it involves in the storage of users' all important data and backup files. Nevertheless, being a type of network application service, cloud storage carries more consumers' perceived risk than other physical products. Therefore, users are more carefully in making their decision on using the CSS.

In this study, the concept of perceived risks is taken into account. Through literature review and market research, the risk reliever strategies, which are possibly taken by users when they make a decision on using the CSS, were carefully chosen. Then, the hierarchical structure of evaluation for this study is constructed based on such risk reliever strategies. Further, FAHP analysis is employed to find out the key factors that have an influence on the users' adoption of the CSS.

The results from FAHP indicate that main influence of first three factors on the users' adoption of the CSS are namely "Acquaintance with the service and brand", "Online word-of-mouth" and "Warranties of data security and privacy". Being a type of newly developed technological service, CSS inevitably encounters the problem that general users might doubt whether it is technically mature enough for commercial use. That is why most users, when considering using CSS, tend to choose the service provided by a brand company they are familiar with or they can trust. In addition, compared to the traditional data access model, CSS proposes a completely new concept of web hard disk. However, in view of the considerably high switching costs, potential users considering using CSS would first try to get more related information from some discussion forums. The CSS potential users would also refer to online user experience. Such information and user experience obtained via Internet together form a main source of reference for the users intending to use CSS. However, since the CSS, as its name suggests, provides a user with an online storage space for the user's important or private files and data, it is doubtlessly an extremely important factor to consider by most users whether the files and data stored in the cloud hard disk are always safe without the risk of being illegally or maliciously accessed.

This study can provide a reference in designing business promotion policies for CSS providers. In future, subsequent studies can be conducted to analyze whether different industrial groups, such as information or electronic industry, are different in the factors being considered by them in their decision making on the use of the CSS.

References

1. Alonso, G., Hagen, C., Agrawal, D., El Abbadi, A., Mohan, C.: Enhancing the Fault Tolerance of Workflow Management System. IEEE Concurrency 8, 74–81 (2000)
2. Bauer, R.A.: Consumer Behavior as Risk Taking. In: 43rd Conf. of the American Marketing Association, pp. 389–398 (1960)
3. Biswas, D., Biswas, A.: The Diagnostic Role of Signals in the Context of Perceived Risks in Online Shopping: Do Signals Matter More on the Web? Journal of Interactive Marketing 18(3), 30–45 (2004)
4. Bowers, K., Juels, A., Oprea, A.: HAIL: A High-Availability and Integrity Layer for Cloud Storage. In: 16th ACM Conf. on Computer and Communications Security (2009)
5. Buckley, J.J.: Fuzzy Hierarchical Analysis. Fuzzy Sets & Systems 17(3), 233–247 (1985)
6. Cox, D.F., Rich, S.V.: Perceived Risk and Consumer Decision Making-The Case of Telephone Shopping. Journal of Marketing Research 1(4), 32–39 (1964)
7. Dowling, G.R., Staelin, R.: A Model of Perceived Risk and Intended Risk-Handling Activity. Journal of Consumer Research 21(1), 119–134 (1994)
8. Ghormley, Y.: Cloud Computer Management from the Small Business Perspective. International Journal of Management & Information Systems 16(4), 349–356 (2012)
9. Kim, W.: Cloud Computing: Today and Tomorrow. Journal of Object Technology 8(1), 65–72 (2009)
10. Kim, I.: Consumers' Rankings of Risk Reduction Strategies in E-shopping. International Journal of Business Research 10(3), 143–148 (2010)
11. Lakhani, F.A., Shah, G., Syed, A.A., Channa, N., Shaikh, F.M.: Consumer Perceived Risk and Risk Relievers in E-shopping in Pakistan. Computer and Information Science 6(1), 32–38 (2013)
12. Mahdavi, I., Fazlollahtabar, H., Heidarzade, A., Mahdavi-Amiri, N., Rooshan, Y.I.: A Heuristic Methodology for Multi-Factors Evaluation of Web-Based E-Learning Systems Based on User Satisfaction. Journal of Applied Sciences 8, 4603–4609 (2008)
13. Marston, S., Li, Z., Bandyopadhyay, S., Zhang, J., Ghalsasi, A.: Cloud Computing- the Business Perspective. Decision Support Systems 51(1), 176–189 (2011)
14. Mitchell, V.W., Greatorex, M.: Risk Perception and Reduction in the Purchase of Consumer Services. The Service Industries Journal 13(4), 179–200 (1993)
15. Nelson, P.: Information and Consumer Behavior. Journal of Political Economy 78(2), 311–329 (1970)
16. Obeidat, M.A., Turgay, T.: Empirical Analysis for the Factors Affecting the Adoption of Cloud Computing Initiatives by Information Technology Executives. Journal of Management Research 5(1), 152–178 (2013)
17. Roselius, T.: Consumer Rankings of Risk Reduction Methods. Journal of Marketing 35(1), 56–61 (1971)
18. Saaty, T.L.: A Scaling Method for Priorities in Hierarchical Structures. Journal of Mathematical Psychology 15(3), 231–281 (1977)
19. Tan, S.J.: Strategies for Reducing Consumers' Risk Aversion in Internet Shopping. Journal of Consumer Marketing 16(2), 163–180 (1999)

A Novel Neural-Fuzzy Guidance Law Design by Applying Different Neural Network Optimization Algorithms Alternatively for Each Step

Jium-Ming Lin[1] and Cheng-Hung Lin[2]

[1] Department of Communication Engineering, Chung-Hua University, 707, Sec 2, Wu-Fu Rd Hsin-Chu, 30012 Taiwan, ROC

[2] Ph. D. Program in Engineering Science, College of Engng., Chung-Hua University, 707, Sec 2, Wu-Fu Rd Hsin-Chu, 30012 Taiwan, ROC

{jmlin,b09306014}@chu.edu.tw

Abstract. In this research, a novel neural-fuzzy guidance law by applying different neural network optimization algorithms alternatively in each step is proposed, such as the Gradient Descent (GD), SCG (Scaled Conjugate Gradient), and Levenberg-Marquardt (LM) methods are applied to deal with those parameter variation effects as follows: target maneuverability, missile autopilot time constant, turning rate time constant and radome slope error effects. Comparing with the proportion navigation (PN) and fuzzy methods are also made; the miss distances obtained by the proposed method are lower, and the proposed acceleration commands are always without polarity changes or oscillation at the final stage.

Keywords: neural-fuzzy guidance law, proportion navigation, turning rate time constant, radome aberration error slope, miss distance.

1 Introduction

Generally, the tactical missile terminal guidance laws are derived on some classical and optimal control techniques [1-2]. However, they were suffered from system parameter variations, such as target maneuverability, time delays of missiles autopilot and turning rate, and radome slope error effect defined in Fig. 1. Although some neural and/or fuzzy methods are also proposed for the guidance system design, the missile turning rate time constant and radome aberration error slope were not taken into account, but they play very important roles in the guidance performances [3-5]. In general, the radome shape must be very sharp to reduce the aerodynamic drag, but the target line-of-sight (LOS) refraction error would be increased as in Fig.1, so do the miss distance for the cases of larger missile turning rate time constants [1-2]. Some fuzzy guidance

N.T. Nguyen et al. (Eds.): ACIIDS 2014, Part II, LNAI 8398, pp. 292–301, 2014.

laws were applied to the missile guidance law design [3-5], but the improvements are limited. In this paper, a novel neural-fuzzy guidance law by applying different neural network optimization algorithms alternatively in each step is proposed, the Gradient Descent (GD) [6], SCG (Scaled Conjugate Gradient) [7], and Levenberg-Marquardt (LM) [6-7] methods are applied to deal with those parameter variation effects, such as target maneuverability, missile autopilot time constant, turning rate time constant and radome slope error effects. The GD algorithm is a standard gradient descent method; all data samples are processed at each step of iteration to determine the steepest descent. If the data samples are linearly separable, then the global minimum point can be reached [2]. Otherwise, the method will never converge as the optimum point does not exist. This method has another weakness; it requires all data samples to process at each step of iteration. The scaled conjugate gradient algorithm [7] is to combine the scaled memoryless BFGS (Broyden-Fletcher-Goldfarb-Shanno) method and the pre-conditioning technique in the frame of gradient method. The preconditioner is also a scaled memoryless BFGS matrix, which is reset when the restart criterion holds.

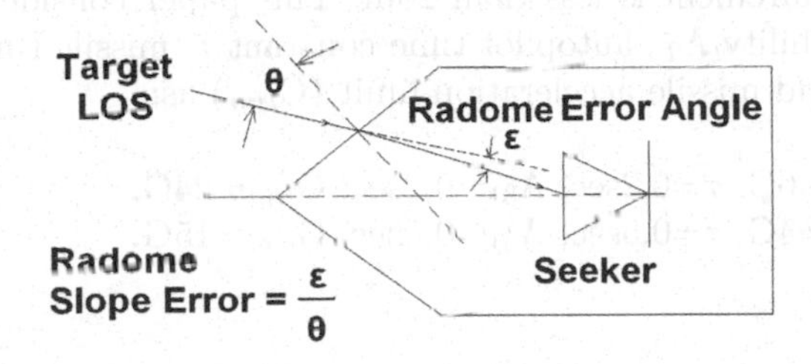

Fig. 1. Missile radome slope error effect

On the other hand, the LM algorithm is an iterative technique that locates the minimum of a multivariate function expressed as the sum of squares of non-linear real-valued functions. It is a standard technique for non-linear least-squares problems, widely adopted in a broad spectrum of disciplines. LM can be thought of as a combination of steepest descent and Gauss-Newton methods. When the current solution is far from the correct one, it behaves like a steepest descent method; this method is slow, but guaranteed to converge.

Comparing with the traditional proportion navigation (PN) and fuzzy methods; we note the miss distances obtained by the proposed method are lower. Besides, the proposed acceleration commands are always without polarity changes or oscillations at the final stage. Thus even there are parameter variation effects; the guidance loop can always track the target without oscillation, and yields better guidance performances. The paper is organized as: the first section is introduction. The second one is problem formulation of guidance system.

The next one is PD-type fuzzy guidance law design and simulation discussion. The proposed method is given in section 4. The last part is the conclusion.

2 Problem Formulation of Missile Terminal Guidance System

In general, the two-degree-of-dimension (2D) terminal engagement geometry of target and missile can be shown in Fig. 2. The target (missile) initial coordinate and velocity are respectively as (X_{TO}, Y_{TO}), and V_{TO}, $((X_{MO}, Y_{MO})$ and $V_{MO})$. Let the target (missile) initial heading angle and maneuver acceleration be respectively as θ_{TO} (θ_{MO}) and A_{TO} (A_{MO}). Fig.3 shows the block diagram by using neural-fuzzy guidance law in which the seeker tracking and stabilization loop gain are 10 and 100 respectively. R is the radome aberration error slope. The velocity of missile and target are 600 and 400 m/sec, respectively. The initial range of missile to target is 10km. The initial aspect angles of target and missile are θ_{TO} (30°) and θ_{MO} (135°), respectively. The target maneuver is assumed to be a unit-step lateral acceleration with amplitude A_T closed to the missile. The miss distance requirement is less than 20m. This paper considers two cases of target maneuverability A_T, autopilot time constant τ, missile turning rate time constant (A_{31}), and missile acceleration limit (G_{lim}) as:

(1)Case 1: A_T=6G, τ=0.2sec, A_{31}=0.2sec, G_{lim}=24G.
(2)Case 2: A_T=4G, τ=0.5sec, A_{31}=0.5sec, G_{lim}=15G.

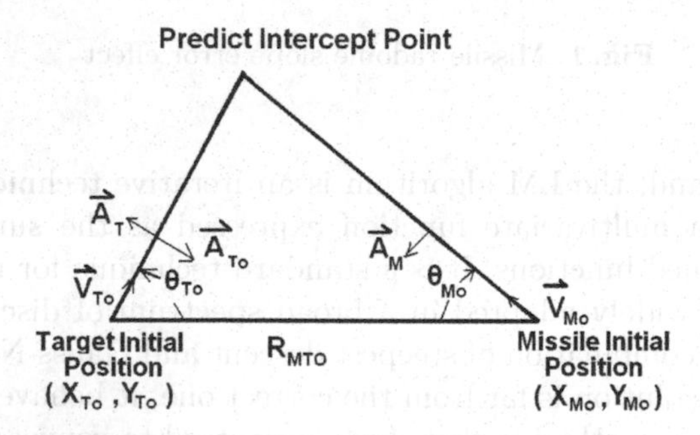

Fig. 2. Trajectories of intercept for target and missile

3 PD-Type Fuzzy Guidance Law

To reduce the miss distance this paper applied an intelligent PD-type fuzzy controller [3-5] firstly. The cross reference rules for the seeker tracking error E and ΔE (deviations of present E and the previous E) and outputs of PD-type

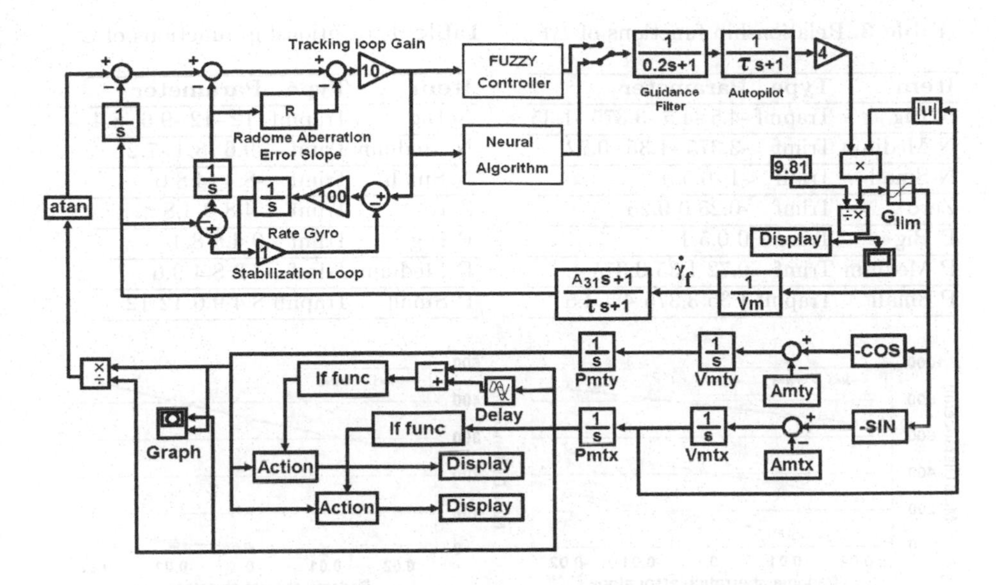

Fig. 3. Block diagram by using neural-fuzzy guidance law

fuzzy controllers are defined in Table 1, where NB, NM, NS, ZE, PS, PM, and PB respectively stand for negative big, negative middle, negative small, zero, positive small, positive middle, and positive big. Then the relationship functions of E, $\triangle$E and U (control input) are defined and listed in Tables 2-4. The guidance performances obtained by using fuzzy controllers are analyzed by simulation. By trial-and-error the proportion and derivative gains are respectively set to 10 and 0.115 to speed up the response of guidance loop. To reduce the gain of the derivative part is to reduce the noise amplification problem. Figs. 4-7 show the performance (miss distance vs. radome aberration error slope) by using the traditional PN and PD-fuzzy guidance laws. Note that both the traditional PN and PD-fuzzy guidance laws cannot meet the miss distance requirement for some cases.

Table 1. Rules of PD fuzzy controller

E,$\triangle$E	NB	NM	NS	ZE	PS	PM	PB
NB	NB	NB	NM	NM	NS	NS	ZE
NM	NB	NM	NM	NS	NS	ZE	PS
NS	NM	NM	NS	NS	ZE	PS	PS
ZE	NM	NS	NS	ZE	PS	PS	PM
PS	NS	NS	ZE	PS	PS	PM	PM
PM	NS	ZE	PS	PS	PM	PM	PB
PB	ZE	PS	PS	PM	PM	PB	PB

Table 2. Relationship functions of E

Item	Type	Parameter
N Big	Trapmf	-1 -1 -0.75 -0.3
N Medium	Trimf	-0.75 -0.3 -0.15
N Small	Trimf	-0.15 -0.1 0
Zero	Trimf	-0.05 0 0.05
P Big	Trimf	0 0.1 0.15
P Medium	Trimf	0.15 0.3 0.75
P Small	Trapmf	0.3 0.75 1 1

Table 3. Relationship functions of $\triangle$E

Item	Type	Parameter
N Big	Trapmf	-4.5 -4.5 -3.375 -1.35
N Medium	Trimf	-3.375 -1.35 -0.72
N Small	Trimf	-1 -0.5 0
Zero	Trimf	-0.25 0 0.25
P Big	Trimf	0 0.5 1
P Medium	Trimf	0.72 1.35 3.375
P Small	Trapmf	1.35 3.375 4.5 4.5

Table 4. Relationship functions of $\cup$

Item	Type	Parameter
N Big	Trapmf	-12 -12 -9.6 -8.4
N Medium	Trimf	-9.6 -8.4 -7.2
N Small	Trimf	-8.4 -4.8 0
Zero	Trimf	-4.8 0 4.8
P Big	Trimf	0 4.8 8.4
P Medium	Trimf	7.2 8.4 9.6
P Small	Trapmf	8.4 9.6 12 12

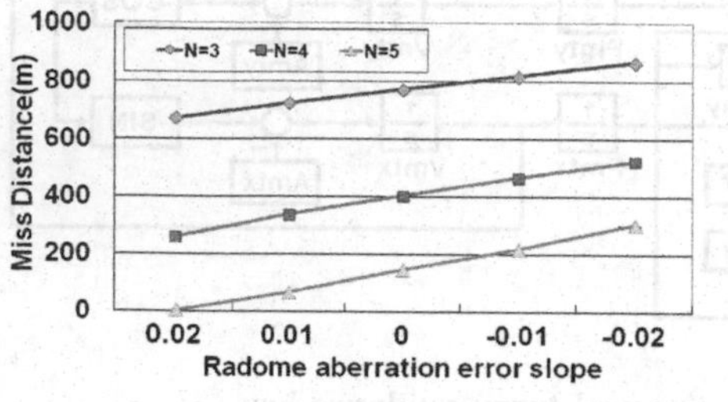

Fig. 4. PN law performance (Case 1)

Fig. 5. PN law performance (Case 2)

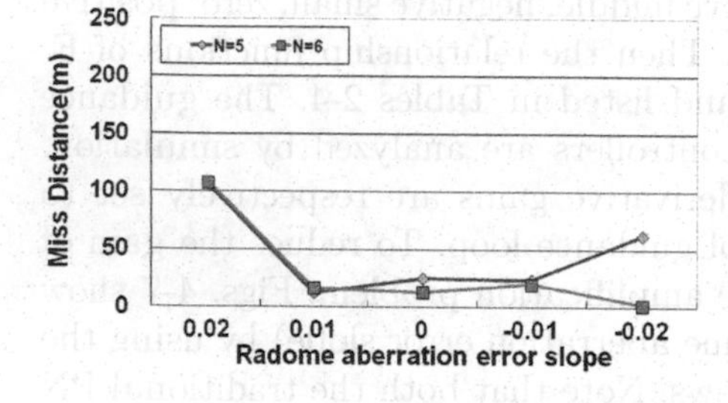

Fig. 6. PD-fuzzy performance (Case 1)

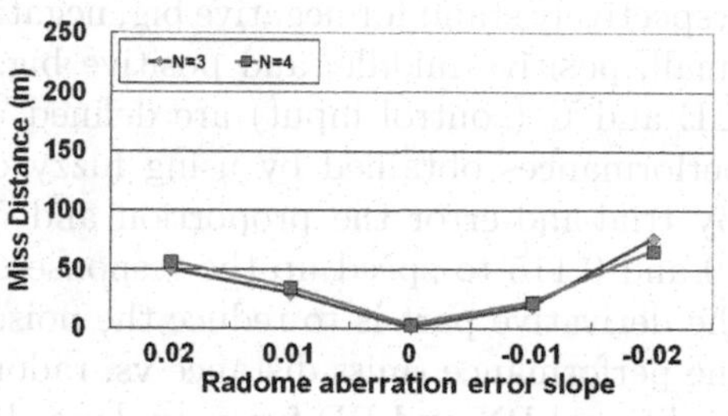

Fig. 7. PD-fuzzy performance (Case 2)

4 Neural-Fuzzy (PD) System Design

Thus the neural-fuzzy (PD) controller is applied for the design with two inputs and four hidden layers. The optimization algorithms such as GD, LM, and SCG in MATLAB toolbox are applied to train the neural network parameters in Fig.3. The fuzzy relationship functions for training the neural network are as listed in Tables 1-4.

4.1 Case 1: Using Only One Neural Optimization Algorithm

Figs. 8 and 9 respectively show the miss distances and acceleration commands of case 1 by using only one neural optimization algorithm. Note the results can

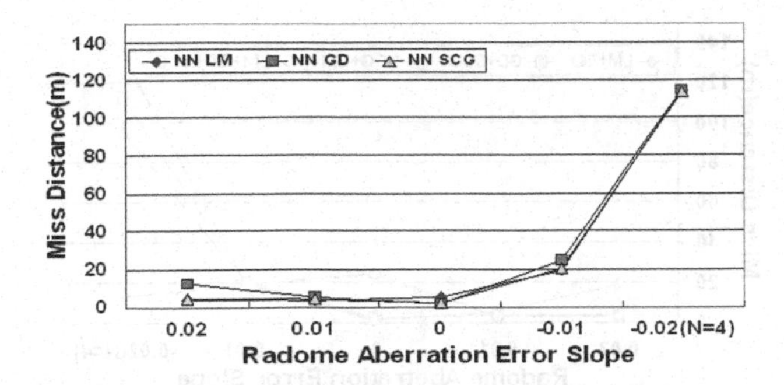

Fig. 8. Miss distance with only one neural optimization algorithm (Case 1)

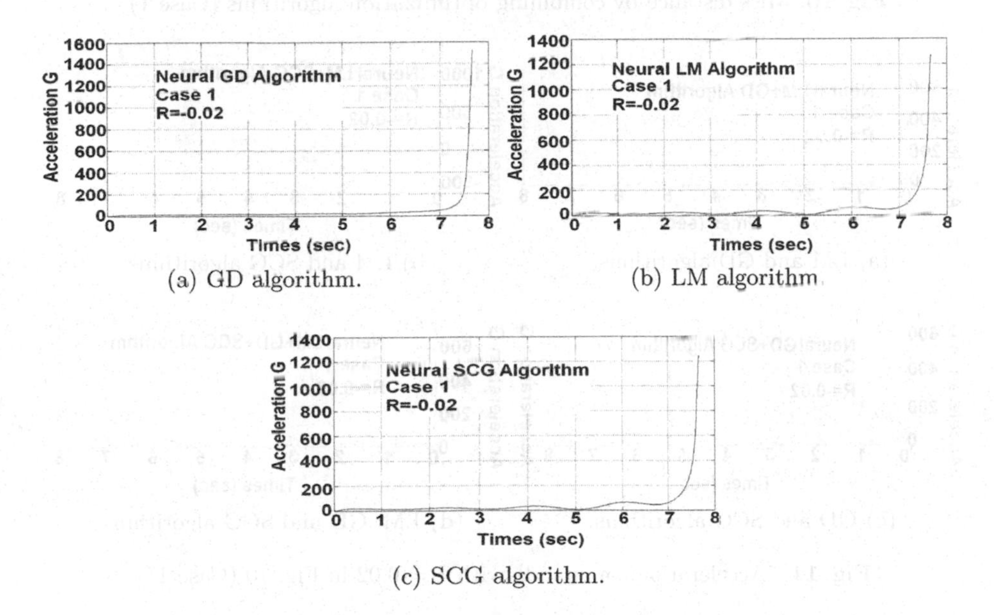

(a) GD algorithm.

(b) LM algorithm.

(c) SCG algorithm.

Fig. 9. Accelerations commands with R=-0.02 in Fig. 8 (Case 1)

meet the requirement except the case with R=-0.02 (N=4). By the way, note the acceleration command of GD algorithm is better as shown in Fig. 9(a) without any oscillation.

4.2 Case 1: Combinations of Neural Optimization Algorithms

Fig.10 shows the results of case 1 by using various combinations of neural algorithms alternatively for each step. The combination sequence is without any influence on the miss distance, because the methods are applied in alternative manner for each step. Note the miss distances obtained by combining both LM and GD algorithms are better for all the cases of radome aberration error slope. The other methods cannot meet the requirement for the case with R=-0.02.

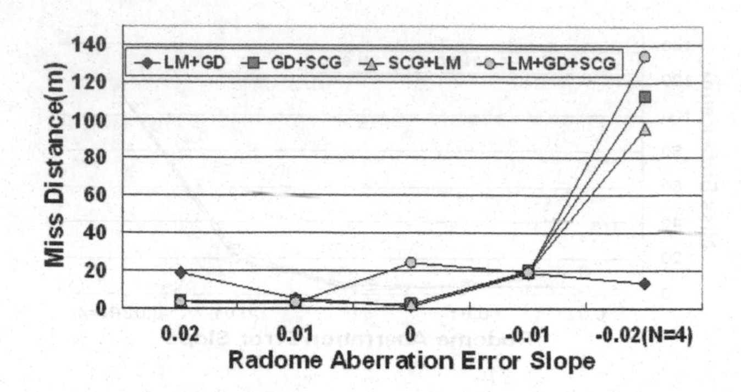

Fig. 10. Miss distance by combining optimization algorithms (Case 1)

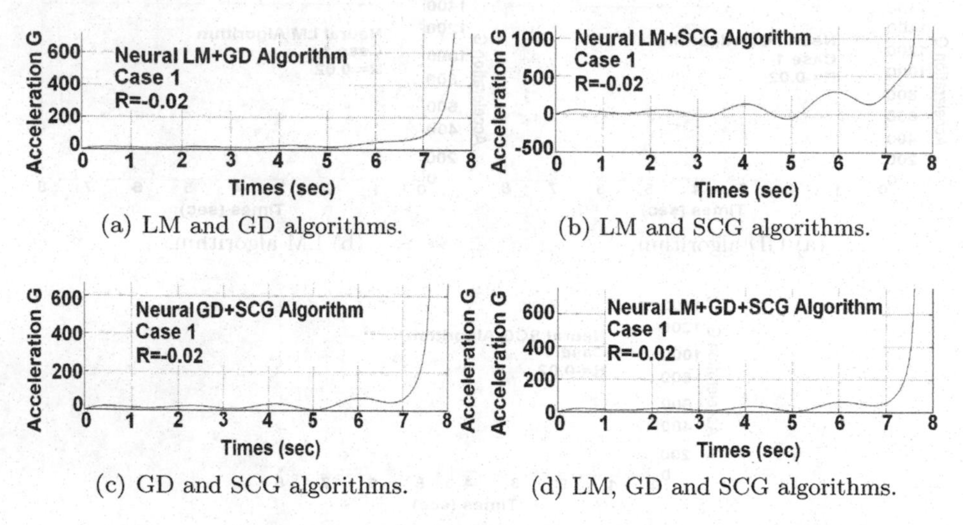

(a) LM and GD algorithms. (b) LM and SCG algorithms.

(c) GD and SCG algorithms. (d) LM, GD and SCG algorithms.

Fig. 11. Acceleration commands with R=-0.02 in Fig. 10 (Case 1)

The acceleration commands in Fig. 11 coincide with the results in Fig. 10, because only the acceleration command by combining LM and GD algorithms is better without oscillation as shown in Fig. 11(a), while the others not.

4.3 Case 2: Using Only One Neural Optimization Algorithm

Figs.12 and 13 (R=0.02) show the results by using only one neural optimization algorithm for Case 2. Note the miss distances by using GD algorithm only can meet the requirement for all the values of R. On the other hand, the acceleration command by using either LM or SCG algorithm is with polarity changing effect at the final stage as shown in Fig. 13(b) or 13(c), so the guidance performances are not good and the miss distances cannot meet the requirement. While by using GD algorithm only, the polarity of the acceleration command is without changing effect at the final stage as shown in Fig. 13 (a), and thus the guidance

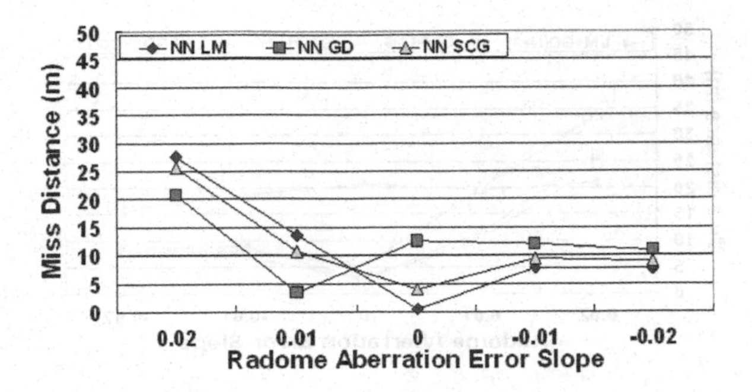

Fig. 12. Miss distances with only one optimization algorithm (Case 2)

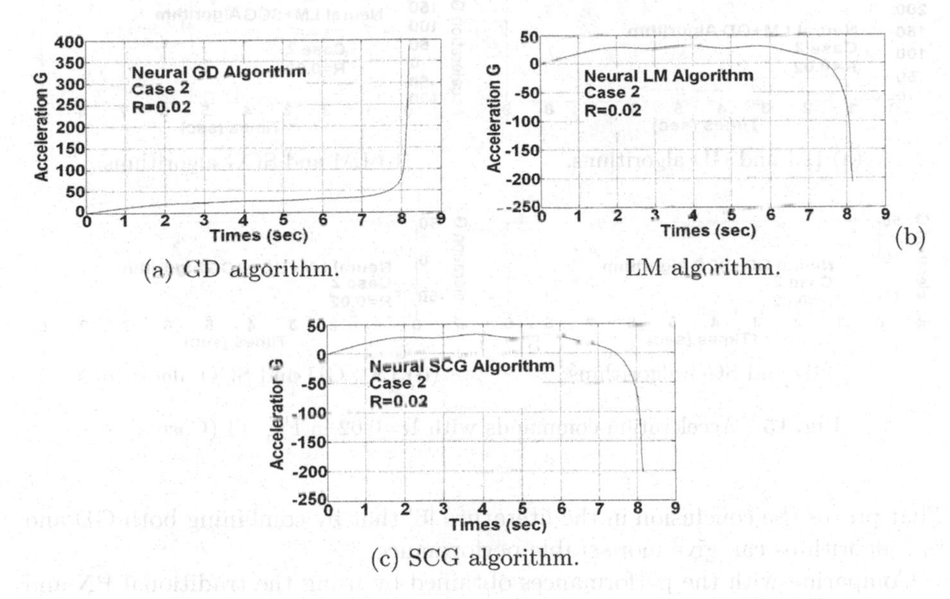

(a) GD algorithm. (b) LM algorithm.

(c) SCG algorithm.

Fig. 13. Acceleration commands with R=0.02 in Fig. 12 (Case 2)

loop can always track the target without oscillation and yields better guidance performances.

4.4 Case 2: Combinations of Neural Optimization Algorithms

Figs.14 and 15 (R=0.02) show the results by using various combinations of neural optimization algorithms for Case 2. Note the miss distances by using LM and GD algorithms only can meet the requirement for all the values of R. Besides, the polarity of the acceleration command by using this algorithm only is without changing effect at the final stage as shown in Fig. 15(a) for R=0.02, thus the guidance tracking performance is better and more stable, so do the performance.

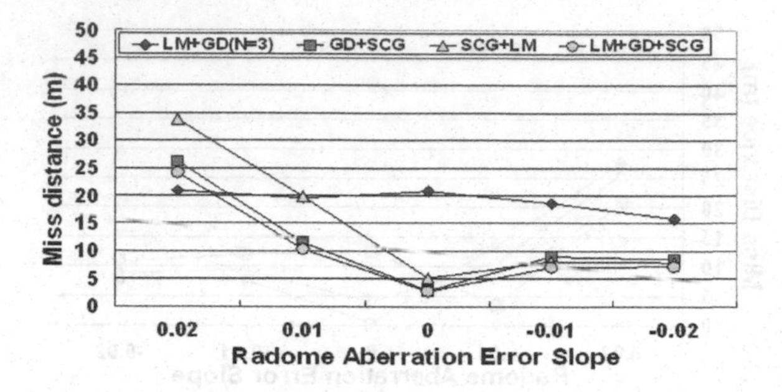

Fig. 14. Miss distances using various combinations of neural algorithms (case 2)

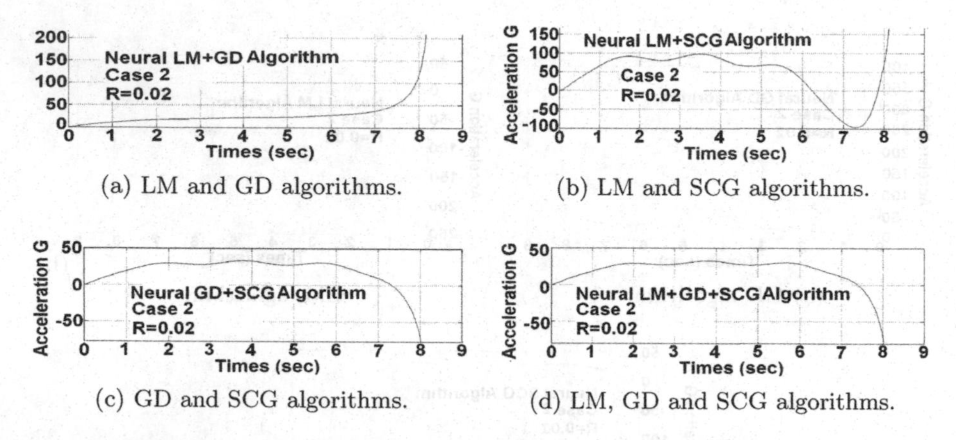

(a) LM and GD algorithms. (b) LM and SCG algorithms.

(c) GD and SCG algorithms. (d) LM, GD and SCG algorithms.

Fig. 15. Acceleration commands with R=0.02 in Fig. 14 (Case 2)

That proves the conclusion in the literature [6] that by combining both GD and LM algorithms can give more stable performance.

Comparing with the performances obtained by using the traditional PN and the fuzzy controllers respectively as in Figs. 4-5 and 6-7, we can see the miss distances obtained by using the proposed neural-fuzzy controllers are lower in general. Besides, the controllers obtained by combining different neural methods are better than those by using only one neural algorithm. Because the guidance loops can always track the target without oscillations and yield better guidance performances. Moreover, the GD and LM algorithms are alternatively applied for each step, so the proposed method would not increase the computation time.

5 Conclusion

This research applies the proportion navigation, fuzzy and neural-fuzzy methods for terminal guidance law design of a surface-to-air missile. Both the missile turning rate time constant and radome aberration error slope are taken into consideration. Comparing with the traditional PN, PD-fuzzy, and the guidance laws obtained by applying both Gradient Descent and Levenberg-Marquardt neural optimization algorithms alternatively for each step are also made; we note the miss distances obtained by the proposed novel method are lower, and the acceleration commands are without polarity changes or oscillation at the final stage. Thus the guidance loop tracking performance is very good. By the way since the GD and LM algorithms are alternatively applied for each step, so the proposed method would not increase the computation time.

Acknowledgments. This work was supported by the grant of National Science Council of the Republic of China with the project number NSC-101-2622-E-216-001-CC3, 101-2221-E-216-006-MY2, and 101-2221-E-216-019-.

References

1. Nesline, F.W., Zarchan, P.: Missile Guidance Design Tradeoffs for High-Altitude Air Defense. Journal of Guidance, Control, and Dynamics 6, 207–212 (1983)
2. Jinho, K., Jungsoon, J.: Nonlinear Model Inversion Control for Bank-to-Turn Missile. In: AIAA Guidance, Navigation and Control Conference, Baltimore, MD (1995)
3. Mishra, S.K., Sarma, I.G., Swamy, K.N.: Performance Evaluation of Two Fuzzy-Logic-Based Homing Guidance Schemes. Journal of Guidance, Control, and Dynamics 17, 1389–1391 (2006)
4. Lin, C.L., Hung, H.Z., Chen, Y.Y., Chen, B.S.: Development of an Integrated Fuzzy Logic-Based Missile Guidance Law Against High Speed Target. IEEE Trans. on Fuzzy Systems 12, 157–169 (2004)
5. Chen, B.S., Chen, Y.Y., Lin, C.L.: Nonlinear Fuzzy H∞ Guidance Law with Saturation of Actuators Against Maneuvering Targets. IEEE Trans. Control Syst. Technol. 10, 769–779 (2002)
6. Mitchell, T.M.: Machine Learning. McGraw-Hill (1997)
7. Andrei, N.: Scaled Memoryless BFGS Preconditioned Conjugate Gradient Algorithm for Unconstrained Optimization. Optimization Methods and Software, 561–571 (2007)

Detection of Bus Routes Number in Bus Panel via Learning Approach

Chun-Ming Tsai[1,*] and Zong-Mu Yeh[2]

[1] Department of Computer Science, University of Taipei,
No. 1, Ai-Kuo W. Road, Taipei 100, Taiwan
cmtsai2009@gmail.com
[2] Department of Mechatronic Technology, National Taiwan Normal University,
Taipei 106, Taiwan
zongmu@ntnu.edu.tw

Abstract. Detection of bus route number is a very important issue for assisting visually impaired people to take the bus. This paper proposes an intelligent approach to detect bus route numbers to help visually impaired people to "see" the bus route number of a moving bus. Current e-bus stations in Taipei, Taiwan, are not very robust in bad weather. For visually impaired people, it is very difficult know which bus route number is approaching the bus station. Some past research used RFID, GIS, sound, and image-based methods to help visually impaired people to "see" bus route number. However, the development of bus route number reading devices for visually impaired people in real time is still at an early stage. In this paper, we propose an efficient and effective learning-based method to detect the bus route number in displayed on the bus façade panel of the moving bus. Experimental results show that the proposed method can reduce time complexity and achieve high bus route number detection rates.

Keywords: Bus route number detection, Moving bus, Learning approach, Cascaded rule-based localization, visually impaired people.

1 Introduction

Every metropolitan community in the world has a public bus network, and this system is a very important resource for visually impaired people to be able get around a city independently. For example, in Taipei, Taiwan, there are many bus routes for people to use, and there has been some effort to assist visually impaired people to use it. In particular, Taipei and New Taipei city has developed the Taipei e-bus system [1]. This system uses e-bus stations to present the information about the bus route number. However, this system is affected by the climate. If it rains, this system information may be out of order.

* This paper is supported by the National Science Council, R.O.C., under Grants NSC 101-2221-E-133-004-.

N.T. Nguyen et al. (Eds.): ACIIDS 2014, Part II, LNAI 8398, pp. 302–311, 2014.

In Taipei, as in most systems, the bus route number is displayed on the lighted panel on the front of the bus. Every bus system has many different companies' buses and many different routes, each with their own bus route numbers, indicating the route and destination of the particular bus. The bus user knows which of the different numbers indicates the correct bus for the place where the user needs to go. Usually, the bus user waits for his/her bus at the bus station. Any given station will normally be served by many routes, and the rider must know whether an approaching bus is the one the rider needs. However, if he/she is a visually impaired people, it is very difficult for them to know which bus route number is coming. If the e-bus stations work properly, it is fortunate, because then the visually impaired person can hear the voice from the e-bus station announced the approaching route number. However, if the e-bus station is out of order, they must rely on help from other people. If no people can help them, they have no choice but to try by themselves to identify the number. They can show the driver a card with the desired bus route number written on it in advance. However, if the bus driver cannot see the card, the visually impaired rider cannot know whether to get on the bus.

The key task therefore is to have a device that will detect the bus route number so visually impaired people will always know which bus is coming. There have been several devices, using a glasses-mounted camera, which are designed to help visually impaired people to "see" bus route number using an alternative sense such as RFID [2], and RFID combined with GIS [3] and sound. However, the development of bus route number reading devices for visually impaired people is still at an early research stage.

A primary image-based detection system is proposed by Pan et al. [4] to assist visually impaired people to independently travel by bus and obtain the route information at a bus station. Their system includes bus detection and bus route number recognition. For bus detection, HOG-based feature [5] and cascaded SVM [6] are used to detect the bus façade. In the bus route number recognition, a scene text extraction algorithm [7] is used to localize and recognize the text information. Their system can achieve high accuracy of bus region detection. Furthermore, the text information of the bus route number can be successfully retrieved by their scene text extraction algorithm. However, their bus dataset is captured by the cell phone, so for their system to be used by a visually impaired people individual, the person must know how to use a cell phone. Furthermore, their system is an image-based system. For real-time application, a video-based system should be implemented. Another problem is that the accuracy of detecting text information from the detected bus has not been presented in their article. For effectively assisting visual impaired people, high accuracy of both text detection and recognition should be achieved.

Tsai and Yeh [8] proposed a text detection method to detect the bus route number in the text region on the bus façade panel. Their method includes moving bus detection, bus panel detection, and text detection. However, their method only detected the text region on the bus façade panel and did not extract the bus route number.

In this paper, we will propose an efficient and effective learning-based method to detect the bus route number in the bus façade panel of the moving bus. The proposed method is not equipped for bus route number recognition and concentrates only on bus route number detection.

2 Detection of Bus Route Number

Helping the visually impaired people with a glasses camera mounted on the ear to know which route number of the bus is coming is a challenging task. When a bus approaching the bus station is moving from far to near, the apparent size of the moving bus varies from small to large. Further, the bus route number is on the top of the bus façade panel, which is covered by glass, causing non-uniform illumination on the bus façade panel. These conditions make the detecting the bus route number in the bus façade panel difficult. In order to solve above-mentioned non-uniform illumination problem, a learning approach based on YCbCr color space is proposed to detect the bus route number.

2.1 Learning Stage

The YCbCr color space is widely used in skin and face detection [9]. The Y, Cb, and Cr components represent the luminance information, the difference between the blue component and the luminance component, and the difference between the red component and the luminance component, respectively.

The transformation used to convert from RGB to YCbCr color space [10] is shown as follows:

$$\begin{bmatrix} Y \\ Cb \\ Cr \end{bmatrix} = \begin{bmatrix} 0 \\ 128 \\ 128 \end{bmatrix} + \begin{bmatrix} 0.299 & 0.587 & 0.114 \\ -0.168935 & -0.331665 & 0.50059 \\ 0.499813 & -0.418531 & -0.081282 \end{bmatrix} \begin{bmatrix} R \\ G \\ B \end{bmatrix}. \tag{1}$$

To obtain the corresponding cluster of bus route number, a supervised learning method is proposed to learn the cluster parameters for bus route number in YCbCr color space. The proposed supervised learning to compute the Y cluster of the bus route number is described as follows:

1. Take a video containing 60 frames as training examples.
2. Use GrabCut [11] to label the bus route number in each training frame image.
3. Compute the total luminance (Y) histogram of all training examples.
4. Use a Gaussian smoothing filter [12] to smooth the total Y histogram.
5. Employ the average differences to determine the major peaks [12].
6. Find the maximum peak (M) with the maximum population. The maximum peak is the distribution of the bus route number.
7. Find the minimum level one (V_1) between the maximum peak and the former peak.
8. Find the minimum level two (V_2) between the maximum peak and the latter peak.
9. Use confidence interval (CI) [13] concept, the lower (L_Y) endpoint and the upper (U_Y) endpoint of the 95% confidence Y interval for the bus route number are defined as follows:

$$Y(L_Y) = \sum_{i=V_1}^{L_Y} n_i = 0.05 \times \sum_{i=V_1}^{M} n_i , \tag{2}$$

$$Y(U_Y) = \sum_{i=M+1}^{U_Y} n_i = 0.95 \times \sum_{i=M+1}^{V_2} n_i \,. \tag{3}$$

Here, n_i is the number of pixels at level i. $Y(L_Y)$ and $Y(U_Y)$ are the zeroth cumulative moments of the luminance histogram from $V_1 th$ up to the L_Yth level and from $(M+1)th$ to U_Yth level, respectively.

10. Similar to steps 3~9, the lower (L_{Cb}) endpoint and the upper (U_{Cb}) endpoint of the 99% confidence Cb interval for the bus route number are defined as follows:

$$Cb(L_{Cb}) = \sum_{i=V_1}^{L_{Cb}} n_i = 0.01 \times \sum_{i=V_1}^{M} n_i \tag{4}$$

$$Cb(U_{Cb}) = \sum_{i=M+1}^{U_{Cb}} n_i = 0.99 \times \sum_{i=M+1}^{V_2} n_i \tag{5}$$

11. Similar to steps 3~9, the lower (L_{Cr}) endpoint and the upper (U_{Cr}) endpoint of the 99% confidence Cr interval for the bus route number are defined as follows:

$$Cr(L_{Cr}) = \sum_{i=V_1}^{L_{Cr}} n_i = 0.01 \times \sum_{i=V_1}^{M} n_i \tag{6}$$

$$Cr(r_{Cb}) = \sum_{i=M+1}^{U_{Cr}} n_i = 0.99 \times \sum_{i=M+1}^{V_2} n_i \tag{7}$$

2.2 Thresholding Stage

Let $\mathbf{x} = [Y, Cb, Cr]^T$ be a pixel in a color video frame. The classification rule is as follows: $\mathbf{x}$ is bus route number color if it satisfies the following three tests:

$$206 \le Y \le 254, \tag{8}$$

$$98 \le Cb \le 148, \tag{9}$$

$$115 \le Cr \le 202. \tag{10}$$

Let $\mathbf{X} = [x_{ij}]_{W \times H}$ be the input color video frame. The output of detection of bus route number color is a binary map $\mathbf{B} = [b_{ij}]_{W \times H}$:

$$b_{ij} = \begin{cases} 1 & if \quad x_{ij} \quad is \quad bus \quad route \quad number \quad color \\ 0 & otherwise \end{cases}. \tag{11}$$

Each set of connected 1's in **B** is a potential candidate of bus route number. However, even with highly accurate detection of the bus route number, two sources of error remain to be addressed: (i) background pixels, i.e. non-color of bus route number such as taxi color or scenery, can have bus route number colors and this leads to false detection; (ii) some bus route number regions are affected by the illumination and this leads to broken detection. To solve these problems, a cascaded localization method is proposed to locate the bus route number and is described in the following Section.

Figure 1 shows an example of the bus route number detection. The original image of frame #26 and the images of thresholding results are shown in Fig. 1(a) and 1(b), respectively. From Fig. 1(b), the bus route number "15" is detected. However, many noises are also detected.

(a) (b)

Fig. 1. Example of the bus route number detection. (a) The image of frame #26. (b) The image of thresholding result.

3 Cascaded Bus Route Number Localization

After thresholding the bus route number, connected component labeling [14] is applied to detect the candidates for the bus route number. These candidates are represented by the bounding-boxes (BBs), which may be gray sky, advertising, taxi, bus façade body, start text, end text, traffic marking, or bus route number, etc. Figure 2(a) shows the example of applying the connected component labeling standard to label the bus route number candidates in the image of Fig. 1(b). From Fig. 2(a), there are 222 bounding-box candidates that have to be labeled. Most of these candidates are noises and should be removed.

In order to locate the bus route number in the candidate bounding-boxes, a cascaded localization method is proposed to remove the non-bus route number candidates, to merge concentrated broken BBs, and to locate the bus route number. The proposed localization method includes thirteen cascaded rules and is described as follows.

(1) Removing large width BBs: Because the bus is moving from the far to the near. The characteristic feature of bus route numbers is the *widths* and these widths are varied from small to large. When the bus is still at a distance, the width of the bus route number is small. On the contrary, when the bus is moving close to the observer, the width of the bus route number is large. Furthermore, when the bus is moving closest to the visually impaired people, the width of the bus route number is the largest. Thus, the *largest width* of the bus

route number is used to remove as noise bounding-boxes with larger width than that value. This *removing large width BBs rule* is defined as: If the width in a bounding-box is greater than a predefined value (T_{LW1}, set as 35 for frame image with width is 640), this bounding-box is removed.

(2) Merging closed BBs: After applying the above-mentioned rule, the bus route number is broken. In order to obtain the complete bus route number, a BB-based closing [15] is used to merge them. In this merging closed BBs rule, in order to vertically merge the adjacent broken BBs, the vertical dilation and erosion constants are considered. This *merging closed BBs rule* is defined as: The vertically dilation constants are set as T_{MVD}; a BB-based dilation operation uses this dilation constant to dilate the broken BBs, a geometry-based BB merging operation is used to merge the dilation broken BBs, and a BB-based erosion operation uses the erosion constant (=T_{MVD}) to erode the merged BBs. Herein, the merged result depends on the size of the dilation and erosion constants. In order to obtain the optimal result, a supervised learning method is applied to set dilation and erosion constants as $T_{MVD} = 7$.

(3) Removing large height BBs: Similar to *removing large width BBs rule*, the height of the bus route number when the moving bus is closest the largest possible height of the bus route number. This *largest height* of the bus route number is used to remove the noise bounding-boxes with larger height. This *removing large height BBs rule* is defined as: If the height in a bounding-box is greater than a predefined value (T_{LH}, set as 60 for frame image with height 480), this bounding-box is removed.

(4) Removing small height BBs: After applying above-mentioned rules, there are still remaining many small noise BBs. Thus, the *small height* of the bus route number is used to remove the noise bounding-box with small height. This *removing small height BBs rule* is defined as: If the height in a bounding-box is smaller than a predefined value (T_{SH}, set as 6 for frame image with height 480), this bounding-box is removed.

(5) Merging closed BBs and removing isolated small BBs: The bus route number is covered by glass. This glass affects the appearance of the bus route number. The reason for this is the non-uniform illumination on the bus façade panel. These conditions result in the bus route number detection process produce broken BBs. Thus, the *distance* between two horizontal BBs and the *width* of the BB are used to merge the closed BBs and remove the isolated small BBs. This *merging closed BBs and removing isolated small BBs rule* is defined as: If the distance between a testing BB and the other BBs is smaller than a predefined value (T_{dl}, set as 6 for frame image with width 640), these two BBs are closed and merged; else if the width of the testing BB is smaller than a predefined value (T_{sw}, set as 5 for frame image with width 640), this testing BB is an isolated small BB and is removed.

(6) Removing large width BBs: After applying above-mentioned rules, some additional BBs with large width are produced. These large width BBs should be removed. This removing large width BBs rule is defined as rule1 in addition to the predefined value, T_{LW2}, set as 50 for frame image with width is 640.

(7) Removing small-area BBs: In the both horizontal side of the bus route number, many texts are displayed to tell the passenger this bus starts at one place and ends its route at another place. Due to the illumination, the texts are broken after connected component labeling. These text BBs affect the detection of the bus route number and need to be removed. In addition, when the bus is moving at a distance, the area of the bus route number is small. Thus, this *small area* is used as the basis to remove the BB with small area. This *removing small-area BBs rule* is defined as: If the area of a BB is smaller than a predefined value (T_{SA}), this BB is removed. T_{SA} is set as 119 for frame image with 640*480.

(8) Merging horizontal and near BBs: The bus route numbers in the Taipei system considered here has up to three digits. In order to group these digits to form a complete region for bus route number, the *distance* between two horizontal and near BBs and the *ratio* of the merged BB are used to merge the horizontal and near BBs. This *merging horizontal and near BBs rule* is defined as: If the distance between a testing BB and its horizontal distance to the other BBs is smaller than a predefined value (T_{d2}, set as 10 for frame image with width 640) and the ratio of the merged BB is between T_{rL1} and T_{rH1}, these two BBs are merged. The T_{rL1} and T_{rH1} are set as 0.8 and 0.9, respectively, for two digits in frame image with 640*480.

(9) Removing illegal ratio BBs: An approaching bus is moving from far to near. The ratio of bus route numbers is approximately constant. Thus, the *ratio* of the bus route number is used to remove the illegal BBs. This *removing illegal ratio BBs rule* is defined as: If the ratio in a bounding-box is smaller than a pre-learning value (T_{rL2}) or greater than a pre-learning value (T_{rH2}), this BB is removed. The T_{rL2} and T_{rH2} are set as 0.73 and 1.01, respectively, for two digits in frame image with 640*480.

(10) Removing large area BBs: When the bus is moving at its closest to the visually impaired people, the area of the bus route number is the largest. Thus, the *largest area* of the bus route number is used to remove the noise BBs with large area. This *removing large area BBs rule* is defined as: If the area in a bounding-box is greater than a predefined value (T_{LA}, set as 3000 for frame image with area is 640*480), this BBs is removed.

(11) Removing BB with minor red color: After applying above-mentioned rules, some noise BBs are still remaining. By observing these remaining noise BBs, we found the red color in the bus number pixels to be greater than the green and blue color. Furthermore, the red color is the majority in the BBs that are in fact part of the bus number. In order to remove the noise BBs, the *red percent* is used. This *removing BB with minor red color* is defined as: If red percent in a BB is smaller than a pre-learning value (T_{RP}, set as 0.588), this BBs is removed.

(12) Adjusting boundaries of BBs: The bus route number is covered by glass. Furthermore, when light shines onto the glass, the illumination of the bus route number is non-uniform. Thus, after applying above-mentioned rules, some boundaries of the remaining BBs are not corrected. Herein, a scan line color differentiation algorithm [16] is used to adjust the four boundaries of the BBs.

(13) Locating the BB of the bus route number: After applying all above-mentioned rules, the remaining BBs are the bus route number, start text, end text, and traffic marking. In order to locate the BB of the bus route number, the *maximum area* of the BBs is used. The BB with the maximum area is the bus route number.

After applying all above-mentioned rules for Fig. 2(a), the image of the bus route number localization is shown in Fig. 2(b). From this figure, the bus route number "15" has been located.

Fig. 2. Example of the bus route number localization. (a) The image of the labeled result. (b) The image of the localization result.

4 Experimental Results

The proposed bus route number detection method was implemented in a Microsoft Visual C# 2012 Windows-based application on a Intel(R) Core(TM) i7-3667U CPU @ 2.00 GHz Apple MacBook Air, carried out on a video clip with 252 frame images. This video clip is captured by a glasses camera with 640 x 480 pixel resolutions, simulated a visually impaired person "seeing" a moving bus at a bus station. The video clip has been divided into training set and testing set consisting of 60 and 192 frame images, respectively.

To demonstrate the performance of the proposed bus route number detection on the moving buses, the predefined values of the proposed method are obtained from the 60 frame training set via learning approach. The proposed method used these training predefined values to detect the bus route number in the training and testing sets, respectively. Table 1 shows the performance of the proposed bus route number detection method. The number of original bus route number in the training and testing sets are 60 and 192, respectively. The number of correct bus route number detections by the proposed method in the two sets are 59 and 187, respectively, i.e., detected rates for the proposed method in two sets are 98.33% and 97.39%, respectively. The reason for the error detected BB is the detected BB is not the bus route number. The correct bus route number has not been detected.

Table 1. Performance of the proposed bus route number detection method

Video clip	Original frames	Original bus route number	Detected bus route number	Detected rate
Training set	60	60	59	98.33%
Testing set	192	192	187	97.39%

Table 2 shows the execution time performance of the proposed method. The sizes of training and testing sets are 640 x 480 and 640 x 480, respectively. The numbers of the frame for training and testing are 60 and 192, respectively. The total execution times for the training and the testing videos are 8.951 and 28.839 seconds, respectively, and the FPSs are 6.7 and 6.6, respectively. From this table, the execution time of the thresholding step is the major time consumer. Overall, these times are fast enough to make the system useful to a person waiting for a bus.

Table 2. Execution times performance of the proposed method

Video clip	frames	binary	CCL	located	whole	FPS
Training set	60	7.265(s)	0.391(s)	1.295	8.951(s)	6.7
Testing set	192	23.111(s)	1.751(s)	3.977	28.839(s)	6.6

These experiments show that the proposed method can detect the bus route number in the moving bus. Furthermore, the size of the execution program is only 36KB, so the proposed method can be run on mobile devices.

5 Conclusions

This study proposes an intelligent, efficient, and effective bus route number detection method that includes a supervised learning, thresholding, and cascaded rule-based localization operations. Experimentally, the bus route number of the moving objects in a video clip can be detected by the proposed method. The experimental results show that the computation and the detection of the proposed method are efficient and effective, respectively. In the future, the detected bus route number will be recognized and translated into voice to form the full system to "see" the bus route number for helping the visually impaired peoples.

Acknowledgements. The author would like to express his gratitude to Dr. Jeffrey Lee and Walter Slocombe, who assisted editing the English language for this article.

References

1. Taipei e-bus System, http://www.e-bus.taipei.gov.tw/new/english/en_index_6_1.aspx
2. Noor, M.Z.H., Ismail, I., Saaid, M.F.: Bus detection device for the blind using RFID application. In: International Colloquium on Signal Processing & Its Applications, pp. 247–249 (2009)

3. Mustapha, A.M., Hannan, M.A., Hussain, A., Basri, H.: UKM campus bus identification and monitoring using RFID and GIS. In: IEEE Student Conference on Research and Development, pp. 101–104 (2009)
4. Pan, H., Yi, C., Tian, Y.: A Primary Travelling Assistant System of Bus Detection and Recognition for Visually Impaired People. In: IEEE Workshop on MAP4VIP, in conjunction with ICME 2013 (2013)
5. Dalal, N., Triggs, B.: Histograms of Oriented Gradients for Human Detection. In: Computer Vision and Pattern Recognition, vol. 1, pp. 886–893 (2005)
6. Chang, C.C., Lin, C.J.: LIBSVM: a Library for Support Vector Machine (2001), http://www.csie.ntu.edu.tw/~cjlin/libsvm
7. Yi, C., Tian, Y.: Text String Detection from Natural Scenes by Structure-based Partition and Grouping. IEEE Trans. on IP 20(9), 2594–2605 (2011)
8. Tsai, C.M., Yeh, Z.M.: Text detection in bus panel for visually impaired people 'seeing' bus route number. In: ICMLC, pp. 1234–1239 (2013)
9. Kukharev, G., Novosielski, A.: Visitor identification elaborating real time face recognition system. In: Proc. 12th WSCG, pp. 157–164 (2004)
10. Color Conversion (RGB to YCbCr), http://msdn.microsoft.com/en-us/library/ff635643.aspx
11. Rother, C., Kolmogorov, V., Blake, A.: Grabcut - interactive foreground extraction using iterated graph cuts. ACM Trans. on Graphics 23(3), 309–314 (2004)
12. Tsai, C.M., Lee, H.J.: Binarization of color document images via luminance and saturation color features. IEEE Trans. on IP 11(2), 434–451 (2002)
13. Confidence interval, http://en.wikipedia.org/wiki/Confidence_interval
14. Chang, F., Chen, C.J., Lu, C.J.: A linear-time component-labeling algorithm using contour tracing technique. CVIU 93(2), 206–220 (2004)
15. Tsai, C.-M.: Intelligent Post-processing via Bounding-Box-based Morphological Operations for Moving Objects Detection. In: Jiang, H., Ding, W., Ali, M., Wu, X. (eds.) IEA/AIE 2012. LNCS (LNAI), vol. 7345, pp. 647–657. Springer, Heidelberg (2012)
16. Tsai, C.M.: Intelligent region-based thresholding for color document images with highlighted regions. Pattern Recognition 45(4), 1341–1362 (2012)

Performance of an Energy Efficient Bandwidth Allocation for Wireless Communication Systems

Yung-Fa Huang[1,*], Che-Hao Li[1], Hua-Jui Yang[1], and Ho-Lung Hung[2]

[1] Department of Information and Communication Engineering,
Chaoyang University of Technology, Taichung 41349, Taiwan (R.O.C)
yfahuang@mail.cyut.edu.tw,
james200405@gmail.com

[2] Department of Electrical Engineering, Chien-Kuo Technology University,
No.1, Chien Shou Rd., Changhua, Taiwan (R.O.C)
hlh@ctu.edu.tw

Abstract. In this paper, an improved energy efficient bandwidth expansion (IEEBE) scheme is proposed to assign the available system bandwidth for users in wireless communication systems. When the system bandwidth is not fully loaded, the available bandwidth can be energy efficiently allocated to users. Simulation results show that energy efficiency of proposed IEEBE scheme outperforms the traditional same bandwidth expansion (SBE) scheme with sub-optimal bandwidth expansion coefficients (BEC). Thus, the proposed IEEBE can effectively assign the system bandwidth and improve energy efficiency.

Keywords: Bandwidth resource allocation, energy efficiency, same bandwidth expansion, sub-optimal bandwidth expansion coefficients.

1 Introduction

The global warming issues become the most important. Thus, how to reduce Carbon oxide gas emissions attracted all over world to work out [1]. Due to the global growth of mobile users, the transmission data traffic increases so much in wireless communication networks in recent years. Thus, it is need to provide a technology to reduce energy consumption in mobile communications [1].

Orthogonal frequency division multiplexing (OFDM) can provide a higher data transmission [2], and has been widely used in mobile communication systems [3-4]. In order to enhance the system capacity, the OFDM based orthogonal frequency division multiple access (OFDMA) technique [5] technique is adopted to be as the multiple access scheme in 4G mobile communication systems [6-8].

In OFDMA systems, different users use the same super frame assigned to the sub-channel. But the different users occupied different bandwidth. In the full load (bandwidth usage rate of 100%) or the low load (bandwidth utilization below 50%) cases, with maintaining the user transmission rate, while providing higher bandwidth to the user, the user can reduce the transmission signal power. But the bandwidth is limited,

* Corresponding author.

N.T. Nguyen et al. (Eds.): ACIIDS 2014, Part II, LNAI 8398, pp. 312–322, 2014.

how to allocate the remaining bandwidth to the user and to reduce the user's energy consumption is an important issue studied in this paper.

2 Energy Efficiency with Bandwidth Expansion Models

The capacity of a communication system can be obtained by Shannon-Hartley theory as [11]

$$C = B\log_2(1+\frac{E}{N_o}) \tag{1}$$

where the required bandwidth of the user is denoted by B. The signal-to-noise ratio (SNR) of received signal is $\Gamma = E/N_o$, where E is the received signal energy and N_o is the power spectral density of the added white Gaussian noise (AWGN). To improve energy efficiency, the bandwidth of users is re-allocated to be α times of required bandwidth, α> 1. It is called by bandwidth expansion mode (BEM) [9], where α is called bandwidth expansion coefficient (BEC). After the bandwidth expansion, we obtain [9][10][12]

$$\alpha(B\log_2(1+\Gamma^{BEM})) = B\log_2(1+\Gamma) \tag{2}$$

where $\Gamma^{BEM} = E^{BEM}/N_o$, E^{BEM} is the required consumption energy. From (2), we obtain Γ^{BEM} by

$$\Gamma^{BEM} = \sqrt[\alpha]{1+\Gamma} - 1 \tag{3}$$

Therefore, from (3), the required energy E^{BEM} can be obtained with the known N_o, E and α.

After bandwidth expansion, the total energy reduce factor (ERF) E_r^U for N users can be obtained by

$$E_r^U\,(\%) = (1 - \frac{E_U^{BEM}}{E_U}) \times 100\,\% \tag{4}$$

where $E_U = \sum_{i=1}^{N} E_i$ is the total energy consumption for users, E_i is the energy consumption of the ith user. Both $E_U^{BEM} = \sum_{i=1}^{N} E_i^{BEM}$ and E_i^{BEM} are the total energy consumption for users and the energy consumption of the ith user, respectively, after bandwidth expansion. To investigate the relationship between the ERF and BEC, α, we define the ERF by

$$E_r = E - E^{BEM} \tag{5}$$

where E is the energy consumption of a user and E^{BEM} is the energy consumption for the user after bandwidth expansion. Fig. 1 shows that the relationship between ERF and BEC with N_o=1nJ for various SNR. From Fig. 1, it is observed that ERF is increasing with SNR. It is intuitive that the higher SNR users consume more energy, the BEM will benefit more efficiency.

Furthermore, in Fig. 1, the ERF are E_r =156.4293nJ and E_r =181.5161nJ with α=1.4 and α=1.8, respectively, for SNR=23dB. Moreover, the ERF are E_r =181.5161nJ and E_r =189.3976nJ with α=1.8, and α=2.2, respectively, for SNR=23dB. Therefore, it is mean that the less energy efficiency the higher BEC.

In the BEM systems, there are total N users, the total bandwidth is B(Hz). With available bandwidth can be allocated to users, we can equals the BEC for users by α_S, called same bandwidth expansion (SBE). Thus, the BEC α_S, can be obtained by $\alpha_S = B/\sum_{i=1}^{N} B_i$ where B_i is the required bandwidth of the ith user. Thus, the allocated bandwidth of the ith user will be $B_i^{BEM} = \alpha_i \times B_i$. The total energy consumption for users of SBE is obtained by

$$E_{U,S}^{BEM} = \sum_{i=1}^{N} (\sqrt[\alpha_S]{1+E_i} - 1) \tag{6}$$

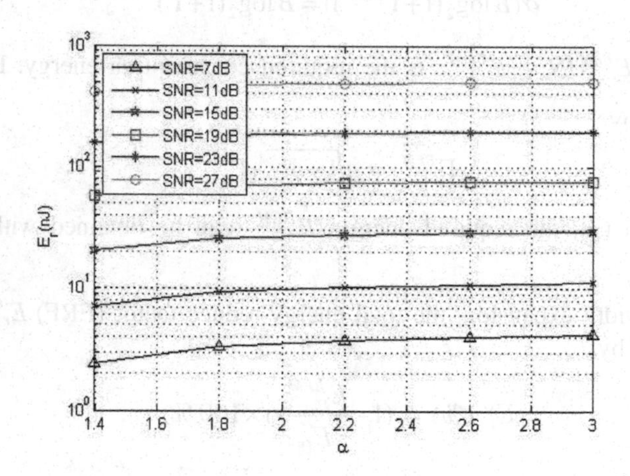

Fig. 1. The relationship between E_r and BEC with N_o=1nJ for various SNR

Then, the ERF of SBE can be expressed by

$$E_{r,S}^{U}(\%) = (1 - \frac{E_{U,S}^{BEM}}{E_U}) \times 100\% \tag{7}$$

Moreover, to perform more energy efficiency the available bandwidth can be adaptively allocated to users with different BEC α_i for the ith user, respectively. When we perform the optimal bandwidth allocation (OBA) with the sub-optimal BEC for sub-optimal energy efficiency, it is called improved energy efficient bandwidth expansion (IEEBE). The IEEBE optimization algorithms can be expressed by

$$\begin{Bmatrix} \text{minimize} & E_{U,IEEBE}^{BEM} \\ \text{subject to} & \sum_{i=1}^{N} \alpha_i B_i = B \end{Bmatrix} \tag{8}$$

where the total consumption energy can be obtained by

$$E_{U,IEEBE}^{BEM} = \sum_{i=1}^{N} \sqrt[\alpha_i]{1+E_i} \text{ -1} . \tag{9}$$

The α_i is the sub-optimal BEC for the ith user, B_i is the allocated bandwith for the ith user. Thus the ERF of IEEBE can be obtained by

$$E_{r,IEEBE}^{U}(\%) = (\frac{E_U - E_{U,IEEBE}^{BEM}}{E_U}) \times 100\% \tag{10}$$

The ERF ratio of IEEBE and SBE can be obtained by

$$E_{r,I,S}^{U}(\%) = (1 - \frac{E_{U,IEEBE}^{BEM}}{E_{U,SBE}^{BEM}}) \times 100\% \tag{11}$$

In the two-user system, we develop the sub-optimal BEC in this section. Assume that total bandwidth is B=20MHz, The required bandwidth for user A and user B is $B_A = 6\text{MHz}$ and $B_B = 4\text{MHz}$, respectively.

The BECs for two users are α_A and α_B. Then, the allocated bandwidth can be $B_A^{BEM} = \alpha_A \times 6\text{MHz}$ and $B_B^{BEM} = \alpha_B \times 4\text{MHz}$ after BEM where B_A^{BEM} and B_B^{BEM} are the allocated bandwidth for user A and B, respectively.

Thus with SNR=15dB, by using SBE, the BECs can be obtained by $\alpha_S = 2$. Then the ERF can be obtained from (4a) by E_r=85.14% as shown in Table 1. However, with IEEBE the BECs can be obtained from the IEEBE algorithms in (8) by α_A=1.92, and α_B=2.135. Thus, the allocated bandwidth of IEEBE is $B_A^{BEM} = 1.92 \times 6\text{MHz} = 11.52\text{MHz}$ and $B_B^{BEM} = 2.12 \times 4\text{MHz} = 8.48\text{MHz}$ for user A and B respectively. The ERF can be obtained (4) by E_r =85.27% as shown in Table 1a. Therefore, it is easily seen that the IEEBE can be more energy efficient than SBE.

We can compare the ERF for different BEC of user A as shown in Fig. 2. From Fig. 2, it is observed that with SBE α_A=2, E_r is increasing to the maximum of IEEBE, α_A=1.91. Furthermore, the performance of IEEBE for different total bandwidth environment is investigated as the results in Tables 1b and 1c. With the assumption of B=15MHz, SNR=15dB, $B_A = 6\text{MHz}$ and $B_B = 4\text{MHz}$, the IEEBE outperform the SBE with ERF=0.34%. From (11), in comparison between IEEBE and SBE, the energy reduction gain can be obtained by $E_{r,O,S}^{U} = 1.20\%$ as shown in Table 1b.

With the assumption of B=15MHz, SNR=15dB, $B_A = 8\text{MHz}$ and $B_B = 2\text{MHz}$, the IEEBE outperform the SBE with ERF=1.64%. From (11), in comparison between IEEBE and SBE, the energy reduction gain can be obtained by

$E_{r,O,S}^{U} = 11.01\%$ as shown in Table 1b. Thus, when the required bandwidth between the two users is higher, the improved ERF of IEEBE becomes more higher.

Table 1. Comparisons of Energy efficiency for SBE and IEEBE with *SNR*=15dB, for total bandwidth: (a) B=20MHz, $B_A = 6\text{MHz}$, $B_B = 4\text{MHz}$, (b) B=15MHz, $B_A = 6\text{MHz}$, $B_B = 4\text{MHz}$, (c) B=15MHz, $B_A = 8\text{MHz}$, $B_B = 2\text{MHz}$.

(a)

	B_A^{BEM}	B_B^{BEM}	α_A	α_B	E_r^U (%)
SBE	**12MHz**	**8MHz**	**2**	**2**	**85.14%**
IEEBE	**11.52MHz**	**8.46MHz**	**1.92**	**2.12**	**85.27%**

(b)

	B_A^{BEM}	B_B^{BEM}	α_A	α_B	E_r^U (%)
SBE	**9MHz**	**6MHz**	**1.5**	**1.5**	**70.88%**
IEEBE	**8.64MHz**	**6.36MHz**	**1.44**	**1.59**	**71.22%**

(c)

	B_A^{BEM}	B_B^{BEM}	α_A	α_B	E_r^U (%)
SBE	**16MHz**	**4MHz**	**2**	**2**	**85.10%**
IEEBE	**14.64MHz**	**5.36MHz**	**1.83**	**2.68**	**86.74%**

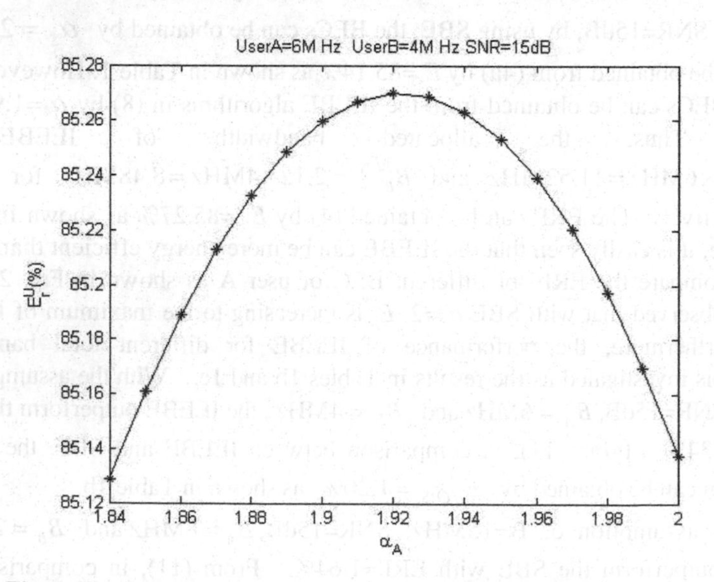

Fig. 2. The ERF comparison for different BEC of user A, α_A with SNR=15dB

3 Simulation Results

From the first investigation in previous section, improving energy efficiency of the bandwidth re-allocation is related with the required bandwidth and the SNR of users. Therefore, in this section, the effectiveness of IEEBE on the system environments is further studied.

At first, the simulation environment is shown in Table 2, where $M = B_A / B_B$, M=1/4, 1/3, 1/2, 1, 2, 3, 4 and let B_A=1.2, 1.5, 2,3,4,4.5, and 4.8 MHz. Then, the BEC $\alpha_{A,I}$ of user A can be found by proposed IEEBE of (8) for different total bandwidth B. Because we fix the total required bandwidth of two users $B^U = B_A + B_B$, the BECs of SBE α_S can be obtained as proportional to the total bandwidth B. With the reallocation algorithms, the BECs $\alpha_{A,I}$ of user A vs. BECs α_S of SBE can be obtained as shown in Fig. 3. From Fig. 3, it observed that with M=1 the $\alpha_{A,I}$ are the same as α_S. That is no need to relocate the bandwidth when the required bandwidth of users are the same. Moreover, when M <1 and smaller, the $\alpha_{A,I}$ becomes higher. But on contrary, when M >1 and larger, the $\alpha_{A,I}$ becomes lower.

The performance of the improvement of proposed IEEBE over SBE is depicted in Fig. 4 for M=1/8, ¼, 1/2, 1, 2, 4, 8, and SNR=10dB. From Fig. 4, it is obvious that the improvement of proposed IEEBE over SBE $E^U_{r,I,S}$ are higher than 18% for 5> α_S> 1.5 for M=8 and 1/8. Moreover, when M=4 and 1/4, the proposed IEEBE outperforms the SBE with 10% higher of $E^U_{r,I,S}$ for 5> α_S> 1.5.

Table 2. Simulation parameters of M, B_A and B_B

B_A (MHz)	B_B (MHz)	M
0.67	5.33	1/8
1.2	4.8	1/4
2	4	1/2
3	3	1
4	2	2
4.8	1.2	4
5.33	0.67	8

Previous results are performed with the same SNR for all users. However, when the SNRs of users are different, we define the SNR ratio of SNR_A to SNR_B by

$$n = \frac{SNR_A}{SNR_B} \tag{12}$$

where SNR_A and are the SNR of user A and user B, respectively. The simulation parameters of n, SNR_A and SNR_B are shown in Table 3.

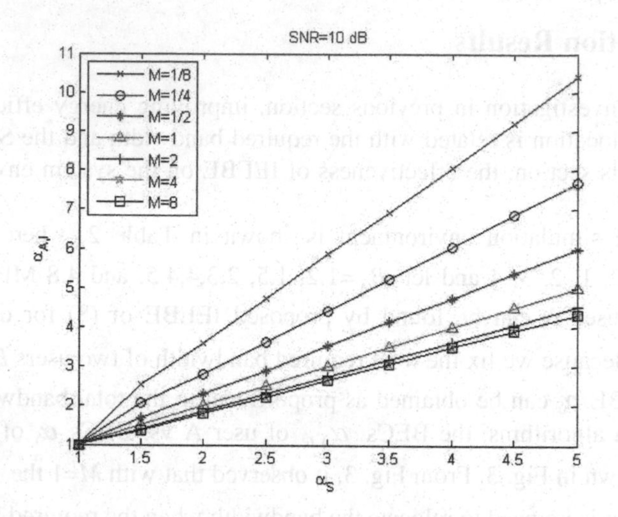

Fig. 3. Relationships of $\alpha_{A,I}$ to α_S with different *M* and SNR=10dB

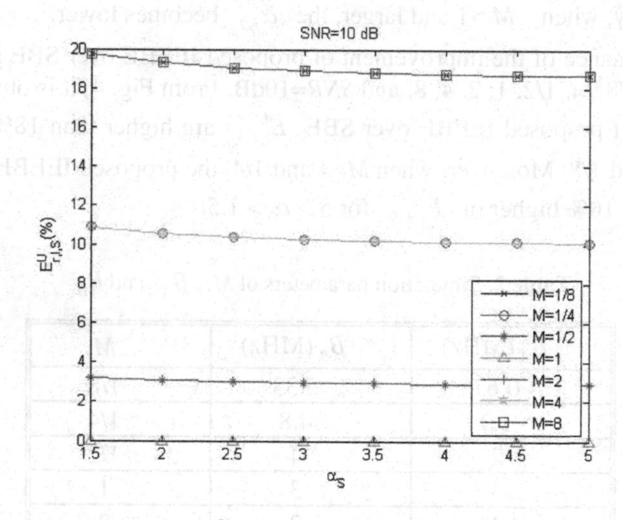

Fig. 4. Relationships of $E^U_{r,I,S}$ to α_S with different *M* and *SNR*=10dB

The simulation results of relationships of $\alpha_{A,I}$ to α_S with different *n* and *M*=1 ($B_A = B_B$ =3MHz) are depicted in Fig. 5. From Fig. 5, it is observed that when *n*>1 the $\alpha_{A,I}$ is becoming higher than that of *n*=1. That is to say when SNR_A is higher than SNR_B, user A can have more energy efficiency gain than user B. On the contrary, when *n* <1, the $\alpha_{A,I}$ is becoming smaller than that of *n*=1. The improvement of ERF of IEEBE to SBE, $E^U_{r,I,S}$ is depicted in Fig. 6 for *n* with *M*=1. From Fig. 6, it is

Table 3. Simulation parameters of n, SNR_A and SNR_B

SNR_A	SNR_B	n
1dB	10dB	1/8
4dB	10dB	1/4
7dB	10dB	1/2
10dB	10dB	1
10dB	7dB	2
10dB	4dB	4
10dB	1dB	8

observed that with n=1/8 and 8, the improvement of ERF of IEEBE to SBE, $E^U_{r,I,S}$ is the highest of 24% with α_S =1.5. However, when the n=1/4 or 4 and 1/2 or 2, the improvement of ERF, $E^U_{r,I,S}$ becomes smaller.

After we investigate the performance of energy efficiency improvement of the proposed IEEBE algorithms on the scenarios of same SNR and same required bandwidth for two users, we further combine the previous results as shown in Figs. 7 and 8. The sub-optimal BECs for users A are depicted in Fig. 7 for different M and n with $\alpha_S = 2$. From Fig. 7, when M=1, $n \leq 1$ ($SNR_A < SNR_B$), the sub-optimal BECs obtain $\alpha_{A,I} < 2$. However, when M increases, the $\alpha_{A,I}$ begins to increase. Moreover, when n=2, 3, M=1, the $SNR_A > SNR_B$, the $\alpha_{A,I}$ increases more. When M increases, the $\alpha_{A,I}$ begins to decrease and $\alpha_{A,I} < 2$.

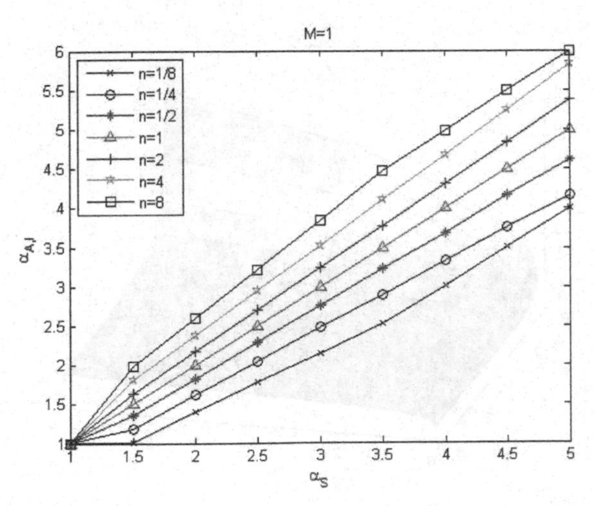

Fig. 5. Relationships of $\alpha_{A,I}$ to α_S with different n and M=1

The performance of energy efficiency improvement of the proposed IEEBE, $E^U_{r,I,S}$ is shown in Fig. 8 with $\alpha_S = 2$. From Fig. 8, it is seen that while M >2, $E^U_{r,I,S} > 10\%$. Especially with n=1/8, and M=4, $E^U_{r,I,S}$ approaches the highest 45%. However, with n=2, 4, $E^U_{r,I,S}$ is lower than 10%. When n=8 and M=1, $E^U_{r,I,S}$ >10%. Moreover, $E^U_{r,I,S}$ <10% for M>2. It is observed that with two regions: a. 1/8<n≤1 and M=1, b. 2<n≤ 8 and M>2, the $E^U_{r,I,S}$ is less than 10%.

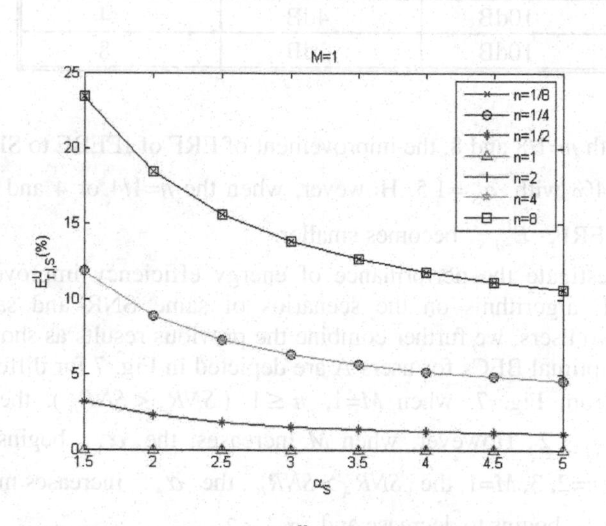

Fig. 6. Improvement of $E^U_{r,I,S}$ for *n and* α_S with *M*=1

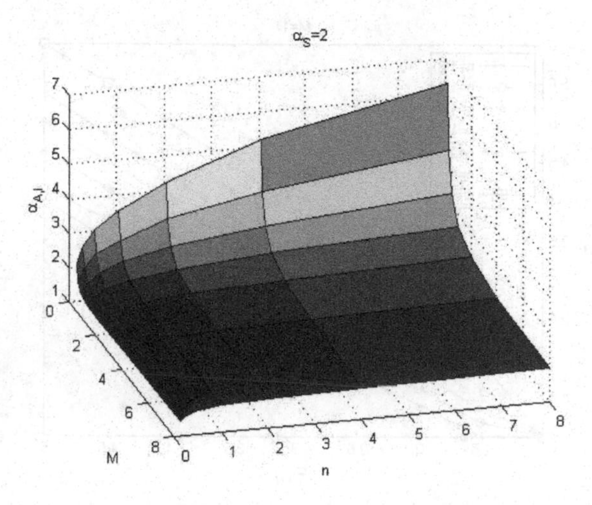

Fig. 7. The 3-D relationship of $\alpha_{A,I}$, with *M* and *n* for $\alpha_S = 2$

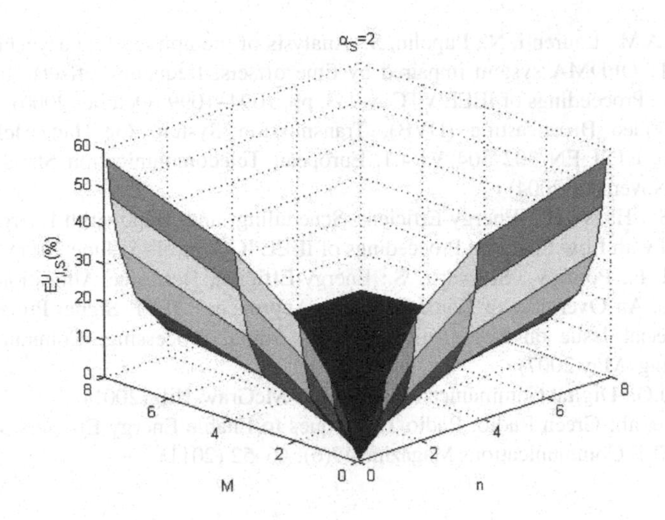

Fig. 8. The 3-D relationship of $E^U_{r,O,S}$ with M and n for $\alpha_S = 2$

4 Conclusion

In this study, we propose an improved bandwidth allocation algorithm to improve the energy efficiency for wireless communication systems. Simulation results show that the proposed IEEBE outperforms the SBE with different SNR and different required bandwidth for users. In two users scenario, the ERF improvements of IEEBE to SBE ratio $E^U_{r,I,S}$ can reach 45%. The ERF of IEEBE, $E^U_{r,I}$ achieves 86.5662%, which is 11.6095% higher than the ERF of SBE, $E^U_{r,S}$.

Acknowledgment. This work was funded in part by National Science Council of Taiwan under Grant NSC101-2221-E-324-024.

References

1. Edler, T.: Green Base Stations — How to Minimize CO2 Emission in Operator Networks. In: Ericsson Seminar, Bath Base Station Conf. (2008)
2. Ipatov, V.P.: Spread Spectrum and CDMA: Principles and Applications. Wiley (2005)
3. Lee, J.S., Miller, L.E.: CDMA System Engineering Handbook. Artech House (1998)
4. Adachi, F., Sawahashi, M., Suda, H.: Wideband DS-CDMA for Next-Generation Mobile Communications Systems. IEEE Commun. Mag. 36(9), 56–69 (1998)
5. IEEE Std. 802.16-2004 Standard for Local and Metropolitan Area Networks, Part16: Air Interface for Fixed Broadband Wireless Access Systems, IEEE Std. 802.16 (2004)
6. Dai, X.: Carrier frequency offset estimation and correction for OFDMA uplink. IET Communications 1(2), 273–281 (2007)

7. Tonello, A.M., Laurenti, N., Pupolin, S.: Analysis of the uplink of an asynchronous multi-user DMT OFDMA system impaired by time offsets, frequency offsets, and multi-path fading. In: Proceedings of IEEE VTC, vol. 3, pp. 1094–1099 (October 2000)
8. Digital Video Broadcasting (DVB): Transmission System for Hand-Held Terminals (DVB-H), ETSI EN 302 304 V1.4.1, European Telecommunication Standards Institute (ETSI) (November 2004)
9. Videv, S., Haas, H.: Energy-Efficient Scheduling and Bandwidth-Energy Efficiency Trade-off with Low Load. In: Proceedings of IEEE ICC, pp. 1–5 (June 2011)
10. Meshkati, F., Poor, V., Schwartz, S.: Energy-Efficient Resource Allocation in Wireless Networks: An Overview of Game Theoretic Approaches. IEEE Signal Processing Magazine: Special Issue on Resource-Constrained Signal Processing, Communications and Networking (May 2007)
11. Proakis, J.G.: Digital Communications, 4th edn. McGraw-Hill (2001)
12. Han, C., et al.: Green Radio: Radio Techniques to Enable Energy Efficient Wireless Networks. IEEE Communications Magazine 49(6), 46–52 (2011)

A Hierarchical Road Model for Shortest Route Discovery in Vehicular Networks

Yung-Fa Huang[1], Jia-Yi Lin[1], Chung-Hsin Hsu[1],
Sowat Boonyos[2], and Jyh-Horng Wen[3,*]

[1] Department of Information and Communication Engineering,
Chaoyang University of Technology, Taichung 41349, Taiwan (R.O.C)
yfahuang@mail.cyut.edu.tw, billy10353@yahoo.com.tw
[2] Department of Communications Engineering, National Chung Cheng University
Min-Hsiung, Chia-Yi County 62145, Taiwan (R.O.C.)
boonyos7@gmail.com
[3] Department of Electrical Engineering, Tunghai University, Taiwan (R.O.C.)
jhwen@thu.edu.tw

Abstract. In this paper, a simplified hierarchical route network model (SHRNM) is proposed for the optimal shortest road route discovery of transportation systems in metropolitan area. The Dijkstra algorithms are applied for the optimal path calculation. In our model, the metric of shortest path is the driving time. The four types of roads of urban road, bypass route, expressway and freeway are designed for different driving speeds in the hierarchical model. Thus, the shortest path can easily be found with proposed simple road traffic model. Simulation results show that the driving time of the discovered shortest route is faster than that of the four types of roads.

Keywords: shortest road route discovery, hierarchical route network model, transportation systems.

1 Introduction

In the intelligent transportation systems (ITS) and the wide-spread Global Positioning System (GPS), the optimal shortest path algorithms is one of the important tools to be accepted in the commercial market [1]. The most frequently used Google Map have been adopted to find the shortest road for the travelers. However, the shortest road usually is obtained by the general traffic environments. Therefore, the driving time of traffic lights, traffic jam and the traffic accidents is not considered in the systems. Therefore, the best road path for the drivers can not easily be exactly found by the Google Map [3].

In the optimization algorithms for the shortest path, the Dijkstra [2], A*[3] and the hierarchical path searching methods are used recently [5-6]. The A* algorithm uses the heuristic methods for the calculation on the straight line distance to predict the

---

* Corresponding author.

N.T. Nguyen et al. (Eds.): ACIIDS 2014, Part II, LNAI 8398, pp. 323–331, 2014.

distance between the beginning point and the destination [4]. The A* algorithm is based on the distance between two points are straight line distance. In the uniformly distributed nodes, the performance of A* algorithm exhibits excellent. However, in real metropolitan area, the roads are interested by the buildings to be zigzag type roads.

The hierarchical methods search the path by water diffusion for the neighbor node of the starting nodes. Like the water drops on the surface, the neighbors can be label by the wave number of the diffusion wave. According to the hierarchical number, the shortest path can be efficiently found. However, when the area is large, the efficiency is degraded [6].

Dijkstra algorithm is the most useful for the shortest path algorithm in networking. [2]. However, the complexity is very high for the high density nodes networking to be difficultly implemented [7]. Thus, the simplified Dijkstra algorithm are developed for the vehicle networks [8, 9].

In the nowadays, the road networks in metropolitan are very complexity. To speed up the vehicle traffics and to shorten the arrival time for the drivers, the different speed limit of roads are developed for efficiently transportation systems [9]. Therefore, in this study, a simplified hierarchy road network model (SHRNM) is developed to reduce the computational complexity for the shortest route discovery.

2 Shortest Path Algorithms

The Dijkstra's algorithm [7,8] is a famous algorithm to exhaustively find a global shortest path in computer networks. Therefore, Dijkstra's algorithm is applied to find the shortest path in this paper. An example of the transportation road map is shown in Fig. 1. The weight on the path between two points is the cost to drive through. Therefore, an example of Dijkstra's algorithm to find the shortest path from point a to point g can be briefly described as the followings.

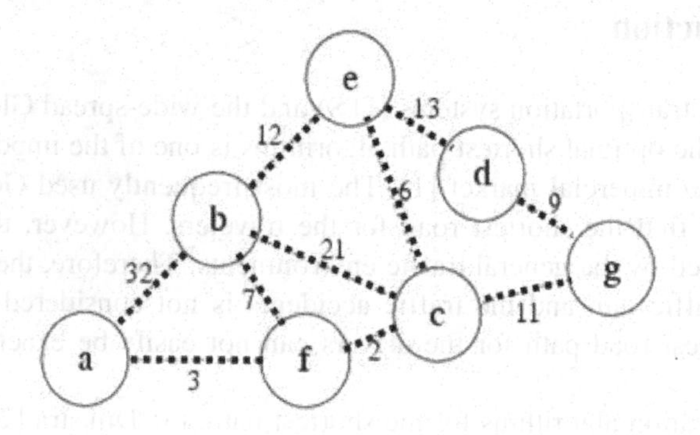

Fig. 1. An network example for Dijkstra's algorithm

At step 0, we list the distance between two points as unlimited because the distances are still not found yet, as shown in Table 1. At step 1, we can obtain the distances from 'a' to 'b' and 'a' to 'f' by 32 and 3 respectively from the weights in Fig. 1 as shown in Table 1.

Table 1. The calculated distance at step 0 and step 1

	a	b	c	d	e	f	g
0	∞	∞	∞	∞	∞	∞	∞
1	0	32	∞	∞	∞	3	∞

At step 2, the distance of a-b is updated by 3+7 with 10. the distance can be obtained by 3+2=5 as shown in Table 2. After step 6 the distances between 'a' point and 'b, c, d, e, f, g' points can be obtained by 10, 5, 24,11, 3, 16, respectively, as shown in Table 2.

Table 2. The calculated distance after step 6

	a	b	c	d	e	f	g
0	∞	∞	∞	∞	∞	∞	∞
1	0	32	∞	∞	∞	3	∞
2	0	10	5	∞	∞	3	∞
3	0	10	5	∞	11	3	16
4	0	10	5	∞	11	3	16
5	0	10	5	24	11	3	16
6	0	10	5	24	11	3	16

3 Hierarchical Vehicle Road Model

In this study, we develop a simplified road model for the shortest route discovery in metropolitan. A simple 9×9 grid map is designed for the proposed SHRNM as shown in Fig. 2. Four types of roads of urban road, bypass route, expressway and freeway are designed for different driving speeds hierarchical models. In Fig. 4, the position of a point is designed for starting point, and the point of Des. Is designed to be destination. Then central grid is designed to be city center. Thus, the starting road is always urban road and the final roads are urban roads, as well. The bypass route is near the city center for short distance bypassing. The expressway is placed to far from the city center for the middle distance travelling. Moreover, the freeway is provided for travelling through whole city, thus is place in the farthest position from city center.

In the proposed SHRNM, the driving time of each segment road is set to a random variable with uniformly distributed in $[t_L, t_H]$, where t_L and t_H are the driving time in light traffic volume and heavy traffic volume, respectively, for the road type. The driving speed limit and the driving time of each road segment are design as shown in Table 3. In Tablc 3, that the driving time of a segment on urban road is set by 1 to 3 minutes is designed to model the variation of traffic volume. Similarly, for other three types of roads, the traffic volumes are set to different parameters as shown in Table 3.

In a simplest model, only traffic volumes of roads are considering. However, to further discover the real shortest route, the traffic signs with the delay time by one to three minutes are added at some urban roads, bypass routes and expressways. The 15 traffic signs are added in the road intersection as shown in Fig. 3. Moreover, to verify the dynamic performance, 6 accidents are randomly generated with the delay driving time of 1 to 3 minutes randomly. The position of 6 accidents are shown in Fig. 3.

Table 3. Parameters in SHRNM

Road type	Speed limit	Driving time of each segment $[t_L, t_H]$
Urban road	50m/hr	[1, 3] min/seg
Bypass route	70km/hr	[1.7, 3.4] min/seg
Expressway	90km/hr	[2.2, 3.5] min/seg
Freeway	11km/hr	[2.3, 3.3] min/seg

4 Simulation Results

The shortest route discovery for a metropolitan is investigated by the proposed SHRNM. In our simulations, the driving time of each segment road is shown in Table 3. We set the driving time of each segment to a random variable with uniformly distributed in $[t_L, t_H]$, where t_L and t_H are the driving time in light traffic volume and heavy traffic volume, respectively, for the road type. The simulation is performed 100 times to discover the shortest route from starting point to destination. We would like to find what types of route of the shortest route. Therefore, the shortest route will be classified to one of four types of road. The route type classification is described as the followings.

A. If the route only passes by urban road, it will be classified to "urban road".
B. Else, if the route only passes by urban road and bypass route, it will be classified to "bypass route".
C. Else, if the route only passes by urban road, bypass route and expressway, it will be classified to "expressway".
D. Else, it will be classified to "freeway".

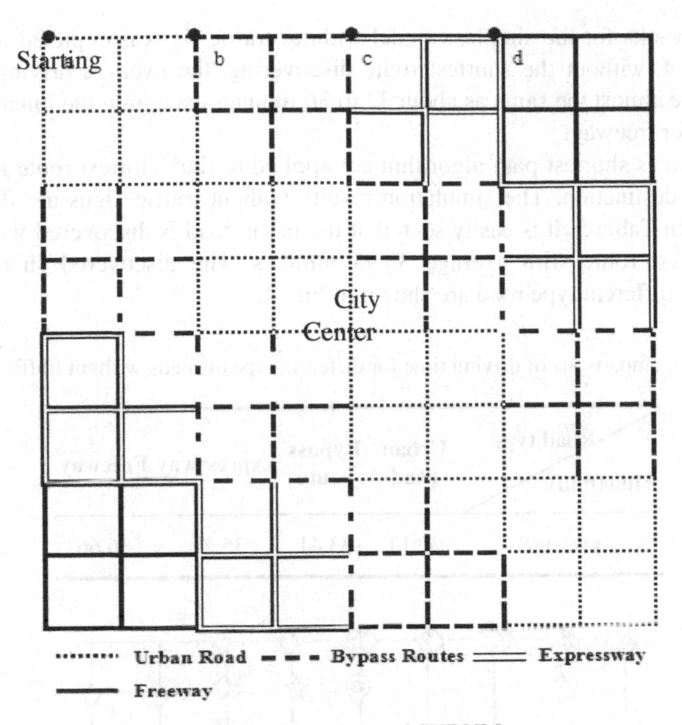

Fig. 2. The proposed SHRNM

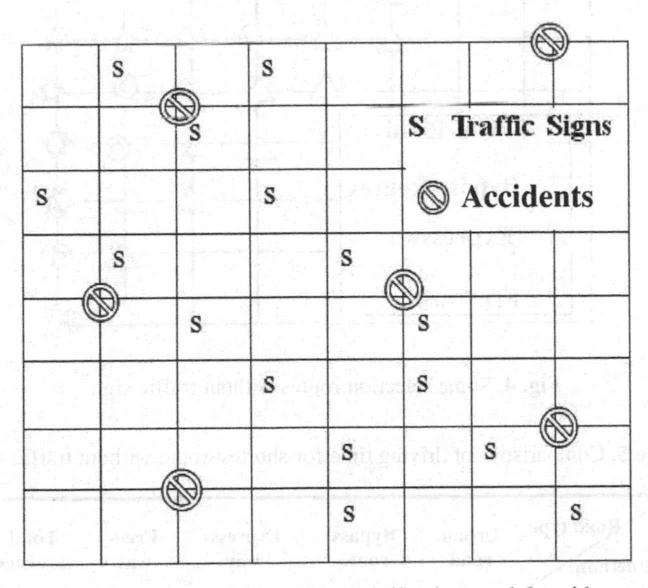

Fig. 3. The location of 15 designed traffic signs and 6 accidents

The test results for the simplest model without traffic signs is depicted in Table 4. From Table 4, without the shortest route discovering, the average driving time for four types are almost the same as about 32 to 36 minutes excepting the longest time of 36 minutes for freeway.

The Dijkstra's shortest path algorithm are applied to find shortest route from starting point to destination. The simulation results without traffic signs are depicted in Table 5. From Table 5, it is easily seen that the urban road is discovered with 39% to be the shortest route with average 30.93 minutes. The discovered shortest route examples for different type road are shown in Fig. 4.

Table 4. Comparisons of driving time for different type of roads without traffic signs

Road type / Time(min)	Urban road	Bypass route	Expressway	Freeway
average	32.14	33.41	35.24	36.66

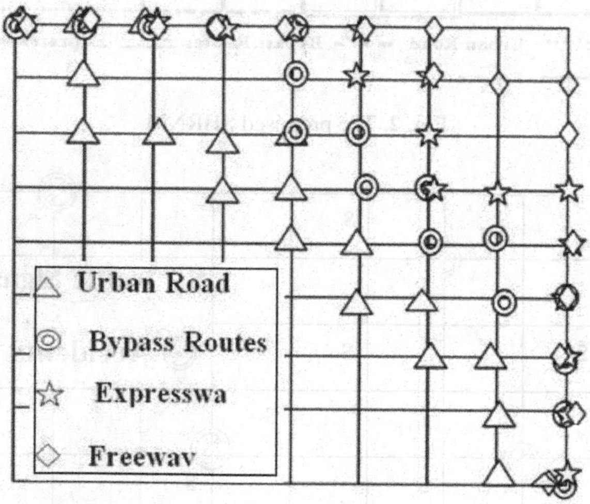

Fig. 4. Some selection routes without traffic sign

Table 5. Comparisons of driving time for shortest route without traffic signs

Road type / Time(min)	Urban road	Bypass route	Express-way	Free-way	Total Average
Average	30.93	31.73	32.12	32.54	31.81
Times	39	27	20	14	

Moreover, when the traffic signs are added, the simulation results of driving time for four type are shown in Table 6. From Tables 4 and 6, it is observed that the traffic signs increase the driving time by about 13 minutes for urban road. However, the traffic signs increase the driving time by 9 minutes for bypass route. It is easily to know the reason is that the traffic signs are mostly added on urban roads. Furthermore, the shortest path algorithm is applied to find shortest route from starting point to destination with traffic signs as depicted in Table 7. From Table 7, it is easily known that discovery rate of the urban road is decreased with 7 % to be the shortest route and the average time is increased to 44.5 minutes. Thus, the discovered shortest route for freeway is the highest with 45% and the driving time is average 35.5 minutes.

Table 6. Comparisons of driving time for different type of roads with traffic signs

Road type / Time(min)	urban road	bypass route	Expressway	Freeway
Average	44.65	42.66	39.76	38.11

Table 7. Comparisons of driving time for shortest route with traffic signs

Road type / Time(min)	Urban road	Bypass route	Expressway	Freeway	Total Average
Average	40.55	37.94	36.73	35.54	38.1
Times	7	12	36	45	

The comparisons of driving time for different types of roads with traffic signs and accidents are further shown in Table 8. From Table 8, when the accidents are added, it is observed that the accidents increase a lot of driving time for urban road to be the longest way among the four types of roads. Thus the urban road is never selected for the shortest path as shown in Table 9. However, the accidents increase the driving time for all four type road. It increase the driving time more for expressway and free way. The shortest path algorithm is applied to find shortest route from starting point to destination with traffic signs and accidents as depicted in Table 9. From Table 9, it is observed that the discovery rate of the urban road is decreased to 0 % to be the shortest route. Moreover, the discovered shortest route rate for freeway is the highest with 52% and the driving time is average 38 minutes.

Table 8. Comparisons of driving time for different type of roads with traffic signs and accidents

Road type / Time(min)	Urban road	Bypass route	Express-way	Free-way
Average	53.6	48.44	43.62	42.31

Finally the comparisons of driving time for four types of roads and the discovered shortest route are depicted in Fig. 5. From Fig. 5, it is observed that the driving time of the discovered shortest route is always faster than that of the four types road. That is to say that if we did not adaptively discover the shortest route and drive all the same type road, it will spend more driving time. On contrary, if we apply the proposed SHRNM, based on the dynamic traffic information, the driving time can be decreased and save the time and money.

Table 9. Comparisons of driving time for shortest route with traffic signs and accidents

Road type / Time(min)	Urban road	Bypass route	Expressway	Freeway	Total Average
Average		44.79	39.91	38.28	39.8
Times	0	10	38	52	

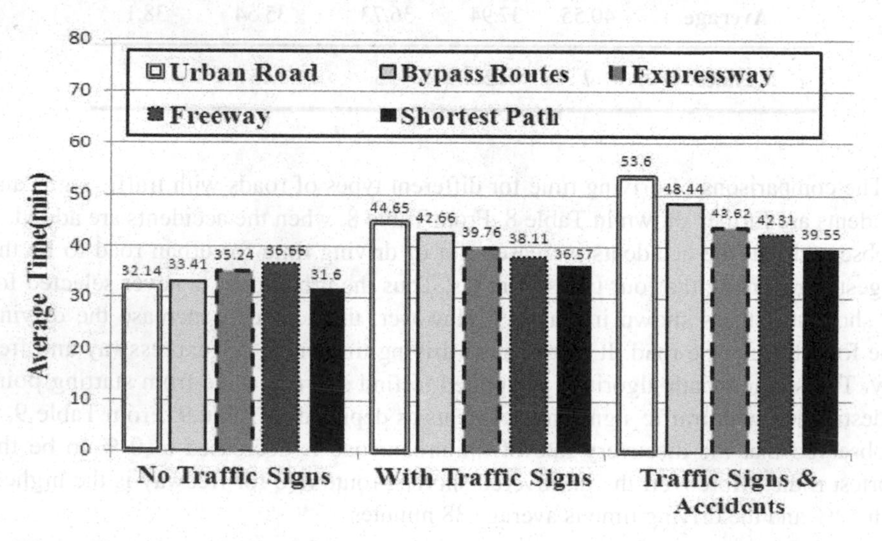

Fig. 5. Comparisons of driving time for four type roads and the discovered shortest route

5 Conclusion

In this paper, an SHRNM is proposed for fast shortest route discovering for vehicular network. Simulation results show that the driving time of the discovered shortest route is always fast than that of the four types road. That is to say that if we did not adaptively discover the shortest route and drive all the same type road, it will spend more driving time. On contrary, if we apply the proposed SHRNM, based on the dynamic traffic information, the driving time can be decreased and save the time and money.

Acknowledgment. This work was funded in part by National Science Council of Taiwan under Grant NSC101-2221-E-324-024.

References

1. Zhang, C., Zhang, F., Ren, J., Ren, J.: Multidimensional Traffic GPS Data Quality Analysis Using Data Cube Model. In: International Conference on Transportation, Mechanical, and Electrical Engineering (TMEE), pp. 307–310 (December 2011)
2. Fan, D., Shi, P.: Improvement of Dijkstra's Algorithm and Its Application in Route Planning. In: Proceedings of Seventh International Conference on Fuzzy Systems and Knowledge Discovery (FSKD), pp. 1901–1904 (2010)
3. Chabini, I., Lan, S.: Adaptations of the A* algorithm for the computation of fastest paths in deterministic discrete-time dynamic networks. Proceedings of IEEE Transactions on Intelli gent Transportation Systems, 60–71 (March 2002)
4. Nordin, N.A.M., Kadir, N., Zaharudin, Z.A., Nordin, N.A.: An Application of the A* Algorithm on the Ambulance Routing. In: Proceedings of IEEE Colloquium on Humanities, Science and Engineering Research (CHUSER), pp. 855–859 (December 2011)
5. Shen, G.J.: An Intelligent Hybrid Forecasting Model for Short-term Traffic Flow. In: Proceedings of the World Congress on Intelligent Control and Automation, pp. 486–491 (July 2010)
6. Mainali, M.K., Mabu, S., Hirasawa, K.: Evolutionary Approach for the Traffic Volume Estimation of Road Sections. In: Proceedings of IEEE International Conference on Systems Man and Cybernetics (SMC), pp. 100–105 (2010)
7. Comer, D.E.: Computer Networks And Internets (M-PIE), 5th edn. Peterson (January 2009)
8. Uppoor, S., Pai, M.M.M., Boussedjra, M., Mouzna, J.: Scalable routing technique using road hierarchy for vehicular networks. In: Proceedings of 2009 9th International Conference on Intelligent Transport Systems Telecommunications (ITST), pp. 403–407 (2009)
9. Zhimin, A.W., Xianfeng, B.L.: The model and algorithm for finding the optimal route in a dynamic road network. In: Proceedings of 2003 IEEE Intelligent Transportation Systems, vol. 2, pp. 1495–1498 (2003)

Factor Analysis as the Feature Selection Method in an Emotion Norm Database

Chih-Hung Wu[1,*], Bor-Chen Kuo[1], and Gwo-Hshiung Tzeng[2]

[1] National Taichung University of Education, Taichung, Taiwan
{chwu,kbc}@mail.ntcu.edu.tw
[2] National Taipei University, Taipei, Taiwan
ghtzeng@mail.ntpu.edu.tw

Abstract. Few studies have focused on integrating affective computing and soft computing technique for human physiology signals recognition in digital content learning system. This study develops a human affective norm (emotion and attention) recognition system for U-learning system. Eight elementary school students (4 male and 4 female) were recruited as participants to see some emotion pictures in international affective picture system (IAPS), and to collect their affective information—attention, meditation, electroencephalography (EEG), electrocardiogram (ECG), and SpO2 for developing the affective norm recognition system. These bio-physiology signals were extracted important features selection (Factor analysis) to serve as the input variables for radial basis function (RBF) neural network model. The results showed that all types of factor analysis did not perform well in our emotion norm database. Factor analysis with covariance extraction has higher accumulative variances than correlation extraction. This study suggested that future research can adopt more nonlinear feature selection methods to develop a high accuracy support vector machine based emotion recognition system.

Keywords: Affective Computing, Radial basis function (RBF) neural network, Emotion and Attention Recognition System, Eye Tracker, EEG (Electroencephalography), ECG (Electrocardiogram), Feature Selection.

1 Introduction

Feature selection has been proven quite necessary as a preprocessing step for classification algorithms [1]. Several well-known feature selection technique were developed as linear principal component analysis (LPCA), kernel principal component analysis (KPCA), linear discriminant analysis (LDA), generalized discriminant analysis (GDA), nonparametric weighted feature extraction (NWFE) and kernel-based NWFE etc. We summarize the application of feature selection methods in EEG/ECG. According the result, KPCA and LDA are the main feature selection methods in EEG. In addition, LPCA and LDA are the main feature selection methods in ECG. The GDA,

* Corresponding author.

N.T. Nguyen et al. (Eds.): ACIIDS 2014, Part II, LNAI 8398, pp. 332–341, 2014.

NWFE and kernel-based NWEF are new feature selection methods, but there are no applications in EEG or ECG analysis. GDA is a kernel-based LDA, it was proposed by [2]. Nonparametric weighted feature extraction (NWFE) was proposed by [3] to solve the poor performance of LDA. And, to extend NWFE for nonlinear situations, a kernel-based NWFE [4] was proposed. Suited feature extraction, feature selection methods and classifiers are unknown. Therefore, this study used factor analysis as feature selection method for e-learning emotion recognition. Eight elementary school students (4 male and 4 female) were recruited as participants to see some emotion pictures in international affective picture system (IAPS), and to collect their affective information—attention, meditation, electroencephalography (EEG), electrocardiogram (ECG), and SpO2 for developing the affective norm recognition system. These bio-physiology signals were extracted important features selection (Factor analysis) to serve as the input variables for radial basis function (RBF) neural network.

2 Literature Review

2.1 The Physiological Input Signals for Emotion Recognition

The physiological input signals of eye movement, EEG and ECG will be selected to input our learning state recognition system. Attention and emotion is a complex phenomenon. A single physiological parameter can't evaluate the state of attention and emotion. For including more objective physiological indices when recognize learning state, several techniques need to be combined for classifying the state of attention and emotion based on the past studies. .With the emergence of Electroencephalography (EEG) technology, learner's brain characteristics could be accessed directly and the outcome may well hand-in-hand supported the conventional test, recognize a learner's Learning Style [5]. The arousal state of the brain [6], alertness, cognition, and memory [7, 8] also can be measure. Heart rate variability (HRV) from ECG, has gained widespread acceptance as a sensitive indicator of mental workload [9]. The positive emotions may change the HF components of HRV [10].

2.2 Feature Selection Methods

EEG features extracted through wavelet analysis may provide a more reliable method [11]. Several recent studies still used discrete wavelet transform (DWT) to extractive feature in EEG [12-15]. In addition, wavelet packet decomposition also be used usually [16-19]. Feature selection has been proven quite necessary as a preprocessing step for classification algorithms [1]. But it is difficult to know which method is suited for a given problem. The performance is depend on many factors [20]. According the past studies, we found that kernel principal component analysis (KPCA) [21-24] and LDA [25-28] are the main feature selection methods in EEG. In addition, linear principal component analysis (LPCA) [29-33] and linear discriminant analysis (LDA) [34-36] are the main feature selection methods in ECG. Marvelously, the generalized discriminant analysis (GDA), nonparametric weighted feature extraction (NWFE) and kernel-based NWEF are new feature selection methods, but there are no applications in EEG or ECG analysis. In addition, multi-class support vector machine (MSVM) had been practically used in several problems of pattern recognition, for example, machine vision [37, 38], corporate credit rating [39], alcohol identification [40] and EEG-based mental fatigue measurement [41]. An emerging kernel-based machine-learning technique, the relevance vector machine (RVM) proposed by Tipping (2000), has been successfully applied in many areas. For example, credit risk analysis [42], EEG signal classification for epilepsy diagnosis [43] and canal flow prediction [44]. Fuzzy support vector machine had been practically used in several problems of pattern recognition, for example, human identification [45], product design [46], bankruptcy prediction [47] and road slope collapse [48].

3 Method

3.1 Participants and Procedure

Four male and four female elementary school students in the fifth grade will participate in this study. Participants will be tested individually. On arrival and after attaching physiological sensors, Participants will be asked to rest quietly. Participants will be told that pictures differing in emotional content would be displayed for 5 seconds on a screen in front of them, and that each picture should be viewed during this moment. Each rating will be preceded by a 5 seconds preparatory slide showing the number (1–24) of the next photograph to be presented (baseline period). The photograph will be rated and then be screened for 5 seconds (stimulus period), while the

Fig. 1. Participant in experiment

physiological responses will be recorded. The inter-trial interval was set at 30 seconds permitting recovery from the previous slide on all physiological measures. Slide presentation will be randomized for all subjects. Prior to the onset of the experimental trials, three pictures (highly pleasant-arousing, neutral on both valence/arousal, and highly unpleasant-arousing) will serve as practice stimuli. We collected participant's physiology signals (EEG, ECG, heart beating, SpO2) during the experiment every one second.

3.2 Materials

The aim of emotion norm construction is to determine (calculate) the relations between physiological changes and subjective ratings of International Affective Picture System (IAPS) photographs. A total number of 18 IAPS were used in the experiment. The pictures were divided into three different groups; 6 pleasant, 6 neutral and 6 unpleasant pictures [49]. Twenty-four International Affective Picture System (IAPS) photographs are grouped into 3 sets of 8 photographs: highly pleasant-arousing, neutral on both valence/arousal, and highly unpleasant-arousing ones. IAPS picture sequence: 1750(pleasant), 5480(pleasant), 1740(neutral), 1050(unpleasant), 9584(unpleasant), 6260(unpleasant), 2210(neutral), 7004(neutral), 7330(pleasant), 1300(unpleasant), 7050(neutral), 1670(neutral), 1463(pleasant), 7460(pleasant), 6370(unpleasant), 7150(neutral), 9592(unpleasant), 1710(pleasant) [50].

Table 1. IAPS picture

Affective	IAPS picture number
Neural	7009, 7150, 7170, 7705, 7050, 7004, 7100, 7140, 7500, 7560, 1670, 2190
Pleasant	1710, 7330, 8461, 5480, 1999, 7430, 1463, 1750, 7460, 8490, 7450, 8420
Unpleasant	1050, 9584, 9050, 1201, 1300, 9592, 9911, 1930, 3530, 1302, 3230, 3210

Reference: [49]

4 Database

NeuroSky is comprised of a complex combination of artifact rejection and data classification methods. According the NeuroSky proprietary Attention & Meditation eSense algorithms, NeuroSky can report the wearer's 10 brainwave state (attention score , meditation score, delta1, theta, beta1, beta2, alpha1, alpha2, gamma1, gamma2).

Table 2. Brain wave data

attention	meditation	delta1	theta	beta1	beta2	alpha1	alpha2	gamma1	gamma2
30.00	56	81.580802917	1.526600003	0.444700003	0.920000017	0.639900029	0.17330000	0.216600001	0.10650000
30.00	51	81.580802917	1.526600003	0.444700003	0.920000017	0.639900029	0.17330000	0.216600001	0.10650000
30.00	51	113.480300903	12.122699738	3.572700024	0.800400019	4.807000160	0.88050002	0.548500001	0.38609999
30.00	51	113.480300903	12.122699738	3.572700024	0.800400019	4.807000160	0.88050002	0.548500001	0.38609999
44.00	51	113.480300903	12.122699738	3.572700024	0.800400019	4.807000160	0.88050002	0.548500001	0.38609999
44.00	67	113.480300903	12.122699738	3.572700024	0.800400019	4.807000160	0.88050002	0.548500001	0.38609999
44.00	67	102.461402893	5.898600101	1.733600020	6.022600174	2.160900116	2.03329992	0.661599994	0.71980000
44.00	67	102.461402893	5.898600101	1.733600020	6.022600174	2.160900116	2.03329992	0.661599994	0.71980000
41.00	67	102.461402893	5.898600101	1.733600020	6.022600174	2.160900116	2.03329992	0.661599994	0.71980000
41.00	70	102.461402893	5.898600101	1.733600020	6.022600174	2.160900116	2.03329992	0.661599994	0.71980000
41.00	70	291.343414307	12.582300186	1.400099993	2.613199949	3.210999966	1.44760001	1.377799988	0.70660001
41.00	70	291.343414307	12.582300186	1.400099993	2.613199949	3.210999966	1.44760001	1.377799988	0.70660001
35.00	70	291.343414307	12.582300186	1.400099993	2.613199949	3.210999966	1.44760001	1.377799988	0.70660001
35.00	64	291.343414307	12.582300186	1.400099993	2.613199949	3.210999966	1.44760001	1.377799988	0.70660001

Based on the distribution of frequency zones, when a learner's emotional state is negative, peaceful or positive, the Coherence score will be calculated as 0, 1 and 2, respectively .The value of Heart Rate Artifacts (HRAs) is zero when the human emotion detected is in normal situations, whereas the value of HRAs is one when the human emotion is detected in abnormal situations. The emWave system identifies learner emotional states every 5 s. In this study, identifying the percentages spent in positive or negative emotions was applied to assess the effects of two learning methods on learning emotions. In computing the percentages of positive and negative emotions, the Accumulated Coherence Score (ACS) has the key role, whereas the method for computing ACS based on different coherence states and HRAs is as follows:

$$CV(t)=\begin{cases} -1\text{, if Coherence}(t)=0 \text{ and HRA}(t)=0 \text{ (negative emotion)} \\ +1\text{, if Coherence}(t)=1 \text{ and HRA}(t)=0 \text{ (peaceful emotion)} \\ +2\text{, if Coherence}(t)=2 \text{ and HRA}(t)=0 \text{ (positive emotion)} \end{cases}, CV(0)=0, t=0,1,2,\ldots,m$$

Our developed emotion norm database is shown in Table 4. We collected data from four male and four female elementary school students. Each student was asked to see 20 IAPS pictures. The database included brain wave data, hear beating data, SpO2, and emotion data. There are totally 14972 records in this database.

Table 3. Emwave data

time	emotion	HeartBeat	SPO2
8:35:13	-1	72	98
8:35:14	-1	70	97
8:35:15	-1	70	97
8:35:16	-1	76	97
8:35:17	-1	82	97
8:35:18	-1	95	97
8:35:19	-1	96	97
8:35:20	-1	98	97
8:35:21	-1	96	97
8:35:22	-1	96	97
8:35:23	-1	97	97
8:35:24	-1	98	97
8:35:25	0	98	97

Table 4. Our Developed Emotion Norm Database

Id	**Attention**	**Meditation**	**Delta1**	**Theta**	**Beta1**	**Beta2**	**Alpha1**	**Alpha2**	**Gamma1**	**Gamma2**	**Emwave**	**HeartBeat**	**SPO2**	**Output**
A001	53	61	44.9	2.1	0.6	1.4	1.1	0.2	0.2	0.1	0	87	98	2
A001	51	61	44.9	2.1	0.6	1.4	1.1	0.2	0.2	0.1	0	87	98	2
A001	51	50	44.9	2.1	0.6	1.4	1.1	0.2	0.2	0.1	0	87	98	2
A001	51	50	86.8	11.1	1.4	0.5	1.7	0.8	0.2	0.4	0	87	98	2
A001	51	50	86.8	11.1	1.4	0.5	1.7	0.8	0.2	0.4	0	85	98	2
A001	51	50	86.8	11.1	1.4	0.5	1.7	0.8	0.2	0.4	0	85	98	2
A001	51	57	86.8	11.1	1.4	0.5	1.7	0.8	0.2	0.4	0	85	98	2
A001	51	57	67.9	14.9	2.7	9.4	5.8	1.3	0.8	0.4	0	85	98	2
A001	51	57	67.9	14.9	2.7	9.4	5.8	1.3	0.8	0.4	0	82	98	2

Classification Accuracy

This study uses support vector machine with 10 folds cross-validation. The results are summarized as shown in Table 5. Among four types of factor analysis, correlation extraction in varimax rotation has the best RBF classification accuracy (35.31%). However, all RBF neural network classification accuracies are quiet low and have no significant differences. Table 6 to Table 9 showed the detail accuracy by class.

Table 5. Classification Results of SVM among four types of Factor analysis

Extract	Correlation	Correlation	Covariance	Covariance
Rotation	Varimax	Oblique	Varimax	Oblique
Correct ratio	35.33%	35.31%	34.96%	34.86%
InCorrect ratio	64.67%	64.69%	65.04%	65.14%
Accumulate Variance	65.37%	65.37%	88.37%	88.37%

Table 6. Detail Accuracy by Class (Extract: Correlation; Varimax Rotation)

TP rate	**FP rate**	**Precision**	**Recall**	**F-measure**	**ROC Area**	**Class**
0.27	0.28	0.33	0.27	0.30	0.50	Negative
0.19	0.15	0.38	0.19	0.25	0.53	Neutral
0.60	0.54	0.36	0.60	0.45	0.53	Positive

Table 7. Detail Accuracy by Class (Extract: Correlation; Oblique Rotation)

TP rate	FP rate	Precision	Recall	F-measure	ROC Area	Class
0.26	0.27	0.33	0.26	0.29	0.50	Negative
0.19	0.15	0.38	0.19	0.25	0.53	Neutral
0.60	0.55	0.36	0.60	0.45	0.53	Positive

Table 8. Detail Accuracy by Class (Extract: Covariance; Varimax Rotation)

TP rate	FP rate	Precision	Recall	F-measure	ROC Area	Class
0.22	0.20	0.36	0.22	0.27	0.52	Negative
0.02	0.02	0.38	0.02	0.04	0.51	Neutral
0.80	0.77	0.35	0.80	0.48	0.52	Positive

Table 9. Detail Accuracy by Class (Extract: Covariance; Oblique Rotation)

TP rate	FP rate	Precision	Recall	F-measure	ROC Area	Class
0.22	0.20	0.36	0.22	0.27	0.51	Negative
0.02	0.02	0.35	0.02	0.04	0.51	Neutral
0.80	0.77	0.35	0.80	0.48	0.52	Positive

5 Conclusion and Suggestion

This study designs an experiment to develop an emotion norm database. After that, we use four types of factor analysis as feature selection methods for developing a RBF neural network classifier for emotion recognition system. The results showed that all types of factor analysis did not perform well in our emotion norm database. Factor analysis with covariance extraction has higher accumulative variances than correlation extraction. Based on research results, we found that linear factor analysis performed poor in our emotion norm database. We guess the emotion norm data might more suitable for nonlinear feature selection method, such as NWFE, kernel-based NWEFE. This study suggested that future research can adopt more nonlinear feature selection methods to develop a high accuracy support vector machine based emotion recognition system.

Acknowledgements. Authors thank the National Science Council of Taiwan for support (grants NSC 101-2410-H-142-003-MY2).

References

1. Muller, K.R., Mika, S., Ratsch, G., Tsuda, K., Scholkopf, B.: An introduction to kernel-based learning algorithms. IEEE Transactions on Neural Networks 12, 181–201 (2001)
2. Baudat, G., Anouar, F.: Generalized Discriminant Analysis using a Kernel Approach. Neural Computation 12, 2385–2404 (2000)

3. Kuo, B.-C., Landgrebe, D.A.: Nonparametric Weighted Feature Extraction for Classification. IEEE Trans. Geosci. Remote Sens. 42, 1096–1105 (2004)
4. Kuo, B.-C., Li, C.H., Yang, J.M.: Kernel Nonparametric Weighted Feature Extraction for Hyperspectral Image Classification. IEEE Trans. Geosci. Remote Sens. 47, 1139–1155 (2009)
5. Rashid, N.A., Taib, M.N., Lias, S., Sulaiman, N., Murat, Z.H., Kadir, R.S.S.A.: Learners' Learning Style Classification related to IQ and Stress based on EEG. Procedia - Social and Behavioral Sciences 29, 1061–1070 (2011)
6. Zhang, Q., Lee, M.: Emotion development system by interacting with human EEG and natural scene understanding. Cognitive Systems Research 14, 37–49 (2012)
7. Berka, C., Levendowski, D.J., Cvetinovic, M.M., Petrovic, M.M., Davis, G., Lumicao, M.N., Zivkovic, V.T., Popovic, M.V., Olmstead, R.: Real-Time Analysis of EEG Indexes of Alertness, Cognition, and Memory Acquired With a Wireless EEG Headset. International Journal of Human-Computer Interaction 17, 151–170 (2004)
8. Berka, C., Levendowski, D.J., Lumicao, M.N., Yau, A., Davis, G., Zivkovic, V.T., Olmstead, R.E., Tremoulet, P.D., Craven, P.L.: EEG correlates of task engagement and mental workload in vigilance, learning, and memory tasks. Aviat Space Environ. Med. 78, B231–B244 (2007)
9. Lin, T., Imamiya, A., Mao, X.: Using multiple data sources to get closer insights into user cost and task performance. Interacting with Computers 20, 364–374 (2008)
10. von Borell, E., Langbein, J., Després, G., Hansen, S., Leterrier, C., Marchant-Forde, J., Marchant-Forde, R., Minero, M., Mohr, E., Prunier, A., Valance, D., Veissier, I.: Heart rate variability as a measure of autonomic regulation of cardiac activity for assessing stress and welfare in farm animals — A review. Physiology & Behavior 92, 293–316 (2007)
11. Merzagora, A.C., Bunce, S., Izzetoglu, M., Onaral, B.: Wavelet analysis for EEG feature extraction in deceptive detection. IEEE Proceedings on EBMS 6, 2434–2437 (2006)
12. Murugappan, M., Ramachandran, N., Sazali, Y.: Classification of human emotion from EEG using discrete wavelet transform. Journal of Biomedical Science and Engineering 3, 390–396 (2010)
13. Hasan, O.: Automatic detection of epileptic seizures in EEG using discrete wavelet transform and approximate entropy. Expert Systems with Applications 36, 2027–2036 (2009)
14. Übeyli, E.D., İlbay, G., Şahin, D., Ateş, N.: Analysis of spike-wave discharges in rats using discrete wavelet transform. Computers in Biology and Medicine 39, 294–300 (2009)
15. Cvetkovic, D., Übeyli, E.D., Cosic, I.: Wavelet transform feature extraction from human PPG, ECG, and EEG signal responses to ELF PEMF exposures: A pilot study. Digital Signal Processing 18, 861–874 (2008)
16. Samjin, C.: Detection of valvular heart disorders using wavelet packet decomposition and support vector machine. Expert Systems with Applications 35, 1679–1687 (2008)
17. Ting, W., Guo-zheng, Y., Bang-hua, Y., Hong, S.: EEG feature extraction based on wavelet packet decomposition for brain computer interface. Measurement 41, 618–625 (2008)
18. Yang, B.-H., Yan, G.-Z., Yan, R.-G., Wu, T.: Adaptive subject-based feature extraction in brain–computer interfaces using wavelet packet best basis decomposition. Medical Engineering & Physics 29, 48–53 (2007)
19. Yusuff, A.A., Fei, C., Jimoh, A.A., Munda, J.L.: Fault location in a series compensated transmission line based on wavelet packet decomposition and support vector regression. Electric Power Systems Research 81, 1258–1265 (2011)
20. Abdulhamit, S.: Automatic recognition of alertness level from EEG by using neural network and wavelet coefficients. Expert Systems with Applications 28, 701–711 (2005)

21. Li, W., He, Q.-C., Fan, X.-M., Fei, Z.-M.: Evaluation of driver fatigue on two channels of EEG data. Neuroscience Letters (2012)
22. Liu, J., Zhang, C., Zheng, C.: EEG-based estimation of mental fatigue by using KPCA–HMM and complexity parameters. Biomedical Signal Processing and Control 5, 124–130 (2010)
23. Zhang, C., Zheng, C.-X., Yu, X.-L.: Automatic Recognition of Cognitive Fatigue from Physiological Indices by Using Wavelet Packet Transform and Kernel Learning Algorithms. Expert Systems with Applications 36, 4664–4671 (2009)
24. Zhao, C., Zheng, C., Zhao, M., Tu, Y., Liu, J.: Multivariate autoregressive models and kernel learning algorithms for classifying driving mental fatigue based on electroencephalographic. Expert Systems with Applications 38, 1859–1865 (2011)
25. Boostani, R., Sadatnezhad, K., Sabeti, M.: An efficient classifier to diagnose of schizophrenia based on the EEG signals. Expert Systems with Applications 36, 6492–6499 (2009)
26. Brankačk, J., Kukushka, V.I., Vyssotski, A.L., Draguhn, A.: EEG gamma frequency and sleep–wake scoring in mice: Comparing two types of supervised classifiers. Brain Research 1322, 59–71 (2010)
27. Hsu, W.-Y., Lin, C.-H., Hsu, H.-J., Chen, P.-H., Chen, I.R.: Wavelet-based envelope features with automatic EOG artifact removal: Application to single-trial EEG data. Expert Systems with Applications 39, 2743–2749 (2012)
28. Subasi, A., Ismail Gursoy, M.: EEG signal classification using PCA, ICA, LDA and support vector machines. Expert Systems with Applications 37, 8659–8666 (2010)
29. Ceylan, R., Özbay, Y.: Comparison of FCM, PCA and WT techniques for classification ECG arrhythmias using artificial neural network. Expert Systems with Applications 33, 286–295 (2007)
30. Chawla, M.P.S., Verma, H.K., Kumar, V.: A new statistical PCA–ICA algorithm for location of R-peaks in ECG. International Journal of Cardiology 129, 146–148 (2008)
31. Korürek, M., Nizam, A.: Clustering MIT–BIH arrhythmias with Ant Colony Optimization using time domain and PCA compressed wavelet coefficients. Digital Signal Processing 20, 1050–1060 (2010)
32. Martis, R.J., Chakraborty, C., Ray, A.K.: A two-stage mechanism for registration and classification of ECG using Gaussian mixture model. Pattern Recognition 42, 2979–2988 (2009)
33. Polat, K., Güneş, S.: Detection of ECG Arrhythmia using a differential expert system approach based on principal component analysis and least square support vector machine. Applied Mathematics and Computation 186, 898–906 (2007)
34. Diery, A., Rowlands, D., Cutmore, T.R.H., James, D.: Automated ECG diagnostic P-wave analysis using wavelets. Computer Methods and Programs in Biomedicine 101, 33–43 (2011)
35. Froese, T., Hadjiloucas, S., Galvão, R.K.H., Becerra, V.M., Coelho, C.J.: Comparison of extrasystolic ECG signal classifiers using discrete wavelet transforms. Pattern Recognition Letters 27, 393–407 (2006)
36. Yeh, Y.-C., Wang, W.-J., Chiou, C.W.: Cardiac arrhythmia diagnosis method using linear discriminant analysis on ECG signals. Measurement 42, 778–789 (2009)
37. Li, D., Yang, W., Wang, S.: Classification of foreign fibers in cotton lint using machine vision and multi-class support vector machine. Computers and Electronics in Agriculture 74, 274–279 (2010)
38. Li, X., Nie, P., Qiu, Z.-J., He, Y.: Using wavelet transform and multi-class least square support vector machine in multi-spectral imaging classification of Chinese famous tea. Expert Systems with Applications 38, 11149–11159 (2011)

39. Kim, K.-J., Ahn, H.: A corporate credit rating model using multi-class support vector machines with an ordinal pairwise partitioning approach. Computers & Operations Research 39, 1800–1811 (2012)
40. Acevedo, F.J., Maldonado, S., Domínguez, E., Narváez, A., López, F.: Probabilistic support vector machines for multi-class alcohol identification. Sensors and Actuators B: Chemical 122, 227–235 (2007)
41. Shen, K.-Q., Li, X.-P., Ong, C.-J., Shao, S.-Y., Wilder-Smith, E.P.V.: EEG-based mental fatigue measurement using multi-class support vector machines with confidence estimate. Clinical Neurophysiology 119, 1524–1533 (2008)
42. Li, S., Tsang, I.W., Chaudhari, N.S.: Relevance vector machine based infinite decision agent ensemble learning for credit risk analysis. Expert Systems with Applications
43. Lima, C.A.M., Coelho, A.L.V., Chagas, S.: Automatic EEG signal classification for epilepsy diagnosis with Relevance Vector Machines. Expert Systems with Applications 36, 10054–10059 (2009)
44. Flake, J., Moon, T.K., McKee, M., Gunther, J.H.: Application of the relevance vector machine to canal flow prediction in the Sevier River Basin. Agricultural Water Management 97, 208–214 (2010)
45. Lu, J., Zhang, E.: Gait recognition for human identification based on ICA and fuzzy SVM through multiple views fusion. Pattern Recognition Letters 28, 2401–2411 (2007)
46. Shieh, M.-D., Yang, C.-C.: Classification model for product form design using fuzzy support vector machines. Computers & Industrial Engineering 55, 150–164 (2008)
47. Chaudhuri, A., De, K.: Fuzzy Support Vector Machine for bankruptcy prediction. Applied Soft Computing 11, 2472–2486 (2011)
48. Cheng, M.-Y., Roy, A.F.V., Chen, K.-L.: Evolutionary risk preference inference model using fuzzy support vector machine for road slope collapse prediction. Expert Systems with Applications 39, 1737–1746 (2012)
49. Waters, A.M., Lipp, O.V., Spence, S.H.: The effects of affective picture stimuli on blink modulation in adults and children. Biological Psychology 68, 257–281 (2005)
50. Rantanen, A., Laukka, S.J., Lehtihalmes, M., Seppänen, T.: Heart Rate Variability (HRV) reflecting from oral reports of negative experience. Procedia - Social and Behavioral Sciences 5, 483–487 (2010)

Harmony Search Algorithm Based Nearly Optimal Peak Reduction Tone Set Selection for PAPR Reduction in OFDM System

Jong-Shin Chen[1], Ho-Lung Hung[2], and Yung-Fa Huang[3,*]

[1] Department of Information and Communication Engineering
Chaoyang University of Technology, Taichung 41368, Taiwan
jschen26@cyut.edu.tw

[2] Department of Electrical Engineering, Chien-Kuo Technology University,
No.1, Chien Shou Rd., Changhua, Taiwan
hlh@ctu.edu.tw

[3] Department of Information and Communication Engineering,
Chaoyang University of Technology, Taichung 41349, Taiwan, R.O.C
yfahuang@mail.cyut.edu.tw

Abstract. This paper considers the use of the tone reservation (TR) technique to reduce the peak-to-average power ratio (PAPR) of an orthogonal frequency division multiplexing (OFDM) signal. In the TR scheme, a small number of unused subcarriers called peak reduction carriers (PRCs) are reserved to reduce the PAPR and the goal of the TR scheme is to find the optimal values of the PRCs that minimize the PAPR of the transmitted OFDM signal. The conventional TR technique can provide good PAPR reduction performance for OFDM signals; however, it requires an exhaustive search over all combinations of allowed PRC set, resulting in high complexity. In this paper, the harmony research (HS) algorithm is applied to search the optimal combination of PRC, which can significantly reduce the computational complexity and offers lower PAPR at the same time. Simulation results show that the TR based on HS algorithm can achieve the good tradeoff between PAPR reduction performance and computational complexity.

Keywords: Orthogonal frequency division multiplexing, Harmony search algorithm Peak-to-average power ratio reduction, Tone reservation.

1 Introduction

Orthogonal frequency division multiplexing (OFDM) is a promising modulation technique that can support high data rate transmission in harsh propagation environments. Thus, it is adopted in several wireless standards [1]–[3]. Moreover, it has been regarded as a promising transmission technique for fourth-generation wireless mobile communications. However, due to its multicarrier nature, the major drawback of the

* Corresponding author.

N.T. Nguyen et al. (Eds.): ACIIDS 2014, Part II, LNAI 8398, pp. 342–351, 2014.

OFDM system is the high peak-to-average power ratio (PAPR), which may cause high out-of-band radiation when the OFDM signal is passed through a radio frequency power amplifier. When a high PAPR OFDM signal is passed through a nonlinear power amplifier, it reduces the efficiency of power consumption and causes in-band distortion and undesired spectral spreading [1]. Furthermore, the high PAPR is one of the most important implementation challenges that face designers of OFDM [2]. As a result, the OFDM signal will be clipped when passed through a nonlinear power amplifier at the transmitter end. Clipping degrades the bit-error-rate (BER) performance and causes spectral spreading [4].

To reduce the search complexity, some simplified search techniques have recently been proposed [5]–[12]. However, for all these search methods, either the computational complexity is still high, or the PAPR reduction performance is not good enough. In those methods have been proposed to reduce the PAPR of OFDM signals [6-15], including amplitude clipping [5], tone reservation [7], tone injection [7], selected mapping (SLM) [8], and partial transmit sequences (PTS) [9]. These techniques improve PAPR statistics of an OFDM signal significantly without any in-band distortion and out-of-band radiation. But, they require side information to be transmitted from the transmitter to the receiver in order to let the receiver know what has been done in the transmitter. An efficient method to reduce the PAPR of OFDM signals is tone reservation (TR) [10-15]. In the approach, the transmitter reserves a small number of unused subcarriers for PAPR control. These subcarriers are referred to as peak reduction carriers (PRC) and are designed to be orthogonal with the data subcarriers, thereby causing no distortion. The problem of optimum selection of PRCs' values is a convex optimization problem [11-12] that leads to increased computational complexity and low convergence rates. On the other hand, the PAPR reduction performance and the computational complexity of TR scheme depend on the method of PRC. In other words, there is a trade-off between PAPR reduction performance and computational complexity in TR scheme. The aim of this paper is to demonstrate some suboptimum solutions] to the OFDM PAPR reduction problem, based on the TR concept and by exploiting the intrinsic advantages of Harmony search (HS) [16-18] algorithm.

In this paper, we first propose a new suboptimal PRC set selection scheme based on HS algorithm, which can efficiently solve the optimal values of the PRCs that minimize the PAPR of the transmitted OFDM signal and NP-hard problem. As a result, the HS optimization scheme achieves a nearly optimal PRC set and requires far less computational complexity than the GA based- PRC and conventional TR method. The simulation results show that the performance of the proposed HS method can achieve significant PAPR reduction while maintaining low complexity. This paper is organized as follows: In Section 2, OFDM system and TR scheme are described. Section 3 introduces a new HS-assisted TR scheme for the nearly optimal PRC set and discusses the computational complexity issue. The simulation results are shown in Section 4, and, finally, the concluding remarks are given in Section 5.

2 System Models

A. *OFDM System Model*

Let $X = [X_0, X_1, \cdots, X_{N-1}]^T$ denote an input symbol sequence in the frequency domain, where X_K represents the complex data of the *K*th subcarrier and *N* the number of subcarriers of OFDM signal. The *N* subcarriers are chosen to be orthogonal, that is, $f_m = m\Delta f$, where $\Delta f = \frac{1}{NT}$. The OFDM signal is generated by summing all the *N* modulated subcarriers each of which is separated by $\frac{1}{NT}$. Then the complex OFDM signal in the time domain is expressed as

$$x_n = \frac{1}{\sqrt{N}} \sum_{i=0}^{N-1} X_i e^{j2\pi ni/LN}, \quad 0 \le n \le N-1 \tag{1}$$

where can also be written in matrix form $\mathbf{x} = [x_0, x_1, \cdots, x_{N-1}]^T = \mathbf{DX}$, where $\mathbf{D}$ is the DFT matrix with the (*n,k*) the entry $d_{n,k} = (1/\sqrt{N})e^{j2\pi nk/N})$, where $j = \sqrt{-1}$, and *n* stands for a discrete time index. *L* is the oversampling factor, where *L* = 4, which is enough to provide an accurate approximation of the PAPR [4] and x_n is the *n*th signal component in OFDM output symbol.

B. *Peak-to-Average Power Ratio*

The PAPR of *x* is defined as the ratio between the maximum instantaneous power and its average power, which can be written as [2-3][11],

$$PAPR(x) = 10\log_{10} \frac{\max_{0 \le n \le LN-1} |x_n|^2}{E\left[|x_n|^2\right]}, \tag{2}$$

where $\max|x_n|^2$ is the maximum values of the OFDM signal power, and $E[\cdot]$ is the mathematical expectation. As PAPR is a random variable so it can not be a single value for a system. Generally PAPR is characterized by Complementary Cumulative Distribution Function (CCDF) defined as

$$CCDF(PAPR_0) = P_r[PAPR > PAPR_0] \tag{3}$$

where $PAPR_0$ is any positive value and $P_r[\cdot]$ is probability function.

C. Tone reservation scheme to reduce the PAPR

The objective is to find the time domain signal to be added to the original time domain signal **x** such that the PAPR is reduced. It is known that a randomly generated PRC set performs better than the contiguous PRC set and the interleaved PRC set. The secondary peak minimization problem is known to be nondeterministic polynomial-time (NP)-hard, which cannot be solved for the practical number of tones.

In order to reduce the PAPR of OFDM signal using tone reservation (TR) scheme, some subcarriers are reserved as a peak reduction tones set which is used to generate the peak canceling signal. In the TR-based OFDM scheme, peak reduction tones are reserved to generate PAPR reduction signals. These reserved tones do not carry any data information, and they are only used for reducing PAPR. Specifically, the peak-canceling signal $\mathbf{C}=[C_0,C_1,...,C_{N-1}]^T$ generated by reserved PRC is added to the original time domain signal $\mathbf{X}=[X_0,X_1,...,X_{N-1}]^T$ to reduce its PAPR. The PAPR reduced signal can be expressed as [14-18]

$$Q \ = \mathbf{x}+\mathbf{c}=D(\mathbf{X}+\mathbf{C}) \tag{4}$$

where **X** and **C** denote the signals transmitted on the data tones and reserved tones, respectively. The $\mathbf{c}$ is the time domain due to **C.** The TR technique restricts the data block **X** and frequency domain vector $\mathbf{C}=[C_0,C_1,...,C_{N-1}]^T$ is avoid signal distortion, **X** and **C** are orthogonal with each other,

$$\mathbf{X}_n+\mathbf{C}_n=\begin{cases}\mathbf{X}_n, & n\in M,\\ \mathbf{C}_n, & n\in R,\end{cases} \tag{5}$$

and

$$M\cap R=\varnothing$$

where $M=\{m_0,m_1,...,m_{N-W-1}\}$ represents the index set of the data-bearing subcarriers and $R=\{r_0,r_1,...,r_{W-1}\}$ is the index set of the reserved tones. In addition, R is also referred to as a PRC set and $W<N$ is the number of PRC. Notably, PRCs are not used to transmit information, but rather to generate PAPR reduction signals. In [11], the PAPR of the OFDM signal sequence is defined Q as

$$PAPR(Q)=\frac{\max\limits_{0\le n\le N-1}|x_n+c_n|^2}{\mathrm{E}\left[|x_n|^2\right]} \tag{6}$$

Basically, we can adjust the value of C_n to reduce the peak value of x_n without disturbing the actual data contained in X. Thus, C_n must be investigated to minimize the maximum norm of the time domain signal x_n, The optimization problem is to find optimum c according to

$$c^{(opt)}=\arg\min_{C}\max_{0\le n<N}|x_n+c_n|^2 \tag{7}$$

Since subcarriers in OFDM systems are orthogonal, and the vectors added are restricted by (7), signals in spare subcarriers can be simply discarded at the receiver. Accordingly, in an exhaustive search approach, the computational complexity increases exponentially with the number of PRCs. Then a minima PAPR optimization problem is formulated as

$$\min_{C\in\Pi} \quad E, \\ \text{s.t. } |x_n + c_n|^2 \le E, \text{ for all } n = 0,1,\ldots,N-1, \tag{8}$$

where E represents the peak power of the peak-reduced signal, and Π is the signal space of all possible peak-compensation signals generated from the set of R .To reduce the complexity of the QCQP, Tellado in [11], [15] proposed a simple a simple gradient algorithm to iteratively approach c and updates the vector as follows:

$$\mathbf{c}^{(i+1)} = \mathbf{c}^{(i)} - \beta_i \mathbf{w}\left[\left((k-k_i)\right)_n\right] \tag{9}$$

where $\mathbf{w} = [w_0, w_1, \ldots, w_{N-1}]^T$ is called the time domain kernel, β_i is a scaling factor relying on the maximum peak found at the i th iteration, and $\mathbf{w}[((k-k_i))_N]$ denotes a circular shift of $\mathbf{w}$ to the right by a value k_i , where k_i is calculated by

$$k_i = \underset{k}{\operatorname{argmax}} \left| x_k + c_k^{(i)} \right| \tag{10}$$

For ease of presentation, the time domain kernel is expressed as

$$\mathbf{w} = [w_0, w_1, \ldots, w_{N-1}]^T = \mathbf{DW} \tag{11}$$

where $\mathbf{W} = [W_0, W_1, \ldots, W_{N-1}]^T$, $W_n \in \{0,1\}$ is called the frequency domain kernel whose elements are defined by

$$W_n = \begin{cases} 0\ , & n \in R^C \\ 1\ , & n \in R \end{cases} \tag{12}$$

Then, the optimal frequency domain kernel $\mathbf{W}$ corresponds to the characteristic sequence of the PRC set R and the maximum peak $\mathbf{w}$ of is always because is a {0, 1} sequence. The PAPR reduction performance depends on the time domain kernel $\mathbf{W}$ and the best performance can be achieved when the time domain kernel $\mathbf{w}$ is a discrete impulse because the maximum peak can be cancelled without affecting other signal samples at each iteration. If the maximum number of iterations is reached or the desired peak power is obtained, iteration stops. After G iterations of this algorithm, the peak-reduced OFDM signal is obtained:

$$Q = \mathbf{x} + \mathbf{c}^{(G)} = \mathbf{x} - \sum_{g=1}^{G} \beta_i \mathbf{w}\left[\left((k-k_i)\right)_n\right] \tag{13}$$

The PAPR reduction performance of the TR scheme depends on the selection of the PRC set R. Thus, it is cannot be solved for practical values of N. To find the optimal PRC set, in mathematical form, we require solving the following combinatorial optimization problem:

$$R^* = \arg\min_R \left\| [w_1, w_2, \cdots, w_{N-1}]^T \right\|_\infty \tag{14}$$

where $\|\bullet\|_j$ denotes the j-norm and ∞-norm refers to the maximum values. Here, R^* denotes the global optimum of the objective function, which requires an exhaustive search of all combination of possible PRC set R.

D. Harmony search Algorithm Based PRC set Search for reduced PAPR

Harmony search (HS) algorithm, originally proposed in [16], is a metaheuristic approach which attempts to simulate the improvisation process of musicians. In such a search process, each decision variable (musician) generates a value (note) for finding a global optimum (best harmony). An initial population of musician is randomly generated. Each tone sequences is a vector of length N, and each element of the vector is a binary zero or one depending on the existence of a PRC set at that position (one denotes existence and zero denotes non-existence). The number of the PRC in each binary vector is $M < N$. Denote the s tones as $\{v_1, v_{2,\ldots,} v_s\}$. Then each $\mathbf{v}_u$ is a binary vector of length. For each tones $\mathbf{v}_u$, the PRC set R_u is the collection of the locations whose elements are one. Then the frequency domain kernel corresponding to the PRC set is obtained by (12), and the time domain kernel $W_u = \left[w_0^u, w_1^u, \cdots, w_{N-1}^u \right]$ is obtained by (11). The merit (secondary peak) of the sequence is defined as [14-15]

$$m(\mathbf{V}_u) = \left\| \left[w_1^u, w_2^u, \cdots, w_{N-1} \right]^T \right\|_\infty \tag{15}$$

The T sequences (called elite sequences) with the lowest merits are maintained for the next population generation. The best merit of the sequences is defined as

$$\hat{m} = \min_{1 \le u \le N} m(\mathbf{V}_u) \tag{16}$$

3 Minimize PAPR Using Harmony Search Algorithm-Based PRC Set

According to the above algorithm concept, the HS metaheuristic algorithm consists of the following five steps [16-18]:

Step 1) **Initialization of the optimization problem and algorithm parameters:** In the first step, a possible value range for each design variable of the optimum design problem is specified. A pool is constructed by collecting these values together, from which the algorithm selects values for the design variables. Furthermore, the number of solution vectors in harmony search memory that is the size of the harmony memory matrix, harmony considering rate (HMCR), pitch adjusting rate (PAR) and the maximum number of searches are also selected in this step.

$$\text{Minimize} \quad f(\hat{x})$$

$$subject \quad to \quad x_i \in X_i, \quad i = 1, 2, \cdots, M \tag{17}$$

where $f(\cdot)$ is a scalar objective function to be optimized, $\hat{x}$ is a solution vector composed of decision variables x_i, X_i is the set of possible range of values for each decision variable x_i, that is, $x_i^L \le X_i \le x_i^U$, where x_i^L and x_i^U are the lower and upper bounds for each decision variable, respectively, and M is the number of decision variables.

Step 2. Initialize the harmony memory (HMS). In this step, the ''harmony memory'' matrix shown in Eq. (18) is filled with as many randomly generated solution vectors as the size of the HM (i.e., HMS) and sorted by the values of the objective function, $f(x)$

$$HM = \begin{bmatrix} x_{11} & x_{21} & \cdots & x_{n1} \\ x_{12} & x_{22} & \cdots & x_{n2} \\ \vdots & \vdots & \cdots & \vdots \\ \vdots & \vdots & \cdots & \vdots \\ x_{1,HMS-1} & x_{2,HMS-1} & & x_{n,HMS-1} \\ x_{1,HMS} & x_{2,HMS} & & x_{n,HMS} \end{bmatrix} \tag{18}$$

where $x_{i,j}$ is the value of the *i*th design variable in the *j*th randomly selected feasible solution. These candidate designs are sorted such that the objective function value corresponding to the first solution vector is the minimum.

Step 3. In generating a new harmony matrix, the new value of the *i*th design variable can be chosen from any discrete value within the range of *i*th column of the harmony memory matrix with the probability of HMCR which varies between 0 and 1. In other words, the new value of x_i can be one of the discrete values of the vector $\{x_{i,1}, x_{i,2}, \cdots, x_{i,HMS}\}^T$ with the probability of HMCR. The same is applied to all other design variables. In the random selection, the new value of the *i*th design variable can also be chosen randomly from the entire pool with the probability of $1 - HMCR$. That is,

$$x_i^{new} = \begin{cases} \{x_{i,1}, x_{i,2}, \cdots, x_{i,HMS}\}^T & \text{with probability HMCR} \\ \{x_1, x_2, \cdots, x_{ns}\} & \text{with probability (1-HMCR)} \end{cases} \quad (19)$$

where *ns* is the total number of values for the design variables in the pool. The PAR decides whether the decision variables are to be adjusted to a neighboring value. This operation uses the pitch adjustment parameter PAR that sets the rate of adjustment for the pitch chosen from the harmony memory matrix as follows: Pitch adjusting decision for

$$x_i^{new} \Leftarrow \begin{cases} yes \quad with\ probability\ of\ PAR \\ no \quad with\ probability\ of\ (1\text{-}PAR) \end{cases} \quad (20)$$

Supposing that the new pitch-adjustment decision for x_i^{new} came out to be yes from the test and if the value selected for x_i^{new} from the harmony memory is the *k*th element in the general discrete set, then the neighboring value $k+1$ or $k-1$ is taken for new x_i^{new}.

Step 4. After selecting the new values for each design variable, the objective function value is calculated for the new harmony vector. If this value is better than the worst harmony vector in the harmony matrix, it is then included in the matrix while the worst one is taken out of the matrix.

Step 5. Check stopping criterion. Steps 3 and 4 are repeated until the termination criterion, which is the pre-selected maximum number of cycles, is reached. This number is selected large enough such that within this number of design cycles no further improvement is observed in the objective function.

4 Simulation Results and Discussions

In this section, simulation experiments are conducted to verify the PAPR performance of the proposed IHS -assisted TR method for OFDM systems. The PAPR reduction performance is evaluated in terms of its complementary cumulative distribution function , which is the probability that the PAPR of a symbol exceeds the threshold level . To generate the CCDF of the PAPR, 100,000 quadrature phase-shift keying (QPSK)-modulated OFDM symbols are randomly generated. In addition, the transmitted signal is oversampled by a factor of 4 for accurate PAPR. Figs. 1 and 2 shows the CCDFs for various PRT sets with $W=32$ and $W=64$ in the TR scheme, where the original OFDM is obtained directly from the output of the IFFT operation. The simulation results show that, as expected, the TR scheme can more significantly reduce PAPR compared to that of the original OFDM signal, regardless of PRT sets employed. In addition, the PAPR performance improves with an increase in the number of PRTs for all the PRT sets. However, with the same number of PRTs, PAPR reduction performance of the proposed CE PRT set is superior to those with the consecutive PRT set, the equally spaced PRT set, and the random optimal PRT set for the cases and. When $P_r[PAPR > PAPR_0] = 10^{-4}$, the PAPR reduction of the HIS PRT set with $W=32$

is in fact slightly better than that of the random optimal PRT set with $W = 64$. This signifies that, at the same PAPR reduction, the HIS PRT set requires less reserved tones in the TR scheme. However, the random optimal PRT set requires more reserved tones in the TR scheme to suppress PAPR. In this case, sacrifice of the data rate of the IHS PRT set is lower than that of the random optimal PRT set.

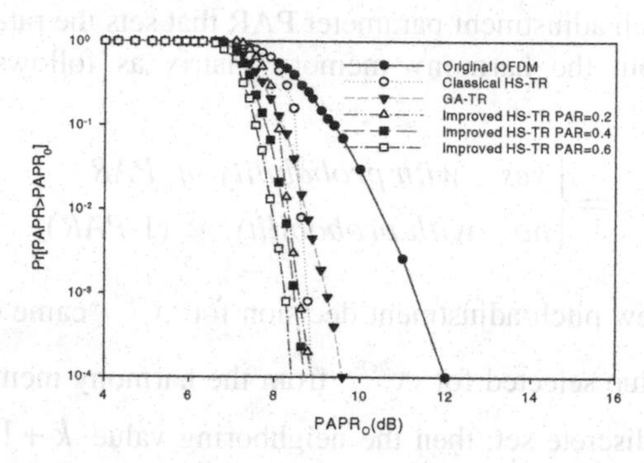

Fig. 1. CCDF of the PAPR for the TR schemes with various PRT sets with N=256 for $W = 32$

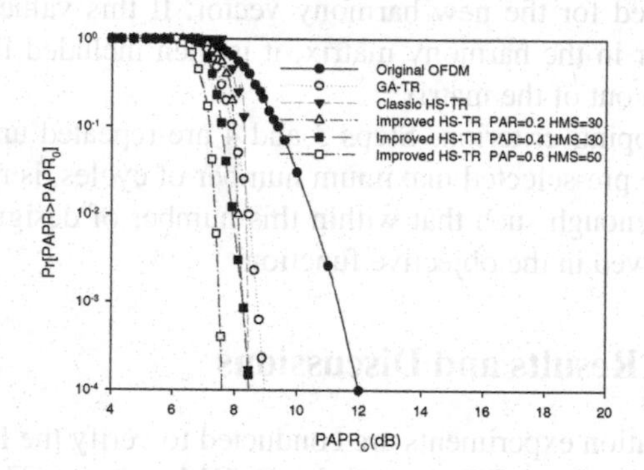

Fig. 2. CCDF of the PAPR for the TR schemes with various PRT sets with N=256 for $W = 64$

5 Conclusion

In this paper, we have developed a reduced-complexity TR scheme for PAPR reduction of OFDM signals, where a new criterion was derived to effectively select samples for peak power calculation in each candidate signal. There is a trade-off between the computational complexity and performance in the PAPR reduction method. To reduce the computational complexity while still obtaining the desirable PAPR reduction, we introduce the IHS method, an effective algorithm that solves various combinatorial optimization problems, to search the optimal phase factors. We have proposed a IHS-based method used to search the values of PRCs in the TR technique to obtain PAPR values below the predefined threshold. The simulations demonstrate that the proposed HS-based TR scheme not only provide better PAPR performance but also enjoys complexity advantages compared with the conventional GA-assisted TR method. Most importantly, the proposed IHS-based TR scheme can achieve all considered target PAPRs; in contrast, the GA-assisted TR method is easily trapped into local solutions and cannot meet the target PAPRs.

Acknowledgements. The works of Jong-Shin Chen were supported in part by the National Science Council (NSC) of Taiwan under grant NSC102-2221-E-324-013.

References

[1] Andrews, J.G., Ghosh, A., Muhamed, R.: Fundamentals of WiMAX, Understanding Broadband Wireless Networking. Prentice-Hall, Englewood Cliffs (2007)

[2] Jiang, T., Wu, Y.: An overview:peak-to-average power ratio reduction techniques for OFDM signals. IEEE Trans. on Broadcasting 54(2), 257–268 (2008)

[3] van Nee, R., Prasad, R.: OFDM for Wireless Multimedia Communications. Artech House, Boston (2000)

[4] Li, X., Cimini Jr., L.J.: Effects of clipping and filtering on the performance of OFDM. In: Proc. Globecom, pp. 1634–1638 (1997)

[5] Tellado, J., Cioffi, J.M.: Peak power reduction for multicarrier transmission. In: Proc. Globecom, pp. 219–224 (1998)

[6] Gross, R., Veeneman, D.: Clipping distortion in DMT ADSL systems. Electron. Lett. 29, 2080–2081 (1993)

[7] Tellado, J.: Multicarrier Modulation With Low PAR: Applications to DSL and Wireless. Kluwer, Norwell (2000)

[8] Breiling, M., Muller, S.H., Huber, J.B.: SLM peak-power reduction without explicit side information. IEEE Commun. Lett. 5(6), 239–241 (2001)

[9] Muller, S.H., Huber, J.B.: OFDM with reduced peak-to-average power ratio by optimum combination of partial transmit sequences. Electron. Lett. 33, 368–369 (1997)

[10] Tellado, J.: Peak to Average Power Reduction for Multicarrier Modulation, Ph.D. dissertation, Stanford Univ. (2000)

[11] Wang, Y., Chen, W., Tellambura, C.: Genetic Algorithm Based Nearly Optimal Peak Reduction Tone Set Selection for Adaptive Amplitude Clipping PAPR Reduction. IEEE Transactions on Broadcasting 58, 462–471 (2012)

[12] Tellado, J., Cioffi, J.M.: PAR reduction in multicarrier transmission system. ANSI Document, T1E1.4 Technical subcommittee, pp. 1–14, no. 97-367 (December 8, 1997)

[13] Li, H., Jiang, T., Zhou, Y.: An Improved Tone Reservation Scheme With Fast Convergence for PAPR Reduction in OFDM Systems. IEEE Transactions on Broadcasting 57(4), 902–906 (2011)

[14] Chen, J.-C., Chiu, M.-H., Yang, Y.-S., Li, C.-P.: A Suboptimal Tone Reservation Algorithm Based on Cross-Entropy Method for PAPR Reduction in OFDM Systems. IEEE Transactions on Broadcasting 57(3), 752–757 (2011)

[15] Lim, D.W., Noh, H.-S., No, J.-S., Shin, D.-J.: Near optimal PRT set selection algorithm for tone reservation in OFDM systems. IEEE Trans. Broadcast. 54(3), 454–460 (2008)

[16] Geem, Z.W., Kim, J.H., Loganathan, G.V.: A new heuristic optimization algorithm: Harmony search. Simulation 76(2), 60–68 (2001)

[17] Geem, Z.W., Kim, S.: Harmony Search Algorithms for Structural Design. Old City Publishing, Philadelphia (2009)

[18] Das, Mukhopadhyay, A., Roy, A., Abraham, A., Panigrahi, B.K.: Exploratory power of the harmony search algorithm: Analysis and improvements for global numerical optimization. IEEE Trans. Syst., Man, Cybern. B, Cybern. 41(1), 89–106 (2011)

Predicting Stock Returns by Decision Tree Combining Neural Network

Shou-Hsiung Cheng

Department of Information Management,
Chienkuo Technology University, Changhua 500, Taiwan
shcheng@ctu.edu.tw

Abstract. This study presents a hybrid models to look for sound financial companies that are really worth making investment in stock markets. The following are three main steps in this study: First, we utilize neural network theory to find out the core of the financial indicators affecting the ups and downs of a stock price. Second, based on the core of the financial indicators coupled with the technology of decision tree, we establish the hybrid classificatory models and the predictable rules that affect the ups and downs of a stock price. Third, by sifting the sound investing targets out, we use the established rules to set out to invest and calculate the rates of investment. These evidences reveal that the average rates of reward are far larger than the mass investment rates.

Keywords: hybrid models, financial indicators, neural network, decision tree.

1 Introduction

The problem of predicting stock returns has been an important issue for many years. Advancement in computer technology has allowed many recent studies to utilize machine learning techniques such as neural networks and decision trees to predict stock returns. Generally, there are two instruments to aid investors for doing prediction activities objectively and scientifically, which are technical analysis and fundamental analysis. Technical analysis considers past financial market data, represented by indicators such as Relative Strength Indicator (RSI) and field-specific charts, to be useful in forecasting price trends and market investment decisions. In particular, technical analysis evaluates the performance of securities by analyzing statistics generated from various marketing activities such as past prices and trading volumes. Furthermore, the trends and patterns of an investment instrument's price, volume, breadth, and trading activities can be used to reflect most of the relevant market information to determine its value [1]. The fundamental analysis can be used to compare a firm's performance and financial situation over a period of time by carefully analyzing the financial statements and assessing the health of a business. Using ratio analysis, trends and indications of good and bad business practices can be easily identified. To this end, fundamental analysis is performed on both historical

N.T. Nguyen et al. (Eds.): ACIIDS 2014, Part II, LNAI 8398, pp. 352–360, 2014.

and present data in order to perform a company stock valuation and hence, predict its probable price evolution. Financial ratios including profitability, liquidity, coverage, and leverage can be calculated from the financial statements [2]. Thus, the focus of this study is not on effects but on causes that should be seeking and exploring original sources actually. Therefore, selective a good stock is the first and the most important step for intermediate- or even long-term investment planning. In order to reduce risk, in Taiwan, the public stock market observation of permit period will disclose regularly and irregularly the financial statements of all listed companies. Therefore, this study employs data of fundamental analysis by using hybrid models of classification to extract. Employing these meaningful decision rules, a useful stock selective system for intermediate- or long-term investors is proposed in this study.

In general, some related work considers a feature selection step to examine the usefulness of their chosen variables for effective stock prediction, e.g. [3]. This is because not all of features are informative or can provide high discrimination power. This can be called as the curse of dimensionality problem [4]. As a result, feature selection can be used to filter out redundant and/or irrelevant features from a chosen dataset resulting in more represen- tative features for better prediction performances [5]. The idea of combining multiple feature selection methods is derived from classifier ensembles [6]. The aim of classifier ensembles is to obtain high highly accurate ones. They are intended to improve the classification performance of a single classifier.The idea of combining multiple feature selection methods is derived from classifier ensembles (or multiple classifiers) [7]. The aim of classifier ensembles is to obtain highly accurate classifiers by combining less accurate ones. They are intended to improve the classification performance of a single classifier. That is, the combination is able to complement the errors made by the individual classifiers on different parts of the input space. Therefore, the performance of classifier ensembles is likely better than one of the best single classifiers used in isolation.

The rest of the paper is organized as follows: In Section 2 an overview of the related works is introduced, while Section 3 presents the proposed procedure and briefly discusses its architecture. Section 4 describes analytically the experimental results. Finally, Section 5 shows conclusions of this paper.

2 Related Works

This study proposes a new stock selective system applying the neural network and decision tree algorithm to verify that whether it can be helpful on prediction of the shares rose or fell for investors. Thus, this section mainly reviews related studies of the association rules, cluster analysis and decision tree.

2.1 Neural Network

The basic element of a neural network is a neuron. This is a simple virtual device that accepts many inputs, sums them, applies a nonlinear transfer function, and generates the result, either as a model prediction or as input to other neurons. A neural network is a structure of many such neurons connected in a systematic way. The neurons in such networks are arranged in layers. Typically, there is one layer for input neurons, one or more layers of the hidden layers, and one layer for output neurons. Each layer is fully interconnected to the preceding layer and the following layer. The connections between neurons have weights associated with them, which determine the strength of influence one neuron has on another. Information flows from the input layer through the processing layers to the output layer to generate predictions. By adjusting the connection weights during training to match predictions to target values for specific records, the network learns to generate better and better predictions.

In this study, a radial basis function network is selected. It consists of three layers: an input layer, a receptor layer, and an output layer. The input and output layers are similar to those of a multilayer perceptron. However, the hidden or receptor layer consists of neurons that represent clusters of input patterns, similar to the clusters in a k-means model. These clusters are based on radial basis functions. The connections between the input neurons and the receptor weights are trained in essentially the same manner as a k-means model. Specifically, the receptor weights are trained with only the input fields; the output fields are ignored for the first phase of training. Only after the receptor weights are optimized to find clusters in the input data are the connections between the receptors and the output neurons trained to generate predictions. Each receptor neuron has a radial basis function associated with it. The basis function used in the study is a multidimensional Gaussian, $\exp\left(-\frac{d_i^2}{2\sigma_i^2}\right)$, where d_i is the distance from cluster center i, and σ_i is a scale parameter describing the size of the RBF for cluster i. You can think of the size of the RBF as the receptive field of the neuron, or how wide a range of inputs it will respond to.

The scale parameter σ_i is calculated based on the distances of the two closest clusters, $\sqrt{\frac{d_1 + d_2}{2}}$, where d_1 is the distance between the cluster center and the center of the closest other cluster, and d_2 is the distance to the next closest cluster center. Thus, clusters that are close to other clusters will have a small receptive field, while those that are far from other clusters will have a larger receptive field.

During output weight training, records are presented to the network as with a multilayer perceptron. The receptor neurons compute their activation as a function of their RBF size and the user-specified overlapping value h. The activation for receptor neuron j is calculated as

$$\sigma_j = e^{\left(-\frac{\|r-c\|^2}{2\sigma_j^2 h}\right)} \tag{1}$$

where r is the vector of record inputs and c is the cluster center vector. The output neurons are fully interconnected with the receptor or hidden neurons. The receptor neurons pass on their activation values, which are weighted and summed by the output neuron,

$$O_k = \sum_j W_{jk} a_j \tag{2}$$

The output weights W_{jk} are trained in a manner similar to the training of a two-layer back-propagation network. The weights are initialized to small random values in the range $-0.001 \leq w_{ij} \leq 0.001$, and then they are updated at each cycle t by the formula

$$w_{jk}(t) = w_{jk}(t-1) + \Delta w_{jk}(t) \tag{3}$$

The change value is calculated as

$$\Delta w_{jk}(t) = \eta(r_k - O_k)a_i + \alpha \Delta w_{jk}(t-1) \tag{4}$$

which is analogous to the formula used in the back-propagation method.

2.2 Decision Tree

ID3 decision tree algorithm is one of the earliest use, whose main core is to use a recursive form to cut training data. In each time generating node, some subsets of the training input tests will be drawn out to obtain the volume of information coming as a test. After selection, it will yield the greatest amount of value of information obtained as a branch node, selecting the next branch node in accordance with its recursively moves until the training data for each part of a classification fall into one category or meet a condition of satisfaction. C4.5 is the ID3 extension of the method which improved the ID3 excessive subset that contains only a small number of data issues, with handling continuous values-based property, noise processing, and having both pruning tree ability C4.5 decision tree in each node use information obtained on the volume to select test attribute, to select the information obtained with the highest volume (or maximum entropy compression) of the property as the current test attribute node.

Let A be an attribute with k outcomes that partition the training set S into k subsets S_j (j = 1,..., k). Suppose there are m classes, denoted $C = \{c_1, \cdots, c_m\}$, and $p_i = \frac{n_i}{n}$ represents the proportion of instances in S belonging to class c_i, where $n = |S|$ and n_i is the number of instances in S belonging to c_i. The selection measure relative to data set S is defined by:

$$Info(S) = \sum_{i=1}^{m} p_i \log_2 p_i \tag{1}$$

The information measure after considering the partition of S obtained by taking into account the k outcomes of an attribute A is given by:

$$Info(S,A) = \sum_{j=1}^{k} \frac{|S_j|}{|S|} Info(S_i) \tag{2}$$

The information gain for an attribute A relative to the training set S is defined as follows:

$$Gain(S,A) = Info(S) - Info(S,A) \tag{3}$$

The Gain(S, A) is called attribute selection criterion. It computes the difference between the entropies before and after the partition, the largest difference corresponds to the best attribute. Information gain has the drawback to favour attributes with a large number of attribute values over those with a small number. To avoid this drawback, the information gain is replaced by a ratio called gain ratio:

$$GR(S,A) = \frac{\mathrm{Gain(S,A)}}{-\sum_{j=1}^{k} \frac{|\mathrm{S_i}|}{|S|} \log_2 \frac{|S_i|}{|S|}} \tag{4}$$

Consequently, the largest gain ratio corresponds to the best attribute.

3 Methodology

In this study, using the rate of operating expenses, cash flow ratio, current ratio, quick ratio, operating costs, operating income, accounts receivable turnover ratio (times), payable accounts payable cash (days), the net rate of return (after tax)net turnover ratio, earnings per share, operating margin, net growth rate, rate of return of total assets and 14 financial ratios, the use of data mining technology on the sample of the study association rules, cluster analysis, decision tree analysis taxonomy induction a simple and easy-to-understand investment rules significantly simplify the complexity of investment rules. The goal of this paper is proposes a straightforward and efficient stock selective system to reduce the complexity of investment rules.

3.1 Flowchart of Research Procedure

The study proposes a new procedure for a stock selective system. Figure 1 illustrates the flowchart of research procedure in this study.

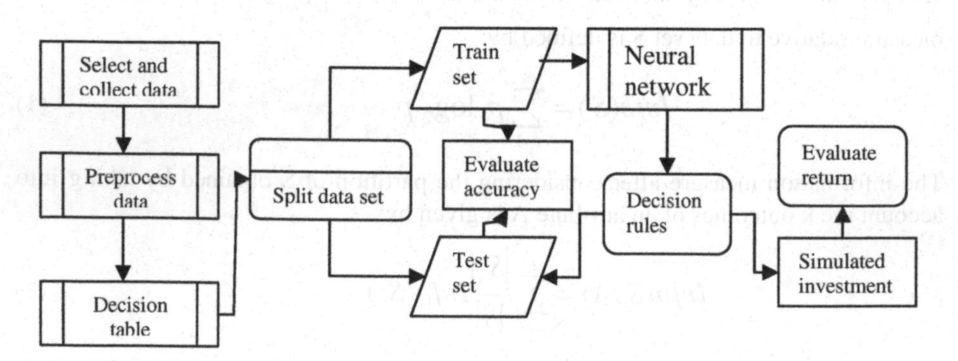

Fig. 1. Flowchart of research procedure

3.2 Flowchart of Research Procedure

This subsection further explains the proposed stock selective system and its algorithms. The proposed stock selective system can be devided into seven steps in detail, and its computing process is introduced systematically as follows:

Step 1: Select and collect the data.

Firstly, this study selects the target data that is collected from Taiwan stock trading system.

Step 2: Preprocess data.

To preprocess the dataset to make knowledge discovery easier is needed. Thus, firstly delete the records that include missing values or inaccurate values, eliminate the clearly irrelevant attributes that will be more easily and effectively pro-cessed for extracting decision rules to select stock. The main jobs of this step includes data integration, data cleaning and data transformation.

Step 3: Build decision table.

The attribute sets of decision table can be divided into a condition attribute set and a decision attribute set. Use financial indicators as a condition attribute set, and whether the stock prices up or down as a decision attribute set.

Step 4: Neural network.

The output weights W_{jk} are trained in a manner similar to the training of a two-layer back-propagation network. The weights are initialized to small random values in the range $-0.001 \le w_{ij} \le 0.001$, and then they are updated at each cycle t by the formula

$$w_{jk}(t) = w_{jk}(t-1) + \Delta w_{jk}(t) \tag{3}$$

The change value is calculated as

$$\Delta w_{jk}(t) = \eta(r_k - O_k)a_i + \alpha \Delta w_{jk}(t-1) \tag{4}$$

which is analogous to the formula used in the back-propagation method.

Step 5: Extract decision rules.

Based on condition attributes clustered in step 5 and a decision attribute (i.e. the stock prices up or down), generate decision rules by decision tree C5.0 algorithm.

Step 6: Evaluate and analyze the results and simulated investment.

For verification, the dataset is split into two sub-datasets: The 67% dataset is used as a training set, and the other 33% is used as a testing set. Furthermore, evaluate return of the simulated investment,

4 Empirical Analysis

4.1 The Computing Procedure

A practically collected dataset is used in this empirical case study to demonstrate the proposed procedure: The dataset for 993 general industrial firms listed in Taiwan stock trading system from 2005/03 to 2012/12 quarterly. The dataset contains 11916 instances which are characterized by the following 14 condition attributes: (i) operating expense ratio(A1), (ii) cash flow ratio(A2), (iii) current ratio(A3), (iv) quick ratio(A4), (v) operating costs(A5), (vi) operating profit(A6), (vii) accounts receivable turnover ratio(A7), (viii) the number of days to pay accounts(A8), (ix) return on equity (after tax) (A9), (x) net turnover(A10), (xi) earnings per share(A11), (xii) operating margin(A12), (xiii) net growth rate(A13), and (xiv) return on total assets growth rate(A14); all attributes are continuous data in this dataset. The computing process of the stock selective system can be expressed in detail as follows:

Step 1: Select and collect the data.
This study selects the target data that is collected from Taiwan stock trading system. Due to the different definitions of industry characteristics and accounting subjects, the general industrial stocks listed companies are considered as objects of the study. The experiment dataset contains 11916 instances which are characterized by 14 condition attributes and one decision attribute.

Step 2: Preprocess data.
Delete the 19 records (instances) that include missing values, and eliminate the 10 irrelative attributes. Accordingly, in total the data of 637 electronic firms that consist of 12 attributes and 6793 instances are included in the RGR dataset. The attributes information of the RGR dataset is shown in Table 1.

Step 3: Build decision table.
The attribute sets of decision table can be divided into a condition attribute set and a decision attribute set. Use financial indicators as a condition attribute set, and whether the stock prices up or down as a decision attribute set.

Step 4: Extract core attributes by neural network
The use of neural network attributes reduction, the core financial attributes can be obtained. The core financial attributes are: (1) return on total assets, (2) the net rate of return after tax, (3) earnings per share, (4) net growth rate, (5) cash flow ratio, (6) operating profit and (7) operating margin.

Step 5: Extract decision rules.
Based on core attributes extracted in step 4 and a decision attribute (i.e. the stock prices up or down), generate decision rules by decision tree C5.0 algorithm. C5.0 rule induction algorithm, is to build a decision tree recursive relationship between the interpretation of the field with the output field data divided into a subset of the and export decision tree rules, try a different part in the interpretation of the data with output field or the relationship of the results. Financial Ratios and Price Change decision tree analysis Table 3, as follows:

Table 1. The partial raw data of the dataset

Case No.	A1	A2	A3	A4	A5	A6	A7	A8	A9	A10	A11	A12	A13	A14	D1
1	0.453	0.550	0.453	0.453	0.453	0.545	0.455	0.620	0.568	0.458	0.575	0.482	0.548	0.582	0.582
2	0.463	0.510	0.463	0.463	0.463	0.503	0.465	0.463	0.500	0.465	0.505	0.468	0.508	0.632	0.632
3	0.365	0.455	0.366	0.366	0.366	0.513	0.367	0.367	0.512	0.368	0.512	0.415	0.602	0.595	0.595
4	0.705	0.648	0.708	0.708	0.708	0.653	0.705	0.710	0.673	0.707	0.673	0.702	0.637	0.620	0.620
5	0.385	0.533	0.385	0.385	0.385	0.498	0.385	0.387	0.470	0.387	0.462	0.412	0.498	0.562	0.562
6	0.618	0.548	0.618	0.618	0.618	0.615	0.617	0.618	0.608	0.617	0.607	0.623	0.595	0.580	0.580
7	0.537	0.551	0.537	0.537	0.537	0.615	0.535	0.533	0.625	0.538	0.630	0.563	0.640	0.597	0.597
8	0.920	0.695	0.920	0.920	0.920	0.718	0.918	0.917	0.728	0.918	0.723	0.877	0.757	0.575	0.575
9	0.545	0.553	0.545	0.545	0.545	0.587	0.543	0.543	0.600	0.543	0.602	0.558	0.583	0.608	0.608
10	0.810	0.668	0.810	0.810	0.810	0.735	0.808	0.808	0.758	0.813	0.760	0.805	0.737	0.657	0.657
11	0.615	0.588	0.615	0.615	0.615	0.638	0.617	0.612	0.635	0.618	0.638	0.630	0.618	0.587	0.587
12	0.083	0.317	0.082	0.082	0.082	0.235	0.083	0.085	0.287	0.083	0.292	0.110	0.247	0.460	0.460
Average	0.542	0.552	0.542	0.542	0.542	0.571	0.542	0.555	0.580	0.543	0.582	0.554	0.581	0.588	0.588

Table 2. The Decision rule set

No.	Decision rule set
1	If the return on total assets growth rate <= 0.070 and the return on total assets growth rate> -0.430 and earnings per share of $ <= 0.900 and the cash flow ratio> -2.800 operating margin <= 20.120 and the business interests <= 184,323 and operating income> -97,782 rose.
2	If the return on total assets growth rate <= 0.070 and the return on total assets growth rate> -0.430 and earnings per share of $ <= 0.900 and the cash flow ratio> -2.800 operating margin> 20.120 shares rose.
3	If the return on total assets growth rate <= 0.070 and the return on total assets growth rate> -0.430 and earnings per share of NT $ 0.900 and the cash flow ratio of> 36.240 and the return on total assets growth rate <= -0.240 then the share price rose.
4	If the growth rate of total assets> 0.070 and return on total assets growth rate <= 2.950 and Cash Flow Ratio <= 13.760 and earnings per share of RMB> -0.820 and net growth rate <= 16.710 shares rose.
5	If the growth rate of total assets> 0.070 and return on total assets growth rate <= 2.950 and the cash flow ratio <= 13.760 per share surplus element> -0.820 and net growth rate of> 16.710 and earnings per share of $ 3.750 pricerise.
6	If the return on total assets growth rate> 0.070 and return on total assets growth rate <= 2.950 and the cash flow ratio of> 13.760 and return on total assets growth rate <= 0.390 and the return on total assets growth rate <= 0.360 shares rose.
6	If the return on total assets growth rate> 0.070 and return on total assets growth rate <= 2.950 and the cash flow ratio of> 13.760 and return on total assets growth rate <= 0.390 and the return on total assets growth rate <= 0.360 shares rose.
7	If the return on total assets growth rate> 0.070 and return on total assets growth rate <= 2.950 and the cash flow ratio of> 13.760 and return on total assets growth rate <= 0.390 and the return on total assets growth rate <= 0.360 shares rose.
8	If the return on total assets growth rate> 0.070 and the return on total assets growth rate> 2.950 share price rose.

Table 3. The Decision rule set

No.	Decision rule set
1	If the return on total assets growth rate <= 0.070 and the return on total assets growth rate <= -0.430 then fell.
2	If the return on total assets growth rate <= 0.070 and the return on total assets growth rate> -0.430 and earnings per share of $ <= 0.900 and the cash flow ratio <= -2.800 fell.
3	If the return on total assets growth rate <= 0.070 and the return on total assets growth rate> -0.430 and earnings per share of $ <= 0.900 and the cash flow ratio> -2.800 operating margin <= 20.120 and the business interests <= 184,323 and business interests <= -97,782 then fell.
4	If the return on total assets growth rate <= 0.070 and the return on total assets growth rate> -0.430 and earnings per share of $ <= 0.900 and the cash flow ratio> -2.800 operating margin <= 20.120 and the business interests of> 184,323 then fell.
5	If the return on total assets growth rate <= 0.070 and the return on total assets growth rate> -0.430 and earnings per share of RMB> 0.900, and the cash flow ratio <= 36.240 fell.
6	If the return on total assets growth rate <= 0.070 and the return on total assets growth rate> -0.430 and earnings per share of RMB> 0.900 cash flow ratio of> 36.240 and return on total assets growth rate> - 0.240 then the share price fell.
7	If the growth rate of total assets> 0.070 and return on total assets growth rate <= 2.950 and the cash flow ratio <= 13.760 and earnings per share of $ <= -0.820 then fell.
8	If the growth rate of total assets> 0.070 and return on total assets growth rate <= 2.950 and the cash flow ratio <= 13.760 and earnings per share of RMB 3.750> -0.820 and net growth rate> 16.710 and earnings per share of $ <= fell.
9	If the growth rate of total assets> 0.070 and return on total assets growth rate <= 2.950 and the cash flow ratio of> 13.760 and return on total assets growth rate <= 0.390 and the growth rate of total return on assets 0.360 fell.

4.2 Simulated Investment

The rules generating from the above statement get down on stock selection from the three quarters former listed companies rise in the rules in year 2008. There are 204 in the first quarter in line with the rise in the rules, the average quarter rate of return of 34.20 %; second quarter 52, thc average quarter rate of return of 17.45%; 12 in the third quarter, the average quarter rate of return of 12.72%.

Further, get down on stock selection from the three quarters former OTC quarter financial report in 97 years. There are 93 in the first quarter in line with the rise in the rules, the average quarter rate of return of 12.88%; second quarter of 38, the average quarter rate of return of 16.18%; 14 in the third quarter, the average quarter rate of return of 7.38%.

5 Conclusion

This paper presents an a stock selective system by using hybrid models of classification. From the results of empirical analysis obtained in this study, some conclusions can be summarized as follows:

(1.) This study presents stock selection method. By the dependence of each company's financial indicators and by the ups and downs of the stock, the use of neural network and decision tree, we can gain a simple set of classification and prediction rules.

(2.) The average rate of return derives from the empirical results show that return on investment on stock price in the research is obvious higher than general market average.

(3.) For the welfare of majority of the investing public, the study will bring a practical and easy to understand stock selection method that can look for quality investment targets in a fast and efficient way.

References

1. Murphy, J.J.: Technical Analysis of the Financial Markets. Institute of Finance, New York (1999)
2. Bernstein, L., Wild, J.: Analysis of Financial Statements. McGraw-Hill (2000)
3. Abraham, A., Nath, B., Mahanti, P.K.: Hybrid intelligent systems for stock market analysis. In: Alexandrov, V.N., Dongarra, J., Juliano, B.A., Renner, R.S., Tan, C.J.K. (eds.) ICCS 2001. LNCS, vol. 2074, pp. 337–345. Springer, Heidelberg (2001)
4. Huang, C.L., Tsai, C.Y.: A hybrid SOFM-SVR with a filter-based feature selection for stock market forecasting. Expert System with Applications 36(2), 1529–1539 (2009)
5. Chang, P.C., Liu, C.H.: A TSK type fuzzy rule based system for stock price prediction. Expert Systems with Application 34(1), 135–144 (2008)
6. Yu, L., Wang, S., Lai, K.K.: Mining stock market tendency using GA-based support vector machines. In: Deng, X., Ye, Y. (eds.) WINE 2005. LNCS, vol. 3828, pp. 336–345. Springer, Heidelberg (2005)
7. Kim, K.J.: Financial time series forecasting using support vector machines. Neurocomputing 55, 307–319 (2003)

Computing Intentions Dynamically in a Changing World by Anticipatory Relevant Reasoning

Jingde Cheng

Department of Information and Computer Sciences, Saitama University,
Saitama, 338-8570, Japan
cheng@ics.saitama-u.ac.jp

Abstract. Intention (and its change, if any) is an indispensable step in the process from belief to action in various types of human behavior. Although the notion of intention has been originally investigated in philosophy and psychology at first, and then was modeled and formalized in Artificial Intelligence, traditional research has focused so much attention on the static properties of intention, but its dynamic properties have received much less attention. However, in a changing world, intention of an agent may be changed before it leads into a real action and the last change and/or intention just before the real action is the most important in almost all cases. In order to provide a computational foundation for various advanced application systems where the key requirement is to accurately grasp intentions of agents/users just before their actions, this position paper proposes a new research direction: Computing intentions dynamically in a changing world by anticipatory relevant reasoning.

Keywords: Intention calculus, Anticipatory computing, Relevant reasoning, Strong relevant logic, Three-dimensional spatio-temporal deontic relevant logic.

1 Introduction

Intention (and its change, if any) is an indispensable step (and often the last step just before action) in the process from belief to action in various types of human behavior. Today, as a key requirement, many advanced application systems require that the systems should accurately grasp intentions of agents/users just before their actions. For examples, in order to provide users with active services and/or personalized services satisfactorily, many advanced service systems in the modern information society require grasping users' intentions before their actions, i.e., to get and/or predict intentions of users beforehand; in a computing anticipatory system consisting of various agents, to get and/or predict agents' intentions before their actions anticipatorily is also a characteristic requirement; and so on.

The notion of intention was originally investigated in Philosophy by Anscombe in 1957 [6]. Fishbein and Ajzen investigated the notion of intention in 1969 from the viewpoint of Social Psychology [1-3], [20-21]. Bratman studied the notion of intention as a planning theory [7-8]. In the area of Artificial Intelligence, Cohen and Levesque first modeled and formalized the notion of intention based on Bratmen's

N.T. Nguyen et al. (Eds.): ACIIDS 2014, Part II, LNAI 8398, pp. 361–371, 2014.

theory [15], and almost work in AI follows from their work [22]. Traditional research has focused so much attention on the static properties of intention, but its dynamic properties have received much less attention. As a result, almost all traditional research approaches on intention are descriptive but not predictive. However, in a changing world, intention of an agent/user may be changed before it leads into a real action and the last change and/or intention just before the real action is the most important in almost all cases. Until now, there is no research focused attention on grasping intentions of agents/users just before their actions.

In order to provide a computational foundation for various advanced application systems where the key requirement is to accurately grasp intentions of agents/users just before their actions, this position paper proposes a new research direction: Computing intentions dynamically in a changing world by anticipatory relevant reasoning.

The rest of this paper is organized as follows: Section 2 discusses the logical basis for reasoning about intention and its change in a changing world, Section 3 presents three-dimensional spatio-temporal deontic relevant logic, Section 4 discusses anticipatory relevant reasoning about intention and its change based on three-dimensional spatio-temporal deontic relevant logic, and concluding remarks are given in Section 5.

2 The Logical Basis for Reasoning about Intention and Its Changes in a Changing World

To represent and reason about intention and its change in a changing world, a logic basis as the criterion for the validity of reasoning is indispensable. It is obvious that not any logic system can serve as the fundamental logic system satisfactorily. The question, "Which is the right logic for reasoning about something?" invites the immediate counter-question "Right for what?" Only if we certainly know what we need, we can make a good choice. The present author considers that the fundamental logic system to underlie representing and reasoning about intention and its change in a changing world must satisfy all the following essential requirements.

First, as a general logical criterion for the validity of reasoning, the logic must be able to underlie relevant and truth-preserving reasoning in the sense of conditional, i.e., for any reasoning based on the logic to be valid, if its premises are true in the sense of conditional, then its conclusion must be relevant to the premises and true in the sense of conditional. Logic is the study of the methods and principles used to distinguish correct reasoning from incorrect reasoning. The notion of a conditional is the heart of logic, because there is no reasoning that does not invoke the notion of conditional.

Second, the logic must be able to underlie ampliative reasoning, i.e., for any reasoning based on the logic to be valid, the truth of conclusion of the reasoning should be recognized after the completion of the reasoning process but not be invoked in deciding the truth of premises of the reasoning. Reasoning is the process of drawing new conclusions from given premises, and therefore, any meaningful reasoning must be ampliative but not circular and/or tautological.

Third, the logic must be able to underlie paracomplete and paraconsistent reasoning, i.e., for any reasoning based on the logic to be valid, the conclusion of the reasoning may not be the negation of a sentence when (even if) the sentence is not a conclusion of the premises of that reasoning, and also may not be an arbitrary sentence when (even if) the premises of that reasoning is inconsistent. In particular, the so-called principle of Explosion that everything follows from a contradiction should not be accepted by the logic as a valid principle. In general, our knowledge about a domain as well as a scientific discipline may be incomplete and/or inconsistent in many ways, i.e., it gives us no evidence for deciding the truth of either a proposition or its negation, and/or it directly or indirectly includes some contradictions. Therefore, reasoning with incomplete and/or inconsistent knowledge is the rule rather than the exception in our everyday lives and almost all scientific disciplines.

Fourth, the logic must be able to underlie spatio-temporal reasoning. In a changing world consisting of various mobile agents in three-dimensional space, any reasoning may somehow depend on notions of time and three-dimensional space. Not only propositions (statements) about mobile agents but also relevant relationships among mobile agents may be dependent on three-dimensional spatial regions and/or points, and may change over time. Some properties of mobile agents, i.e., motion and speed, are intrinsically dependent on both time and three-dimensional space. This naturally requires that the logic must be able to underlie temporal, three-dimensional spatial, and three-dimensional spatio-temporal reasoning.

Finally, the logic must be able to underlie normative reasoning. Because our empirical rules often describe only those ideal situations, when they be used to in actual situations, we need to distinguish between what ought to be done and what is the case. Therefore, a formalisation of normative notions is indispensable. Normative statements make claims about how things should or ought to be, and which actions are right or wrong.

Classical mathematical logic (**CML**), which is based on the classical account of validity and represents the notion of conditional by the notion of material implication, cannot satisfy any of the above essential requirements [4-5], [9-14], [16], [19], [23], [26]. Because any temporal (classical) logic is a classical conservative extension of classical mathematical logic in the sense that it is based on the classical account of validity and it represents the notion of conditional directly or indirectly by the material implication, no temporal (classical) logic can satisfy the essential requirements [10-12]. Those existing spatial logics [16] are classical conservative extensions of classical mathematical logic in the sense that they are based on the classical account of validity and they represent the notion of conditional directly or indirectly by the material implication. Therefore, similar to the case of temporal (classical) logic, these spatial logics also cannot satisfy the essential requirements [14].

Traditional relevant (relevance) logics were constructed during the 1950s in order to find a mathematically satisfactory way of grasping the elusive notion of relevance of antecedent to consequent in conditionals, and to obtain a notion of implication which is free from the so-called 'paradoxes' of material and strict implication [4-5], [19], [23]. As a modification of traditional relevant logics **R**, **E**, and **T**, strong relevant logics **Rc**, **Ec**, and **Tc** rejects all conjunction-implicational paradoxes and disjunction-implicational paradoxes in **R**, **E**, and **T**, respectively [9], [11]. The strong relevant logics can underlie relevant reasoning in the sense of the strong relevance. The three-dimensional spatio-temporal relevant logics [14] are obtained by introducing predicates and axiom schemata about solid-region connection, predicates and axiom schemata about point position, and predicates and axiom schemata about motion of mobile objects into temporal relevant logics. However, the three-dimensional spatio-temporal relevant logics do not provide explicit means to deal with the normative notions and therefore they cannot satisfy the fifth essential requirement.

Therefore, no existing logic system can satisfy all of the essential requirements.

3 Three-Dimensional Spatio-temporal Deontic Relevant Logic

Let $\{r_1, r_2, r_3, \ldots\}$ be a countably infinite set of individual variables, called ***solid-region variables***. Atomic formulas of the form $\boldsymbol{C}(r_1, r_2)$ are read as 'region r_1 connects with region r_2.' Let $\{p_1, p_2, p_3, \ldots\}$ be a countably infinite set of individual variables, called ***point variables***. Let ***TCP*** be an individual constant of point, called ***the central point***. Atomic formulas of the form $\boldsymbol{I}(p_1, r_1)$ are read as 'point p_1 is included in region r_1.' Atomic formulas of the form $\boldsymbol{Id}(p_1, p_2)$ are read as 'point p_1 is identical with p_2.' Atomic formulas of the form $\boldsymbol{Arc}(p_1, p_2)$ are read as 'points p_1, p_2 are adjacent such that there is an arc from point p_1 to point p_2, or more simply, point p_1 is adjacent to point p_2.' Note that an arc has a direction. Atomic formulas of the form $\boldsymbol{Reachable}(p_1, p_2)$ are read as 'there is at least one directed path (i.e., a sequence of arcs such that one connects to the next one) from point p_1 to point p_2.' Atomic formulas of the form $\boldsymbol{NH}(p_1, p_2)$ are read as 'taking ***TCP*** as the reference point, the position of point p_1 is not higher than the position of point p_2, i.e., the length of the vertical line from p_1 to ***TCP*** is not longer than the length of the vertical line from p_2 to ***TCP***. Here, $\boldsymbol{C}(r_1, r_2)$, $\boldsymbol{I}(p_1, r_1)$, $\boldsymbol{Id}(p_1, p_2)$, $\boldsymbol{Arc}(p_1, p_2)$, $\boldsymbol{Reachable}(p_1, p_2)$ and $\boldsymbol{NH}(p_1, p_2)$ are primitive binary predicates to represent three-dimensional geometric relationships between three-dimensional geometric regions and points. Note that here we use a many-sorted language. Let $\{o_1, o_2, o_3, \ldots\}$ be a countably infinite set of individual variables, called ***object variables***. Atomic formulas of the form $\boldsymbol{A}(o_1, p_1)$ are read as 'object o_1 arrives at point p_1.' Atomic formulas of the form $\boldsymbol{NS}(o_1, o_2)$ are read as 'the speed of object o_1 is not faster than the speed of object o_2.' Here, $\boldsymbol{A}(o_1, p_1)$ and $\boldsymbol{NS}(o_1, o_2)$ are primitive binary predicates to represent motion relationships between mobile objects in a three-dimensional geometric space.

The symbols (logical connectives, quantifiers, individual variables, individual constants, solid-region variables, point variables, object variables, predicates, temporal operators, deontic operators), region connection predicates, point position predicates, object movement predicates, axiom schemata, and inference rules are as follows:

Symbols: $\{\neg, \Rightarrow, \wedge, \forall, \exists, x_1, x_2, \ldots, x_n, \ldots,$ ***TCP***, $c_1, c_2, \ldots, c_n, \ldots, r_1, r_2, r_3, \ldots, r_n, \ldots, p_1, p_2, p_3, \ldots, p_n, \ldots, o_1, o_2, o_3, \ldots, o_n, \ldots, p_0^{1}, \ldots, p_0^{n}, \ldots, p_1^{1}, \ldots, p_1^{n}, \ldots, p_2^{1}, \ldots, p_2^{n}, \ldots, p_k^{1}, \ldots, p_k^{n}, \ldots,$ (,), $\boldsymbol{G}, \boldsymbol{H}, \boldsymbol{F}, \boldsymbol{Pa}, \boldsymbol{O}, \boldsymbol{Pe}\}$

Primitive and Defined Logical Connectives: $\Rightarrow$ (entailment), $\neg$ (negation), $\otimes$ (intensional conjunction, $A \otimes B =_{\mathrm{df}} \neg(A \Rightarrow \neg B)$), $\oplus$ (intensional disjunction, $A \oplus B =_{\mathrm{df}}$

$\neg A \Rightarrow B$), $\Leftrightarrow$ (intensional equivalence, $A \Leftrightarrow B =_{df} (A \Rightarrow B) \otimes (B \Rightarrow A)$), $\wedge$ (extensional conjunction), $\vee$ (extensional disjunction, $A \vee B =_{df} \neg(\neg A \wedge \neg B)$), $\rightarrow$ (material implication, $A \rightarrow B =_{df} \neg(A \wedge \neg B)$ or $\neg A \vee B$), $\leftrightarrow$ (extensional equivalence, $A \leftrightarrow B =_{df} (A \rightarrow B) \wedge (B \rightarrow A)$).

Temporal Operators: $\boldsymbol{G}$ (future-tense always or henceforth operator, $\boldsymbol{G}A$ means 'it will always be the case in the future from now that A'), $\boldsymbol{H}$ (past-tense always operator, $\boldsymbol{H}A$ means 'it has always been the case in the past up to now that A'), $\boldsymbol{F}$ (future-tense sometime or eventually operator, $\boldsymbol{F}A$ means 'it will be the case at least once in the future from now that A'), and $\boldsymbol{Pa}$ (past-tense sometime operator, $\boldsymbol{Pa}A$ means 'it has been the case at least once in the past up to now that A'). Note that these temporal operators are not independent and can be defined as follows: $\boldsymbol{G}A =_{df} \neg \boldsymbol{F} \neg A$, $\boldsymbol{H}A =_{df} \neg \boldsymbol{Pa} \neg A$, $\boldsymbol{F}A =_{df} \neg \boldsymbol{G} \neg A$, $\boldsymbol{Pa}A =_{df} \neg \boldsymbol{H} \neg A$.

Deontic Operators: $\boldsymbol{O}$ (obligation operator, $\boldsymbol{O}A$ means "It is obligatory that A"), and $\boldsymbol{Pe}$ (permission operator, $\boldsymbol{Pe}A =_{df} \neg \boldsymbol{O}(\neg A)$, $\boldsymbol{Pe}A$ means "It is permitted that A").

Primitive Binary Predicates: $\boldsymbol{C}(r_1, r_2)$ (connection), $\boldsymbol{I}(p_1, r_1)$ (inclusion), $\boldsymbol{Id}(p_1, p_2)$ (the same point), $\boldsymbol{Arc}(p_1, p_2)$ (arc), $\boldsymbol{Reachable}(p_1, p_2)$ (reachable), $\boldsymbol{NH}(p_1, p_2)$ (not higher than), $\boldsymbol{A}(o_1, p_1)$ (arrives at), and $\boldsymbol{NS}(o_1, o_2)$ (not speedier than).

Defined Binary Predicates: $\boldsymbol{DC}(r_1, r_2) =_{df} \neg \boldsymbol{C}(r_1, r_2)$, $\boldsymbol{DC}(r_1, r_2) =_{df} \neg(\exists p_1(\boldsymbol{I}(p_1, r_1) \wedge \boldsymbol{I}(p_1, r_2)))$ ($\boldsymbol{DC}(r_1, r_2)$ means 'r_1 is disconnected from r_2'), $\boldsymbol{Part}(r_1, r_2) =_{df} \forall r_3(\boldsymbol{C}(r_3, r_1) \Rightarrow \boldsymbol{C}(r_3, r_2))$, $\boldsymbol{Part}(r_1, r_2) =_{df} \forall p_1(\boldsymbol{I}(p_1, r_1) \Rightarrow \boldsymbol{I}(p_1, r_2))$ ($\boldsymbol{Part}(r_1, r_2)$ means 'r_1 is a part of r_2'), $\boldsymbol{PrPart}(r_1, r_2) =_{df} \boldsymbol{Part}(r_1, r_2) \wedge (\neg \boldsymbol{Part}(r_2, r_1))$ ($\boldsymbol{PrPart}(r_1, r_2)$ means 'r_1 is a proper part of r_2'), $\boldsymbol{EQ}(r_1, r_2) =_{df} \boldsymbol{Part}(r_1, r_2) \wedge \boldsymbol{Part}(r_2, r_1)$ ($\boldsymbol{EQ}(r_1, r_2)$ means 'r_1 is identical with r_2'), $\boldsymbol{Overlap}(r_1, r_2) =_{df} \exists r_3(\boldsymbol{Part}(r_3, r_1) \wedge \boldsymbol{Part}(r_3, r_2))$ ($\boldsymbol{Overlap}(r_1, r_2)$ means 'r_1 overlaps r_2'), $\boldsymbol{DR}(r_1, r_2) =_{df} \neg \boldsymbol{Overlap}(r_1, r_2)$ ($\boldsymbol{DR}(r_1, r_2)$ means 'r_1 is discrete from r_2'), $\boldsymbol{PaOverlap}(r_1, r_2) =_{df} \boldsymbol{Overlap}(r_1, r_2) \wedge (\neg \boldsymbol{Part}(r_1, r_2)) \wedge (\neg \boldsymbol{Part}(r_2, r_1))$ ($\boldsymbol{PaOverlap}(r_1, r_2)$ means 'r_1 partially overlaps r_2'), $\boldsymbol{EC}(r_1, r_2) =_{df} \boldsymbol{C}(r_1, r_2) \wedge (\neg \boldsymbol{Overlap}(r_1, r_2))$ ($\boldsymbol{EC}(r_1, r_2)$ means 'r_1 is externally connected to r_2'), $\boldsymbol{TPrPart}(r_1, r_2) =_{df} \boldsymbol{PrPart}(r_1, r_2) \wedge \exists r_3(\boldsymbol{EC}(r_3, r_1) \wedge \boldsymbol{EC}(r_3, r_2))$ ($\boldsymbol{TPrPart}(r_1, r_2)$ means 'r_1 is a tangential proper part of r_2'), $\boldsymbol{NTPrPart}(r_1, r_2) =_{df} \boldsymbol{PrPart}(r_1, r_2) \wedge (\neg \exists r_3(\boldsymbol{EC}(r_3, r_1) \wedge \boldsymbol{EC}(r_3, r_2)))$ ($\boldsymbol{NTPrPart}(r_1, r_2)$ means 'r_1 is a nontangential proper part of r_2'), $\boldsymbol{SA}(p_1, p_2) =_{df} \boldsymbol{NH}(p_1, p_2) \wedge \boldsymbol{NH}(p_2, p_1)$ ($\boldsymbol{SA}(p_1, p_2)$ means 'the position of point p_1 and the position of point p_2 are in the same altitude'), $\boldsymbol{Hi}(p_1, p_2) =_{df} \neg \boldsymbol{NH}(p_1, p_2)$ ($\boldsymbol{Hi}(p_1, p_2)$ means 'the position of point p_1 is higher than the position of point p_2'), $\boldsymbol{SS}(o_1, o_2) =_{df} \boldsymbol{NS}(o_1, o_2) \wedge \boldsymbol{NS}(o_2, o_1)$ ($\boldsymbol{SS}(o_1, o_2)$ means 'the motion of object o_1 and the motion of object o_2 are in the same speed'), $\boldsymbol{Sp}(o_1, o_2) =_{df} \neg \boldsymbol{NS}(o_1, o_2)$ ($\boldsymbol{Sp}(o_1, o_2)$ means 'the motion of object o_1 is faster than the motion of object o_2'), $\boldsymbol{ND}(p_1, p_2, p_3) =_{df} \exists o_1 \exists o_2((\boldsymbol{A}(o_1, p_1) \wedge \boldsymbol{A}(o_2, p_2) \wedge \boldsymbol{Reachable}(p_1, p_3) \wedge \boldsymbol{Reachable}(p_2, p_3) \wedge \boldsymbol{SS}(o_1, o_2)) \Rightarrow \boldsymbol{G}(\boldsymbol{A}(o_1, p_3) \Rightarrow \boldsymbol{A}(o_2, p_3)))$ ($\boldsymbol{ND}(p_1, p_2, p_3)$ means 'the distance of between point p_2 and point p_3 is not more distant than the distance of between point p_1 and point p_3'), $\boldsymbol{SD}(p_1, p_2, p_3) =_{df} \boldsymbol{ND}(p_1, p_2, p_3) \wedge \boldsymbol{ND}(p_2, p_1, p_3)$ ($\boldsymbol{SD}(p_1, p_2, p_3)$ means 'the distance of between point p_1 and point p_3 is equal to the distance of between point p_2 and point p_3'), and $\boldsymbol{Ne}(p_1, p_2, p_3) =_{df} \neg \boldsymbol{ND}(p_1, p_2, p_3)$ ($\boldsymbol{Ne}(p_1, p_2, p_3)$ means 'the distance of between point p_1 and point p_3 is nearer than the distance of between point p_2 and point p_3').

Axiom Schemata:

E1: $A \Rightarrow A$,

E2: $(A \Rightarrow B) \Rightarrow ((C \Rightarrow A) \Rightarrow (C \Rightarrow B))$, E2′: $(A \Rightarrow B) \Rightarrow ((B \Rightarrow C) \Rightarrow (A \Rightarrow C))$,

E3: $(A \Rightarrow (A \Rightarrow B)) \Rightarrow (A \Rightarrow B)$, E3′: $(A \Rightarrow (B \Rightarrow C)) \Rightarrow ((A \Rightarrow B) \Rightarrow (A \Rightarrow C))$,

E3″: $(A \Rightarrow B) \Rightarrow ((A \Rightarrow (B \Rightarrow C)) \Rightarrow (A \Rightarrow C))$,

E4: $(A \Rightarrow ((B \Rightarrow C) \Rightarrow D)) \Rightarrow ((B \Rightarrow C) \Rightarrow (A \Rightarrow D))$, E4′: $(A \Rightarrow B) \Rightarrow (((A \Rightarrow B) \Rightarrow C) \Rightarrow C)$,

E4″: $((A \Rightarrow A) \Rightarrow B) \Rightarrow B$, E4‴: $(A \Rightarrow B) \Rightarrow ((B \Rightarrow C) \Rightarrow (((A \Rightarrow C) \Rightarrow D) \Rightarrow D))$,

E5: $(A \Rightarrow (B \Rightarrow C)) \Rightarrow (B \Rightarrow (A \Rightarrow C))$, E5′: $A \Rightarrow ((A \Rightarrow B) \Rightarrow B)$,

N1: $(A \Rightarrow (\neg A)) \Rightarrow (\neg A)$, N2: $(A \Rightarrow (\neg B)) \Rightarrow (B \Rightarrow (\neg A))$, N3: $(\neg(\neg A)) \Rightarrow A$,

C1: $(A \wedge B) \Rightarrow A$, C2: $(A \wedge B) \Rightarrow B$, C3: $((A \Rightarrow B) \wedge (A \Rightarrow C)) \Rightarrow (A \Rightarrow (B \wedge C))$,

C4: $(\mathbf{L}A \wedge \mathbf{L}B) \Rightarrow \mathbf{L}(A \wedge B)$, where $\mathbf{L}A =_{\mathrm{df}} (A \Rightarrow A) \Rightarrow A$,

D1: $A \Rightarrow (A \vee B)$, D2: $B \Rightarrow (A \vee B)$, D3: $((A \Rightarrow C) \wedge (B \Rightarrow C)) \Rightarrow ((A \vee B) \Rightarrow C)$,

DCD: $(A \wedge (B \vee C)) \Rightarrow ((A \wedge B) \vee C)$,

C5: $(A \wedge A) \Rightarrow A$, C6: $(A \wedge B) \Rightarrow (B \wedge A)$, C7: $((A \Rightarrow B) \wedge (B \Rightarrow C)) \Rightarrow (A \Rightarrow C)$,

C8: $(A \wedge (A \Rightarrow B)) \Rightarrow B$, C9: $\neg(A \wedge \neg A)$, C10: $A \Rightarrow (B \Rightarrow (A \wedge B))$,

T1: $\mathbf{G}(A \Rightarrow B) \Rightarrow (\mathbf{G}A \Rightarrow \mathbf{G}B)$, T2: $\mathbf{H}(A \Rightarrow B) \Rightarrow (\mathbf{H}A \Rightarrow \mathbf{H}B)$,

T3: $A \Rightarrow \mathbf{G}(\mathbf{P}A)$, T4: $A \Rightarrow \mathbf{H}(\mathbf{F}A)$, T5: $\mathbf{G}A \Rightarrow \mathbf{G}(\mathbf{G}A)$,

T6: $(\mathbf{F}A \wedge \mathbf{F}B) \Rightarrow (\mathbf{F}(A \wedge \mathbf{F}B) \vee \mathbf{F}(A \wedge B) \vee \mathbf{F}(\mathbf{F}A \wedge B))$,

T7: $(\mathbf{Pa}A \wedge \mathbf{Pa}B) \Rightarrow (\mathbf{Pa}(A \wedge \mathbf{Pa}B) \vee \mathbf{Pa}(A \wedge B) \vee \mathbf{Pa}(\mathbf{Pa}A \wedge B))$,

T8: $\mathbf{G}A \Rightarrow \mathbf{F}A$, T9: $\mathbf{H}A \Rightarrow \mathbf{Pa}A$, T10: $\mathbf{F}A \Rightarrow \mathbf{F}(\mathbf{F}A)$,

T11: $(A \wedge \mathbf{H}A) \Rightarrow \mathbf{F}(\mathbf{H}A)$, T12: $(A \wedge \mathbf{G}A) \Rightarrow \mathbf{Pa}(\mathbf{G}A)$,

DR1: $\mathbf{O}(A \Rightarrow B) \Rightarrow (\mathbf{O}A \Rightarrow \mathbf{O}B)$, DR2: $\mathbf{O}A \Rightarrow \mathbf{Pe}A$,

DR3: $\neg(\mathbf{O}A \wedge \mathbf{O}\neg A)$, DR4: $\mathbf{O}(A \wedge B) \Rightarrow (\mathbf{O}A \wedge \mathbf{O}B)$, DR5: $\mathbf{Pe}(A \wedge B) \Rightarrow (\mathbf{Pe}A \wedge \mathbf{Pe}B)$

IQ1: $\forall x(A \Rightarrow B) \Rightarrow (\forall xA \Rightarrow \forall xB)$, IQ2: $(\forall xA \wedge \forall xB) \Rightarrow \forall x(A \wedge B)$,

IQ3: $\forall xA \Rightarrow A[t/x]$ (if x may appear free in A and t is free for x in A, i.e., free variables of t do not occur bound in A),

IQ4: $\forall x(A \Rightarrow B) \Rightarrow (A \Rightarrow \forall xB)$ (if x does not occur free in A),

IQ5: $\forall x_1 \ldots \forall x_n(((A \Rightarrow A) \Rightarrow B) \Rightarrow B)$ $(n \geq 0)$,

RCC1: $\forall r_1(\mathbf{C}(r_1, r_1))$, RCC2: $\forall r_1 \forall r_2(\mathbf{C}(r_1, r_2) \Rightarrow \mathbf{C}(r_2, r_1))$,

PRCC1: $\forall p_1 \forall r_1 \forall r_2((\mathbf{I}(p_1, r_1) \wedge \mathbf{DC}(r_1, r_2)) \Rightarrow \neg\mathbf{I}(p_1, r_2))$,

PRCC2: $\forall p_1 \forall r_1 \forall r_2((\mathbf{I}(p_1, r_1) \wedge \mathbf{Pa}(r_1, r_2)) \Rightarrow \mathbf{I}(p_1, r_2))$,

PRCC3: $\forall r_1 \forall r_2(\mathbf{O}(r_1, r_2) \Rightarrow \exists p_1(\mathbf{I}(p_1, r_1) \wedge \mathbf{I}(p_1, r_2)))$,

PRCC4: $\forall r_1 \forall r_2(\mathbf{PaOverlap}(r_1, r_2) \Rightarrow (\exists p_1(\mathbf{I}(p_1, r_1) \wedge \mathbf{I}(p_1, r_2)) \wedge \exists p_2(\mathbf{I}(p_2, r_1) \wedge \neg\mathbf{I}(p_2, r_2)) \wedge \exists p_3(\neg\mathbf{I}(p_3, r_1) \wedge \mathbf{I}(p_3, r_2))))$,

PRCC5: $\forall r_1 \forall r_2(\mathbf{EC}(r_1, r_2) \Rightarrow \exists p_1(\mathbf{I}(p_1, r_1) \wedge \mathbf{I}(p_1, r_2) \wedge \forall p_2(\neg\mathbf{Id}(p_2, p_1) \Rightarrow (\neg\mathbf{I}(p_2, r_1) \wedge \neg\mathbf{I}(p_2, r_2)))))$,

PRCC6: $\forall p_1 \forall r_1 \forall r_2((\mathbf{I}(p_1, r_1) \wedge \mathbf{TPrPart}(r_1, r_2)) \Rightarrow \mathbf{I}(p_1, r_2))$,

PRCC7: $\forall p_1 \forall r_1 \forall r_2((\mathbf{I}(p_1, r_1) \wedge \mathbf{NTPrPart}(r_1, r_2)) \Rightarrow \mathbf{I}(p_1, r_2))$,

RC1: $\forall p_1 \forall p_2(\mathbf{Arc}(p_1, p_2) \Rightarrow \mathbf{Reachable}(p_1, p_2))$,

RC2: $\forall p_1 \forall p_2 \forall p_3((\mathbf{Reachable}(p_1, p_2) \wedge \mathbf{Reachable}(p_2, p_3)) \Rightarrow \mathbf{Reachable}(p_1, p_3))$,

HC1: $\forall p_1(\mathbf{NH}(p_1, p_1))$, HC2: $\forall p_1 \forall p_2 \forall p_3((\mathbf{NH}(p_1, p_2) \wedge \mathbf{NH}(p_2, p_3)) \Rightarrow \mathbf{NH}(p_1, p_3))$,

MC1: $\forall o_1(\mathbf{NS}(p_1, p_1))$, MC2: $\forall o_1 \forall o_2 \forall o_3((\mathbf{NS}(o_1, o_2) \wedge \mathbf{NS}(o_2, o_3)) \Rightarrow \mathbf{NS}(o_1, o_3))$,

DC1: $\forall p_1 \forall p_3(\mathbf{ND}(p_1, p_1, p_3))$,

DC2: $\forall p_1 \forall p_2 \forall p_3 \forall p_4((\mathbf{ND}(p_1, p_2, p_4) \wedge \mathbf{ND}(p_2, p_3, p_4)) \Rightarrow \mathbf{ND}(p_1, p_3, p_4))$.

Inference Rules:
⇒E: from *A* and *A*⇒*B* to infer *B* (Modus Ponens),
∧I: from *A* and *B* to infer *A*∧*B* (Adjunction),
O-necessitation: "if *A* is a logical theorem, then so is ***O****A*" (Deontic Generalization).
TG: from *A* to infer ***G****A* and ***H****A* (Temporal Generalization),
∀I: if *A* is an axiom, so is ∀*xA* (Generalization of axioms).

Various relevant logic systems are defined as follows, where we use 'X | Y' to denote any choice of one from two axiom schemata X and Y:

$\mathbf{T}_{\Rightarrow} =_{df} \{E1, E2, E2', E3 \mid E3''\} + \Rightarrow E$,
$\mathbf{E}_{\Rightarrow} =_{df} \{E1, E2 \mid E2', E3 \mid E3', E4 \mid E4'\} + \Rightarrow E$,
$\mathbf{E}_{\Rightarrow} =_{df} \{E2', E3, E4''\} + \Rightarrow E$,
$\mathbf{E}_{\Rightarrow} =_{df} \{E1, E3, E4'''\} + \Rightarrow E$,
$\mathbf{R}_{\Rightarrow} =_{df} \{E1, E2 \mid E2', E3 \mid E3', E5 \mid E5'\} + \Rightarrow E$,
$\mathbf{T}_{\Rightarrow,\neg} =_{df} \mathbf{T}_{\Rightarrow} + \{N1, N2, N3\}$,
$\mathbf{E}_{\Rightarrow,\neg} =_{df} \mathbf{E}_{\Rightarrow} + \{N1, N2, N3\}$,
$\mathbf{R}_{\Rightarrow,\neg} =_{df} \mathbf{R}_{\Rightarrow} + \{N2, N3\}$,
$\mathbf{T} =_{df} \mathbf{T}_{\Rightarrow,\neg} + \{C1{\sim}C3, D1{\sim}D3, DCD\} + \wedge I$,
$\mathbf{E} =_{df} \mathbf{E}_{\Rightarrow,\neg} + \{C1{\sim}C4, D1{\sim}D3, DCD\} + \wedge I$,
$\mathbf{R} =_{df} \mathbf{R}_{\Rightarrow,\neg} + \{C1{\sim}C3, D1{\sim}D3, DCD\} + \wedge I$,
$\mathbf{Tc} =_{df} \mathbf{T}_{\Rightarrow,\neg} + \{C3, C5{\sim}C10\}$,
$\mathbf{Ec} =_{df} \mathbf{E}_{\Rightarrow,\neg} + \{C3{\sim}C10\}$,
$\mathbf{Rc} =_{df} \mathbf{R}_{\Rightarrow,\neg} + \{C3, C5{\sim}C10\}$,
$\mathbf{TQ} =_{df} \mathbf{T} + \{IQ1{\sim}IQ5\} + \forall I$,
$\mathbf{EQ} =_{df} \mathbf{E} + \{IQ1{\sim}IQ5\} + \forall I$,
$\mathbf{RQ} =_{df} \mathbf{R} + \{IQ1{\sim}IQ5\} + \forall I$,
$\mathbf{TcQ} =_{df} \mathbf{Tc} + \{IQ1{\sim}IQ5\} + \forall I$,
$\mathbf{EcQ} =_{df} \mathbf{Ec} + \{IQ1{\sim}IQ5\} + \forall I$,
$\mathbf{RcQ} =_{df} \mathbf{Rc} + \{IQ1{\sim}IQ5\} + \forall I$.

The minimal or weakest propositional temporal relevant logics are defined as follows: $\mathbf{T_0Tc} = \mathbf{Tc} + \{T1{\sim}T4\} + TG$, $\mathbf{T_0Ec} = \mathbf{Ec} + \{T1{\sim}T4\} + TG$, $\mathbf{T_0Rc} = \mathbf{Rc} + \{T1{\sim}T4\} + TG$. Note that the minimal or weakest temporal classical logic $\mathbf{K_t}$ = all axiom schemata for **CML** + →E + {T1~T4} + TG. Other characteristic axioms such as T5~T12 that correspond to various assumptions about time can be added to $\mathbf{T_0Tc}$, $\mathbf{T_0Ec}$, and $\mathbf{T_0Rc}$ respectively to obtain various propositional temporal relevant logics. Various predicate temporal relevant logics then can be obtained by adding axiom schemata IQ1~IQ5 and inference rule ∀I into the propositional temporal relevant logics. For examples, minimal or weakest predicate temporal relevant logics are defined as follows: $\mathbf{T_0TcQ} = \mathbf{T_0Tc} + \{IQ1{\sim}IQ5\} + \forall I$, $\mathbf{T_0EcQ} = \mathbf{T_0Ec} + \{IQ1{\sim}IQ5\} + \forall I$, $\mathbf{T_0RcQ} = \mathbf{T_0Rc} + \{IQ1{\sim}IQ5\} + \forall I$.

We can obtain the minimal or weakest propositional temporal deontic relevant logics as follows: $\mathbf{DT_0Tc} = \mathbf{T_0Tc} + \{DR1{\sim}DR5\}$ + ***O***-necessitation, $\mathbf{DT_0Ec} = \mathbf{T_0Ec} + \{DR1{\sim}DR5\}$ + ***O***-necessitation, $\mathbf{DT_0Rc} = \mathbf{T_0Rc} + \{DR1{\sim}DR5\}$ + ***O***-necessitation [19]. Other characteristic axioms such as T5~T12 that correspond to various assumptions about time can be added to **DT_0Tc**, **DT_0Ec**, and **DT_0Rc** respectively to obtain various propositional temporal deontic relevant logics. Various predicate temporal deontic relevant logics then can be obtained by adding axiom schemata IQ1~IQ5 and inference rule ∀I into the propositional temporal deontic relevant logics.

Now, we can obtain various three-dimensional spatio-temporal deontic relevant logics by adding axiom schemata about region connection, point position, and motion of mobile objects into the various predicate temporal relevant deontic logics. For examples: **SDT_0TcQ** = **DT_0TcQ** + {RCC1, RCC2, PRCC1~PRCC7, RC1, RC2, HC1, HC2, MC1, MC2, DC1, DC2}, **SDT_0EcQ** = **DT_0EcQ** + {RCC1, RCC2, PRCC1~PRCC7, RC1, RC2, HC1, HC2, MC1, MC2, DC1, DC2}, **SDT_0RcQ** = **DT_0RcQ** + {RCC1, RCC2, PRCC1~PRCC7, RC1, RC2, HC1, HC2, MC1, MC2, DC1, DC2}.

4 Anticipatory Relevant Reasoning about Intention and Its Change Based on Three-Dimensional Spatio-temporal Deontic Relevant Logic

Having the three-dimensional spatio-temporal deontic relevant logics as the fundamental logic systems, we can represent and reason about intention and its change based on the logics.

First, the strong relevant logics provide a logical validity criterion for relevant reasoning in the sense of strong relevance, i.e. for any valid reasoning based on a strong relevant logic, its premises include no irrelevant and unnecessary conjuncts and its conclusion includes no irrelevant or unnecessary disjuncts (Note that the logical validity criterion provided by traditional relevant logics is not necessarily relevant in this sense). Therefore, in the framework of strong relevant logic, if a reasoning is valid, then the strong relevance between its premises and its conclusion can be guaranteed necessarily, i.e. the logics can certainly underlie relevant reasoning in the sense of strong relevance [9], [11]. On the other hand, because the strong relevant logics are free of not only implicational paradoxes but also conjunction-implicational and disjunction-implicational paradoxes, the logical validity criterion provided by strong relevant logics is truth-preserving in the sense of conditional. Note that the logical validity criterion provided by **CML** is truth-preserving only in the sense of material implication; it is not truth-preserving in the sense of conditional. Also note that the logical validity criterion provided by traditional relevant logics is truth-preserving only in the sense of relevant implication; it is not truth-preserving in the sense of conditional. Therefore, in the framework of strong relevant logic, if a reasoning is valid, then the truth of its conclusion in the sense of conditional can be guaranteed necessarily, i.e. the logics can certainly underlie truth-preserving in the sense of conditional [9], [11].

Second, a reasoning based on any of relevant logics including strong relevant logics is ampliative but not circular and/or tautological. This is because the notion of entailment (conditional) that plays the most intrinsic role in any reasoning is represented in relevant logics by a primitive intensional connective satisfying the

Wright-Geach-Smiley criterion, i.e. to come to know the truth of an entailment without coming to know the falsehood of its antecedent or the truth of consequent [4-5], [19], [23].

Third, all relevant logics including strong relevant logics reject the principle of Explosion, and therefore, they are paraconsistent but not explosive [4-5], [19], [23]. All relevant logics can certainly underlie paracomplete and paraconsistent reasoning.

Fourth, the predicates and axiom schemata about solid-region connection and point position provide the means to represent and reason about three-dimensional geometric relationships among three-dimensional geometric regions and points. The temporal operators and related axiom schemata provide the means to represent and reason about those propositions and/or formulas whose truth-values may depend on time. The predicates and axiom schemata about motion of mobile agents provide the means to represent and reason about motion relationships among mobile agents in a three-dimensional geometric space. Therefore, the three-dimensional spatio-temporal relevant logics can also underlie three-dimensional spatial reasoning, temporal reasoning, and three-dimensional spatio-temporal reasoning.

Finally, the three-dimensional spatio-temporal deontic relevant logics provide explicit means to deal with the normative notions, and therefore, they can underlie normative reasoning.

Therefore, the three-dimensional spatio-temporal deontic relevant logics can satisfy all of the essential requirements for the fundamental logic systems to underlie representing and reasoning about intention and its change in a changing world.

An anticipatory reasoning is a reasoning to draw new, previously unknown and/or unrecognized conclusions about some future event or events whose occurrence and truth are uncertain at the point of time when the reasoning is being performed [10], [12], [17-18], [24-25]. By using temporal operators and predicates about motion of mobile agents, one can easily represent and reason about future situations of mobile agents moving in a three-dimensional geometric space.

To represent and reason about intention and its change based on the three-dimensional spatio-temporal deontic relevant logics, for any real case, we have to define some concrete (empirical) facts and predicates to represent agent's concrete beliefs, intentions, actions, and situations of the changing world in which the agent is playing/working, find some concrete (empirical) relationships between them, represent these facts, predicates, and relationships by the language of the three-dimensional spatio-temporal deontic relevant logic, choose a three-dimensional spatio-temporal deontic relevant logic system that is suitable to the real case, and finally take these facts, predicates, and relationships as premises to reason about agent's intention and its change based on the logic system.

5 Concluding Remarks

We have proposed a new approach to represent and reason about intention and its change based on the three-dimensional spatio-temporal deontic relevant logics. The work presented in this paper is only a preliminary theoretical one, and there are many future works. We are working on various case studies to apply our approach to some real cases to confirm the effectiveness of our approach.

Grasping agents/users' intentions in a changing world before their actions, i.e., to get and/or predict intentions of agents/users beforehand, may be a key issue in an application system. It is obvious that the system will have more time to do something if it gets and/or predicts intentions (in particular, the last intention before action) of agents/users as early as possible. Therefore, from the viewpoint of application, the efficiency of anticipatory reasoning about intention and its change is a key requirement as well as its validity and effectiveness. In this paper, we only consider the validity of anticipatory reasoning about intention and its change. There are many interesting and challenging research problems on the effectiveness and efficiency of anticipatory reasoning.

References

1. Ajzen, I.: From Intentions to Actions: A Theory of Planned Behavior. In: Kuhl, J., Beckman, J. (eds.) Action Control: From Cognition to Behavior, pp. 11–39. Springer, Heidelberg (1985)
2. Ajzen, I.: The Theory of Planned Behavior: Reactions and Reflections. Psychology & Health 26, 1113–1127 (2011)
3. Ajzen, I.: The Theory of Planned Behavior. In: Lange, P.A.M., Kruglanski, A.W., Higgins, E.T. (eds.) Handbook of Theories of Social Psychology, vol. 1, pp. 438–459 (2012)
4. Anderson, A.R., Belnap Jr., N.D.: Entailment: The Logic of Relevance and Necessity, vol. I. Princeton University Press, Princeton (1975)
5. Anderson, A.R., Belnap Jr., N.D., Dunn, J.M.: Entailment: The Logic of Relevance and Necessity, vol. II. Princeton University Press, Princeton (1992)
6. Anscombe, G.E.M.: Intention. Basil Blackwell (1957), 2nd edn. (1963). Harvard University Press, Cambridge (2000)
7. Bratman, M.E.: Intentions, Plans, and Practical Reason. Harvard University Press, Cambridge (1987)
8. Bratman, M.E.: Faces of Intentions: Selected Essays on Intention and Agency. Cambridge University Press, Cambridge (1999)
9. Cheng, J.: A Strong Relevant Logic Model of Epistemic Processes in Scientific Discovery. In: Kawaguchi, E., Kangassalo, H., Jaakkola, H., Hamid, I.A. (eds.) Information Modelling and Knowledge Bases XI. Frontiers in Artificial Intelligence and Applications, vol. 61, pp. 136–159. IOS Press, Amsterdam (2000)
10. Cheng, J.: Temporal Relevant Logic as the Logical Basis of Anticipatory Reasoning-Reacting Systems. In: Dubois, D.M. (ed.) Computing Anticipatory Systems: CASYS 2003. AIP Conference Proceedings, vol. 718, pp. 362–375. American Institute of Physics, Melville (2004)
11. Cheng, J.: Strong Relevant Logic as the Universal Basis of Various Applied Logics for Knowledge Representation and Reasoning. In: Kiyoki, Y., Henno, J., Jaakkola, H., Kangassalo, H. (eds.) Information Modelling and Knowledge Bases XVII. Frontiers in Artificial Intelligence and Applications, vol. 136, pp. 310–320. IOS Press, Amsterdam (2006)
12. Cheng, J.: Adaptive Prediction by Anticipatory Reasoning Based on Temporal Relevant Logic. In: Proc. 8th International Conference on Hybrid Intelligent Systems, pp. 410–416. IEEE Computer Society Press, New York (2008)

13. Cheng, J.: Deontic Relevant Logic as the Logical Basis for Representing and Reasoning about Legal Knowledge in Legal Information Systems. In: Lovrek, I., Howlett, R.J., Jain, L.C. (eds.) KES 2008, Part II. LNCS (LNAI), vol. 5178, pp. 517–525. Springer, Heidelberg (2008)
14. Cheng, J.: Qualitative Spatio-temporal Reasoning about Moving Objects in Three-Dimensional Space. In: Kang, L., Cai, Z., Yan, X., Liu, Y. (eds.) ISICA 2008. LNCS, vol. 5370, pp. 637–648. Springer, Heidelberg (2008)
15. Cohen, P.R., Levesque, H.J.: Intention Is Choice with Commitment. Artificial Intelligence 42, 213–261 (1990)
16. Cohn, A.G., Hazarika, S.M.: Qualitative Spatial Representation and Reasoning: An Overview. Fundamenta Informaticae 46(1-2), 1–29 (2001)
17. Dubois, D.M.: Computing Anticipatory Systems with Incursion and Hyperincursion. In: Dubois, D.M. (ed.) Computing Anticipatory Systems. AIP Conference Proceedings, vol. 437, pp. 3–29. American Institute of Physics, Melville (1998)
18. Dubois, D.M.: Mathematical Foundations of Discrete and Functional Systems with Strong and Weak Anticipations. In: Butz, M.V., Sigaud, O., Gérard, P. (eds.) Anticipatory Behavior in Adaptive Learning Systems. LNCS (LNAI), vol. 2684, pp. 110–132. Springer, Heidelberg (2003)
19. Dunn, J.M., Restall, G.: Relevance Logic. In: Gabbay, D., Guenthner, F. (eds.) Handbook of Philosophical Logic, 2nd edn., vol. 6, pp. 1–128. Kluwer Academic, Dordrecht (2002)
20. Fishbein, M., Ajzen, I.: Belief, Attitude, Intention, and Behavior: An Introduction to Theory and Research. Addison-Wesley (1975)
21. Fishbein, M., Ajzen, I.: Predicting and Changing Behavior: The Reasoned Action Approach. Psychology Press (2010)
22. van der Hoek, W., Jamroga, W., Wooldridge, M.: Towards a Theory of Intention Revision. Synthese 155, 265–290 (2007)
23. Mares, E.D.: Relevant Logic: A Philosophical Interpretation. Cambridge University Press, Cambridge (2004)
24. Nadin, M.: Anticipatory Computing. Ubiquity - The ACM IT Magazine and Forum, Views 1(40) (2000)
25. Rosen, R.: Anticipatory Systems - Philosophical, Mathematical and Methodological Foundations. Pergamon Press, Oxford (1985), 2nd edn. Springer, New York (2012)
26. Tagawa, T., Cheng, J.: Deontic Relevant Logic: A Strong Relevant Logic Approach to Removing Paradoxes from Deontic Logic. In: Ishizuka, M., Sattar, A. (eds.) PRICAI 2002. LNCS (LNAI), vol. 2417, pp. 39–48. Springer, Heidelberg (2002)

One Step Ahead towards the Rearrangeability of 4D-Hypercube Interconnection Networks

Ibrahima Sakho and Jean-Pierre Jung

LITA, Université de Lorraine, Ile du Saulcy, BP 794
57045 Metz, France

Abstract. This paper addresses the problem of the nD-hypercube interconnection networks rearrangeability that is the capability of such networks to route optimally arbitrary permutations under queueless communication constraints. To that purpose, the paper exploits the recursive structure of nD-hypercube as two (n-1)D-hypercubes and proposes the k-partitioning paradigm. For n = 4, the paper characterizes permutations that do not admit 1-partitioning. It then exhibits representatives of some of the subclasses of such permutations. Each representative is then proved to admit 2-partitioning and one of its corresponding routing strategies is exhibited. Such routing strategy is obtained as a concatenation of a routing strategy of one of the upstream permutations of the considered permutation and a routing strategy of each of the two downstream permutations induced by the considered upstream permutation.

Keywords: interconnection network, hypercube, network rearrangeability, permutation, routing, bipartite graph, maximum matching, graph partitioning.

1 Introduction

A n-dimensional hypercube, nD-hypercube, is a bidirectional interconnection network (IN) whose graph, say $H^{(n)} = (V, E)$, is a set V of 2^n nodes u = 0, 1, ..., 2^n-1 and a set E of the edges such that a node u is connected to a node v if and only their binary codes ($u_{n-1}u_{n-2} \ldots u_0$) and ($v_{n-1}v_{n-2} \ldots v_0$) differ on one and only one dimension. Fig. 1 illustrates a 4D-hypercube.

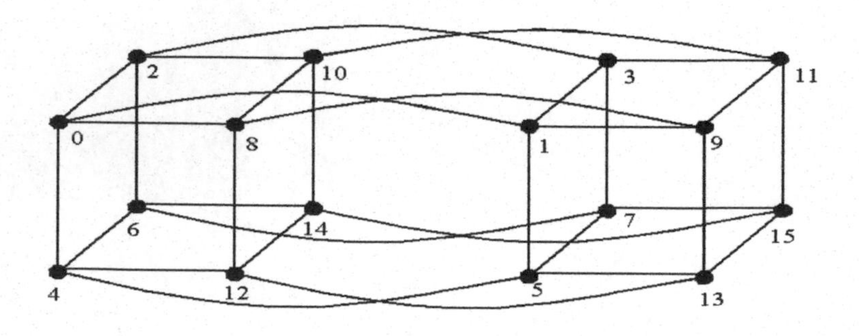

Fig. 1. A 4D-hypercube

N.T. Nguyen et al. (Eds.): ACIIDS 2014, Part II, LNAI 8398, pp. 372–381, 2014.

In this paper we will consider that each node is connected to itself by a sort of internal link. This interconnection logic confers to hypercubes interesting mathematical properties which allow them to match most of the IN performance criteria. For this reason they still constitute very attractive alternative for numerous applications among which parallel computers IN. In this context, its rearrangeability that is its capability to optimally route arbitrary permutations and, if so, a switching strategy to do so are, beyond their theoretical aspects, practical valuable properties to be insured. Unfortunately, this problem is still an open one since at least a quarter of century. If the problem can be tackled with brute force for very low dimension, $n \leq 2$, hypercubes, this is inconceivable for hypercubes with higher dimensions. Indeed, apart from the fact that the number of permutations to be examined is exponential, $2^n!$, 20,922,789,888,000 for $n = 4$ for instance, each permutation would require an exhaustive search for 2^n non-conflicting and optimal routes through the hypercube. To avoid so much effort, efficient alternative approaches have to be developed.

In this paper, we propose the k-partitioning paradigm. It consists in decomposing the permutation in an upstream permutation routable in k steps and two independent downstream permutations routable in n-k steps on two disjoint hypercubes.

The remainder of the paper is organized in four sections. Section 2 surveys the state-of-the-art on the optimal queueless routing of arbitrary permutations. Section 3 presents the mathematical foundations of the k-partitioning paradigm. Section 4, firstly establishes the characterization of non-1-partitionable permutations. Then it analyzes the non-1-partitionability on 4D-hypercubes and exhibits the representatives of some of the non-1-partitionable permutations classes. Section 5 deals with and proves the 2-partitionability of the exhibited classes of non-1-partitionable permutations. It then shows how to devise required non-conflicting routing paths for each class. Section 6 concludes the paper.

2 Problem Formulation and Related Works

Let π be a permutation on a $H^{(n)}$ and a set of 2^n messages of the same size each one located at one node u and destined for the node $\pi(u)$. Routing π under queueless communication constraints consists in conveying all the messages to their respective destination given that each interconnection link may convey no more than one message and each node may host no more than one message. Given the constraints, such routing consists in a sequence of global and synchronous exchanges between neighbor nodes. As the messages have the same size, its complexity is of the order of the required number of exchange steps. Therefore an optimal routing is the one with the minimal exchange steps. For an arbitrary permutation it is well known, from e-cube routing [1], that at least n exchange steps are required.

Optimal routing of arbitrary permutations on nD-hypercubes is, since at least a quarter of century, one of the most challenging open problems in the theory of IN. It has so been extensively well studied and several communication models and routing paradigms have been used to that purpose.

In [2] Szimansky considers the offline routing in circuit-switched and packet switched commutation under all-port MIMD communication model. Under the circuit-switched hypothesis he proves, for $n \leq 3$, that any hypercube is rearrangeable.

He also conjectured that routing can be made on the shortest paths, conjecture for which a counterexample has been given in [3] by Lubiw. Under packet-switched hypothesis he shows that routing can be made in 2n-1 steps. Under the single port MIMD communication model, Zhang in [4] proposes a routing in O(n) steps on a spanning tree of the hypercube. In [5, 6] Hwang et al considered online oblivious routing under buffered all port MIMD communication models. They prove that n steps routing is possible for $n \leq 7$ and later for $n \leq 12$ if local information are used. The better routings under the models viewed above are due to Vöckling [7]. He proves that deterministic offline routing in buffered all port MIMD model can be done in $n+O(\sqrt{n}\log n)$ steps while online oblivious randomized one can be done in $n+O(n/\log n)$ steps.

For the more restrictive models, that is single-port, queueless, and MIMD communication model, the personal communication of Coperman to Ramras, according to Ramras, and the works of Ramras [8] constitute certainly the leading ones. Indeed while Coperman gives computational proof that arbitrary permutations can be routed in 3D-hypercube in 3 steps, Ramras proves that if a permutation can be routed in r steps in rD-hypercube, then for $n \geq r$ arbitrary permutations on nD-hypercubes can be routed in 2n-r steps. Recently, Laing and Krumme in [9] have introduced an approach which simplifies the problem enough to permit a human verification of the possibility of routing in 3 steps arbitrary permutations on 3D-hypercube and computer verification for the 4 steps routing in 4D-hypercube. We also have addressed the problem with a paradigm similar to Laing and Krumme one that we call k-partitioning. However instead of looking explicitly for a partition into 2^k permutations, we look for a partition into two permutations on two disjoint (n-1)D-hypercubes routable in n-k steps. We formally proved in [10] that for $n \leq 3$, arbitrary permutations are routable in n steps.

3 Mathematical Foundations of k-Partitioning

As the routing hypercube rearrangeability requires a sequence of synchronous exchange between neighbor nodes, this one can be realized on one side according to the sequence of perfect matching between hypercube nodes. On the other side, to guarantee the convergence of the messages towards their destination, each matching must split the hypercube in other of lesser dimensions. So, formally, the k-partitioning process comes down to the computation of a perfect matching in bipartite graphs that are graphs $G=(V_1, V_2, E)$ where V_1 and V_2 are disjoint sets of nodes and E, the set of the interconnection between the pairs $\{u, v\}$ of the nodes such that $u \in V_1$, $v \in V_2$.

The bipartite graph associated to a nD-hypercube is the one where V_1 and V_2 are two disjoint copies of the nD-hypercube nodes and E is the set of the pairs $\{u, v\}$ such that $\{u, v\}$ is an interconnection of the hypercube nodes. The perfect matching we are looking for are one-to-one correspondences Γ which associates to each node u of V_1 a node $\Gamma(u)$ of V_2 such that the set of all the pairs $\{u, \Gamma(u)\}$ constitute a set of two-by-two non adjacent edges of E.

With this purpose, let's consider the adjacency matrix M of the hypercube interconnection graph whose rows and columns are indexed by the hypercube nodes and the components are such that $M[u, v] = 1$ (resp. 0) if $\{u, v\} \in$ (resp. $\notin$) E. Observe that a nD-hypercube, as illustrated in Fig. 1 for n = 4 in dimension i = 0, a nD-hypercube suppress the last a nD-hypercube can be viewed as an interconnection, in any of its n dimensions, of two (n-1)D-hypercubes say $H^{(n)}_{0,\,i}$, and $H^{(n)}_{1,\,i}$ where $H^{(n)}_{x,\,i}$ is the restriction of the nD-hypercube to its nodes u such that $u_i = x$. Similarly, the adjacency matrix is anyone of the n (2x2)-blocks matrices whose extra-diagonal blocks, which express the interconnection between its two (n-1)D-hypercubes, are identity matrices and diagonal blocks are the adjacency matrices of its two (n-1)D-hypercubes. Table 1 illustrates the four adjacency matrices of the 4D-hypercubes.

Table 1. Adjacency matrices of 4D-hypercubes

			3	**0**	**1**	**2**	**3**	**4**	**5**	**6**	**7**	**8**	**9**	**10**	**11**	**12**	**13**	**14**	**15**
			2	**0**	**1**	**2**	**3**	**8**	**9**	**10**	**11**	**4**	**5**	**6**	**7**	**12**	**13**	**14**	**15**
			1	**0**	**1**	**4**	**5**	**8**	**9**	**12**	**13**	**2**	**3**	**6**	**7**	**10**	**11**	**14**	**15**
3	*2*	*1*	*0*	**0**	**2**	**4**	**6**	**8**	**10**	**12**	**14**	**1**	**3**	**5**	**7**	**9**	**11**	**13**	**15**
0	**0**	**0**	**0**	1	1	1		1				1							
1	**1**	**1**	**2**	1	1		1		1				1						
2	**2**	**4**	**4**	1		1	1			1				1					
3	**3**	**5**	**6**		1	1	1				1				1				
4	**8**	**8**	**8**	1				1	1	1						1			
5	**9**	**9**	**10**		1			1	1		1						1		
6	**10**	**12**	**12**			1		1		1	1							1	
7	**11**	**13**	**14**				1		1	1	1								1
8	**4**	**2**	**1**	1								1	1	1		1			
9	**5**	**3**	**3**		1							1	1		1		1		
10	**6**	**6**	**5**			1						1		1	1			1	
11	**7**	**7**	**7**				1						1	1	1				1
12	**12**	**10**	**9**					1				1				1	1	1	
13	**13**	**11**	**11**						1				1			1	1		1
14	**14**	**14**	**13**							1				1		1		1	1
15	**15**	**15**	**15**								1				1		1	1	1

In dimension i the indexes of the rows and the columns of the matrix are the boldfaced numbers of the column and the row numbered i in shadowed italic font.

Given a permutation π on a nD-hypercube $H^{(n)}$, $x \in \{0, 1\}$ and $0 \le i \le n-1$, let:

– r be its optimal routing steps,
– $S_{x,i}$ (resp. $D_{x,i}$) be the set of the nodes u of $H^{(n)}$ such that $\pi_i(u)$ (resp. u_i) = x; in fact $D_{x,i}$ is the set of the nodes of $H^{(n)}_{x,i}$,

– $G_{x,i,k}$ be the bipartite graph $(S_{x,i}, D_{x,i}, E^k)$, k = 0, 1, 2, …, r-1 where $\{u, v\} \in E^k$ if and only if there is a path of length less than or equal k, from $u \in S_{x,i}$ to $v \in D_{x,i}$, and a path of length less than or equal r-k from v to $\pi(u)$,
– $M_{x,i,k}$ be the adjacency matrix of $G_{x,i,k}$,
– $\Gamma_{x,i,k}$ be a maximum matching of $G_{x,i,k}$.

Definition 1. π is said to be k-partitionable, for k < n, in dimension i if there is a permutation $\Gamma = (\Gamma_{0,i,k}, \Gamma_{1,i,k})$ on $H^{(n)}$ such that α (resp. β) which associates $\pi(u)$ to $\Gamma(u)$ such that $\Gamma_i(u) = 0$ (resp. 1) is a permutation on the (n-1)D-hypercube $H^{(n)}_{0,i}$ (resp. $H^{(n)}_{1,i}$). α and β are then downstream permutations and Γ upstream permutation of π.

Definition 2. π is said to be k-partitionable if there is a dimension for which it is k-partitionable. Otherwise it is non-k-partitionable.

Obviously, permutations such that $S_{x,i} = D_{x,i}$, or $S_{x,i} \cap D_{x,i} = \varnothing$ k-partitionable with k = 0 and 1 respectively. Thus in the sequel we will consider only permutations such that $S_{x,i} \cap D_{y,i} \neq \varnothing$ for y = 0, 1.

4 Characterization of Non-1-Partitionable Permutations

The k-partionability is the guaranty for a permutation to be optimally routed, first in k steps and then in n-k steps as two distinct permutations on two disjoint (n-1)D-hypercubes. In this intention, it is essential to identify the classes of non-1-partitionable permutations. In [10] we proved the following characterization.

Proposition 1. A permutation is non-1-partitionable in dimension i of a hypercube if and only if one of the adjacency matrices $M_{x,i,1}$ contains a null column.

It can be easily proved, consequently to Proposition 1, that:

Proposition 2. A necessary and sufficient condition for a permutation π to be non-1-partitionable in dimension i, is that there is at least a node v of $H^{(n)}_{x,i}$ such that for any node u, neighbor of v, $\pi_i(u) = x$.

In the sequel, a node v of $H^{(n)}_{x,i}$ whose neighbors u, including itself, are such that $\pi_i(u) = x$ will be called a x-node.

Now let's examine how the characterization from the proposition 2 comes in various forms for n = 4. Let π be a non-1-partitionable permutation on $H^{(4)}$. Then for any dimension i, there is a x-node. As it may exist more than one x-node per dimension, we have first to determine the maximum number of x-nodes admissible per dimension then, in case of several x-nodes, to study their compatibility in the sense that they can lead to a permutation.

It can be observed that with the labelling induced by Proposition 2, $H^{(4)}_{x,i}$ admits at most two x-nodes. Thus there are two non-1-partitionability models in any dimension: the 1-x-node 1 and the 2-x-nodes models. In this study we restrict ourselves to the second model. As $H^{(4)}_{x,i}$ is a 3D-hypercube, three situations may happen according to the distance between the x-nodes. Without loss of generality, we can consider the case of the 1-nodes in $H^{(4)}_{0,0}$. Let node 0 be one of the 1-nodes.

Case 1. The 1-nodes are distant of 1. Let node 2 be the second 1-node. By definition of a 1-node, nodes 0, 2 and all their neighbors, the half of $H^{(4)}$ nodes have to be labeled with 1. Consequently, the other nodes have to be labeled with 0. Fig. 2 illustrates this case. We can observe that in each of the 3D-hypercubes $H^{(4)}{}_{x,k}$, $\pi_i(u) = \underline{\pi_i(u)}$.

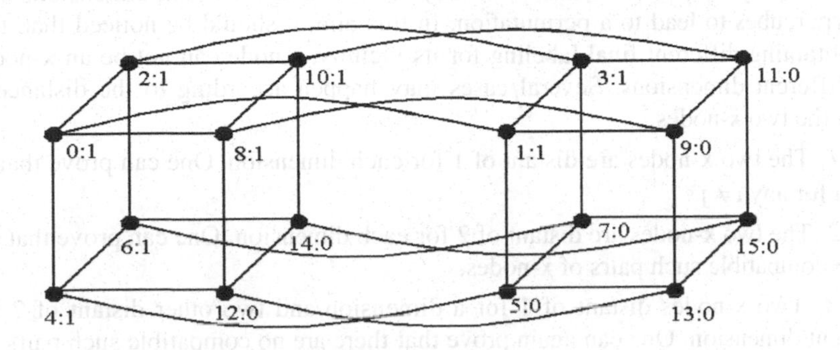

Fig. 2. The labelling of $H^{(4)}$ nodes according to the 1-nodes 0 and 2

Case 2. The 1-nodes are distant of 2. Let node 10 be the second 1-node. The resulting labelling is illustrated in Fig. 3.

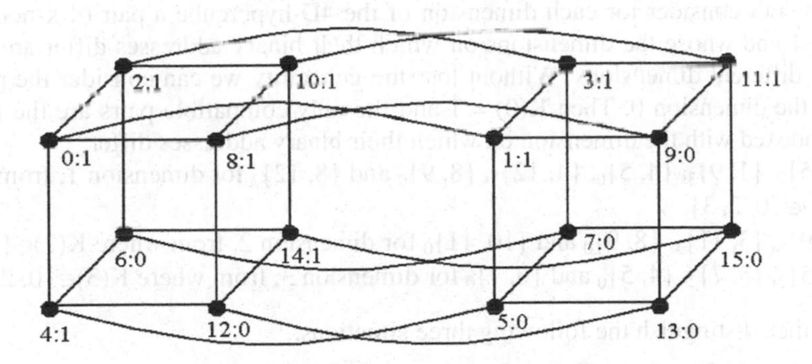

Fig. 3. The labelling of $H^{(4)}$ nodes according to the 1-nodes 0 and 10

As previously, the labeling is consistent. Furthermore in each of the 3D-hypercubes $H^{(4)}{}_{x,k}$, $\pi_i(u) = \underline{\pi_i(u)}$.

Case 3. The 1-nodes are distant of 3. Node 0 being one the 1-node in $H^{(4)}{}_{0,0}$, the only choice for the second 1-node is node 14. The resulting labeling is obviously not consistent.

This analysis can be summarized by the following proposition.

Proposition 3. If a permutation on $H^{(4)}$ admits two x-nodes in $H^{(4)}{}_{x,i}$ then :

a) it also admits two x-nodes in $H^{(4)}{}_{x,i}$,

b) both x (resp. x)-nodes are distant of 1 or 2 in $H^{(4)}{}_{x,i}$ (resp. $H^{(4)}{}_{x,i}$),

c) each x-node remains a x-node in a different $H^{(4)}{}_{y,k}$ where k is a dimension in which the binary addresses of the x-nodes differ.

In the sequel, we will consider arbitrary permutations having two x-nodes in each dimension of the 4D-hypercube and we will denote K(i) the set of the dimensions k for which the binary addresses of the x-nodes in the dimension i of the 4D-hypercube differ. Now we have to study how these two consistent models of non-1-partitionability self combine then how they mingle in the different dimensions of the 4D-hypercubes to lead to a permutation. In this aim, it should be noticed that, to allow obtaining different final labeling for its vicinity, a node can not be an x-node in two different dimensions. Several cases may happen according to the distance between the two x-nodes.

Case 1. The two x-nodes are distant of 1 for each dimension. One can prove that K(i) ≠ K(j) for any i ≠ j.

Case 2. The two x-nodes are distant of 2 for each dimension. One can prove that there are no compatible such pairs of x-nodes.

Case 3. Two x-nodes distant of 1 for a dimension and two other distant of 2 for a different dimension. One can again prove that there are no compatible such pairs of x-nodes.

We can conclude this part of our analysis by the fact that if a permutation on a 4D-hypercube admits two x-nodes for each dimension, they are distant of 1. Moreover for two different dimensions the dimensions on which the binary address of its x-nodes differ are different too.

Now let's consider for each dimension of the 4D-hypercube a pair of x-nodes distant of 1 and whose the dimensions on which their binary addresses differ are different for different dimensions. Without loss the generality we can consider the pair {0, 2} for the dimension 0. Then K(0) = 1 and the only compatible pairs are the followings, indexed with the dimension on which their binary addresses differ:

– $\{1, 5\}_2$, $\{1, 9\}_3$, $\{4, 5\}_0$, $\{4, 12\}_3$, $\{8, 9\}_0$ and $\{8, 12\}_2$ for dimension 1, from where K(1)∈ {0, 2, 3}
– $\{1, 9\}_3$, $\{3, 11\}_3$, $\{8, 9\}_0$ and $\{10, 11\}_0$ for dimension 2, from where K(2)∈ {0, 3}
– $\{1, 5\}_2$, $\{3, 7\}_2$, $\{4, 5\}_0$ and $\{6, 7\}_0$ for dimension 3, from where K(3)∈ {0, 2}.

We then distinguish the following three situations.

– K(1) = 0, K(2) = 3 and K(3) = 2: which leads for the triplets of dimensions (1, 2, 3) to one of the triplets of pairs ({4, 5}, {1, 9}, {3, 7}) and ({8, 9}, {3, 11}, {1, 5}).
– K(1) = 2, K(2) = 3 and K(3) = 0: which leads for the triplets of dimensions (1, 2, 3) to one of the triplets of pairs ({1, 5}, {3, 11}, {6, 7}) and ({8, 12}, {1, 9}, {4, 5}).
– K(1) = 3, K(2) = 0 and K(3) = 2: which leads for the triplets of dimensions (1, 2, 3) to one of the triplets of pairs ({1, 9}, {10, 11} , {3, 7}) and ({4, 12}, {8, 9}, {1, 5}).

Considering the labels induced for each node of the 4D-hypercube by the above pairs of 1-nodes we easily verify that they lead to permutations isomorphic to:

- $\pi^{(1)}$ = (3, 11, 13, 15, 1, 10, 9, 8, 7, 6, 5, 14, 0, 2, 4, 12)
- $\pi^{(2)}$ = (7, 15, 9, 13, 3, 14, 11, 10, 5, 4, 1, 12, 2, 6, 0, 8)

The above analysis can then be summarized by the following proposition.

Proposition 4. A permutation on $H^{(4)}$ admits two x-nodes for each dimension if and only if it is isomorphic to one of the permutations $\pi^{(1)}$ and $\pi^{(2)}$.

5 Optimal Routing of Non-1-Partitionable Permutations

In this section we deal with the optimal routing of $\pi^{(1)}$ and $\pi^{(2)}$. We will restrict our analysis mainly to $\pi^{(1)}$, the reasoning being the same for $\pi^{(2)}$.

In the sequel an upstream permutation which can not be routed in k steps on a nD-hypercube or whose at least one of induced downstream permutations can not be routed in n-k steps on a (n-1)D-hypercube will be called T-permutation.

5.1 On the 2-Partitionability of the Permutations

Let π be a non-1-partitionable permutation on a hypercube $H^{(4)}$. Let us examine if there is a dimension i such that the bipartite graphs $G_{x,i,2}$ admit a perfect matching. With this purpose we have to consider, without loss of generality, the adjacency matrices $M_{x,i,2}$ for x = 0, 1 of the bipartite graphs of a representative permutation of each class. We proved in [10] that such permutations are isomorphic to (3, 2, 1, 7, 0, 6, 5, 4), or (6, 2, 1, 7, 0, 3, 5, 4) for 3D-hypercube.

Let's consider the adjacency matrices of the bipartite graphs $G_{x,i,2}$, x = 0, 1. For i=0, for instance these are the matrices of Table 2.

Table 2. The adjacency matrices of the bipartite graphs $G_{x,i,2}$, x = 0, 1 associated to $\pi^{(1)}$

	0	2	4	6	8	10	12	14
12	**1**	0	1	1	1	1	1	0
14	0	**1**	1	1	1	0	1	1
5	1	0	0	**1**	0	0	1	0
7	0	1	**1**	0	0	0	0	1
9	1	0	0	0	0	**1**	1	0
11	0	1	0	0	**1**	1	0	1
13	0	0	1	0	1	0	0	**1**
15	0	0	0	1	0	1	**1**	1

	1	3	5	7	9	11	13	15
1	1	1	0	1	1	1	1	0
3	0	1	1	1	1	1	0	1
0	1	1	1	0	1	0	0	0
2	1	0	0	1	0	1	0	0
4	1	0	1	1	0	0	1	0
6	0	1	1	0	0	0	0	1
8	1	0	0	0	0	1	1	0
10	0	1	0	0	1	0	0	1

We can observe, and this is the case for the other dimensions, that the two matrices are similar modulo a rows and columns permutation. Indeed, $M_{1,0,2}[u, v] = M_{0,0,2}[u, v]$ $\forall u \in S_{0,0}$, $v \in D_{0,0}$. Thus in the sequel, for each dimension i we will consider only the bipartite graph $G_{0,0,2}$.

Considering in $M_{0,0,2}$ only the boldfaced 1 which constitute a permutation matrix of the same order and then a perfect matching of $G_{0,0,2}$, we can conclude, from the similarity with $G_{1,0,2}$ that in each dimension, $\pi^{(1)}$ admits $\Gamma^{(0)}$ = (5, 1, 7, 3, 13, 6, 15, 4, 11, 10, 9, 8, 0, 14, 2, 12) as an upstream permutation. Thus its 2-partitionability question comes down to the existence in at least one dimension of an upstream permutation which is not a T-permutation. Unfortunately, such problem is generally not decidable and may require a prohibitive and often futile exhaustive research as this is the case with $\Gamma^{(1)}$ = (6, 7, 1, 3, 0, 15, 5, 13, 11, 10, 9, 2, 4, 14, 8, 12).

Now, we have to consider the 2-steps-routability on two distinct 3D-hypercubes of the downstream permutations, say $\alpha^{(i)}$ and $\beta^{(i)}$, induced by at least one of the 2-steps-routable upstream permutations $\Gamma^{(i)}$. With this purpose, for each dimension of the 4D-hypercube, renumber the nodes of each of its 3D-hypercube from 0 to 7 according to their ascending number in their 3D-hypercube. We obtain for instance form $\Gamma^{(0)}$ the downstream permutations $\alpha^{(0)} = (0, 2, 4, 5, 7, 3, 6, 1)$ and $\beta^{(0)} = (5, 7, 1, 6, 2, 3, 0, 4)$ which are routable in 2 steps on their respective 3D-hypercube. In conclusion, $\pi^{(1)}$ admits at least an upstream permutation which is not a T-permutation. Therefore it is 2-partitionable and it can be routed in at most 4 steps on 4D-hypercube.

5.2 Routing of the Permutations

In this subsection we deal with the routing of the two representatives of the considered non-1-partitionable permutations. We restrict our analysis to $\pi^{(1)}$. The data exchange steps required to do so follows from the analysis which has led to its 2-partitionability. Let's consider the upstream permutation $\Gamma^{(0)}$ and its corresponding downstream ones $\alpha^{(0)}$ and $\beta^{(0)}$. As $\Gamma^{(0)}$ is routable in 2 steps by 1-partitioning let's consider one of the perfect matching of its corresponding bipartite graphs, for example in the dimension 1, we have (1, 0, 3, 2, 12, 7, 14, 5, 10, 11, 8, 9, 4, 15, 6, 13).

Similarly, as $\alpha^{(0)}$ and $\beta^{(0)}$ are routable in 2 steps by 1-partitioning let's consider one of the perfect matching of their corresponding bipartite graphs. For example in the dimension 0, we have respectively (2, 0, 6, 1, 5, 7, 4, 3) and (1, 5, 3, 2, 0, 7, 4, 6).

In restoring the 3D-hypercubes nodes numbers to their original ones in the 4D-hypercube, the perfect matching of $\Gamma^{(0)}$, $\alpha^{(0)}$ and $\beta^{(0)}$ exhibited above lead to the routing table of Table 5.

Table 3. A routing table for $\pi^{(1)}$

Source	**0**	**1**	**2**	**3**	**4**	**5**	**6**	**7**	**8**	**9**	**10**	**11**	**12**	**13**	**14**	**15**
Transition	1	0	3	2	12	7	14	5	10	11	8	9	4	15	6	13
	5	1	7	3	13	6	15	4	11	10	9	8	0	14	2	12
	7	3	5	11	9	2	13	12	10	14	1	15	4	6	0	8
Destination	**3**	**11**	**13**	**15**	**1**	**10**	**9**	**8**	**7**	**6**	**5**	**14**	**0**	**2**	**4**	**12**

By a similar reasoning the routing table of the table 6.

Table 4. A routing table for $\pi^{(2)}$

Source	**0**	**1**	**2**	**3**	**4**	**5**	**6**	**7**	**8**	**9**	**10**	**11**	**12**	**13**	**14**	**15**
Transition	4	3	10	1	5	13	2	7	0	9	11	15	8	12	6	14
	6	11	8	5	7	15	2	3	0	1	9	13	10	14	4	12
	7	15	9	13	3	14	10	11	4	5	1	12	2	6	0	8
Destination	**7**	**15**	**9**	**13**	**3**	**14**	**11**	**10**	**5**	**4**	**1**	**12**	**2**	**6**	**0**	**8**

This study can be summarized by the following proposition.

Proposition 5. Any permutation on $H^{(4)}$ which admits two x-nodes in each dimension is routable in 4 steps.

6 Conclusion and Future Works

This paper promoted the k-partitioning paradigm to contain the algorithmic complexity of the hypercube rearrangeability problem. This paradigm is based on the recursive structure of hypercube interconnection networks that it exploits by reducing each routing step of a given permutation on nD-hypercube to routing two different permutations each one on a distinct (n-k)D-hypercube. For the still open case of 4D-hypercube, non-1-partitionable permutations have been characterized and some classes, the ones with two x-nodes for each dimension, exhibited. For each representative of these latter, an optimal routing strategy has been devised.

To decide of the rearrangeability of 4D-hypercube, this study has to be completed with two other of the same vein. The first concerns how the 1-x-node models of non-1-partitionability may consistently self-combine and mingle with the 2-x-nodes models. The second concerns the structure of the resulting permutations and their 4-steps-routability. Simulations are currently under way on each aspect of these studies. The results let think that the concerned permutations are 2-partitionable too. Consequently 4D-hypercube would be rearrangeable too.

References

1. Draper, J.T., Ghosh, J.: Multipath e-cube algorithms (MECA) for adaptive wormhole routing and broadcasting in k-ary n-cubes. In: International Parallel Processing Symposium, pp. 407–410 (1992)
2. Szymanski, T.: On the permutation capability of a circuit switched hypercube. In: International Conference on Parallel Processing, pp. I-103–I-110. IEEE Computer Society Press, Silver Spring (1989)
3. Lubiw, A.: Counter example to a conjecture of Szymanski on hypercube routing. Informations Processing Letters 35, 57–61 (1990)
4. Zhang, L.: Optimal bounds for matching routing on trees. In: 8th Annual ACM-SIAM Symposium on Discrete Algorithms, New Orleans, pp. 445–453 (1997)
5. Hwang, F., Yao, Y., Grammatikakis, M.: A d-move local permutation routing for d-cube. Discrete Applied Mathematics 72, 199–207 (1997)
6. Hwang, F., Yao, Y., Dasgupta, B.: Some permutation routing algorithms for low dimensional hypercubes. Theoretical Computer Science 270, 111–124 (2002)
7. Vöcking, B.: Almost optimal permutation routing on hypercubes. In: 33rd Annual ACM-Symposium on Theory of Computing, pp. 530–539. ACM Press (2001)
8. Ramras, M.: Routing permutations on a graph. Networks 23, 391–398 (1993)
9. Laing, A.K., Krumme, D.W.: Optimal Permutation Routing for Low-dimensional Hypercubes. Networks 55, 149–167 (2010)
10. Jung, J.P., Sakho, I.: Towards Understanding Optimal MIMD Queueless Routing of Arbitrary Permutations on Hypercubes. Int. J. of Supercomp., doi: 10.1007/s11227-011-0574-8

Bi-level Sensor Planning Optimization Process with Calls to Costly Sub-processes

Frédéric Dambreville

Lab-STICC UMR CNRS 6285, ENSTA Bretagne,
2 rue François Verny, 29806 Brest Cedex 9, France
lcns@fredericdambreville.com
http://www.fredericdambreville.com

Abstract. While there is a variety of approaches and algorithms for optimizing the mission of a sensor, there are much less works which deal with the implementation of several sensors within a human organization. In this case, the management of the sensors is done through at least one human decision layer, and the sensors management as a whole arises as a bi-level optimization process. The following hypotheses are considered as realistic: Sensor handlers of first level plans their sensors by means of elaborated algorithmic tools based on accurate modelling of the environment; Higher level plans the handled sensors according to a global observation mission and on the basis of an approximated model of the environment and submit its plan to a costly assessment by the first level. This problem is related to the domain of experiment design. A generalization of the Efficient Global Optimization method (Jones, Schonlau and Welch) is proposed, based on a rare event simulation approach.

Keywords: sensor management, Efficient Global Optimization, rare event simulation, non-Gaussian/non-linear models, experiment design.

1 Introduction

The main background of this paper is the optimal planning of sensors in the context of an acquisition mission. Typically, the acquisition mission may result in the localisation of a target, with the final purpose of intercepting this target. In this work, we focus especially on dealing with the modelling errors of the sensor planning problem. Then, the question of interest is: *how to spend resources optimally in order to reduce the model errors, and how does that affect the sensor planning problem?*

Sensor planning, especially in order to localize a target, has been thoroughly studied in the literature. First works in this domain track back to the works of Koopman during World War II [1, 2]. This seminal works has been extended in various manner, so as to take into account motion models [3, 4], or reactive behaviours of the target [5, 6]. Sensor planning now deals with the general domain of search and surveillance [7, 8]. The combination of multiple sensors with their constraints is addressed by some works and in various application contexts:

N.T. Nguyen et al. (Eds.): ACIIDS 2014, Part II, LNAI 8398, pp. 382–391, 2014.

optimizing the performance of a sensor network [9, 10]; optimizing the tasks-to-sensors affectation in the context of an intelligence collection process [11–14]. Another major issue in sensor planning is also to maximize the positive effect of subsequent data processing in regards to mission objectives. For example, entropic-based criterion is used in order to take into account optimal post-processing (data fusion) of the collected information[15]. A more direct approach has also been addressed by means of Partially Observable Markov Decision Processes [16, 17]. From this last point of view, sensor planning is clearly related to the domain of robotic.

Thus, a variety of approaches have been investigated for many contexts of the sensor planning. Nevertheless, there is not as much works dedicated to the question of modelling the sensor planning. In their inspiring work[18, 19], Koopman addressed initially this formalisation, priorly to sensor planning problem. Le Cadre studied various practical case of use of the model of Koopman, and deduced related parametrization of the models[20]. Whatever, it appears that a minimal effort is necessary for acquiring a good estimation of the parameters modelling our sensor planning. In the case of a reproducible scenario, it is possible to learn such parameters.

However, there are cases where a prior learning of the parameters is clearly impossible. Such cases hold typically when the planning team has a limited control on the sensors, and relies on sub-processes or on sub-teams in order to implement the sensors or compute their performance parameters. Especially, military organizations are characterized by a hierarchical structure, where decisions are made through at least two human-driven levels. In practice, *accurate models of the sensors and mission contexts are only available to the first, close-to-sensor, level.* The coordination level only works on the basis of approximated models; Accurate parameters learning or acquisition are generally not possible at this level, since the request to sub-processes resources are costly *and* restricted.

The following hypotheses are thus considered as realistic:

- Sensor handlers of the first level plans their sensors by means of elaborated algorithmic tools based on accurate modelling of the environment,
- Higher level plans globally a variety of sensors according to a global observation mission on the basis of an approximated model; at this stage, the global plan has to be confirmed by the first level,
- In order to assess the global plans and to enhance their accuracy, higher level may request the first level. Each request to the first level is costly: it implies communication procedures, as well as the parametrization and execution of the algorithmic tools by the sensors handlers.

This bi-level problem is formalized very generally as the maximization of a function, defined with a *prior* model noise. Each actual evaluation of the function increases the knowledge about the function, and subsequently the efficiency of the maximization. The issue is to optimize the sequence of value to be evaluated, in regards to the evaluation costs: this problem is related to the domain of *experiment design.* Jones, Schonlau and Welch proposed a general method, the

Efficient Global Optimization (EGO) [21, 22], for solving this problem in the case of additive functional Gaussian law. In this work, a generalization of the EGO is proposed, based on a rare event simulation approach. This simulated approach makes possible the implementation of non-Gaussian functional law, and even of simulated functional law. It is applied to the aforementioned bi-level sensor planning.

In the first section 2 of this paper, we propose a general formalisation of the sensor planning with experiment sub-processes, and its description as an abstract problem. In section 3, a rare-event simulation approach is proposed for solving this bi-level sensor planning. Section 4 presents a scenario and numerical results. Section 5 concludes.

2 Sensor Planning with Experiment Sub-processes

2.1 A Model-Noised Sensor Planning

The purpose of this paper is to solve the planning of a set of sensors in order to answer to a set of requests and on the basis of a noisy prior knowledge of the environment. More precisely, the problem is characterized as follows:

- M requests characterized by their locations $z[m]$ with $m = 1 : M$,
- K sensors with indices $k = 1 : K$:
  - Starting/ending location of sensor k is $s[k]$,
  - Maximum autonomy of sensor k (maximum cumulative cost performed by k) is $\gamma[k]$,
- A noisy map μ, which describes the difficulty of the ground: from this map is computed the minimum cost $c[z_1, z_2; \mu]$ for moving from z_1 to z_2 . The exact map is not known from the planner: only the prior law p_μ on μ is known,
- A true map $\widehat{\mu}$ which is unknown to the planner and only known by the monitoring teams which process the sensors,
- Moving constraints: a sensor travels from starting/ending point through some request locations and back to starting/ending point. The trip of sensor k is denoted $\tau[k] = s[k]z[m_1^k] \cdots z[m_{i_k}^k]s[k]$. Moreover the cumulative cost for a trip is smaller than the sensor autonomy:

$$\text{If } \tau[k] \text{ is a valid trip, then } C(\tau[k]; \mu) \leq \gamma[k] \,, \tag{1}$$

where:

$$C(\tau[k]; \mu) = c[s[k], z[m_1]; \mu] + c[z[m_1], z[m_2]; \mu] + \cdots + c[z[m_{i_k}^k], s[k]; \mu] \,. \tag{2}$$

- Criterion to maximize:

Priority 1: Maximize the number of requests performed by the sensors:

$$G[\tau; \mu] = \text{card}\left(\{m = 1 : M \,/ \exists k = 1 : K,\, m \in \tau[k] \text{ and } \tau[k] \text{ is valid}\}\right) \,, \tag{3}$$

Priority 2: Minimize the global cost:

$$C[\tau;\mu] = \sum_{k=1:K} C(\tau[k];\mu)\,. \tag{4}$$

Since the environment is known with noise, the global plan τ does not always fit the actual constraints, and *has to be evaluated by the first planning level.* It is assumed that the proposed plan is truncated by the first planning level, in order to fit the accurate models:

$$\tau[k] \text{ is truncated to: } \widehat{\tau}[k] = s[k]z[m_1^k]\cdots z[m_{n_k}^k]s[k]\,, \tag{5}$$

where:

$$n_k = \arg\max\{n = 1 : i_k/C(s[k]z[m_1^k]\cdots z[m_n^k]s[k];\widehat{\mu}) \le \gamma[k]\}\,. \tag{6}$$

The first planning level also provides an actual evaluation of the (truncated) plan in regards to the true map:

$$\widehat{G}[\tau] = G[\widehat{\tau};\widehat{\mu}] \text{ and } \widehat{C}[\tau] = C[\widehat{\tau};\widehat{\mu}]\,. \tag{7}$$

These information imply an improvement of the knowledge of the map, but it is costly. In order to optimize the plan, the global planner has to tune between the optimization of the plan and the actual evaluation requests.

2.2 The Theoretical Problem

A refined theoretical formulation of the planning problem of section 2.1 is now made with the following meaning: f is the evaluation of the plan, x is a plan, ν is the noisy map. Then, the optimization problem is characterized by means of a noisy criterion function:

$$f : (x,\nu) \in X \times N \mapsto f(x,\nu)\,, \tag{8}$$

where:

$$x \in X \text{ is a parameter to be optimized}\,, \tag{9}$$

$$\nu \in N \text{ is a model noise}\,, \tag{10}$$

$$p_\nu \in \mathcal{P}(N) \text{ is a known probabilistic noise prior}\,. \tag{11}$$

and by an unknown *actual* model noise:

$$\widehat{\nu} \in N \text{ is the actual value of the model noise}\,. \tag{12}$$

The noise on f is a model noise and *it is always possible to evaluate the actual criterion $f(\cdot,\widehat{\nu})$ for any specific actual parameter $\widehat{x}$.* Then the purpose is not to optimize a mean criterion, but rather to choose a *good* sequence of actual parameters $\widehat{x}$ so as to approximate an optimum for the actual criterion $f(\cdot,\widehat{\nu})$. It comes that each evaluation of the actual criterion is costly, while, in comparison,

the evaluation of the modelled criterion $f(\cdot,\nu)$ is considered free for any noise hypothesis ν. Since each evaluation of the actual criterion provide also some knowledge about the actual model noise $\widehat{\nu}$, the issue is to balance optimally between actual evaluation and model-based optimization, so as to find a near optimal solution to the actual criterion.
So as to deal with this problem, Welch proposed[21] the famous Efficient Global Optimization method, which is based on an iterative optimization maximizing the Expected Improvement. More precisely, Welch considered the case of a (spatial) Gaussian noise combined with a linear model, and derived exact computation of the sequence. Our main contribution is to extend Welch algorithm to any cases by means of simulation approaches. Rare event simulation methods are quite instrumental here.

From a general point of view, Welch approach takes the form of the following recursive computation:

[Expected Improvement Maximization (EIM)]
1. Set $n = 0$,
2. Repeat:
   (a) Compute $\widehat{x}_{n+1}$, the next candidate for an actual evaluation:

$$\widehat{x}_{n+1} \in \arg\max_{x\in X} \int_{\nu\in N} p_\nu[n](\nu) f[n](x,\nu)\,\mathrm{d}\nu\;, \tag{13}$$

$$\text{where:}\quad p_\nu[n](\nu) = p_\nu\big(\nu \mid \forall k = 1:n,\; f(\widehat{x}_k,\nu) = \widehat{y}_k\big)\;, \tag{14}$$

$$f[n](x,\nu) = \max\left\{ f(x,\nu), \max_{k=1:n} \widehat{y}_k \right\}\;. \tag{15}$$

   (b) Request the actual evaluation of $\widehat{x}_{n+1}$ $\widehat{y}_{n+1} = f(\widehat{x}_{n+1},\widehat{\nu})$,
   (c) Set $n \leftarrow n+1$,
   until the convergence of $(\widehat{x}_{1:n},\widehat{y}_{1:n})$ is sufficient.
   [Output:] The sequence $(\widehat{x}_{1:n},\widehat{y}_{1:n})$ and model noise estimation $p_\nu[n]$.

The function $f[n] - \max_{k=1:n}\widehat{y}_k$ evaluates the improvement of f at step n. The conditional probability $p_\nu[n]$ is the posterior knowledge of ν obtained after the n first measurements.

From these considerations, it appears that we need to:

- Evaluate the conditional probability $p_\nu[n]$,
- Compute the optimal parameter $\widehat{x}_{n+1}$.

We will see that both tasks are performed by rare event simulations. We will also propose a combined approach for performing these tasks at same time.

3 Rare Event Simulation

A rare event is an event with very small probability. In this section, the following notations are considered:

- Ω is a probabilistic space,
- $p_\omega \in \mathcal{P}(\Omega)$ is a probabilistic distribution on Ω,
- $\phi : \omega \in \Omega \mapsto \mathbb{R}$ is a measurable function,

We are considering events the form $\phi^{-1}([\gamma, +\infty[)$. Then $\phi^{-1}([\gamma, +\infty[)$ is a rare event, if $\int_{\omega\in\Omega} I[\phi(\omega) \geq \gamma] p_\omega(\omega)\, d\omega \ll 1$, where $I[\text{true}] = 1 - I[\text{false}] = 1$. The two following subsection explain how conditional sampling and optimization may be solved by simulating a rare event. Third section explains the *cross-entropy method* as a general process for rare event simulation.

3.1 Conditional Sampling and Rare Event Simulation

It is possible to sample the conditional law $p_\omega(\omega|\phi(\omega) \geq \gamma)$ by sampling the law p_ω and rejecting all samples such that $\phi(\omega) < \gamma$. In the case where $\phi^{-1}([\gamma, +\infty[)$ is a rare event, a direct Monte-Carlo approach is not feasible, but dedicated rare event simulation approaches may be used.

As a specific case, the simulation of conditional law $p_\nu[n]$ may be approximated by $p_\omega(\omega|\phi(\omega) \leq \epsilon)$, where $\omega = \nu$ and:

$$\phi(\nu) = \sum_{k=1:n} \left(f(\widehat{x}_k, \nu) - \widehat{y}_k\right)^2 . \tag{16}$$

In such approach, *the threshold ϵ is a measure of the quality of the conditional sampling.*

3.2 Optimization and Rare Event Simulation

Actually, the set of maximizers of a function could be defined as a limit of rare events. More precisely:

$$\arg\max_{\omega\in\Omega} \phi(\omega) = \bigcap_{\gamma<\max\phi(\Omega)} \phi^{-1}([\gamma, +\infty[) . \tag{17}$$

Then, it comes naturally that the optimization of a function may be obtained by simulating an arbitrarily rare event. Especially, the Cross-Entropy simulation method presented subsequently has been applied to the optimization of functions. Such approaches may be compared to population-based metaheuristics (*eg.* genetic algorithm).

3.3 The Cross-Entropy Method

The cross-entropy method (CE) has been pioneered by Rubinstein [23], and was initially settled for the simulation of rare event. It is based on a recursive importance sampling driven by a family of sampling densities:

$$\pi(\cdot|\Theta) = \left(\pi(\cdot|\theta)\right)_{\theta\in\Theta} \tag{18}$$

Without loss of generality, it is assumed that there is $\theta^o \in \Theta$ such that $p_\omega = \pi(\cdot|\theta^o)$. By denoting N_t the number of samples ω_t^i generated at step t, $R_t : \varphi \in \mathbb{R} \mapsto R_t(\varphi) \in [0,1]$ a selective function for the samples (typically, a quantile-based selection) and $\alpha_t \in]0,1]$ a smoothing parameter, the CE simulation may be defined as follows:

[CE simulation]

1. Set $t = 0$ and $\theta_0 = \theta^o$,
2. Repeat until the convergence is sufficient:
   (a) Generate the samples $\omega_t^i \in \Omega$, for $i \in \{1 : N_t\}$, according to the probabilistic density function (pdf) $\pi(\cdot|\theta_t)$,
   (b) Compute the evaluations $\phi(\omega_t^i)$ of the samples for $i \in \{1 : N_t\}$,
   (c) Compute the selective parameters:
   $$\rho_t[0] = 1 - \alpha_t \; , \tag{19}$$
   $$\rho_t[i] = \alpha_t \frac{R_t(\phi(\omega_t^i))}{\sum_{i=1:N_t} R_t(\phi(\omega_t^i))} \times \frac{\pi(\omega_t^i|\theta^o)}{\pi(\omega_t^i|\theta_t)} \; , \quad \text{for all } i \in \{1 : N_t\} \; , \tag{20}$$
   (d) **[Update]** Update the importance sampler by maximizing the cross-entropy with the selected samples:
   $$\theta_{t+1} \in \arg\max_{\theta\in\Theta} \int_\Omega \left(\rho_t[0]\pi(\omega|\theta_t) + \sum_{i=1:N_t} \rho_t[i]\delta[\omega = \omega_t^i] \right) \log\left(\pi(\omega|\theta)\right) d\omega \; ,$$
   where: $\delta[\omega = \omega_t^i]$ is the Dirac distribution on ω_t^i, (21)
   (e) Set $t \leftarrow t + 1$,

   [Output:] The importance sampler $\pi(\cdot|\theta_t)$ and likelihood ratio $\dfrac{p_\omega}{\pi(\cdot|\theta_t)}$.

The criterion for convergence may be, as in the classical CE [23], achieved when a sufficient ratio of samples is within the rare event $\phi^{-1}([\gamma, +\infty[)$.

4 Practical Implementation and Numerical Results

4.1 Definition of the Maps, Costs and Plans

The map is defined by mapping from a vector parameter to a matrix of practicability level: $Z \in \mathbb{R}^p \mapsto \mu[Z] \in \mathbb{R}_+^{[0,1]\times[0,1]}$. In practice, Z combines the positions of threats and $\mu[Z]$ is computed as distances from these threats. The practicability level $\mu[Z](x)$ infers a local cost to any sensor which moves on position z: the cost $c[z,z']$ of a trajectory $z \to z'$ is obtained by integrating the local cost. The threat vector Z is known with a Gaussian noise, the model noise of our problem.

4.2 Generating Laws for Conditional Sampling and Optimization

The map is sampled as a real-valued vector. The plan is obtained by sampling mission to sensor mapping and a priority order between the mission. Both are easily derived from a real-valued vector by mean of a surjective discrete mapping. Gaussian laws are thus considered for both conditional sampling and optimization. The update step (21) is quite easy for such laws family and is typically derived from the empirical mean and covariance.

4.3 Numerical Results

Settings. The considered scenario is characterized by 5 sensors and 20 missions:

- Sensor position: $(1,1)$, $(1,1)$, $(9,1)$, $(9,1)$, $(5,1)$
- Sensor autonomy: 1, 2, 1, 2, 2
- Missions: 20 missions chosen uniformly on $[1,10] \times [1,10]$

and by a map of threats characterized by 4 threats:

- Theoretical threat position, with noise $\nu_i \sim N(0, \mathrm{diag}(2,2))$:
$$\mu = \{(2,3) + \nu_1, (5,4) + \nu_2, (4,7) + \nu_3, (2,3) + \nu_4\}$$
- Actual threat position: $\widehat{\mu} = \{(1,1), (4,6), (3,7), (1,4)\}$.

The cost inferred by the map is computed as follows:

- The local cost $c[z] = 1/(1 + d(z,T)^2)$ decrease with the distance to the set of threats,
- The cost of a path $z \to z'$ is computed by integrating on the interval $[z, z']$, *ie.* $C[z, z'; \mu] = \int_{\omega \in [z,z']} c[\omega]\, \mathrm{d}\omega$.

A Sequence of Run. First at all, a reference plan is optimized on the basis on the true map:

iter	1	20	40	60	80	100	120	140	160	180	200
opt	11.6	13.3	14.3	14.8	15.1	15.5	16	16.6	17.3	17.7	17.9
$\int$ *samp*	100	$2K$	$4K$	$6K$	$8K$	$10K$	$12K$	$14K$	$16K$	$18K$	$20K$

In this table, *iter* is the number of iteration in the CE algorithm, *opt* is the reached maximal value and $\int$ *samp* is the cumulative number of generated samples. This optimization required 200 iteration and 20000 samples for convergence. The following optimized plan is sampled be means of the sampler $\pi(\cdots|\theta_{200})$ after the last iteration:

Sensor	planned trajectory	corrected trajectory
0	$\{3\}$	$\{\}$
1	$\{2,10,12,16\}$	$\{2,10,12,16\}$
2	$\{4,0,18\}$	$\{4,0\}$
3	$\{13,6,5,15,7,8\}$	$\{13,6,5,15,7,8\}$
4	$\{17,1,9,19,14,11\}$	$\{17,1,9,19,14,11\}$

Now, the following sequence of actual evaluations is obtained by applying the CE-based EIM algorithm:

ϵ	NaN	0.07	0.96	0.72	0.8	2.6
y	15.7	17	17.7	17.7	17.1	NaN
$\widehat{y}$	14.9	13.9	11.9	15.92	15.93	NaN

The quality of the conditional estimation is evaluated by means of the feasible threshold ϵ, as defined in section 3.1, obtained after CE convergence. It is of course undefined at step 0 (there is no conditioning). The estimation is good at step 1, rather good from step 2 to step 4. The estimation is bad at step 5. The optimized actual evaluation $\widehat{y}$ is 14.9 at step 0 and it is 15.93 at step 4, and only 4 evaluation is needed in order to reach 15.92. At step 5, the conditioning is bad, and the CE-based EIM fails.

Obviously, the failure of CE-based EIM is related to the quality of the conditional estimation. In fact, the conditional law is obtained from nonlinear constraints. As a consequence, it is multi-modal and cannot be efficiently sampled by means of Gaussian sampler. In order to enhance the algorithm, it will be necessary to consider mixtures of laws, or multi-modal law by construction.

5 Conclusion

In this paper we considered the problem of planning a set of sensors in the presence of model noises. The following hypotheses were considered: the planner only knows the law of the model noise, but he can request an actual but costly evaluation of a solution. In such case, each actual evaluation of the criterion function increases the knowledge about the model, and subsequently the efficiency of the plan optimization. The issue was to optimize the sequence of value to be evaluated, in regards to the evaluation costs. In our work, we defined a generalization of the Efficient Global Optimization (EGO) algorithm, based on a rare event simulation approach. The results are promising, and the algorithm produced good plans while requesting quite few sub-process calls. It appeared that this optimization was limited by the Gaussian approximation of potentially multi-modal conditional law. Future works will consider mixtures of laws for approximating the conditional law.

References

1. Koopman, B.O.: The theory of search. iii. the optimum distribution of searching effort. Operations Research 5(5), 613–626 (1957)
2. de Guenin, J.: Optimum distribution of effort: An extension of the Koopman basic theory. Operations Research 9, 1–7 (1961)
3. Washburn, A.R.: Search for a moving target: The FAB algorithm. Operations Research 31(4), 739–751 (1983)
4. Brown, S.S.: Optimal search for a moving target in discrete time and space. Operations Research 28(6), 1275–1289 (1980)

5. Iida, K., Hohzaki, R., Furui, S.: A search game for a mobile target with the conditionally deterministic motion defined by paths. Journal of the Operations Research of Japan 39(4), 501–511 (1996)
6. Dambreville, F., Le Cadre, J.-P.: Search game for a moving target with dynamically generated informations. In: Int. Conf. on Information Fusion (Fusion 2002), Annapolis, Maryland, pp. 243–250 (July 2002)
7. Frost, J.R.: Principle of search theory. Technical report, Soza & Company Ltd. (1999)
8. Haley, K.B., Stone, L.D.: Search Theory and Applications. Plenum Press, New York (1980)
9. Chakrabarty, K., Iyengar, S.S., Qi, H., Cho, E.: Grid coverage for surveillance and target location in distributed sensor networks. IEEE Transactions on Computers 51, 1448–1453 (2002)
10. Jayaweera, S.K.: Optimal node placement in decision fusion wireless sensor networks for distributed detection of a randomly-located target. In: IEEE Military Communications Conference, pp. 1–6 (2007)
11. Le Thi, H.A., Nguyen, D.M., Pham, D.T.: A DC programming approach for planning a multisensor multizones search for a target. Computers & Operations Research (July 2012) (online first)
12. Simonin, C., Le Cadre, J.-P., Dambreville, F.: A hierarchical approach for planning a multisensor multizone search for a moving target. Computers and Operations Research 36(7), 2179–2192 (2009)
13. Janez, F.: Optimization method for sensor planning. Aerospace Science and Technologie 11, 310–316 (2007)
14. Nguyen, D.M., Dambreville, F., Toumi, A., Cexus, J.C., Khenchaf, A.: A column generation method for solving the sensor management in an information collection process. Submitted to Optimization (October 2012)
15. Céleste, F., Dambreville, F., Le Cadre, J.-P.: Optimized trajectories for mobile robot with map uncertainty. In: IFAC Symp. on System Identification (SYSID 2009), Saint-Malo, France, pp. 1475–1480 (July 2009)
16. Tremois, O., Le Cadre, J.-P.: Optimal observer trajectory in bearings-only tracking for maneuvering sources. Sonar and Navigation 146(1), 1242–1257 (1997)
17. Dambreville, F.: Cross-entropic learning of a machine for the decision in a partially observable universe. Journal of Global Optimization 37, 541–555 (2007)
18. Koopman, B.O.: The theory of search, part i. kinematic bases. Operations Research 4(5), 324–346 (1956)
19. Koopman, B.O.: The theory of search, part ii. target detection. Operations Research 531, 503–531 (1956)
20. Le Cadre, J.-P.: Approximations de la probabilité de détection d'une cible mobile. In: Actes du Colloque GRETSI, Toulouse (September 2001)
21. Jones, D.R., Schonlau, M.J., Welch, W.J.: Efficient global optimization of expensive black-box function. J. Glob. Optim. 13(4), 455–492 (1998)
22. Marzat, J., Walter, E., Piet-Lahanier, H.: Worst-case global optimization of black-box functions through Kriging and relaxation. J. Glob. Optim. (2012)
23. De Boer, P.T., Kroese, D.P., Mannor, S., Rubinstein, R.Y.: A tutorial on the cross-entropy method. Annals of Operations Research 134 (2002)
24. Hu, J., Fu, M.C., Marjus, S.I.: A model Reference Adaptive Search Method for Global Optimization. Oper. Res. 55, 549–568 (2007, 2008)

DC Programming and DCA for Portfolio Optimization with Linear and Fixed Transaction Costs

Tao Pham Dinh[1], Viet-Nga Pham[2], and Hoai An Le Thi[3]

[1] Laboratory of Mathematics, National Institute for Applied Sciences - Rouen, 76801 Saint Etienne du Rouvray, France

[2] Departement of Mathematics, Hanoi University of Agriculture, Trau Quy, Gia Lam, Ha Noi, Viet Nam

[3] Laboratory of Theorical and Applied Computer Science LITA EA 3097, University of Lorraine, Ile du Saulcy-Metz 57045, France

pham@insa-rouen.fr,
pvnga@hua.edu.vn,
hoai-an.le-thi@univ-lorraine.fr

Abstract. In this work, a single-period portfolio selection problem which consists of minimizing the total transaction cost subject to different types of constraints on feasible portfolios was considered. The total transaction cost function is separable and discontinuous. This problem is nonconvex and very hard to solve. First, by using additional binary variables, we transform it into a mixed zero-one program and then investigate a DC (Difference of Convex functions) programming framework for designing solution methods. Two approaches are developed: DCA (DC Algorithm) and a combination of DCA and Branch & Bound technique. Computational experiments are reported to demonstrate high efficiency and computational inexpensiveness of DCA, which provides good approximate global solutions.

Keywords: portfolio selection, separable transaction cost, DC programming, DCA, Branch and Bound.

1 Introduction

In 1952, Markowitz proposed the mean-variance's model [11] which is a basis for the development of many portfolio selection techniques. Based on this model, many researchers considered portfolio optimization problems taking into account real features such as transaction costs, cardinality constraints, shortselling, buy-in threshold constraints, etc,... The appearance of these features makes these problems nonconvex and very difficult to solve in most of the cases, see e.g., [3–6].

In [10], Lobo et al. studied two alternative models for the problem of single-period portfolio optimization. The first model deals with maximizing the expected return, taking the nonconvex transaction cost into account, and subject to

N.T. Nguyen et al. (Eds.): ACIIDS 2014, Part II, LNAI 8398, pp. 392–402, 2014.

different types of constraints on the feasible portfolios. They proposed a heuristic method for solving this model. The authors confirmed that their heuristic method can be adapted to solve the second model which consists of minimizing separable and discontinuous transaction costs subject to feasible portfolio constraints.

In this work we are interested in the second model introduced in [10]. We consider a slightly modified model where the constraints include shortselling constraints, limit on expected return, limit on variance, and diversification constraints. The considered transaction cost is also assumed to be separable. This function is discontinuous making the program nonconvex.

We investigate a deterministic approach for its solution methods to this problem based on DC programming and DCA. DC programming and DCA were first introduced by Pham Dinh Tao in 1985 in their preliminary form and have been extensively developed since 1994 by Le Thi Hoai An and Pham Dinh Tao in their common works. DCA has been successfully applied to many large-scale nonconvex programs in various domains of applied sciences, to become now classic and popular (see e.g. [7–9, 12, 13, 1] and references therein). We first introduce binary variables and then rewrite the initial problem as an equivalent mixed 0-1 programming problem. By using penalty techniques in DC programming, this program is reformulated as a DC program, which can be handled by a DCA. For evaluating the quality of solutions provided by DCA, we propose a hybridization algorithm that combines DCA and a Branch-and-Bound (BB) scheme. Lower bounds of the optimal value are obtained by solving convex relaxation subproblems.

The rest of the paper is organized as follows. In the next section, we describe the portfolio problem and its mathematical formulation. Section 3 deals with the reformulation of this problem as a mixed zero-one program and the transformation of the resulting problem into a DC program by penalty techniques. A combined DCA-Branch-and-Bound algorithm is also proposed in the same section. Numerical simulations are reported in section 4 and some conclusions are included in the last section.

2 Problem Description and Mathematical Formulation

An investment consists of holding in some or all of n assets. The current holding is $w = (w_i, \ldots, w_n)^T$, the current wealth is then $\mathbf{1}^T w$. The amount transacted in asset i is x_i, with $x_i > 0$ for buying, $x_i < 0$ for selling and $x = (x_1, \ldots, x_n)^T$ is portfolio selection. After transactions, the adjusted portfolio is $w + x$. This adjusted portfolio $w + x$ is held for a fixed period of time. At the end of that period, the return on asset i is the random variable a_i. Let $a = (a_1, \ldots, a_n)$. We assume that $\mathbf{E}(a)$ and $\mathbf{Var}(a)$ are known.

The wealth at the end of the period is a random variable, $W = a^T(w + x)$. We consider the problem of minimizing the total transaction costs subject to portfolio constraints:

$$\min\{\phi(x) : \mathbf{E}(W) \geq r_{\min}, w + x \in \mathcal{S}\} \tag{1}$$

where the portfolio constraint set S can be defined from the following convex constraints:

1. *Shortselling constraints:* Individual bounds s_i on the maximum amount of shortselling allowed on asset i are

$$w_i + x_i \geq -s_i, \quad i = 1, \ldots, n. \tag{2}$$

   If shortselling is not permitted, the s_i are set to zero. Otherwise, $s_i > 0$.
2. *Variance:*The standard deviation of the end period wealth W is constrained to be less than $\sigma_{\max}$ by the convex quadratic inequality

$$\mathbf{Var}(W) = (w + x)^T \Sigma (w + x) \leq \sigma_{\max}^2 \tag{3}$$

   ((3) is a *second-order cone constraint*).
3. *Diversification constraints:* Individual diversification constraints limit the amount invested in each asset i to a maximum of p_i,

$$w_i + x_i \leq p_i, \quad i = 1, \ldots, n. \tag{4}$$

   Alternatively, we can limit the fraction of the total wealth held in each asset,

$$w_i + x_i \leq \lambda_i \mathbf{1}^T (w + x), \quad i = 1, \ldots, n. \tag{5}$$

In this study, the transaction costs $\phi(x)$ is defined by

$$\phi(x) = \sum_{i=1}^{n} \phi_i(x_i), \tag{6}$$

where ϕ_i is the transaction cost function for asset i. Let β_i be the common fixed costs associated with buying and selling asset i. The functions ϕ_i are given by

$$\phi_i(x_i) = \begin{cases} 0, & x_i = 0 \\ \beta_i + \alpha_i |x_i|, & x_i \neq 0. \end{cases} \tag{7}$$

The function ϕ is nonconvex, unless the fixed costs are zero.

We develop below a zero-one approach based on DC programming and DCA for solving (1) with S being defined in (2), (3), (5) and ϕ given in (6), (7).

3 Solving (1) by a Zero-One Approach

Recall that the problem (1) is the following (C being the feasible convex set of portfolio cosntraints)

$$\omega = \min \left\{ \phi(x) = \sum_{i=1}^{n} \phi_i(x_i) : x \in C, \; l_i^0 \leq x_i \leq u_i^0, \forall i = 1, \ldots, n \right\}$$

3.1 A Mixed Zero-One Formulation

Suppose that lower bounds l_i^0 and upper bounds u_i^0 for x_i are calculated. We introduce n binary variables y_i such that $y_i = 0$ if and only if $x_i = 0$, and $y_i = 1$ if $x_i \neq 0$. Then $\phi_i(x_i) = (\beta_i + \alpha_i|x_i|)y_i$, $\forall i = 1, \dots, n$, and we have

$$\forall i,\ l_i^0 \leq x_i \leq u_i^0 \Longleftrightarrow y_i \in \{0,1\} \text{ and } l_i^0 y_i \leq x_i \leq u_i^0 y_i. \tag{8}$$

Each ϕ_i is now considered as a function of two variables x_i, y_i and we can replace $\phi_i(x_i)$ by $\varphi_i(x_i, y_i) := (\beta_i + \alpha_i|x_i|)y_i$.

The mixed 0-1 programming formulation of (1) is

$$\omega = \min \left\{ \varphi(x,y) := \sum_{i=1}^{n} \varphi_i(x_i, y_i) : x \in C, y_i \in \{0,1\}, l_i^0 y_i \leq x_i \leq u_i^0 y_i\ \forall i \right\} \tag{Q01}$$

3.2 DC Programming and DCA for Solving (Q01)

DC Programming and DCA [12, 13, 9, 1]. For a function $\theta : \mathbb{R}^n \to \mathbb{R} \cup \{+\infty\}$ lower semicontinuous proper convex function, the subdifferential of θ at $x_0 \in \text{dom} f := \{x \in \mathbb{R}^n : \theta(x) < +\infty\}$, denoted by $\partial\theta(x_0)$, is defined by

$$\partial\theta(x_0) := \{y \in \mathbb{R}^n : \theta(x) \geq \theta(x_0) + \langle x - x_0, y\rangle, \forall x \in \mathbb{R}^n\},$$

and the conjugate θ^* of θ is

$$\theta^*(y) := \sup\{\langle x, y\rangle - \theta(x)\ :\ x \in \mathbb{R}^n\}, \quad y \in \mathbb{R}^n.$$

A general DC program is that of the form:

$$\alpha = \inf\{F(x) := G(x) - H(x)\ :\ x \in \mathbb{R}^n\} \quad (P_{dc}) \tag{9}$$

where G, H are lower semicontinuous proper convex functions on $\mathbb{R}^n$. Such a function F is called a DC function, and $G - H$ a DC decomposition of F while G and H are the DC components of F. Minimizing the DC function F on a nonempty closed convex C set can be recast into the standard form (9) by changing g in $g + \chi_C$ where χ_C is the indicator function of C defined by $\chi_C(x) = 0$ if $x \in C$, and $+\infty$ otherwise.

DC duality associates a primal DC program with its dual [12, 13, 9]

$$\alpha := \inf\{h^*(y) - g^*(y) : y \in \mathbb{R}^n\} \quad (D_{dc}) \tag{10}$$

which is also a DC program with the same optimal value.

A point x^* is called a *critical point* of $G - H$, or a generalized Karush-Kuhn-Tucker point (KKT) of (9) if

$$\partial H(x^*) \cap \partial G(x^*) \neq \emptyset. \tag{11}$$

DC programming and DCA have been introduced by Pham Dinh Tao in their preliminary form in 1985 and extensively developed by Le Thi Hoai An and Pham Dinh Tao since 1994 (see [12, 13, 9, 1], and the references therein). These theoretical and algorithmic tools have been successfully applied by researchers and practitioners to model and solve their nonconvex programs from different filds of Applied Sciences, especially in the large scale setting.

Based on local optimality conditions and duality in DC programming, DCA consists in constructing two sequences $\{x^k\}$ and $\{y^k\}$ (of trial solutions of the primal and dual DC programs respectively) which are improved at each iteration:

Generic DCA scheme

Initialization: Let $x^0 \in \mathbb{R}^n$, $k \longleftarrow 0$.
Repeat
 – Calculate $y^k \in \partial H(x^k)$
 – Calculate $x^{k+1} \in \partial G^*(y^k) = \arg\min\{G(x) - \langle x, y^k\rangle : x \in \mathbb{R}^n\} \quad (P_k)$
 – $k \longleftarrow k+1$
Until convergence of $\{x^k\}$.

It is worth noting that DCA works with the convex DC components G and H but not the DC function F itself (see [7, 9, 12, 13]). Moreover, a DC function F *has infinitely many DC decompositions* which have crucial impacts on the performance (speed of convergence, robustness, efficiency, globality of computed solutions,...) of DCA.

Convergence properties of DCA and its theoretical basis can be found in [7, 9, 12]. For instant, it is important to mention that (for the sake of simplicity, we omit here the similar dual part)

- DCA is a descent method (the sequences $\{G(x^k) - H(x^k)\}$ is decreasing) without linesearch but with global convergence
- If the optimal value α of problem (9) is finite and the sequence $\{x^k\}$ is bounded then every limit point x^* of the sequence $\{x^k\}$ is a critical point of $G - H$.
- DCA has a linear convergence for general DC programs.
- DCA has a finite convergence for polyhedral DC programs (when either G or H is polyhedral convex).

The next subsection is devoted to the development of DCA applied on $(Q01)$.

DC Algorithm for the Problem $(Q01)$. Firstly we claim that the objective function of $(Q01)$ is a DC function on $\mathbb{R}^n \times [0,1]^n$. Indeed, we have

$$\varphi_i(x_i, y_i) = \left(\frac{\alpha_i}{2}(|x_i| + y_i)^2\right) - \left(\frac{\alpha_i}{2}(x_i^2 + y_i^2) - \beta_i y_i\right)$$

Let $\theta_i(x_i, y_i) = \frac{\alpha_i}{2}(|x_i| + y_i)^2$ and $\kappa_i(x_i, y_i) = \frac{\alpha_i}{2}(x_i^2 + y_i^2) - \beta_i y_i$, then θ_i, κ_i are convex functions on $\mathbb{R} \times [0,1]$. Let $\theta(x,y) = \sum_{i=1}^{n} \theta_i(x_i, y_i)$ and $\kappa(x,y) = \sum_{i=1}^{n} \kappa_i(x_i, y_i)$, then $\theta - \kappa$ is a DC decomposition of φ.

Furthermore, let

$$A := \{(x,y) \in \mathbb{R}^n \times [0,1]^n : x \in C,\ l_i^0 y_i \le x_i \le u_i^0 y_i,\ \forall i = 1,\dots,n\}$$

and define the function $p : \mathbb{R}^n \times \mathbb{R}^n \longrightarrow \mathbb{R}$, $p(x,y) := \sum_{i=1}^{n} y_i(1-y_i)$, then p is a nonnegative concave (quadratic) function on the convex set and the feasible set of $(Q01)$ is

$$\{(x,y) \in A : y_i \in \{0,1\}, \forall i\} = \{(x,y) \in A : p(x,y) = 0\} = \{(x,y) \in A : p(x,y) \le 0\}.$$

Thus $(Q01)$ becomes

$$\min\{\varphi(x,y) = \theta(x,y) - \kappa(x,y) : (x,y) \in A,\ p(x,y) \le 0\}. \tag{12}$$

Note that the objective function of (12) is DC and (12) contains a reverse convex constraint: $p(x,y) \le 0$. In order to overcome these difficulties when solving (12), we propose to use penalty techniques in DC programming [14, 2].

With a scalar $t > 0$, we define a penalty function F_t of φ on $\mathbb{R}^n \times \mathbb{R}^n$ by

$$F_t(x,y) = \varphi(x,y) + tp(x,y) = \theta(x,y) - (\kappa(x,y) - tp(x,y)) \tag{13}$$

then the penalized problem for (12) can be

$$\min\{F_t(x,y) = \theta(x,y) - (\kappa(x,y) - tp(x,y)) : (x,y) \in A\} \tag{14}$$

or in a natural DC form

$$\min\{G(x,y) - H(x,y) : (x,y) \in \mathbb{R}^n \times \mathbb{R}^n\}$$

where $G(x,y) = \theta(x,y) + \chi_A(x,y)$ and $H(x,y) = \kappa(x,y) - tp(x,y)$.

Applying DCA to solve (14) leads to compute $(z^k, v^k) \in \partial H(x^k, y^k)$ at each iteration k, and then solve the convex program

$$\min\{G(x,y) - \langle (z^k, v^k), (x,y)\rangle : (x,y) \in \mathbb{R}^n \times \mathbb{R}^n\}$$

to obtain $(x^{k+1}, y^{k+1}) \in \partial G^*(z^k, v^k)$. This convex program is equivalent to the following

$$\min\{\theta(x,y) - \langle (z^k, v^k), (x,y)\rangle : (x,y) \in A\} \tag{15}$$

A subgradient of H, $(z^k, v^k) \in \partial H(x^k, y^k)$, is computed by

$$(z^k, v^k) \in \partial H(x^k, y^k) \Longleftrightarrow \begin{cases} z_i^k = \alpha_i x_i^k,\ \forall i = 1,\dots,n, \\ v_i^k = (\alpha_i + 2t) y_i^k - (\beta_i + t),\ \forall i = 1,\dots,n. \end{cases} \tag{16}$$

We describe the DCA applied on (14) as follows.

Algorithm 1 (DCA for 0-1 model).

- **Initialization:**
  Let $(x^0, y^0) \in \mathbb{R}^n \times [0,1]^n$ and ε be a small enough positive number.
  Iteration $k \longleftarrow 0$.

- **Repeat:**
  - ◇ Calculate $z_i^k = \alpha_i x_i^k$ and $v_i^k = (\alpha_i + 2t)y_i^k - (\beta_i + t)$, $\forall i = 1, \ldots, n$.
  - ◇ Solve (15) to obtain (x^{k+1}, y^{k+1}).
  - ◇ $k \longleftarrow k+1$
- **Until:** $|F(x^{k+1}, y^{k+1}) - F(x^k, y^k)| \leq \varepsilon$ or $\|x^{k+1} - x^k\| + \|y^{k+1} - y^k\| \leq \varepsilon$.

3.3 A Combined DCA-Branch and Bound Algorithm

To evaluate the globality of solutions computed by DCA in Algorithm 1, we propose to solve $(Q01)$ or its equivalent problem (12) by a BB algorithm. The subdivision of the combined algorithm is performed in the way that either $y_i = 0$ or $y_i = 1$.

For lower bounding, we solve the relaxation problem $(R_k cp)$ of ϕ on $C \cap R_k$ at iteration k in the BB scheme.

Suppose that at iteration k in the BB scheme, we have two sets of indices $I_k, J_k \subset \{1, \ldots, n\}$ such that $y_i = 0 \ \forall i \in I_k$, $y_j = 1 \ \forall j \in J_k$. The corresponding DC objective function at this iteration is

$$\varphi^k(x, y) = \sum_{i \notin I_k \cup J_k} \varphi_i(x_i, y_i) + \sum_{j \in J_k} (\alpha_j |x_j| + \beta_j) \tag{17}$$

The penalty function of φ^k at iteration k, with $t > 0$, $F_t^k(x, y) = \varphi^k(x, y) + tp(x, y)$ is a DC function with DC components

$$G^k(x, y) = \sum_{i \notin I_k \cup J_k} \theta_i(x_i, y_i) + \sum_{j \in J_k} (\alpha_j |x_j| + \beta_j)$$

$$H^k(x, y) = \sum_{i \notin I_k \cup J_k} (\kappa_i(x_i, y_i) + t y_i (y_i - 1))$$

The penalty problem of (12) at iteration k can be given by

$$\min\{F_t^k(x, y) : (x, y) \in A, \ y_i = 0, \ \forall i \in I_k, \ y_j = 1, \ \forall j \in J_k\} \tag{18}$$

Solving (18) by DCA leads to determine two sequences $\{(x^s, y^s)\}$, $\{(z^s, v^s)\}$ in $\mathbb{R}^n \times \mathbb{R}^n$ satisfying

$$(z^s, v^s) \in \partial H^k(x^s, y^s) \Longleftrightarrow \begin{cases} z_i^s = \alpha_i x_i^k, \ \text{if } i \notin I_k \cup J_k, \\ z_i^s = 0, \ \text{if } i \in I_k \cup J_k, \\ v_i^s = (\alpha_i + 2t) y_i^k - (\beta_i + t), \ \text{if } i \notin I_k \cup J_k, \\ v_i^s = 0, \ \text{if } i \in I_k \cup J_k, \end{cases} \tag{19}$$

and

$$(x^{s+1}, y^{s+1}) \in \partial (G^k)^*(z^s, v^s) \Longleftrightarrow (x^{s+1}, y^{s+1}) \text{ solves the convex program}$$

$$\min\{G^k(x,y) - \langle (z^s, v^s), (x,y)\rangle \ : \ (x,y) \in A, y_i = 0, i \in I_k, y_j = 1, j \in J_k\} \quad (20)$$

If solving the penalty problem (18) by DCA provides $(\overline{x}, \overline{y})$ as a solution then $\varphi(\overline{x}, \overline{y})$ is an upper bound for ω.

Therefore, we can describe below a combined algorithm for solving (12).

Algorithm 2 (DCA-BB for 0-1 model).

- **Initialization:**
  Compute the first bounds $[l_i^0, u_i^0]$ for variables x_i and the first rectangle $R_0 = \prod_{i=1}^{n} [l_i^0, u_i^0]$.
  $I_0 := \emptyset$, $J_0 := \emptyset$, iteration $k \longleftarrow 0$.
  The optimal value μ_0 of the relaxation problem (R_0cp) of (P) provides the first lower bound for ω.
- **Iteration k:**
  We apply DCA described in Algorithm 1 inside BB algorithm. DCA is used in the first time at the end of the first iteration of BB scheme (iteration 0). And then, in the BB process, we restart DCA when the current upper bound is updated. More precisely, the DCA used inside the BB algorithm is carried on as follows:

  1. Construct the current relaxation problem (R_kcp) of $(Q01)$ at the node k. Solve (R_kcp) to obtain a lower bound for ω at this node and a solution $x^{R_k} \in C \cap R_0$.
  2. If $\phi(x^{R_k})$ is smaller than the current upper bound then construct the penalty problem of the form (18) and launch DCA for solving it.
  3. Let $(\widetilde{x}^1, \widetilde{y}^1)$ be the solution obtained by DCA. Let ϵ be a sufficiently small positive number. For each $i \notin I_k \cup J_k$, if $\widetilde{y}_i^1 \leq \epsilon$ then add the constraints $x_i = 0$ into the set of constraints of (R_kcp) and add this index i into I_k. Name the new problem (P'_{re}).
  4. Solve (P'_{re}).
     - If (P'_{re}) provides a solution then launch DCA for solving the corresponding penalty problem of $(Q01)$ (constructed with the new I_k) to obtain $(\widetilde{x}^2, \widetilde{y}^2)$. Update the upper bound, the best current solution known so far by comparing $\phi(x^{R_k})$ with $\phi(\widetilde{x}^1)$ and $\phi(\widetilde{x}^2)$ then return to the BB algorithm.
     - If (P'_{re}) is infeasible, update the upper bound, the best current solution and return to the BB algorithm.
  5. Continue the BB process until the convergence.

Numerical experiments in the next section show the efficiency of the proposed algorithms.

4 Computational Results

The algorithms are coded in C and run on a PC equipped with Window 7 Intel(R) Core(TM) i5-2540M CPU 2.60GHz, 8.00 Go RAM. To solve the convex programs, we use CPLEX solver version 12.4.

We have tested the proposed algorithms on the set of data used in [10]. The portfolio selection consists of $(n-1)$ risky and one riskless assets (the riskless asset corresponds to the nth-asset in the portfolio decision). The mean and covariance of $(n-1)$ risky assets were estimated from daily closing prices of S&P 500 stocks (for the tests with $n \leq 101$, we chose the first $(n-1)$ stocks, alphabetically by ticker, with a full year of data from January 9, 1998 to January 8, 1999; for $n > 101$, the first $(n-1)$ stocks were chosen with the data from January 01, 2005 to January 01, 2007). The mean of riskless asset is set to be 0.1.

The results presented in Table 1 have been computed using the values

$$\begin{aligned}
&w_i = 1/n, \ \forall i = 1, \dots, n \\
&\alpha_i = 0.01, \ \forall i = 1, \dots, n-1, \quad \alpha_n = 0 \\
&\beta_i = 0.1/(n-1), \ \forall i = 1, \dots, n-1, \quad \beta_n = 0 \\
&s_i = 5\beta_i, \ \forall i = 1, \dots, n-1, \quad s_n = 0.5 \\
&\lambda_i = 0.5, \ \forall i = 1, \dots, n.
\end{aligned}$$

We have tested Algorithm 1 (denoted DCA for 0-1 model), Algorithm 2 (denoted DCA-BB for 0-1model) and the BB algorithm without DCA (denoted BB). The tolerance ε for stopping DCA is equal to 10^{-8}. The stopping criteria of the BB algorithm (with DCA or without DCA) is either the CPU time (in seconds) is greater than 1 hour or the difference between the best upper bound and the best lower bound is smaller than $\epsilon := 10^{-8}$. In our numerical tests, the hybrid algorithm and the BB algorithm always provide an ϵ-optimal solution. In Table 1, the number of iterations for each algorithm as well as the ϵ-optimal values found by BB, "DCA-BB for 0-1 model" and the CPU time are reported.

Comments on the Numerical Results. From numerical results we observed that

- "DCA for 0-1 model" provides usually a good approximation of the optimal solution within a very short running time (less than 5 seconds) and the number of iterations "DCA for 0-1 model" is less than 5.
- The combined algorithm "DCA-BB for 0-1 model" (in which the number of restarting DCA is less than 10) provides the same optimal values in comparison with the classical BB algorithm (BB) within a bit larger CPU time when $n \leq 161$. However, in the last four cases, when $n = 171, 181, 191, 201$, respectively, we can observe the performance of DCA when combining it with BB: it greatly reduces the number of iterations of the BB process and the computation time of BB (without DCA) is really more expensive.

Table 1. Minimize Transaction costs

	BB			DCA-BB for 01 model				DCA for 01 model		
n	iter	Opt.val	CPU	iter	rest.	Opt.val	CPU	iter	valDCA	CPU
11	51	0.104149	2.699	50	4	0.104149	2.684	2	0.104544	0.140
21	35	0.089724	1.077	84	4	0.089724	3.447	4	0.089742	0.593
31	31	0.086269	1.685	30	2	0.086269	1.981	4	0.087762	0.390
41	41	0.085362	3.260	40	3	0.085362	4.557	4	0.086853	0.889
51	51	0.084845	7.005	50	3	0.084845	7.035	3	0.086337	0.514
61	70	0.084487	11.732	100	2	0.084487	15.418	3	0.085977	0.983
71	159	0.084228	27.877	242	4	0.084228	35.100	4	0.085719	0.827
81	181	0.083878	35.631	428	4	0.083878	76.767	3	0.085530	1.029
91	3803	0.083601	956.046	3802	4	0.083601	786.803	3	0.085382	1.045
101	4877	0.083683	1573.743	4876	3	0.083683	1109.761	4	0.085208	1.310
111	189	0.084263	51.995	1116	3	0.084263	305.000	2	0.085251	1.560
121	316	0.084130	106.190	1403	4	0.084130	430.733	2	0.085169	1.794
131	298	0.083862	101.924	561	3	0.083862	225.102	2	0.085097	2.153
141	495	0.083775	181.563	627	5	0.083775	272.205	2	0.085039	2.262
151	364	0.083678	139.508	666	4	0.083678	297.383	2	0.084989	2.793
161	612	0.083593	281.261	628	5	0.083593	329.627	2	0.084945	2.730
171	721	0.083516	519.326	661	5	0.083516	389.767	2	0.084905	2.715
181	986	0.083447	788.477	874	5	0.083447	590.711	2	0.084869	3.291
191	1135	0.083386	982.932	915	6	0.083386	680.646	2	0.084837	4.337
201	1643	0.083329	1781.377	1348	5	0.083329	981.273	2	0.084807	4.103

5 Conclusion

In this work, we have proposed solution methods for solving a hard portfolio selection problem where the total transaction cost function is nonconvex, discontinuous. By introducing binary variables, the problem can be rewritten as a nonconvex mixed zero-one program, and then transformed into a DC program in the continuous framework due to penalty techniques in DC programming. We developed DCA, BB and the combined DCA-BB for solving the resulting DC program. In our numerical simulations with real data, CPLEX solver was used for minimizing convex quadratically constrained quadratic programs.

The computational aspects of the proposed approaches show high efficiency, computational inexpensiveness of DCA, which provides good approximate global solutions, and also the positive influence of DCA on BB algorithm, especially for large-scale problems.

References

1. Le Thi, H.A.: DC Programming and DCA, `http://lita.sciences.univ-metz.fr/~lethi`
2. Le Thi, H.A., Pham Dinh, T., Huynh, V.N.: Exact penalty and Error Bounds in DC programming. Journal of Global Optimization 52(3), 509–535 (2012)

3. Kellerer, H., Mansini, R., Speranza, M.G.: Selecting Portfolios with Fixed Costs and Minimum Transaction Lots. Annals of Operations Research 99, 287–304 (2000)
4. Konno, H., Wijayanayake, A.: Mean-absolute deviation portfolio optimization model under transaction costs. Journal of the Operation Research Society of Japan 42(4), 422–435 (1999)
5. Konno, H., Wijayanayake, A.: Portfolio optimization problems under concave transaction costs and minimal transaction unit constraints. Mathematical Programming 89(B), 233–250 (2001)
6. Konno, H., Yamamoto, R.: Global Optimization Versus Integer Programming in Portfolio Optimization under Nonconvex Transaction Costs. Journal of Global Optimization 32, 207–219 (2005)
7. Le Thi, H.A.: Contribution à l'optimisation non convexe et l'optimisation globale: théorie, algorithmes et applications. Habilitation à Diriger de Recherches. Université de Rouen, France (1997)
8. Le Thi, H.A., Pham Dinh, T.: A continuous approach for globally solving linearly constrained quadratic zero-one programming problems. Optimization 45(1-2), 12–28 (2001)
9. Le Thi, H.A., Pham Dinh, T.: The DC (Difference of convex functions) Programming and DCA Revisited with DC Models of Real World Nonconvex Optimization Problems. Annals of Operations Research 133, 23–46 (2005)
10. Lobo, M.S., Fazel, M., Boyd, S.: Portfolio optimization with linear and fixed transaction costs. Annals of Operations Research 157, 341–365 (2007)
11. Markowitz, H.: Portfolio selection. The Journal of Finance 7(1), 77–91 (1952)
12. Pham Dinh, T., Le Thi, H.A.: Convex analysis approach to d.c. programming: Theory, Algorithms and Applications. Acta Mathematica Vietnamica (dedicated to Professor Hoang Tuy on the occasion of his 70th birthday) 22(1), 289–355 (1997)
13. Pham Dinh, T., Le Thi, H.A.: A d.c. optimazation algorithm for solving the trust region subproblem. SIAM Journal of Optimization 8(2), 476–505 (1998)
14. Le Thi, H.A., Pham Dinh, T., Le, D.M.: Exact penalty in DC programming. Vietnam Journal of Mathematics 27(2), 169–178 (1999)
15. Rockafellar, R.T.: Convex Analysis. Princeton University Press, Princeton (1970)

A Filter Based Feature Selection Approach in MSVM Using DCA and Its Application in Network Intrusion Detection

Hoai An Le Thi, Anh Vu Le, Xuan Thanh Vo, and Ahmed Zidna

Laboratory of Theoretical and Applied Computer Science
UFR MIM, University of Lorraine, Ile du Saulcy, 57045 Metz, France
{hoai-an.le-thi,anh-vu.le,xuan-thanh.vo,ahmed.zidna}@univ-lorraine.fr

Abstract. We develop a filter based feature selection approach in Multi-classification by optimizing the so called Generic Feature Selection (GeFS) measure and then using Multi Support Vector Machine (MSVM) classifiers. The problem is first formulated as a polynomial mixed 0-1 fractional programming and then equivalently transformed into a mixed 0-1 linear programming (M01LP) problem. DCA (Difference of Convex functions Algorithm), an innovative approach in nonconvex programming framework, is investigated to solve the M01LP problem. The proposed algorithm is applied on Intrusion Detection Systems (IDSs) and experiments are conducted through the benchmark KDD Cup 1999 dataset which contains millions of connection records audited and includes a wide variety of intrusions simulated in a military network environment. We compare our method with an embedded based method for MSVM using $l_2 - l_0$ regularizer. Preliminary numerical results show that the proposed algorithm is comparable with $l_2 - l_0$ regularizer MSVM on the ability of classification but requires less computation.

Keywords: Feature Selection, Filter approach, Generic Feature Selection measure, MSVM, DCA.

1 Introduction

Feature selection consists of choosing a subset of available features that capture the relevant properties of the data. In supervised pattern classification, a good choice of features is fundamental for building compact and accurate classifiers. Feature selection is often applied to high-dimensional data prior to classification learning. The main goal is to select a subset of features of a given data set while preserving or improving the discriminative ability of a classifier. Generally speaking, feature selection can be classified into three categories: filter approaches, wrapper approaches, and embedded approaches.

Wrapper methods exploit a machine learning algorithm to evaluate the usefulness of features. Filter methods rank the features according to some discrimination measure and select features having higher ranks without using any learning algorithm (it utilizes the underlying characteristics of the training data

N.T. Nguyen et al. (Eds.): ACIIDS 2014, Part II, LNAI 8398, pp. 403–413, 2014.

to evaluate the relevance of the features or feature set by some independent measures such as distance measure, correlation measures, consistency measures [12]). The wrapper approach is generally considered to produce better feature subsets but runs much more slowly than a filter. In contrast to the filter and wrapper approaches, the embedded approach of feature selection does not separate the learning from the feature selection part. It integrates the selection of features in the model building. For instance, for feature selection in classification, an embedded method uses a machine learning algorithm to search a classifier that uses as few features as possible while a filter method selects the features by optimizing a measure criterion and then finds a classifier defined on the selected features.

In this paper we propose a filter approach for feature selection in MSVM. The starting point of our work is the GeFS measure studied in [2] that is a common model of several feature selection measures, for instance the correlation-feature-selection (CFS) measure [6] and the minimal-redundancy-maximal-relevance (mRMR) measure [13]. Considering the GeFS measure, we follow [2] to formulate the feature selection problem as a polynomial mixed 0-1 fractional programming (PM01FP) problem. Based on the Chang's method [2,3] this PM01FP problem is reformulated equivalently as a mixed 0-1 linear programming (M01LP) problem. We follow [4] to refine the last model in a MP01LP problem with smaller size. Our main contributions concern with the use of DCA for solving the M01LP problem and the application of the proposed method for Intrusion Detection Systems (IDSs) which play a vital role of detecting various kinds of attacks. Due to the computational efficiency, filter approaches are usually utilized to select features from high-dimensional data sets, such as IDSs. A major challenge in the IDS feature selection process is to choose appropriate measures that can precisely determine the relevance and the relation between features of a given data set. Hence the correlation feature selection (CFS) measure seems to be suitable and is often used to feature selection and classification in IDSs. Here we consider the GeFS measure whose CFS is an instance. We apply our algorithm for optimizing the GeFS measure, from which a subset of features is selected. Finally, l_2-MSVM is used to classification on the selected features. Experiments are conducted through the benchmark KDD Cup 1999 dataset which contains millions of connection records audited and includes a wide variety of intrusions simulated in a military network environment. We compare our method with an embedded based method for MSVM using $l_2 - l_0$ regularizer [9]. Preliminary numerical results show that the proposed algorithm is comparable with $l_2 - l_0$ -MSVM but requires less computation.

2 Optimizing Generic Feature-Selection Measure: A PM01FP Formulation

2.1 Generic Feature-Selection Measure and the Feature Selection Problem

Let $x = (x_1, ..., x_n)$ be the vector with binary values x_i indicating the appearance ($x_i = 1$) or the absence ($x_i = 0$) of the feature f_i, for $i = 1, \ldots, n$. A generic feature-selection measure used in the filter model is a function $GeFS(x)$ defined as ([4])

$$GeFS(x) = \frac{a_0 + \sum_{i=1}^{n} A_i(x)x_i}{b_0 + \sum_{i=1}^{n} B_i(x)x_i}, x \in \{0,1\}^n. \quad (1)$$

where a_0, b_0 are constants; $A_i(x), B_i(x)$ are affine functions of variables $x_1, \ldots, x_n$:

$$A_i(x) = a_{i0} + \sum_{j=1}^{n} a_{ij}x_j, \quad B_i(x) = b_{i0} + \sum_{j=1}^{n} b_{ij}x_j,$$

The feature selection problem is to find $x \in \{0,1\}^n$ that maximizes the function $GeFS(x)$ ([4]):

$$\max_{x\in\{0,1\}^n} \quad GeFS(x) = \frac{a_0 + \sum_{i=1}^{n} A_i(x)x_i}{b_0 + \sum_{i=1}^{n} B_i(x)x_i}, \quad (2)$$
$$s.t. \ b_0 + \sum_{i=1}^{n} B_i(x)x_i > 0.$$

Correlation Feature Selection Measure is a Special Case of GeFS: The Correlation Feature Selection (CFS) measure evaluates subsets of features on the basis of the following hypothesis: *"Good feature subsets contain features highly correlated with the classification, yet uncorrelated to each other"* [6]. The following quation gives the merit of a feature subset S consisting of k features:

$$Merit_{S_k} = \frac{k\overline{r_{cf}}}{\sqrt{k + k(k-1)\overline{r_{ff}}}}$$

Here, $\overline{r_{cf}}$ is the average value of all feature-classification correlations, and $\overline{r_{ff}}$ is the average value of all feature-feature correlations. The CFS criterion is defined as follows:

$$\max_{S_k} \left[\frac{r_{cf_1} + r_{cf_2} + \ldots + r_{cf_k}}{\sqrt{k + 2(r_{f_1f_2} + .. + r_{f_if_j} + .. + r_{f_kf_1})}} \right]. \quad (3)$$

This problem takes the form

$$\max_{x\in\{0,1\}^n} \left[\frac{(\sum_{i=1}^{n} a_i x_i)^2}{\sum_{i=1}^{n} x_i + \sum_{i\neq j} 2b_{ij}x_ix_j} \right], \quad (4)$$

where $a_i := r_{cf_i}$ and $b_{ij} := r_{f_if_j}$.
It is obvious that the CFS measure is an instance of the GeFS measure. We denote this measure by $GeFS_{CFS}$.

2.2 A Mixed 0-1 Linear Programming Formulation

By introducing an additional positive variable, denoted by y, one consider the following problem equivalent to (2):

$$\min_{x\in\{0,1\}^n} \quad -a_0 y - \sum_{i=1}^{n} A_i(x) x_i y \tag{5a}$$

$$s.t.\ b_0 y + \sum_{i=1}^{n} B_i(x) x_i y = 1,\ y > 0. \tag{5b}$$

Proposition 1. *Suppose that $m \leq a(x,\xi) \leq M$ and $x \in \{0,1\}$. Then, the mixed term $a(x,\xi)x$ can be represented via a continuous variable z by one of two following ways*

$$\begin{array}{ll} \min_z & z \\ s.t. & \begin{cases} z \geq mx, \\ z \geq M(x-1) + a(x,\xi), \end{cases} \end{array} \quad \text{and} \quad \begin{cases} mx \leq z \leq Mx, \\ z \leq m(x-1) + a(x,\xi), \\ z \geq M(x-1) + a(x,\xi). \end{cases} \tag{6}$$

Proposition 2. *A term $-A_i(x)x_i y$ from* (5a) *and a term $B_i(x)x_i y$ from* (5b) *can be represented via a continuous variables z_i and v_i as follows*

$$\begin{array}{ll} \min_{z_i} & z_i \\ s.t. & \begin{cases} z_i \geq -Cx_i, \\ z_i \geq C(x_i - 1) - A_i(x)y, \end{cases} \end{array} \quad \text{and} \quad \begin{cases} -Cx_i \leq v_i \leq Cx_i, \\ v_i \leq C(1-x_i) + B_i(x)y, \\ v_i \geq C(x_i - 1) + B_i(x)y, \end{cases} \tag{7}$$

where C is a large positive number.

We substitute each term $x_j y, (j = 1, \ldots, n)$ that will appear in (7) by new variables t_j satisfying constraints from Proposition 1. By applying these technique, the problem (5) is equivalent to a mixed 0-1 linear program given below

$$\begin{aligned} \min_{x\in\{0,1\}^n, y, z, v, t} \quad & -a_0 y - \sum_{i=1}^{n} z_i \\ s.t. \quad & y > 0, \text{ and for all } i = 1, \ldots, n \\ & t_i \geq C(x_i - 1) + y,\ t_i \leq y,\ 0 \leq t_i \leq Cx_i, \\ & z_i \geq -Cx_i,\ z_i \geq C(x_i - 1) - a_{i0} y - \sum_{j=1}^{n} a_{ij} t_j, \\ & -Cx_i \leq v_i \leq Cx_i, \\ & C(x_i - 1) \leq v_i - b_{i0} y - \sum_{j=1}^{n} b_{ij} t_j \leq C(1 - x_i). \end{aligned} \tag{8}$$

The total number of variables for the M01LP problem will be $4n+1$. Therefore, the number of constraints on these variables will also be a linear function of n. As we mentioned above, with Chang's method [2,3] the number of variables and constraints depends on the square of n. Thus this new method actually improves Chang's method by reducing the size of the problem.

3 Solving the M01LP Problem by DCA

3.1 DC Programming and DCA

DC Programming and DCA were introduced by Pham Dinh Tao in their preliminary form in 1985. These theoretical and algorithmic tools are extensively developed by Le Thi Hoai An and Pham Dinh Tao since 1994 to become now classic and increasingly popular. DCA is a continuous primal dual subgradient approach based on local optimality and duality in DC programming for solving standard DC programs which take the form

$$(P_{dc}) \quad \alpha = \inf\{f(x) := g(x) - h(x) : \, x \in \mathbb{R}^n\}, \tag{9}$$

with $g, h \in \Gamma_0(\mathbb{R}^n)$. Such a function f is called a DC function, and $g - h$, a DC decomposition of f, while the convex functions g and h are DC components of f.

The main idea of DCA is simple: each iteration of DCA approximates the convex function h by its affine minorant defined by $y^k \in \partial h(x^k)$, and solves the resulting convex program.

$$\begin{aligned} y^k &\in \partial h(x^k) \\ x^{k+1} &\in \arg\min_{x \in \mathbb{R}^n} \{g(x) - h(x^k) - \langle x - x^k, y^k \rangle\}. \; (P_k) \end{aligned}$$

The construction of DCA involves DC components g and h but not the function f itself. Moreover, a DC function $\mathcal{F}$ *has infinitely many DC decompositions which have crucial impacts on the qualities* (speed of convergence, robustness, efficiency, globality of computed solutions,...) of DCA. Hence, for a DC program, each DC decomposition corresponds to a different version of DCA. DCA is so a philosophy rather than an algorithm. For each problem we can design a family of DCA based algorithms. To the best of our knowledge, DCA is actually one of the rare algorithms for nonsmooth nonconvex programming which allow to solve large-scale DC programs. DCA was successfully applied to a lot of different and various nonconvex optimization problems to which it quite often gave global solutions and proved to be more robust and more efficient than related standard methods ([14,15,16] and the list of reference in [7]).

3.2 From Combinatorial Optimization to DC Programming

Let $D \neq \emptyset$ be a bounded polyhedral convex set in $\mathbb{R}^n$ and let $J \subset \{1, \ldots, n\}$. Consider now the M01LP problem in the form

$$(M01LP) \quad \min\{\langle c, x \rangle : x \in D, x_i \in \{0,1\} \forall i \in J\}.$$

For solving (M01LP) by DCA we first reformulate it in a continuous optimization problem.

Let $K := \{x \in D : 0 \leq x_i \leq 1, \forall i \in J\}$ and define $p(x) = \sum_{i \in J} x_i(1 - x_i)$. Clearly, p is a concave function with nonnegative values on K and

$$\{x \in D : \, x_i \in \{0,1\}\} = \{x \in K : p(x) = 0\} = \{x \in K : p(x) \leq 0\},$$

the M01LP problem is equivalent to

$$\min\{\langle c,x\rangle : x\in K,\, p(x)\leq 0\}.$$

Using the exact penalty theorem proved in [8,11] we can reformulate equivalently the M01LP problem as, for any $t > t_o$,

$$\min\left\{\langle c,x\rangle + t\sum_{i\in J} x_i(1-x_i) : x\in K\right\}. \tag{10}$$

3.3 DCA for Solving (10)

By introducing the indicator function χ_K ($\chi_K = 0$ if $x \in K$, $+\infty$ otherwise) we rewritte the last problem as

$$\min\{\mathcal{F}(x) := \chi_K(x) + \langle td + c, x\rangle - tx^TQx : x\in\mathbb{R}^n\}. \tag{11}$$

where $d\in\mathbb{R}^n$ satisfying $d_i = 1$ if $i\in J$ and 0 if $i\notin J$. Since K is a polyhedral convex set, χ_K is a polyhderal convex function. On the other hand, the quadratic function $-\langle td+c,x\rangle + tx^TQx$ is clearly convex. Hence (11) can be written in a DC form

$$\alpha = \inf\{\mathcal{F}(x) := g(x) - h(x) \;:\; x\in\mathbb{R}^n\}, \tag{12}$$

with the following natural DC decomposition

$$g(x) := \chi_K(x); \quad h(x) := -\langle td+c,x\rangle + tx^TQx. \tag{13}$$

By the definition of h, we have

$$y^k = \nabla h(x^k) = -(td+c) + 2tQx^k. \tag{14}$$

Finding x^{k+1} consists in solving the following convex program after removing useless constants:

$$x^{k+1} \in \arg\min_{x\in\mathbb{R}^n}\{g(x) - \langle x, y^k\rangle\}$$

which is in fact a linear program

$$\{\langle -(td+c) + 2te^J, x\rangle : x\in K\} \tag{15}$$

Algorithm 1 describes the DCA for solving problem (12).

Algorithm 1. DCA applied to (12)

Initilization: Let $x^0\in\mathbb{R}^n$ be a guess, $k\leftarrow 0$

Repeat:

- Set $y^k := -(td+c) + 2te^J$.
- Solve the linear program (15) to obtain x^{k+1}.
- $k\leftarrow k+1$.

Until: Convergence of $\{x^k\}$.

The convergence of this DCA is proven by the following theorem:

Theorem 1. *(Convergence properties of* **DCA***)*

(i) DCA generates a sequence $\{x^k\}$ contained in the vertex set of K (denoted $V(K)$) such that the sequence $\{\langle e, x^k\rangle + tp(x^k)\}$ is decreasing.
(ii) The sequence $\{x^k\}$ converges to a critical point $x^* \in V(K)$ of (12) after a finite number of iterations.
(iii) Moreover, the point x^* is almost always a local minimizer of Problem (12).
(iv) For a sufficiently large number t , if at an iteration r we have $x^r \in \{0,1\}^n$, then $x^k \in \{0,1\}^n$ for all $k \geq r$.

Proof. This is consequence of the convergence properties of general DC programs, those of polyhedral DC programs, as well as the transportation of local minimizers in DC programming ([14,15]).

It is worth to note that, intesrestingly, with this suitable DC decomposition, although our algorithm works on a continuous domain, it constructs a sequence of integer solution (see iii) in the above theorem). Such an original property is important for large scale setting: in ultra large problems (it is often the case of IDSs), if we stop the algorithm before its convergence, we get always a feasible solution of the M01LP program. Moreover, DCA enjoys interesting convergence properties: it converges, after a finitely many iterations, to a local solution in almost cases.

4 Numerical Experiments

We will use the $GeFS_{CFS}$ measure mentioned in Sect. 2.1 to evaluates subsets of features. Solving the problem (2) (or equivalently (8)) results in a subset of selected features. To perform classification task, we use Multi-class Support Vector Machines (MSVM) proposed in [17] that uses a l_2-norm regularization. This approach will be referred to $GeFS_{CFS}$ when mentioning the feature selection task and l_2-MSVM (or l_2 for short) when indicating the classification task.

We compare our approach with an embedded feature selection in MSVM called l_2l_0-MSVM (or l_2l_0 for short) that is considered in [9,10]. In this approach, l_2l_0 regularizer has been utilized and the concave approximation has been imposed on the l_0-norm. The approximate optimization problem is a DC program and was solved by DCA. This approach has been already shown by the authors to be more efficient than several other competitors [9].

The environment used for the experiments is Intel Core^{TM} I5 (2.8 Ghz) processor, 4 GB RAM. The CPLEX solver library (ILOG CLPEX 11.2) for C^{++} is used to solve the linear programs in Algorithm 1 and convex quadratic programs in l_2-MSVM and l_2l_0-MSVM.

4.1 Datasets and Evaluation Criteria

The performances of the comparative methods are evaluated through the KDD Cup 1999 dataset. This original dataset contains more than 4 millions connection records of intrusions which are simulated in a military network environment. These records are extracted from the sequence of TCP packets to and from some IP addresses.

Each element includes 41 attributes. An other set, equal to 10% of the original dataset (Ten-Percent set), containing 494,021 records is considered. The attributes are devised into 4 groups. The first group includes 9 features of individual TCP connections. The second and the third group consist of, respectively, 13 features which describe the content within a connection suggested by domain knowledge and 9 traffic features. The final group contains 10 host based features [10,1].

We first construct the training datasets. Three training datasets with the size of 5,000; 10,000 and 50,000 records are generated from the Ten-Percent set. In these sets, the percentages of samples in each category are maintained as in the original dataset. In the Table 1, we summarize the statistics of the used training datasets.

Table 1. Statistics of training datasets

Dataset	#Samples	#Attributes	#Categories	Intrusion samples
SET05	5,000	41	5	80.34%
SET10	10,000	41	5	80.32%
SET50	50,000	41	5	80.31%

The test sets are separated into two types. The first type are three above training datasets. In the second type, the test sets having different sizes are disjoint with the training sets. As above, we maintain the percentages of samples in each category as those in the original dataset [10].

We use five main evaluation criteria in IDSs for these methods. The first criterion is Classification Error (CE) which is often used to evaluate the quality of classifiers. The second criterion is ACTE (Average Cost per Test Example) defined via a Cost Matrix of each misclassification. We use the Cost Matrix published by the KDD CUP'99 competition organizers. The third criterion is True Positive or Detection Rate (DeR) which is defined by the rate of the number of intrusions detected as intrusions (regardless of intrusion type) over the total number of intrusions. The fourth criterion is Diagnosis Rate (DiR) that is the rate of the number intrusions correctly classified over the total number of intrusions. The final criterion is False Positive (FP) which is the rate of the number of normal connections identified as the intrusions over the total number of normal connections [10].

4.2 Experiment Results

In the Table 2, we show the number of full-set features and the number of selected features by the $GeFS_{CFS}$ and the l_2l_0-MSVM [9,10].

In the Table 3, we report the results via the five above criteria on traning sets. The Table 4 presents the CPU time (in seconds) of training and testing processes on each set. The results on test sets is presented in Table 5.

Table 2. Full-set features and the number of selected features

Dataset	*Ful-set*	$GeFS_{CFS}$	l_2l_0-*MSVM*
SET05	41	13	15
SET10	41	13	20
SET50	41	15	29

Table 3. The performance of l_2-MSVM and l_2l_0-MSVM for IDSs on the training sets

Data	CE (%)		ACTE		DeR (%)		DiR (%)		FP(%)	
set	l_2	l_2l_0	l_2	l_2l_0	l_2	l_2l_0	l_2	l_2l_0	l_2	l_2l_0
SET05	1.24	0.03	0.0248	0.0038	98.80	99.93	98.66	99.90	0.813	0.711
SET10	1.27	0.09	0.0252	0.0021	98.78	99.93	98.67	99.93	1.016	0.152
SET50	0.87	0.09	0.0173	0.0020	99.07	99.94	99.03	99.93	0.477	0.172

Table 4. The CPU time (s) of two approaches. Time of preparing data is time for calculating correlation coefficients used in CFS measure

Data set	Toltal times of Trainning and Feature Selection					Testing time	
	$GeFS_{CFS}$				l_2l_0		
	preparing data	DCA(*M01LP*)	l_2	Total		l_2	l_2l_0
SET05	18.330	0.062	3.032	**21.424**	67.765	**0.093**	0.296
SET10	48.859	0.063	5.117	**54.039**	139.854	**0.171**	0.514
SET50	866.503	0.070	66.144	**932.717**	1169.550	**1.101**	2.038

Table 5. The performance of l_2-MSVM and l_2l_0-MSVM for IDSs on the test sets

Test	CE (%)		ACTE		DeR (%)		DiR (%)		FP(%)	
set	l_2	l_2l_0	l_2	l_2l_0	l_2	l_2l_0	l_2	l_2l_0	l_2	l_2l_0
The training set SET05										
5,000	1.30	0.46	0.0252	0.0078	98.80	99.83	98.78	99.70	0.915	1.118
10,000	1.30	0.54	0.2560	0.0086	98.69	99.75	98.51	99.64	0.508	1.270
50,000	1.26	0.55	0.2514	0.0086	98.71	99.81	98.57	99.71	0.559	1.625
100,000	1.25	0.53	0.0248	0.0083	98.73	99.82	98.58	99.72	0.589	1.569
The training set SET10										
10,000	1.34	0.24	0.0269	0.0052	98.74	99.85	98.62	99.79	1.169	0.356
50,000	1.25	0.25	0.0247	0.0047	98.78	99.88	98.69	99.83	1.016	0.569
100,000	1.24	0.24	0.0245	0.0046	98.81	99.80	98.71	99.84	1.051	0.574
494,021	1.23	0.22	0.0244	0.0042	98.82	99.90	98.72	99.86	1.048	0.561
The training set SET50										
50,000	0.92	0.17	0.0178	0.0034	99.06	99.89	99.01	99.88	0.586	0.335
100,000	0.91	0.16	0.1786	0.0033	99.06	99.90	99.00	99.88	0.564	0.335
494,021	0.90	0.14	0.0177	0.0029	99.07	99.96	99.02	99.90	0.557	0.310
4898431	0.38	0.10	0.0060	0.0017	99.71	99.96	99.67	99.95	0.577	0.335

Through these results, we observe that: both l_2-MSVM and l_2l_0-MSVM provide good quality solutions for very large sizes of test sets. They give very good results via all of the criteria, especially, for the two most important criteria of the network intrusion detection techniques: ACTE and FP. Overall, the l_2l_0-norm gives better results on classification but the filter approach selects a smaller number of features and its running time is shorter.

5 Conclusion

In this paper, we have studied the feature selection problem by means of the GeFS measure as a polynomial mixed 0-1 fractional programming (*PM01FP*) problem. We use the improved Chang's method to reformulate this PM01FP problem into a mixed 0-1 linear programming (M01LP) problem. DCA, an efficient algorithm in nonconvex programming framework is investigated to solve this M01LP problem via an exact penalty technique. Finally, the classification task is performed by using l_2-MSVM on reduced dataset with selected features. Experimental results investigated to intrusion detection systems show that the proposed algorithm is comparable with $l_2 - l_0$ regularizer MSVM on the ability of classification but requires less computation.

References

1. KDD Cup 1999 data set (1999), `http://www.sigkdd.org/kddcup/index.php?section=1999&method=data`
2. Chang, C.-T.: On the polynomial mixed 0-1 fractional programming problems. European Journal of Operational Research 131(1), 224–227 (2001)
3. Chang, C.-T.: An efficient linearization approach for mixed integer problems. European Journal of Operational Research 123, 652–659 (2000)
4. Nguyen, H.T., Franke, K., Petrovic, S.: Towards a generic feature-selection measure for intrusion detection. In: International Conference Pattern Recognition, pp. 1529–1532 (2010)
5. Nguyen, H.T., Franke, K., Petrovic, S.: Reliability in a feature-selection process for intrusion detection. In: Dai, H., Liu, J.N.K., Smirnov, E. (eds.) Reliable Knowledge Discovery, pp. 203–218. Springer US (2012)
6. Hall, M.: Correlation Based Feature Selection for Machine Learning. Doctoral Dissertation, University of Waikato, Department of Computer Science (1999)
7. Le Thi, H.A.: DC Programming and DCA, `http://lita.sciences.univ-metz.fr/~lethi`
8. Le Thi, H.A., Pham Dinh, T., Le, D.M.: Exact penalty in DC programming. Vietnam Journal of Mathematics 27(2), 169–178 (1999)
9. Le Thi, H.A., Nguyen, M.C.: Efficient Algorithms for Feature Selection in Multi-class Support Vector Machine. In: Nguyen, N.T., van Do, T., Thi, H.A. (eds.) ICCSAMA 2013. SCI, vol. 479, pp. 41–52. Springer, Heidelberg (2013)
10. Le, A.V., Le Thi, H.A., Nguyen, M.C., Zidna, A.: Network Intrusion Detection Based on Multi-Class Support Vector Machine. In: Nguyen, N.-T., Hoang, K., Jędrzejowicz, P. (eds.) ICCCI 2012, Part I. LNCS (LNAI), vol. 7653, pp. 536–543. Springer, Heidelberg (2012)

11. Le Thi, H.A., Pham Dinh, T., Huynh, V.N.: Exact penalty and Error Bounds in DC programming. Journal of Global Optimization 52(3), 509–535 (2012)
12. Chen, Y., Li, Y., Cheng, X.-Q., Guo, L.: Survey and Taxonomy of Feature Selection Algorithms in Intrusion Detection System. In: Lipmaa, H., Yung, M., Lin, D. (eds.) Inscrypt 2006. LNCS, vol. 4318, pp. 153–167. Springer, Heidelberg (2006)
13. Peng, H., Long, F., Ding, C.H.Q.: Feature selection based on mutual information: Criteria of max-dependency, max-relevance, and min-redundancy. IEEE Transactions on Pattern Analysis and Machine Intelligence 27(8), 1226–1238 (2005)
14. Pham Dinh, T., Le Thi, H.A.: Convex analysis approach to dc programming: Theory, algorithms and applications. Acta Mathematica Vietnamica 22(1), 289–357 (1997)
15. Pham Dinh, T., Le Thi, H.A.: Dc optimization algorithms for solving the trust region subproblem. SIAM J. Optimization 8, 476–505 (1998)
16. Pham Dinh, T., Le Thi, H.A.: Recent advances on DC programming and DCA. To appear in Transactions on Computational Collective Intelligence, 37 pages. Springer (2013)
17. Weston, J., Watkins, C.: Support Vector Machines for Multi-Class Pattern Recognition. In: Proceedings - European Symposium on Artificial Neural Networks, ESANN 1999, pp. 219–224. D-Facto public (1999)
18. Le Thi, H.A., Pham Dinh, T.: DC programming in Communication Systems: challenging problems and methods. Vietnam Journal of Computer Science, invited issue, 21 pages (2013), doi:10.1007/s40595-013-0010-5

Building a Trade System by Genetic Algorithm and Technical Analysis for Thai Stock Index

Monruthai Radeerom

Faculty of Information Technology, Rangsit University, Pathumtani, Thailand 12000
mradeerom@yahoo.com, m.radeerom@rsu.ac.th

Abstract. Recent studies in financial markets suggest that technical analysis can be a very useful tool in predicting the trend. Trading systems are widely used for market assessment. This paper employs a genetic algorithm to evolve an optimized stock market trading system. Our proposed system can decide a trading strategy for each day and produce a high profit for each stock. Our decision-making model is used to capture the knowledge in technical indicators for making decisions such as buy, hold and sell. The system consists of two stages: elimination of unacceptable stocks and stock trading construction. The proposed expert system is validated by using the data of 5 stocks that publicly traded in the Thai Stock Exchange-100 Index from the year 2010 through 2013. The experimental results have shown higher profits than "Buy & Hold" models for each stock index, and those models that included a volume indicator have profit better than other models. The results are very encouraging and can be implemented in a Decision- Trading System during the trading day.

Keywords: Computational intelligence, Genetic Algorithms, Stock Index, Technical Analysis.

1 Introduction

In recent financial markets, the use of automatic trading methods, which are often referred to as algorithmic trading, is expanding rapidly. Many works are found in applications of computational intelligence methodologies in finance [1]. Evolutionary computation, such as genetic algorithm (GA) [2], is promising in these methodologies, because of their robustness, flexibility and powerful ability to search.

Various works have been done on automated trading using evolutionary computation (e.g. [3], [4], [5] and [6]). These methods are mainly based on technical analysis, which is one of the two basic approaches in trading methods. Technical analysis is an attempt for the forecast of the future direction of prices by analyzing past market data, such as price and volume.

There is a large body of GA work in the computer science and engineering fields, but little work has been done concerning business related areas. Latterly, there has been a growing interest in GA use in financial economics, but so far there has been little research concerning automated trading. According to Allen and Karjalainen [7], genetic algorithm is an appropriate method to discover technical trading rules.

N.T. Nguyen et al. (Eds.): ACIIDS 2014, Part II, LNAI 8398, pp. 414–423, 2014.

Fern´ andez-Rodr´ıguez et al. [8] by adopting genetic algorithms optimization in a simple trading rule provides evidence for successful use of GAs from the Madrid Stock Exchange. Some other interesting studies are those by Mahfoud and Mani [9] that presented a new genetic-algorithm-based system and applied it to the task of predicting the future performances of individual stocks; by Neely et al. [10] and by Oussaidene et al. [11] that applied genetic programming to foreign exchange forecasting and reported some success.

The aim of this study is to show how genetic algorithms, a class of algorithms in evolutionary computation, can be employed to improve the performance and the efficiency of computerized trading systems. It is not the purpose here to provide the theoretical or empirical justification for the technical analysis. We demonstrate our approach in a particular forecasting task based on emerging stock markets.

The paper is organized as follows: Section 2 presents the background about the Technical Analysis and genetic programming. Section 3 presents the GA decision-making model (GATradeTool); Section 4 is devoted to experimental investigations and the evaluation of the decision-making model, and models the structure. The main conclusions of the work are presented in Section 5, with remarks on future directions.

2 Literature Reviews

2.1 Technical Analysis

Our approach to analyze financial markets is the technical analysis based on the past changes of prices and volume. The technical analysis needs various indicators for trading. They are evaluated from past stock data. Generally, technical indicators have several parameters. For example, moving average has a parameter, namely period, which is used as the denominator of averaging calculation. Various derived indicators, such as 10-days moving average, 50-days moving average, etc., are defined with the parameter.

Relative Strength Index (RSI)

The Relative Strength Index (RSI) is a momentum oscillator used to compare the magnitude of a stock's recent gains to the magnitude of its recent losses, in order to determine the overbought or oversold conditions. The calculation formula used is

$$RSI = 100 - \frac{100}{1 + \frac{\sum(positive\,change)}{\sum(negative\,change)}} \tag{1}$$

Where RS=Average gains/Average losses.

Fig. 1 (a) shows a RSI index in 14 periods time. After that, we can evaluate buy, hold and sell signals by experience trader. Sell zone was RSI more than 70 and buy zone was RSI below 30. Thus, hold zone was between 30 and 70. Sell, buy and hold zone shown Fig 1(b).

Because the index prices and technical indicators are not in the same scale, the same maximum and minimum data are used to normalize them. Normalization can be used to reduce the range of the data set to values appropriate for inputs to the system. The max is derived from the maximum value of the linked time series; similarly the minimum is derived from the minimum value of the linked time series. The maximum and minimum values are from the training. Testing and validation data sets must be same scale. The outputs of the system will be rescaled back to the original value according to the same formula.

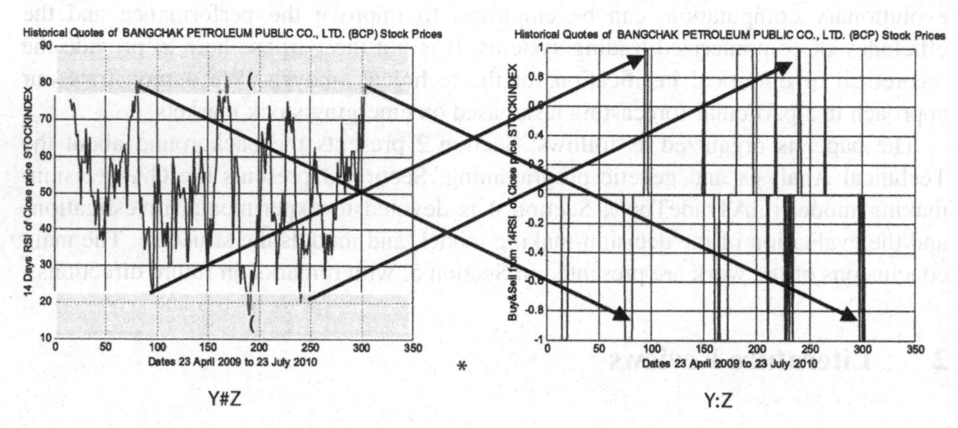

Fig. 1. (a) The 14 periods RSI index (RSI14 (t)) calculated by close price(t) (b) The Buy (1), Hold (0), Sell (-1) evaluated by RSI14(t)

After normalization, input scale is between -1 and 1. The normalization and scaling formula is

$$y = \frac{2x-(\max+\min)}{(\max-\min)}, \tag{2}$$

Where x is the data before normalizing, y is the data after normalizing.

For NFs trading system, the expected returns are calculated considering the stock market. That is, the value obtained on the last investigation day is considered the profit. The trader's profit is calculated as

$$\text{Profit(n)} = \text{Stock Value(n)} - \text{Investment value} \tag{3}$$

Where n is the number of trading days.

And the Rate of Return Profit (RoRP) is

$$\text{RoRP} = \frac{\text{Profit(n)}}{\text{Investment value}} \times 100 \tag{4}$$

Moving Average Convergence Divergence (MACD)

The MACD indicator constitutes one of the most reliable indicators within the market. It is a trend following the momentum indicator that exhibits the relation between two distinct moving averages. Essentially, it defines two lines: the MACD line that corresponds to the difference between a 26-week and 12-week EMA and a trigger line that corresponds to an EMA of the MACD line. The difference between the former lines allows us to obtain a histogram that can be easily analyzed and offers us perspectives on price evolution.

$$MACDt(s,l) = EMA(s)t\text{-}EMA(l)t \tag{5}$$

where S=12, l=26

2.2 Genetic Programming

Genetic Programming is a branch of genetic algorithms. The difference between them is the way of representing the solution. Genetic programming creates computer programs as the solution whereas genetic algorithms create a string of numbers that represent the solution. Here the one-dimensional vector is called the chromosome and the element in it is a gene. The pool of chromosomes is called the population.

Genetic Programming uses these steps to solve problems.

i. Generate a population of random polynomials.
ii. Compute the fitness value of each polynomial in the population based on how well it can solve the problem.
iii. Sort each polynomial based on its fitness value and select the better one.
iv. Apply reproduction to create new children.
v. Generate new population with new children and current population.
vi. Repeat step ii – vi until the system does not improve anymore.

The final result that we obtain will be the best program generated during the search. We have discussed how these steps are implemented in our work in the next subsections.

Initial Population Generation

The initial population is made of randomly generate programs. We have used the traditional grow method of tree construction to construct the initial population. A node can be a terminal (value) or function (set of functions +, -, x, /, exp) or variable. If a node is a terminal, a random value is generated. If a node is a function, then that node has its own children. This is how a tree grows.

Fitness Evaluation

After the initial random population is generated, individuals need to be assessed for their fitness. This is a problem specific issue that has to answer “how good or bad is this individual?” In our case, fitness is computed by $\sum_{k=1}^{l}(g(p_k) - f_k)^2$ where k is the day in past, p is the past data, f is the present data and l is the length of the section found by concordance measures.

Crossover and Mutation
In a crossover, two solutions are combined to generate two new off spring. Parents for the crossover are selected from the population based on the fitness of the solutions. Mutation is a unary operator aimed to generate diversity in a population and is done by applying random modifications. A randomly chosen subtree is replaced by a randomly generated subtree. First, a random node is chosen in the tree, and then the node as well as the subtree below it is replaced by a new randomly generated subtree.

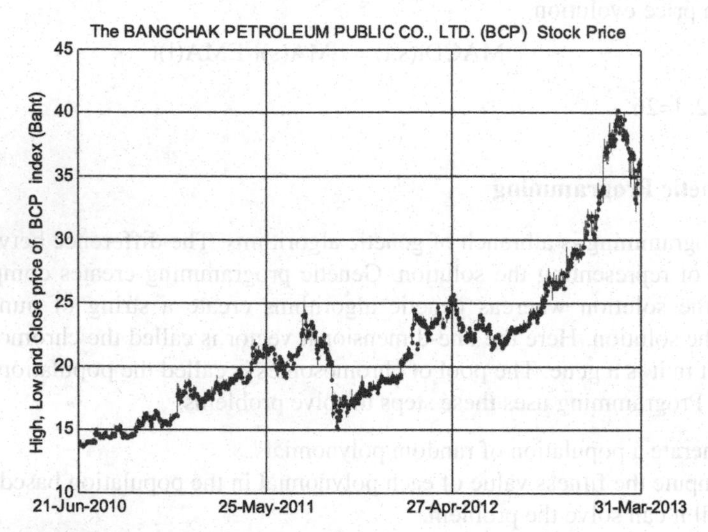

Fig. 2. Historical Quotes of Bangchak Petroleum public Co., Ltd. (BCP) Stock Prices

3 Methodology for The Intelligence Decision Trading System

Many stock market traders use conventional statistical techniques for decision-making in purchasing and selling [7]. Popular techniques use fundamental and technical analysis. They are more than capable of creating net profits within the stock market, but they require a lot of knowledge and experience. Because stock markets are affected by many highly interrelated economic, political and even psychological factors, and these factors interact with each other in a very complex manner, it is generally very difficult to forecast the movements of stock markets (see Fig 2). Fig. 2 shows historical quotes of Bangchak Petroleum Public Co., Ltd. (BCP) stock prices. It is a high nonlinear system. In this paper, we are working on one-day decision making for buying/selling stocks. For that we are developing a decision-making model, besides the application of an intelligence system.

3.1 Architecture

Many technical indicators have been proposed, but it is hard to select optimal indicators for actual algorithmic trading. Furthermore, it is also difficult to determine parameters for the selected indicators. In this paper, GAs are applied for this problem. The same trading system is used in both the training and test phase. We investigate fifteen trading systems. Each system consists of a combination of five technical indicators. We encode technical indicators and their parameters on chromosomes in GAs and apply the GA search for acquiring appropriate combinations of technical indicators and their parameters.

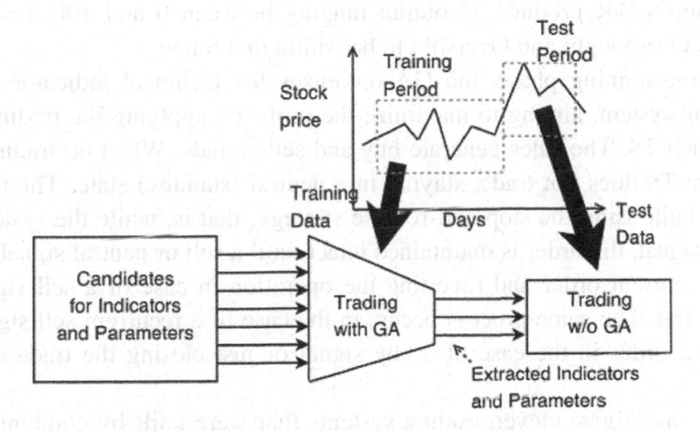

Fig. 3. Flow of trading using GA

The flow of our trading method is shown in Fig. 3. First, we apply the GA using stock price data for a predetermined period. This is the training phase to extract a set of effective technical indicators and their parameters. Next, we try an automated trading with the obtained set of technical indicators and their combined parameters using another stock price dataset. This is the testing phase to examine the performance of the selected indicators and their parameters. This process was executed with the 15 trading systems. Note that we do not apply the GA in the testing phase. It is important to prevent the overfitting in the training phase for improvement of the performance in the test.

3.2 Technical Indicators (TI)

Technical indicators are used by financial technical analysts to predict market movements, and many analysts use the standard parameters of these indicators. However, the best set of values for the parameters may vary greatly for each Thai Stock Index. Therefore, our approach uses GA to optimize these parameters considering the past time series history in order to maximize the profit. In this paper, we use five popular technical indicators: (i) Exponential Moving Average (EMA), (ii Moving Average Convergence/Divergence (MACD), (iii) Relative Strength Index (RSI), (iv) Williams's %R (W) and (v) On Balance Volume. All of them correspond on close price and volume. The TA-Lib MATLAB toolbox [20] is used to generate the technical indicators for each Thai Stock Index.

3.3 Trading Strategy

To analyze the proposed method, fifteen trading systems (TS) were created using the five technical indicators listed above. We define four TSs shown in Table I. The eleven remaining TSs are defined as a combination of the systems of Table I. In the GA search, the parameters assume only the integer values within the search ranges listed in Table I. Table I also shows the output variables of the technical indicators. The RSI and Williams's %R produce an output ranging between 0 and 100, restricting the parameters Overbought and Oversold to be within that range.

During the training phase, the GA optimizes the technical indicator's values of each trading system, aiming to maximize the profit by applying the trading rules defined by each TS. The rules generate buy and sell signals. When no trading signal is emitted, the TS does not trade, staying in a neutral (standby) state. The trading systems were built using the stop-and-reverse strategy, that is, while the system is emitting a buy signal, the order is maintained intact until a sell or neutral signal is emitted, closing the current order and reverting the operation in case of a sell signal or just staying neutral. The same process occurs in the case of a recurrent sell signal output, reverting the order in the case of a buy signal or just closing the trade and staying neutral.

We also investigate eleven trading systems that were built by combining the signals generated from the five technical trading indicators. All the TSs considered are listed in Table 1.

Table 1. Input and output of Decision Trading System

TS	TS Rules	TS	TS Rules	TS	TS Rules
GA1	EMA	GA6	EMA+RSI	GA11	EMA+RSI+W
GA2	MACD	GA7	EMA+W	GA12	MACD+RSI+W
GA3	RSI	GA8	MACD+RSI	GA13	EMA+RSI+W+OBV
GA4	William(W)	GA9	MACD+W	GA14	MACD+RSI+W+OBV
GA5	EMA+MACD	GA10	RSI+W	GA15	EMA+MACD+RSI+W+OBV

3.4 Genetic Algorithm

The method of Fig. 3 was implemented using the genetic algorithm of the MATLAB Global Optimization Toolbox [19], which has an implementation to handle Mixed Integer Optimization Problems [20]. This is used in the experiments to cope with the chromosome integer restrictions. The best profit in 50 executions of the training phase was chosen as the best fitted solution and then validated against the test data. The algorithm stops if there is no improvement in the objective function after at least 50 generations. The maximum number of generations chosen was 100 and two elite children are guaranteed to survive to the next generation.

4 Results and Discussion

4.1 Setup

The model realization could be run having different groups of stocks (like Banking group, Energy group, etc.), indexes or other groups of securities. For that we are using market orders, as it allows simulating buying stocks when the stock exchange is nearly closed. All the experimental investigations were run according to the above presented scenario and were focused on the estimation of Rate of Return Profit (RoRP). At the beginning of each realization, the starting investment is assumed to be 1,000,000 Baht (Approximately USD 29,412). The data set, including 10 approved stock indexes in the Stock Exchange of Thailand (SET) index, has been divided into two different sets: the training data and test data. The stock index data is from April 23, 2010 to March 31, 2013 totaling 674 records. The first 539 records are training data, and the rest of the data, i.e., 135 records, will be test data. Moreover, the data for stock prices includes the buy-sell strategy, closing price and its technical data. Consequently, max-min normalization can be used to reduce the range of the data set to appropriate values for inputs and output used in the training and testing method.

4.2 Stage 1: Elimination of Unacceptable Stocks

In this stage, the stocks that are not preferred by investors are eliminated. The unacceptable stocks are those that have a negative price to earnings ratio (P/E) or a negative shareholder's equity value. For this reason, investors generally do not prefer to invest in these stocks. In this study, the data that are taken into consideration in this stage are the 1-year-data that precedes the investment date. Investors may also eliminate some stocks according to their preferences or specific knowledge about those stocks. This stage reduces the burden on the stock evaluation stage, and prevents the system from suggesting unacceptable stocks to user.

4.3 Stage 2: Stock Trading Construction and Results

After developing the intelligence trading system, we were given 1,000,000 baht for investment at the beginning of the testing period. The decision to buy and sell stocks is given by the proposed intelligence output. We translated the produced RoRP results that verify the effectiveness of the trading system. Table 1 shows a Financial Simulation Model for calculating profit in our trading strategy. For example, results of Model GA13 of Historical Quotes of Bangchak Petroleum public Co., Ltd. (BCP) Stock Prices within training days and testing day are shown in Figure 4 and 5, respectively. The Profit of BCP on training periods is 80.3% and 50.2 % on testing periods.

Table 2. Example of Financial Simulation Model in trading strategy

Stock	GA13 : Buy & Sell (BCP)				Stock	Buy & Hold : Buy & Sell (BCP)			
Index	Action		# of Shared	Cash(Baht)	Index	Action		# of Shared	Cash(Baht)
69.50	1	STAY	-	400,000	69.50	1	STAY	-	300,000
69.25	0	Hold	-	400,000	69.25	1	Buy	4,332	-
68.00	0	Hold	-	400,000	68.00	0	HOLD	4,332	
67.25	0	Hold	-	400,000	67.25	0	HOLD	4,332	
67.00	0	Hold	-	400,000	67.00	0	HOLD	4,332	
67.00	1	Buy	5,970	-	67.00	0	HOLD	4,332	
67.25	0	Hold	5,970	-	67.25	0	HOLD	4,332	
68.50	0	Hold	5,970	-	68.50	0	Hold	4,332	-
69.50	0	Hold	5,970	-	69.50	0	Hold	4,332	-
75.50	0	Hold	5,970	-	75.50	0	Hold	4,332	-
77.25	0	Hold	5,970	-	77.25	0	Hold	4,332	-
77.75	0	Hold	5,970	-	77.75	0	Hold	4,332	-

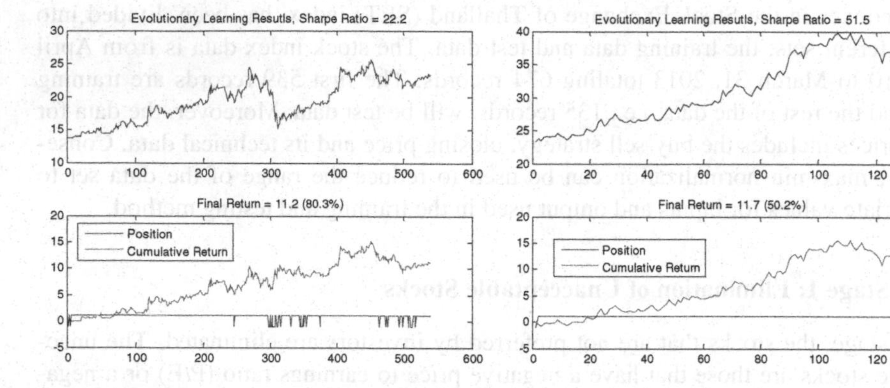

Fig. 4. Comparison profit between Possible Rate of Return Profit (Possible RoRP) and Profit from our proposed Trading System in Training Days of BCP Stock Index

Fig. 5. Comparison profit between Possible Rate of Return Profit (Possible RoRP) and Profit from our proposed Trading System in Testing Days of BCP Stock Index

Our experimental results show that GATradeTool can improve digital trading by quickly providing a set of near optimum solutions. Concerning the effect of different GA parameter configurations, we found that an increase in population size can improve performance of the system. The parameter of the crossover rate does not seriously affect the quality of the solution.

Finally, it would be interesting for further research to test a series of different systems in order to see the correlation between a genetic algorithm and system performances. At a time of frequent changes in financial markets, researchers and traders can easily test their specific systems in GATradeTool by changing only the function that produces the trading signals

5 Conclusion

This paper presented the decision-making model based on the application of GA. The model was applied in order to make a one-step forward decision, considering historical data of daily stock returns. The experimental investigation has shown a scenario of

Intelligence Trading System based on Technical Analysis and GA to make a trading strategy, achieving more stable results and higher profits when compared with Buy and Hold strategy. Some models that included volume indicators have profit better than other models. For future work, several issues could be considered. Other techniques, such as support vector machines, particle swarm algorithms, etc. can be applied for further comparisons. This method should also be used to combine different signals that capture market dynamics more effectively (say a bear, bull, or sideways market), or to analyze other stock index groups, other stock exchanges or other industries for comparisons.

References

1. Brabazon, A., O'Neill, M.: An introduction to evolutionary computation in finance. IEEE Computational Intelligence Magazine, 42–55 (November 2008)
2. Goldberg, D.E.: Genetic Algorithms in Search, Optimization and Machine Learning. Addison-Wesley (1989)
3. de la Fuente, D., Garrido, A., Laviada, J., Gomez, A.: Genetic algorithms to optimise the time to make stock market investment. In: Proc. of Genetic and Evolutionary Computation Conference, pp. 1857–1858 (2006)
4. Hirabayashi, A., Aranha, C., Iba, H.: Optimization of the trading rule in foreign exchange using genetic algorithms. In: Proc. of the 2009 IASTED Int'l Conf. on Advances in Computer Science and Engineering (2009)
5. Dempster, M.A.H., Jones, C.M.: A real-time adaptive trading system using genetic programming. Quantitative Finance 1, 397–413 (2001)
6. Matsui, Sato, H.: A comparison of genotype representations to acquire stock trading strategy using genetic algorithms. In: Proc. of Int'l Conf. on Adaptive and Intelligent Systems, pp. 129–134 (2009)
7. Allen, F., Karjalainen, R.: Using genetic algorithms to find technical trading rules. Journal of Financial Economic 51, 245–271 (1999)
8. Fernández-Rodríguez, F., González-Martel, C., Sosvilla-Rivero, S.: Optimisation of Technical Rules by Genetic Algorithms: Evidence from the Madrid Stock Market, Working Papers 2001-14, FEDEA (2001), `ftp://ftp.fedea.es/pub/Papers/2001/dt2001-14.pdf`
9. Mahfoud, S., Mani, G.: Financial forecasting using genetic algorithms. Journal of Applied Artificial Intelligence 10(6), 543–565 (1996)
10. MATLAB Global Optimization Toolbox (May 2012), `http://www.mathworks.com/products/global-optimization/`
11. MATLAB Mixed Integer Optimization Problems (May 2012), `http://www.mathworks.com/help/toolbox/gads/bs1cibj.html#bs1cihn`
12. Neely, C., Weller, P., Ditmar, R.: Is technical analysis in the foreign exchange market profitable? A genetic programming approach. In: Dunis, C., Rustem, B. (eds.) Proceedings of the Forecasting Financial Markets: Advances for Exchange Rates, Interest Rates and Asset Management, London (1997)

A Study of Shift Workers for Role Conflict Effect: Quality of Life-Emotional Exhaustion as Mediator

YuhShy Chuang, Jian-Wei Lin, Jhih-Yu Chin, and Ying-Tong Lin

Chien Hsin University of Science and Technology International Business Dept.
Yuhshy60@gmail, jwlin@uch.edu.tw,
{siaojhij,sea0717}@gmail.com

Abstract. Globalization of the competitive market and the diverse needs of society have produced today's work patterns, which differ from the agricultural age and produce a twenty-four hour industrial society. Shift work has become one of the social work patterns which, while shift workers often need to adjust to different times of the day and change their daily routine of life to adjust to the long-term impact of shifts on work performance. These changes have reduced life quality, leading to negative emotions and physiological and mental fatigue. This study on Semiconductor technology conducted an employee questionnaire survey using 250 questionnaires, and 226 questionnaires were recovered for a rate of 94.1%. The study found that Semiconductor technology industry employees present a significant effect for Role Conflict to Emotional Exhaustion, Emotional Exhaustion and a negative impact on Quality of Life, and Role Conflict insignificantly influences Quality of Life. A full mediating affect of Emotional Exhaustion between Role Conflict and Quality of Life is supported.

Keywords: Shift Work, Emotional Exhaustion, Role Conflict, Quality of Life.

1 Introduction

In response to growing competition in the global marketplace and changing social demands, a 24-hour work pattern has emerged in industrial society, which leads to shift work, an employment practice that allows for both adjustable and flexible work hours. Although shift work has produced tremendous economic effects, it has also had many mental and physical impacts on shift employees. Over the long term, the practice of shift work may affect employees' job performance and quality of life, or even result in employees' experiencing negative emotions and psychological fatigue, as they withstand multiple physical and psychological effects (Knutsson & Bøggild, 2000).

Without timely regulation of emotions, the employee may suffer from excessive pressure, start to develop indifferent and negative attitudes toward work, and be vulnerable to emotional exhaustion (Halbesleben & Bowler, 2007). Over time, shift work, already a characteristic of modern society, will continue to be extensively used, and thus, a growing number of employees will be required to engage in shift work. Hence, the issue of shift work and its impact on the shift worker is worthy of in-depth research.

N.T. Nguyen et al. (Eds.): ACIIDS 2014, Part II, LNAI 8398, pp. 424–433, 2014.

2 Literature Review

Role Conflict. Role conflict appears when an individual is pulled between two or more role expectations, and while meeting one of these expectations, he or she must act in conflict with the other/s (Jackson & Schuler.1985). According to the definition of role conflict proposed by Kahn et al. (1964), of all the set of roles that an individual plays, each may present different demands for the individual, and consequently, impose different role pressure and role expectations on that individual. That role conflict becomes a form of mental pressure or emotional distress, which occurs in a situation where an individual's ability falls short of satisfying all demands, when facing two or more conflicting role expectations. Babakus et al. (1999) have pointed out that employees with a higher level of role ambiguity or role conflict are also more likely to experience emotional exhaustion.

Emotional Exhaustion. Emotional exhaustion is a feeling of depleted feelings, vigor, and energy. Emotion exhaustion is also a state that tends to make a worker feel over-extended and exhausted in capability, as well as sensing helplessness at work, and emotional burnout. Meanwhile, the worker may suffer from anxiety, stress, and depression, along with other negative emotions, and feel weary about doing his or her work and unable to maintain devotion and commitment to work (Maslach, 1993; Maslach & Jackson, 1981; Freudenberger, 1974). Wu (2009) points out that there is a positive correlation between the level of role conflict and emotional exhaustion. Role conflict can have an indirect effect on an individual's health due to emotional exhaustion. Emotional exhaustion refers to a situation when an employee at work feels an exhaustion of emotional resources from extreme consumption, the cause of which may attribute to the depletion of that employee's emotional energy after a long period of enduring excessive stress.

Quality of Life. Quality of life emphasizes that an objective assessment of one's status of life should be added to an individual's subjective feelings about life to have a more comprehensive assessment of the total quality of life. Quality of life, an inherent condition of each person and an ongoing and dynamic process of personal sentiment, is interconnected with other factors, such as body, soul, social culture, and environment, as well as the forming of relationships of mutual influence using these factors (Anne Hickey, 2005; Kleinpell & Ferrans, 2002; Register & Herman, 2006). The pursuit of a positive and healthy life, however, can not be solely achieved through satisfaction with physical-level matters, such as medical care and one's life, as satisfaction at a spiritual level is the key for experiencing an enhanced quality of life.

3 Methodology

3.1 Role Conflict and Emotional Exhaustion

Role conflict arises when an individual perceives conflicts stemming from incompatible expectations and demands for two or more roles in the collection of roles that an individual plays (Singh, 1998). An employee whose job duties primarily involve simple and routine tasks is less likely to experience emotional exhaustion, as that employee has fewer opportunities to be exposed to the sources of stress that can cause emotional exhaustion (Cordes, 1997). Role conflict also has a positive effect on emotional exhaustion (Schwab & Iwanicki, 1982). Thus, this study proposes the following research hypothesis:

H1: Role conflict has a positive and significant correlation with emotional exhaustion.

3.2 Emotional Exhaustion and Quality of Life

An absolute relationship exists between recreational activities and quality of life. In addition to alleviating mental stress, proper physical recreational activities provide people benefits during activity participation and thus improve people's quality of life (Iwasaki & Mannell, 2000a). Bergner (1989) and Coleman(1993) studied the influence of recreational activities on quality of life and found that recreational activities can enhance an individual's quality of life, life satisfaction, well-being, and self-efficacy. Thus, this study proposes the following hypothesis:

H2: Emotional exhaustion has a significantly negative influence on quality of life.

3.3 Role Conflict and Quality of Life

Iwasaki and Mannell (2000a) propose that recreational activities can help people when handling various types of stress. Bergner (1989) discussed the effects of recreational activities on quality of life and concluded that recreational activities can enhance quality of life, life satisfaction, well-being, and self-performance, as well as help people alleviate stress, self-anxiety, and depression. Ongoing participation in recreational activities also contributes to a person's positive mindset and in turn affects that person's physical health in positive ways Therefore, this study proposes the following hypothesis:

H3: Role conflict has a positive and significant influence on quality of life.

With reference to these three research hypotheses, it can be inferred that role conflict has a significant influence on quality of life through emotional exhaustion. Therefore, this study proposes the following hypothesis:

H4: Emotional exhaustion has a mediating effect on the influence of role conflict on quality of life.

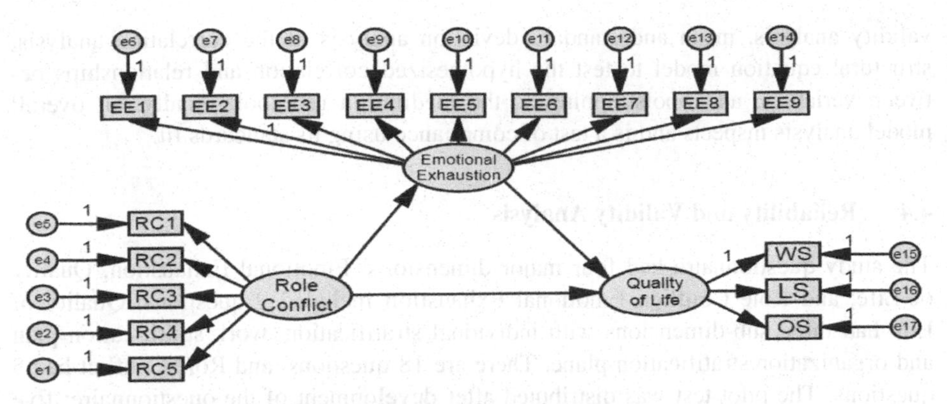

Fig. 1. The research structure concept

4 Data Analysis and Results

4.1 Sample Information

The study took convenience sampling from Hsinchu Science Park shift worker questionnaires. The test session occurred periodically from Aug 1st to Aug 31st about four weeks periodic. After the convenient random sample was distributed to employee who work in Science Park at a Science Semiconductor company and total 250 surveys were delivered and 250 surveys were returned. Eliminate the 24 unusable questionnaires were eliminated, leaving 226 samples useable. The total returned rate became 94.1.8%.

4.2 The Operational Definition

The operational definition of **Role Conflict,** when referring to shift workers is the conflict to roles stress or expectation for two or more statuses. Role Conflict was assessed using Li and Shani's developed role conflict 5-item scale (1991).

The operational definition of **Emotional Exhaustion** refers to shift workers tired from physical and emotional depletion. Emotional Exhaustion was assessed using Maslach and Jackson's (1981) 9-item scale.

The operational definition of **Quality of Life** refers to what shift workers face in life that makes them feel satisfied. Quality of Life was assessed using Cumming and Huse's 18 -item scale (1985) And three dimensions as: Individual stratification, work stratification plane, and the organization stratification plane.

All of the questionnaires applied Likert's scale 7-point scale and a response scale consisting of seven points, ranging from 1 to 7 (Strongly agree).

4.3 Data Analysis

The statistical analysis software package uses IBM SPSS 20.0 and AMOS 20.0 as the analysis technology. The statistic methods used for this study include a reliability and validity analysis, mean and standard deviation analysis of the correlation analysis, structural equation model to test the hypothesized correlation and relationships between variables, and bootstrapping as the meditation test tool. Finally, an overall model analysis inspects and is a test of compliance using its standards fit.

4.4 Reliability and Validity Analysis

The study questionnaire had four major dimensions: Emotional Exhaustion, Quality of Life, and Role Conflict. Emotional Exhaustion included 9 questions; Quality of Life had three sub-dimensions with individual stratification, work stratification plan and organization stratification plane. There are 18 questions, and Role Conflict had 5 questions. The pilot test was distributed after development of the questionnaire; five experts reviewed and provided word correction. A total 32 of questions became the final survey. After the survey sample analysis, a reliability analysis of the questionnaire total dimensions of reliability test was greater than 0.7, and for each scale the total dimensions Cronbach α values were> 0.7, the Quality of Life slightly lower than 0.7 (Cronbach's α0.654) , Emotional Exhaustion Cronbach α was 0.860, and Role Conflict Cronbach α was 0.863. The three dimensions all presented a high degree of reliability.

4.5 Construct Validity

4.5.1 Convergent Validity Analysis

Confirmatory factor analysis (CFA) enabled us to test how well the measured variables represented the construct. Scholars Fornell and Larcker (1981) indicated the best CR value should be above 0.6. After a CAF examination of those dimensions, the CR values are almost greater than 0.7 and shows this scale has a good reliability. The result of AVE offered the following information: The Quality of Life dimensions for AVE values were 0.56, 0.30, and 0.47; Emotional Exhaustion at all observed variables had AVE values at 0.42; Role Conflict scale for AVE was 0.57. The construct variables were close to the suggested AVE values being greater than 0.5 (Hair et al, 2009).

4.5.2 Discriminant Validity Analysis

Discriminant validity verifies the different dimensions and suggests they should have differentiation that is less statistically; further, the questions in each dimension should not be highly related. In this study the confidence interval method (bootstrap) was used to exam the correlation coefficient between the dimensions. If the test number did not include 1, then that dimension completely related (Torzadeh, Koufteros, & Pflughoeft, 2003). The bootstrapping sample for 2000 times with a confident interval of 95%. Three different methods can identify the result of discriminant validity: Estimates value ±2 standard error, Bias-corrected, and the Percentile method. The results show that all dimension numbers did not include 1 and there is a discriminant between variables as shown in Table 1.

Table 1. The Four Dimensional Scale of Discriminant Validity Analysis

Parameters		Estimate	Estimates ±2 SE		Bias-corrected		Percentile Method	
			Lower	Upper	Lower	Upper	Lower	Upper
Quality of Life	work <--> individual	0.765	0.617	0.913	0.643	0.93	0.633	0.917
	work <--> organization	0.703	0.519	0.887	0.499	0.872	0.515	0.876
	individual <--> organization	0.317	0.155	0.479	0.147	0.464	0.153	0.469

4.6 Overall Model Fit Indices

This study focused on the correlation of shift workers to Role Conflict, Quality of Life, and Emotional Exhaustion as its dimensions measurement. The first phase of the dimensions value indicated the second phase adjusted model. The initial confirmatory factor analysis indicators were not entirely satisfactory in this study. Therefore, the adjusted parameter was released offered by the modification indices.

The indices showed that released and deleted Emotional Exhaustion total of the two questions dimension will reduce the chi-square value. The final adjust model fit as follows: The partial model fit results are acceptable for the absolute model fit and the parsimony and incremental model fit were also good. The result of the model fit is shown in Table 2.

Table 2. Overall Model Fit for Construct Structure

Model fit	Evaluation Indicators	Range	Evaluation Criterion	Preliminary Validation of the Model	Result	Acceptable
Absolute Fit Index	χ2/df	Less is better	<5	4.436	3.933	Good
	GFI	0-1	>0.9	0.782	0.823	Acceptable
	AGFI	0-1	>0.9	0.713	0.756	Not Good
	RMR	-	as less as better	0.24	0.200	Acceptable
	SRMR	0-1	<0.05	0.0863	0.0769	Acceptable
	RMSEA	0-1	<0.08	0.124	0.114	Acceptable
Parsimony Fit Index	PGFI	0-1	>0.5	0.593	0.597	Acceptable
	PCFI	0-1	>0.5	0.665	0.694	Good
	PNFI	0-1	>0.5	0.628	0.659	Good
Incremental Fit Index	NFI	0-1	>0.9	0.736	0.796	Acceptable
	NNFI/TLI	0-1	>0.9	0.742	0.804	Acceptable
	CFI	0-1	>0.9	0.780	0.838	Acceptable

4.7 Hypotheses Test

In the unstandardized estimates model, the test results show that the correlation between Role Conflict and Emotional Exhaustion estimated value was 0.431, standard error was 0.065 、CR had a 6.655、P value =***, and the standardized path coefficient = 0.702. The results show that Role Conflict and Emotional Exhaustion have a positive effective significance. Thus **H1 is accepted.**

In the unstandardized estimates model, the test results show that the correlation between Emotional Exhaustion and Quality of Life estimated value was -0.572, standard error was 0.127 、CR was -4.518 、P value =***, and the standardized path coefficient =-0.593. This result shows that Emotional Exhaustion and Quality of Life have a positive effective significance. Thus **H2 is accepted.**

In the unstandardized estimates model, the test results show that the correlation between Role Conflict and Quality of Life estimated value was0. 056, standard error was 0.066、CR was 0. 841、P value was =0.400, and the standardized path coefficient =-0.094. The results show that Role Conflict and Quality of Life did not have effective significance. Thus H3 was not accepted.

4.8 Mediating Effect Test

In this study, the bootstrapping method was used for testing the mediating effect for judgment value that under a (1-α) 100% confidence interval (95% CI). The results between the highest and lowest range does not include zero mean, namely, that the α result reached statistically significant level (Cheung & Lau, 2008).However, this study tested Emotional Exhaustion a mediator between Role Conflict and Quality of Life, and the result shows that the total effect and the Indirect Effects of Z scored higher than 1.96.

Table 3. The Relationship Between Role Conflict and Quality of Life Mediated by Emotional exhaustion

Variable	Estimate	Product of Coefficients		Bootstrapping			
				Bias-Corrected 95% CI		Percentile 95% CI	
		SE	Z	Lower	Upper	Lower	Upper
Total Effects							
Role Conflict→ Quality of Life	-0.191	0.060	3.18	-0.315	-0.084	-0.312	-0.081
Indirect Effects							
Role Conflict→ Quality of Life	-0.247	0.063	3.92	-0.416	-0.153	-0.393	-0.144
Direct Effects							
Role Conflict→ Quality of Life	0.056	0.069	0.81	-0.079	0.197	-0.084	0.193

Moreover, the Bias-Corrected and Percentile exam value of the highest and lowest scores does not include the Zero after the bootstrapping exam. The direct effect of the Z score of 0.81 is lower than 1.96; also the Bias-Corrected and Percentile exam value of the highest and lowest score included the Zero after the bootstrapping exam. **Thus, H4 is complete mediation.** The result of the mediation effect is in seen in Table 3.

5 Conclusion and Discussion

Santavirta et al (2007) indicate that role conflict is perceived when the protagonist receives two or more role expectations and does not know how to satisfy them. Shirom (2003) suggests that emotional exhaustion is accompanied by physical and mental fatigue. When the resources available at work fail to meet work requirements, employees feel depletion of emotional resources, which leads to emotional exhaustion.

This study found that role conflict has a positive and a significant influence on emotional exhaustion, and role conflict causes mental and physical fatigue and further leads to emotional exhaustion. Hideto, Yasushi & Kouichi (2005) indicate that shift workers have a more inferior quality of life than day shift workers, implying that an accelerated level of occupational stress under the practice of shift work results in a dropped quality of life. As discussed, occupational stress has a positive and significant influence on emotional exhaustion.

This study found that emotional exhaustion has a negative and significant influence on the quality of life, implying that a higher level of emotional exhaustion is accompanied by a lower quality of life, and vice versa. As an individual plays a variety of roles in diverse environments, role conflicts arise when these roles conflict with the individual (Santavirta et al., 2007). Thus, to ensure a certain quality of work and minimize the impacts of role conflict, one should relieve stress in a timely way and participate in more recreational activities to maintain a healthy and sound mental state (Bergner, 1989).

This study suggests that emotional exhaustion has a mediating effect on the influence of role conflict on the quality of life, while role conflict has an indirect, yet significant, influence on quality of life when a person is emotionally exhausted. In other words, emotional exhaustion has a complete mediating effect on the influence of role conflict on quality of life. With respect to workers in the technology industry, workers' following the shift practice of work are more likely to become victims of occupational stress and experience a lowered personal quality of life due to such issues as life patterns and job performance. To cope with this predicament, enterprises should allow employees to work on a fixed time shift and maintain their routine sleep patterns. In addition, enterprises could consider providing extra rewards to employees who do volunteer to work the night shift for a long time. On the other hand, employees should relax through proper recreational activities to release anxiety and stress and thus enhance their quality of life, elevate their work performance, and reinforce organizational commitment to the worker.

References

1. Hickey, A., Barker, M., McGee, H., O'Boyle, C.: Measuring health-related quality of life in older patient populations. Pharmacoeconomics 23(10), 971–993 (2005)
2. Babakus, E., Cravens, D.W., Johnston, M., Moncrief, W.C.: The role of emotional exhaustion in sales forces attitude and behavior relationships. Academy of Market Science 27(1), 58–70 (1999)
3. Bergner, M.: Quality of life, health status, and clinical research. Medical Care 27(3), 148–156 (1989)
4. Coleman, D.: Leisure based social support, leisure dispositions and health. Journal of Leisure Research 25, 250–361 (1993)
5. Cordes, C.L., Dougherty, T.W.: A review and an integration of research on job burnout. Academy of Management Review 18, 621–656 (1993)
6. Cordes, C.L., Dougherty, T.W., Blum, M.: Patterns of burnout among managers and professionals: A comparison of models. Journal of Organizational Behavior 18, 685–701 (1997)
7. Fornell, Larcker: Structural equation models with unobservable variables and measurement errors. Journal of Marketing Research 18(2), 39–50 (1981)
8. Freudenberger, H.J.: Staff Burnout. Journal of Social Issues 30(1), 159–165 (1974)
9. Hair Jr., J.F., Anderson, R.E., Tatham, R.L., Black, W.C.: Multivariate Data Analysis, 5th edn. Prentice-Hall, Englewood Cliffs (1998)
10. Halbesleben, J.R.B., Bowler, W.M.: Emotional exhaustion and job performance: The mediating role of motivation. Journal of Applied Psychology 92(1), 93–106 (2007)
11. Hideto, H., Yasushi, S., Kouichi, S.: Three-shift system increase job-related stress in Japanese workers. J. Occup. Health 47, 397–404 (2005)
12. Iwasaki, Y., Mannell, R.C.: Hierarchical dimensions of leisure stress coping. Leisure Sciences 22, 163–181 (2000a)
13. Jackson, S.E., Schuler, R.S.: A meta-analysis and concept critique of research on role ambiguity and role conflict in work setting. Organizational Behavior and Human Decision Processes 36, 16–78 (1985)
14. Kahn, R.L., Wolfe, D.M., Quinn, R.P., Snock, J.: Organizational Stress: Studies in Role Conflict and Ambiguity. John Willy, New York (1964)
15. Kleinpell, R.M., Ferrans, C.E.: Quality of life of elderly patients after treatment in the ICU. Research in Nursing & Health 25, 212–221 (2002)
16. Knutsson, A., Bøggild, H.: Shif work and cardiovascular disease: Review of disease mechanisms. Rev. Environ. Health 15, 359–372 (2000)
17. Maslach, C.: Burnout: A multidimensional perspective. In: Schaufeli, W.B., Maslach, C., Marek, T. (eds.) Professional Burnout: Recent Developments in Theory and Research, pp. 19–32. Taylor & Francis, New York (1993)
18. Maslach, C., Jackson, S.E.: The measurement of experienced burnout. Journal of Occupational Behavior 2, 99–113 (1981)
19. Register, M.E., Herman, J.: A middle range theory for generative quality of life for the elderly. Advances in Nursing Science 29(4), 340–350 (2006)
20. Santavirta, N., Solovieva, S., Theorell, T.: The association between job strain and emotional exhaustion in a cohort of 1028 Finnish teachers. British Journal of Educational Psychology 77, 213–228 (2007)
21. Schwab, R.L., Iwanicki, E.F.: Perceived Role Conflict, Role Ambiguity, and teacher burnout. Educational Administration Quarterly 18, 60–74 (1982)

22. Shirom, A.: Job-related burnout. In: Quick, J.C., Tetrick, L.E. (eds.) Handbook of Occupational Health Psychology, pp. 245–265. American Psychological Association, Washington, DC (2003)
23. Singh, Jagdip: Striking a Balance in Boundary-Spanning Positions: An Investigation of Some Unconventional Influences of Role Stressors and Job Characteristics on Job Outcomes of Salespeople. Journal of Marketing 62(3), 69–87 (1998)
24. Torzadeh, G., Koufteros, X., Pflughoeft, K.: Confirmatory Analysis of Computer Self-Efficacy. Structural Equation Modeling 10(2), 263–275 (2003)
25. Van Sell, M., Brief, A.P., Schuler, R.S.: Role conflict and role ambiguity: Integration of the literature and directions for future research. Human Relations 34(1), 43–71 (1981)
26. Wu, C.H.: Role conflicts, emotional exhaustion and health problems: a study of police officers in Taiwan. Stress and Health 25(3), 259–265 (2009)

How Product Harm Recalls Affect Customer Brand Equality and Trust and Customer Perceived Value

Lung Far Hsieh and Ping Chuan Lee

Chung Yuan Christian University, Taiwan
lungfar@cycu.edu.tw, Supercute.apple@gmail.com

Abstract. Being capable and doing well as a core strategy are very important for business corporations that desire good products, quality, marketing share and cooperative social responsibility (CSR) .International business has developed with the deepening of globalization, and CSR covers a wide range of discussion. Customers are to awakening the attention of business brand equation, trust and businesses to corporate social responsibility (CSR).

This paper applies a SEM structural equation model as the analysis technique for product harm recall impact for brand equity, trust, and customer perceived value correlation. Corporate social responsibility does not play a moderation role between brand equity perceived under the bootstrapping test. There is no correlation between customer trust and perceived value, but customer trust has a positive correlation with corporate social responsibility (CSR).

Keywords: Product harm recall, brand equity, customer perceived value, customer trust, CSR, Structural equation model.

1 Introduction

Being capable and doing well are very important core strategies for business corporations that include good products, quality, marketing share, and cooperate social responsibility (CSR).International business develop via the deepening of globalization, so the topic of CSR covers a wide range of discussion . Moreover, customers are awakening to business brand equation, trust and businesses -corporate social responsibility (CSR). Product-harm crises often produce product recalls that have a significant impact on a firm's sale, brand reputation, stock price and financial value. Moreover, company product-harm crises also cause negative publicity that substantially affects future purchase intentions considering consumer complaints.

Business operations involving business activities and achievements of all pro-social organizations not only strictly affect the maximizing of shareholder profits, but also "corporate social responsibility. That responsibility "is accepted for Taiwan enterprises and regarded as a model. Although t many researchers have sought CSR initiatives to understand the relationship between motivation, behavior and the consequences for consumers in brand recognition under corporate social responsibility and has an impact on perceived value and trust (e.g., Martins, 2005; Waddock &Graves, 1997).

N.T. Nguyen et al. (Eds.): ACIIDS 2014, Part II, LNAI 8398, pp. 434–443, 2014.

2 Literature Review

From a marketing perspective, the enterprises from their economic interests focus on corporate social responsibility practice linked to positive consumer product marketing, brand assessment, brand choice and brand recommendations (Brown & Dacin, 1997; Drumwright, 1994; Handelman & Arnold, 1999; Osterhus, 1997). Many scholars studies conclude that corporate social responsibility (CSR), makes an important contribution in that: (1)corporate social responsibility affects consumer behavior and play an important role in everyday life more than economic or rational considerations;(2)the additional effect of the halo effect, unrelated to the consumer on the judgment of daily consumption of corporate social responsibility, such as evaluating new products. Corporate social responsibility plays an important role in ongoing consumer behavior and the impact of consumer purchases (Folkes, V.1984). Therefore, the following hypothesis is offered:

H1: Corporate Social Responsibility has a significant effect on consumer value perceptions

Dodds, Monroe (1985) and Monroe,& Krishnan(1984) tested perceived quality and purchase intention, brand under an empirical model. The results show that prices will positively affect perceived quality, but there is a negative relationship between perceived value and purchase intention. In other words, consumers believe that higher price lead to better quality product perception. Higher perceived quality and purchase intention will also increase. Therefore, trust will result in greater perceived value and a positive effect. The second hypothesis thus is as follows:

H2: Consumers trust relationship and consumer perceived value have a positive and significant relationship

Assessing the implementation of corporate social responsibility lets the average consumer not only to consider their own CSR expectations, but also considerations relative to a Company's unique corporate culture and whether the product is consistent with personal moral beliefs, values to generate consumer trust and loyalty(Endacott, 2003). Menon, & Kahn's(2003)study found that consumers support corporate social responsibility of products under good advertising and believe product quality and are willing to pay a higher price to buy those products. Therefore, when the company takes more CSR actions for their products, the more will consumer value perception and trust relationship be linked. Therefore, the third hypothesis is offered as follows:

H3: The mediation effect of corporate social responsibility between the consumer relationship of trust and perceived value relationship is significant.

Monroe and Krishnan(1985) and Dodds, et al (1985) point out that brand equity achieved through brand name(brand name) or corporate symbol (company symbol)has a positive and direct impact on customer perceived value of products.

Morgan (2000) points out that brand equity is the key to consumer perception toward the brand value for the individual. Further empirical research, Yoo et al(2000) found that brand equity will positively influence the value of the customer and the company's perceived value based on the impact of brand equity and its dimensions. Aaker(1991) proposed a brand equity model for sale price and quality impact of social values and emotional response. Based on this literature, a fourth hypothesis is as follows:

H4: Brand equity has a significant positive impact on perceived value.

Brown and Dacin's(1997)study points out that a company's corporate social responsibility to participate in social activities has a positive impact on its perceived benefits, interests, and stakeholders. The active participation incorporate social responsibility relates to consumer perceived value. It has a negative impact on enterprise responsibility for product failures and brand equity for competing products (Jones, 2005). However, the implementation of corporate social responsibility will also help consumers to agree to fulfill the expectations of stakeholders and enable more value for brand equity. Based on the above theory, this study proposes a fifth hypothesis as follows:

H5: Corporate social responsibility plays a mediator role between brand equity and perceived value.

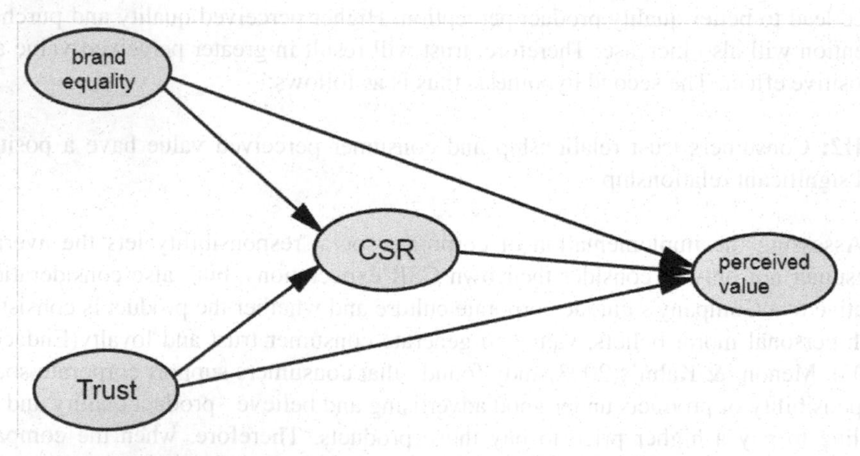

Fig. 1. The Research Concept Model for CSR as Mediator

3 Methodology

In this study, the operational definitions include corporate social responsibility, which is the consumer perception of organizations response to economic, legal, ethical, and spontaneous responsibility. Economic responsibility is the corporate responsibility to maintain the performance of economic growth and produce products that meet consumer expectations. The legal responsibility of the enterprise abides by national business regulations and compliance to sustain economic growth. Moral responsibility is abiding by the social ethical standards and complying with social expectations. Spontaneous responsibility for law and ethics, corporate voluntary implementation bear the responsibility for consumer. The instrument was developed by Maignan & Ferrell (2000). There are total of 18 questions on four dimensions, including Economic, Legal, Ethic and Spontaneous concepts.

Trust refers to the consumer's feeling about the enterprise and its produced products. The products of a company are reliable based on the information provided, the service is trustworthy and the customer believes that the enterprise has the ability to meet these demands, thus achieving commitment to the customer. There are two dimensions, namely reliability and capability, and the survey instrument has eight questions developed by Walter Muller and Helfert (2003).

4 Data Analysis and Results

In this study convenience sampling was used. The students from University that were under junior and senior levels were the sampling target for random sample questionnaire survey. The thrust of this study is the plasticizer events from large enterprises of the plasticizer crisis that was reported on by newspapers and magazines. This event reported on perception of consumer awareness of corporate social responsibility and trust, and product impact as the background for the study case. A total of 250 questionnaires were issued after a random convenience sampling and a total of 213questionnaires were returned, leaving 188questionnaires as valid samples.

4.1 Reliability and Validity Analysis

The questionnaire addressed total four dimensions: Corporate social responsibility, brand equity, trust, and perceived value. Corporate Social Responsibility (CSR) has four dimensions with a total of 18 questions; there are four dimensions of brand equity with total 15 questions, while trust has two dimensions and a total of eight questions; perception of value has four dimensions of total 21 questions. In this study, Cronbach's α and a confirmatory factor analysis assessed the limits and validity, Nunnally (1978) pointed out that the terms of basic research for a reliability coefficient should be above .70 . All questionnaire dimensions showed greater than the reliability Cronbach's α test 0.7 or above. For Corporate Social Responsibility, he Cronbach's alpha value is 0.95; for brand equity, the Cronbach's alpha value is 0.94 and for trust, it is 0.869. The perceived value of Cronbach's alpha value is 0.962. All of the dimensions showed a high degree of validity and reliability.

4.2 Analysis of Convergent Validity

Confirmatory factor analysis dimensions as convergent validity is based on the average number of variables extracted amount (average variance extracted, AVE) as a benchmark. The AVE is mainly the dimensions in the calculation constructs surface of each observed variable (the measured items) the explanatory power of the average variance. If the average variance extracted was higher (the AVE ≧0.5), then the dimensions of the higher reliability and convergent validity. For the corporate social responsibility of all observed variables, each dimension AVE values were 0.562, 0.629, 0.60, and 0.634. The AVE value of the perceived value is 0.758, 0.803, and 0.724. Trust Scale dimensions and their AVE value are 0.585and0.499. The brand equity scale AVE value is 0.693, 0.830, 0.557, and 0.697. All of the various dimensions for recommended AVE values were greater than 0.5 (Hair et al, 2009)

4.3 Overall Model Fit

The scale confirmatory factor analysis model fit indicators were χ 2value of 376.81(df= 59); χ2/df(chi-square value /degrees of freedom)value is 6.387, other fit indexes: GFI= .754, AGFI =0.621, NFI =0.798, CFI= 0.823, all with moderate were lower than the evaluation value of 0.9,but not nearly to the ideal value. Overall were unable to achieve the adaptation index of 0.9 or more, while RMSEA=0.170 and

Table 1. The Overall Model Fit for Construct Structure

Evaluation Indicators	Scope	Evaluation Standard	Model Validated	Research Result	Ideal Evaluation Value Conformity
χ2/df	Less is better	<5	6.387	5.592	Acceptable
GFI	0-1	>0.9	0.754	0.808	Acceptable
AGFI	0-1	>0.9	0.621	0.688	Not good
NFI	0-1	>0.9	0.798	0.835	Acceptable
CFI	0-1	>0.9	0.823	0.859	Acceptable
RMR	Less is better	<0.05	0.094	0.082	Acceptable
RMSEA	Less is better	<0.05	0.170	0.157	Not good

RMR =0.094, SRMR= 0.092, those values only close to the minimum tolerance standard, according to the comprehensive judgment of the goodness of the fit index of the initial confirmatory factor analysis for indicators and are less than ideal.

Therefore, it is needed to release parameters and adjustments based on the modification indices provided by the AMOS program. The results found in the dimensions of brand equity, brand loyalty factor loading 0.773, proposed the deletion of the modification indices to reduce the chi-square value of143. This study is based on the recommendations dimensions to delete the higher score various and after the release of the dimensions recomputed for the model fit analysis. Therefore, according to data provided by the modification indices of the index mode correction, the model fit indicators are as follows: χ 2value of268.434(df= 48); χ 2/ df(chi-square value /degrees of freedom)is5.59,and other indicators of model fit: GFI=.808, AGFI =0.688, NFI =0.835, CFI= 0.859, the overall fit indicators should be greater than 0.9 but remain above the minimum acceptable range, and RMSEA =0.157andRMR =0.082, also close to the standard range, the results in the overall fit indicators, such as shown in Table1.

4.4 Bootstrapping Effect Analysis of the Mediating Test

When using the Bootstrap method in this study to test the mediating effect (Cheung &Lau, 2008), the judgment for (1-α) 100% confidence interval (usually set at 95% CI) lies between the highest and lowest range and does not contain zero. This finding indicates that the α level of each statistic is significant. According to results of the study, brand equity to perceived value is estimated as a value of 0.882, the Z value> 1.96 and presents significantly, while the bootstrap in the Bias-correct and percentile and 95% confidence interval of high standard and low standard value does not contain zero, indicating the establishment of the brand equity for the total effect of perceived value. However, the value of the indirect effect of the bootstrap method in two detection methods is zero, which points out that the indirect effect does not exist. Direct effect from the three tests under high and low values do not contain zero confirms the direct effect of the existence, while the mediator effect is not supported. These results are shown in Table 2.

CSR as the trust and perceived value of the mediator and verification statement results shows that trust under the total effect of perceived value in the coefficient is relative to the product of the Z value <1.96, high and low values of the bootstrap method in the two detection methods that are containing zero. The effect does not exist. Indirect effects and direct effects are not only showing zero detection value, but also the indirect effects do not exist. Therefore, the CSR as mediator between trust and perceived value does not exist. These results are shown in Table 3.

Table 2. The CSR Mediator Test for Brand Equity and Perceived Value

Variable	Estimated Value	Multiplied Coefficient: Product of the Coefficients		Bootstrapping: Bias-Corrected 95% CI		Percentile 95% CI	
		SE	Z	Lower	Upper	Lower	Upper
Total Effects							
Brand Equity → Perceived Value	0.882	0.218	4.04	0.645	1.638	0.592	1.407
Indirect Effects							
Brand Equity → Perceived Value	0.012	0.227	0.052	-0.013	0.156	-0.089	0.067
Direct Effects							
Brand Equity → Perceived Value	0.817	0.352	2.32	0.619	1.1771	0.576	1.451

Table 3. The CSR Mediator Test for Trust and Perceived Value

Variable	Estimated Value	Multiplied Coefficient: Product of Coefficients		Bootstrapping: Bias-Corrected 95% CI		Percentile 95% CI	
		SE	Z	Lower	Upper	Lower	Upper
Total Effects							
Trust →Perceived Value	-0.083	0.256	0.324	-0.910	0.232	-0.688	0.295
Indirect Effects							
Trust →Perceived Value	-0.039	0.342	0.114	-0.228	0.087	-0.178	0.159
Direct Effects							
Trust →Perceived Value	-0.044	0.489	0.089	-1.017	0.325	-0.783	0.408

Hypothesis Testing

The trust to corporate social responsibility estimated value is 0.476, standard error=0.162, the CR value is 2.934(P = 0.003) significant in the model test results. Brand equity to corporate social responsibility points to an estimated value of-0.143, standard error=0.126, the CR value of-1.137(P =.255) showing it is not significant. Corporate social responsibility and perceived value estimate= - 0.082, standard error=0.11, the CR value of-0.745(p = 0.255) showed as not significant.

Trust to perceived value estimate is-0.044, standard error=0.202, CR value of-0.218(P = 0.828) showing as not significant. Brand equity to perceived value estimated value is 0.871, standard error=0.159, so the CR value of 5.459(P = 0.001) is significant.

5 Conclusion and Discussion

The globalization wave has reached many countries. International companies with strong capital, technology, and individuals now enter the development market in countries and battle for market share. Both the supply chain and supplier relationships are an essential tools for strategic competitiveness. Suppliers and downstream information integration issues with the concept of corporate social responsibility is mature and gradually expanding. However, corporate social responsibility will no longer be limited to operating brands and their image from enterprises. The suppliers of raw materials will also be a critical partner and be pulled into the circle of trust and business strategies to enhance international trade supply chain requirements.

The Taiwan plasticizer event indicates the broad supply chains problem on the surface. Whether this product harm recall reaction to a health issue event affects consumer trust in business brand image produces a deep impact on product value perception and customer trust, moreover lower customer brand equality. For consumers, the research find out on corporate social responsibility in corporate trust and brand equity of products' perceived value for the plasticizer incident value perception test as the mediator role does not affect and even not relative as important linkage. The result suggests that consumers require corporate social responsibility that is higher than the average of the other dimensions, especially in terms of legal liability and spontaneous responsibility being higher than ethical responsibility and economic responsibility. Corporate responsibility in the minds of consumers should be consistent with national law and companies' self-motivation to produce CSR. Therefore, enterprises should be consistent in their social responsibility to fulfill consumers' expectations.

However, under corporate social responsibility associated with the perceived value and trust of consumers, the perception and trust of the consumer is substantial, and accepted in that direct responsibility, especially for price dimensions, does higher affect than other dimensions. Moreover, the trust dimensions for consumers still indicate that enterprises have the ability to solve any product harm recovery and handle product returns. The result shows that the event affected the reliability of product perception, product reliability and value under the slow move and rejection of responsibility since the event as reported by the press was unable to resolve the product harm issues and crisis management for large enterprise access to consumer trust.

The company that faces its corporate social responsibility and recalls products immediately will have a positive significant impact on the trust structure. Corporate social responsibility for the perceived value of trust and brand equity cannot render the mediator effect; consumers do not support the outcome of the corporate social responsibility for the association of perception of trust and product value and need more work out to retrieve customer confidence and trust.

References

1. Aaker, D.A.: Managing Brand Equity. Free Press, New York (1991)
2. Brown, T.J., Dacin, P.A.: The company and the product: corporate associations and consumer product responses. Journal of Marketing 61, 68–84 (1997)
3. Chang, T.Z., Wildt, A.R.: Price, product information, and purchase intention: An empirical study. Journal of the Academy of Marketing Science 22(1), 16–27 (1994)
4. Cheung, G.W., Lau, R.S.: Testing mediation and suppression effects of latent variables: Bootstrapping with structural equation models. Organizational Research Methods 11(2), 296–325 (2008)
5. Dodds, W.B., Monroe, K.B.: The effect of brand and price information on subjective product evaluations. Advances in Consumer Research 12, 85–90 (1985)
6. Doney, Cannon: An Examination of the Nature of Trust in Buyer-Seller Relationships. Journal of Marketing, 35–52 (April 1997)
7. Drumwright, M.: Socially responsible organizational buying: Environmental concern as a non-economic buying criterion. Journal of Marketing 58(3), 1–19 (1994)
8. Endacott, R.W.J.: Consumers and CRM: a national and global perspective. Journal of Consumer Marketing 21(3), 183–189 (2003)
9. Erickson, G.M., Johansson, R.I.: The role of price in multi-attribute product evaluation. Journal of Consumer Research 12, 195–199 (1985)
10. Folkes, V.: Consumer reactions to product failure: An attribution approach. Journal of Consumer Research 10(4), 398–409 (1984)
11. Gale, B.T., Wood, R.C.: Managing customer value: Creating quality and service that customers can see. The Free Press, New York (1994)
12. Hair Jr., J.F., Anderson, R.E., Tatham, R.L., Black, W.C.: Multivariate Data Analysis, 5th edn., p. 14. Prentice-Hall, Englewood Cliffs (1998)
13. Handelman, J., Arnold, S.: The role of marketing actions with a social dimension: Appeals to the institutional environment. Journal of Marketing 63(3), 33–48 (1999)
14. Richard, J.: Finding sources of brand value: developing a stakeholder model of brand equity. J. Brand Manag. 13(1), 10–32 (2005)
15. Maignan, I., Ferrell, O.C.: Measuring corporate citizenship in two countries: The case of the United States and France. Journal of Business Ethics 23(3), 283–297 (2000)
16. Martins, L.L.: A Model of the Effects of Reputational Rankings on Organizational Change. Organization Science 16, 701–720 (2005)
17. Menon, S., Kahn, B.E.: Corporate sponsorships of philanthropic activities: when do they impact perception of sponsor brand? Journal of Consumer Psychology 13(3), 316–327 (2003)
18. Mohr, L.A., Webb, D.J.: The effects of corporate social responsibility and price on consumer responses. Journal of Consumer Affairs 39(1), 121–147 (2005)
19. Monroe, K.B., Krishnan, R.: The effect of price on subjective product evaluations. In: The Perception of Merchandise and Store Quality. D. C. Heath, MA (1984)
20. Morgan, R.P.: A consumer-orientated framework of brand equity and loyalty. International Journal of Market Research 42(1), 65–78 (2000)
21. Nunnally, J.C.: Psychometric theory, 2nd edn. McGraw-Hill, New York (1978)
22. Osterhus, T.: Pro-social consumer influence strategies: When and how do they work? Journal of Marketing 61(4), 16–29 (1997)
23. Parasuraman, A.: Reflections on gaining competitive advantage through customer value. Journal of the Academy of Marketing Science 25, 154–161 (1997)

24. Peloza, J.: Using corporate social responsibility as insurance for financial performance. California Management Review 48(2), 52–72 (2006)
25. Petrick, J.F.: Development of a Multi-Dimensional Scale for measuring the perceived value of a service. Journal of Leisure Research 34(2), 119–134 (2002)
26. Porter, L.W., Steers, R.M., Mowday, R.T., Boulian, P.V.: Organizational commitment, job-satisfaction, and turnover among psychiatric technicians. Journal of Applied Psychology 59, 603–609 (1974)
27. Ring, P.S., Van de Ven, A.H.: Structuring Cooperative Relationships between Organizations. Strategic Management Journal (13), 483–498 (1992)
28. Rundle-Thiele, S., Ball, K., Gillespie, M.: Raising the bar: from corporate social responsibility to corporate social performance. Journal of Consumer Marketing 25(4), 245–253 (2008)
29. Sheth, J.N., Newman, B.I., Gross, B.L.: Why we buy what we buy: a theory of consumption values. Journal of Business Research 22(2), 159–170 (1991)
30. Suri, R., Kohli, C., Monrce, K.B.: The effects of perceived sarcity on consumers' processing of price information. Academy of Marketing Science 35(1), 89–105 (2007)
31. Walter, A., Muller, T.A., Helfert, G., Ritter, T.: Functions of industrial supplier relationships and their impact on relationship quality. Industrial Marketing Management 32, 159–169 (2003)
32. Yoo, B., Donthu, N., Lee, S.: An examination of selected marketing mix elements and brand equity. Journal of the Academy of Marketing Science 28(2), 195–211 (2000)

Using Chance Discovery in Recruiting Strategies to Explore Potential Students in Taiwan

Feng-Sueng Yang[1], Ai-Ling Wang[2], and Ya-Tang Yang[3]

[1] Department of Information Management, Aletheia University No. 32, Zhenli Street, Tamsui Dist., New Taipei City, 25103, Taiwan, R.O.C.
fsyang@mail.au.edu.tw

[2] English Department, Tamkang University, Tamsui, New Taipei City, Taiwan, R.O.C.
wanga@mail.tku.edu.tw

[3] Department of Business Administration, Chihlee Institute of Technology, Banciao, New Taipei City, Taiwan, R.O.C.
tammy@mail.chihlee.edu.tw

Abstract. The issue of decreasing birth rate has caused a big challenge for the Ministry of Education in Taiwan. The universities are looking for solutions to the problem. Current recruiting strategies include college admission fairs, college admission seminars, campus visits, and press release. How to find the potential students has become a big challenge. The present study aims at exploring the potential source of the Department of Information Management of a university via the application of Chance Discovery and KeyGraph and the operation of frequency and association parameter value and sensitivity analysis. Findings of the study showed that students from Taipei and New Taipei City, living in a distance within 15 Km to the university are basic students of the university. Furthermore, the Zip code 970, 608, 636 and high schools that have a distance around 63 Km to the university may also be potential students of the university.

Keywords: Chance Discovery, KeyGraph, decreasing birth rate, recruiting activities.

1 Introduction

Due to the phenomenon of "decreasing number of children" in Taiwan, the Ministry of Education (MOE) is facing a serious problem. In the three years to come, higher education in Taiwan will be confronting its first challenge. It will be especially apparent in private colleges or universities, who are eager to look for solutions to the problems of recruiting students. Decreasing number of children means the decrease in birth rate and, consequently, the decrease in young population and the entire population of Taiwan. This phenomenon is going to affect the entire social structure and the economical development of Taiwan. The entire population of Taiwan was slightly over 300,000 between 1989 and 1997, and it dropped to 270,000 in 1998. This will first cause serious problems for colleges or universities in the 2016 academic year. In 2000, the population of Taiwan slightly raised to 300,000. However, from the next

N.T. Nguyen et al. (Eds.): ACIIDS 2014, Part II, LNAI 8398, pp. 444–453, 2014.

year on, the population of Taiwan dropped dramatically. In 2010, as it turned out, there were only about 160,000 people in Taiwan, approximately half of the population in the 1990s [15]. This phenomenon has worried the Taiwanese government, and they have been looking for solutions to the serious problem.

Currently, the common practice of promoting the school and recruiting students includes nation-wide college fairs, seminars for high school students, face-to-face talks with high school students, advertisements in mass media. How effective these unfocused recruiting strategies can be is really doubtful. The author argues that recruiting strategies should be decided based on the attributes of individual departments or programs. It may be more effective to target your recruiting campaign at those highly potential students. Generally speaking, high school students do not really understand a college or its departments, and their decision is probably influenced by the ranking of schools, their parents, teachers, or relatives. They are not really sensitive to schools or departments. Aside from those high-achieving and low-achieving students, who have less schools or departments to choose from, it does not make a real difference for those falling at the intermediate level to choose department A or department B. It live schools some room for promoting themselves. However, it is a great and rewarding challenge for universities to explore their sources of students.

The purpose of this study is to explore the sources of students for the information department in a university via applying the theories of Chance Discovery and KeyGraph and use of frequency and association value in sensitivity analysis. The sections that will follow are literature review that include an overview of the common practice of recruiting students in Taiwan, and theories and applications of Chance Discovery and KeyGraph, methodology and procedure, findings, and conclusion and discussions.

2 Literature Review

2.1 Studies on Higher Education's Recruiting Strategies

Generally speaking, previous studies on recruiting strategies of higher education were based on the application of marketing, and questionnaire surveys and case studies were mostly used [8] [10]. However, the quality of the questionnaire, the willingness, involvement, and reliability of the respondents may greatly affect the findings of the survey [2]. When the distribution of potential students appears to have unique characteristics in a particular group, using the strategies of market segmentation may be helpful for universities to respond to a loosely distributed student market and to students' needs. Many researchers now is resorting to the strategies of market segmentation and offering alternatives to respond to the issues of recruiting students [1] [3] [5]. In order to solve the potential problems caused by questionnaire survey, some researchers make use of the techniques of detecting clusters in data mining to explore the patterns of student distribution and the characteristics of a particular group so as to provide the school with recruiting strategies that meet the requirement of the student market [12].

As far as marketing strategies are concerned, universities in the U.S. resorted to commercial ad, development of public relations, fund raising, recruiting management, etc. The president of a university is always busy recruiting students. Business terms, such as customers, shareholders, Niche marketing, brand marketing, appeared in academia, and it is a sign that higher education is transforming into business enterprise [11]. In Taiwan, the common practice for colleges and universities to recruiting students include college recruiting fairs, campus visits, workshops for potential students, TV commercials, newspaper or magazine advertisements, road signs, audio or video tapes, reports, and propagandas [14].

2.2 Theories and Applications of Chance Discovery and KeyGrraph

Chance Discovery was first proposed by Ohsawa [16] [17] as a new issue of research. Ohsawa considered "chance" an event or a scenario critical to decision making existing in a dynamic environment. These events or scenarios are rare and are easily ignored. From the perspective of information management, data that are rare and are structurally important can be considered "chances." Chance Discovery is mainly a combination and extension of information interception, social networks, and small world. Generally speaking, the tool KeyGraph proposed by Ohsawa [17] and others is currently used in the study of Chance Discovery.

KeyGraph is a tool used to present strategic information in a visualized interface to decision makers. It is a tool used to do data mining and, based on the frequencies and positions of words or events appearing in a database to calculate the association values between and among words or events. The results are then presented to decision makers in a figure form. The steps as to how KeyGraph calculates are illustrated as follows [7]:

Here we assume that a document, D , is composed of sentences and each sentence is composed of words. The main steps of the KeyGraph algorithm can be outlined as follows.

Document Preprocessing, Which Consists of Two Tasks:

1. Document compaction: insignificant words from the document are removed using a user-supplied list of words and word stems. Word stems are used to reduce related words to the same root. For example, words like 'innovate', 'innovates' and 'innovating' are reduced to 'innovate' using a method proposed by Porter [18].

2. Phrase construction: here preference is given to longer phrases with higher frequency. A subset of l_{phrase} words are chosen from the document and all possible phrases out of those words are constructed. A phrase that occurs with the highest frequency in the document is retained.

It should be noted that in this study we don't use any preprocessing. After preprocessing, if any, the document D is reduced to D', which consists of unique terms $w_1, w_2, \cdots, w_l$, where a term w_i refers to either a word or a phrase.

Extracting High-Frequency Terms:
Terms in D' are sorted by their frequencies in the document D and top n_{nodes} (high-frequency terms) are retained. These high-frequency terms are represented as nodes in a graph G. A set of the high-frequency terms is denoted by N_{hf}.

Extracting Links:
Links represent co-occurrence-term-pairs that often occur in the same sentence. A measure for co-occurrence of terms w_i and w_j is defined as

$$\text{assoc}(w_i, w_j) = \sum_{s \in D} \min\left(|w_i|_s, |w_j|_s\right)$$

where w_i and w_j are elements of the set N_{hf}, and $|w_i|_s$ is the number of times a term w_i occurs in a sentence s.

The assoc values are computed for all pairs of high-frequency terms in N_{hf}. The term-pairs are sorted according to their assoc values and top $N_{\text{hf}} - 1$ tightly associated term-pairs are taken to be the links. The links between term-pairs are represented by the edges in G.

Extracting Key Terms:
Key terms are terms that connect clusters of high-frequency terms together. To measure the tightness with which a term w connects a cluster, the following function is defined:

$$\text{key}(w) = 1 - \prod_{g \subset G} \left[1 - \frac{\text{based}(w,g)}{\text{neighbors}(g)}\right]$$

where g is a cluster, and

$$\text{based}(w, g) = \sum_{s \in D} |w|_s |g - w|_s$$
$$\text{neighbors}(g) = \sum_{s \in D} \sum_{w \in s} |w|_s |g - w|_s$$
$$|g - w|_s = \begin{cases} |g|_s - |w|_s & \text{,if } w \in g, \\ |g|_s & \text{,if } w \notin g, \end{cases}$$

where $|g|_s$ is the number of times a cluster g occurs in a sentence s.

Qualitatively, $\text{key}(w)$ gives a measure of how often a term w occurs near a cluster of high-frequency terms.

The *key* values are computed for all the terms in Dassoc values and n_{key} top *key* terms are taken as *high-key terms*. These high-key terms are added as nodes-if they are not already present-in G and are elements of a set K_{hk}.

Extracting Key Links:
For each high-frequency term $w_i \in N_{\text{hk}}$ and each high-key term $w_j \in N_{\text{hk}}$, $\text{assoc}(w_i, w_j)$ is calculated. Links touching w_j are sorted by their assoc values for each high-key term $w_j \in N_{\text{hk}}$. A link with highest assoc values connecting w_j to two or more clusters is chosen as a key link. Key links are represented by edges-if they are not already present-in G.

Extracting Keywords: nodes in G are sorted by the sum of assoc values associated with the key links touching them. Terms represented by nodes of higher values of these sums than a certain threshold are extracted as keywords for the document D.

The theory of Chance Discovery has been widely used by researchers. Hong [9] used Grounded Theory as basis to integrate with domain knowledge and the interactive procedure of data mining to develop Qualitative Chance Discovery Model (QCDM), and this model has been used in the Optical Industry in Taiwan to explore the potential chances. Li et al. [13], on the other hand, used the techniques of Chance Discovery to explore the rare and important features or techniques of patented DVD and provided the industries with information for developing patent items. Furthermore, the theory of Chance Discovery is also used in analysis of bankruptcy of banks [6] and are applied to Bulletin Board Services [19].

The researcher of the present study made use of the features of KeyGraph, applied the operation of key value, and operated Sensitivity Analysis to solve the problem of KeyGraph caused by different settings of parameter value [20]. In this study, the researcher analyzed the data of students admitted to the Department of Information Management. The following section to present the methodology of the present study.

3 Methodology

3.1 Data Collection and Data Analysis

The author collected data from the statistical information of the students admitted to the Department of Information Management of AU, ranging from the 2008 to 2013 academic year, including students' demographic information, levels of students' scores based on the five categories assigned in that year, the distance between their high schools and AU. There were 275 students' information collected for the present study. Individual student's information included gender, ID number, the high school from which he or she graduated and the distance between the high school and AU, address and zip code, finally the scores he or she earned in the five main subjects they took on the entrance exam, namely Chinese, English, Math, Social Science, and Natural Science. The author then programmed a KeyGraph formula to obtain the key values and Red Nodes (meaning the potential opportunities).

3.2 Sensitivity Analysis

It is essential to set parameter values of Threshold of Frequency (TF), Association rule, Threshold of Association (TA), Number of double-circled black nodes, Number of red nodes, and Threshold of Frequency of KeyGraph. Double cycle black nodes stand for High Frequency and High Key Value Terms and Red nodes stand for Key words or Terms. There is no standardized way to set parameter values, and it is also possible that different settings of parameter values can generate different KeyGraph [20]. A common practice is to use the principles of Try and Error to obtain the best result. The researcher used Sensitivity Analysis in the hope to obtain the best results. Sensitivity Analysis answers "what if" questions: "If we make a slight change in one or more aspects of the model, does the optimal decision change?" If so, the decision is said to be sensitive to these small changes, and the decision maker may wish to reconsider more carefully those aspects to which the decision is sensitive [4].

The parameter values settings at the beginning were as follows: Threshold of Frequency=2, Association rule=Initial association, Threshold of Association=5, Number of black nodes= 6, Number of red nodes= 6, Threshold of Frequency of KeyGraph=3. The researcher adjusted the values of sensitivity for the Threshold of Frequency from 2 to 100 and the Threshold of Association from 5 to 50 or 90. The researcher stopped the task in case the same red nodes appeared. It is a sign of stability. Table 1 is an example of Sensitivity Analysis, using Threshold of Frequency=15.

In this case, TF=2, and TA ranges from 5 to 50 with an interval of 5. At the point of 50 in TA, the 6 key terms remain unchanged. However, at the point of TF=15, the situation is not stable until TA=80. At the point of TF=100, the 6 Key Terms are unchanged no matter how TA are changed. The fact is that the 6 Key Terms are unchanged from TF=60 on no matter how TA are changed. Therefore, in the present study, the potential chances were decided by the 6 stable key terms of TF.

Table 1. Example of Sensitivity Analysis

TF	TA	Key term 1	Key term 2	Key term 3	Key term 4	Key term 5	Key term 6
15	5	242	HS248	dis_24	114	251	dis_27
15	10	242	HS248	dis_24	114	251	dis_27
15	15	HS118	HS218	241	HS222	dis_24	HS227
15	20	HS118	HS218	dis_27	dis_24	241	HS354
				…			
15	75	HS248	HS118	242	247	HS232	dis_24
15	80	HS248	HS118	242	247	dis_24	HS232
15	85	HS248	HS118	242	247	dis_24	HS232
15	90	HS248	HS118	242	247	dis_24	HS232

The procedure to calculate is to first look for the points of potential chances, and then calculate the frequency of each key term and the frequencies are ranked. The next step is to calculate the key value of each potential point and rank the key value. The difference between the two values is termed "ranking difference". Those key terms that have large ranking difference are in low frequency and high key value, and they can be considered those with greater potential chances. Findings of the study are presented in the following section.

4 Findings

4.1 Frequency of Demographic Data

Frequency of demographic data is shown in Fig. 1.

This chart shows that the percentages of males and females admitted to the department are 58.91 and 41.09 respectively. The highest frequency is "Total_Level 4", and it takes 80%. Birthplace codes A and F have a total of 68%. The highest frequency of distance between the high school and AU is 15 Km.

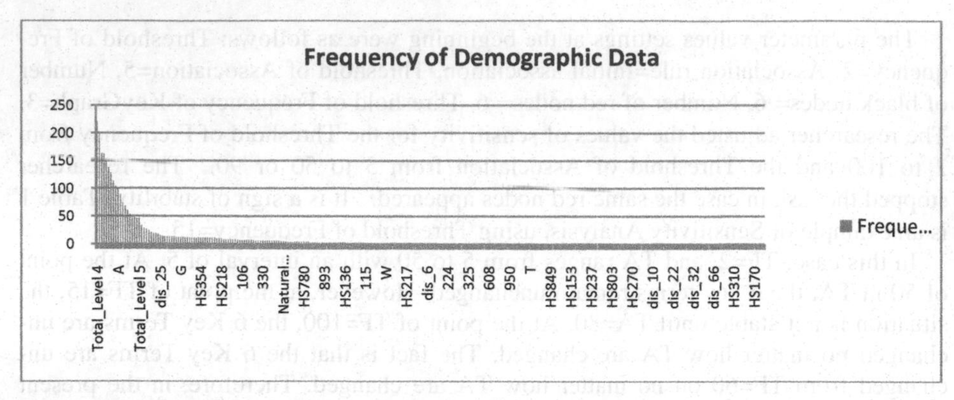

Fig. 1. Frequency of Demographic Data

4.2 K-Means Cluster Analysis

The researcher first coded the 275 pieces of data and came up with 6 codes, namely distance (dis) , five levels for Chinese (11, 12, 13, 14, 15), five levels for English (21, 22, 23, 24, 25, five levels for math (31, 32, 33, 34, 35), five levels for social science (41, 42, 43, 44, 45), five levels for natural science (51, 52, 53, 54, 55), and five levels for the total grades (61, 62, 63, 64, 65). K-means cluster analysis was then done to look for possible relationships between the distance between the high school and AU and possible clusters based on students' scores of different subjects.

Table 2. K-means Cluster Analysis

	Cluster		
	1	2	3
dis	500	21	221
1 Chinese	15	14	14
2 English	24	24	24
3 Math	34	34	34
4 Social Science	44	44	43
5 Natural Science	55	54	54
6 Total Scores	64	64	64
Total No.	2	250	23

Findings of the study (as shown in Table 2) showed that the sources of students of the department can be roughly divided into 3 clusters, and most of the students mainly belong to the 2nd cluster, meaning their grades of each subject fall on Level 4, school distance is around 21 Km. The next highest cluster are students living in Central

Taiwan and the school distance is around 221 Km. and they had an average higher scores in social science (Level 3). There are only 2 students who belong to the 3rd cluster, who are overseas Chinese from mainland China, and they had low scores in Chinese and Natural Science (Level 5).

K-means Cluster Analysis mainly aims at exploring possible clusters. However, it is not to make sense as to how the chart means. The researcher used the theory of Chance Discovery to analyze the KeyGraph data to explore important information.

4.3 Analysis of Key Value and Key Terms

Based on the procedures stated above, key terms are presented in Table 3. The value of Ranking Difference has its threshold set for 10 and above. Because "dis_63" is the first item shown in the chart and the following items are all names of high schools, these high frequency schools do not meet the principles and requirements of KeyGraph, 10 is chosen as the threshold value. Zip code 970 (Hualien City) has the highest ranking deviation, meaning low frequency and high key term. It is a potential term. Zip code 320 (Jungli City), 249(Bali District), and 247(Luzhou District) are currently not really the sources of AU students, these areas are close to AU and are potential sources of AU students. Findings of study can provide the recruiting department with important information on key high schools to focus on, mainly in Chiayi City, Yunlin County, Taoyuan County, Hsinchu City, Taipei and New Taipei City. It is crucial to develop academic relationship and collaboration with those potential high schools. The appropriate distance between the high school and AU is 63 Km, roughly around Hsinchu City.

Table 3. Ranking Difference between Frequency and Key value

Ranking Difference	Key term	Key value
163	970	0.109134
	…	
34	320	0.072109
32	HS335	0.104901
29	249	0.087884
	…	
19	247	0.081877
19	HS150	0.115871
19	HS129	0.102742
	…	
10	dis_63	0.036143

5 Conclusion and Discussions

In an era of decreasing birth rate, academic institutions are looking for ways to mitigate the impact caused by the challenge. Generally speaking, the academia applies traditional ways of mass marketing to recruiting students. However, as the author argues, different recruiting strategies should be applied to different schools because of their different academic attributes and different historical backgrounds. It has become an important issue for school administrators and teachers to look for potential students and to provide these students with one-on-one promotion. It is an interesting issue as to how we can discover those potential students.

The author of the present study applied KeyGraph used for chance discovery to the discovery of potential students and their shared characteristics. Findings of the study showed that this method can indeed help discovery of potential students. This method not only focused on the levels of frequency, but also took the relationships or connections between two schools into consideration. Take AU as an example, in addition to being physically closed to AU, some high schools may be potentially the source of prospective students because of their religious attributes. Just after the completion of this study, one of the high schools made an alliance with AU. It is apparent from the fact that the method can be helpful for schools to discover the sources of potential students. In the future studies, the author will look for more pieces of evidence to ensure the validity and reliability of the method.

References

1. Adler, K.: Degree upgrades: a new service, a new market and a new strategy for higher education. Journal of Marketing for Higher Education 9(11), 11–23 (1998)
2. Berry, M.J.A., Linoff, G.S.: Data Mining Techniques for Marketing, Sales and Customer Support. John Wiley & Sons, Inc., New York (1997)
3. Bonnici, J.L., Reddy, A.C.: Breaking away from the pack: positioning the marketing discipline through a triangular analysis. Journal of Marketing for Higher Education 4(1), 107–119 (1993)
4. Clemen, R.T., Reilly, T.: Making Hard Decisions with DecisionTools. Duxbury/Thomson Learning, CA (2001)
5. Coccari, R.L., Javalgi, R.G.: Analysis of students' needs in selecting a college or university in a changing environment. Journal of Marketing for Higher Education 6(2), 27–39 (1995)
6. Goda, S., Ohsawa, Y.: Estimation of Chain Reaction Bankruptcy Structure by Chance Discovery Method - with Time Order Method and Directed KeyGraph. Journal of Systems Science and Systems Engineering 16(4), 489–498 (2007)
7. Goldberg, D.E., Sastry, K., Ohsawa, Y.: Discovering Deep Building Blocks for Competent Genetic Algorithms Using Chance Discovery via KeyGraphs. IlliGAL Report, No. 2002026 (2002)
8. Hall, M.C., Elliott, K.M.: Strategic planning for academic departments: a model and methodology. Journal of Marketing for Higher Education 4(2), 295–307 (1993)
9. Hong, C.F.: Qualitative Chance Discovery-Extracting Competitive Advantages. Information Sciences 179(11), 1570–1583 (2009)

10. Ko, T.H.: Admissions Strategy and Status of Taiwan Vocational Education in Viet Nam. Quarterly Journal of Technological and Vocational Education 1(4), 7–13 (2011) (in Chinese)
11. Krip, D.L.: Shakespeare, Einstein, and the Bottom Line-The Marketing of Higher Education. Harvard University Pr. (2004)
12. Lee, W.I., Shih, C.C.: The Research of University Recruiting Strategies Positioning by the Application of Data Mining. Journal of National University of Tainan 37(2), 47–66 (2003) (in Chinese)
13. Li, Y.R., Wang, L.H., Hong, C.F.: Extracting the Significant-rare Keywords for Patent Analysis. Expert Systems with Applications 36(3-1), 5200–5204 (2009)
14. Liu, H.H., Wei, C.: A Study of Recruiting Strategies of Sports and Leisure Department. Sports Research Review (100), 21–28 (2009) (in Chinese)
15. Ministry of Education in Taiwan (2013), https://stats.moe.gov.tw
16. Ohsawa, Y., McBurney, P. (eds.): Chance Discovery. Advanced Information Processing Series. Springer, Berlin (2003)
17. Ohsawa, Y., Benson, N.E., Yachida, M.: KeyGraph: Automatic Indexing by Co-occurrence Graph Based on Building Construction Metaphor. In: Proceedings of Advanced Digital Library Conference (IEEE ADL 1998), pp. 12–18 (1998)
18. Porter, M.F.: An algorithm for suffix stripping. Automated Library and Information Systems 14(3), 130–137 (1980)
19. Seo, Y., Iwase, Y., Takama, Y.: KeyGraph based BBS for Online Chance Discovery. In: IEEE International Conference on Systems, Man and Cybernetics, pp. 1754–1758 (2006)
20. Takama, Y., Iwase, Y.: Scenario to Data Mapping for Chance Discovery Process. In: Braham, A., Dote, Y., Furuhashi, T., Köppen, M., Ohuchi, A., Ohsawa, Y. (eds.) Soft Computing as Transdisciplinary Science and Technology. Advance in Soft Computing, pp. 470–479. Springer (2004)

A Fees System of an Innovative Group-Trading Model on the Internet

Pen-Choug Sun[1,*], Rahat Iqbal[2], and Shu-Huei Liu[3]

[1] Department of Information Management, Aletheia University No. 32, Zhenli Street, Tamsui Dist., New Taipei City, 25103, Taiwan, R.O.C.
{au1159,au1076}@au.edu.tw
[2] Department of Computer and Network Systems, Coventry University, Priory Street, Coventry, CV1 5FB, UK
aa0535@coventry.ac.uk

Abstract. A new Core Broking Model (CBM) has been introduced to e-markets, involving joint-selling of multiple goods and offering volume discount for group-buying coalitions. It is a core-based model, maintaining the stability of coalitions and using physical brokers to resolve group-trading problems in e-markets. After a survey has been made on the present commission of brokers and the current fees systems of some popular e-marketplaces, a fees system for the CBM has been set up and was suggested to the model. The fees system consists of four kinds of fees: final value fee, handling fee, session fee and online store fee. It is evaluated and discussed at the end of this paper.

Keywords: Fees system, Brokers, Broking system, Commission, E-Markets.

1 Introduction

E-commerce is “the best approach for organizations to manage their activity as an integral part of their approach to developing and sustaining customer relationship” [1]. The growth of e-commerce was at one time more than 20% per year all over the world. “Much of the retail sector's overall growth in both the US and the EU over the next five years will come from the Internet” [2], so there will be unceasing large potential profits for traders in Internet e-commerce. When traders get together on the Internet and work out deals, they seize every opportunity to maximize their own profits. Forming coalitions is an effective way of striving to achieve their goals. Therefore, concepts and algorithms for coalition formation have been investigated in Multi-Agent Systems, Computer Science and Economics communities [3, 4, 5]. Issue of coalition problems is increasingly popular to researchers in e-commerce, “where business transactions take place via telecommunications networks” [6].

E-market is a new trading model for Internet e-commerce. It gives product information and trading mechanisms to traders and provides them opportunities to execute transactions. Buying or selling goods in e-markets has become an essential aspect of the daily lives of many people. It is understandable why coalition problems in e-markets attract so much attention from both academics and practitioners. A coalition

N.T. Nguyen et al. (Eds.): ACIIDS 2014, Part II, LNAI 8398, pp. 454–464, 2014.

with stability is in a condition when every member has the incentive to remain in it [7]. The core is the stable set of profits that no coalition can improve upon [8]. However, it is incapable of dealing with large coalitions [9].

A new core-based model called Core Broking Model (CBM) has been built [10]. The CBM combines coalitions in e-markets into a bigger coalition [11]. It inherits from the core to ensure the stability of coalitions, but makes many improvements [12] to have incentive compatibility, distributed computing, and less computational complexity [13] and can be adopted in e-markets [14]. It highly depends on physical brokers, who are "persons who buy and sell goods or assets for others" [15]. They carry out transactions and play a wide variety of roles in trading process. They may be salespersons and fill orders for their clients. They may also serve as trading advisers to their customers providing services like market research or trade recommendations. Commission is an effective way to encourage brokers to fulfill their duties. A commission agreement is used in the new model.

The CBM consists of a site, which is a broking system defining as "the business or service of buying and selling goods or assets for others" [15]. Many online shopping sites such as eBay are such systems providing services to their users and executing transactions. Their main revenues come from the users. A financial reward can be an incentive to these sites to provide a good service to their users. People are willing to pay for the services, because they are tailored to their needs. The sellers pay for the privilege of coming to the e-markets and finding potential customers there. These sites keep themselves afloat with the fees from sellers. Fees systems of shopping sites are explored here, so that a reasonable fees system can be set up for this new model.

Here, a fees system is proposed to provide guidelines for the CBM about how the brokers and the site may be paid. To construct a suitable fees system, in section 3, there are some investigations into commissions for brokers and fees systems in 15 popular shopping websites. In the last section, the proposed fees system is evaluated.

2 Core Broking Model (CBM)

This core-based model involves joint-selling of multiple goods, offering volume discount for group-buying coalitions in e-markets. Several providers conduct bundle selling together, while, many buyers are for the amount discounts. The descriptions of

- **Core-brokers:** the initiators of the group-trading projects.
- **Projects:** Each project has several sessions of group-trading in e-markets.
- **Providers:** The core-brokers invite them to provide products and services for customers.
- **Market-brokers:** who play the role of team members to help with the projects.
- **E-markets:** may be any existing online shopping avenues such as eBay or the market-brokers' own sites on which they can post projects and find customers.
- **Buyers:** the market-brokers' clients, who have been attracted to the projects.
- **The Core Broking System (CBS)** consists of three components as follows:
  - **CBS Website:** A place where core-brokers and market-brokers meet together. Market-brokers may come here and search for the projects which interest them. Members of the site can report and open a case for a problematic transaction in its resolution centre.

– **Project Subsystem:** a system specially designed to assist the core-broker in managing all the necessary tasks to assure quality outcomes.
– **Market Subsystem:** the market-brokers can use it to perform transactions for a session on a project; purchase electronic coupons from the core brokers and sell them to their clients.

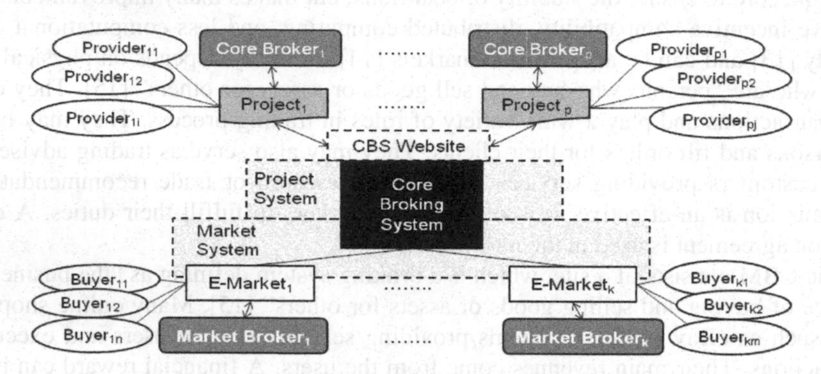

Fig. 1. The Structure of the CBM

The system flow chart of the model shown in Fig. 2 is used to explain the process of the CBM. A brief description for the six stages in the process is as follows:

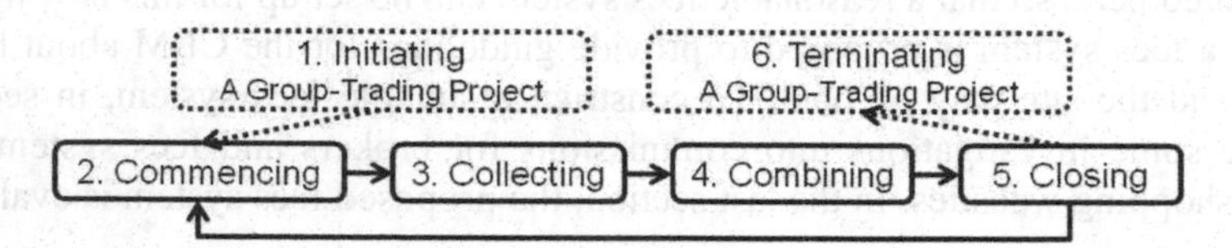

Fig. 2. System Flow Chart of the CBM

1. **Initiating** – A core broker setups a proposal for a group-trading project, settle the project with some providers and lists the project on the CBS website.
2. **Commencing** – After recruiting several market brokers, the core broker officially begins sessions of group trading in the group-trading project.
3. **Gathering** – The market brokers attract buyers to their websites, combine the orders of buyers into market-orders and submit the market-orders to the core broker.
4. **Combining** – The core broker checks the stability of the coalitions, combines the coalitions together, decides the final prices for the items and sends acceptance notices to the market brokers.
5. **Closing** – When the buyers have paid for their purchases, they receive coupons. Finally, the brokers close the deal with their clients and the benefits of the participants are calculated.
6. **Terminating** –The core broker backups the data tables and analyse the transactions in the sessions of the group-trading project for the future use.

To ensure a healthy level of competition, the new model adopts brokers to prevent price makers as would occur in 'monopolies', giving them "exclusive possessions or control of the supply of or trade in a commodity or service" [15] or 'cartels', acting as a formal arrangement between producers and manufacturers who agree to fix prices, marketing, and production [16]. Both core brokers and market brokers play important roles in the CBM. They make possible the collaboration between the members of coalitions. The core-broker is like a project manager. On the other hand, the market-brokers act like salesmen in the CBM.

The core-brokers initiate projects on the CBS website. The market-brokers promote products of the project on the appropriate shopping sites and form buyers' coalitions there. The core-brokers invite providers to perform joint-selling to increase the 'competitive advantage' [17]. They provide information to the market-brokers for them to promote the product. Each session has a starting and an ending date. The suggested duration for a session is usually one week. The iterative process is looping between stage 2 and stage 5. At the end of a session, the core brokers may choose to have a new session of trading or stop the project for good.

The model manages the orders on a First Comes First Serves (FCFS) basis. If an order is submitted late, the required items may no longer be available, especially when the supply of products falls short of demand. If the actual quantity is less than the quantity ordered, the brokers have to decide how to distribute the items fairly. There are many ways to do so. Other than FCFS, another option is to use the Shapley value, which is decided by the original amount ordered by each customer. An agreement may also be reached amongst the customers on the distribution over the conflicting issues through a multiple stage negotiation process [18].

When buyers pay for the orders, they receive electronic coupons from the brokers. Each coupon has a unique ID to ensure that one coupon may be redeemed only once. No extra fee for shipping is charged, when they claim the real products and services from the providers printed on the coupons. Alternative payment methods are used including bank transfers, PayPal and utility & debit cards. The results of 248 surveys in Tulsa indicated that most sellers do not feel any risk associated with transactions with a reliable payment mechanism [19]. Bank transfers are regarded as secure and are a common and efficient way of making payments today. PayPal is an alternative safe way but it involves additional cost [20]. According to the assumptions in this paper, the providers and the core-brokers receive money by bank transfer, while customers pay for their items via PayPal.

A prototype of the CBS was developed in C# under the development environment of Visual Studio 2008. The database in the CBM is designed so that the core-brokers and market-brokers can manage and store all the data they need to fulfil their tasks in the group-trading.

3 Commissions and Fees Systems in Current E-Markets

The fees are normally calculated on the basis of a percentage of the sale price. The percentage is negotiable. For instance, in the area of home buying and selling, brokers usually charge the homeowner 5% to 7% commission [21], but commissions may range widely between 1.5% and 12% in practice. Likewise, online broking sites take various commissions. However most of them seem to have standard and non-negotiable fees systems. A good fees system is essential for group-trading sites, therefore a survey of the fees for online shopping sites is needed in order to construct a

Table 1. Fees for 15 Shopping Sites

Company	*Final Value Fees*	*Online Store Fees*	*Insertion Fees*
Amazon	8.05%, 11.5%, 17.25% £0.86 + £0.16 ~£1.32	£28.75	£0.06
Atomic Mall	1~6% of TV	£0~ £12.30	0
Blujay	0	0	0
Bonanzle	1.5~ 3.5% of FOV + £0.31	0	0
CQout	1.8~5.4%	£3.41~£9.25	0
Craigslist	£6.20	0	0
eBay	1.5~10%	£14.99, £49.99, £349.99	£0~£7.95 or 3%+£1
eBid.net	3%	£4~£8	0
eCrater	0	0	0
Etsy	3.5%	0	£0.20
GoAntiques	2~10%	£24.20~£49.62	0
iOffer	1.5~5%	0	£0~£12.40
OnlineAuction.com	0	£5	£0~£6.20
Ruby Lane	0	£12.40	£0.19
TIAS.com	2% or10%	£24.80	0

reasonable fees system for the new model. The sites in Table 1 are selected because they "had traction, had a substantial number of users" [22]. Generally speaking, there are three types of fees which sellers are normally charged: online store fees (OSF), insertion fees and final value fees (FVF).

Opening an online store is one of the most economic ways for sellers to setup a business. The sellers do not need to spend money on renting a warehouse, but they need Uniform Resource Locators (URLs) for their online shops to "provide means of locating the resource by describing its primary access mechanism" [23]. When the sellers pay a monthly fee for an online store on these sites, they can have URLs for their stores, to which they can easily direct their customers from around the world.

Many of the sites provide special features to allow sellers to customize their professional-looking store with logos, images and colors of their choice. Some marketplaces help sellers with promotional tools, so that they can advertise their goods and bring in more buyers. EBay charge the highest OSF, which is £349.99 for an anchor shop and £49.99 for a featured shop. A GoAntiques online shop at £49.62 comes next. And it is then followed by Amazon's £28.75 and TIAS.com's online shop at £24.80. The lowest OSF is £5 on OnlineAuction.com. The average OSF is about £24.50 excluding the above two over-exaggerated fees.

An insertion fee may be charged whenever a seller lists an item on a shopping site. All the marketplaces charge a fixed amount of money, except eBay. In nine out of the fifteen shopping venues, listing an item is free. On the eBay site, the insertion fee is zero when the starting price of an auction-style item is less than £0.99. The concept of an insertion fee is perhaps similar to the fee for a newspaper advertisement, but newspaper companies do not charge a fee when the products are sold. It does not seem fair to ask the seller to pay for insertion fees and FVF at the same time. Most sellers are discontented with this [24], especially when they get no profit at all out of an unsold item, but they still have to pay the extra expense of an insertion fee.

Most of the sites calculate the FVF using a certain percentage of the final selling price handling fees or sales taxes. Amazon has the highest percentage, at 17.25% a possible reason as to why many of its sellers may have complaints [25]. The second highest percentage is 10%, which is charged by eBay, GoAntiques and TIAS.com.

And it is then followed by Atomic Mall, at 6%, CQout, at 5.4% and iOffer, at 5%. The average percentage of these 15 sites is around 7.5%. Amazon and eBay are consumers' top two websites for shopping this year in the UK [26], but they seem to charge the sellers the highest percentages of the final selling price.

In Atomic Mall, the final fee is up to 6% of the Total Value (TV), which is the total price of an item including shipping fees and handling costs. The FVF on some sites like Bonanzle is based on the Final Offer Value (FOV), which is the sale price minus a shipping fee exemption of up to £6.20. For example, assume that a seller sells a £472 item after shipping. The total shipping for the item is £10, of which only £6.20 is deducted. The FOV is £472–£6.20=£465.80. The fee for the first £310 is £0.31+£310×3.5%=£11.36 while the fee for the remaining amount is £155.80×1.5%=£2.34. So the FVF is £11.36+£2.34=£13.70. The final value fee for a seller derived from the schemes based on TV or FOV usually turns out to be more than the fee calculated from the final selling price of an item.

In summary, the above survey of the fees systems reveals that these markets' average percentage of FVF is around 7.5% of the final selling price of an item. Their average OSF is about £24.50 and their range of insertion fees is from £0 to £35.00.

4 A Fees System of the CBM

A fees system of the CBM including the commission for brokers and how the members of the CBS site pay their fees is constructed here. There are four kinds of fees: FVF, handling fee, session fee and OSF. Session fee is the only fee the core-brokers need to pay for a project. It is suggested that a session fee of £30.00, every time they list a session in a project on the site. An OSF of £24.50 is a suggested monthly fee for market-brokers, who wish to open an online store on the CBS site including purchase of domain name, server use, maintenance etc.

A FVF is paid by the providers, 5.5% of final selling price. When a FVF is received, it is then divided into 3 portions. The core-broker takes 2%, the market-broker gains 3% and the CBS site keeps 0.5%. For example, a core-broker CB has four market-brokers: MB1, MB2, MB3 and MB4, who order 100, 80, 120 and 50 items respectively. Assume that the retail price of an item from a provider is £100 and the buyers get 40% discount. The final selling price for one unit of the item is £100×(1–40%)=£60. The total number of items is 100+80+120+50=350. The provider earns £60×350×(1–5.5%)=£198,450. The FVF is £60×350=£21,000, and CB, MB1, MB2, MB3, MB4 and the CBS site gain £420, £180, £144, £216, £90 and £105 respectively.

A handling fee from the buyers rewards the brokers, because they let the buyers get better discounts. A suggested handling fee is 15% of the extra discount, after each of the brokers has processed the orders. For example, an item' retail price is £100 and its

Table 2. Handling Fees

Customer	*Quantity*	*Original Discount*	*Extra Discount*	*Payment*	*Handling Fee*	*Total Payment*
C1	10	3%	£370	£600	£55.50	£655.50
C2	33	12%	£924	£1980	£138.60	£2118.60
C3	49	15%	£1225	£2940	£183.75	£3123.75
C4	28	9%	£868	£1680	£130.20	£1810.20

discount is 40%. In Table 2, market-broker MB3 has four customers C1, C2, C3 and C4. Customer C1 orders 10 items and has a discount of 3% at the beginning. C1 gains extra discount £100×(40%–3%)=£370. So his handling fee is £370×15%=£55.50. The total payment for customer C1 is £100×10×(1–40%)+£55.50=£655.50.

The final discounts of the customers are due to the contributions of two brokers. Table 3 shows how the handling fees from the customers are distributed to be commissions for the brokers. When MB3 put the orders together, the market order has 21% discount. Customer C1's discount increases 21%–3%=18%, so he pays £100×10×18%×15%=£27 to reward MB3. When core-broker CB combines all the orders together, the final discount becomes 40%. C1 pays another £100×10×(40%–3%–18%)×15%=£28.50 to CB for services. This means C1's handling fee £55.50 is divided into two portions, £27.00 and £28.50. The commission that MB3 gets from this fee, is £27.00+£44.55+£44.10+£50.40=£166.05, while core-broker CB gains £28.50+£94.05+£139.65+£79.80=£342.00 from MB3's clients.

Table 3. Commissions

Customer	*MB3 (21%)*	*Handling Fee for MB3*	*CB (40%)*	*Handling Fee for CB*
C1	18%	£27.00	19%	£28.50
C2	9%	£44.55	19%	£94.05
C3	6%	£44.10	19%	£139.65
C4	12%	£50.40	19%	£79.80

With the fees system, the brokers and the CBS site may earn money from the execution of group-trading. Brokers get their commission out of FVF and handling fee. They are in return of their valuable efforts, such as setting the projects in motion, keeping a watchful eye on the process of group-trading, ensuring that what is on offer to the buyers will be attractive to them, and most of all, combining the orders to bring high benefits to the traders. Certainly, their earnings need to be good enough to cover the payments of session fee and OSF. The fees, FVF, session fee and OSF, keep the CBS site running and provide on-going services to the brokers.

5 Evaluation of the Fees System in the CBM

The aim of the new model is not only to bring lower prices for buyers but also to create higher profits for service providers. It is essential to prove that both the buyers and the providers may get more benefits in the CBM than they can gain in a traditional market (will be referred to as TM) even when the benefits have excluded the fees for brokers and websites. Therefore, the proposed fees system must be judged here to see whether it functions properly in the CBM. Although prototype of CBS site has been developed in C#, at this stage, a real-world test would be inappropriate; further it would be unlikely to produce the data set necessary for a rigorous testing. Therefore, a simulation system was developed using the scenario of a travel agent and was used to produce outputs from the TM and the CBM.

Core-broker Ben created group-trading project S1, 'Summer Time around the Midlands' , which integrates the products from the three providers offering inexpensive hotel rooms and low car rentals for economical travel in the Midlands. By offering volume wholesale discounts, customers may form groups to purchase items. Coupons can be chased and sent to the providers on them and exchanged into hotel rooms or car for the buyers to travel around the Midlands in the UK.

The simulation system was written in C# in the Visual Studio 2008 development environment. The results in this paper were produced in it on a common personal computer with Windows Vista. The system contains a Test Case Generator (TCG), a TM Simulator (TMS) and a CBM Simulator (CBMS). The TMS is based on the core concept and aims to find a core of the coalition in a TM. The CBMS is built to the pattern of the CBM and aims to find a bigger core of coalitions in multi-e-markets. The data generated by the TCG were put into the TMS and the CBMS at the same time. By examining the outputs of the TMS and CBMS, a comparison between the TM and the CBM and the results were made. According to the results of the simulation system in Fig. 3, the net discounts of the customers and the net profits for the providers in the CBM are not less than the ones in the TM. The net benefits of the traders in the CBM have had the brokers' commission deducted from them. The brokers do bring higher benefit to both the customers and the providers in the multi e-markets of the CBM. The results of the scenario of demanding customers, who desire to get high discounts, definitely may convince the traders to joint in the CBM.

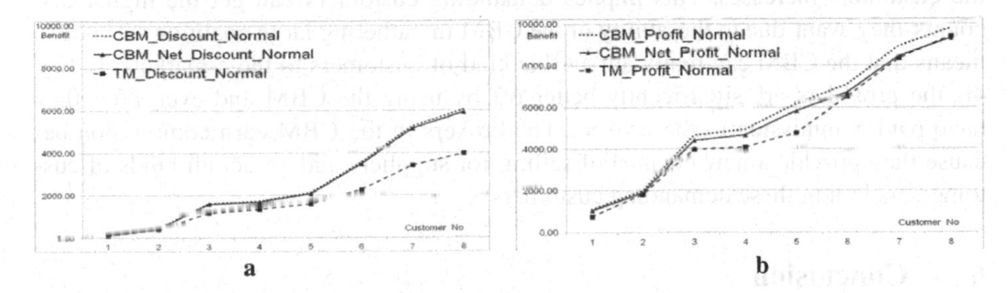

Fig. 3. Normal Customers' Discounts and their Providers' Profits

According to the results of the simulation system in Fig. 3, the net discounts of the customers and the net profits for the providers in the CBM are all higher than the ones in the TM. The net benefits of the traders in the CBM have had the brokers' commission deducted from them. The brokers do bring higher benefit to both the customers and the providers in the multi e-markets of the CBM. The results of the scenario of demanding customers, who desire to get high discounts, definitely may convince the traders to joint in the coalitions of the CBM.

It is quite common in an e-market nowadays to have many demanding customers. It is also very difficult to attract such customers to the TM. It is important to show that CBM can effectively allure such buyers to e-markets and bring more profit to providers. It is crucial that the simulation system provides evidence to show that there are far more discounts in the CBM for customers than in the TM. For this purpose, the TCG created test data for a scenario of demanding buyers, who only buy items at an extremely low price. Within the order detail table, there is a special field called 'expected discount'. This field allows customers to place the orders without committing to buy the items. The CBM will wait for the final discount to be settled and decide whether the purchase should go ahead or be dropped, by comparing the final and expected discount. For instance, product Ca's discount for customers is between 0% and 40%. The TCG generates data randomly ranging from 19% to 40% and puts in this particular field of the orders of the demanding buyers.

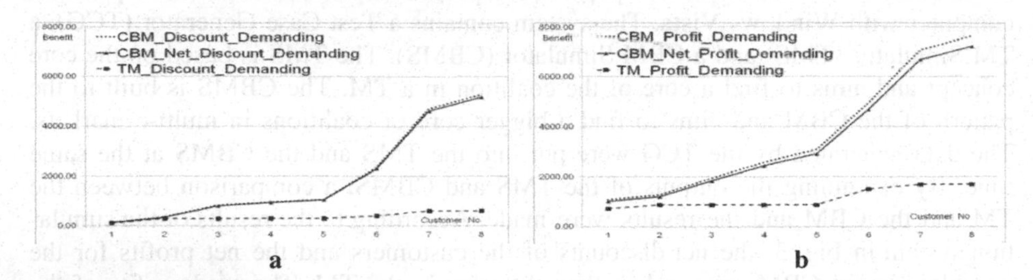

Fig. 4. Demanding Customers' Discounts and their Providers' Profits

With the demanding customers, the graph in Fig. 4a shows that the discount stays low in the TM, but the net discount in the CBM is getting higher when the number of the customers increases. This implies demanding customers can get the higher discounts they want due to the ability of the CBM in gathering large coalitions. This also means that the CBM can really attract this kind of customers to the e-markets. In Fig. 4b, the providers are significantly better off by using the CBM and even after they have paid commission to the brokers. The brokers in the CBM earn commission because they provide a new channel of selling for suppliers and attract all kinds of customers including these demanding customers.

6 Conclusion

It is usual that if the customers get better discounts, then the service providers receive less profit, or the other way round, but the CBM has considered the interests of both the customers and providers and makes all of them better off than the TM. The simulation also shows that the traders may gain higher benefits in the CBM in both scenarios: the one with normal buyers, and the one with demanding buyers. In this way, the new model definitely has chances to attract as many as buyers to the e-markets.

With the proposed fees system, the traders are proven to have better benefits in the CBM than they gain in the TM, even after part of the benefits goes to the brokers as commission. Besides, the process of the CBM allows brokers to merge smaller coalitions in different e-markets into a bigger coalition and gives buyers higher discounts. This can really encourage the customers go through with their purchases and bring larger total profits to the providers. As a result, a wise trader will definitely choose to join the group-trading in the CBM rather than stay in the core of a TM.

Core brokers may set up different rates of fees, if they feel it is necessary, before a session of group trading in the project starts. Through the aid of the simulation system, an experienced core broker usually can figures out reasonable rates without problems. They may find out some useful solutions and set up special rates for their projects by consulting providers or other brokers in the CBS site.

The distribution of discounts for buyers in a real e-market is normally distributed in some contexts. However, the data in the expected discount fields of the orders of the normal buyers, which was generated randomly by the TCG and used in the simulation system, is no way near to this distribution. To simulate natural e-markets using data in such a distribution is one of future tasks in the CBM. To work out a negotiable fees system will be another target for future research. The rates of commission in the CBM's fees system for the brokers are fixed here. In practice, they are fixed for the duration of the session, but all the fees are subject to negotiation.

References

1. Abdulghader, A.A., Dalbir, S., Ibrahim, M.: Potential E-Commerce Adoption Strategies for Libyan Organization. IJICT 1(7), 321–328 (2012)
2. Forrester Research, Inc. 'Forrester Forecast: Double-Digit Growth for Online Retail in the US and Western Europe', Cambridge, Mass. (2010), `http://www.forrester.com/ER/Press/Release/0,1769,1330,00.html` (February 3, 2011)
3. Moulin, H.: Cooperative Microeconomics: A game theoretic Introduction. Princeton University Press, Princeton (1995)
4. Sandholm, T., et al.: Coalition structure generation with worst case guarantees. Artificial Intelligence 111, 209–238 (1999)
5. Turban, E., et al.: Electronic Commerce: A Managerial Perspective. Prentice Hall, New Jersey (1999)
6. Lieberman, H.: Autonomous Interface Agents. In: Pemberton, S. (ed.) Human Factors in Computing, pp. 67–74 (1997)
7. Shehory, O., Kraus, S.: Feasible Formation of coalitions among autonomous agents in non-super-additive environments. Computational Intelligence 15(3), 218–251 (1999)
8. Gillies, D.: Some theorems on n-person games. Unpublished PhD thesis, Princeton University (1953)
9. Sun, P., et al.: Core-based Agent for Service-Oriented Market. In: Lee, T., Zhou, M. (eds.) Proceedings of 2006 IEEE International Conference on Systems, Man, and Cybernetics (SMC 2006), The Grand Hotel, Taipei, Taiwan, October 8-11, pp. 2970–2975. The IEEE Inc., Piscataway (2006)
10. Sun, P., et al.: A Core Broking Model for E-Markets. In: Proceedings of the 9th IEEE International Conference on e-Business Engineering (ICEBE 2012), Zhejiang University, Hangzhou, China, September 9-11, pp. 78–85. IEEE Press (2012)
11. Sun, P., et al.: Extended Core for E-Markets. In: Isaias, P., White, B., Nunes, M.B. (eds.) Proceedings of IADIS International Conference WWW/Internet 2009, Rome, Italy, November 19-22, pp. 437–44. IADIS Press (2009)
12. Sun, P.: A Core Broking Model for E-Markets. Unpublished PhD thesis, Coventry University (2011)
13. Sun, P., et al.: Evaluations of A Core Broking Model from the Viewpoint of Online Group Trading. In: Proceedings of the IEEE International Conference on Industrial Engineering and Engineering Management (IEEM 2012), Hong Kong Convention and Exhibition Centre, Hong Kong, December 10-13, pp. 1964–1968. IEEE Press (2012)

14. Sun, P., Yang, F.: An Insight into an Innovative Group-Trading Model for E-Markets Using the Scenario of a Travel Agent. In: Bădică, C., Nguyen, N.T., Brezovan, M. (eds.) ICCCI 2013. LNCS (LNAI), vol. 8083, pp. 582–592. Springer, Heidelberg (2013)
15. The New Oxford Dictionary of English, 3rd edn. Oxford University Press, Oxford (2010)
16. Sullivan, A., Sheffrin, S.M.: Economics: Principles in Action. Pearson Prentice Hall, Upper Saddle River (2003)
17. Porter, M.: Competitive advantage: Creating and Sustaining Superior Performance. Free Press, New York (1985)
18. Chao, K., et al.: Using Automated Negotiation for Grid Services. IJWIN, 141–150 (2006)
19. Leonard, L.: Attitude Influencers in C2C E-commerce: Buying and Selling. Journal of Computer Information Systems 52(3), 11–17 (2012)
20. PayPal 'Transaction Fees for Domestic Payments', `https://www.paypal.com/uk/cgi-bin/webscr?cmd=_display-receiving-fees-outside` (May 15, 2011) (n. d.)
21. Kokemuller, N.: What Percentage Do Realtors Make on Commission? eHow Contributor, May 26 (2011), `http://www.ehow.co.uk/info_8494252_percentage-do-realtors-make-commission.html` (July 11, 2011)
22. Steiner, I.: Seller's Choice: Merchants Rate Ecommerce Marketplaces. AuctionBytes.com, January 24 (2010), `http://www.auctionbytes.com/cab/abu/y210/m01/abu0255/s02` (April 24, 2011)
23. Berners-Lee, T., et al.: Uniform Resource Identifier (URI): Generic Syntax. Internet Society. Request for Comments: 3986, STD 66 January 1 (2005), `http://labs.apache.org/webarch/uri/rfc/rfc3986.html` (July 1, 2011)
24. Steiner, I.: Seller's Choice Marketplace Ratings: eBay. EcommerceBytes.com, January 24 (2010), `http://www.auctionbytes.com/cab/abu/y210/m01/abu0255/s09` (April 25, 2011)
25. Steiner, I.: Seller's Choice Marketplace Ratings: Amazon. EcommerceBytes.com, January 24 (2010), `http://www.auctionbytes.com/cab/abu/y210/m01/abu0255/s03` (July 19, 2011)
26. Sillitoe, B.: Amazon & eBay - the top websites for UK shoppers. RetailGazette Daily Retail News, June 2 (2011), `http://www.retailgazette.co.uk/articles/20133-amazon-ebay-the-top-websites-for-uk-shoppers` (July 21, 2011)

Non Dominated Sorting Genetic Algorithm for Chance Constrained Supplier Selection Model with Volume Discounts

Remica Aggarwal[1] and Ainesh Bakshi[2]

[1] Department of Management, Birla Institute of Technology & Science, Pilani, India
[2] Department of Computer Science, Rutgers New Brunswick, USA

Abstract. This paper proposes a Stochastic Chance-Constrained Programming Model (SCCPM) for the supplier selection problem to select best suppliers offering incremental volume discounts in a conflicting multi-objective scenario and under the event of uncertainty. A Fast Non-dominated Sorting Genetic Algorithm (NSGA-II), a variant of GA, adept at solving Multi Objective Optimization, is used to obtain the Pareto optimal solution set for its deterministic equivalent. Our results show that the proposed genetic algorithm solution methodology can solve the problems quite efficiently in minimal computational time. The experiments demonstrated that the genetic algorithm and uncertain models could be a promising way to address problems in businesses where there is uncertainty such as the supplier selection problem.

Keywords: Supplier selection, Chance constrained approach, Incremental quantity discount model, Genetic algorithms.

1 Introduction

Today businesses run not only on a higher level of performance but also on maintaining cordial relationship with the clients. For a manufacturing firm , particularly , suppliers acts as a backbone of the business as the right choice of suppliers reduces costs, increases profit margins, improves component quality and ensures timely delivery and therefore choosing a few but superior suppliers becomes a very important strategic decision. Choice of suppliers depends on several dimensions such as price, delivery, quality, capacity etc. Selection of suppliers becomes more challenging when suppliers offer interesting deals such as better price and quality, incremental or all units quantity discounts etc. to attract and retain buyers for a longer period, to motivate them to procure larger quantities and to reduce cost of transportation per commodity. These deals and discounts could be based on the quantity of each product ordered from a supplier or based on the total value of all products ordered from a supplier. Therefore in general, the supplier selection problem is a multi-criteria problem based on joint consideration of purchasing cost, quantity discounts, order size restrictions, product quality and service levels, supplier capacity and lead time. Selecting an optimal supplier under such multiple conflicting criteria often gets complicated even for

N.T. Nguyen et al. (Eds.): ACIIDS 2014, Part II, LNAI 8398, pp. 465–474, 2014.

an experienced purchase manager because competing suppliers have different levels of achievement under these criteria.

However, in solving for practical supplier selection and purchasing plans, businesses are faced with some uncertain factors. In the event of uncertainty such as uncertain or fluctuating demand, changing supplier's capacity, supplier's unreliable lead time and varying quality, selection of suppliers becomes even more complex. Although the researches have been made in the field of selecting suppliers under uncertainty or under discount pricing and lot size restrictions, no research to the best of our knowledge has integrated quantity discounts model with uncertainty and lot size restrictions at a time.

Various multi-objective optimization techniques have already been established to solve a variety of deterministic supplier solution problems. However, when uncertain factors are explicitly considered in a supplier selection problem, it is hardly treated since the traditional mathematical modeling is difficult to deal with the uncertain programming problem. In such cases, where risk is a major part of the decision, it is imperative to capture the risk factors into a mathematical model. Chance-constrained programming that was pioneered by Charnes and Cooper [4] is one approach that can handle the uncertainty of the problem. Nature inspired / Evolutionary algorithms can be used as an alternative to obtain a global optimal solution to solve such Multi Objective Optimization problems. One of the most popular of these techniques is Genetic Algorithms (GA). GA is a stochastic search method for optimization problems based on the mechanics of natural selection and natural genetics (i.e., the principle of evolution—survival of the fittest). A Fast Non-dominated Sorting Genetic Algorithm (NSGA-II) [7], a variant of GA, adept at solving Multi Objective Optimization, is used to obtain the Pareto optimal solution set for our problem statement.

In this paper, an attempt has been made to present a mathematical model for supplier selection supplying multiple products to a buyer under uncertain scenario incorporating incremental quantity discounts on lot sizes of multiple products to be supplied by multiple suppliers. Uncertainty in model parameters such as capacity, lead time and demand uncertainty is handled through the Chance constraints approach [4-6]. These uncertainties are captured by probability distribution of capacity, demand, cost and lead time. A stochastic chance-constrained programming model for the supplier selection problem is transformed into the deterministic equivalent mathematical programming model which is then solved using fast non dominated genetic algorithm (NSGAII). Our results show that the proposed genetic algorithm solution methodology can solve the problems quite efficiently in minimal computational time. The experiments demonstrated that the genetic algorithm and uncertain models could be a promising way to address problems in businesses where there is uncertainty such as the supplier selection problem.

The paper is structured as follows. Section 2 presents a review of the relevant literature on supplier selection and use of genetic algorithms to solve such problems. Section 3 formulates the Stochastic Chance-Constrained Programming Model (SCCPM) for the supplier selection problem with uncertain cost, quality and lead time. Section 4 explains the SVGAII algorithm to solve the formulated problem.

A case example has been presented in section 5 to validate the proposed model. Details of how SVGA applied to the case problem is presented in section 6 .Section 7 conclude the paper with future directions for research.

2 Literature Review

Although the process of supplier selection has been studied extensively, the problem of supplier selection under multi-supplier with quantity discounts has received attention quite recently. A weighted fuzzy multi-objective model for the supplier selection under price breaks or quantity discounts environment in a supply chain is proposed by [2]. Xia and Wu [16] propose a multi-objective optimization problem, where one or more buyers order multiple products from different vendors in a multiple sourcing network using using business volume discounts. Ebrahim et al. [8] uses a scatter search algorithm for supplier selection and order lot sizing under multiple price discount environment.

Alonso Ayuso et al. [1] proposed a two-phase stochastic program where they considered plant selection, product allocation and supplier selection under uncertain costs, product prices and demand. Burke et al. [3] also developed a supplier selection model with demand uncertainty and unreliable suppliers. These researches has not explored uncertainties related to demand, capacity and lead time at a time and also lack the concept of quantities discounts on the part of supplier.

Application of genetic algorithms is gaining momentum in a variety of industrial applications [9,10,11,12,15]. Rezaei and Davoodi [12] who formulated a fuzzy mixed integer programming model of a multi-period inventory lot sizing problem with supplier selection. The problem is converted to equivalent crisp decision making problems and solved by using a genetic algorithm that determines what items to order in what quantities from each supplier in which periods. But this paper however lacks the concept of uncertainty which prevails in the real business scenario.

Present paper integrates the concept of incremental quantities discounts and lot size restrictions offered by multiple suppliers in the event of uncertain demand, supplier's capacity and lead time. A stochastic chance-constrained programming model for the supplier selection problem is transformed into the deterministic equivalent mathematical programming model which is then solved using fast non dominated genetic algorithm (NSGAII) [7].

3 Problem Formulation

This section formulates a Stochastic Chance-Constrained Programming Model (SCCPM) for the supplier selection problem. Following **notations** are used to formulate the model.

j 1,2,...,J suppliers
k 1,2,...K products

x_{jk} 1 if supplier j is assigned as supplier of product k, 0 otherwise

h_{jk} The amount of product *k* shipped from supplier *j*

D_k Demand for product *k*

Cap_{jk} Capacity at supplier *j* for product *k*

m_{jk} middle order quantity to be supplied from supplier j of product k

u'_{jk} k[th] product unit cost incurred on buyer from j[th] supplier if order quantity $\leq m_{jk}$

u''_{jk} k[th] product unit cost incurred on buyer from j[th] supplier if order quantity $> m_{jk}$

u_{jk} Aggregate unit cost incurred when order quantity $\geq m_{jk}$

F_j Fixed cost of operating with supplier *j*

Q_{jk} Percentage of good quality items of product *k* procured from supplier *j*

L_{jk} Lead time of product *k* from supplier *j*

q_j Minimum order quantity to be supplied from supplier *j*

L_{jk} Lead time of product *k* from supplier *j*

μ_{ljk} Mean lead time of product *k* from supplier *j*

α_k Level of probability that units supplied satisfies the demand of kth product

α_{jk} Level of probability that $h_{jk} \leq cap_{jk} x_{jk}$

α_l Level of risk for the calculated value of lead time to be greater than aspired level

$F_{D_k}^{-1}$ Constant inverse probability distribution function for random demand

$F_{cap_{jk} x_{jk}}^{-1}$ Constant inverse probability distribution function for random capacity

$F_{A_1}^{-1}(\alpha_l)$ Constant inverse probability distribution function for random lead time

3.1 Stochastic Chance-Constrained Programming Model (SCCPM)for the Supplier Selection Problem with Uncertain Cost, Quality and Lead Time

A stochastic chance-constrained programming model for the supplier selection problem integrating incremental quantity discounts and lot size restrictions can be written as:

$$\text{Minimize } Z_1 = \sum_{j=1}^{J} \sum_{k=1}^{K} u_{jk} h_{jk} x_{jk} + \sum_{j=1}^{J} \sum_{k=1}^{K} F_j x_{jk} \tag{1}$$

$$\text{Maximize } Z_2 = \sum_{j=1}^{J} \sum_{k=1}^{K} Q_{jk} h_{jk} x_{jk} \tag{2}$$

$$\text{Minimize } Z_3 = \sum_{j=1}^{J} \sum_{k=1}^{K} L_{jk} h_{jk} x_{jk} \tag{3}$$

Subject to

$$P\left(\sum_{j=1}^{J} h_{jk} \geq D_k\right) \geq \alpha_k, \quad \forall k \in K \tag{4}$$

$$P\left(h_{jk} \leq cap_{jk} x_{jk}\right) \geq \alpha_{jk}, \quad \forall j \in J, k \in K \tag{5}$$

$$P\left(\sum_{j=1}^{J} \sum_{k=1}^{K} \mu_{l_{jk}} h_{jk} x_{jk} \geq A_1\right) \geq \alpha_l \tag{6}$$

$$\sum_{k=1}^{K} h_{jk} x_{jk} \geq q_j, \forall j \in J, \forall k \in K \tag{7}$$

$$x_{jk} \in [0,1] \ \forall j \in J, \forall k \in K \tag{8}$$

$$h_{jk} \geq 0, \forall j \in J, \forall k \in K \tag{9}$$

$$u_{jk} = u'_{jk} \quad if \ h_{jk} \leq m_{jk} \tag{10}$$

$$u_{jk} = u_{jk} = u'_{jk} * m_{jk} + (h_{jk} - m_{jk}) u''_{jk}, \ if \ h_{jk} > m_{jk} \tag{11}$$

Objective function Z_1 minimizes the total cost incurred by the buyer. Objective function Z_2 maximizes the total quality of purchased products. Objective function Z_3 minimizes the total lead time. Constraint given by equation (4) ensures that shipments from the suppliers cover the entire demand for each product. Equation (5) restricts the amount shipped from the suppliers to their capacity. Constraint (6) ensures the lead time of suppliers .Constraint (7) provides the lot size restriction on different products. Constraints in equation (8) and equation (9) enforce binary and non-negativity condition in the decision variables respectively. Constraints (10) and (11) indicates the incremental quantities discount provided by the jth supplier for kth product based on the number of components ordered.

3.2 Stochastic Chance-Constrained Programming Model (SCCPM) for the Supplier Selection with Deterministic Equivalents

Stochastic chance-constrained programming model (SCCPM) for the supplier selection with its deterministic equivalents is described from equation (12) to equation (17) as follows [4]:

$$\text{Minimize } Z_1 = \sum_{j=1}^{J} \sum_{k=1}^{K} u_{jk} h_{jk} x_{jk} + \sum_{j=1}^{J} \sum_{k=1}^{K} F_j x_{jk} \tag{12}$$

$$\text{Maximize } Z_2 = \sum_{j=1}^{J} \sum_{k=1}^{K} Q_{jk} h_{jk} x_{jk} \tag{13}$$

$$\text{Minimize } Z_3 = \sum_{j=1}^{J} \sum_{k=1}^{K} \mu_{l_{jk}} h_{jk} x_{jk} \tag{14}$$

Subject to

$$\sum_{j=1}^{J} h_{jk} = F_{D_k}^{-1}(\alpha_k) \ \forall k \in K \tag{15}$$

$$h_{jk} \leq F_{cap_{jk} x_{jk}}^{-1}\left(1 - \alpha_{jk}\right) \ \forall j \in J, k \in K \tag{16}$$

$$\sum_{j=1}^{J} \sum_{k=1}^{K} \mu_{1_{jk}} h_{jk} x_{jk} \geq F_{A_1}^{-1}(\alpha_1) \tag{17}$$

Constraints (7)-(11)

4 Solution Methodology: Fast Non Dominated Sorting Genetic Algorithm (NSGAII)

In GA terminology, a solution vector is called an individual or a chromosome. In the modern implementations of GA, chromosomes can be real numbers, strings, matrices and other data structures. For the purpose of Multi Objective Optimization, GA provide a Pareto set of final solutions to the given problem. The set of all feasible non-dominated solutions in X is referred to as the Pareto optimal set, and for a given Pareto optimal set, the corresponding objective function values in the objective space is called the Pareto front. For many problems, the number of Pareto optimal solutions is enormous, sometimes even infinite. Thus the eventual goal of a multi-objective optimization algorithm, such as ours, is to closely approximate solutions in the Pareto optimal set. Therefore, a more pragmatic approach is followed in which the best known Pareto set is investigated as much as possible.

Pareto-ranking approaches explicitly utilize the concept of Pareto dominance in evaluating fitness or assigning selection probability to solutions. Initially, a random parent population P_0 is created. The population is sorted based on the non-domination. Each solution is assigned a fitness equal to its non-domination level. At first, the usual binary tournament selection, recombination, and mutation operators are used to create an offspring population Q_0 of size N. A combined population $\mathbf{R}_t = \mathbf{P}_t \cup \mathbf{Q}_t$ is formed. The population $\mathbf{R}_t$ is of size 2N and is sorted according to non-domination thereby ensuring non domination. Solutions belonging to the best non dominated set $\mathbf{F}_1$ are of best solutions in the combined population and must be emphasized more than any other solution in the combined population. If the size of $\mathbf{F}_{t+1}$ is smaller than N we definitely choose all members of the above set for the new population $\mathbf{P}_{t+1}$. The remaining members of the population $\mathbf{P}_{t+1}$ are chosen from subsequent non dominated fronts in the order of their ranking. Thus, solutions from the set $\mathbf{F}_2$ are chosen next, followed by solutions from the set $\mathbf{F}_3$ and so on. Crowding distance approaches (Rajagopalan et al. [11]) aim to obtain a uniform spread of solutions along the best11 known Pareto front without using a fitness sharing parameter.

Step 1. Rank the population and identify non-dominated fronts $\mathbf{F}_1, \mathbf{F}_2, \ldots \mathbf{F}_R$ .For each front *j=1 ... R* repeat Steps 2 and 3.

Step 2. For each objective function k, sort the solutions in F jF_j in the ascending order. Let l = $|F_j|$ and x $_{[i,k]}$ represent the ith solution in the sorted list with respect to the objective function k. Assign $cd_k(x_{[1,k]}) = \infty$ ∞ and $cd_k(x_{[l,k]}) = \infty$ ∞ and

for i =2 … l assign

$$cd_k(x_{[i,k]}) = [z_k(x_{[i+1,k]}) - z_k(x^k_{[i-1,k]})]/[z_k^{max} - z_k^{min}] \quad (18)$$

Step 3. To find the total crowding distance cd(x) of a solution x, sum the solution crowding distances with respect to each objective, i.e. $cd(x) = \sum_k cd_k(x)(x)$.

In the NSGA-II, this crowding distance measure is used as a tie-breaker as in the selection phase that follows. Randomly select two solutions x and y; if the solutions are in the same non dominated front, the solution with a higher crowding distance wins. Otherwise, the solution with the lowest rank is selected.

Deb also proposed the constrain-domination concept and a binary tournament selection method based on it, called a constrained tournament method. A solution x is said to constrain-dominate a solution y if either of the following cases are satisfied:

Case 1: Solution x is feasible and solution y is infeasible.
Case 2: Solutions x and y are both infeasible; however, solution x has a smaller constraint violation than y.
Case 3: Solutions x and y are both feasible, and solution x dominates solution y.

In the constraint tournament method in NSGA-II, first non-constrain-dominance fronts $F_1 F_2 \ldots F_R$ F $_R$ are identified by using the constrain-domination criterion. In the constraint tournament selection, two solutions x and y are randomly chosen from the population. If solutions x and y are both in the same front, then the winner is decided based on crowding distances of the solution.

5 Numerical Illustration : (5*3 Test Problem; Three Objectives)

Consider Five suppliers (S1, S2, S3, S4, S5) supplying three different products (P1, P2, P3) to the buyer. Stochastic data corresponding to the capacity, minimum order, middle order quantities, demand, unit cost, fixed cost, quality levels and stochastic lead time are given from Table 1 to Table 6. All data are randomly generated. The reliability level for the capacity is set at $\alpha_{jk} = 0.95 \ \forall j = 1,2,\ldots 5; k = 1,2,3$ meaning that at least 95% of demand should be met. The risk level is set at $\alpha_k, \alpha_l = 0.05$.

We begin with developing the NSGA-II algorithm framework and subsequently approach the given problem from a GA viewpoint. From the equations formulated above it is inferred that there are 15 independent real variables (h_{jk}) and 15 binary variables (x_{jk}) involved in the optimization problem, the values of which need to be determined. The feature vector for each individual in the population thus consists of 30 variables. The problem is to be optimized with respect to 3 objective functions and is subject to 6 constraints. The 15 intermediate variables (u_{jk}) are introduced in the problem formulation according to the given data. The genetic algorithm is initialized with a random starting population of 100 individuals. The probability for crossover is 0.75 and that for mutation is 0.05. The termination condition is set at 250 generations.

Table 1. Stochastic capacity data and lot size restriction (in units)

	P1	**P2**	**P3**	**Min. order**
S 1	N(50,6.25)	N(45, 5)	N(100,25)	100
S 2	N(90, 20)	N(100,25)	N(20,1)	100
S 3	N(70,12)	N(50,6.25)	N(150,56)	100
S 4	N(80, 6)	N(200,100)	N(50,6.25)	100
S 5	N(70,12)	N(100,25)	N(70,12.25)	100

Table 2. Stochastic demand data (in units)

P1	P2	P3
N(210,36)	N(250,49)	N(250,64)

Table 3. Fixedcost data (in dollars)

S 1	**S 2**	**S 3**	**S 4**	**S 5**
100	200	150	150	120

Table 4. Unit cost data (in dollars)

	Without discount				**With discount**			
	Ranges	P1	P2	P3	Ranges	P1	P2	P3
S 1	$30 \leq h \leq 50$	15	10	8	h>50	12	9	6
S 2	$30 \leq h \leq 60$	10	8	10	h>60	8	7	9
S 3	$30 \leq h \leq 50$	6	5	9	h>50	5	4	7
S 4	$30 \leq h \leq 60$	5	9	7	h>60	4	8	6
S 5	$30 \leq h \leq 60$	7	7	10	h>60	5	6	8

Table 5. Quality data (% of good items)

	P1	**P2**	**P3**
S 1	0.95	0.95	0.93
S 2	0.95	0.97	0.99
S 3	0.9	0.9	0.9
S 4	0.9	0.93	0.9
S 5	0.9	0.92	0.97

Table 6. Stochastic lead time data (days)

	P1	**P2**	**P3**
S 1	N(10,6)	N(9, 5)	N(1,0)
S 2	N(5, 2)	N(2,1)	N(8,1)
S 3	N(8,2)	N(3,1)	N(9,2)
S 4	N(3,1)	N(4,2)	N(6,2)
S 5	N(8,2)	N(2,1)	N(4,1)

Table 7. Results in terms of cost, quality and lead time

	NSGA-II (Population1)	NSGA-II (Population2)	NSGA-II (Population3)
Cost	8736.04	9023.806	8709.357
Quality	687.4363	689.433	685.5609
Lead time	3671.174	3742.611	3652.361

Table 8. Number of units supplied from each supplier

NSGA II Population1				NSGA II Population 2				NSGA II Population 3			
x_{11}	1	h_{11}	26.2	x_{11}	1	h_{11}	26.79	x_{11}	1	h_{11}	24.07
x_{12}	1	h_{12}	38.2	x_{12}	1	h_{12}	40.93	x_{12}	1	h_{12}	38.25
x_{13}	1	h_{13}	62.4	x_{13}	1	h_{13}	59.59	x_{13}	1	h_{13}	62.48
x_{21}	1	h_{21}	78.4	x_{21}	1	h_{21}	80.08	x_{21}	1	h_{21}	78.42
x_{22}	1	h_{22}	45.8	x_{22}	1	h_{22}	45.95	x_{22}	1	h_{22}	45.84
x_{23}	0	h_{23}	0.00	x_{23}	1	h_{23}	0.00	x_{23}	0	h_{23}	0.00
x_{31}	0	h_{31}	0.00	x_{31}	0	h_{31}	0.00	x_{31}	0	h_{31}	0.00
x_{32}	1	h_{32}	37.2	x_{32}	1	h_{32}	35.22	x_{32}	1	h_{32}	37.22
x_{33}	1	h_{33}	101.9	x_{33}	1	h_{33}	104.9	x_{33}	1	h_{33}	101.8
x_{41}	1	h_{41}	54.9	x_{41}	1	h_{41}	49.64	x_{41}	1	h_{41}	54.97
x_{42}	1	h_{42}	45.9	x_{42}	1	h_{42}	45.89	x_{42}	1	h_{42}	45.98
x_{43}	1	h_{43}	40.8	x_{43}	1	h_{43}	40.40	x_{43}	1	h_{43}	40.05
x_{51}	1	h_{51}	59.9	x_{51}	1	h_{51}	63.66	x_{51}	1	h_{51}	60.89
x_{52}	1	h_{52}	91.9	x_{52}	1	h_{52}	92.00	x_{52}	1	h_{52}	91.98
x_{53}	1	h_{53}	58.2	x_{53}	1	h_{53}	58.99	x_{53}	1	h_{53}	58.17

6 Interpretation of Results

Results in terms of objective function values for cost, quality and lead time corresponding to different Pareto optimal solutions obtained from SVGAII is given in Table 7. Although many optimal solution were obtained from the algorithm (which is the distinct advantage of this algorithm), three have been shown. Although these solutions are not necessarily optimal and are almost near-optimal, it can be possible for decision maker select one of them that matches with the real world condition.

7 Conclusions and Future Directions

Fast non dominated Sorting NSGA II algorithm has been used to solve the multi-objective optimization model related to supplier selection with incremental quantities discount and lot size restrictions under uncertainties in demand, capacity and lead time associated with supplier. Model is validated using a case problem. The problem can be extended to include the cases of multiple buyers, business volume discounts and all unit discounts as well. Other costs such as transportation costs and variable costs can also be included as a part of total costs.

References

1. Alonso-Ayuso, A., Escudero, L.F., Garin, A., Ortuno, M.T., Perez, G.: An approach for strategic supply chain planning under uncertainty based on stochastic 0-1 programming. Journal of Global Optimization 26(1), 97–124 (2003)
2. Amid, A., Ghodsypour, S.H., O'Brien, C.: A weighted additive fuzzy multi-objective model for the supplier selection problem under price breaks in a supply Chain. Int. J. Production Economics 104, 394–407 (2007)
3. Burke, G.J., Geunes, J., Romeijnb, H.E., Vakharia, A.: Allocating procurement to capacitated suppliers with concave quantity discounts. Operations Research Letters 36(1), 103–109 (2008)
4. Charnes, A., Cooper, W.: Chance-constrained programming. Management Science 5, 73–79 (1959)
5. Charnes, A., Cooper, W.: In management models and industrial applications of linear programming, vols. 1-2. Wiley, New York (1961)
6. Charnes, A., Cooper, W.: Deterministic equivalents for optimizing and satisfying under chance constraints. Operations Research 11, 18–39 (1963)
7. Deb, K., Pratap, A., Agarwal, S., Meyarivan, T.: A fast and elitist multi-objective genetic algorithm: NSGA-II. IEEE Transactions on Evolutionary Computation 6(2), 182–197 (2002)
8. Ebrahim, M., Razmi, J., Haleh, H.: Scatter search algorithm for supplier selection and order lot sizing under multiple price discount environment. Advances in Engineering Software 40, 766–776 (2009)
9. Lu, H., Yen, G.G.: Rank-density-based multi-objective genetic algorithm and benchmark test function study. IEEE Transactions on Evolutionary Computation 7(4), 325–343 (2003)
10. Oh, K.J., Kim, T.Y., Min, S.: Using genetic algorithm to support portfolio optimization for index fund management. Expert Systems with Applications 28(2), 371–379 (2005)
11. Rajagopalan, R., Mohan, C.K., Mehrotra, K.G., Varshney, P.K.: Evolutionary Multi-objective crowding algorithm for path computations. In: Proc. Fifth International Conference on Knowledge based Computer Systems, pp. 46–65 (2004)
12. Rezaei, J., Davoodi, M.: Genetic algorithm for inventory lot-sizing with supplier selection under fuzzy demand and costs. In: Ali, M., Dapoigny, R. (eds.) IEA/AIE 2006. LNCS (LNAI), vol. 4031, pp. 1100–1110. Springer, Heidelberg (2006)
13. Sawik, T.: Single vs. multiple objective supplier selection in a make to order environment. Omega 38(3/4), 203–212 (2010)
14. Shiromaru, I., Inuiguchi, M., Sakawa, M.: A fuzzy satisfying method for electric power plant coal purchase using genetic algorithms. European Journal of Operations Research 126, 218–230 (2000)
15. Vergara, F.E., Khouja, M., Michalewicz, Z.: An evolutionary algorithm for optimizing material flow in supply chains. Computers and Industrial Engineering 43, 407–421 (2002)
16. Xia, Wu: Supplier selection with multiple criteria in volume discount Environments. Omega 35, 494–504 (2007)

Improving Military Demand Forecasting Using Sequence Rules

Rajesh Thiagarajan, Mustafizur Rahman, Greg Calbert, and Don Gossink

Defence Science and Technology Organization, Edinburgh SA 5111, Australia
firstname.lastname@dsto.defence.gov.au

Abstract. Accurately forecasting the demand of critical military stocks is a vital step in the planning of a military operation. A large number of stocks in the military inventory do not specify their consumption and usage rates per platform (e.g., ship). As a result, the demand of many critical military stocks may be under or over estimated leading to undesired operational impacts. To address this, we propose an approach to improve the platform-based military demand modelling process by deriving the demand of items with no usage models from the demands of correlated items with known per-platform usage rates. We adopt a data mining approach using sequence rule mining to automatically determine demand correlations by assessing frequently co-occurring usage patterns. Our experiments using our approach in a military operational planning system indicate a considerable reduction in the prediction errors across several categories of military supplies.

1 Introduction

Identifying and accurately forecasting demand for supply items are vital steps in the planning of a military operation as they facilitate effective decision making. Such demand forecasts, referred to as Advanced Demand Information (ADI) models in the supply chain literature, have been shown to be beneficial in several aspects of supply chain management; see [1–3]. In a military context, ADI of supply items should be accurately modelled on the basis of a their usage and Rate of Effort(ROE) by military platforms such as ships and aircraft. Consider the logistics planning of a week long military operation O, which involves a ship with a ROE set to 5 hours of sailing per day. Assuming the ship consumes 300 litres of diesel an hour, the diesel ADI model for this operation would consist of a daily demand for 1500 litres of diesel over 7 days resulting in a total demand of 10500 litres of diesel.

Not all stock items in the military inventory have platform-based ROE usage models (referred as ROEM henceforth) that can be used to derive accurate ADI models. A large number of stocks do not have ROEM mainly due to the fact that most stocks in the military inventory are managed through demands aggregated across a number of platforms. For example, demand for a lubricant at a military base over a time period is generally managed by aggregating its demand from a number of military platforms that require the lubricant during that period.

N.T. Nguyen et al. (Eds.): ACIIDS 2014, Part II, LNAI 8398, pp. 475–484, 2014.

While the aggregate demand for supply items at different military bases are captured as a part of the historic demand data, the per-platform usage details are not recorded. Therefore, automatic generation of ROEM is difficult. The absence of ROEM, however, leads to a lack of ADI models for many stock items. This inhibits effective and comprehensive operational planning as the demand for critical stocks may be under or over estimated leading to undesirable operational or cost impacts.

To address this gap, we propose an approach to improve the military demand modelling process by deriving unknown ADI models from known ADI models. In our approach, ADI models of items with no ROEM are derived from the ADI models of correlated supply items with known ROEM. Consider the planning of operation O, which as discussed above, consists of an ADI model of diesel for a ship, but lacks the ADI models of other critical stocks such as lubricants that are essential to the ship's operation. In our approach, the ADI model of a lubricant required for the ship can be derived from the ship's diesel ADI model because typically there would a surge in lubricant usage as a part of the ship's maintenance routine after a surge in diesel consumption during the operation.

A key concern of our demand modelling approach is the identification of correlation between demand for supply items. With over a million unique stock items in the military inventory, manually identifying correlation in demand between items is infeasible. Therefore, we adopt a data mining approach using sequence rule mining to determine frequently co-occurring usage between supply items from historic demand data. Our approach combines the results from sequence rule mining with time series regression analysis to derive correlated ADI models. We illustrate the effectiveness of our approach by predicting unknown ADI models in a military operational planning system. We also study the possibility of combining an auto-regressive demand prediction technique with correlated ADI models. Our experimental evaluation indicates that incorporating demand correlations in the ADI forecasting process considerably reduces the prediction errors across several categories of military supplies, improving the accuracy of the demand planning process.

The rest of the paper is organised as follows. Section 2 provides an overview of the related work followed by a brief discussion on the existing demand modelling process in Section 3. Our approach to extend the demand modelling process with sequence rule mining is presented in Section 4. We illustrate our ADI generation approach within a military operational planning system in Section 5 and discuss the results from our experimental evaluation in Section 6. Section 7 summarises our contribution and future work.

2 Related Work

A large body of work that deals with the advantages ADI models have in several aspects of supply chain management exists, such as [1–3]. The approaches in [1, 2] emphasise the utility of ADI models, which are established through market research, inputs from sales managers and advance (but imperfect) orders, to effectively realign manufacturing processes according to the ADI models. Existing works such as [3] have shown that military planning systems are a promising source of ADI models particularly when ROEM exist. However, the problem of generating ADI models for items with no ROEM has not received as much attention. To the best of our knowledge, there are no studies that address the problem of missing ADI model generation in the military context. We address this gap by extending a military operational planning system that generates a few ADI models with an approach to infer a large number of new ADI models based on the existing ones.

In the absence of ADI models, demand prediction techniques that use auto-correlation may be used to estimate the demand for critical military supplies for impending military operation based on their past usage history [4]. However, even items with continuous non-intermittent demands may be under or over estimated if the impending military operation is different to its past occurrences. In our approach, we complement an existing auto-correlated demand prediction technique with ADI correlation and show the resultant reduction in the prediction errors across several categories of military supplies.

3 The Military Demand Modelling Process

Accurate ADI models are essential to undertake effective decision making at the planning stage of a military operation. The work in [3] shows that a military operational planning system can be used to estimate the ADI models for an operation. The planning system allows a logistics planner to specify the force elements including platforms and personnel to be used in an operation, locations and routes in the operation, a schedule of activities including resource allocations, and the supply chains used to sustain the operation [3]. Apart from troop and equipment movements, the logistics plan of an operation consists of ADI models to effectively procure and distribute supplies to the operation. The planning system utilises pre-specified ROEM to generate the ADI models. The ADI models generated by the planning system facilitate a variety of decision support tasks including assessing plan feasibility, logistics sustainability, and risks or weak points in the operational plan.

4 Demand Modelling by Inferring Correlated ADI Models

The ideal solution to the lack of ADI models would be to establish ROEM for all stock items. However, such an effort would be an enormous, data-intensive and time-consuming undertaking with high costs due to the size of the military inventory. As mentioned earlier, automatic generation of ROEM is difficult as the historic demand data does not record per-platform usage details. Therefore, an approach to increase the number of ADI models available for analyses in the absence of ROEM is required.

In response, we propose an approach to extend the existing demand modelling process with new correlated ADI models inferred from known ADI models. Our approach builds on the basic premise that the demand for certain supply items are correlated in the context of a military operation. The correlation in demands may arise from a range of factors including complementary relationships (e.g., a gear lubricant oil and an engine oil), part-of relationships (e.g., an engine and a fuel pump), dependence (e.g., a gun and ammunition), and operational circumstances (e.g., an operation at a tropical region requires both anti-malarial drugs and repellents). As a result, in some cases, when the demands of 2 supply items are correlated and if only one item's ADI model is generated by the military operational planning system then their correlation may potentially be leveraged to infer the missing ADI model.

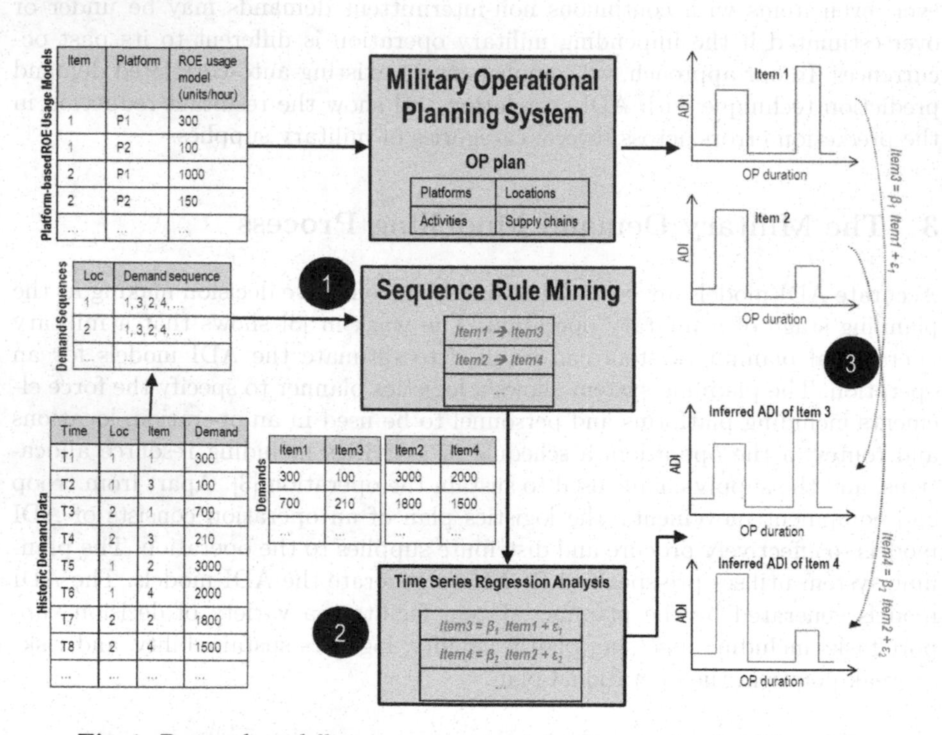

Fig. 1. Demand modelling process extended with correlated ADI models

Our demand modelling approach presented in Figure 1 extends the basic process discussed in Section 3 with following three additional steps.

1. Identification of correlation in demands using the historic demand data
2. Modelling the correlation in demands
3. Generating inferred ADI models on the basis of correlated demand models

The 3 steps in the extended demand modelling approach are detailed below.

4.1 Sequence Rules to Identify Correlated Demands

The key challenge of our demand modelling approach is the identification of correlation in demands. Specifically, the challenge is to determine frequently co-occurring usage between supply items from the historic demand data. We adopt a data mining approach to address this issue because the size of the military inventory makes manual discovery of correlations infeasible. Widely used data mining methods such as association rule mining [5] and all-pairs correlation techniques [6] are unsuitable to this problem mainly because the historic demand data is not a market basket database. Although conversion to a market basket database using time period based sampling is possible (e.g., all demands on a day can be treated as one market basket transaction), previous works like [6] have noted that such methods are vulnerable to false-positive and false-negative correlations because they do not take into account possible lags between the demands of correlated items. For example, unless the whole duration of the operation O discussed in Section 1 is considered as a single market basket transaction, the lagged correlation between demand for diesel and the lubricant for post-operational maintenance would be ignored.

To address this, we transform the identification of correlation in demands as a sequence rule mining problem, where the goal is to discover sequential rules from a database of sequences [7]. A sequence database is of the form $S = \{s_1, \ldots, s_n\}$, where each sequence s_i consists of chronologically ordered itemsets $\{\{i_1^1, \ldots, i_x^1\}, \ldots, \{i_1^m, \ldots, i_y^m\}\}$. The sequence rule mining process over S returns a set of sequence rules of the form $X \rightarrow Y$, where $X \subset s_i$ and $Y \subset s_i$ are disjoint itemsets such that Y occurs after X in S with a certain support and confidence [7]. While users are allowed to set a minimum support *minsup* value to filter infrequent rules, in most application domains, however, it is difficult to ascertain an optimal *minsup* before the mining process. Therefore, we adopt the rule mining algorithm in [7] that allows the users to efficiently search for *top-k* sequence rules with a certain minimum confidence. The algorithm works by a process called RuleGrowth, where small sequence rules are recursively expanded by adding frequent itemsets to the left and right side of the rules. The process continuously updates a *top-k* rules set when new rules with higher support are found.

Step 1 in Figure 1 shows the adoption of a sequence rule mining process in our extended demand modelling process to identify correlated demands. Firstly, a sequence database is generated from the historic demand data. Each sequence is formed by chronologically ordering the demand transactions at a military base. The rationale behind location-based sequence formation is that if the demand for two items co-occur frequently (with or without lag) across a number of military bases then it is likely that their demands are correlated. The demand sequences table in Figure 1 shows the location-based sequence database generated from the historic demand data. For example, the demand sequence at location 1 consists of items $\{1, 3, 2, 4, \ldots\}$ that occur at times $\{T1, T2, T5, T6, \ldots\}$. The demand sequences are provided as input to the sequence rule mining process [7] to generate sequence rules that are above a user-specified confidence threshold across all locations. Figure 1 shows the sequence rules generated $\{Item1 \rightarrow Item3, Item2 \rightarrow Item4, \ldots\}$. The rule $Item1 \rightarrow Item3$ implies that if a military base demands $Item1$ then it is likely that in the future it will be followed by a demand for $Item3$, indicating a potential correlation in the demands of items 1 and 3. Note that, unlike all-pairs correlation [6], the sequence rules also capture correlations between several items (e.g., $\{safety\_goggles, safety\_gloves\} \rightarrow \{safety\_boots\}$).

In some data mining domains (e.g., retail) it is sufficient to just identify potential correlations between items as this may already be enough to effectively adapt marketing tactics. In the context of ADI model generation, however, modelling the characteristics of a correlation is equally critical because it is important to quantify how the demand for one or more supply items impact the demand of a correlated item. To this end, we adopt time series regression analysis to model the correlated demands.

4.2 Time Series Regression to Model Correlated Demands

While a sequence rule indicates a potential correlation, in order to quantify the relationship between items it is important to consider the demand quantities of the correlated items over time. Time period based sampling (e.g., daily demand) can be used to represent the chronologically ordered demands of an item as a time series. Time series regression analysis, a well known statistical method to model the relationship between time series variables, is one way to model the correlated demands. A linear regression model of the form $Y_t = \beta_1 X_t^1 + \ldots + \beta_n X_t^n + \epsilon$ models the relationship between a dependent variable Y at time t and independent variables $\{X_t^1, \ldots, X_t^n\}$ at t using the regression coefficients $\beta_1, \ldots, \beta_n$, and an error term ϵ.

The problem of selecting suitable independent variables X from a large pool of potential variables is a common issue in regression analysis. Methods such as stepwise regression [8], which sequentially add/remove variables to a regression model based on a scoring criteria, are suitable if the pool of independent variables is small but do not scale well to larger pools. In the context of ADI model generation, the sequence rules generated in step 1 can be used to substantially reduce the pool of independent variables available to the regression analysis step. All items in the antecedent part of a sequence rule are considered as a part of the pool of independent variables to predict the unknown ADI models of items in the subsequent part of the rule. Step 2 of Figure 1 shows the regression analysis process conducted, leveraging the sequence rules to generate linear (or non-linear) models of the correlated demands. For every item in a demand correlation identified by a sequence rule, a time period based sampling is performed to generate the item's demand time series. For example, the demands table in step 2 of Figure 1 shows the time series creation of items $\{1, 3\}$ and $\{2, 4\}$. The regression analysis results include a linear model $Item3 = \beta_1 Item1 + \epsilon_1$ quantifying the demand for $Item3$ from the demand for $Item1$. A user-defined threshold for the coefficient of determination R^2 [9] is used to filter poorly correlated models.

Recall the issue of false-negatives and false-positives with time period based sampling raised in Section 4.1. The problem was due to the inability to deal with lagged correlations. We address this issue in step 2 by allowing lagged variants of known ADI models to be part of the pool of independent variables available to the regression analysis. For example, if the monthly demand of a lubricant l_t is dependent on a truck's ROE (i.e., diesel demand) over the last 6 months $\{d_{t-1}, \ldots, d_{t-6}\}$ then it can be modelled as $l_t = \beta_1 d_{t-1} + \ldots + \beta_6 d_{t-6} + \epsilon$.

4.3 Generating Inferred ADI Models

The final step in our extended demand modelling process is to infer new ADI models by applying the data from the known ADI models on the linear models formulated in step 2. It is important to note that both step 1 and 2 may be performed off-line prior to the planning stage. The results from step 2 can be used to infer new ADI models as shown in step 3 of Figure 1, where the ADI of $Item3$ and $Item4$ are derived from the ADI of $Item1$ and $Item2$.

5 Correlated Demand Models in Military Contingency Planning

To illustrate of our extended demand modelling approach, we have implemented a software system that infers new ADI models from the ones generated by a military operational planning system. Consider the planning of a military contingency operation. The mission's logistics plan using the military operational planning system would consist of ADI models of items with ROEM. Based on this initial plan, our implementation offers the planner a set of new ADI models that can be derived based on a library of correlated demand models. The planner can select the new ADI models as required depending on the mission's circumstances. The library of correlated demand models used in our system is developed prior to the planning process. As discussed in Section 4.1, the identification of demand correlation is transformed into a sequence rule mining problem that is solved using the *top-k* rule algorithm [7] in the Sequential Rule Mining Framework (SPMF) [10]. The linear models quantifying the correlated demands are generated using the *dynlm* package in the statistical package *R* [11].

We conducted a pilot study on using the extended demand modelling approach to plan for a military contingency. Our initial analysis indicates that the military operation planning system can only generate ADI models for some frequently used military consumables. The military consumables under consideration include fuels and lubricants (e.g., engine oil), food (e.g., combat ration packs), clothing (e.g., combat boots), cleaning products (e.g., engine cleaning brush), building products (e.g., aircraft sealant), and packaging (e.g., fuel drums). Our pilot study shows that inferring new ADI models based on sequence rule mining increases the available ADI models 5 fold across all the above mentioned categories of military consumables.

To validate the new ADI models generated using our extended demand modelling process, our pilot study included a blind experiment, where known ADI models are assumed to be unknown and are inferred using the demand correlation process. To quantify the accuracy of a predicted ADI model, we use the Root

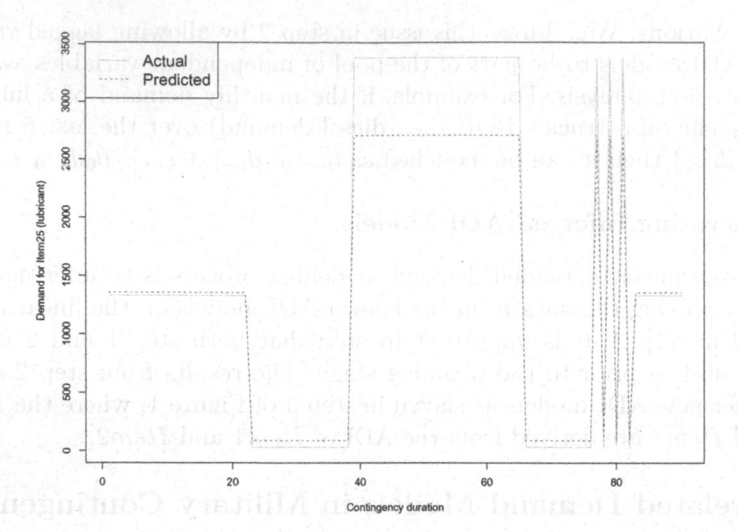

Fig. 2. Inferring a known ADI model

Mean Square Error (RMSE) metric [12] normalised as $RMSE/(max(ADI_a) - min(ADI_a))$ (referred as NRMSE henceforth), where ADI_a is the actual ADI model. Figure 2 shows the results from the blind test conducted on the ADI generation of a lubricant (*Item*25) for a military contingencies. The grey line shows the ADI model of *Item*25 as generated by the military operational planning system based on its ROEM. The red dotted line shows the ADI model inferred based on the demand correlations identified using the sequence rule mining process. Figure 2 shows that the inferred ADI model of *Item*25 closely matches (NRMSE is 7%) the actual ADI model generated by the military operational planning system, illustrating the validity of ADI model inference process.

With only a small number of known ADI models available for validation, large scale blind experimentation is difficult. Our future work includes extending this pilot study to a full-scale validation user study involving military logistics planners and subject matter experts.

6 Combining Auto-correlation and ADI Correlation

Due to the lack of ROEM, demand prediction techniques are a common part of the military operational planning process. In these techniques, the demand of an item for an upcoming operation (e.g., a training exercise) is auto-correlated with the item's usage in the past occurrences of the operation [4]. The demand of an item based on auto-correlation may be assumed to be its ADI model over the operational period. Auto-correlation techniques such as Simple Exponential Smoothing (SES) [13] and Autoregressive Integrated Moving Average models (ARIMA) [14] have been shown to be useful in the prediction of recurrent non-intermittent demand of military supplies [4]. Nevertheless, the retrospective demand-based predictions are only effective if the impending operation closely

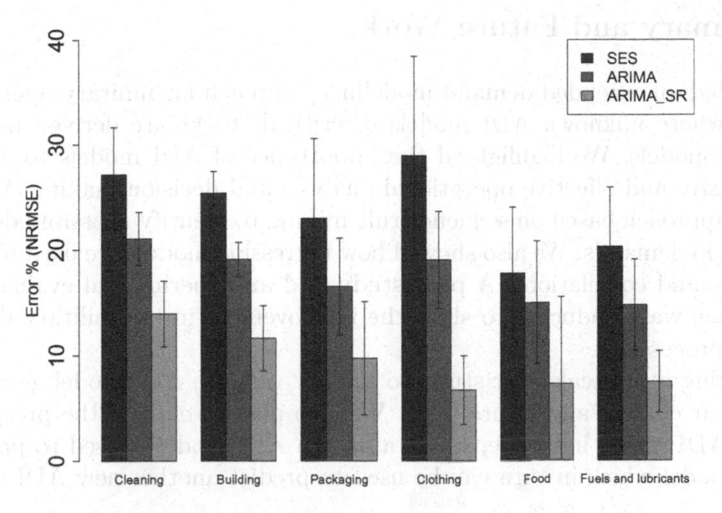

Fig. 3. Average Demand Prediction Errors

resembles its previous occurrences. To address this, we study the effects of combining auto-correlation with the ADI correlation.

To improve the ADI model generation using demand prediction, we incorporate the correlated demand models within the ARIMA prediction technique to produce a combined prediction. Specifically, when predicting an unknown ADI model, we make the correlated ADI models discovered using the sequence rule mining process outlined in Section 4.1 as external regressors in the ARIMA model, referred to as ARIMA_SR. As a result, when an ARIMA_SR model is used to predict an item's unknown ADI model then the prediction is not only based on the item's past usage but also on its correlation with other known ADI models. For example, if the diesel ADI model in the operation O discussed in Section 1 is added as an external regressor to the ARIMA_SR model predicting the ADI of a lubricant then its ADI would be predicted not only on the basis of the lubricant's usage in the past occurrences of O but also on the basis of the correlation that exists with diesel ADI.

To evaluate, we used the SES, ARIMA and ARIMA_SR techniques to predict the demand of items across several categories of military consumables mentioned in Section 5. We used the demand data from military training exercises conducted over the last 20 years in these experiments. The first 10 years of data from the training exercises were used to train the demand predictors, while the remaining data was used for testing the prediction accuracy. The results from our experiments presented in Figure 3 shows that on average the ARIMA_SR model is the best technique for ADI model generation with a relatively low NRMSE across all categories. When the standard deviations shown as error bars in Figure 3 are taken into consideration, the ARIMA_SR technique clearly outperforms the SES and ARIMA techniques in terms of NRMSE in 4 out of the 6 categories of military consumables. Therefore, combining correlated ADI models discovered using sequence rule mining with auto-correlated demand prediction improves the demand modelling process.

7 Summary and Future Work

We presented an extended demand modelling approach for military operational planning, where unknown ADI models of critical stocks are derived from existing ADI models. We highlighted the importance of ADI models to conduct comprehensive and effective operational analysis and decision making. We presented an approach based on sequence rule mining to identify (possibly delayed) correlation in demands. We also showed how regression models are used to quantify the demand correlations. A pilot study and an experimental evaluation of our approach was conducted to show the improvement in the military demand modelling process.

Conducting a full-scale user study to further validate ADI model generation is part of our current and future work. We also plan to explore the prospect of transitive ADI model inference, where a known ADI model is used to predict a new ADI model which in turn can be used to predict another new ADI model.

References

1. Benjaafar, S., Cooper, W.L., Mardan, S.: Production-inventory systems with imperfect advance demand information and updating. Naval Research Logistics (NRL) 58(2) (2011)
2. Karaesmen, F.: Value of advance demand information in production and inventory systems with shared resources. In: Handbook of Stochastic Models and Analysis of Manufacturing System Operations, vol. 192 (2013)
3. Thiagarajan, R., Mekhtiev, M.A., Calbert, G., Jeremic, N., Gossink, D.: Using military operational planning system data to drive reserve stocking decisions. In: 29th IEEE International Conference on Data Engineering (ICDE) Workshops (2013)
4. Downing, M., Chipulu, M., Ojiako, U., Kaparis, D.: Forecasting in airforce supply chains. International Journal of Logistics Management 22(1) (2011)
5. Agrawal, R., Imieliński, T., Swami, A.: Mining association rules between sets of items in large databases. SIGMOD Rec. 22(2) (1993)
6. Xiong, H., Zhou, W., Brodie, M., Ma, S.: Top-k correlation computation. INFORMS Journal on Computing 20(4) (2008)
7. Fournier-Viger, P., Tseng, V.S.: Tns: mining top-k non-redundant sequential rules. In: ACM Symposium on Applied Computing (SAC) (2013)
8. Wilkinson, L.: Tests of significance in stepwise regression. Psychological Bulletin 86(1) (1979)
9. Myers, R.H.: Classical and modern regression with applications, vol. 2 (1990)
10. Fournier-Viger, P., Gomariz, A., Soltani, A., Gueniche, T.: SPMF: Open-Source Data Mining Platform (2013), `http://www.philippe-fournier-viger.com/spmf/`
11. Zeileis, A.: dynlm: Dynamic Linear Regression (2013), R package version 0.3-2
12. Willmott, C.J.: On the validation of models. Physical Geography 2(2) (1981)
13. Gardner, E.: Exponential smoothing: The state of the art - part II. International Journal of Forecasting 22(4) (2006)
14. Box, G., Jenkins, G., Reinsel, G.: Time Series Analysis: Forecasting and Control (2008)

Agent-Based System for Brokering of Logistics Services – Initial Report

Lucian Luncean[1], Costin Bădică[2], and Amelia Bădică[2]

[1] Romanian-German University of Sibiu, Romania
[2] University of Craiova, Romania

Abstract. The competition between logistics companies leads to their continuous concern of improving their competitiveness in the market by increasing the quality of provided logistics services and by reducing logistics costs. The key to solving the above problems is the efficient management of logistics information. Therefore, building a smart logistics services supply system is imperative to each logistics company. In this paper we propose a multi-agent system for smart brokering of logistics services and we provide a preliminary analysis and design sketch of the system.

1 Introduction

With the explosive development of the Internet applications, entrepreneurs have identified new opportunities to grow their businesses with help from the virtual environment and the services that it provides. Also, the interest in increasing the degree of automation of business activities is continuously growing (e.g. by providing (semi-)automated support for negotiation between providers-sellers and consumers-buyers of online services). This can be achieved using software agents that enable dynamic trading between business partners. In this context, considerable effort on the analysis, design and implementation of agent-based support for automation of activities in e-Commerce transactions has been made by recent e-Commerce research [1].

Both providers of freight and cargo owners are continuously seeking new transport opportunities. Their common desire triggered the emergence of the new business model of *freight transportation exchanges*. This model includes, additionally to the online announcement of transportation opportunities, the provisioning of matchmaking services that facilitate the connection of the owners of goods with the freight transportation providers and their appropriate contracting. The businesses of freight transportation exchanges can be defined as virtual logistics platforms that operate in the domain of freight transport, exploiting the opportunities provided by the requirements of goods that need to be transported, as well as of the availability of free vehicles.

Freight transportation exchanges are used by companies that operate in transport and logistics sectors, and also by companies that aim at acquiring and/or providing goods that are necessary as inputs or outputs to carry out their business processes.

Typically, freight transportation exchanges are supported by online platforms[1]. The owners of these platforms are inviting users to register and post their availability of

[1] Popular examples are: www.bursadetransporturi.ro, www.europeancargo.ro, www.eurofreightexchange.com, www.easycargo.ro, ro.trans.eu, www.timocom.com, a.o.

N.T. Nguyen et al. (Eds.): ACIIDS 2014, Part II, LNAI 8398, pp. 485–494, 2014.

freight transport requests or freight vehicles provisions in public directories. Then, potential customers can easily browse, search and manually inspect through these directories in order to determine appropriate offers that suit their business needs. The available information contains real-time transport and goods postings in numerous and diverse fields of activity. Therefore, otherwise said, these online platforms facilitate the meeting of tenderers and recipients of transport services on a freight transportation exchange to contract and make exchanges of cargos between the holders of goods.

While certainly very useful, based on our initial analysis, the use of existing online platforms for freight transportation exchanges is heavily based on human-driven processes (i.e. manual browse, search and inspect of information), and thus, in our opinion, this process has several drawbacks:

1. The human operator intervention is always needed to link carriers to holders of goods.
2. Transport application forms are not standardized by a certain structured document format. Consequently, sometimes either important information is missing, or irrelevant information is provided (i.e. information overload).
3. A filtering mechanism for discarding posted ads that do not have any relation with goods transport is missing.
4. Transport capabilities of the owners of transport vehicles are not properly highlighted. Consequently, the cost incurred by customers for extracting useful information may sometimes be too high.
5. There is no monitoring system for the management of the lifetime of an application form. For example, a post might have a preset expiration time or it might become invalid or obsolete at some point in time (e.g. when the announced provision is not available anymore).
6. There is lack of a control model of virtual companies (e.g. reputation mechanisms or audits).

Agents are defined as computer systems situated in an environment that are able to achieve their objectives by: (i) acting autonomously, i.e. by deciding themselves what to do, and (ii) being sociable. Agents are able to flexibly combine reactive and intentional behavior in order to successfully meet the design objectives of many contemporary computer applications [2]. Multi-agent systems are especially suitable for application environments that are dynamic, partly accessible and unpredictable, thus also addressing the requirements of the logistics domain [3,4].

Therefore, in this paper we propose and introduce the initial design of a multi-agent system for freight brokering services. This paper presents the results of our preliminary analysis of using multi-agent systems for the development of brokering systems for the delivery of smart logistics services, trying to address some of the drawbacks of existing solutions, while highlighting some of the advantages of the agent-based approach.

The paper is outlined as follows. Following Section 1, where we briefly introduce and motivate our research, in Section 2 we proceed to an overview of research works related to the subject of our paper. Section 3 describes our proposed initial system architecture, introducing the system goals as well as initial use cases of our proposed system, while in Section 4 we present in details a use case of our system. Finally, in Section 5 we present our conclusions and point to future works.

2 Related Works

According to our analysis, there is a quite rich research literature on the use of multi-agent systems for modeling and development of domain-independent or domain-specific brokering services, as well as for design of e-business applications in logistics sector.

Authors of paper [5] propose a sound taxonomy of domain-independent middle-agents based on formal modeling using process algebras and temporal logic. In particular, they present the formal details of a *Broker* agent, highlighting its differences from *Matchmaker* and *Front-agent* types of middle-agents. Our proposed *FBAgent* type follows quite closely the behavioral model of *Broker* discussed in [5]. While the research results presented in [5] are theoretically sound and provide a solid foundation for our design, our proposal discussed in this paper goes one step forward by introducing a practical domain-specific multi-agent system for the logistics sector.

An agent-based domain-independent brokering service in also presented in paper [6]. However, here the focus of the research is set on the semantic capabilities of the *Broker* using rules and ontologies, rather than on its formal behavioral features. So, in some sense this work is orthogonal and complementary to [5]. We plan to incorporate some of these ideas into our own work by focusing on the domain-specific semantics in the logistics sector [7].

A new agent and cloud-based model for logistics chain is proposed in research paper [8]. Here, the author introduces the SMART cooperation model that allows companies to collaborate during a logistics process, thus contributing to the overall growth of the "system intelligence". Compared to our work, the goal of [8] is broader in scope and maybe more ambitious. Nevertheless, there are similarities with our approach, especially in the use of agent communication languages and interaction protocols for the design and implementation of agent collaboration in the system.

Another research that, similarly to us, proposes the use of agent-based negotiation for improving the logistics management system is [4]. However, there the focus is on the whole supply chain, rather than on brokering of logistics services, as in our work. Nevertheless, similarly to us, the use of multi-agent systems is motivated by the interactivity, responsiveness, and social characteristics of the domain.

Finally, other related works in the area of multi-agent domain-specific brokering services can be mentioned, for example paper [9] in the domain of resource brokering for Grid computing and paper [10] in the e-insurance domain.

3 Preliminary System Design

We propose a multi-agent system that provides an intelligent logistics brokerage service. We are focusing on the transport activity for the purpose of efficient allocation of the existing resources to the transport applications. Our initial analysis revealed that one of the desirable goals is to provide an intelligent service that is able to create an optimal transport route or policy such that a vehicle does not move without cargo or at least, that the movement without cargo is kept at a minimum, on the road segments between two loading points. Our agent-based system can benefit from the existing mathematical optimization methods developed for the freight brokerage industry [11].

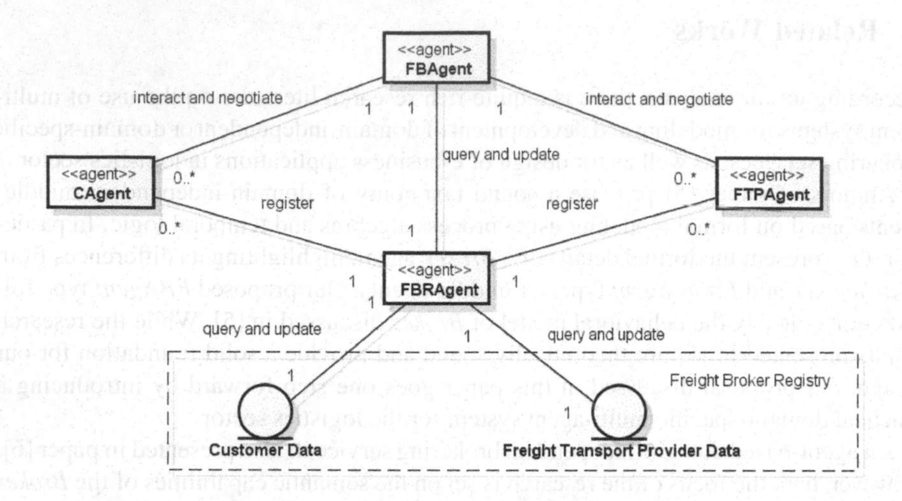

Fig. 1. System diagram

Users of our system, covering both customers and freight transport providers, are represented by software agents playing appropriate roles in the multi-agent system. We identified four types of agents in our system representing both external users of the system, as well as internal system components with specific responsibilities:

- *Customer Agent – CAgent* represents a customer of the transport brokering service.
- *Freight Transportation Provider Agent – FTPAgent* represents a provider of a transport resource.
- *Freight Broker Agent – FBAgent* represents the transport brokering service.
- *Freight Broker Registry Agent – FBRAgent* manages the registry of requests of customers and transport providers.

The system diagram illustrating the types of agents and their acquaintance relations is illustrated in Fig. 1. The diagram is using UML class diagrams, while agent types are represented using *agent* stereotype. Agent acquaintance relations are represented as UML associations [12].

Each agent has a set of specific objectives and in order to achieve them it needs appropriate information from human users and from other agents. In what follows we describe the specific objectives of each agent, as well as sample information needed to achieve them. The detailed formalization of the information needs of each agent type is the subject of ongoing research and it is outside the scope of the present paper.

3.1 Customer Agent

CAgent is responsible with resolving of requests issued by the customer of the transport service and it has three objectives: (i) to manage the registration of new customers of the transport brokering service; (ii) to issue transport requests to *FBAgent* based on customer input; (iii) to acquire (via subsequent interaction and negotiation with *FBAgent*) a convenient transport contract for the human user that represents the customer.

In order to achieve its objectives, *CAgent* needs certain pieces of information. Firstly, *CAgent* needs information from the customer user that describes his or her request of a transport service, including: date of request, point of dispatch, point of destination, proposed date of charging, proposed date of discharge, weight, volume, special transport constraints, lifetime. Secondly, *CAgent* must interact with *FBRAgent* for registration and update of customer information. Thirdly, *CAgent* must interact with *FBAgent* by issuing a new transport request and performing negotiation of an appropriate transport deal for the customer. During negotiation, *CAgent* might need confirmations or updated information about the transport request from the customer. Finally, *CAgent* might succeed or fail in finding an appropriate transport deal for its customer request. If successful, it will return to the customer a suitable transport deal including negotiated dates of charge and discharge, as well as other specific details dependent on the request.

3.2 Freight Transportation Provider Agent

FTPAgent is responsible with resolving of provisions of availability of freight transport resources and it has two objectives: (i) to manage the registration of new freight transport providers; (ii) to negotiate the terms of contract for providing transport.

In order to achieve these objectives *FTPAgent* needs certain pieces of information. Firstly, *FTPgent* needs information from the freight transport provider (a) about the vehicle description, including: type of vehicle (truck, tan pit, van), vehicle dedicated for special type of freight (logs, cars, animals, etc.), type of fuel (petrol, diesel), carrying capacity, length, width, height, as well as special characteristics like: canvas vehicle, hydraulic tailgate vehicle (hydraulic lift), and (b) about the transport resource availability, including location, date and time of availability. Secondly, *FTPAgent* must interact with *FBRAgent* for registration and update of freight transport provider information. Thirdly, *FTPAgent* must interact and negotiate with *FBAgent* a potential contract for the availability of the freight transport resource to a certain customer transportation request issued by *CAgent*.

3.3 Freight Broker Agent

FBAgent is responsible with the mediation between customer transport requests and freight transport provisions. This agent has a *broker* functionality [5] and it has the following list of objectives: (i) to record the transport requests issued by *CAgent* agents; (ii) to acquire from the *FBRAgent*, on request, a list of available vehicles that are capable to perform a given transport request; (iii) to negotiate with providers of transportation *FTPAgent* the possibility of carrying out a requested transport; (iv) to negotiate convenient terms and conditions (for example, a lower price and/or a convenient date and time); (v) to collect historical data about past transactions and perform a post-action analysis for improving the decision making processes.

In order to achieve these objectives *FBAgent* needs certain pieces of information. Firstly, *FBAgent* needs information from the *CAgent* about the transport request, including vehicle description and point of dispatch/undispatch. Secondly, *FBAgent* needs information from the *FTPAgent* about the freight transport resource, including vehicle description as well as location and date of availability. Thirdly, *FBAgent* queries *FBRAgent* to determine a list of potential freight providers matching a given transport request or a list of potential travel requests matching a given freight transport resource provision. Finally, *FBAgent* mediates between customer transport requests issued by *CAgent* and freight transport provisions issued by *FTPAgent*, possibly using negotiation, to determine convenient matches for both parties.

3.4 Freight Broker Registry Agent

FBRAgent is responsible with the management of the registry of requests of customers and transport providers. These registries are functioning following the "yellow pages" or *matchmaker* model [5]. *FBRAgent* has the following list of objectives: (i) to provide a list of provided transport resources that match the requirements of the customer transport request received from *FBAgent* and that are available or will become available on short-term; (ii) to provide a list of matching customer transport requests that meet the requirements of a new transport resource provision received from *FBAgent*; (iii) to record information about customers and their transport requests received from the *FBAgent* into the *Customer data* database; (iv) to record information about transport providers and their available transport resources received from the *FBAgent* into the *Freight Transportation Provider Data* database.

FBRAgent interacts with following agents to acquire the information needed for the achievement of its objectives: with *CAgent* and *FTPAgent* for receiving registration requests and with *FBAgent* for receiving forwarded transportation requests from *CAgent* and forwarded transport provisions requests from *FTPAgent*.

4 Use Cases

The functionality of our system was inspired by the operation of the *freight broker* business model [11]. Following [3], a freight broker coordinates transportation arrangements between transport customers (usually shippers and consignees) and transport resource providers or carriers. Their revenue model is based on transaction fees or commissions. For a successful and sustainable operation, freight brokers create and manage a vast network of customers. Their value proposition is the convenient and efficient matchmaking of freight and/or transport availability according to various customer specifications that allows customer agents (*CAgent* and *FTPAgent* in our case) agents to find the best terms and conditions for fulfilling their specific transport needs, including for example price, dispatch/undispatch points, duration, and safety.

In our opinion a key capability of a freight broker is to be able to negotiate with both the transport providers and customers to efficiently choose among the numerous delivery segments that define a certain transport task such that each segment is covered and the vehicles do not travel without cargo on the delivery segments. A *delivery segment* is defined as the distance traveled by the vehicles of a transport provider between two locations in order to deliver the goods from address *A* to address *B*.

A freight broker that deals with arranging transport shipments between suppliers and customers is always keen to develop its business by identifying opportunities and quality of transport services, but also seeks solutions to reduce costs in order to maximize profit. Moreover, the freight broker is directly concerned, in addition to the above mentioned aspects, about increasing portfolio of customers and transport providers.

To support those goals, the freight broker acquires a location on the Internet, and sets up a Web application called from this point on the e-Marketplace to communicate more effectively with transportation providers and customers. This e-Marketplace allows the registration in the system of transport providers and customers, primarily as actors representing legal persons or individuals. Moreover, each actor may subsequently register availability of certain vehicles, respectively certain transport applications. By creating the e-Marketplace, the company aims at the efficient automation of allocating a freight request to a transport resource resulting from a process of negotiation that occurs between the freight broker, customer and supplier of transport. Transport providers can register for this e-Marketplace with several transport vehicles, so basically, allocating a transport request is made to a transport vehicle owned by a specific transport provider. While the e-Marketplace is mainly perceived as a user-friendly interface by the clients of the freight broker, the actual functionality behind this interface is provided by our proposed multi-agent system that is the focus of this paper.

We identified three types of actors representing the users of our system: the *Freight Broker*, the *Freight Transport Provider* and the *Customer*. Each has the private goal of its own business development to optimize costs and increase service quality and quantity. Each type of user is represented in the system by a suitable agent type, as follows: *CAgent*, *FBAgent*, and *FBRAgent* [13]. In order for the system to be functional our system incorporates a passive entity called *Freight Broker Registry* inspired by *Client Information Center* from [9]. The *Freight Broker Registry* acts as a repository comprising the *Customer Data* and *Freight Transport Provider Data* data bases shown on Fig. 1. The *Freight Broker Registry* stores descriptions of transportation resources and requests for transportation resources, including: (i) meta-data describing each *Customer* (including contact information, delivery history and reputation data); (ii) meta-data describing each *Freight Transport Provider* (including contact information, number of vehicles and "home" location of each vehicle); (iii) information about requests for transportation resources; and (iv) information describing available transportation resources.

According to this analysis, we identified two main use cases that underlie the development of our proposed multi-agent system for the mediation of logistics services provision: (i) to allow customers to make transport requests; (ii) to allow transport providers to join a freight broker. In what follows we describe in detail the agent interactions during a sample scenario based on the first use case of our system. The UML sequence diagram [12] that concisely describes those interactions is presented in Fig. 2. The detailed description of the second use case was omitted because of lack of space.

The scenario involves a customer represented by *aCAgent* that is already registered and logged into our system. Let us assume that *aCAgent* is interested to contract and purchase freight services and thus he prepares and submits an application form to the *Freight Broker* represented in the system by his *aFBAgent*.

Following the interactions described in Fig. 2, the scenario proceeds as follows:

- The user *aCustomer* prepares and submits an application form via his *aCAgent* personal agent. We assume here that the customer is already registered and represented

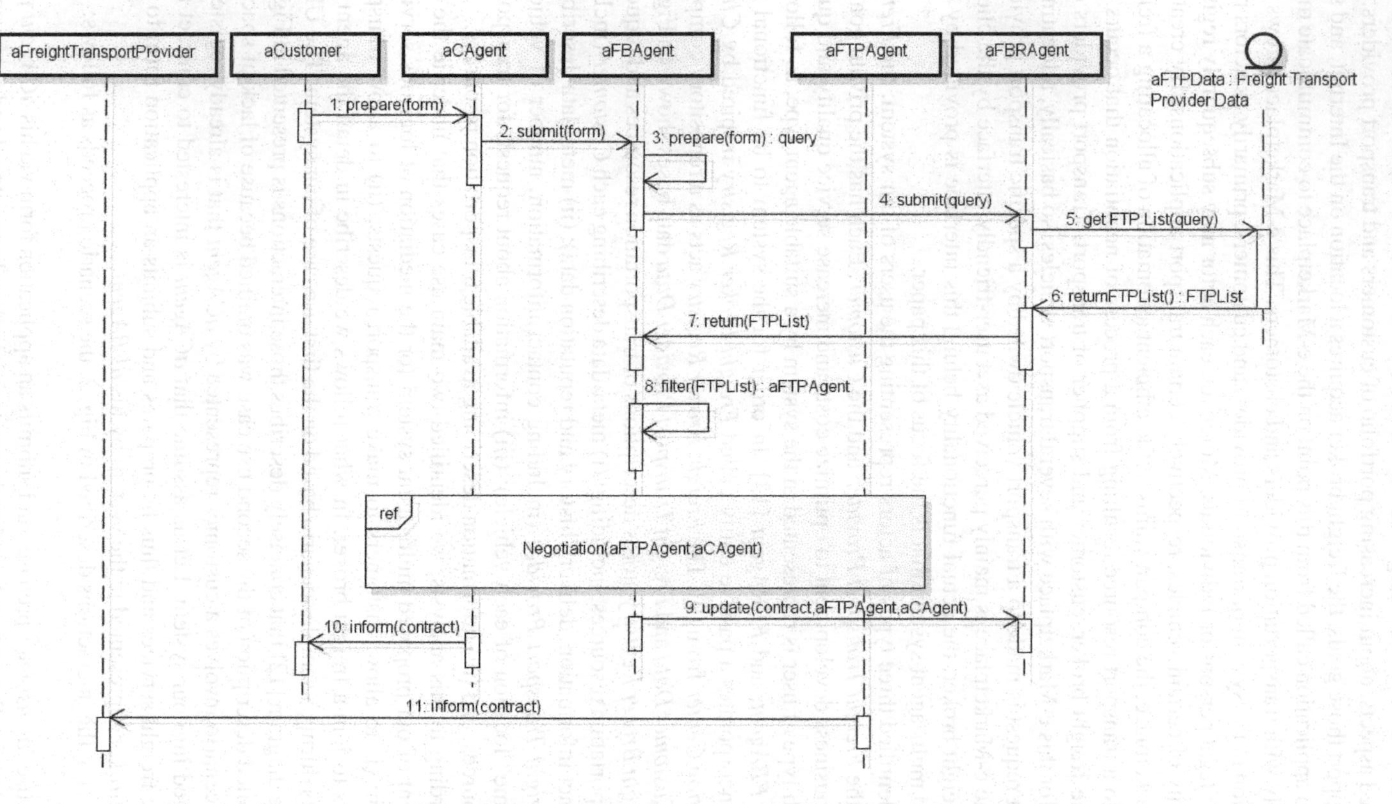

Fig. 2. Agent interactions for resolving the transport request submitted by a customer

in the system by *aCAgent* (for technical details on the coupling of the external Web-based interface with the multi-agent system the interested reader can consult reference [14]).
- *aCustomer* forwards the application form, represented in serialized form (for example, using an XML-based or other structured representation), to *aFBAgent* that represents the *Freight Broker*.
- *aFBAgent* processes the form and determines (message 3) and submits (message 4) a query to the *aFBRAgent*.
- *aFBRAgent* queries (message 5) its database *Freight Transport Provider Data* and determines (message 6) a list *FTPList* of suitable freight transport providers that meet the requirements of the *aCustomer*.
- *aFBRAgent* returns the list of transport providers to the *aFBAgent* (message 7).
- *aFBAgent* is using an internal decision component to filter out less promising freight transport providers (message 8), finally keeping a single most promising one, denoted by *aFTPAgent*.
- *aFBAgent* is coordinating a negotiation between *aCAgent* and the chosen *aFTPAgent* to determine the terms and conditions for the freight transport contract. This is represented by the *Negotiation(aCustomer,aFTPAgent)* interaction box.
- Finally, *aCAgent* and *aFTPAgent* send the contract information representing the negotiation outcome to the customer and the freight transport provider (messages 10 and 11), while *aFBAgent* updates the *aFBRAgent* with the status of the transport request and transport provision (message 9). The request was fulfilled, so it must be discarded from the registry, while the provision should be marked as not available during the activity of carrying out the transport.

Please note that the diagram does not model the situation when the customer request cannot be immediately fulfilled. This can happen either if there is no matching transport resource provision recorded in *Freight Broker Registry* or the negotiation between the *aCAgent* and the chosen *FTPAgent* fails. In the first case the request can be simply memorized in the *Freight Broker Registry* until an appropriate resource is discovered or the request becomes obsolete. In the second case the decision and negotiation processes of the *aFBAgent* can be retried, either immediately or slightly later, after a certain duration of time. The underlying agent interactions were not included on the diagram from Fig. 2 because of lack of space, but nevertheless, they were taken into consideration during the system design.

5 Conclusion and Future Works

In this paper we formulated our initial proposal of a multi-agent system that aims to improve the competitiveness of logistics companies by increasing the quality of provided logistics services and by reducing logistics costs. During the preliminary system design we identified two main use cases of the system, as well as a set of agent types with specific goals and functionalities for the efficient management of logistics information. For each use case we proposed a sequence diagram that describes the interaction protocols between the agents. Our initial analysis and design provide a good basis for producing a more detailed design and implementation of the proposed multi-agent system.

As future works, on short term, we would like to perform a more detailed analysis and specification of the information requirements for our system. In particular, we will clearly identify and formalize the information required by each agent type to achieve its tasks, and the information that must be exchanged between agents, as well as between human users and agents as inputs and results. On long term we are interested in the implementation and experimentation with the proposed multi-agent system on realistic logistic scenarios.

References

1. Dobriceanu, A., Biscu, L., Bădică, A., Bădică, C.: The design and implementation of an agent-based auction service. International Journal of Agent-Oriented Software Engineering 3(2/3), 116–134 (2009)
2. Wooldridge, M.: An Introduction to MultiAgent Systems, 2nd edn. John Wiley & Sons (2009)
3. Bowersox, D.J., Closs, D.J.: Logistical Management – The Integrated Supply Chain Process. McGraw-Hill (1996)
4. Guanghai, Z., Runhong, Y.: Multi-agent supply logistics intelligent management system based on negotiation. Computer Science Applications and Education 3(1), 476–481 (2013)
5. Bădică, A., Bădică, C.: Fsp and fltl framework for specification and verification of middle-agents. Applied Mathematics and Computer Science 21(1), 9–25 (2011)
6. Antoniou, G., Skylogiannis, T., Bikakis, A., Doerr, M., Bassiliades, N.: Dr-brokering: A semantic brokering system. Knowledge-Based Systems 20(1), 61–72 (2007)
7. Scheuermann, A., Hoxha, J.: Ontologies for intelligent provision of logistics services. In: Proceedings of the Seventh International Conference on Internet and Web Applications and Services (ICIW 2012), pp. 106–111 (2012)
8. Kawa, A.: Smart logistics chain. In: Pan, J.-S., Chen, S.-M., Nguyen, N.T. (eds.) ACIIDS 2012, Part I. LNCS (LNAI), vol. 7196, pp. 432–438. Springer, Heidelberg (2012)
9. Wasielewska, K., Ganzha, M., Paprzycki, M., Drozdowicz, M., Petcu, D., Bădică, C., Attaoui, N., Lirkov, I., Olejnik, R.: Negotiations in an agent-based grid resource brokering system. In: Iványi, P., Topping, B. (eds.) Trends in Parallel, Distributed, Grid and Cloud Computing for Engineering, pp. 355–374. Saxe-Coburg Publications (2011)
10. Nogueira, L., Oliveira, E.: A multi-agent system for e-insurance brokering. In: Kowalczyk, R., Müller, J.P., Tianfield, H., Unland, R. (eds.) Agent Technology Workshops 2002. LNCS (LNAI), vol. 2592, pp. 263–282. Springer, Heidelberg (2003)
11. Silver, J.L.: Optimization tools for the freight brokerage industry. Massachusetts Institute of Technology. Engineering Systems – Master's thesis (2003), `http://hdl.handle.net/1721.1/28574`
12. Fowler, M.: UML Distilled: A Brief Guide to the Standard Object Modeling Language, 3rd edn. Addison-Wesley Professional (2003)
13. Gawinecki, M., Vetulani, Z., Gordon, M., Paprzycki, M.: Representing users in a travel support system. In: Proceedings of the Fifth International Conference on Intelligent Systems Design and Applications (ISDA 2005), pp. 393–399. IEEE Computer Society (2005)
14. Ilie, S., Bădică, C., Bădică, A., Sandu, L., Sbora, R., Ganzha, M., Paprzycki, M.: Information flow in a distributed agent-based online auction system. In: Burdescu, D.D., Akerkar, R., Bădică, C. (eds.) Proc. 2nd International Conference on Web Intelligence, Mining and Semantics (WIMS 2012), p. 42. ACM (2012)

Cloud Community in Logistics e-Cluster

Arkadiusz Kawa and Milena Ratajczak-Mrozek

Poznań University of Economics,
al. Niepodległości 10, 61-875 Poznań, Poland
{arkadiusz.kawa,milena.ratajczak}@ue.poznan.pl

Abstract. In Europe there are a lot of transportation and logistics companies. They provide different services, depending on their sizes and scope. The largest logistics service providers use advanced ICT solutions enabling them to capture, transform and interchange data. However, others, especially small and medium enterprises (SMEs), do not have enough money to invest in new ICT solutions which allow to monitor data e.g. related to the inbound freight flows.

In this paper, the authors propose a solution which is based on cloud community and e-cluster. It gives SMEs a better possibility to automatically capture information and exchange it within a particular enterprises' network. Also, it helps to build up business relations and, finally, to better match demand and supply capacity data. The main idea of the project and the proposal to optimize business operations in the context of the usage of cloud computing facilities are presented.

Keywords: cloud community, e-cluster, logistics service providers.

1 Industry Clusters and e-Clusters as a Network Structure

Business environment is based on formal and informal interactions (including transactions, trade exchange) hence business actors are interlinked and form different kinds of network structures. Those network structures are strictly or loosely defined sets of collaborating entities (nodes, actors) linked by so called network relationships (ties, arcs).

An important type of a network structure is business (industry) cluster. Such clusters are „(…) geographic concentrations of interconnected companies and institutions in a particular field. Clusters encompass an array of linked industries and other entities important to competition [15]". Actors within clusters include, e.g. suppliers of components, providers of specialised infrastructure, research & development institutions, universities and governmental institutions [15]. Within cluster usually there is a so called broker acting as a flagship company/institution and the task integrator. The integrator is the main entity that is actively creating the network in a strategic manner but only has strategic control over those aspects of its partners' business systems which are dedicated to the network [10, 16]. The actors within cluster are independent and interconnected by cooperation, commonalities and/or complementarities. This in turn should generate value added and lead to a competitive advantage both for involved actors as well as the whole economy (mainly region).

N.T. Nguyen et al. (Eds.): ACIIDS 2014, Part II, LNAI 8398, pp. 495–503, 2014.

Traditionally, within the original concept of a cluster the importance of geographic proximity is stressed. Cluster actors must be located geographically close to each other to gain competitive advantage [3, 13]. However, Internet technology and other information and communication technologies enable the cooperation and creation of virtual relationships even between the widely dispersed actors [1, 11]. The notion that geographical proximity is not a vital factor for collaboration within clusters led to the identification of e-clusters [2, 6]. Those e-clusters are defined as "digital enterprise communities enabled by one or more intermediaries and based on a new type of electronically enabled inter-organizational system" [5]. The e-clusters use information and communication technologies as a digital platform for cooperation [6].

2 The Idea and Benefits of Cloud Computing

Cloud computing is "a model for enabling ubiquitous, convenient, on-demand network access to a shared pool of configurable computing resources (e.g., networks, servers, storage, applications, and services) that can be rapidly provisioned and released with minimal management effort or service provider interaction" [14]. In cloud computing a pool of abstracted, virtualized, dynamically-scalable, managed computing power, storage, platforms, and services are delivered on demand to external customers over the Internet [7].

Essential characteristics of cloud computing include [14]: on-demand self-service (the use of computing capabilities, such as server time and network storage, automatically), broad network access (the availability of capabilities over the network and accessed through standard mechanisms), resource pooling (the provider's computing resources serve multiple consumers using a multi-tenant model, with location independence of the provided resources), rapid flexibility (the elastic, often automatically and unlimited provision of capabilities) and measured service (the monitoring, controlling and reporting of resource usage providing transparency for both the provider and customer).

Depending on whether the cloud is made available to the general public or refers only to internal data centers of a business or other organization, public and private clouds are distinguished [4]. There are also community clouds when the cloud infrastructure is shared by several organizations and supports a specific community [14]. Such of a cloud may be used by e-cluster members and in this way serving a supporting and strengthening role for cooperation and acting as a center hub of a network structure.

Unlike the private data centers public clouds offer appearance of infinite computing resources on demand, elimination of an up-front commitment by cloud users, ability to pay for use of computing resources on a short-term basis as needed, economies of scale due to very large data centers, usually higher utilization by multiplexing of workloads from different organizations and simplifying operation and increasing utilization via resource virtualization [4]. Community cloud is in this context a solution "in between" of private and public clouds and their advantages.

The basic advantage resulting for companies from the use of cloud computing is cost reduction. Cloud computing allows provisioning and consuming IT capabilities on a need and pay by use basis ("pay as you go" method), according to the actual user demands. It eliminates the need to purchase licenses and pay for software installation and administration and helps the IT systems to be more agile [17].

Cloud computing allows also more flexible use of resources and operations, an asset-free provision of technological resources [17], giving the access to almost unlimited resources on demand. It saves "overprovisioning for a service whose popularity does not meet their predictions (...) or underprovisioning for one that becomes wildly popular, thus missing potential customers and revenue" [4].

The cloud platform works as a so-called trusted third party that imposes a clear and established security model for all participants, provides a reliable source of information on such issues as time stamping, non-repudiation of certain actions, etc. Moving some actions of e-cluster to the cloud improves cooperation security, because it does not require trust in the partner but trust in the platform.

3 Problems with Traditional Systems Used in the Logistics Industry

IT systems commonly used by logistics companies are mostly limited to the management of operations within a single organization but less frequently to cooperation between contractors who are closer or further upstream and downstream. This is due to several difficulties that limit the development of logistics clusters.

Companies do not want to completely get rid of the previously used, often old, but very specialized systems, which, in their view, are tailored to their specific business. For this reason, they are combined and, as a result, precisely customized systems are created. This raises a number of problems with data consistency, their repetition or lack thereof. Customized systems do not ensure uniform data synchronization and standardization in the business network. If one of the companies makes a change in its system, the systems of the other partners are not even informed about these changes.

Most systems operate on the principle of one-way communication and the information is "pushed", which means that if one company is to receive the information concerned, another one must send such information. For this reason, companies have a difficult task and cannot analyze the suppliers' systems to adapt their activities to the needs of the entire network.

The lack of synchronization and database updating is also a strong concern. Every time new information is placed in one database or any category changes, amendments must be made in each of the linked databases, because in many cases there is still no integration mechanism that understands these changes and automatically adapts to them.

One of the most significant disadvantages is the relatively high cost of implementation and maintenance of a traditional system. This is not only financial expenditure incurred for the right software, but also for electronic devices. This eliminates a large portion of small enterprises from the electronic information exchange process. For example, any company using EDI (Electronic Data Interchange) standards has to bear fixed costs of software licenses and fees for technical support which is supposed to ensure that the system functions properly regardless of the type of the IT solutions used in the enterprise, consistency of data exchange protocols and a uniform rate of their transmission.

Another problem is the relatively high cost of providing an adequate level of mass customization. For example, it is difficult to monitor individual processes in the supply chain by different users. Each monitored product may be different, used differently; furthermore, contact methods may vary. There may be thousands or even millions of tracked products and their users, which, in the face of the high specificity of these products, raises a number of problems associated with effective monitoring and sharing of information.

It is also worth noting that customers using services of different logistics service providers must adapt to the tools offered by them each time. In the case of one-off collaboration, searching for the carrier, they browse the Web, shopping catalogs, etc. The logistics sector lacks solutions that integrate services from different providers in one place.

4 Cloud Community in e-Cluster

A remedy for the problems concerning cooperation in logistics clusters presented in the previous section is the aforementioned concept of cloud computing, in particular cloud community in e-cluster.

It is worth to note that, thanks to the Internet technology, the electronic flow of information between companies in the global economic network can be faster and easier. Due to the low cost of Internet access, participation in such a network is not limited to large companies, but also includes small and medium enterprises. Thanks to that, all suppliers, distributors, manufacturers and retailers can work together more closely and effectively than before.

The use of the concept of cloud community in e-clusters makes it possible to create an electronic logistics platform (cf. Section 5) which is targeted at a limited set of organizations [19]. This platform provides, inter alia, cooperation of logistics companies, especially to gain access to data about logistics services and supply capacities. It allows to work together in the changing conditions and access resources; software and information are provided to computers and other devices on demand.

Most importantly, thanks to the idea of cloud community it is possible to derogate from the concept of rigid links, traditionally built on the basis of long-term contracts, and replace them with e-clusters, which are adapted to changing trends, customer preferences and high competitiveness. Such e-clusters are often configured and maintained only for specific operations that must be performed in a very short period of time or have a very specific nature [8]. E-clusters adapt to the changing needs with extraordinary flexibility - potential partners "are drawn into" the e-cluster to perform specific tasks. After completing the order, this particular cluster dies. Of course, this cluster can be re-engaged when a similar contract appears.

The use of cloud computing in e-cluster allows to achieve transparency of most processes taking place in the network. The scope of management is expanded, which enables managers to make operational decisions on the basis of the information received from various areas where previously information was missing. It gives the possibility to monitor and control transactions irrespective of the geographical location.

5 Proposed Use of Cloud Community in Logistics e-Cluster

The idea of the research proposal starts from the assumption that through cooperation companies should rationalize their logistics processes, obtain cost savings and reduce empty shipments. At the moment, companies are not activating collaboration as they are traditionally managed like small "family enterprises". This limits their ability to get potential opportunities offered by collaboration with other actors operating in the market.

The presented approach is based on the Logical project [21] and other European research projects [9, 18, 20]. The main idea of the Logical project is the creating of logistics platform which provides complex services to customers and simplifies management and administration of joint contracts. It gives more transparency and comfort to customers, shipment tracking, a fulfillment forecast, a change to pull-processes [21]. In order to make it possible, a cloud computing platform is to enable an exchange of information in real time between the entities involved in the transport (transport user, logistics service provider, coordinator). The cooperation between the companies reduces transportation costs through the optimization of the vehicle and complete elimination of "empty runs" and, at the same time, it may even improve customer service. It is possible thanks to the access to and matching of demand and supply capacity data [12].

The described logistics platform will run on "a pay as you go" principle. Contrary to ERP (Enterprise Resource Planning) and SCM (Supply Chain Management), this platform will not require significant expenditure on the design, construction and maintenance of the system and applications supporting production planning, logistics and sales.

Easy configuration of resources, capabilities and competences of actors happens in an e-cluster in order to achieve a certain objective but within the community network. We assume, in accordance with the e-cluster definition, that the actors forming the community are dispersed geographically, or, to be more specific, their seats may be located in any place. E-cluster is perceived by the final user as a single coherent whole, even though it is composed of independent entities.

The logistics e-cluster may also optimize its activities through group purchases of services, thus aggregating the sum of needs for basic logistics services such as freight transport, storage or handling demands. This optimization consists in achieving the economies of scale while maintaining the principles of sustainable use of resources. For example, single companies transport small volumes with varying frequency, which does not give them appropriate bargaining power towards logistics operators. To maximize customer satisfaction a small company increases the frequency of deliveries, but frequent deliveries in small amounts cause a rise in the transport cost and a reduction in the rate of the cargo area utilization. The accumulation of public procurement and implementation of full load supplies contributes, in turn, to a reduction of transport costs, but does not provide them with an acceptable level of customer service. Group purchases of transportation services make the realization of the transport needs of several companies possible on one vehicle. In addition, this form of shopping may lead to a common reduction of personnel costs related to the organization of the transport by automating certain processes (route planning, reporting, billing, communication). For the customer, faster transport means faster inventory turnover, and, consequently, a lower maintenance cost.

An important way to improve the time and cost of transport is to use multi-modal transportation. With a variety of means of transport belonging to companies representing independent logistics providers, the cloud computing platform selects the optimum solution. Often, the use of multi-modal terminals and freight train connections for transnational freight transport could be intensified if matching of sufficient intermodal transport demands and related service offers could be facilitated and accelerated. The efficiency and sustainability of multimodal transport will be raised by establishing cooperation between freight villages, airports, ports, other logistic centers and networks based upon interconnection of their logistics data clouds and intercloud computing services [21].

Including the fleet tracking module (Track & Trace), one can also monitor the company's vehicle in real time, thus reducing costs. Thanks to that, the logistics service provider can freely and rationally coordinate the operation of its business. An agent observes the way of the consignment all the time and on that basis selects the optimal route; in case of a loss of the shipment they can take appropriate action. Track & Trace can also get complete statistical information about the quality of the services provided and, if necessary, the company may seek to raise their standard. With Track & Trace, logistics service providers have a clearly defined system of accountability for the delivery and know where the consignment is at every moment.

In addition, thanks to integration with the European transport exchanges, companies have access to the best offers for both cargo and freight. Looking through them and placing new ones can be done automatically by the cloud computing platform. For example, the searching process takes place on the basis of the criteria put into the system by the forwarding agent.

Another example is the shipment realization order process, which is mostly realized with the use of cloud computing. The preparation of and sending the offer as well as receiving feedback are done automatically in a cloud and are based on the data collected in the previous processes. The forwarding actor's task is to approve the prepared order and pass it on to the realization stage. Also, the coordination of the transportation process takes place entirely without interference of the forwarding agent. Changes of the transport status are made on the basis of the information received from the carrier automatically and are available to the client [21].

6 Processes Optimization in Cloud Community

Cloud computing is expected to contribute to the optimization of enterprises in business processes as well as in IT solutions. From the perspective of a business process, the optimization can be carried out in the field of finance, administration, marketing, human resource management and logistics.

In the case of the logistics processes optimization in e-cluster we can present the following criteria:

- General measures
    - Direct product cost
    - Direct product profitability
    - Share of fixed cost in the total cost
    - Use of the "pull" process
    - Use of the "push" process
    - Access to other enterprise resources

  - Process transparency
  - Total logistics services cost
  - Customer service level
  - Use of existing transport infrastructure (including storage, transport and transshipment)
  - Consumption of enterprise resources
  - Achieve quality in logistical services (reliability, supplier loyalty)
  - Cash-to-cash cycle time
  - Personnel costs associated with logistics support orders
  - Degree of concentration on core business
  - Value added services
- Warehouse and inventory measures
  - Stock structure
  - Economic stock
  - Average Stock
  - Buffer stock
  - Stock rotation
  - Stock cost
  - Stock carrying costs
  - Completion waiting time
  - Replenishment lead time
  - Stock-out risk
- Transport measures
  - Transport time
  - Total deliver costs
  - Product shipped per delivery
  - Quantity per shipment
  - Occupancy rate of transport means
  - Time work rate of transport means
  - Share of transport costs in the value of the cargo
  - Share of various modes of transport in the transport of goods
  - Using collective (sustainable) modes of transport (multi-modal co-operation).
  - Average time of loading and unloading
  - Just in time delivery
  - Delivery reliability
- Order fulfillment measures
  - Order processing time
  - Perfect order fulfillment
  - Delivery lead time
  - Operational efficiency
  - Cost to schedule product deliveries
  - Average days per schedule change
  - Reliability shipment quality
  - Matching demand/capacities for segmented partner networks

The proposed criteria first need to be given the right priorities and only then can be gradually used in a e-cluster. Of course, it is a long-term process, but thanks to it, information is gathered concerning among other things bottlenecks, cost reduction possibilities, the most effective cluster actors, the most effective solutions. This in turn leads to strengthening the cooperation potential and whole e-cluster optimization.

It should be remembered that a solution once prepared and implemented must be constantly improved. Measuring individual indicators such as the rate of information flow, security cost system, time associated with obtaining access to data etc. enables continuous monitoring of individual components of the logistics e-cluster.

7 Conclusions and Further Research

Many small and medium transportation and logistics companies do not have enough money and other resources to invest in advanced ICT solutions enabling them to capture, transform, monitor and interchange data which in turn leads to their relatively lower competitive advantage. The possible solution for this problem may be a cooperation within e-cluster supported by community cloud. Cooperation gives an ability to obtain resources which are at partners' disposal or even resources unavailable thus far to any of the parties. Moreover such, resources resulting from collaboration are quite often hard to duplicate. The community cloud strengthens the aforementioned benefits of cooperation. It gives small and medium companies better possibility to automatically capture information and exchange it within a particular enterprises' network allowing provisioning and consuming IT capabilities on a need and pay by use basis. Such a cloud helps also to build up business relations and serves a supporting role for cooperation, acting as a center hub of a network structure.

It is worth to emphasize that there are some problems concerning cooperation within e-cluster resulting from the use of cloud computing. They include: a breach of trust and the resulting hostile use of the information that is fundamental to a company, communication problems (at the internal level of a company and the entire e-cluster), restriction on freedom of making decisions and shared responsibility for decisions and actions. There is the fear of losing independence and control over the company associated with both e-clusters and IT community clouds. There is also a lot of concern about the security and privacy of the data which may sabotage the willingness of cooperation. Many managers are not comfortable about their data located outside the company. Those problems needs special attention and have to be challenged in further research. They include both proper IT solutions securing data flow and management matters concerning inter-organizational trust and business relationships.

Acknowledgements. The paper was written with financial support from the National Center of Science [Narodowe Centrum Nauki] – the grant of the no. DEC-2012/05/D/HS4/01138. The research was also supported by the grant of the National Center of Science, no. DEC-2011/03/D/HS4/03367.

References

1. Adebanjo, D., Kehoe, D., Galligan, P., Mahoney, F.: Overcoming the barriers to e-cluster development in a low product complexity business sector. International Journal of Operations & Production Management 26(8), 924–939 (2006)
2. Adebanjo, D., Michaelides, R.: Analysis of Web 2.0 enabled e-clusters: A case study. Technovation 30, 238–248 (2010)
3. Anderson, G.: Industry clustering for economic development. Economic Development Review 12(2), 26–32 (1994)
4. Armbrust, M., Fox, A., Griffith, R., Joseph, A.D., Katz, R., Konwinski, A., Lee, G., Patterson, D., Rabkin, A., Stoica, I., Zaharia, M.: A View of Cloud Computing. Communications of the ACM 53(4) (2010)
5. Brown, D.H., Lockett, N.: Engaging SMEs in e-commerce: the role of intermediaries within eClusters. Electronic Markets 11(1), 52–58 (2001)
6. Cecil, P., Castleman, T., Parker, C.: Knowledge management for SME-based regional clusters. Collaborative Electronic Commerce Technology and Research (2004), http://www.collecter.org/coll2004S
7. Foster, I., Zhao, Y., Raicu, I., Lu, S.: Cloud Computing and Grid Computing 360-Degree Compared. In: Grid Computing Environments Workshop, GCE 2008, Austin, Texas, USA, pp. 1–10 (2008)
8. Fuks, K., Kawa, A., Wieczerzycki, W.: Dynamic Configuration and Management of e-Supply Chains Based on Internet Public Registries Visited by Clusters of Software Agents. In: Mařík, V., Vyatkin, V., Colombo, A.W. (eds.) HoloMAS 2007. LNCS (LNAI), vol. 4659, pp. 281–292. Springer, Heidelberg (2007)
9. Hajdul, M.: Model of coordination of transport processes according to the concept of sustainable development. LogForum 3(21), 45–55 (2010)
10. Jarillo, J.C.: Strategic networks. Creating the bordless organization. Butterworth Heinemann, Oxford (1995)
11. Karaev, A., Koh, L., Szamosi, L.: The cluster approach and SME competitiveness: a review. Journal of Manufacturing Technology Management 18(7), 818–835 (2007)
12. Kawa, A.: SMART logistics chain. In: Pan, J.-S., Chen, S.-M., Nguyen, N.T. (eds.) ACIIDS 2012, Part I. LNCS (LNAI), vol. 7196, pp. 432–438. Springer, Heidelberg (2012)
13. Ketels, C.: The Development of the Cluster Concept - Present Experience and Further Developments. In: NRW Conference on Clusters, Duisburg, Germany (2003)
14. Mell, P., Grance, T.: The NIST Definition of Cloud Computing. Recommendations of the National Institute of Standards and Technology (2012), http://www.csrc.nist.gov/publications/nistpubs/800-145/SP800-145.pdf
15. Porter, M.E.: Clusters and the New Economics of Competition. Harvard Business Review, 77–90 (November-December 1998)
16. Ratajczak-Mrozek, M.: Global Business Networks and Cooperation within Supply Chain as a Strategy for High-Tech Companies' Growth. Journal of Entrepreneurship, Management and Innovation 1, 35–51 (2012)
17. Subhankar, D.: From outsourcing to Cloud computing: evolution of IT services. Management Research Review 35(8), 664–675 (2012)
18. http://www.co3-project.eu
19. http://www.gartner.com/it-glossary/community-cloud
20. http://www.i-cargo.eu
21. http://www.project-logical.eu

Models of an Integrated Performance Measurement System of Intelligent and Sustainable Supply Chains

Blanka Tundys[1], Andrzej Rzeczycki[1], Magdalena Zioło[1], and Jarosław Jankowski[2]

[1] University of Szczecin, Faculty of Management and Economics of Services, Szczecin, Poland
{blanka.tundys,andrzej.rzeczycki,magdalena.ziolo}@wzieu.pl
[2] Faculty of Computer Science, West Pomeranian University of Technology, Szczecin, Poland
jjankowski@wi.zut.edu.pl

Abstract. Changes in the nature of supply chains require the dedication of the individual measures and their alignment with the latest trends in supply chain management. The concept presented in the article is to identify a dedicated and integrated performance measurement system for intelligent and sustainable supply chains. The economic conditions and trends, both in transport, shipping, logistics, as well as structural and organizational changes within the supply chain, need to be revised and adapted to a standard measurement in order to evaluate their performance.

Keywords: sustainable supply chains, diffusion of information, performance measurement.

1 Introduction

The management of an innovative and modern supply chain requires the use of appropriate logistics performance. It is assumed that logistics performance is multi-dimensional, reflecting multiple stakeholders and interests. The possible desired outcomes are numerous and range from customer satisfaction over environmental responsibility, to overall cost-effectiveness. Logistics performance is predominantly measured with soft perceptual indicators given the difficulty of obtaining hard performance measures, and is a result of two different variables. On the one hand, it is influenced by the performance of logistics processes performed in-house under the direct responsibility of the logistics service provider's (LSP) customer. On the other hand, it is affected by the performance of outsourcing arrangements in which the customer has delegated logistics and the accompanying responsibility to a logistics service provider. Logistics performance may be defined as the extent to which goals such as cost efficiency, profitability, social responsibility, on-time delivery, sales growth, job security and working conditions, customer satisfaction, keeping promises, flexibility, "fair" prices for inputs, low loss and damage and product availability are achieved [5]. The aim of this paper is to indicate what is relevant and important of appropriate indicators and performance metrics for intelligent and sustainable supply chain.

N.T. Nguyen et al. (Eds.): ACIIDS 2014, Part II, LNAI 8398, pp. 504–514, 2014.

2 Literature Review and Methodology

The changing economic conditions and market trends, as well as structural and organizational changes within the supply chain, have led to a new understanding of the concept of supply chain, and so we can talk about smart and sustainable supply chains. Management of these chains requires adjustments in the standard assessment measures, logistics, by focusing on the importance of individual measures. In the past, the classic approach of the supply chain did not include interest in the environmental aspects, external costs and the impact of the supply chain on the environment. However, today, environmental issues are becoming increasingly important and are a major problem in most global supply chains [11]. The trend associated with the implementation of the principles of sustainable development is defined as "the development that meets the needs of the present without compromising the ability of future generations to meet their own needs" [17]. Reference to the sustainability of the supply chain needs to take into account all dimensions of sustainable development and relate them to the functioning of the supply chain in the context of meeting the needs of customers and the economy and the competitiveness of the chain [15]. In addition to the environmental aspects, the trend in supply chain management has to consider the terms of the use of information technology. The basis of its operation is the use of advanced information technology (POS, EDI, ERP), which is more effectively manage to the supply chain. The definition of supply chains is not sufficient enough to explain how to improve the efficiency of processes. The models indicate that the gauges and indicators are critical to the effective functioning of these chains. Up to now, measurements of the effectiveness of supply chains were presented at a general level, including new character strings. These related mostly to the application of modern tools and metrics to dedicated sectors. Example: [3], [6], [12]. In order to build the model of indicators and measures [8] take four points of view, they are considered: time, cost, flexibility and quality, in terms of strategic, tactical and operational areas [7]. Conceptual studies focus primarily on adaptation measures of logistics performance, comparative comparison of individual elements and the creation of the base models. Sustainable supply chain is the process of integrating a strategic approach, which aims to improve long-term economic efficiency and customer satisfaction taking into account social, environmental and economic. Supply chain management is categorized into three main aspects of sustainable development, the environmental and social criteria must be met by the supply chain (business units), and the competitiveness of the chain, which will help to meet the needs of the customer [2], [15]. The essence of intelligent supply chain is to be used in managing the processes and flows of advanced information technology. The main and characteristic features include: continuous monitoring of the processes occurring, smooth and uninterrupted flow of information, standardization of processes and the use of appropriate technology. In addition to the use of appropriate information technology, essentials of process in intelligent supply chain is integration and communication. Intelligent supply chains use ERP systems as well as VMI, CRP,

FRM, JIT, ECR, EDI VAN or CPFR [9]. Both chains have common and different elements. It may be noted that in a sustainable chain, emphasis is placed on material flows and their impact on the environment, but also on other aspects of classically understood sustainable development. In this case, the flow of capital and information is important as it allows performance of operations and logistics processes in an efficient manner, but it can be said that in this case the physical flows and their protection and organization plays the most important role. The intelligent supply chains are so created, that the physical flow does not guarantee they full efficiency, they need good flow of information's based on IT.

The appropriate categorization of performance measures and their significance in relation to specific supply chains can help to improve their implementation and application. Choosing the right measure of performance is extraordinary difficult, due to of the lack of a uniform system of measurement, which is dedicated for specified supply chains. It also indicates that if you cannot measure something, you cannot manage it. Measuring performance management can provide a wealth of information feedback, so that it will be possible to monitor and better control the efficiency of processes, as well as common errors. For the purposes of this study, logistics performance measures will be defined in four dimensions: cost, quality, time and flexibility [13]. Some researchers limited their studies of the measures as financial indicators (the most common) and non-financial, while others limited their studies to the sharing of measures and indicators for quantitative and qualitative [4]. Creating and evaluation of the measurment system for supply chain should be: full (measurement of all pertinent aspects), universal (allow for comparison under various operating conditions), measurable (data required are measurable) and consistent (measures consistent with organization goals) [1]. Many authors in a variety of ways classified measures of logistics performance. Their characteristics and description of selected classification give a chance to create a framework for a system of supply chain assessment measures. Table 1 presents selected techniques and ways of developing indicators and measures of logistics performance. As you can see, in both areas, the tools are also different, and largely depend on the individual's choice.

The objective is to make a comparative analysis of the efficiency of intelligent and sustainable supply chains. This analysis requires taking a look at the functioning of these chains and their development trends. In order to address these issues, we have utilized an exploratory research methodology, based on a literature review and some initial case studies. Building a research methodology the authors took into account the logic diagram of the research process (for O. Lange). For the purposes of the above studies the authors have adopted a detailed diagram of the research process.

3 Models and Conditions of Its Functioning Literature Review

The literature broadly describes the problem of measuring the performance of supply chains in many different facets. The diversity in the perception relates to the criteria for classification and selection of key metrics and indicators for the analysed supply chains. The basis of differentiation is very often the industry. Nevertheless, it can be

Table 1. Selected techniques of performance measures for supply chain (Source: Own elaboration based on [10], [12])

Author	Logistics performance measurement/metrics
Beamon 1996	Output, - customer responsiveness, quality, and the quantity of final product produced. *Sales. Profit. Fill rate. On-time deliveries. Backorder/stockout. Customer response time. Manufacturing lead time. Shipping errors. Customer complaints.*
	Flexibility; Reductions in the number of backorders. Reductions in the number of lost sales. Reductions in the number of late orders. Increased customer satisfaction. Ability to respond to and accommodate demand variations, such as seasonality.
	Ability to respond to and accommodate periods of poor manufacturing performance (machine breakdowns). Ability to respond to and accommodate periods of poor supplier performance. Ability to respond to and accommodate periods of poor delivery performance. Ability to respond to and accommodate new products, new markets, or new competitors. *Volume flexibility Delivery flexibility Mix flexibility New product flexibility*
	Resource: Total cost; Distribution costs; Manufacturing cost; Inventory; ROI
Model SCOR	SCOR is developed as a cross-industry standard for supply chain management. It uses a process reference model to explain a supply chain. The process reference model is a combination of business-process reengineering, benchmarking and best practices analysis. The process reference model is aimed at providing a framework for performance measures and best practices for standard processes.
Cagnazzo et al. (2010)	Balanced models consider the presence of both financial and non-financial indicators. In these models several separate performance measures which correspond to diverse perspectives (financial, customer, etc.). Quality models - these are frameworks in which a great importance is attributed to Quality Questionnaire-based models - these are frameworks based on questionnaire Hierarchical models - balanced models consider the presence of both financial and non-SCPM models that are strictly hierarchical (or strictly vertical), characterised by cost and non-cost performance on different levels of aggregation are classified as hierarchical models. Support models - Frameworks that do not build a performance measurement system but help in the identification of the factors that influence performance indicator are classified as support models.

Author	Logistics performance measurement/metrics
Chan, Qi, 2003	Based on fuzzy set theory; For identifying appropriate performance measures, for each process and sub processes of the supply chain, corresponding measures are grouped into processes and measures hierarchy
Kaplan, Norton (1992)	Supply chain balanced scorecard: Learning and growth perspective, Business process perspective, Customer perspective, Financial perspective
Neely, 2002	Performance Measurement System (PMS) as a balanced and dynamic system that enables support of decision-making processes by gathering, elaborating and analyzing information.
Tangen (2004)	Performance be defined as the efficiency and effectiveness of action: PM is defined as the process of quantifying the efficiency and effectiveness of action and as a metric used to quantify the efficiency and/or effectiveness of an action; PMS is defined as the set of metrics used to quantify the efficiency and effectiveness of an action.
Gunasekaran (2004)	Strategic Level: *Top-level management decisions, investigation of broad based policies, corporate financial plans, competitiveness and adherence to organizational goals.* Tactical Level: *Resource allocation, measuring performance of goals set in strategic level, feedback on the mid-level management decisions.* Operational Level: *Data analysis, evaluation of decision of low level managers and workers, ensuring that goals set in tactical level are met.*
Kleijnen, Smits (2003)	Gave an example of one division of a large multinational company's evaluation of logistic al performance. The company followed the following five performance metrics: % of order delivered on time (Fill rate) % of order delivered as negotiated (Confirmed Fill rate) Difference between "on time" and "as negotiated" order delivery (Response delay) Total work in process (WIP) expressed as a % of the total sales. If this is less then there will be loss of business (Stock). Difference between actual delivery and delivery as negotiated (Delay).
Bititci et al. (1997)	Defined SCPMS as the reporting process that gives feedback to employees on the outcome of actions

assumed that, although the details differ, it is based mostly on the general functioning of the supply chain strategies such as smart, lean, agile, etc. Thus, the base has common metrics and indicators for each strategy. In the case of intelligent and sustainable supply chain should be take into account selected indicators, which are presented in Table 2.

The selection of indicators has been made assuming the need to monitor the performance of the chain of decision-making at different levels (strategic, tactical and operational) and is based on the different areas of functioning (according to the table these are cost, time, quality, and flexibility). Below they were designated typical common links between the measures and indicators in different researched area and levels of decision-making, Then have been designated weight (importance) indices for the analysed chains - intelligent (ISC) and sustainable (SSC). Discussed relations in researched areas of: logistics cost in supply chain are presented in diagram 1, time on diagram 2, quality on diagram 3, and flexibility on diagram 4. As the basis for the relationships was indicated the high degree of correlation between the measures and indicators. In accordance with the accepted principles of selection of key performance indicators (KPIs), their number should hover around twenty 20-here of indicators, and this also applies to financial measures and indicators, which do not include this analysis (not directly related to logistics efficiency), so the number must be reduced to about fifteen. Thus, the choice of measures and indicators should be to maximize the level of significance (x) by the following function:

$$f_{(x)} \rightarrow max \sum_{i=1}^{i} x_i \,; x \in \{S; T; O\}, i \leq 15 \qquad (1)$$

After analysis, it can be concluded that for intelligent supply chain the most important key performance measures include the costs of internal subsystems and information flow. For the sustainable supply chain the most important measures are in particular external costs and their main factors which are transport costs. In terms of quality in the sustainable supply chain forefront is to measure the timeliness and availability of deliverability, together with the delays at the operational level. In the case of intelligent supply chain may be the same set of indicators, but also the measures based on the order lead time and speed of information flow and reactive time. In the other of quality area of the supply chain, intelligent are characterized by a greater significance for measures of the computerization degree and accuracy of communication, and in the case of the sustainable supply chain the most important are: availability of goods and the completeness of delivery (also very important in ISC), and also the set of indicators on waste management. In the last discussed area for the intelligent supply chain the key measures are flexibility and delivery capability in conjunction with the indicator of spare capacity and the ability to obtain information from the client. The most important aspects of sustainable chain-based flexibility are the accuracy of forecasting and risks of the process, but their aggregate marks are lower than in other areas of analysis.

Table 2. Selected performance indicators of supply chains (Source: Own elaboration based on [1], [7], [8], [10], [12], [14], [16])

Category	Subcategory (based on)	Name of indicator/meter	The method of calculating	Subcategory (based on)	Name of indicator/meter	The method of calculating
Cost	total costs	Global supply chain costs	$C_G = \sum (C_s; C_p; C_d; C_e)$	the subprocesses	Costs indicator of information flows	$Ici = \frac{C_{if}}{n_d}$
	The main processes	Supply costs	$C_s = \sum (C_o; C_i; C_t; C_{if})$		Transport costs indicator	$Ict = \frac{C_t}{n_t}$
		Production costs	$C_p = \sum (C_{pm}; C_t; C_i; C_{if})$		Material management costs indicator	$Ics = \frac{C_m}{n_m}$
		Distribution costs	$C_d = \sum (C_{cs}; C_t; C_i; C_{if})$		Orders costs indicator	$Ico = \frac{C_o}{n_o}$
		C_o - orders costs; C_t - transport costs; C_{if} – costs of information flows; C_i – inventory costs; C_{pm} – production management costs; C_{cs} – customer service costs			Costs indicator of innovative processes	$Icp = \frac{C_{ip}}{n_{ip}}$
	External	External costs C_{ef} - environmental fees; C_{cc} - certification costs; C_{gi} - investments in green technologies	$C_e = \sum (C_{ef}; C_{cc}; C_{gi})$	Cif – costs of information flows; nd – number of deliveries; Ct- tranport costs; nt – number of transports; Cm – material management costs; nm – storage area in square meters; Co - orders costs; no – number of orders; Cip - costs associated with the implementation of innovative processes; nip - number of innovation processes		
Time	Lead time	Order Lead Time	Time from customer order received to customer order delivered.	speed of reaction	Speed of reaction d_b - total deliveries supplied ahead of schedule; d_n - total number of deliveries	$Rs = \frac{d_b}{d_n} x100\%$
		Order Handling Time	Time from customer order received to sales order created.		Readiness of delivery p_a - products available for immediate shipment p_n - total number of products	$Rd = \frac{p_a}{p_n} x100\%$
		Manufacturing Lead Time	Time from sales order created to production finished (ready for delivery).		Flow rate of information	The time flow of information from the order notification by the customer to the launch of order processing
		Production lead time	Time from start of physical production of first submodule/part to production finished (ready for delivery).			
		Delivery lead time	Time from production finished to customer order delivered.			
		The degree of timetable implementation T_r – realized timetable; T_p – planed timetable	$T_d = \frac{T_r - T_p}{T_p} x100\%$	reliability	Delays index d_d - deliveries delayed ; d_n - total number of deliveries	$Di = \frac{d_d}{d_n} x100\%$
	punktuality	Delivery assurance d_t - number of deliveries within the time limit; d_n - total number of deliveries	$Da = \frac{d_t}{d_n} x100\%$		Downtime index p_d - total production downtime; p_t - total production volume	$Dti = \frac{p_d}{p_t}$
	rhythmicity	Periodicity n_d - number of days during the considered period ; t_d - average time of one delivery	$Pd = \frac{n_d}{t_d} x100\%$		Average delay time t_{dn} - n-th delivery delay time; n_d - number of delays	$Adt = \frac{t_{d1} + t_{d2} + \cdots + t_{dn}}{n_d}$

Table 2. (*Continued*)

Quality	consumer satisfaction	Product availability g_a - number of immediately available goods; g_t - total number of goods	$Pa = \frac{g_a}{g_t} x100\%$	infrastructure	Degree of informatization P_{IT} - number of processes using information technology P_n - total number of processes	$TI = \frac{P_{IT}}{P_n} x100\%$
		Complaints indicator d_c - number of challenged deliveries; d_n - total number of deliveries	$Ic = \frac{d_c}{d_n} x100\%$		Technical infrastructure of work V_a - value of assets; n_e - number of employees	$TI = \frac{V_a}{n_e}$
		Completeness of deliveries d_k - number of deliveries supplied in full; d_n - total number of deliveries	$Cd = \frac{d_k}{d_n} x100\%$		Environmental performance P_e - number of ecological processes; P_t - total number of processes	$Ep = \frac{P_e}{P_t} x100\%$
		Openness of information I_s - number of customer satisfying information about the order status; I_t - total number of customer information about the order status	$Oi = \frac{I_s}{I_t} x100\%$	processes	Process improvement rate P_{va} - number of value added processes P_{ev} - expected number of processes that add value to the customer	$Pi = \frac{P_{va}}{P_{ev}} x100\%$
		Accuracy of communication M_{ef} - number of error-free messages; M_t - total number of messages	$Ac = \frac{M_{ef}}{M_t} x100\%$		Precision of planning P_e - number of processes planned and executed; P_p - number of processes planned	$Pp = \frac{P_e}{P_p} x100\%$
flexibility	risk	Accuracy of forecasts P_f - number of processes executed in accordance with the forecast; P_t - total number of processes	$Af = \frac{P_f}{P_t} x100\%$	Customer service	Flexibility of supply n_{fn} - number of fulfilled customer's special needs n_{sn} - number of submitted customer's special needs	$F = \frac{n_{fn}}{n_{sn}} x100\%$
		Indicator of spare production capacity p_p - number of products produced in a given period; p_m - maximum number of products possible to produce in a given period	$SPC = \frac{p_p}{p_m} x100\%$		Delivery ability t_{ds} - delivery time specified by the customer; t_{dr} - delivery time how can be realized	$Ad = \frac{t_{ds}}{t_{dr}}$
		Risk to perform processes P_u - number of unexecuted processes; P_t - total number of processes	$Rp = \frac{P_u}{P_t} x100\%$	market specificity	Seasonality index n_{sp} - number of seasonal products n_p - total number of products	$Is = \frac{n_{sp}}{n_p} x100\%$
		Informational ability r_c - number of customer responses; q_c - number of queries relating to customer preferences	$Ia = \frac{r_c}{q_c} x100\%$		Nature of the product P_{AL} - number of processes performed in strategy agile/lean; P_t - total number of processes	$Pn = \frac{P_{AL}}{P_t} x100\%$

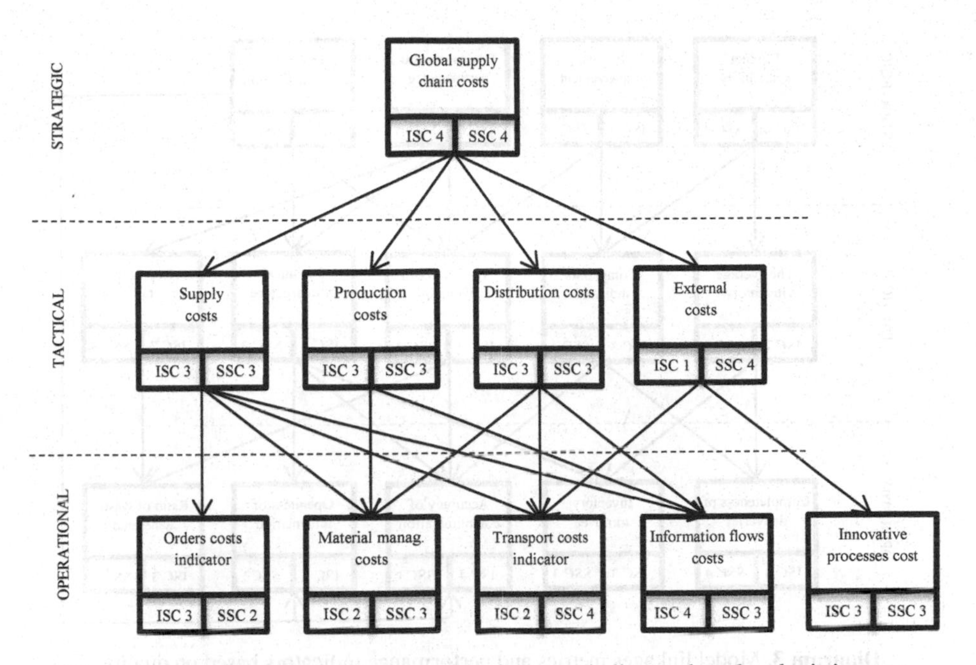

Diagram 1. Model linkages metrics and performance indicators based on logistic costs

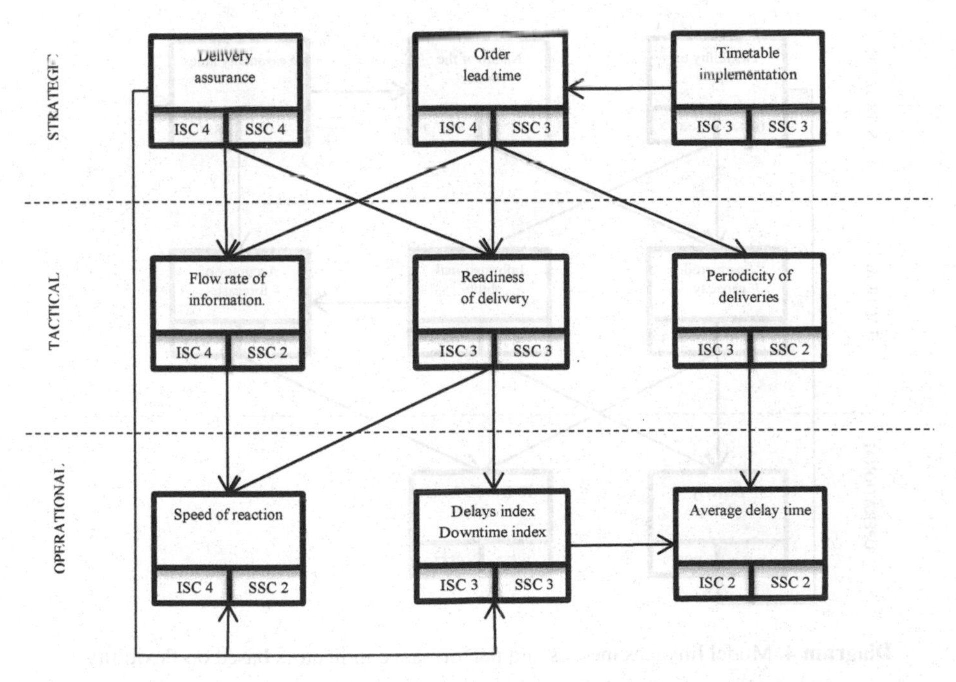

Diagram 2. Model linkages metrics and performance indicators based on time

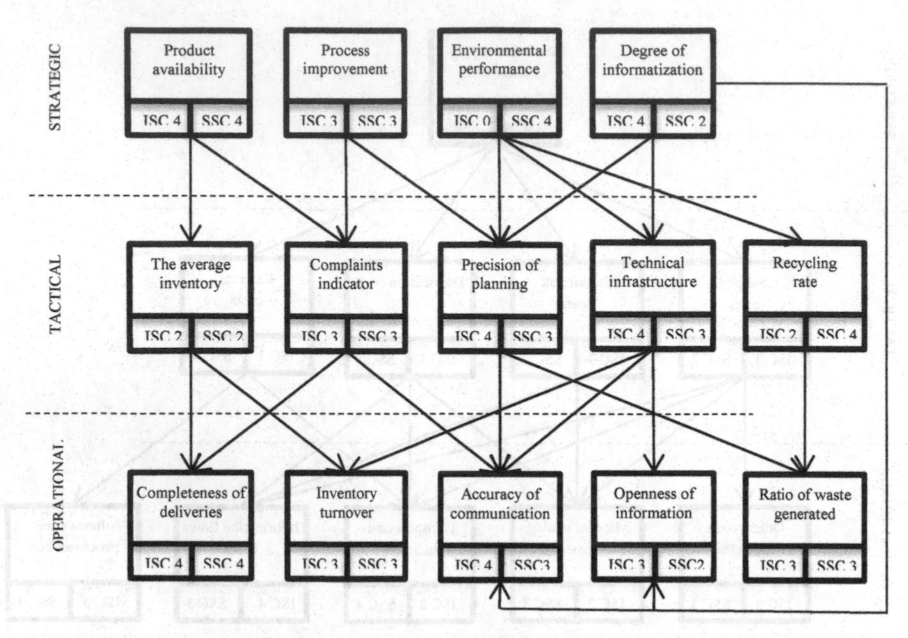

Diagram 3. Model linkages metrics and performance indicators based on quality

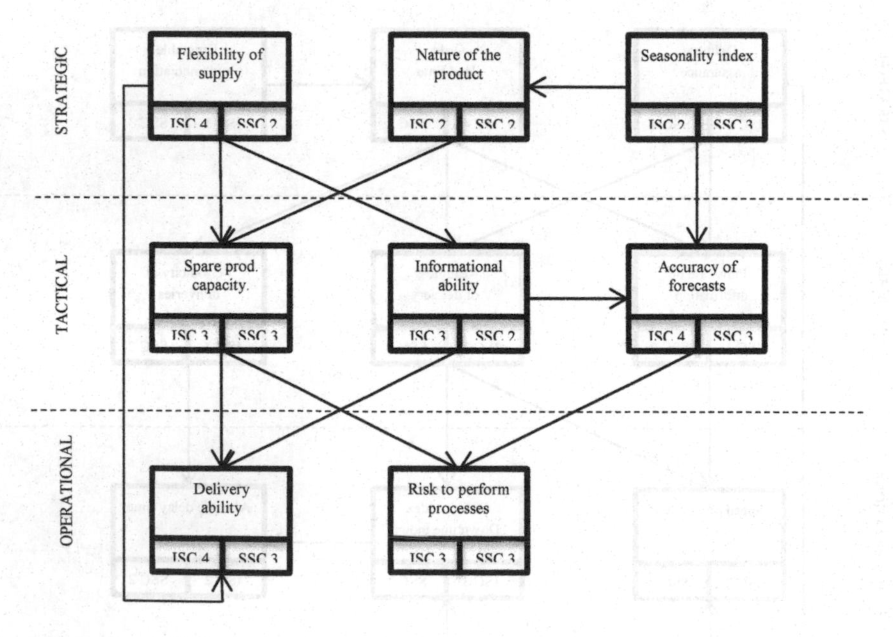

Diagram 4. Model linkages metrics and performance indicators based on flexibility

Rationalizing the selection of measures and indicators should therefore consider the omission of the area and increase the number of measures in other areas. Further work on the model selection of metrics should focus on: an indication of the relationship between the categories of time, cost and quality, combined with the prospect of a financial performance chain and verifying the theoretical assumptions about the extent of the correlation between the measures and indicators that occur between levels of decision-making chains and the significance of indicators in relation to the objectives/strategies chains.

Summary

Supply chain performance measurement is a complex process that requires not only looking through the prism of one company, but at the same time to link them to various levels of government. Determining the situation model facilitates and accelerates decision making. The main issue, however, is to determine the degree of generalization, which determines the fit of models to the market situation. In the authors' opinion, the selection of indicators and models of indicators relating to supply chain strategy provides a good balance between the general model for the chain and industry models. The presented models in intelligent and sustainable chains and allow the capture key differences between these approaches. Model implemented on the level of industry will detect their course much more but it is worth considering the relationship of growth effects to the workload realize that goal. Further work on the model selection of metrics and performance indicators presented chains should focus on identifying the relationship between the categories of time, cost and quality, combined with the prospect of financial performance of the chain. Another direction involves verifying the theoretical assumptions of the extent of correlation between the measures and indicators that occur between levels of decision-making chains and significance of indicators in relation to the objectives/strategies chains.

References

1. Beamon, B.M.: Measuring supply chain performance. International Journal of Operations & Production Management 19(3), 275–292 (1999)
2. Beamon, B.M.: Sustainability andthe Future of Supply Chain Management. Operations and Supply Chain Management: An International Journal 1(1), 4–18 (2008)
3. Bhagwat, R., Sharma, M.K.: Performance measurement of supply chain management: A balanced scorecard approach. Computers & Industrial Engineering 53, 43–62 (2007)
4. Chan, F.T.S., Qi, H.J.: An innovative performance measurement method for supply chain management. Supply Chain Management: An International Journal 8(3), 209–223 (2003)
5. Chow, G., Heaver, T.D., Henriksson, L.E.: Logistics Performance: Definition and Measurement. International Journal of Physical Distribution & Logistics Management 24(1), 17–28 (1994)
6. Erkan, T.E., Baç, U.: Supply chain performance measurement: a case study about applicability of SCOR model in a manufacturing industry firm. International Journal of Business and Management Studies 3(1), 381–390 (2011)
7. Gunasekaran, A., Patel, C., McGaughey, R.E.: A framework for supply chain performance measurement. Int. J. Production Economics 87, 333–347 (2004)

8. Gunasekaran, A., Kobu, B.: Performance measures and metrics in logistics and supply chain management: a review of recent literature (1995–2004) for research and applications. International Journal of Production Research 45(12), 2819–2840 (2007)
9. Hwang, H.J., Seruga, J.: An Intelligent Supply Chain Management System to Enhance Collaboration in Textile Industry. International Journal of u- and e- Service, Science and Technology 4(4), 1–16 (2011)
10. Khare, A., Saxena, A., Tewari, P.: Supply Chain Performance Measures for gaining Competitive Advantage: A Review. Journal of Management and Strategy 3(2), 25–32 (2012)
11. Kuik, S.S., Nagalingam, S.V., Amer, Y.: Challenges in implementing sustainable supply chain within a collaborative manufacturing network. In: 8th International Conference on Supply Chain Management and Information Systems, Hong Kong, pp. 662–669 (2010)
12. Kurien, G.P., Qureshi, M.N.: Study of performance measurement practices in supply chain management. International Journal of Business, Management and Social Sciences 2(4), 19–34 (2011)
13. Neely, A., Gregory, M., Platts, K.: Performance measurement system design. International Journal of Operations & Production Management 15(4), 80–116 (1995)
14. Pomiar funkcjonowania łańcuchów dostaw, red. D. Kisperska-Moroń, Wydawnictwo Akademii Ekonomicznej w Katowicach, Katowice (2006)
15. Seuring, S., Müller, M.: From a literature review to a conceptual framework for sustainable supply chain management. Journal of Cleaner Production 16, 1699–1710 (2008)
16. Twaróg, J.: Mierniki i wskaźniki logistyczne, Instytut Logistyki i Magazynowania, Poznań (2003)
17. World Commission on Environment and Development, Our Common Future. Oxford University Press, Oxford (1987)

Implementation of Quaternion Based Lifting Scheme for Motion Data Editor Software

Mateusz Janiak[1], Agnieszka Szczęsna[2], and Janusz Słupik[2]

[1] Polish-Japanese Institute of Information Technology, Bytom, Poland
Mateusz.Janiak@pjwstk.edu.pl

[2] The Silesian University of Technology, Institute of Informatics, Gliwice, Poland
{Agnieszka.Szczesna,Janusz.Slupik}@polsl.pl

Abstract. Motion analysis is rapidly developing area of research. Due to availability of cheaper hardware with reasonable accuracy for motion acquisition in form of various motion controllers, dedicated mainly for gaming, there are rising new research groups interested in this topic. Proposed solutions for motion analysis have a wide range of application in medicine, sport, entertainment and security. Although motion analysis is one of the most important domains of our everyday life, there are still no good tools supporting knowledge exchange and experiments in this field. In this paper we want to introduce a possibility of implementing specialised wavelet analysis in form of the lifting scheme for motion data in quaternion representation in a data flow processing framework available in Motion Data Editor (MDE) software developed at Polish-Japanese Institute of Information Technology (PJWSTK) in Bytom (Poland). We want to show how easily custom solutions can be introduced to this general purpose data processing software. Usage of this software saves time by concentrating on rapid prototyping of new algorithms and performing experiments, skipping creation of similar solutions for various data types and algorithms.

Keywords: motion analysis, software architecture, data flow, data processing, lifting scheme, quaternions, wavelets.

1 Introduction

Nowadays new techniques are being developed and improved for motion analysis. Among them we can find tools dedicated to motion data comparison and compression. Additionally, many algorithms for motion segmentation, recognition and classification are proposed. Unfortunately, there are no tools that support motion analysis and processing. Two available products: commercial Vicon Polygon [16] and open-source Mokka [1], provide only motion data browsing and visualisation, with no data processing features. Moreover, users can not extend any of those applications to fit their particular needs, in example introducing custom data types and algorithms. To address this problem a new software called MDE was developed at PJWSTK in Bytom (Poland). This is a general purpose data processing tool, with dedicated extensions for motion analysis and medical

N.T. Nguyen et al. (Eds.): ACIIDS 2014, Part II, LNAI 8398, pp. 515–524, 2014.

applications. In this paper we want to present briefly the main features of MDE and show how they affect research work in the domain of motion analysis. As an example, the lifting scheme for multiresolution data analysis has been implemented in the MDE software. We present how particular tests can be performed and how to collect their results for noise reduction and data compression of motion data.

1.1 Motion Data Editor

The MDE software is developed at PJWSTK in Bytom (Poland). Initially it was designed to support medics in diagnosis of various motion dysfunctions. Since then it has been reorganised and refactored to become a mature tool for a general purpose data processing and analysis. The main power of the MDE software is its architecture dedicated to a well defined data processing pipeline.

MDE architecture supports flexible application extension with custom data types, their dedicated operations and algorithms. This is achieved with a specialised plug-in system allowing to fit MDE to particular users' needs, standardized their solutions around one tool and save time on implementation of similar functionality operating on different data types.

Architecture. In the MDE architecture are objects responsible for particular stages of data processing pipeline. In the first step user browse for data that to process. This is realised by *Source* objects. They are responsible for delivering data in containers (usually various file formats) to local hard disk, if required. Later, containers are unpacked with help of *Parser* objects. Loaded data are normalised and wrapped with help of *ObjectWrappers*. This is a completely new approach to uniform data handling in statically and strongly typed C++ programming language with just a small memory and performance overhead. This technique combines generic programming and run-time type information (RTTI). It outperforms functionality of different *variant* types. It allows simple, yet efficient data exchange between all application modules. After loading the data to application user might view the data with help of *Visualiser* objects, presenting data graphically at various perspectives. *Visualisers* allow to present and compare more data on a single scene. *Services* allow users to introduce new functionality to application. This allows users to introduce new algorithms to MDE or functionality completely unconnected with data analysis and processing. All those objects can be introduced to MDE through a dedicated plug-in system.

Features. As a cross-platform software, MDE provides an abstraction layer for most system specific functionality like file system management and threading. Additionally, a dedicated hierarchical *Log* system was designed to simplify notifying about application status. Dedicated managers with transactional mechanisms offer thread-safe memory operations. To control and monitor amount of threads used in application a concept of *ThreadPool* has been introduced. *Threads* are expected to perform simple operations, mainly waiting for some events and actions to schedule their processing. For heavy computations an idea of *Jobs* and *JobManager* is proposed. *Job* represents particular computations that are scheduled to *JobManager*. *JobManager* is responsible to run *Jobs* utilizing optimally available computational resources (specified number of worker threads depending on available central processing unit (CPU)). Analysed data is very often indexed with time - this is especially the case for various physical measurements. To address problem of uniform handling data indexed with time a dedicated and generic *DataChannel* type was introduced. It allows to create illusion of data continuity in the time domain with help of various inter- and extrapolation methods. To handle efficiently memory used by the loaded data a lazy initialization mechanism is proposed.

1.2 Data Flow

The concept of processing data in form of a well defined pipeline is widely applied in CPU architecture [12,11,9,3,4]. Each instruction execution is divided into separate stages realised in dedicated modules. This allows to execute several instructions in parallel on a single CPU. To allow users simple composition of complex data processing algorithms, the concept of data flow processing framework was introduced to MDE in form of a dedicated service. It is based on a graph structure extended with elements called *Pins*. *Pins* are attached to nodes, providing nodes input and output for data. Nodes can be connected together only through compatible *Pins*. Created processing model can be saved and restored for later usage. Among nodes three groups can be defined: sources, processors and sinks (Figure 1).

Source nodes deliver new data for processing in data flow. Processing nodes consume data from attached input pins and provides new data through output pins. Sinks are used to save results of processing for further analysis. Data are

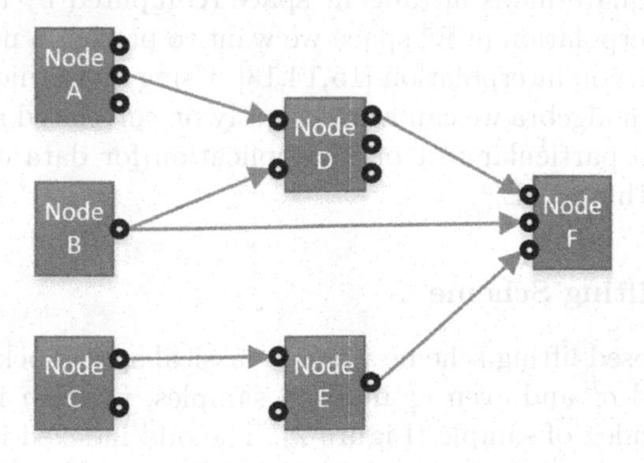

Fig. 1. Example of data flow model

considered to leave the model either when they are processed by sink nodes or when corresponding output pins are not connected. Such approach allows simple utilisation of available computational resources by handling processing logic of each node by a separate thread from *ThreadPool* and scheduling node's processing in form of a *Job* to *JobManager*. With such approach users have to implement processing functions and encapsulate their signatures within proper nodes, where functions' inputs and outputs are represented by particular *Pins* with the given data types.

1.3 Visual Data Flow

Based on data flow processing model a graphical programming environment was created to simplify creation of processing pipelines. Using drag'n'drop mechanism users can easily choose interesting functions and compose them together creating more specific algorithms. Those algorithms can be saved and restored for further usage. It is also possible to group several basic nodes connected together and replace them with a single virtual node. This helps to keep scene clear and maintain good overview of composed algorithm. While creating processing pipeline user is guided visually which pins can be connected together by verifying their data types compatibility.

2 Lifting Scheme and Experiments in MDE

Multiresolution tools are widely used for motion data analysis. They usually operate on Euler angles representation for rotations (orientations) [7,8,5,6,2]. In our work we are developing new algorithms based on the second generation wavelets constructed by the lifting scheme, that work on quaternion representation for rotations. Based on our previous work, where we have proposed a lifting scheme for quaternions in tangent space (computed by logarithm) using classical Bezier interpolation in $\mathbb{R}^3$ space we want to present a new approach using *SQUAD* quaternion interpolation [15,14,13]. Using quaternion lifting scheme based on quaternion algebra we can work directly on correlated motion data. We decided to perform particular test of its application for data compression and noise reduction within MDE.

2.1 *SQUAD* Lifting Scheme

To construct proposed lifting scheme we use classical split block, where samples are divided on odd o_i^j and even e_i^j indexed samples, where j is the resolution level and i is the index of sample (Figure 2). The odd indexed input value after one step of lifting scheme is replaced by the difference (detail value) between the odd value and its prediction. The even indexed samples are updated, so that coarse-scale samples have the same average value as the input samples. Based on even indexed samples we propose prediction and update blocks for forward transform as follows:

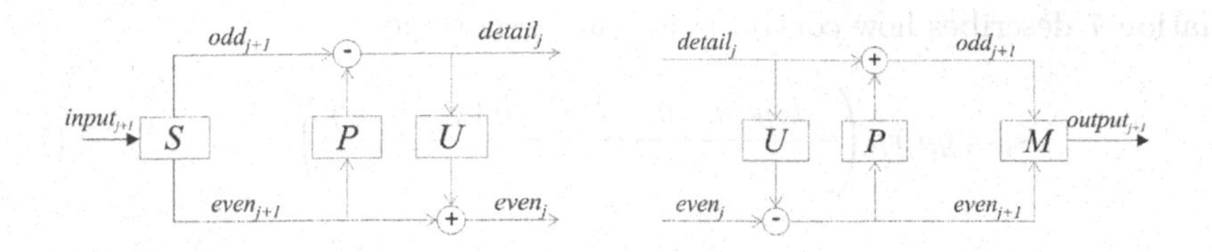

(a) forward lifting scheme (b) inverse lifting scheme

Fig. 2. The lifting scheme

- predict
  SQUAD interpolation uses four nearby even indexed samples to predict odd sample according to the equation 5.

$$o_i^j = o_i^{j+1} * squad\left(e_i^{j+1}, e_{i+1}^{j+1}, e_{i+2}^{j+1}, e_{i+3}^{j+1}, 0.5\right)^{-1} \tag{1}$$

- update
  The update step is obtained in order to preserve the equality (average of the signal) as

$$e_{i+1}^j = \left(o_{i+1}^j\right)^{0.5} * e_{i+1}^{j+1} \tag{2}$$

For the inverse transform the following equations are given:

- undo-predict

$$o_i^{j+1} = o_i^j * squad\left(e_i^{j+1}, e_{i+1}^{j+1}, e_{i+2}^{j+1}, e_{i+3}^{j+1}, 0.5\right) \tag{3}$$

- undo-update

$$e_i^{j+1} = \left(o_i^j\right)^{-0.5} * e_i^j \tag{4}$$

Applying Bezier curves interpolation idea to quaternions we obtain SQUAD interpolation (equation 5) [10]. It requires four quaternions for interpolation, two of them are used as a interpolation range, with other two used to generate control points ensuring smooth and differentiable interpolation curve. Quaternions q_{i-1} and q_i are used as key frames and quaternions q_{i+1} and q_{i+2} are control points.

$$squad\left(q_{i-1}, q_i, q_{i+1}, q_{i+2}, t\right) = slerp\left(slerp\left(q_{i-1}, q_i, t\right), slerp\left(q_{i+1}, q_{i+2}, t\right), 2t\left(1-t\right)\right) \tag{5}$$

and

$$slerp\left(q_a, q_b, t\right) = q_a\left(q_a^* q_b\right)^t \tag{6}$$

Equation 7 describes how control points are generated.

$$s_i = q_i exp\left(-\frac{log\left(q_i^{-1}q_{i+1}\right) + log\left(q_i^{-1}q_{i-1}\right)}{4}\right) \tag{7}$$

2.2 Applications

We want to apply presented lifting scheme for data compression and noise reduction. In case of data compression we propose lossy algorithm. It is based on removing of wavelet quaternion detail coefficients of the specific (see 2.5) number of the highest resolutions. The detail coefficient is the difference between the odd indexed sample and its prediction. The reconstruction introduces differences to original signal because part of the data was removed. For noise reduction we are using selective soft threshold α according to the following rule:

$$q_d = \begin{cases} q_d & \text{if } \alpha < v < 1 - \alpha \\ [1,(0,0,0)] & \text{otherwise} \end{cases} \tag{8}$$

where $v = \frac{arccos(a)}{\pi}$ and detail is a quaternion $q_d = a + bi + cj + dk$. Such approach eliminates meaningless rotations (close to 0°, changing unnoticeable body orientation) and significant rotations (close to 360°, moving body close to its initial orientation). As rotation quaternions are always represented as a cosine of half desired rotation angle, therefore α values were limited to the given range $(0, 0.5)$.

As a measure of quality for the presented applications we introduce a distance measure for quaternion signals. For two normalised quaternions q_a and q_b we can define the distance as:

$$d_q\left(q_a, q_b\right) = \left|arccos\left(Re\left(q_a q_b^{-1}\right)\right)\right| \tag{9}$$

For two quaternion signals Q_A and Q_B with N quaternion samples, we can define the distance between them as:

$$D(Q_A, Q_B) = \frac{1}{N}\sum_{i=1}^{N}\left(d_q\left(q_{ai}, q_{bi}\right)\right)^2 \tag{10}$$

2.3 Test Data

For tests we have chosen motion of a left knee of healthy male, age 26. Recorded data samples count was equal to 812, but for experiments first 512 samples were used. Data were recorded with 100Hz frequency which gives recording duration approximately equal to 5 seconds - long enough to capture several steps. Figure 3 presents knee rotations in time as Euler angles for clarity. Angle values are truncated to the range of $\langle -180°; 180°)$. All results are also presented in this form. Time axis presents values always in seconds.

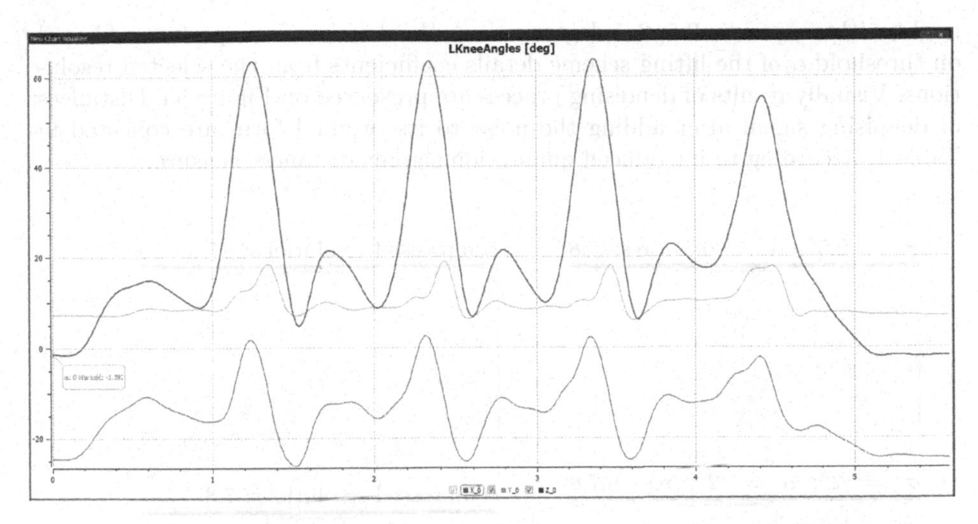

Fig. 3. Euler angles of test data - left knee of 26 years old and healthy man

2.4 Implementation

To verify presented lifting scheme we have implemented it within data flow processing framework for MDE software. The lifting scheme was decomposed to forward and inverse transforms, were each of them was modelled as an independent processing node. For testing a denoising process a dedicated source node was proposed allowing to set noise levels introduced to motion data (the σ value). For compression process two processing nodes were proposed - one for compression and other for decompression. Also signals differences was encapsulated in processing node. Based on such nodes several processing pipelines were created (Figure 4).

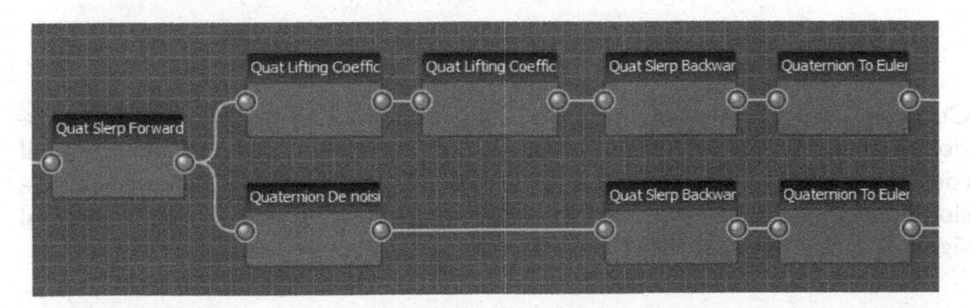

Fig. 4. MDE visual data flow environment for testing lifting scheme algorithms

2.5 Test Results

Noise Reduction. To test denoising, we have added three levels of white Gaussian noise for rotations in Euler angles representation, independently for each angle, with σ equal to: 0.5, 2, 5 degrees. Next, the denoising was performed based on threshold α of the lifting scheme details coefficients from the selected resolutions. Visually results of denoising process are presented on Figure 5a. Distances of denoising signal after adding the noise to its original form are collected in Table 1, according to introduced quaternion signals distance measure.

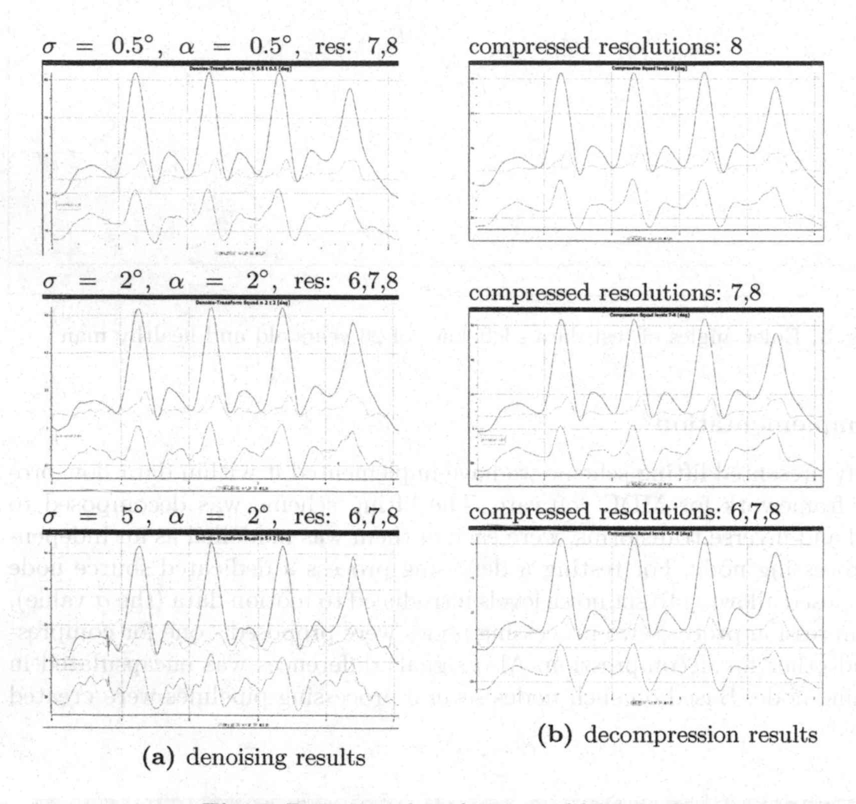

(a) denoising results

(b) decompression results

Fig. 5. Processing based on squad lifting scheme

Compression. Testing compression abilities according to proposed compression method, three levels of compression were proposed, based on removed detail coefficients from resolution levels 6, 7, and 8 (Figure 5b). To compare decompression quality, the distance of reconstructed signal after decompression to original signal is measured (Table 2).

Table 1. Noise reduction results

Noise (σ[°])	Threshold angle (°)	De-noise details resolutions	Distance (radians)
0.5	0.5	7, 8	1.6
2	2	6, 7, 8	5.6
5	2	6, 7, 8	13.3

Table 2. Compression results

Levels of removed details resolution	Distance (radians)
8	0.006
7, 8	0.08
6, 7, 8	1.3

3 Summary

Proposed novel approach to motion data analysis with multiresolution tools for quaternions provide very promising results in the field of data compression and noise reduction. Presented technique opens new directions for motion analysis tools based on the quaternion lifting scheme. Test results were obtained with help of MDE software, proving that it can be easily adopted for custom research solutions, increasing their quality and standardising them for cooperation with other built-in functionality. This makes data analysis simpler and faster. Additionally, users can automatically utilise all available computational resources with help of delivered plug-in. The MDE software, through the idea of rapid prototyping, provides a new, very fast, and easy to use framework for testing various algorithms.

Presented motion data decomposition based on lifting scheme and proposed SQUAD prediction block can be used to create more reliable and descriptive motion processing tools, used further for motion classification, recognition and prediction.

Acknowledgment. This work was supported by projects NN 516475740 from the Polish National Science Centre and PBS I ID 178438 path A from the Polish National Centre for Research and Development.

References

1. Mokka Motion kinematic & kinetic analyzer, `http://b-tk.googlecode.com/svn/web/mokka/index.html`
2. Beaudoin, P., Poulin, P., van de Panne, M.: Adapting wavelet compression to human motion capture clips. In: Proceedings of Graphics Interface 2007, pp. 313–318. ACM (2007)
3. Fog, A.: The microarchitecture of intel, amd and via cpus. In: An Optimization Guide for Assembly Programmers and Compiler Makers. Copenhagen University College of Engineering (2011)
4. Hennessy, J.L., Patterson, D.A.: Computer Architecture: A Quantitative Approach. Elsevier (2012)
5. Hsieh, C.-C.: B-spline wavelet-based motion smoothing. Computers & Industrial Engineering 41(1), 59–76 (2001)
6. Hsieh, C.-C.: Motion smoothing using wavelets. Journal of Intelligent and Robotic Systems 35(2), 157–169 (2002)

7. Lee, J., Shin, S.Y.: A coordinate-invariant approach to multiresolution motion analysis. Graphical Models 63(2), 87–105 (2001)
8. Lee, J., Shin, S.Y.: General construction of time-domain filters for orientation data. IEEE Transactions on Visualization and Computer Graphics 8(2), 119–128 (2002)
9. Shen, J.P., Lipasti, M.H.: Modern processor design: fundamentals of superscalar processors, vol. 2. McGraw-Hill Higher Education (2005)
10. Shoemake, K.: Quaternion calculus and fast animation, siggraph course notes (1987)
11. Šilc, J., Robič, B., Ungerer, T.: Processor Architecture: From Dataflow to Superscalar and Beyond; with 34 Tables. Springer (1999)
12. Smith, J.E., Pleszkun, A.R.: Implementing precise interrupts in pipelined processors. IEEE Transactions on Computers 37(5), 562–573 (1988)
13. Szczesna, A., Slupik, J., Janiak, M.: Motion data denoising based on the quaternion lifting scheme multiresolution transform. Machine Graphics & Vision 20(3), 238–249 (2011)
14. Szczesna, A., Slupik, J., Janiak, M.: Quaternion lifting scheme for multi-resolution wavelet-based motion analysis. In: The Seventh International Conference on Systems, ICONS 2012, pp. 223–228 (2012)
15. Szczęsna, A., Słupik, J., Janiak, M.: The smooth quaternion lifting scheme transform for multi-resolution motion analysis. In: Bolc, L., Tadeusiewicz, R., Chmielewski, L.J., Wojciechowski, K. (eds.) ICCVG 2012. LNCS, vol. 7594, pp. 657–668. Springer, Heidelberg (2012)
16. Vicon Motion Systems. Polygon User Manual

Rough Set Based Classifications of Parkinson's Patients Gaits

Andrzej W. Przybyszewski[1,2], Magdalena Boczarska[2], Stanisław Kwiek[3], and Konrad Wojciechowski[2]

[1] University of Massachusetts Medical School, Dept Neurology, 65 Lake Av., Worcester, MA 01655, USA
Andrzej.Przybyszewski@umassmed.edu
[2] Polish-Japanese Institute of Information Technology, Koszykowa 86, 02-008 Warszawa, Poland
{przy,konrad.wojciechowski}@pjwstk.pl, m.boczarska@gmail.com
[3] Dept. Neurosurgery, Medical University of Silesia, Central University Hospital, Medykow 14, 40-752 Katowice, Poland
skwiek@csk.katowice.pl

Abstract. Motion capture (MoCap) technology becomes recently often used in neurological applications, especially for diagnosis of gait abnormalities. In this paper we present several different approaches to compute important features of gait abnormalities. This is a continuation of our previous experimental results concerning examination of Parkinson's disease (PD) with bilateral subthalamic nucleus stimulation (DBS) patient in the MoCap laboratory. At first, we calculate mean changes of the gait as effects of medication and DBS. We present these changes as phase plots suggesting different dynamics in different patients. In the second part, we apply AI approach related to application of the Rough Set Theory in order to generate decision rules for all our patients and all experiments. We have tested these rules by comparing training and test sets.

Keywords: MoCap, Deep Brain Stimulation (DBS), reducts, information table, decision rules.

1 Introduction

There were already many studies using MoCap measurements for diagnosis of human gait abnormalities related to neurological diseases as presented in references [1-4]. In these papers several different indexes were proposed and verified on experiments with neurological patients. They found that these indices might to be useful in diagnosis of neurological gait abnormalities, but different groups used different MoCap platforms and therefore algorithms for processing MoCap data were not always consistent. Also some indices were specific for patients with different neurological disorders. In our previous work, we have computed indexes for neurological gait abnormalities for PD patients with DBS [5]. We have found a strong influence of the medication and DBS on the decomposition index of knee and hip and hip and ankle. Therefore in the

N.T. Nguyen et al. (Eds.): ACIIDS 2014, Part II, LNAI 8398, pp. 525–534, 2014.

present work we have concentrated analysis on the dynamics of the hip movements during the gait. However, the present approach is different as we proposed to use not only statistical analysis of certain indexes, but also AI approach base on the rough set theory. This new approach not only summarizes actual measurements but also gives some strong predictions that might be better than standard indexes [1-4], which also predict effects of different therapies for PD patients. As effects of medications and DBS are very different in different patients making predictions is very difficult task and we present here only the preliminary data.

2 Methods

Our experiments were performed on 12 Parkinson Disease (PD) patients who have undergone the surgery based on implanting Deep Brain Stimulator (DBS) for improving their motoric skills. Dr. Kwiek performed surgeries in all patients taking part in our tests on in the Dept. of Neurosurgery Medical University of Silesia (MUS) in Katowice. They were qualified for surgery and observed postoperatively in the Dept. of Neurology MUS [6,7]. Both mentioned above medical departments as well as Polish-Japanese Institute of Information Technology (PJIIT) in Bytom are collaborating, as the group of Silesian Interdisciplinary Centre for Parkinson's Disease Treatment. All experiments were performed in MoCap lab of PJIIT. PD patients performed normal walking under four experimental conditions defined by pharmacological medication and subthalamic nucleus (STN) electrical stimulation (DBS): session S1 was related to MedOFFStimOFF, session S2: MedOFFStimON, session S3: MedONStimOFF, and S4: MedONStimOFF.

In the kinematic movement recording set-up were used 10-cameras and 3D motion capture system (Vicon). The 3D position of the patient was analyzed based on 39 reflective markers (tracked at 100 FPS) placed on major body segments: 4 on Head, 5 on Torso, 14 on left and right side of upper limbs and 16 on left and right sides of lower body.

The structure of data is an important point of our analysis. It is represented in the form of information system or a decision table. We define after Pawlak [8] an information system as $S = (U, A)$, where U, A are nonempty finite sets called the *universe of objects* and the *set of attributes*, respectively. If $a \in A$ and $u \in U$, the value $a(u)$ is a unique element of V (where V is a value set). The *indiscernibility relation* of any subset B of A or $I(B)$, is defined [8] as follows: $(x, y) \in I(B)$ or $xI(B)y$ if and only if $a(x) = a(y)$ for every $a \in B$, where $a(x) \in V$. $I(B)$ is an equivalence relation, and $[u]_B$ is the equivalence class of u, or a *B-elementary granule*. The family of all equivalence classes of $I(B)$ will be denoted $U/I(B)$ or U/B. The block of the partition U/B containing u will be denoted by $B(u)$. Having in discernibility relation we define the notion of reduct $B \subset A$ is a reduct of information system if $IND(B) = IND(A)$ and no proper subset of B has this property. In case of decision tables decision reduct is a set $B \subset A$ of attributes such that it cannot be further reduced and $IND(B) \subset IND(d)$. Decision rule is a formula of the form $(a_{i1} = v_1) \wedge ... \wedge (a_{ik} = v_k) \Rightarrow d = v_d$, where $1 \leq i_1 < ... < i_k \leq m$, $v_i \in Va_i$. Atomic subformulas $(a_{i1} = v_1)$ are called conditions. We say that rule r is applicable to object, or alternatively, the object matches rule, if its attribute values satisfy the rule. With the rule we can connect some numerical characteristics such as matching and support. In order to replace the original attribute a_i with new, binary attribute which tells as whether actual attribute value for an object is greater or lower than c (more in [9]), we define c as a cut. By cut for an attribute $a_i \in A$, such that Va_i is an ordered set we will denote a value $c \in Va_i$. Template of A is a propositional formula $v_i \in Va_i$. A generalized template is the formula of the form $\wedge(a_i \in T_i)$ where $T_i \subset Va_i$. An object satisfies (matches) a template if for every attribute a_i $(a_i = v_i)$ where $a_i \in$ A. The template is a natural way to split the original information system into two distinct sub-tables. One of those sub-tables consist of the objects that satisfy the template, the second one of all others. Decomposition tree is defined as a binary tree, whose every internal node is labeled by some template and external node (leaf) is associated with a set of objects matching all templates in a path from the root to a given leaf [10].

We will distinguish in the information system two disjoint classes of attributes: condition and decision attributes. The system S will be called a decision table $S = (U, C, D)$ where U are objects, C and D are condition and decision attributes [8].

3 Results

Recordings in four sessions: S1: MedOFFStimOFF, S2: MedOFFStimON, S3: MedONStimOFF, S4: MedONStimON were performed in all PD patients. The mean for all patients UPDRS III were improving with sessions, S1: 53+/- 4 (SE), S2: 35+/-6, S3: 22+/-3.5, S4: 18+/-3. Mean duration of three consecutive steps were similar between sessions: S1: 3.9+/- 0.2s (SE), S2: 3.6+/-1.6s, S3: 3.6+/-1.4s, S4: 3.5+/-1.2s. These values are similar to slow walk of the healthy person. In this study, we have limited our analysis to x-direction changes in the hip angles for left and right legs during three consecutive steady steps of all PD patients.

A mean of the maximum x-direction hip angles extension (swing phase) for left (L) and right (R) sides were symmetric and improved non-significantly between sessions, S1: L: 29+/-3 deg (SE), R: 29+/-3 deg (SE), S2: L:32+/-3 deg, R: 33+/-3 deg, S3: L :34+/-3 deg, R: 36+/-3 deg, S4: 35+/-4 deg R: 36+/-3 deg. We also found non-significant improvements for the x direction hip angle flexion (stand phase) between sessions. However, we have observed more significant improvements in the maximum velocity of the x-direction hip angles extension (velocity in the swing phase): S1: L: 123+/-8.5 deg/s, R: 124+/-9.5 deg/s; S2: L: 142+/-6 deg/s, R: 140+/-8.4 deg/s; S3: L: 170+/-6.5 deg/s, R: 169+/-9 deg/s; S4: L: 173+/-6 deg/s, R: 174+/-9 deg/s; and hip angle flexion speed (velocity in the stand phase): S1: L: 71+/-85 deg/s, R: 75+/-5 deg/s; S2: L: 82+/-6 deg/s, R: 93+/-6 deg/s; S3: L: 108+/-7 deg/s, R: 127+/-8 deg/s; S4: L: 120+/-9 deg/s, R: 120+/-9 deg/s.

Notice that the most significant increase in velocities was between sessions S2 and S3, so it is an effect of medications. On the basis of mean values for all our patients we can say that medication as well as DBS are improving patients' UPDRS and (hip) movements velocities. The L-DOPA medications as well as DBS are well-established methods so one would expect such results. However, individual patients are very different and even in our small patients populations we have observed significant variability of the medication and stimulation effects. Therefore, we would like to learn, if we can group effects of medication and DBS therapies of individual patients into several categories?

We have tried two different methods; the first one was related to the dynamical system analysis and the second to the machine learning approach. In our first method, we have compared phase plots for individual patients in four sessions S1 to S4.

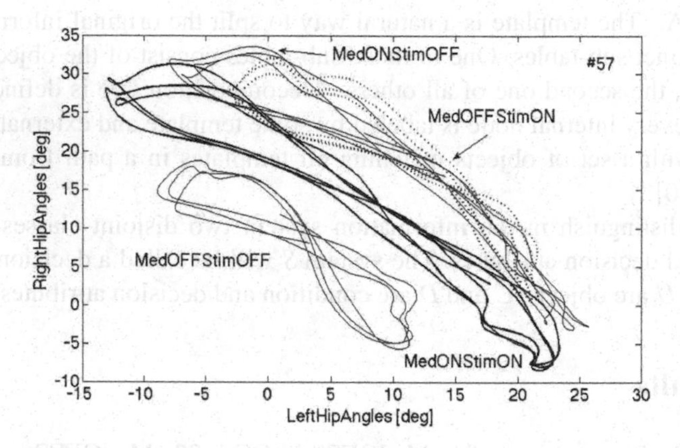

Fig. 1. Phase plots of the right against left x direction hip angles during the gait. Stimulation and medication extend trajectories and shift them up and to the right

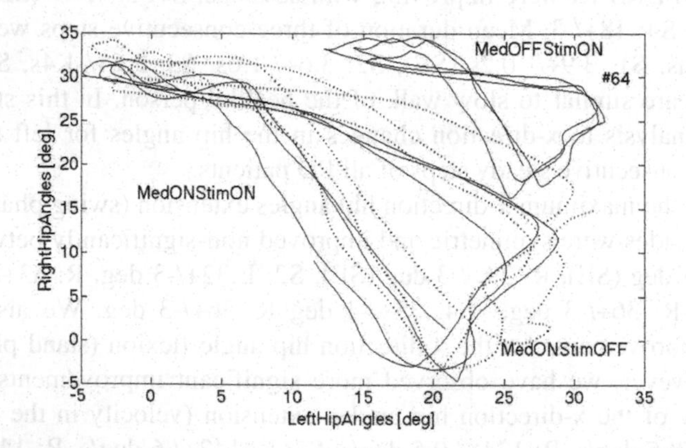

Fig. 2. Phase plots of the right against left x hip angles during walking. Notice a shift down and to the left related to the medication with extent of amplitudes and shift down as effect of DBS during MedON.

We have plotted the movement trajectories in the phase space as changes of the right hip x- angles as a function of the left hip angles changes during three steps stable walk. We have found different types of attractor changes as effect of medication and stimulations, as it is demonstrated on the following figures.

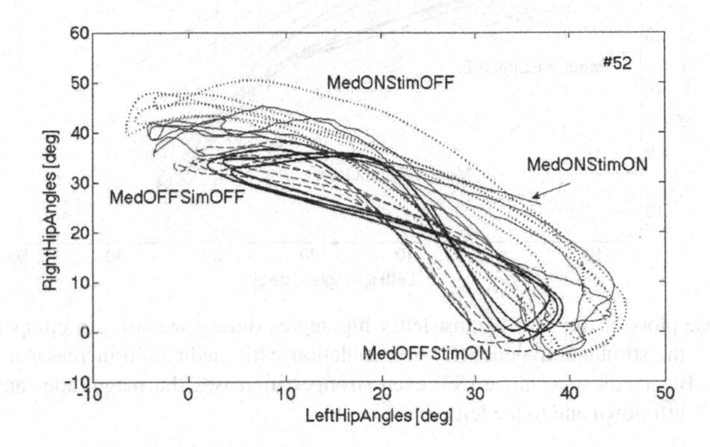

Fig. 3. Phase plots of the right against left x hip angles during walking. Notice that for this patient effects of stimulation and medications are relatively small. Stimulation alone (S2) does not introduce significant changes in comparison to the control (S1). A significant changes in trajectories' amplitude with shift up and to the right are effects of the medication (S3, S4).

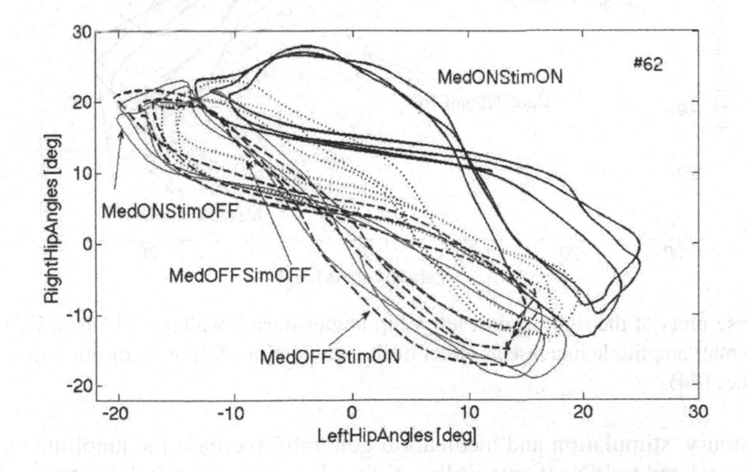

Fig. 4. Phase plots of the right against left x hip angles during walking. Notice very similar trajectories during sessions S1, S2 and S3. In contrast, interaction of medication and stimulation (S4) strongly shifts trajectories up and into the right, but without changes in amplitudes.

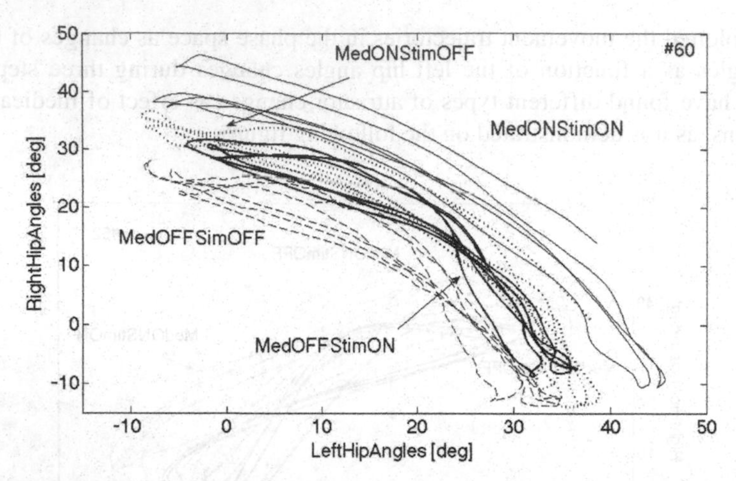

Fig. 5. Phase plots of the right against left x hip angles during walking. In comparison to the control (S1), the stimulation alone (S2) or stimulation with medication increases magnitude of trajectories. But medication alone (S3) even stronger increases the magnitude and introduce trajectories' shift down and to the left .

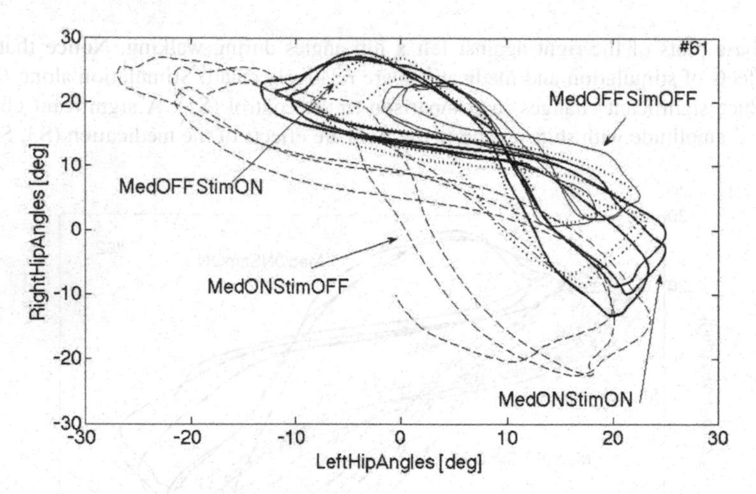

Fig. 6. Phase plots of the right against left x hip angles during walking. Notice a shift up with relatively small amplitude increase as effect of the stimulation (S2) or medication (S3) alone or both together (S4).

In summary, stimulation and medication generally increase the amplitude and shift trajectories related to PD patients walk activity. It is not mainly related to patients gait speed, as mean gait durations were similar in all sessions. These plots might give basis for the dynamical model of the gait in different sessions but as demonstrated, in different patients changes of the particular trajectory are difficult to predict, as they are effects of the system complexity and basal ganglia regulatory numerous loops interactions.

3.1 Rough System Approach

As described above we have used the RSES 2.2 (Rough System Exploration Program) [9] in order to find regularities in our data. At first our data was placed in the decision table as originally proposed by Pawlak [8]. In the row there are following attributes: P# - patient#, S# Session #, t-time, mxaL/mxaR/mnaL/mnaR – max/min Left/Right hip x-direction angles, mxVaL/mxVaR/mnVaL/mnVaR – max/min Left/Right hip x-direction velocity, and UPDRS III as measured by the neurologist in the last column. There are data from two out of 12 patients in the table below:

Table 1. A part of the decision table

P#	S#	t	mxaL	mxaR	mnaR	mnaL	mxVaL	mxVaR	mnVaL	mnVaR	UPDR
52	1	390	38.6	35.9	3.31	-0.65	1.14	1.55	-0.62	-0.65	52
52	2	390	36.2	36.0	-0.14	-2.06	1.30	1.65	-0.89	-0.85	23
52	3	330	44.8	47.9	-5.1	-3.79	2.12	2.25	-1.38	-1.58	13
52	4	400	43.5	42.9	-3.8	-3.04	1.66	1.82	-1.0	-1.21	27
53	1	320	17.7	17.0	-6.9	-6.33	1.49	1.32	-0.70	-0.80	53
53	2	305	23.2	23.3	-1.47	-2.93	1.45	1.50	-0.75	-0.94	23
53	3	290	32.6	24.3	-6.77	-13.99	1.85	1.87	-1.58	-1.79	10
53	4	320	2 6.4	20.3	-8.70	-12.81	1.51	1.59	-1.27	-1.32	8

The last column represents a decision attribute then we can write each row a decision rule as following:

('Pat'=52)&('Sess'=1) &('time'=390)& ('mxaL' =38.6)& … =>('UPDRS'=53) (1)

We read this rule as following: if for patient #52 and session S1 and time of his/her three steps 3.9 s and max hip x-direction angle equal 38.6 deg and … then his/her UPDRS III for this session is 53.

Therefore we obtain 46 decision rules directly from our measurements, as two from our 12 patients did not have all four sessions. The main purpose of our analysis is to reduce these rules and to find regularities in our data. There are many possible steps as described in [9], below we will give some examples. At first, we would like to make rules shorter and find that they apply to more than one case, e.g.:

('Pat'=60)=>('UPDRS'=9[2]) 2 (2)

('mnVaL'=-0.6756)=>('UPDRS'=32[2]) 2 (3)

it reads that Pat# 60 obtained UPDRS=9 in two sessions (eq. 2) and that min velocity of the left hip equal -0.6756 (- is related to the direction of gait) was related to UPDRS=32 in two cases (eq.3). In order to make rules more effective RSES can find optimal linear combinations of different attributes like:

*'mxVaL'*0.594+'mxVaR'*(-0.804)* (4)

*'mx_aL'*0.046+'mn_aL'*(-0.587)+'mn_aR'*0.807* (5)

and these linear combinations may be added as an additional attributes. Also we can use discretization procedure [9] that divides attributes values into non-overlapping parts:

('Pat'="(58.5,Inf)")&('Sess'="(2.5,3.5)"/"(3.5,Inf)")&('mnVaL'="(-0.9803, Inf)") => ('UPDRS'=32[3]) 3 (6)

That reads that for patients' numbers above 58.5 and in sessions S3, S4 min hip velocity is -0.9803 or above then UPDRS equals 32 in three cases (eq. 6).

As we have demonstrated above rules determining possible UPDRS are important but from patient and doctor points of view, the first message should be if the therapy (medication and/or DBS) is effective. In order to find it, we need to correlate our measurements with the session number that is related to the specific procedure. In this case the session number will be the decision attribute. In this case, we can obtain the following more general rules e.g.:

('UPDRS'=52/53/43/56/87/45/58/30/60)=>('Sess'=1[11]) 11 (7)

('UPDRS'=23/13/43/22/39/28/24/81/48/42)=>('Sess'=2[11]) 11 (8)

('time'=440/305/280/365/310)=>('Sess'=2[6]) 6 (9)

that means that session S1 (MedOFFStimOFF) is related to high UPDRS in 11 cases (eq. 7), in session S2 (MedOFFStimON) UPDRS are generally smaller in 11 patients (eq. 8) and in this session (S2) the duration of three steps is between 2.8 and 4.4 s in 6 cases (eq. 9). We can also find rules in which the duration of three steps are similar as in (eq. 10):

('time'=350)&('Pat'=56/57/62/59)=>('Sess'=4[4]) 4 (10)

Another important issue is how values of different attributes are changing in different sessions and patients. More variability is related to better attribute. Below there are two examples for: UPDRS and max hip left angles velocity.

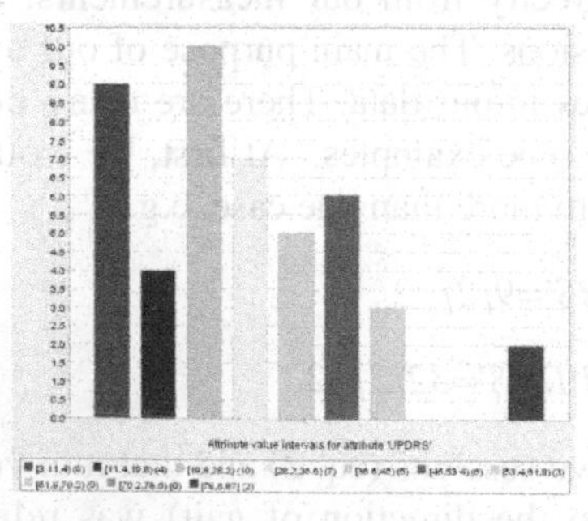

Fig. 7. Statistic for UPDRS

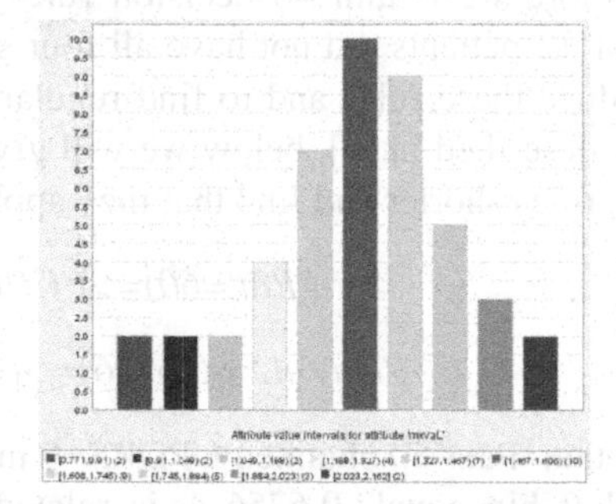

Fig. 8. Statistic for max velocity L. hip

But the main purpose of the ML approach is related to demonstration that proposed rules are enough universal to predict results from new patients on the basis of already measured patients (*test-and-train scenario* –[9]). In order to perform such test, we have divided our data set into two parts: one 60% of our data was training set, and another 40% was set that we have tested. We have removed decision attributes from the test-set and compared them with attributes values obtained from our rules. We have used several different algorithms in order to find rules from training-set. The exhaustive algorithm [9] gave the best results described as the confusion matrix:

Table 2. Confusion matrix for different session numbers (S1-S4)

		Predicted				
		2	3	4	1	ACC
Actual	2	2	0	0	1	0.66
	3	1	0	1	2	0.0
	4	1	3	1	0	0.2
	1	0	1	1	2	0.5
	TPR	0.5	0.0	0.33	0.4	

TPR: True positive rates for decision classes, ACC: Accuracy for decision classes: Coverage for decision classes: 0.75, 1.0, 1.0, 0.66 and global coverage=0.8421, and global accuracy=0.3125. A global accuracy was above 30% that means that we probably need to use more rules as for example combinations of many attributes or/end extend number of measured attributes for our analysis. However problem with this approach is that its results depend on which part of our measurements was taken as training and which part was tested. In order to test in exhaustive manner or all different possibilities we have divided our experimental randomly set into 9 subsets:

Table 3. Confusion matrix for the UPDRS as the decision attribute

		Predicted					
		50,69.5	-Inf,29.5	42.5,50	34,42.5	69.5,Inf	29.5, 34
Actual	50, 69.5	0.67	0.0	0.0	0.0	0.0	0.0
	-Inf, 29.5	0.0	1.67	0.0	0.11	0.11	0.0
	42.5,50	0.0	0.0	0.11	0.0	0.0	0.0
	34,42.5	0.0	0.11	0.0	0.0		
	69.5, Inf	0.0	0.11	0.0	0.0	0.0	0.0
	29.5, 34	0.0	0.0	0.0	0.0	0.0	0.22
	TPR	0.44	0.72	0.11	0.0	0.0	0.22

TPR: True positive rates for decision classes, ACC: Accuracy for decision classes: 0.44, 0.72, 0.11, 0, 0, 0.22. Coverage for decision classes: 0.44, 0.602, 0.11, 0.11, 0.11, 0.167 and global coverage=0.6, and **global accuracy=0.917**. UPDRS decision classes: (50, 69.5), (-Inf, 29.5), (42.5, 50), (34, 42.5), (69.5, Inf), 29.5, 34).

4 Conclusions

We have presented comparison of the classical dynamical systems, and rough set (RS) approaches to process the MoCap data from PD patients in four different treatments. We have plotted effects of the medication and brain stimulation on individual patients gait trajectories. As these effects are strongly patient's dependent they could not give enough information to predict new patient's behavior. The RS approach is more universal as it gives general rules and predictions that cover individual patients reactions to different treatments as demonstrated for the UPDRS predictions.

Acknowledgements. This work was supported by projects NN 518289240 and UMO-2011/01/B/ST6/06988 from the Polish National Science Centre. We thank Marek Kulbacki for his help with the database.

References

1. Zifchock, R.A., Davis, I., Higginson, J., Royer, T.: The symmetry angle: a novel, robust method of quantifying asymmetry. Gait Posture 27(4), 622–627 (2008)
2. Mian, O.S., Schneider, S.A., Schwingenschuh, P., Bhatia, K.P., Day, B.L.: Gait in SWEDDs patients: Comparison with Parkinson's disease patients and healthy controls. Movement Disorders 26, 1266–1273 (2011)
3. Earhart, G.M., Bastian, A.J.: Selection and coordination of human locomotor forms following cerebellar damage. J. Neurophysiol. 85, 759–769 (2001)
4. Lewek, M.D., Poole, R., Johnson, J., Halawa, O., Huang, X.: Arm swing magnitude and asymmetry during gait in the early stages of Parkinson's disease. Gait Posture 31, 256–260 (2010)
5. Stawarz, M., Kwiek, S.J., Polanski, A., Janik, L., Boczarska, M., Przybyszewski, A., Wojciechowski, K.: Algorithms for computing indexes of neurological gait abnormalities in patients after DBS surgery for Parkinson Disease based on motion capture data. Machine Graphics & Vision 20, 299–317 (2011)
6. Kwiek, S.J., Kłodowska-Duda, G., Wojcikiewicz, T., Ślusarczyk, W., Kukier, W., Bazowski, P., et al.: Simultaneous targeting and stimulation of STN and VIM in tremor predominant PD patients. Pro's and Cons. Acta Neurochir. 148, 36 (2006)
7. Kwiek, S.J., Boczarska-Jedynak, M., Świat, M., Kłodowska-Duda, G., Ślusarczyk, W., Kukier, W., Błaszczyk, B., Doleżych, H., Szajkowski, S., Suszyński, K., Kocur, D., Bażowski, P., Opala, G.: DBS for Parkinson's disease treatment. Experience and results interdisciplinary Silesian Centre for Parkinson's Disease Treatment in Katowice. Neurol. Neurochir. Pol. 44(supl.1), S16–S17 (2010)
8. Pawlak, Z.: Rough sets: Theoretical aspects of reasoning about data. Kluwer, Dordrecht (1991)
9. Bazan, J., Son, N.H., Trung, T., Nguyen, S.A., Stepaniuk, J.: Desion rules synthesis for object classification. In: Orłowska, E. (ed.) Incomplete Information: Rough Set Analysis, pp. 23–57. Physica – Verlag, Heidelberg (1998)
10. Bazan, J., Szczuka, M.S.: RSES and RSESlib - A Collection of Tools for Rough Set Computations. In: Ziarko, W.P., Yao, Y. (eds.) RSCTC 2000. LNCS (LNAI), vol. 2005, pp. 106–113. Springer, Heidelberg (2001)

Feature Selection of Motion Capture Data in Gait Identification Challenge Problem

Adam Świtoński[1], Henryk Josiński[2], Agnieszka Michalczuk[1], Przemysław Pruszowski[2], and Konrad Wojciechowski[1]

[1] Polish-Japanese Institute of Information Technology, Aleja Legionów 2, 41-902 Bytom, Poland
{aswitonski,amichalczuk,kwojciechowski}@pjwstk.edu.pl
[2] Silesian University of Technology, ul. Akademicka 16, 41-100 Gliwice, Poland
{henryk.josinski,przemyslaw.pruszowski}@polsl.pl

Abstract. The method of discovering robust gait signatures containing strong discriminative properties is proposed. It is based on feature extraction and selection of motion capture data. Three different approaches of feature extraction applied to Euler angles and their first and second derivates are considered. The proper supervised classification is preceded by specified selection scenario. On the basis of the obtained precision of person gait identification, analyzed feature sets are assessed. To examine proposed method database containing 353 gaits of 25 different males is used. The results are satisfactory. In the best case the recognition accuracy of 97% is achieved. On the basis of classification which takes into consideration only the data of the specified segments, the ranking is constructed. It corresponds to the evaluation of individual features of the joint movements.

Keywords: gait identification, motion capture, supervised learning, feature selection, feature extraction, biometrics

1 Introduction

In a motion capture acquisition positions of attached markers on human body are tracked by calibrated multicamera system. Thus basic motion data representation contains time sequences of global 3D coordinates of the markers. It can be transformed to skeletal model representation with a kinematical chain of a tree-like structure. The root object of the tree usually corresponds to lower part of a spine and subsequent nodes point to following joints. Thus every pose is described by joints rotations and global translation of human body in respect to specified reference frame. The rotation are coded by three Euler angles or unit quaternions by default.

There are two basic approaches of skeleton model estimation [1]. If the markers are located in specified anatomical points of a human body, the simple set of geometric transformations is sufficient. However such a method is very sensitive even to slight markers displacements. That is a reason why preliminary matching of markers and

N.T. Nguyen et al. (Eds.): ACIIDS 2014, Part II, LNAI 8398, pp. 535–544, 2014.

skeleton segments usually is carried out by clustering techniques. It requires a special set of movements to be performed by a human, prior to proper acquisition, which should take into consideration all degrees of freedom of assumed skeleton model.

However direct applications of motion capture acquisition in deployed person gait identification systems is problematic. It removes one of the most important advantages of gait identification - non-awareness of recognized human and what is more the acquisition is time consuming - man has to put special suit with attached makers and perform prepared set of movements for the sake of skeleton calibration purposes. Thus, markerless motion capture can be utilized. The problem of skeleton estimation by such an acquisition in most cases leads to nonlinear high dimensional parametric optimization. The skeleton configuration which matches image data in best way has to be determined. Most often used technique to explore configuration spaces are particle filters. For instance in [2] particle swarm optimization is applied and the obtained mean markers displacement in respect to reference to Vicon motion capture is 50mm. The price which has to be paid for much convenient acquisition is worse precision of measurements. Thus application of motion capture in development phase of gait identification system is justified. It allows to focus on classification stage without influence of acquisition noise and to obtain best possible results. If the classification is satisfactory the second stage has to be completed - proper choice and parameters tuning of markerless motion capture.

The paper presents the method of gait classification based on feature extraction and selection of motion capture data. The final recognition is carried out by supervised classification. To examine proposed approach gait data collected in human motion capture laboratory is utilized. On the basis of obtained classification accuracies features spaces are explored and their subsets are evaluated. It is the main contribution of the paper which relates to discovered features corresponding to the most valuable individual gait properties.

2 Related Work

As described in previous section, motion data is represented as time sequence of pose parameters, markers positions or silhouettes. There are three basic approaches to classification of such a data: feature extraction, dynamic time warping and Hidden Markow Models.

In the first approach, features of time sequences are calculated and motion descriptors are constructed. On the basis of obtained feature vectors, subsequent classification is carried out - for instance machine learning can be applied. In [3] generic extraction is proposed which utilizes tensor reduction technique by multilinear principle component analysis. The final recognition is performed by selected distance function. In [4] and [5] four types of features sets and supervised classification are used for motion capture data. In [6] all body parameters are transformed into the frequency domain and first two lowest Fourier components are chosen. Afterwards PCA reduction is applied.

Dynamic time warping (DTW), originally introduced to spoken word recognition [7], is normalization technique which matches poses of compared motions on the basis of monotonic transformation. It makes them faster or slower in subsequent time instants to obtain the lowest total distance of corresponding poses. DTW is utilized to estimate motion dissimilarity which is a crucial challenge in nearest neighbors classification scheme. In [8] DTW is used for binary relational features, in [9] different distance metrics of Euler angles and unit quaternions spaces are examined for DTW classification of gait motion capture data and in [10] DTW is based on linearly and nonlinearly reduced silhouettes.

In Hidden Markow Model (HMM) approaches motion is modeled as Markow chain. Models are hidden, because it is stated assumption that poses depend on states which are unknown for an observer. A single model usually is taken for every class and its parameters contains probability transition matrix between states in subsequent time moments and probability distribution of poses in specified states. In the training phase parameters of HMM for every class are calculated and in classification stage model with the greatest probability is determined. Exemplary HMM based methods are presented in [11], [12] and [13].

3 Collected Database

To validate proposed method and to discover most valuable individual features, motion database in PJWSTK Human Motion Laboratory[1] was collected. It contains 353 gaits of males at the age from 25 to 35 years old - students and staff of PJWSTK university. The gait route was specified to be a straight line of about 5 meters long. Example data is visualized in Fig. 1.

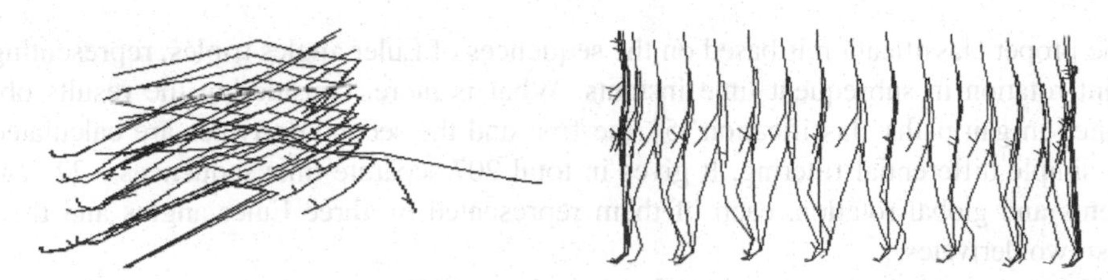

Fig. 1. Example collected gait

In acquisition Vicon Blade software was applied with default skeleton model containing 22 bone segments as presented in Fig. 2a. Complete pose description is defined in 72 dimensional space. There are 23 rotations coded by three Euler angles - additional one is associated with global skeleton rotation The pose space also contains a three dimensional vector related to global translation.

Gait can be defined as coordinated cyclic combination of movements which results in human locomotion [14]. What is more, because requirement of calibration process of Vicon Blade software, every gait starts and terminates with T-pose, which is quite untypical during normal gait. Thus to remove such a data and to detect

[1] `http://hml.pjwstk.edu.pl/en/`

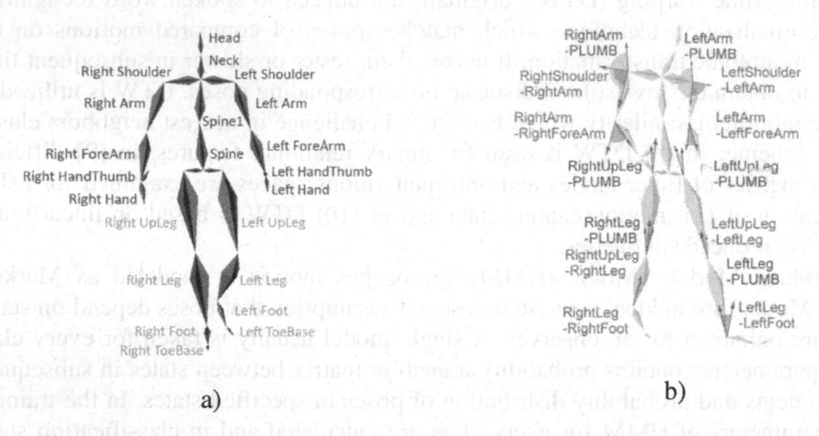

Fig. 2. Skeleton model a) upper and lower body segment names, b) custom angles

representative main cycle containing two adjacent steps distance tracking between two feet is carried out, as is presented in [5].

To avoid learning and recognition on the basis of gait location, which seems to be strongly related to place of acquisition, the global translation vector is removed from motion data and it is not taken into consideration by subsequent feature extraction procedures.

4 Feature Extraction

The proper classification is based on the sequences of Euler angles triples, representing joint rotation in subsequent time instants. What is more, considering the results obtained in gait paths classification [5], the first and the second derivates are calculated by simple differential filtering. It gives in total 207 separate time sequences - 22 segments and global rotation, each of them represented by three Euler angles and their first two derivates.

Example time sequences of randomly selected five persons, reduced to detected main cycle, are visualized in Fig. 3 and in Fig. 4. The charts present Euler angles and their first and second derivates in respect to local coordinate system RX-RY-RZ. What is more rotational data is transformed into representation of angles between two adjacent segments by calculating arc cosine of dot products of unit vectors related to segment orientation in 3D space. For instance in case of left knee joint it is the angle between LeftUpLeg and LeftLeg segments from Fig. 2. It allows to visualize aggregated dependency among joint orientation, its angular velocity and acceleration. as shown in Fig. 3d and Fig. 4. Different scales on charts in Fig. 3d and Fig. 3a are caused by opposite meaning of presented data. In case of raw Euler angles supplied by motion capture acquisition, the pose is specified in respect to given reference frame and after above described transformation, angles correspond to direct rotation which has to be performed to match both adjacent segments.

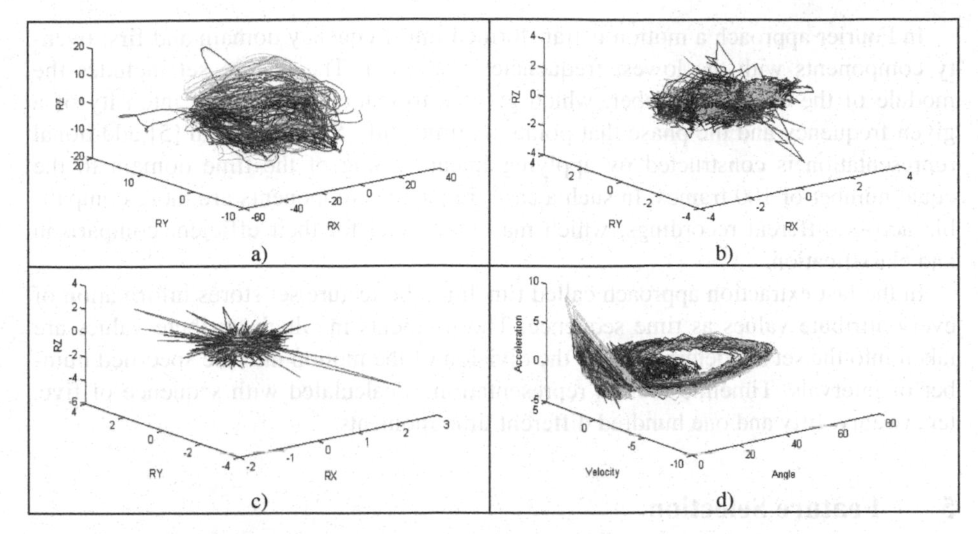

Fig. 3. Left hip joint time sequences: a) Euler angles, b) angular velocities, c) angular accelerations, d) single angle of rotation

Data of different humans in Fig. 3 and in Fig. 4 are marked by separate colors. It is very difficult to manually point discriminative properties of presented time series. What is more problem is probably more challenging and efficient classification requires simultaneous analysis of multiple joints movements.

Similar to methods presented in [5] and [4], three types of feature sets are applied in extraction:

- Statistical
- Fourier transform
- TimeLine

In the first approach two most basic statistics - mean value and variance of every considered pose attribute are calculated. Such a simple feature extraction does not take into consideration dependencies in time domain and analyzes only the emerging values of sequences.

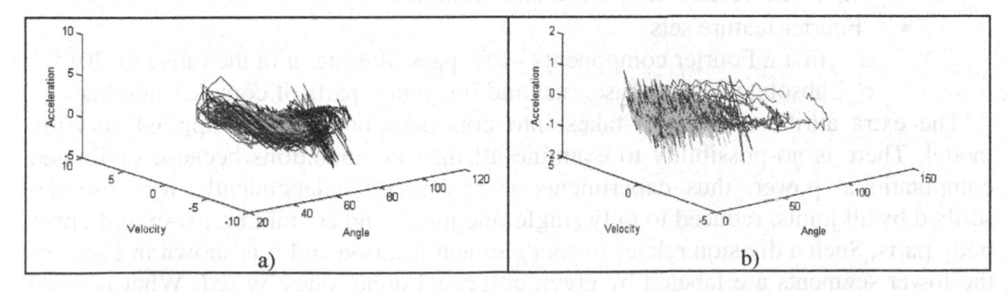

Fig. 4. Left knee and left arm joints time sequences a) left knee, b) left arm

In Fourier approach a motion is transformed into frequency domain and first twenty components with the lowest frequencies are taken. The feature set includes the module of the complex number, which gives information of the total intensity of a given frequency and the phase that points its time shift. Similar like in [5] additional representation is constructed by applying linear scaling of the time domain to the equal number of 100 frames. In such a case frequency components are more compatible across different recordings, which may be crucial for their efficient comparison and classification.

In the last extraction approach called timeline, the feature set stores information of every attribute values as time sequence. The moments in which attribute values are taken into the set are determined by the division of the motion into the specified number of intervals. Timeline motion representation is calculated with sequence of five, ten, twenty, fifty and one hundred different time moments

5 Feature Selection

The proposed feature spaces are high dimensional. In case of most simple statistical extraction feature space contains 414 attributes and for the most complex timeline approach with 100 intervals 20700 separate features are determined. The hypothesis of useless attributes without discriminative, individual properties can be stated in such features spaces. It is proven that irrelevant or distracting attributes to a dataset often confuses machine learning systems [15], which may result in worse classification accuracy.

To verify the hypothesis of useless features and to discover the most valuable ones, attribute selection is carried out prior to proper classification. Ii is based on the manually proposed scenarios with feature subsets evaluated on the basis of obtained identification efficiency. At the current stage, we have not used automatic selection techniques because of the complexity of the problem and to get more interpretable results.

The prepared selection scenarios are as follows. In all cases we selected every possible combination of attributes associated with:

- axis RX, RY and RZ of the local coordinate system.
- rotation, angular velocity and acceleration
- statistical feature sets: mean and variance
- Fourier feature sets
  - first n Fourier components - low pass filtering, n in the range (1,20),
  - absolute value, phase, real and imaginary parts of complex number

The extra attribute selection takes into consideration joints of applied skeleton model. There is no possibility to examine all their combinations because of limited computational power, thus experiments were iterated independently for pose described by all joints, reduced to only single one joint and containing lower and upper body parts. Such a division relates to root element location and it is shown in Fig. 2a - the lower segments are labeled by green color and upper ones by red. What is more the custom combination of angles, containing selected data of direct relationships between adjacent segments and their orientation to vertical axis is investigated as presented in Fig. 2b. The combination was prepared according to our subjective expectations to keep individual gait features. Motion data in this case is aggregated into absolute values of angles without distinction of three basic rotations.

6 Experiments, Results and Conclusions

The experiments were iterated for all previously described feature sets and their selection scenarios. What is more additional variants are generated by applied prefiltrations containing main cycle detection, linear scaling of the time domain into given number of frames and also by (α, β, χ) Euler angles transformation which determines aggregated total angle between two adjacent segments according to equation (1). The number 100 of frames used in linear scaling was adjusted to roughly approximated mean duration of the main cycles.

$$angle(\alpha, \beta, \chi) = \arccos(\cos\alpha \cdot \cos\beta \cdot \cos\chi - \sin\alpha \cdot \sin\beta \cdot \sin\chi) \quad (1)$$

Because of numerous experiments were required, applied classifiers have to be characterized by low computational requirements. That is a reason why k nearest neighbors (kNN) [15] and naive Bayes [15] classifiers are chosen. The number k of considered neighbors for kNN is in the range <1,10> and for Naive Bayes parametrical estimation of normal distribution and non parametrical kernel based one are used. To split collected gait instances into training and testing parts, leave one out validation is utilized [15].

Obtained results are shown in Fig. 5, 6, 7, 8 and 9.which present best achieved accuracy of identification expressed by percent of correctly classified gaits in respect to specified feature extraction and selection approaches and applied prefiltration.

The results are satisfactory. Best obtained precision of recognition is 97.1%, which means only 10 misclassified gaits of 353. It belongs to first five Fourier components calculated for whole set of Euler angles and their first derivates of gaits reduced to detected main cycle.

According to expectations, both main cycle detection and linear scaling improve performance of feature extraction and subsequent classification – the former noticeable, the latter slightly, as shown in Fig . 5 and 6. To achieve maximum possible accuracy of identification. analysis of complete joint description containing three Euler angles is necessary. Its simplification to single rotational angle removes some individual features. This probably explains unexpected worse results obtained by custom angles from Fig 2b.

The most precise classification is carried out on the basis of Fourier extraction. Beside the case of custom angles. timeline extraction is much worse. Simple statistical features are suitable for recognition with 95% accuracy, which is even better in comparing to much more complex timeline features.

First and second derivates corresponding to angular velocity and acceleration still represent strong individual features. They allow to recognize gaits with 95.4% and 95.9% precision, respectively, which is only less than 1% worse in comparing to raw Euler angles. If the derivates are combined together in a single gait signature accuracy increases to 96.3%.

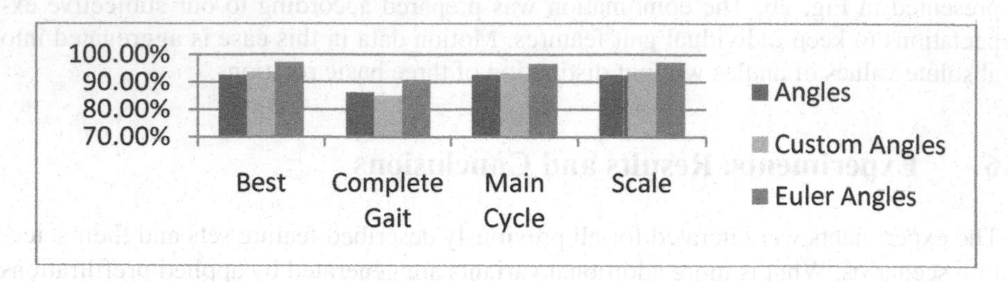

Fig. 5. Classification results in respect to different prefiltrations and parameters of a skeleton data

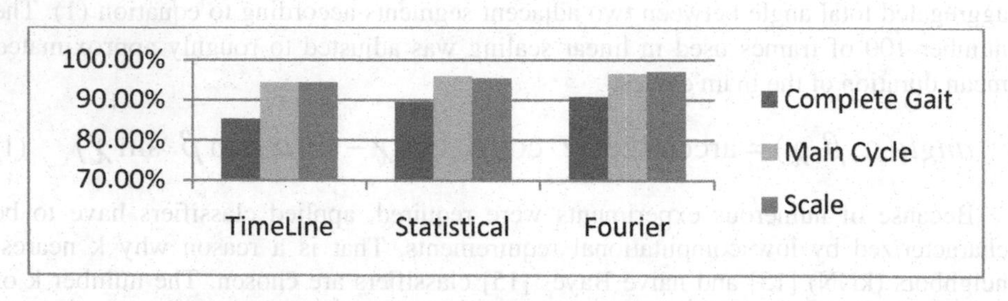

Fig. 6. Classification results in respect to a different extraction approaches and prefiltrations

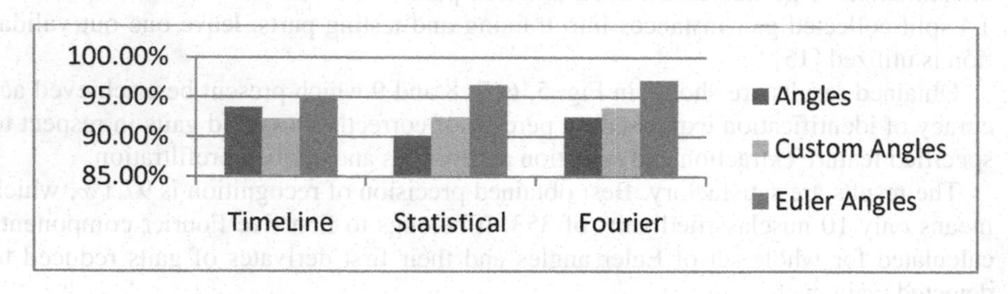

Fig. 7. Classification results in respect to different extraction approaches and parameters of a skeleton data

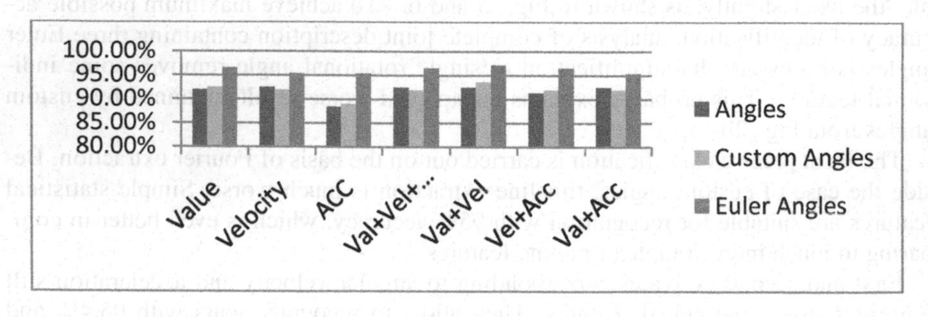

Fig. 8. Classification results in respect to derivates and parameters of a skeleton data

It is sufficient to persist only first five Fourier components to obtain maximum accuracy. Subsequent ones are more influenced by the acquisition noise, which slightly worsens the results. The single first Fourier component corresponding to mean value is efficient only for Euler angles and not for their derivates. It can be explained by individual features of a human posture which are reflected by skeleton Euler angles. while derivates correspond to the joint movements.

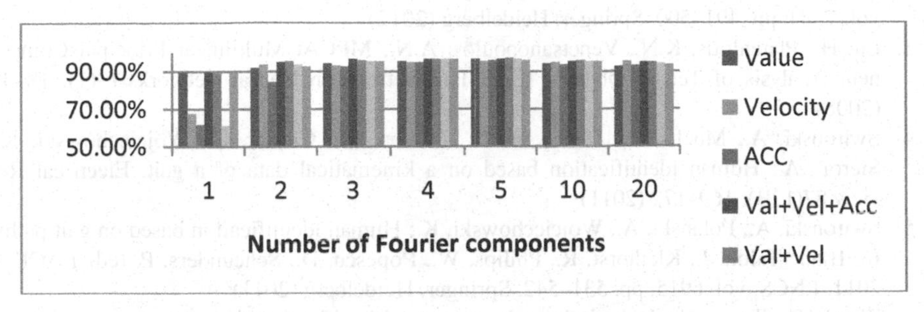

Fig. 9. Classification results in respect to selected number of Fourier components and derivates

On the basis of classification performed by taking into consideration only the data of specified joints, the ranking is constructed as shown in Table 1. Very similar high precision of identification corresponds to complete set of lower and upper body segments. According to expectations, the joints which are active during gait as for in stance, hip, ankle, and shoulder are top ranked. Bit surprisingly classification based on knee joints is much less efficient It also partially explains why results obtained for the custom angles feature extraction, which considers knee movements instead of hip ones, were much worse than expected. High positions of spine segments are probably caused by a human posture abnormalities.

Table 1. Joint ranking

Prec.	Segment
92,49%	UP
91,91%	LeftUpLeg
91,33%	DOWN
90,75%	Spine
88,44%	RightUpLeg
86,13%	Spine1
81,50%	RightFoot
79,77%	LeftShoulder

Prec.	Segment
78,61%	LeftFoot
76,88%	RightShoulder
75,72%	root
71,68%	LeftArm
70,52%	RightArm
61,85%	LeftLeg
57,80%	RightLeg
57,23%	Neck

Prec.	Segment
57,23%	LeftHand
56,07%	Head
53,18%	RightHand
52,60%	LeftForeArm
47,98%	RightForeArm
45,09%	LeftToeBase
38,15%	RightToeBase

Acknowledgement. This work was supported by project NN 516475740 from the Polish National Science Centre.

References

1. Janik, Ł., Polański, K., Wojciechowski, K.: Models and algorithms for human skeleton estimation form 3D marker trajectories. Machine Graphics and Vision (2013)
2. Krzeszowski, T., Kwolek, B., Michalczuk, A., Świtoński, A., Josiński, H.: View independent human gait recognition using markerless 3D human motion capture. In: Bolc, L., Tadeusiewicz, R., Chmielewski, L.J., Wojciechowski, K. (eds.) ICCVG 2012. LNCS, vol. 7594, pp. 491–500. Springer, Heidelberg (2012)
3. Lu, H., Plataniotis, K.N., Venetsanopoulos, A.N.: MPCA: Multilinear Principal Component Analysis of Tensor Objects. IEEE Transactions on Neural Networks 19(1), 18–39 (2008)
4. Świtoński, A., Mucha, R., Danowski, D., Mucha, M., Cieślar, G., Wojciechowski, K., Sieroń, A.: Human identification based on a kinematical data of a gait. Electrical Review 87(12B), 169–172 (2011)
5. Świtoński, A., Polański, A., Wojciechowski, K.: Human identification based on gait paths. In: Blanc-Talon, J., Kleihorst, R., Philips, W., Popescu, D., Scheunders, P. (eds.) ACIVS 2011. LNCS, vol. 6915, pp. 531–542. Springer, Heidelberg (2011)
6. Zhang, Z., Troje, N.: View-independent person identification from human gait. Neurocomputing 69 (2005)
7. Sakoe, H., Chiba, S.: Dynamic programming algorithm optimization for spoken word recognition. IEEE Transaction on Acoustics, Speech and Signal Processing (1978)
8. Müller, M., Röder, T., Clausen, M.: Efficient content-based retrieval of motion capture data. ACM Transactions on Graphics 24(3) (2003)
9. Świtoński, A., Michalczuk, A., Josiński, H., Wojciechowski, K.: Dynamic Time warping in gait classification of motion capture data. In: International Conference on Signal Processing, Pattern Recognition and Applications, Venice (2012)
10. Świtoński, A., Michalczuk, A., Josiński, H., Wojciechowski, K.: Gait human identification by Dynamic Time Warping applied to reduced video recordings data. In: International Conference on Computer Science and Engineering, Kuala Lumpur (2013)
11. Cheng, M., Ho, M., Huang, C.: Gait analysis for human identification through manifold learning and HMM. Pattern Recognition 41 (2008)
12. Josiński, H., Świtoński, A., Michalczuk, A., Kostrzewa, D.: Feature extraction and HMM-based classification of gait video sequences for the purpose of human identification. Vision Based Systems for UAV Applications (2013)
13. Aqmar, M.R., Shinoda, K., Furui, S.: Efficient model training for HMM-based person identification by gait. In: 2012 4th Asia-Pacific Signal and Information Processing Association Annual Summit and Conference (2012)
14. Boyd, J.E., Little, J.: Biometric Gait Identification. Advanced Studies in Biometrics, Berlin (2005)
15. Witten, I.H., Frank, E., Hall, M.A.: Data mining - Practical Machine Learning Tools and Technique. Morgan Kaufman, Burlington (2011)

Motion Data Editor Software Architecture Oriented on Efficient and General Purpose Data Analysis

Marek Kulbacki, Mateusz Janiak, and Wojciech Knieć

Polish-Japanese Institute of Information Technology,
Koszykowa 86, 02-008 Warszawa, Poland
{kulbacki,mjaniak,wkniec}@pjwstk.edu.pl
http://www.pjwstk.edu.pl

Abstract. We present the architecture of the Motion Data Editor (MDE) real-time development framework for multi-modal motion data management, visualization and analysis. There is an emerging need for such tools due to the capability of recording and computing large number of motion modalities with high precision, synchronized in time domain. Such tools should be efficient, easy to use and flexible to apply to various data types and algorithms. Proposed Motion Data Editor (MDE) is dedicated to general data processing, and supports most of the common functionalities during data analysis. We discuss the most important functional requirements and present selected elements of system architecture: core data types, functionality and processing logic elements. The MDE with Human Motion Laboratory (HML) and cloud based Human Motion Database (BDR) [3] constitute collaborative environment for acquisition and analysis of multi-modal synchronized motion data for medical research and entertainment.

Keywords: software architecture, c++ , generic programming, variant type, data flow, continues integration, plug-ins, data analysis, data processing.

1 Introduction

There are many solutions dedicated to general data processing, but most of them are either too specialised and limited for particular applications (biomechanics, medical imaging), or general enough, but offering poor efficiency. This forces the development of dedicated applications that fit particular research projects needs, where very often similar functionality is shared among different applications and only data types with algorithms are different. This leads to lack of compatibility between many tools, error propagation during data processing and other disadvantages. We propose an universal solution, offering flexibility in managed data types with efficiency in data processing, standardizing most common operations and functionalities present during data analysis.

N.T. Nguyen et al. (Eds.): ACIIDS 2014, Part II, LNAI 8398, pp. 545–554, 2014.

1.1 Data Processing Procedure

Figure 1 presents generalized data processing pipeline. We distinguished five independent steps:

1. **Browse** – search for data available for analysis and processing;
2. **Load** – data extracting by allocating required resources and normalize data;
3. **Process** – operating on data with various algorithms;
4. **View** – data observing at different perspectives, making decisions about further analysis and processing steps;
5. **Save** – storing results for further analysis, potentially share them with other users.

Such pipeline, or its parts, can be found in almost any kind of data processing oriented software, although depending on the application purpose and internal realisation, those steps can differ significantly. Most of underlying functionalities are shared across applications, encapsulated with different user interface (UI). Providing general architecture, dedicated to such processing pipeline should allow to apply it for any kind of data and algorithms, reusing those common functionalities. This allows to limit costs and time, focusing on developing and testing essential analysis tools instead of creating once again similar software or looking for new tools for particular application and learn them from scratch for this single case.

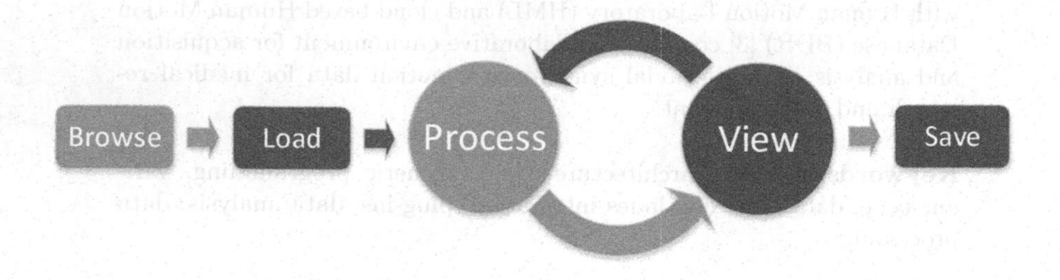

Fig. 1. Generalized data processing pipeline

1.2 Requirements for Multi-modal Data Processing Software

Before we present the architecture and functionalities of MDE software, we want to introduce basic requirements, that tools oriented on data processing should provide. Table 1 presents the most important features of such applications, guaranteeing standardization for performed experiments and developed analysis algorithms. This should lead to easier knowledge exchange between team members, allow to reuse already developed algorithms in other approaches increasing overall productivity.

Table 1. Data processing software functional requirements

Requirement	Description
Support for any data type	Various applications require using specific data types, users should be free in defining and using their custom data types
Unified time management for time indexed data	Shifting, scaling, splitting or merging in time domain should be provided ensuring their efficiency
Standardized and efficient data management	Loading large data sets requires careful memory management and fast data access
Generalized data loading and normalization	Data from various sources (containers) must be extracted and converted before analysis
Support data viewing	Data can be viewed at different perspectives nearby other data types and instances
Expandable with user solutions	Software must be flexible for new, user specific functionalities
Utilization of available computing resources	Efficient data processing should use all available computing resources to limit time required for various tests and experiments
Simple data processing pipelines composition	Tool should provide easy mechanism for creating complex processing pipelines and allow to reuse them in the future or store as an independent processing components

2 Motion Data Editor

Motion Data Editor (MDE) is a software developed at Polish-Japanese Institute of Information Technology (PJWSTK) in Bytom (Poland). Originally it was designed to support clinicans in viewing medical data for diagnosis of various human movement disorders. It has been re-organised and re-factored to become a general purpose data processing software.

Architecture. Architecture of MDE is designed to support data processing pipeline presented in Figure 1. Additionally, proposed functionality and logic realise most common operations during data analysis. Figure 2 presents an overview of MDE architecture.

Introduced data types were proposed to create an abstraction layer for uniform data management for strongly typed C++ programming language [5], standardize time index data and system specific operations.

Data Types. Among various data types especially two should be described in details:

- ObjectWrapper,
- DataChannel.

ObjectWrapper type is a generalised approach to variant type in C++.
DataChannel is a generic approach to data indexed with time.

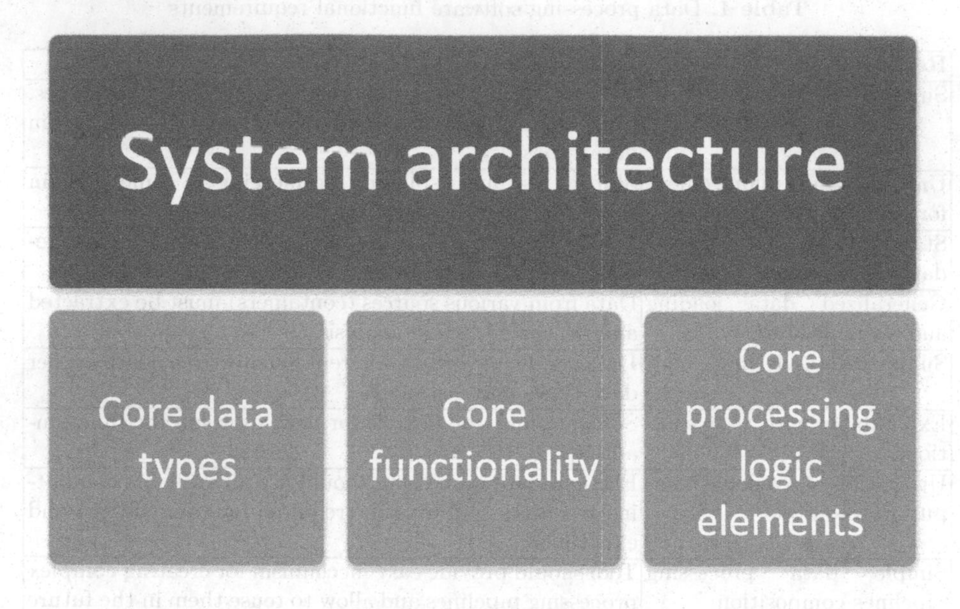

Fig. 2. MDE architecture

ObjectWrapper. To provide uniform data management in strongly typed C++ programming language a dedicated *ObjectWrapper* type has been proposed. It is based on generic programming [18] and *RTTI* mechanism (*typeid* and *type_info*) offering runtime type information about encapsulate data with its hierarchy. To make *ObjectWrapper* applicable for any valid data type in C++ it is based on policies [1,11,10], allowing to customize pointer type storing the data, cloning functionality and for the inherited types their hierarchy information. It simplifies querying data about particular types. Additionally *ObjectWrapper* offers the functionality of lazy initialization, acquiring resources for encapsulated data on data extraction and meta-data information. The whole application architecture is based on *ObjectWrapper* functionality.

DataChannel. Dedicated type was designed for uniform time indexed data management. *DataChannel* provides functionalities for treating discrete time data as continuous in time domain by introduction of specialized interpolation methods. It also allows to extend data with time property, saving memory when data is used in different time contexts. Data access according to time is optimized for channels with equally spread time samples. Moreover, when access to data for time values outside of the *DataChannel* is required, it is possible to use various built-in extrapolation techniques:

- **exception.** An exception is thrown on querying time index values outside of the *DataChannel* time range;
- **border copy.** First or last values are returned, depending on queried time index value;
- **periodic.** Queried time index value is truncated to mimic data periodic behaviour.

2.1 Functionalities

MDE provides many built-in functionalities supporting data processing and analysis procedure. They offer out-of-the box solutions for common operations, usually implemented from scratch in each dedicated software fit for particular need. Among many features, several are worth to be mentioned briefly:

- threads management with thread pool concept [9];
- optimal computing resources utilization with job manager approach [19];
- system status logging through dedicated hierarchical log mechanism;
- standardized file system operations;
- dedicated plug-in system [6];
- standardized UI [2].

All those components introduce an universal abstraction level for operating specific operations, making MDE software a cross-platform tool.

2.2 Processing Logic Elements

Based on presented data types and functionalities the architecture of MDE was decomposed to independent elements supporting particular stages of discussed data processing pipeline. The main goal was to provide a flexible system, easily expandable with users' custom functionalities and data types, handling any kind of data source in an uniform manner. Figure 3 presents five processing logic elements: Parser, Visualizer, DataSource, Service and Plug-in.

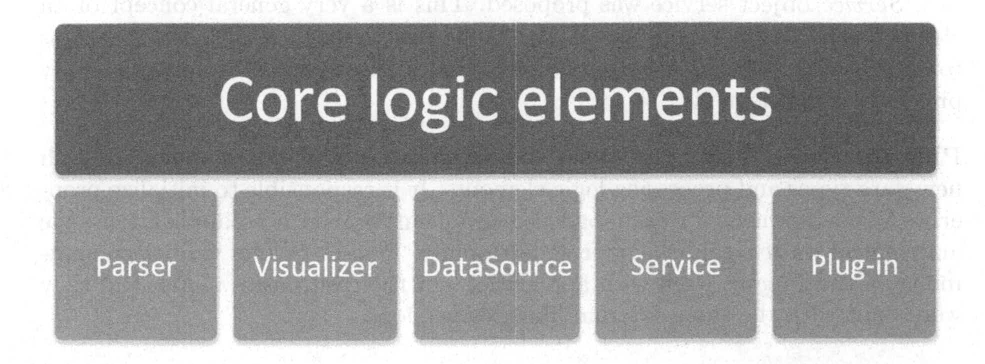

Fig. 3. Processing logic elements

DataSource. To allow users to load data from various sources a dedicated element *DataSource* was introduced. It is responsible for connecting to the source and browsing the content of the source with its specific perspective to present user data for potential analysis. Once user chooses the data of interest, *DataSource* delivers the data in its specific container, unpack the container, finally load the converted and normalised data to application. As most of the data are stored in form of files or provided with some kind of streams, MDE offers a dedicated mechanisms for automated handling such types of containers. They are based on functionalities of *Parsers.*

Parsers. *Parser* objects are propose to extract particular data types from a given file format or stream. Sometimes only a portion of data stored in a container is required, therefore it might be more efficient to skip other data knowing the container format structure and properties. As an example one can point out a video file, where audio and image can be used independently, considering speech analysis and image processing. Therefore proposing two *Parsers* for video file format seems to be reasonable, giving user greater flexibility in choosing data of interest from containers. With *Parsers* user can extend MDE to handle new data containers.

Visualizers. Various data types can be observed from many perspectives. To simplify procedure of viewing the data and standardize it, *Visualizer* objects are proposed. Their main task is to manage presented data in form of data series, where user can add and remove more data to scene of the *Visualizer* to compare them visually, depending on *Visualizer* capabilities. The same data can be viewed in many *Visualizers* showing its various perspectives (i.e. 2D plot and 3D scene). With help of *Visualizers* user can introduce new data perspectives for any kind of handled data types in MDE.

Service. Although MDE provides many useful functionalities for the most common operations in general data processing pipeline, it may happen that in particular specific cases some additional and specialised tasks are repeated in this process. To allow users application extension with completely new functionalities, *Service* object service was proposed. This is a very general concept of an element capable to handle any kind of new functions in MDE, having access to almost all application resources and having highest privileges among already presented logic elements.

Plug-in. *Plug-in* system simplifies and standardize extending application with new data types and processing logic elements. It is responsible to initialize properly the environment for each loaded component, embed it to application logic and control its general behaviour. Additionally, *Plug-in* system verifies compatibility of the *Plug-in* itself with application - if their interfaces match and they were built with the same external libraries versions.

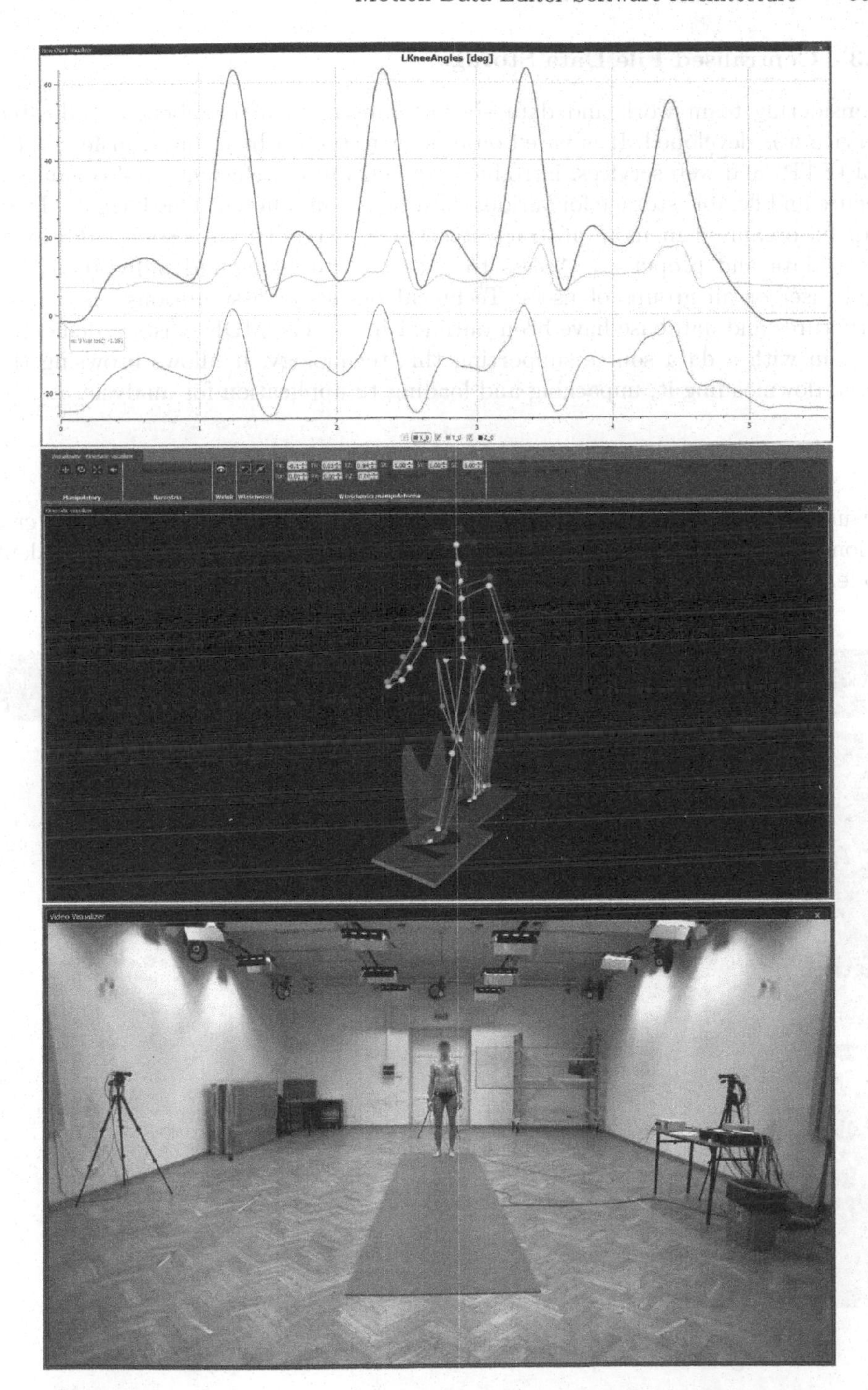

Fig. 4. Types of visualizers from MDE (from top to bottom): 2D Visualizer with motion trajectory, 3D Visualizer with kinematic and GRF data, Video Visualizer with front camera view

2.3 Centralised File Data Storage

Considering team work and data sharing among team members a dedicated service was developed. It is based on a dedicated data base, file transfer protocol (FTP) and web services, introducing a centralised, efficient, well organised, secure and flexible storage for various data stored in different files formats. Data can be organised in more abstract structures, extended with some additional meta data and properties. Access to data can be configured individually for each user or all groups of users. Technical details of assumptions, used data structures and database have been outlined in [3]. For MDE exists a dedicated plug-in with a data source supporting this technology. It allows browsing the data, downloading it, unpacking and loading to application for analysis.

2.4 Visual Programming - Data Flow Processing Framework

Visual programming concept was developed to simplify process of software creation by manipulating graphically logic blocks, instead of writing their equivalent code.

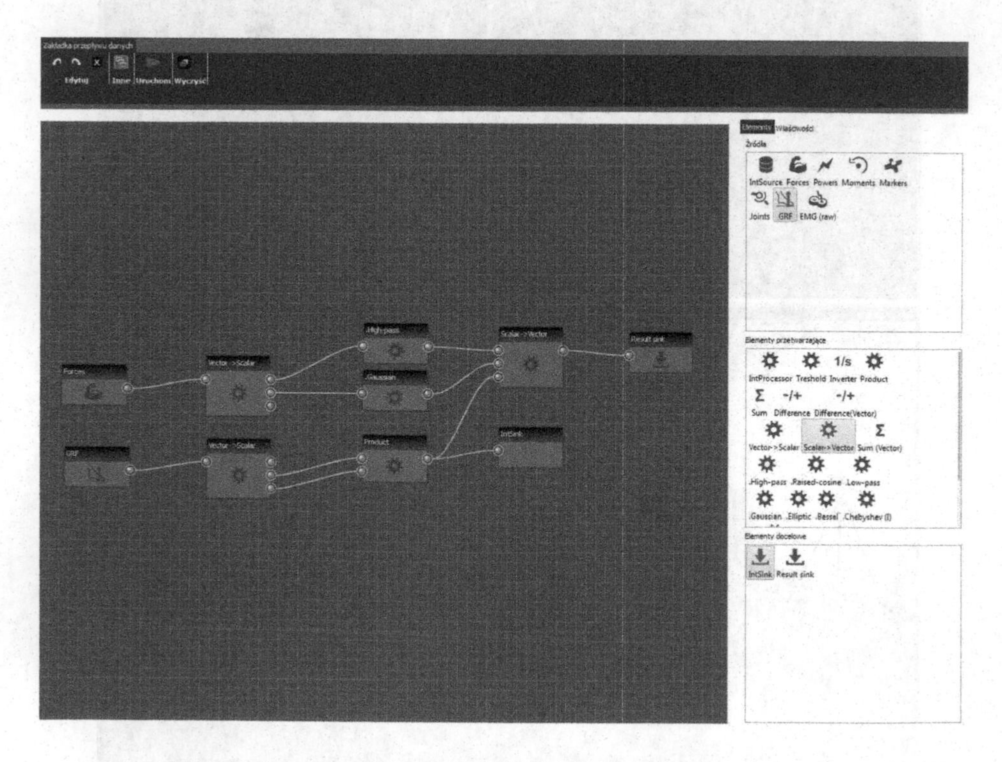

Fig. 5. An example of Visual Data Flow enviroment for MDE

As the final element of our system architecture we decided to provide users with similar visual programming environment, despite processing model and logic, to simplify and speed up process of data flow [4,7,8] creation. It allows users to dynamically create new data flows without need of writing new fragments of code and compilation procedure, based on delivered nodes functionality (see in fig.5). They can now define and launch data flows without any knowledge about programming. Visual data flow environment supports user in creating data flow by presenting graphically available nodes. It guides users, how particular nodes can be connected according to basic model rules and data types compatibility. Required and dependent pins are marked graphically with different styles to point out user places in the model, where connections are still required to make model complete. In the end any model verification failures are also presented to the user, with detailed description of elements and actions leading to fix those problems.

3 Conclusion

We see a great potential in presented modular system especially visual programming for multi-modal spatio-temporal data processing, therefore many new ideas were introduced for its possible applications. MDE is used with success for such research areas like presented in [15,16,17]. Additionally, we see a great potential in applying MDE for work presented in [12,13,14]. Two major solutions cover topics of utilizing GPU computational power and scheduling and distributing work in clusters environment. Because of modular structure, and good separation from external libraries, we are going to fast migrate our system into cloud based scalable environment for multi-modal big data processing, with clients based on mobile solutions.

Acknowledgments. Thanks to Piotr Gwiazdowski, Mariusz Szynalik, Marek Daniluk, Michał Szafarski, Michał Sachajko, Norbert Bastek, Krzysztof Kaczmarski, Piotr Habela, Wiktor Filipowicz and many others for help on this work. This work was supported by projects NN 518289240 and NN 516475740 from the Polish National Science Centre.

References

1. Alexandrescu, A.: Modern C++ Design: Generic Programming and Design Patterns Applied, 1st edn. Addison-Wesley (February 2001)
2. Blanchette, J., Summerfield, M.: C++ GUI Programming with Qt 4, 2nd edn. Prentice Hall Open Source Software Development Series. Prentice Hall (February 2008)
3. Filipowicz, W., Habela, P., Kaczmarski, K., Kulbacki, M.: A generic approach to design and querying of multi-purpose human motion database. In: Bolc, L., Tadeusiewicz, R., Chmielewski, L.J., Wojciechowski, K. (eds.) ICCVG 2010, Part I. LNCS, vol. 6374, pp. 105–113. Springer, Heidelberg (2010)

4. Hennessy, J.L., Patterson, D.A.: Computer architecture: a quantitative approach. Elsevier (2012)
5. Lippman, S., Lajoie, J., Moo, B.: C++ Primer. Addison-Wesley Publishing Company (2012)
6. Reddy, M.: API Design for C++, 1st edn. Morgan Kaufmann (February 2011)
7. Shen, J.P., Lipasti, M.H.: Modern processor design: fundamentals of superscalar processors, vol. 2. McGraw-Hill Higher Education (2005)
8. Smith, J.E., Pleszkun, A.R.: Implementing precise interrupts in pipelined processors. IEEE Transactions on Computers 37(5), 562–573 (1988)
9. Smith, J.M.: Elemental Design Patterns, 1st edn. Addison-Wesley Professional (April 2012)
10. Sutter, H.: Exceptional C++: 47 Engineering Puzzles, Programming Problems, and Solutions. Addison-Wesley Professional (November 1999)
11. Sutter, H., Alexandrescu, A.: C++ Coding Standards: 101 Rules, Guidelines, and Best Practices, 1st edn. Addison-Wesley Professional (November 2004)
12. Szczęsna, A.: The multiresolution analysis of triangle surface meshes with lifting scheme. In: Gagalowicz, A., Philips, W. (eds.) MIRAGE 2007. LNCS, vol. 4418, pp. 274–282. Springer, Heidelberg (2007)
13. Szczęsna, A.: Designing lifting scheme for second generation wavelet-based multiresolution processing of irregular surface meshes. In: Proceedings of Computer Graphics and Visualization (2008)
14. Szczęsna, A.: The lifting scheme for multiresolution wavelet-based transformation of surface meshes with additional attributes. In: Bolc, L., Kulikowski, J.L., Wojciechowski, K. (eds.) ICCVG 2008. LNCS, vol. 5337, pp. 487–495. Springer, Heidelberg (2009)
15. Szczęsna, A., Słupik, J., Janiak, M.: Motion data denoising based on the quaternion lifting scheme multiresolution transform. Machine Graphics & Vision 20(3), 238–249 (2011)
16. Szczęsna, A., Słupik, J., Janiak, M.: Quaternion lifting scheme for multi-resolution wavelet-based motion analysis. In: The Seventh International Conference on Systems, ICONS 2012, pp. 223–228 (2012)
17. Szczęsna, A., Słupik, J., Janiak, M.: The smooth quaternion lifting scheme transform for multi-resolution motion analysis. In: Bolc, L., Tadeusiewicz, R., Chmielewski, L.J., Wojciechowski, K. (eds.) ICCVG 2012. LNCS, vol. 7594, pp. 657–668. Springer, Heidelberg (2012)
18. Vandevoorde, D., Josuttis, N.M.: The Complete Guide, 1st edn. Addison-Wesley Professional (November 2002)
19. Williams, A.: C++ Concurrency in Action: Practical Multithreading, 1st edn. Manning Publications (2012)

4GAIT: Synchronized MoCap, Video, GRF and EMG Datasets: Acquisition, Management and Applications

Marek Kulbacki, Jakub Segen, and Jerzy Paweł Nowacki

Polish-Japanese Institute of Information Technology,
Koszykowa 86, 02-008 Warsaw, Poland
{kulbacki,segen,nowacki}@pjwstk.edu.pl

Abstract. Presented is the 4GAIT, a group of multimodal, high quality Reference Human Motion Datasets. The described in this article Multimodal Human Motion Lab provides a comprehensive environment for multimodal data acquisition, management and analysis. Introduced is a proposed PJIIT Multimodal Human Motion Database (MHMD) model for multimodal data representation and storage, along with Motion Data Editor – a toolkit for multimodal motion data management, visualization and analysis. As an example a group of three synchronized, multimodal, motion datasets using MHMD model: 4GAIT-HM, 4GAIT-Paarkinson and 4GAIT-MIS are described.

Comparing to the currently available Human Motion Datasets, 4GAIT offers multiple data modalities presented within a unified model, higher video resolution, and larger volume of motion data for specific medical tasks, which better fulfills the needs of medical studies and human body biomechanics research.

Keywords: multimodal acquisition, reference human motion, clinical gait analysis.

1 Introduction

Human motion, or motion capture, data describes articulated human body motion in three dimensions. The usual form of the description is a time series consisting of vectors of joint angles plus root position, with respect to a defined skeleton. Human motion data is used in biomechanics, sports, human factors and ergonomics, entertainment, robotics, computer vision research, gait analysis, biometrics, human-computer interfaces, multiple subfields of medicine, such as orthopedics, orthotics, prosthetics, characterization and diagnosis of movement impairments and rehabilitation, and many other applications.

Human motion databases have been constructed by research and industrial centers to serve as reference sets for research studies or as human motion repositories for applications, such as animation in video games or film. These databases usually contain metadata such as the names and descriptions of actions, gender of actors or parameters of the acquisition configuration, and additional signal

N.T. Nguyen et al. (Eds.): ACIIDS 2014, Part II, LNAI 8398, pp. 555–564, 2014.

information associated with the motion data. The nature of such additional signal information depends on the needs of applications for which the database is intended. Audio data is useful for studies of expressions, behavior or emotions, video from single or multiple cameras is needed in computer vision research, force measurements are used in studies related to sport, biomechanics, gait, ergonomics or in rehabilitation, electromyography (EMG) signals may be required in study or diagnosis of neuromuscular diseases and movement disorders, research in prosthetics, biomechanics and robotics and in rehabilitation. If additional signal information is associated with the motion data, the issue of signal synchronization appears. This issue is not usually present in case of metadata. The additional signal information is most useful if the additional signals are synchronized with the motion data signal. The signal synchronization should be as precise as possible, since more precise synchronization leads to more accurate modeling of relations between the signals, for example the relation between the signal from a sensor that measures forces between a foot and the floor and the acceleration of a body part computed from the motion data. The signal synchronization should be provided by the data acquisition hardware.

This article describes a system for acquisition of motion data and other signals that is used in Human Motion Laboratory (HML) of the Polish Japanese Institute of Information Technology (PJIIT) in Bytom, Poland, and databases created using this system. Many publicly accessible motion databases existed before the HML system was built. The list below is representative of such motion database, although it is not exhaustive.

1. **CMU Mocap database** [1] is an extensive motion data set used in many research projects. It contains more than six hours of full body motion data in 2605 motion clips, organized into a hierarchy of activities and actions that were recorded from 144 subjects. The activities include single person motions and interactions with environment, and two people interactions.
2. **CMU Multi-Modal Activity Database (CMU-MMAC)** [2] contains multimodal measurements of subjects performing tasks related to cooking in a specially built kitchen. The database contains associated video with maximum resolution 1024 x 768, audio, and signals from accelerometers attached to the body. The main dataset consists of data from 43 subjects cooking five recipes, the anomalous dataset consists of data from three subjects cooking five recipes.
3. **CMU Motion of Body (MoBo) database** [3] is focused on human gait. It contains four walking actions, described as slow, fast, inclined, and carrying a ball, which are performed on a treadmill by 25 subjects. The data contains only video recorded from six cameras placed around the subject. The main goal of this dataset was research on biometric identification of humans from their gait characteristics.
4. **IEMOCAP Database** [4] from University of Southern California is oriented towards human communication and expression of emotions. Its motion data is recorded from face, head, hands and torso. The dataset contains a

large number of emotions performed by ten trained subjects. The motion data is associated with audio recordings.
5. **HumanEva Database** [5] main objective is to provide ground-truth data computer vision research, specifically, for testing and evaluation of pose estimation and motion tracking methods. Associated signal data are video sequences synchronized with the motion capture data. The video resolution is relatively low, the highest resolution is 659x694. The dataset contains six actions performed by four subjects wearing natural clothing.
6. **HDM05 MoCap Database** at the Hochschule der Medien [6] consists of 1457 motion clips that represent 100 action classes, performed by five actors. The set for each action class contains between 10 and 50 clips. The goal of the dataset was research on analysis, synthesis and classification of motion.
7. **SMILE lab Human Motion Database** [7] consists of five subsets: the praxicon containing around 350 actions of the same subject, the cross-validation dataset of 70 actions performed by 50 subjects, the generalization dataset that contains samples of nine action classes each sample of the same action performing it differently, the compositionality dataset containing complex actions each composed of two or three simple actions, and the interaction dataset consisting of 150 actions performed by two interacting subjects. The goal of the database is research on analysis and synthesis of motion.
8. **Korea University Gesture (KUG) database** [8] contains actions performed by 20 actors of different gender and age. The set of actions contains 14 typical, normal motions such as sitting on a chair and walking, 10 atypical actions such as falling, and 30 gestures representing answers or instructions such as yes, no, pointing, or selecting a number. The goal of the database was to serve as reference in research on gesture recognition and people tracking. The motion data is associated and synchronized with video data from multiple stereoscopic cameras looking at the action from different directions.
9. **Human Identification at Distance (HID) database** [9] at Georgia Tech contains gait data recorded from 20 subjects indoor and outdoor. The goal of the database is gait recognition and identification of people from characteristics of their walk. The motion data is associated with video data recorded from different distances and angles.
10. **ICS Action Database** [10] at the University of Tokyo contains 25 action classes with five motion samples representing each class. The data samples are annotated on a per-frame basis, such that a frame can be a part of more than one action class. The goal of the database is to serve as reference for research on human action segmentation and recognition.

The creation of Human Motion Laboratory at PJIIT and the subsequent activity in building human motion databases was motivated by a plan to work along multiple aspects of human motion, collaborating with other institutions in multidisciplinary groups in three main directions: medical, entertainment, and research in computer vision, biometrics, and motion analysis and synthesis. This objective imposed a tough set of demands on the acquisition system. The vision research required video data of high quality synchronized with motion data, the medical research needed additional measurement modalities such as force and EMG, also synchronized with motion. The entertainment applications, such as animation for videogames, demanded highest quality motion capture data. None of the available databases possessed the required combination of modalities or the quality of video and motion capture that was possible to achieve at the time when the project has begun, and the obvious solution was to build a new lab – the HML described in the next section.

2 Laboratory

A multimodal laboratory for motion analysis – Human Motion Lab allows for simultaneous measurement of a number of patient motion parameters. It enables acquiring of motion data through simultaneous and synchronous measurement and recording of motion kinematics, muscle potentials, ground reaction forces and video streams. Owing to that there occurs the opportunity for spatio-temporal correlation between the video sequences, values of angles, forces, moments of forces, powers in selected joints (one or a few), potentials of muscles determining their activity and ground reaction forces.

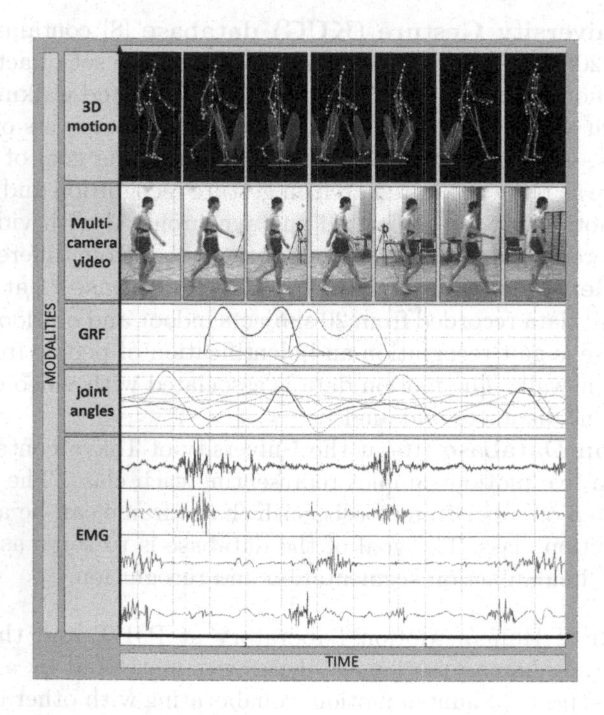

Fig. 1. An example of trial from dataset 4GAIT-HM. Simultaneous visualization of data from mocap, video stream, ground reaction forces, trajectories of joint angles and electromyographic signals with visible contractions of muscles.

HML takes measurements used in orthopedics and rehabilitation and cooperates with the major medical centers in Poland, developing the laboratory's research methods and work techniques.

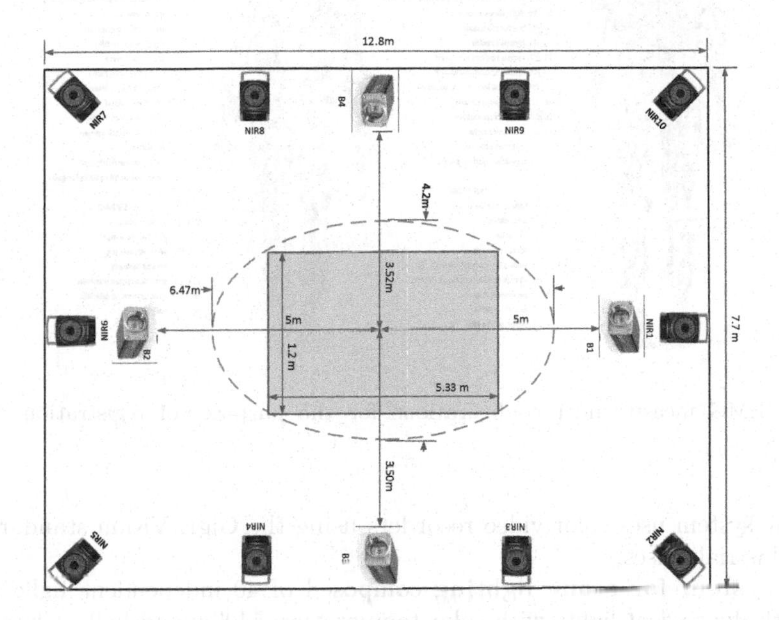

Fig. 2. Camera setup for 4GAIT data acquisition

Basic equipment of the laboratory includes:

- **Vicon's Motion Kinematics Acquisition and Analysis System** equipped with 10 NIR cameras with the acquisition speed of 30 to 2000 frames per second at full frame resolution of 4 megapixels and 8-bit grayscale.
- **Noraxon's Dynamic Electromyography (EMG) System** allowing for 16-channel measurement of muscle potentials with non-gel electrodes in compliance with the SENIAM guidelines. For the purposes of registration 4GAIT datasets containing EMG , have been defined four measurement configurations (see fig. 3).
- **Kistler's Ground Reaction Force (GRF) Measurement System** used for measuring ground reaction forces with two dynamometric platforms with measurement ranges adjusted to gait analysis research. The system measures forces with equal accuracy on the entire surface of the platform in a measurement range not smaller than 5 times the body of an adult person for dynamic function research. The system has a 6-meter path masking two platforms situated in the middle of its length.
- **A system for simultaneous multi-camera video image recording equipped with Basler's cameras** that allows for simultaneous image recording from all the cameras in Full HD and lossless video recording.

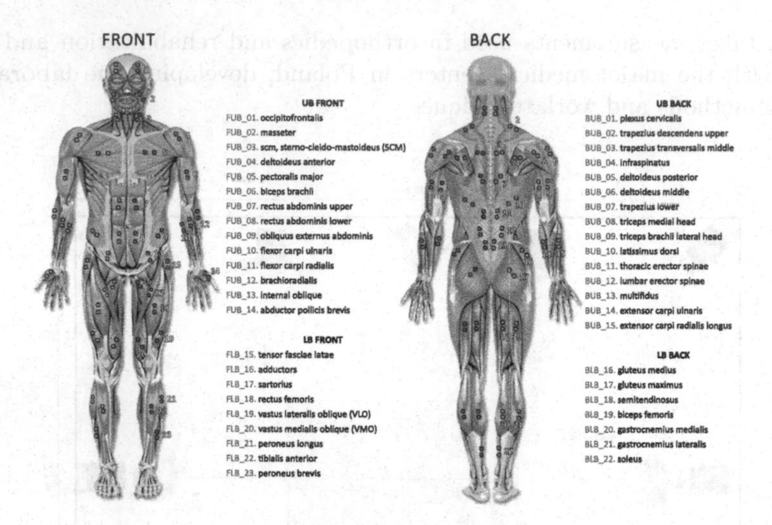

Fig. 3. EMG measurement configurations for the purposes of registration 4GAIT datasets

The system uses color video recorders using the GigE Vision standard and industrial lenses.

- **A system for scene lighting** composed of 40 independent light points with fluorescent lights with color temperature 5400K and ballast frequency > 35000 Hz.

Hardware Synchronization
MX-Giganet Lab is responsible for hardware synchronisation during data acquisition from base systems. Our MX-Giganet Lab is equipped with Analog ADC Option Card for connecting up to 64 analog channels in simple Vicon MX T-series system with a single primary MX Giganet.The analog ADC card is a 64-channel device for generating 16-bit offset binary conversions from analog sources.The maximum rate at which we can sample data via the ADC card is 192000 samples/second (192 kHz) for one channel configuration and 3000 samples/second (3kHz) for 64 channels configuration. So for simultaneous data acquisition from GRF and EMG, sample rate is equal to 3 kHz. To incorporate a Basler GigE camera in a T-Series system for hardware based synchronization of simultaneous data acquisition our four Basler GigE Cameras are connected by four Powered Sync Outputs in MX Giganet Lab.

3 Databases

The PJIIT Multimodal Human Motion Database (MHMD) consists various datasets of time synchronized and calibrated data from: optical high resolution motion capture system, high definition cameras from multiple views, two dynamometric platforms and dynamic electromiography system. In this paper we focus on 4GAIT type datasets, each containing a human gait data with registered up to four types of modalities at the same time. Each presented dataset has the same uniform structure with the following hierarchy **Subject** → **Session** → **Trial** → **Segment**. Subject identifies a subject of motion. Session describes measurement sessions for a given subject. Trial identifies trials which may be done by a subject during a single session. Each instance of multimodal trial is composed of one C3D file containing (kinematic and kinetic motion parameters) up to 4 video streams and configuration files. Segment describes a part of the trial sequence distinguished based on some criteria of interest. At each level of this hierarchy it should be possible to assign instances to categories and to equip them with custom attributes. Technical details of assumptions, used data structures and database have been outlined in [11].

Table 1. The summary and origin of presented datasets

Dataset Name	Modalities	Summary of the project
4GAIT-MIS	4 video streams (25 fps, 1920x1080), motion capture (100 fps)	Novel commercializable technology and industrial grade software for a scalable cloud-based video surveillance system with advanced video analysis functions that include object or person tracking over extended range and time, identification of people, activities and behavior, and automatic learning.
4GAIT-Parkinson	4 video streams (25 fps, 1920x1080), motion capture(100 fps), EMG(EMG_LB_3), GRF	Diagnosis of Parkinson's disease, the correlation of UPDRS with motion descriptors. Evaluation of the drug influence and stimulation.
4GAIT-HM	4 video streams (25 fps, 1920x1080), motion capture(100fps), EMG (EMG_UB_1, EMG_LB_3, EMG_FB_1), GRF	System with a library of modules for advanced analysis and an interactive synthesis of human motion.

4GAIT-MIS. The main goal of this dataset is research on biometric reidentification of humans from their gait characteristics.The dataset contains walking actions which are performed by 33 subjects aged 25-35 years. The actors are dressed in skin-tight red, green, orange or blue T-shirts and black mocap trousers. Each new recording session is preceded by an empty scene. Video cameras are positioned at the same height in the greatest possible distance from the center of the stage (see fig. 1).

4GAIT-Parkinson. The main goal of this dataset is research on gait recognition and identification of PD patients symptoms and stage of the disease progression. In the future multimodal tests should lead to automation of patients'

Fig. 4. An example of trial from dataset 4GAIT-MIS. Simultaneous visualization of data from mocap and four video streams.

disease stage assesment without doctor participation and replace subjective UPDRS measure by "golden neurologist". Participants performed 12 tasks under four experimental conditions called sessions. Medication ON/OFF means that patient was with/without drugs during the session. Stimulation ON/OFF means that patient was with stimulator turned on/turned off during the particular session. Each session groups the following tasks: sway, gait/tandem gait, turnover, walk at normal/fast speed, heel to toe walk in a straight line, sensomotoric test related to tracking light spots, pulling back test, arizing from chair and leg agility test. During the sway task participants stood motionless on force platform with hands lowered at both sides and feet together, once with eyes open (EO), once closed, (EC). The experiment has been performed two times for EO and EC. Duration of one test is 30 s; between attempts was a rest period. The scope of interest has been marked on the paper covering the platform, to ensure the same position between successive attempts. In the case of EO study, patients looked at the sign "x", located 1.6 m above ground and 2 m in front of them. This sign was surrounded by a black circle with a radius of 18 cm. The final dataset contains 1781 trials (including 803 trials of gait) grouped in 12 tasks which were performed by 18 subjects - 14 males and 4 females.

4GAIT-HM. The main goal of this dataset was to investigate the relationship between the kinematics and kinetics of 34 patients with diagnosed coxarthrosis (10 subjects), knee arthrosis, degenerative spinal disorders (19 subjects), stroke (4 subjects) and arthritis (1 subject). The final dataset exceed 1144 trials of gait which were performed by 23 males and 11 females.

4 Applications

Datasets 4GAIT presented in this paper, still growing. The frequent and regular measurements are collected and organized in the cloud based **Human Motion Database**. A client application software, **Motion Data Editor (MDE)**, enables access to datasets in natural way, visualization and processing of data from any number of multimodal measurements simultaneously and synchronously. It supports dozens of industrial formats used for medical data storage. The application has modular structure, that enables simple extensions of its capabilities with a dedicated plugin system. MDE uses an intuitive data flow approach involving data sources, processing modules and data visualization sinks. Processing modules perform appropriate data manipulations such as filtering and transforming data. MDE allows the user to easily swap and extend modules. The developed system demonstrates features that benefit medical informatics applications. It provides centralized storage for medical data and associated descriptions, allows users to exchange data, giving them a possibility to consult and discuss specific cases. Users are able to filter the data and manage filters themselves. This stage supports browsing, viewing, processing and comparing data records. More information can be found at http://hm.pjwstk.edu.pl/mde. During last 3 years the research on motion analysis have been conducted. Eventually one group of researchers utilized and second group extended existing datasets. In [12] the 4GAIT-HM dataset was extended by 90 gaits coming from 15 different patients – 7 of them with diagnosed coxarthrosis on the basis of an earlier different examinations and 8 of them without coxarthrosis. This dataset has been extensively used by [13], [14], [15], [16] for gait based re-identification. At the same time 4GAIT-MIS has been used by 3D inference research in [17], [19] and multiscale processing performance for motion capture in [18].

Acknowledgments. Thanks to Andrzej Przybyszewski, Dawid Pisko, Kamil Wereszczyński, and many others for help on this work. This work was supported by projects NN 518289240 and UMO-2011/01/B/ST6/06988 from the Polish National Science Centre.

References

1. CMU Mocap Database (2001), `http://mocap.cs.cmu.edu`
2. Tech. report CMU-RI-TR-08-22, Robotics Institute, Carnegie Mellon University (July 2009), `http://kitchen.cs.cmu.edu/index.php`
3. Gross, R., Shi, J.: The CMU motion of body MOBO database. Technical report, CMU (2001)
4. Busso, C., Bulut, M., Lee, C.-C., Kazemzadeh, A., Mower, E., Kim, S., Chang, J., Lee, S., Narayanan, S.: IEMOCAP: Interactive emotional dyadic motion capture database. Journal of Language Resources and Evaluation 42(4), 335–359 (2008)
5. Sigal, L., Black, M.: HumanEva: Synchronized video and motion capture dataset for evaluation of articulated human motion. Technical Report CS-06-08, Brown Univ. (2006)

6. Müller, M., Röder, T., Clausen, M., Eberhardt, B., Krüger, B., Weber, A.: Documentation: Mocap database HDM05. Technical report CG-2007-2, Universitt Bonn (2007)
7. Guerra-Filho, G., Biswas, A.: The human motion database: A cognitive and parametric sampling of human motion. Image and Vision Computing 30(3), 251–261 (2012)
8. Hwang, B.-W., Kim, S., Lee, S.-W.: A full-body gesture database for automatic gesture recognition. In: Proc. of Int. Conf. on Automatic Face and Gesture Recognition, pp. 243–248 (2006)
9. Georgia Tech Human Identification at Distance Database, `http://www.cc.gatech.edu/cpl/projects/hid/`
10. ICS Action Database, The University of Tokyo (2003-2009), `http://www.ics.t.u-tokyo.ac.jp/action/`
11. Filipowicz, W., Habela, P., Kaczmarski, K., Kulbacki, M.: A Generic Approach to Design and Querying of Multi-purpose Human Motion Database. In: Bolc, L., Tadeusiewicz, R., Chmielewski, L.J., Wojciechowski, K. (eds.) ICCVG 2010, Part I. LNCS, vol. 6374, pp. 105–113. Springer, Heidelberg (2010)
12. Świtoński, A., Stawarz, M., Boczarska-Jedynak, M., Sieroń, A., Polański, A., Wojciechowski, K.: The effectiveness of applied treatment in Parkinson disease based on feature selection of motion activities. Electrical Review 88, 103–111 (2012)
13. Świtoński, A., Mucha, R., Danowski, D., Mucha, M., Polański, A., Cieślar, G., Wojciechowski, K., Sieroń, A.: Diagnosis of the motion pathologies based on a reduced kinematical data of a gait. Electrical Review 87(12b) (2011) ISSN 0033-2097
14. Świtoński, A., Polański, A., Wojciechowski, K.: Human Identification Based on the Reduced Kinematic Data of the Gait. In: 7th International Symposium on Image and Signal Processing and Analysis (2011)
15. Świtoński, A., Polański, A., Wojciechowski, K.: Human identification based on gait paths. In: Blanc-Talon, J., Kleihorst, R., Philips, W., Popescu, D., Scheunders, P. (eds.) ACIVS 2011. LNCS, vol. 6915, pp. 531–542. Springer, Heidelberg (2011)
16. Świtoński, A., Mucha, R., Danowski, D., Mucha, M., Cieślar, G., Wojciechowski, K., Sieroń, A.: Human identification based on a kinematical data of a gait. Electrical Review (12b), 169–172 (2011)
17. Krzeszowski, T., Kwolek, B., Wojciechowski, K.: Model-based 3D human motion capture using global-local particle swarm optimizations. In: Burduk, R., Kurzyński, M., Woźniak, M., Żołnierek, A. (eds.) Computer Recognition Systems 4. AISC, vol. 95, pp. 297–306. Springer, Heidelberg (2011)
18. Jabłoński, B., Kulbacki, M.: Multiscale Processing Performance for Motion Capture. MGV 20(3), 251–266 (2011)
19. Kwolek, B., Krzeszowski, T., Wojciechowski, K.: Real-time multi-view human motion tracking using 3D model and latency tolerant parallel particle swarm optimization. In: Gagalowicz, A., Philips, W. (eds.) MIRAGE 2011. LNCS, vol. 6930, pp. 169–180. Springer, Heidelberg (2011)

VMASS: Massive Dataset of Multi-camera Video for Learning, Classification and Recognition of Human Actions

Marek Kulbacki, Jakub Segen, Kamil Wereszczyński, and Adam Gudyś

Polish-Japanese Institute of Information Technology,
Koszykowa 86, 02-008 Warsaw, Poland
{mk,js,kw,agudys}@pjwstk.edu.pl

Abstract. Expansion of capabilities of intelligent surveillance systems and research in human motion analysis requires massive amounts of video data for training of learning methods and classifiers and for testing the solutions under realistic conditions. While there are many publicly available video sequences which are meant for training and testing, the existing video datasets are not adequate for real world problems, due to low realism of scenes and acted out human behaviors, relatively small sizes of datasets, low resolution and sometimes low quality of video.

This article presents **VMASS**, a dataset of large volume high definition video sequences, which is continuously updated by data acquisition from multiple cameras monitoring urban areas of high activity. The VMASS dataset is described along with the acqusition and continuous updating processes and compared to other available video datasets of similar purpose. Also described is the sequence annotation process. The amount of video data collected so far exceeds 4000 hours, 540 million frames and 2 million recorded events, with 3500 events annotated manually using about 150 event types.

1 Introduction

The intelligent event recognition is one of the most important issues of computer vision [2–7]. To construct Intelligent Video Analytics systems massive video datasets with annotations are necessary, as training and test data for machine learning processes. The following criteria can help to assess usefulness of a video dataset towards a particular task: realism, quality described by multiple parameters, variety of stored events and actions, actors, scenes and external conditions, and the annotation method.

Realism of dataset is a property of recording not pre-arranged events in a not pre-arranged scene, where subjects are incidental and not conscious of being recorded. This property relates to the degree of how well the recorded data represents the real world. System trained on non-realistic dataset usually fail in the real world. Few existing datasets satisfy this criterion, since most rely on a small number of actors performing pre-arranged actions [1, 4] or use movie scenes [2]. The datasets that possess this property are [5], CAVIAR [8] and VIRAT [3].

N.T. Nguyen et al. (Eds.): ACIIDS 2014, Part II, LNAI 8398, pp. 565–574, 2014.

Quality of a video dataset can be represented numerically by parameters: image resolution, image or video compression level and the number of frames per second (FPS). Within the bounds of present technology, a good quality video dataset will have Full HD resolution (1920x1080), compression corresponding to JPEG 10-20% level without motion compensation and at least 18 FPS. An example of a good quality video dataset is VIRAT [3].

Variety of conditions relates to how broad are the ranges of conditions or parameters within which the system operates or represented objects. It includes the quantity of events or actions, the number of actors and scenes (in background object appearance and activity aspects), the external conditions such as as season, time of day, weather, illumination or stability of the background. Lack of variety in existing datasets is usually represented by short recording time, small number of pre-arranged scenes and actions performed by few actors, or clear weather and good illumination.

Annotation method effects the usefulness of the dataset as training and test data. Multiple types of annotations are needed: (1) Annotation of time segments; (2) Annotation of objects as used by by KTH [1], Weitzmann [4] or TRECVID [9]; (3) Multiple annotations of objects as in Virat [3]; (4) Hierarchical annotations, for example an action divided into smaller structures (the authors don't know of its use before **VMASS**).

Calibration of a view in a video dataset, or camera calibration, concerns the information that makes possible to relate geometrically different views, different cameras and assess true object sizes. While calibration may not be necessary for a dataset with a single view, such as most of the existing datasets, it becomes critical with video segments from different cameras or multiple views. Obtaining the calibration information under minimal operator's involvement is not trivial, but even more difficult is a continuous computing of the calibration parameters with cameras in motion. **VMASS** contains calibration parameters for every view, that are computed in real time during the acquisition.

Size of a video dataset can be expressed by several parameters such as: (1) size of data (GB); (2) length of the recording; (3) number of frames. Existing available datasets range from several to under 100 hours, while **VMASS** contains more than 4.000h of video data.

2 VMASS Dataset

The Polish-Japanese Institute of Information Technology (PJIIT) in its Bytom branch conducts research in the areas of human motion analysis and video based activity recognition. As most projects are based on machine learning techniques, large amounts of video data are needed for the training and testing. The VMASS system and the acquired massive video dataset were build to fulfill this need. The following subsections describe properties of the VMASS system within the guidelines of the criteria introduced in Section 1.

2.1 Realism

The cameras used for vision data acquisition were located in the center of Bytom, Poland. The acquisition system collects video data, outdoor, recording real situations without any control of the scene or events. This classifies it as a high Realism system according to the criteria described in Section 1.

Fig. 1. A typical VMASS video frame as an example for data realism

2.2 Quality

The four video cameras used for the acquisition of video dataset are High Definition IP auto-dome cameras made by Axis Communications, model Q6035E. Technical details of this equipment are:

- Image sensor resolution: p1080 (1920 x 1080)
- Zoom scale: 1x - 20x
- Pan / Tilt position setting precision: about 0.01 deg.
- Frame rate in various resolutions: 1920x1080 MJPEG: 25fps; 1024 x 768: 50 fps; 640 x 480: 50 fps.

An important part of acquisition process is the efficient delivery of video data to a storage center. This task is being realized by optical and 6th category level

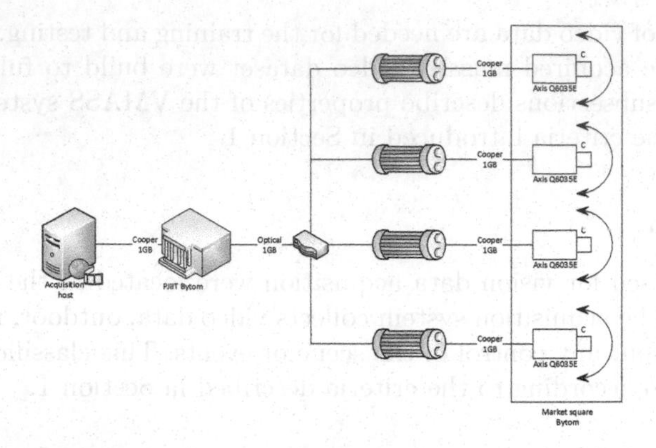

Fig. 2. VMASS video acquisition network

copper network connecting all cameras with CVL with guaranteed bandwidth on level 1GBps.

The quality of the acquired data depends on image resolution, frame-rate and inversely on the compression level. The video parameters of the recorded sequences are: resolution 1920 x 1080, frame-rate 18-27 fps, JPEG compression on 10-13%, video compression: no compression of motion areas (see detailed description in next section). Six to eight hours of video per day are recorded from each camera.

2.3 Variety of Conditions

Variety of conditions in the VMASS system is illustrated by examples in Figure 3.

Fig. 3. Variety of conditions in the VMASS

The range of conditions used in the VMASS dataset is shown in Table 1.

Table 1. Variety of conditions in the VMASS dataset

Condition type	**Range**	**Examples**
event	1 – 150	walk, talking, bicycle, run, run away, photograph, play, sit, downfall, get out of building, carry luggage, waiting, smoke, phone,...
actor	1 – 40	man, woman, pair, threesome, group, crowd, dog, luggage, car, truck, excavator, ...
scene	1 – 52	market square in bird-eye like view, church entrance, supermarket entrance, fountain, open-air cafe, road crossing, trees, street, building entrance, ...
External conditions		
seasons	1 – 4	spring, summer, fall, winter
day times	daylight	morning, noon, evening
weather	any	sunny, windy snowy, rainy, cloudy (many level of cloud cover),...
illumination	visible	depending on weather conditions, daytime, season and so on.
background stability	0 - moderate motion	stable,camera in motion, appearing and disappearing objects in background, birds, leafs, raindrops, etc existing in background.

2.4 Annotation Method

The annotation method of VMASS can be described as Hierarchical, Parallel, Multidimensional and Flexible (HPMF).

Hierarchical means that each annotation can have sub-annotations and each of them sub-sub-annotations and so on (e.g. event "meeting" can have sub-annotations: "greetings", "chat", "walk away" and "greetings" could have sub-annotations: "waving", "handshake", "kiss"). This annotation tree could be as deep as needed.

Parallel means that on the same time and the same object different set of annotations can be defined. (e.g. given person could attend an event "meeting" with person A and event "jostle" with person B; in the same time other two persons can attend event "play"). It means also, that one event could contain more than one tag of each dimension.

Multidimensional means that one event/action could be annotated in several perspectives (called dimensions) with different tags. For each dimension there is defined other tag set. Such annotations are 4-dimensional (action, behavior, actor and scene).

Flexible means that depth of annotation hierarchy is unlimited, dimension count and tags dictionary for each dimension can be modified.

A software system for adding the annotations to video sequences called Annotation Editor has been developed at PJIIT Bytom. Figure 4 shows a snapshot of the Annotation Editor window.

Fig. 4. Screenshots from acquisition application

On the left side of the screen starting from the top:time-line (white); Annotations (blue bars mean not selected and red bar means selected); Sub-annotation of selected annotation - green bars below selected activity.

Rectangular windows: Red window - Selected motion areas (could more then one); Blue window - motion areas not connected with selected activity; Green window - motion areas connected with selected activity.

2.5 Comparison to Other Datasets

Results of comparison between VMASS and other published datasets are summarized in Tables 2, 3, 4 and 5.

Table 2. Realism

	KTH	Wiezmann	HOHA 1	TRECVID	VIRAT	VMASS
incidental events, scenes, actors	No	No	Partial	Yes	Yes	Yes
realistic environment	No	No	No/Partial	Partial	Partial	Yes

Table 3. Quality

	KTH	Wiezmann	HOHA 1	TRECVID	VIRAT	VMASS
resolution (wxh)	160x120	180x144	540x240	720x576	1920x1080	1920x1080
frame rate	N/A	N/A	N/A	N/A	N/A	18-27 fps
frame compression level	N/A	N/A	N/A	N/A	N/A	10-13% (JPEG)
video compression	N/A	N/A	N/A	N/A	N/A	lossless MJPEG
zoom	fixed	fixed	fixed	fixed	N/A	1-20 (optical)
actors height in pixels	80-100	60-70	100-120	20-200	20-180	20-1000

Table 4. Variety of conditions

	KTH	Wiezmann	HOHA 1	TRECVID	VIRAT	VMASS
count of events and action types	6	10	8	10	23	150(e), 400(a)
avg. count of samples per class	100	9	85	3-1670	10-1500	5-1200
count of scene types	N/A	N/A	many	5	17	over 50
seasons	N/A	N/A	N/A	N/A	N/A	4
weather	N/A	N/A	N/A	N/A	N/A	many
illumination	N/A	N/A	N/A	N/A	N/A	many
background stability	Yes	Yes	cam. motion	Yes	cam. motion	varying[1]

Table 5. Annotation system and data size

	KTH	Wiezmann	HOHA 1	TRECVID	VIRAT	VMASS
annotation type	1 per obj	1 per obj	1 per obj	1 per obj	multiple	HPMF
dataset size [h]	few seq.	few seq.	few seq.	few seq.	28	4.000
dataset size [million frames]	N/A	N/A	N/A	N/A	2	540
approximate recorded events count [thousands]	0.6	0.09	0.68	10	23	3.000

[1] Stable, camera motion, appearing and disappearing objects in background, birds, leafs, raindrops, etc existing in background.

3 Overview of Implementation of the VMASS System

The main parts of the data collection process are video acquisition, compression, camera view registration, storage and annotation. The first three parts of the process are automatic, the fourth part - annotation is manual. A video stream acquired from a camera is a sequence of JPEG compressed images. Each frame is additionally compressed using a method of object selective compression, and stored in a compressed form. Each distinct camera view is registered with a map with a reference ground plane. To add annotations to the sequence, the stored video scenes are retrieved and processed manually using the Annotation Editor. The object selective compression method and the view registration are described in the following subsections.

3.1 Object Selective Compression

Object selective compression allows the system to store moving objects at higher accuracy and non-changing background at a lower accuracy. For each frame the regions containing moving objects are stored as new information while the background representation is taken from the previous frame, disregarding minor background changes. The moving objects are preserved at their acquisition quality while the background is stored less accurately since minor changes in the background are omitted, which results in a high degree of compression. The process of object selective compression consist of: (1) Identifying background and storing it as B; (2) Detecting motion rectangles for each frame; they will be called *crumbles* and marked with a letter C; (3) Storing the sequence as: $BCCCCCBCCCB$; (4) Rebuilding a frame on demand as $B_l + C_i$ where i is the index of demanded frame and B_l is last B before C_i.

This process uses the background subtraction method based on Adaptive Gaussian Mixture Model proposed by P. KaewTraKulPong and R. Bowden [12] and Z. Zivkovic [13]. These models focus on one pixel color change between consecutive frames and basing on this change try to estimate most probable background color of this point using an idea that this color is the one which stay

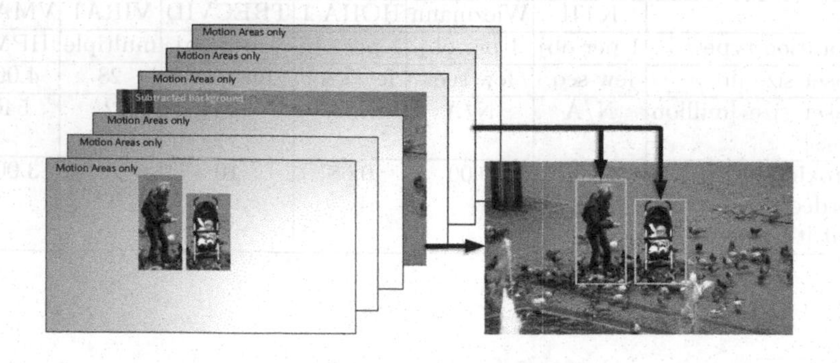

Fig. 5. Object selective compression

longer and more static on given pixel. Once background is known, silhouettes are retrieved, basing on observation differences between background and given frame. For eliminated fragmentation of retrieved object, mixture of filtering and thresholding methods are used. When silhouettes are retrieved for each frame, motion areas could be discovered, basing on calculating motion history, described by Bradsky and Davis in 2002 in [14].

3.2 Camera View Registration

The view registration, which is the mapping of the image elements that lie on the ground plane to positions on the reference frame, is done from known camera calibration parameters. A camera calibration method has been developed as a part of PJIIT-MIS project. It updates the calibration parameters when the camera pan and tilt angles change. The camera calibration is specified by two matrices: intrinsic and extrinsic.

Intrinsic Camera Parameters. Camera intrinsic parameters are: focal length f, skew s, coordinates of principal point $P = (p_u, p_v)$, scale factors in horizontal and vertical directions k_u,k_v. The intrinsic parameters are computed using the Hartley's algorithm [10].

Extrinsic Camera Parameters. The extrinsic parameters are given by the camera pose, which consists of its orientation angles and the position. Pose is represented by translation vector $T_{3\times1}$ and rotation matrix $R_{3\times3}$. Pose is obtained from the correspondence between coordinates of a scene object and its image. The algorithm used by the calibration module is a version of the POSIT method for coplanar points (Oberkampf et. Al, 1996) [11].

Maintaining the Calibration. The initial camera pose estimation (or its extrinsic calibration) is calculated from a set of image points and their corresponding points on the reference plane. Such initial calibration is done for a small set of camera poses. The calibration parameters for the other poses are computed recursively starting from this initial set. For a rotating camera image stitching is used to obtain a sequence of camera poses, following Hartley's algorithm [10]. For each pair of successive images a set of common points is found that allows computing the pose of the second image from the pose of the first image.

4 Conclusion

A new approach and a system VMASS for constructing massive video datasets with annotations has been presented. Such datasets are needed for training and testing methods of classification and recognition of human activities. VMASS dataset represents a significant improvement comparing to the existing datasets with a similar purpose, due to its realism, image quality, variety of actions and situations, hierarchical annotation protocol and a very high volume of stored data. The **VMASS** dataset has become a essential resource for computer vision projects at PJIIT. We expect it will be useful to others and we plan to make it publicly available.

Acknowledgments. This work was supported by projects NN 516475740 from the Polish National Science Centre and PBS I ID 178438 path A from the Polish National Centre for Research and Development.

References

1. Schuldt, C., Laptevand, I., Caputo, B.: Recognizing human actions: A local SVM approach. In: ICPR (2004)
2. Laptev, I., Perez, P.: Retrieving actions in movies. In: ICCV, pp. 1–8 (2007)
3. Oh, S., Hoogs, A., Perera, A., Cuntoor, N., Chen, C.-C., Lee, J.T., Mukherjee, S., Aggarwal, J.K., Lee, H., Davis, L., Swears, E., Wang, X., Ji, Q., Reddy, K., Shah, M., Vondrick, C., Pirsiavash, H., Ramanan, D., Yuen, J., Torralba, A., Song, B., Fong, A., Roy-Chowdhury, A., Desai, M.: A large-scale benchmark dataset for event recognition in surveillance video. In: CVPR 2011, pp. 3153–3160 (2011)
4. Gorelick, L., Blank, M., Shechtman, E., Irani, M., Basri, R.: Actions as Space-Time Shapes. PAMI 29(12), 2247–2253 (2007)
5. Liu, J., Luo, J., Shah, M.: Recognizing realistic actions from videos "in the Wild". In: CVPR 2009 (2009)
6. Weinland, D., Ronfard, R., Boyer, E.: Free viewpoint action recognition using motion history volumes. CVIU 104(2), 249–257 (2006)
7. Ke, Y., Sukthankar, R., Hebert, M.: Volumetric Features for Video Event Detection. IJCV 88(1) (2010)
8. Fisher, R.B.: The PETS04 Surveillance Ground-Truth Data Sets (2004)
9. Smeaton, A.F., Over, P., Kraaij, W.: Evaluation campaigns and TRECVid. In: MIR 2006 (2006)
10. Hartley, R.I.: Self-Calibration from Multiple Views with a Rotating Camera. In: Eklundh, J.-O. (ed.) ECCV 1994, Part I. LNCS, vol. 800, pp. 471–478. Springer, Heidelberg (1994)
11. Oberkampf, D., DeMenthon, D.F., Davis, L.S.: Iterative Pose Estimation Using Coplanar Feature Points. Computer Vision and Image Understanding 63(3), 495–511 (1996)
12. KaewTraKulPong, P., Bowden, R.: An improved adaptive background mixture model for real-time tracking with shadow detection. In: Proc. 2nd European Workshop on Advanced Video-Based Surveillance Systems (2001)
13. Zivkovic, Z.: Improved Adaptive Gaussian Mixture Model for Background Subtraction. In: International Conference Pattern Recognition, UK (August 2004)
14. Davis, J.W., Bradski, G.R.: Motion Segmentation and Pose Recognition with Motion History Gradients. Machine Vision and Applications (2002)

New Tools for Visualization of Human Motion Trajectory in Quaternion Representation

Damian Pęszor[1], Dominik Małachowski[1], Aldona Drabik[2], Jerzy Paweł Nowacki[2], Andrzej Polański[1], and Konrad Wojciechowski[2]

[1] The Silesian University of Technology, Institute of Informatics
[2] Polish-Japanese Institute of Information Technology
Damian.Peszor@pjwstk.edu.pl

Abstract. The paper presents the results of research on the development of a framework and new tools for perceptually oriented visualization of human motion, in particular, a gait. Presented in this paper are new tools for visualizing motion and gait of a human. Their implementation is based on the following principles: i) translational motion component is omitted and time-varying orientations of individual parts of the body are represented by the trajectories of quaternions, ii) the trajectories of quaternions are visualized using a maps: $S^3 \to \mathbb{R}^3$ implemented as: orthogonal projection, stereographic projection, Hopf transformation. This paper describes only the basic functionalities of the tool, the rest are easy to elicit from the main application screen. Rotations of a rigid body represented by the "3D one" about an axis defined by selected vector are presented as an example of using the tool and to facilitate understanding of motion visualization. Finally the exemplary gait visualisations based on three different maps for healthy subjects and patients with impaired movement are presented. Trial carried out using data from the gait laboratory HML show that the best approach to qualitative and quantitative analysis is based on using orthogonal projection. Other types of maps like Hopf and stereographic are difficult to interpret and may be useful in more specific cases.

1 Introduction

In the case of human motion, there are two types of models of human body: i) the skeleton and ii) skeleton-less. The skeleton model is represented by a system of articulated rigid bodies The skeleton-less model is represented by a set of distinguished points and its motion through time. In the rest of this work a skeleton model is used. Each of the rigid bodies of skeleton model is parameterized by the mass and inertia tensor and corresponds to the anatomical body segments like shoulder, arm, foot, hand. Similarly, the number of degrees of freedom for body part connection is equal to the number of degrees of freedom of the respective anatomy joint. In standard models, the number of rigid bodies depends on the assumed model resolution, 22 for Vicon Blade model and 24 for Vicon Nexus model. Specialized models for certain parts of human body do represent those parts with higher resolution, good example of those is clinically

N.T. Nguyen et al. (Eds.): ACIIDS 2014, Part II, LNAI 8398, pp. 575–584, 2014.

tested Oxford Foot Model which uses 3 ridig bodies for foot and 1 for tibia [1]. Configuration of articulated rigid bodies can be described alternatively by: i) definition of position (translation and orientation) of the frame of each body w.r.t. the frame of the world or alternatively w.r.t. the frame of this body in the selected initial configuration (e.g. T pose), ii) definition of the hierarchy tree of bodies and the definition of orientation of the frame body of child w.r.t. current frame. Translation for the frame of child to parent frame is constant due to the assumption of rigidity of bodies. Translation and orientation of the body which is the root of a tree is determined w.r.t. the frame of the world. Regardless of convention of configuration description the key element of visualization is to visualize change in time orientations of individual rigid bodies forming the skeleton. From a formal point of view, the orientation of a rigid body can be parameterized alternatively and equivalently by: i) a 3×3 skew-symmetric matrix, via exponentiation; the 3×3 skew-symmetric matrices are the Lie algebra of SO(3), and this is the exponential map readily applicable in Lie theory, ii) Euler angles (θ, φ, ψ), representing a product of rotations about the z, y and z axes, iii) Tait-Bryan angles (θ, φ, ψ), representing a product of rotations about the x, y and z axes, iv) axis angle pair $(\mathbf{n}, \theta)$ of a unit vector representing an axis, and an angle of rotation about it, v) a quaternion q of length 1; the components of which are also called Euler-Rodrigues parameters. The aim of the current work [16,17,18,19,20,21] was to test a wide variety of parameterizations for usability in problems of human motion analysis and classification, including their sensitivity to some of the characteristics features of motion. In this paper quaternion parameterization of orientation was assumed and the goal was to implement and test various techniques of quaternion trajectory visualization and perceptual diagnosis of abnormalities of movement. Tested quaternion visualization techniques uses alternatively: i) orthogonal projection of S^3 parallel to the real axis ii) stereographic projection and iii) Hopf map. The developed new tools require experience in the interpretation of the observed trajectory obtained as a result of mapping $S^3 \to \mathbb{R}^3$. For this reason, a separate component of the tool is developed that allows to simulate the trajectory of elementary rotation and to show their visualization.

The literature on visualizing quaternions and more generally quaternions trajectories and fields is extensive, the basic positions are [4] and [5]. Despite of this, diversity and utility of tools implemented on the basis of theoretical concepts of visualization are rather restricted in terms of their funcionalities and perceptual values offered. Most of these like *Meshview* [6], *Quaternion Rotation Demo* [7], *Quaternion - Maps* are authored by A. J. Hanson and serve rather for demonstration than for practical use. Another program of Hanson *Quaternion Demonstrator*, [8], allows the visualization of quaternions maps. J.C. Hart program, [9], [10], allows the visualization of one quaternion simultaneously. There exist also *Quaternion visualization tool in Matlab environment*, [12]. Regardless of the critical evaluation of existing programs visualizing quaternions and quaternions trajectories many of theoretical and graphical ideas included in them were used to create new tools presented in this work.

2 Motion Data Acquisition

Gait data used to test the proposed approach and implemented tools have been obtained in the multimodal laboratory-Human Motion Laboratory (HML) of Polish-Japanese Institute of Information Technology http://www.hml.pjwstk.edu.pl A huge database contains the records of gait motion obtained from healthy patients with coxartroza and movements of people with Parkinson's disease. HML measurements equipment are: 1) Vicon's Motion Kinematics Acquisition and Analysis System equipped with 10 NIR cameras with the acquisition speed of 100 to 2000 frames per second at full frame resolution of 4 megapixels and 8-bit grayscale and acquisition space of 12x7x4 meters. 2) Noraxon's Dynamic Electromyography (EMG) System allowing for 16-channel measurement of muscle potentials with non-gel electrodes in compliance with the SENIAM guidelines. 3) Kistler's Ground Reaction Force (GRF) Measurement System used for measuring ground reaction forces with two dynamometric platforms with measurement ranges adjusted to gait analysis research. The system has a 6-meter path masking two platforms situated in the middle of its length. 4) A system for simultaneous multi-camera video image recording equipped with 4 Basler's cameras that allows for simultaneous image recording from all the cameras in Full HD and lossless video recording. The system uses color video recorders using the GigE Vision standard and industrial lenses. The multimodal motion data available in C3D, AMC/ASF, BVH formats.

3 Theoretical Issues

3.1 Quaternions

Lets define three distinguished coordinate vectors $(0,1,0,0)$, $(0,0,1,0)$ and $(0,0,0,1)$ named i, j and k, respectively. The vector (w,x,y,z) written as $q = w + xi + yj + zk$ represents quaternion in algebraic notation. The number w is the real part and x, y and z are called the i, j and k parts, respectively. Beside algebraic notation we use trygonometric $q = \|q\|(\cos\frac{\theta}{2} + \mathbf{n}\sin\frac{\theta}{2})$ and exponential $q = \|q\|\exp(\mathbf{n}\theta)$, $\mathbf{n} = \mathbf{0} + \mathbf{n_1 i} + \mathbf{n_2 j} + \mathbf{n_3 k}$ where $n = (n_1, n_2, n_3)$ defines axis of rotation and θ angle of rotation. Detailed information concerning quaternions can be found in [2,3,11]

The set of unit length quaternions, viewed as points in $\mathbb{R}^4$, is the $3D$-sphere S^3. Each nonzero quaternion q has a multiplicative inverse, denoted as q^{-1} and conjugate denoted as $\overline{q}$.

The set S^3 with the operation of quaternion multiplication satisfies the axioms of a group. The set of rotations in 3-space, with the operation of composition, is also a group, called $SO(3)$. Each rotation R in $SO(3)$ can be realized by quaternion $q, -q \in S^3$

3.2 Orthogonal Projection

Orthogonal projection $S^3 \to \mathbb{R}^3$ is a basic tool for quaternion trajectory visualisation. Lets represent quaternion as:

$$\boldsymbol{q} = (w, x, y, z) = (w, \boldsymbol{v}) \tag{1}$$

where:

$$(w)^2 + (x)^2 + (y)^2 + (z)^2 = 1 \tag{2}$$

Depending on value of w, the set of vectors v can be divided into following subsets; if $w > 0$: north hemisphere, if $w < 0$: south hemisphere and if $w = 0$: equator. Each point of $\mathbf{S^3}$ sphere is projected to a point (x, y, z) in a ball with radius of 1. Value of w can be deduced from $\boldsymbol{v}$ and subset.

$$(w) = \pm\sqrt{1 - \boldsymbol{v} \cdot \boldsymbol{v}} \tag{3}$$

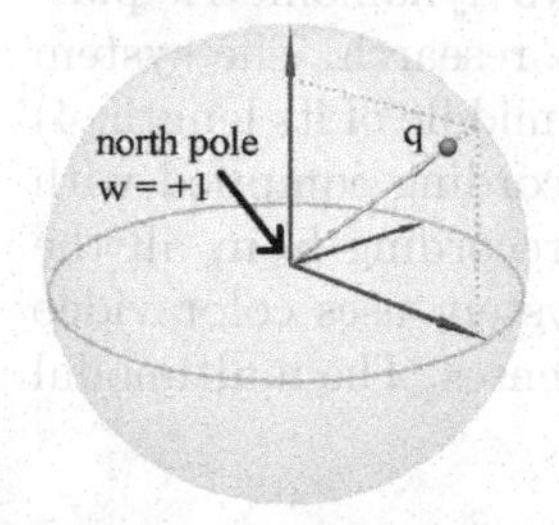

(a) Ball North hemisphere
$x^2 + y^2 + z^2 < 1$
$0 < w \leq 1$

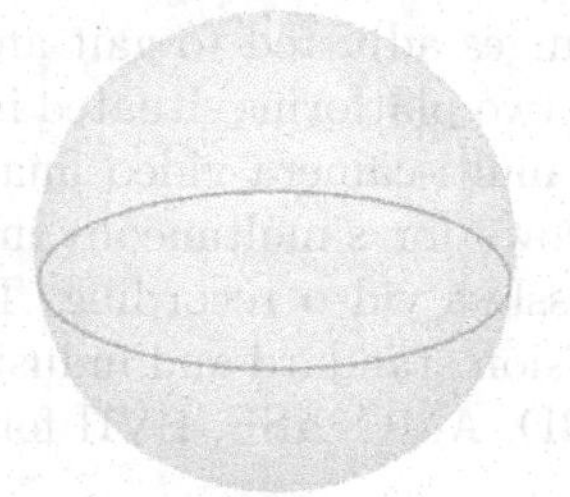

(b) Sphere Equator
$x^2 + y^2 + z^2 = 1$
$w = 0$

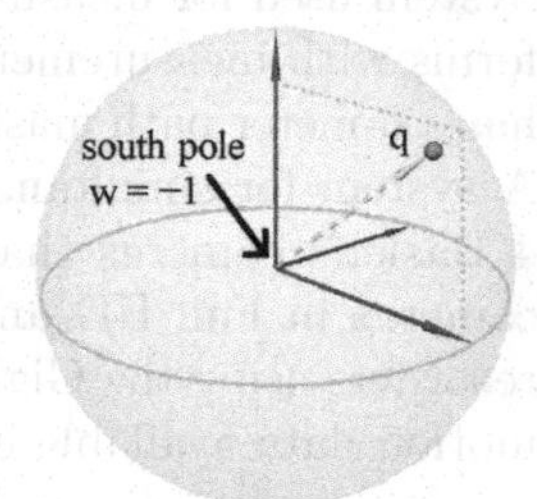

(c) Ball
South hemisphere
$x^2 + y^2 + z^2 < 1$
$-1 \leq w < 0$

Quaternions have a property of double-covering orientation space. Joints in human body; however, have a limited range of rotations. Therefore, for purpose of visualition, all quaternions from southern hemisphere are reflected onto northern.

$$\boldsymbol{q_{vis}} = \begin{cases} \boldsymbol{q} & w \geq 0 \\ -\boldsymbol{q} & w < 0 \end{cases} \tag{4}$$

Relation between angle of rotation θ and distance from center of ball t is non-linear and can be described as:

$$t \in < 0; 1 > \tag{5a}$$

$$\theta = 2 \arcsin t \tag{5b}$$

In order to linearize and make visualization more intuitive, following transformation is applied:

$$\boldsymbol{q_{vis}} = (w,x,y,z) = \left(\cos\frac{\theta}{2}, \boldsymbol{n}\sin\frac{\theta}{2}\right) \tag{6a}$$

$$\boldsymbol{P} = \hat{\boldsymbol{n}}\frac{2\arccos(w)}{\Pi} \tag{6b}$$

where $\boldsymbol{P}$ is transformed position of $\boldsymbol{q_{vis}}$ in visualization space and $\hat{\boldsymbol{n}}$ is normalized $\boldsymbol{n}$ vector.

By applying above formula relation between angle and the distance of P from the centre of $\mathbb{R}^3$ becomes linear. This distance is equal to distance on S^3 from $\boldsymbol{q_{vis}}$ to the north pole divided by Π.

3.3 Stereographic Projection

Stereographic projection $S^3/(1,0,0,0) \to \mathbb{R}^3$ considered in this paper as alternative tool for quaternion trajectory visualisation is defined as follows:

$$(w,x,y,z) \to (\frac{x}{1-w}, \frac{y}{1-w}, \frac{z}{1-w}) \tag{7}$$

3.4 Hopf Fibration

The Hopf fibration, named after Heinz Hopf, is a useful tool in mathematics and physics and has many physical applications such as quantum information theory [13], magnetic monopoles [14] and rigid body mechanics [15]. In context of this paper a key issue is connection of the Hopf fibration with rotations of $3D$-space. Let S^3 and S^2 be spheres in $\mathbb{R}^4$ and $\mathbb{R}^3$ respectively The Hopf fibration is the mapping h from S^3 to S^2 defined as follows:

$$h(w,x,y,z) = (w^2 + x^2 - y^2 - z^2, 2(wz + xy), 2(xz - wy)), \tag{8}$$

under assumption that $q = (w,(x,y,z))$ is quaternion in vector notation.

It can be easily verified that the squares of the three coordinates on the right hand side of (8) sum to 1, so that the image of h is a subset (indeed the whole) of S^2.

Hopf fibration map can be also defined in terms of rotation by quaternions. Let $P_0 = (1;0;0)$, be a distinguished point on S^2. Let for any given point (w,x,y,z) on S^3, $q = w+xi+yj+zk$ be the corresponding unit quaternion. The quaternion q defines a rotation R_q of 3-space. Then the Hopf fibration is defined by:

$$q \to R_q(P_0) = qi\overline{q} \tag{9}$$

Consider once again the point $P_0 = (1,0,0)$ in S^2. One can easily check that the set of points $C = (cost, sint, 0, 0), t \in \mathbb{R}$ or in other representation; a rotation about axis defined by vector $(1,0,0)$, map to $(1;\ 0;\ 0)$ via the Hopf fibration h. In this case set C is the entire set of points that map to $(1,0,0)$ via h. In other words, C is the preimage set $h^{-1}(1,0,0)$. The set C is the unit circle in a plane in $\mathbb{R}^4$. In general for any point P in S^2, the preimage set $h^{-1}(P)$ is a circle in S^3 which is called the fiber of the Hopf map for P. Let us note that the converse statement is not true. For example the circles in S^3 defined as $C = (cost, 0, sint, 0)$, $C = (cost, 0, 0, sint)$ and representing rotations about axis defined by vectors $(0,1,0)$,$(0,0,1)$ are mapped to circles in $\mathbb{R}^2$.

4 General Tool Description

Tool developed for quaternion visualization was implemented using .NET framework and is based on Windows Forms technology. There are two distinct applications of which each presents different frontend, however both utilize same implementation of all transformations to achieve its goals of visualizing quaternions that can be selected using GUI. Visualizations are shown in separate panels, each providing controls for specific visualization. All visualizations can be saved to PNG file at any given time using current orientations of coordinate systems selected in each visualization panel.

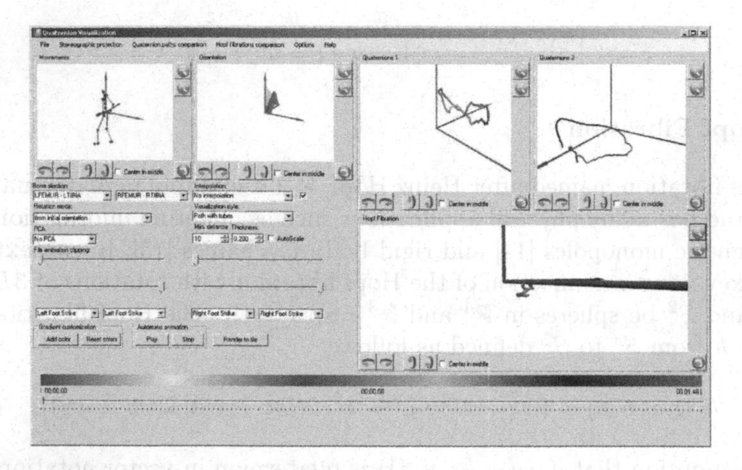

Fig. 1. Main screen of QuaternionVisualization application

First application, QuaternionVisualization has ability to load skeleton-based animation files containing information about rotation of each segment of human body around its parent joint defined by skeleton. Additional data, like time markers related to specific events (e.g. beginning and end of each step) can also be loaded from C3D files. In each frame there is a visualization of quaternion in time using techniques mentioned. Data from animation file can be clipped using arbitrary values for point in time in which to begin and end visualization, or it can be selected from events provided by loaded C3D files. Two different segments can be selected at once in order to compare them on separate or combined screen.

Second application, QuaternionVisualizationLearning, exists to prove correctness of implementation and to show how each type of visualization works for typical data. User can select axis of rotation and an angle and create a list of such rotations that will be visualized together as one path. User can also customize visualization to suit his needs by selecting visualization style, visibility of objects, thickness of lines, etc. User can select specific quaternion using slider to see which part of visualization is based on selected quaternion. A shape of **1** is used to show orientation and rotation axis of provided data.

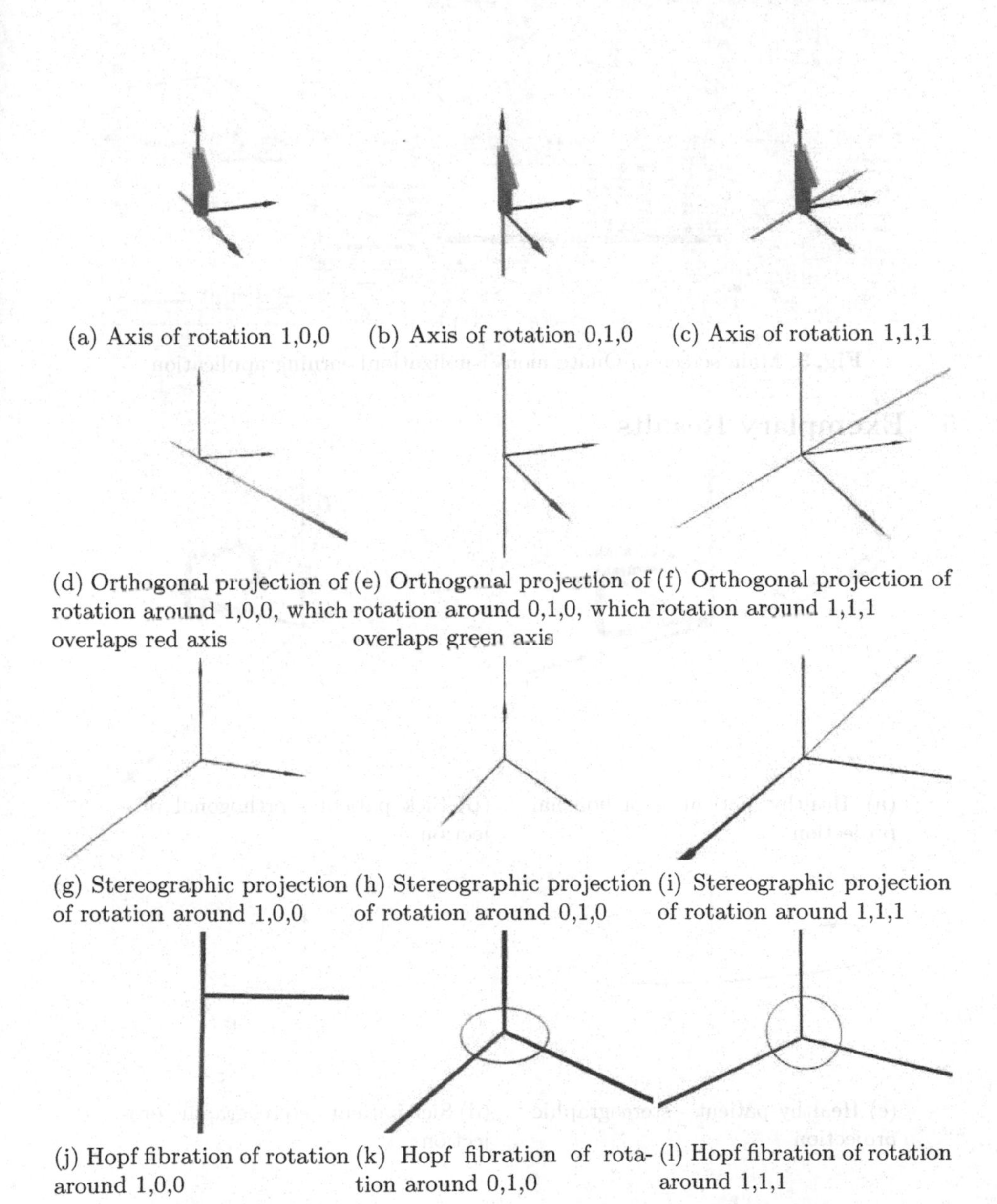

(a) Axis of rotation 1,0,0 (b) Axis of rotation 0,1,0 (c) Axis of rotation 1,1,1

(d) Orthogonal projection of rotation around 1,0,0, which overlaps red axis (e) Orthogonal projection of rotation around 0,1,0, which overlaps green axis (f) Orthogonal projection of rotation around 1,1,1

(g) Stereographic projection of rotation around 1,0,0 (h) Stereographic projection of rotation around 0,1,0 (i) Stereographic projection of rotation around 1,1,1

(j) Hopf fibration of rotation around 1,0,0 (k) Hopf fibration of rotation around 0,1,0 (l) Hopf fibration of rotation around 1,1,1

Fig. 2. Sample visualizations using QuaternionVisualizationLearning application

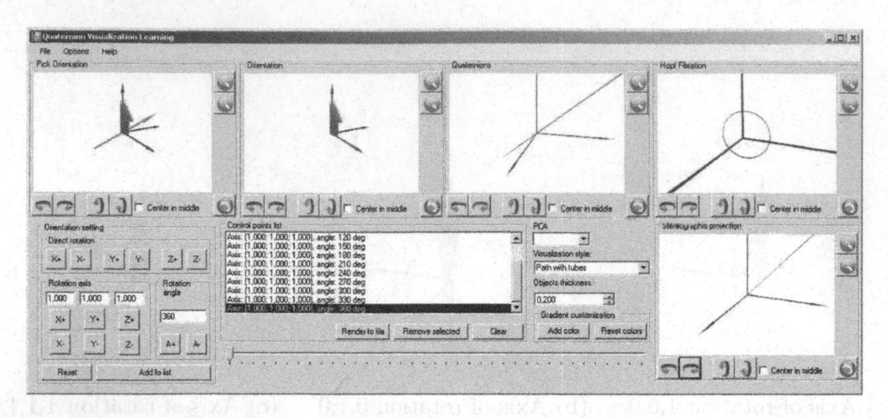

Fig. 3. Main screen of QuaternionVisualizationLearning application

5 Exemplary Results

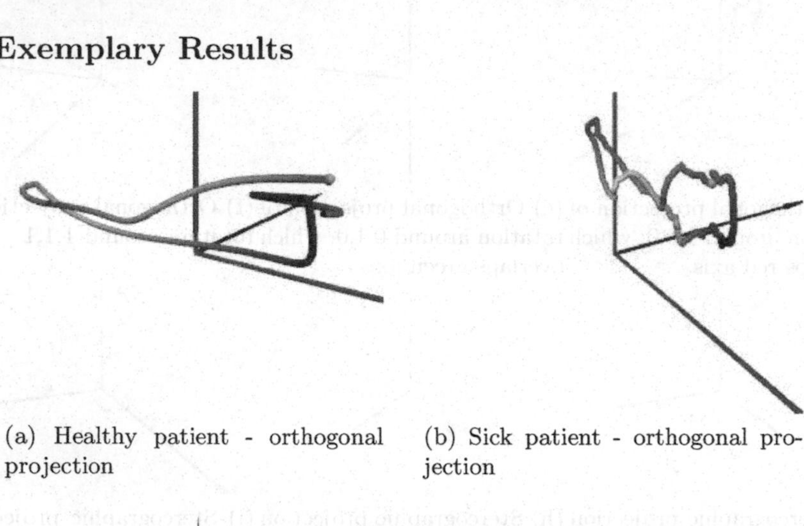

(a) Healthy patient - orthogonal projection

(b) Sick patient - orthogonal projection

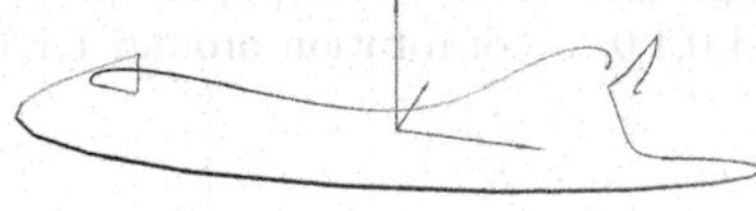

(c) Healthy patient - stereographic projection

(d) Sick patient - stereographic projection

(e) Healthy patient - Hopf fibration

(f) Sick patient - Hopf fibration

Fig. 4. Exemplary results collected from healthy and sick patients

6 Conclusions and Further Works

Visualizations underlying the following comparison and summary refer to the case of hierarchical representation of pose as a ordered set of child body - parent body rotations. Other cases discussed in the first chapter are also acceptable and supported by the software, but their interpretation is more difficult because of the possible change of orientation of the entire skeleton. Visualizations are made based on the following three different techniques of mapping S^3 sphere onto $\mathbb{R}^3$: i) the orthogonal projection which maps the northern hypersphere of S^3 sphere on the B^3 ball, ii) the stereographic projection which maps the northern hypersphere of S^3 sphere to $\mathbb{R}^3$, and finally iii) Hopf map S^3 to S^2. Orthogonal projection is easiest to perceive because in this case the quaternion encoding rotation is visualized by the vector defined rotation axis. The length of this vector is scaled linearly or non-linearly by the size of the angle of rotation. Such visualization is perceptually clear in the case of the regular gait in which the axes of rotation of each rigid body is approximately constant. The variability of the axes may be an important prerequisite for diagnostic reason. For example, in the case of the knee, due to the anatomical structure of this joint axis of rotation performs precession. The shape and size of the surface generated by the vector of rotation axis in some cases approximated by a cone of precession are important for diagnosis. The disadvantage of visualization techniques based on the orthogonal projection is the difficulty in visualizing the time. In the developed software time is visualized by color of the trajectory point, plus a separate time line. The difficulty of visualizing time causes difficulties in reading, even qualitative, angular velocity. Although there is a possibility of visualizing the points of trajectory by markers or numbers, experiments performed have shown that this leads to decreasing perceptual quality of visualization. As seen in fig. 4, one could assess health and sick patient's movement based on shakiness of trajectory. Interpretation of the stereographic projection is based on knowledge of the images of selected arcs on the sphere S^3. Of particular interest is the visualization based on the Hopf map. As shown in section 2 rotation with respect to the axis defined by the vector (1,0,0) maps to a point (1,0,0). Consequently, the "scatter" around the point (1,0,0) on the sphere S^2 may be a measure how much the actual rotation about actual axis is different from the ideal rotation with respect to axis defined by vector (1,0,0). Still though, analysis of the utility of visualization based on Hopf map technology requires further research. Therefore, one objective of the work is wide dissemination of software in order to obtain opinions of its usefulness.

Acknowledgement. This work was supported by projects NN 516475740 and NN 518289240 from the Polish National Science Centre.

References

1. Carson, M.C., Harrington, M.E., Thompson, N., O'Connor, J.J., Theologis, T.N.: Kinematic analysis of a multi-segment foot model for research and clinical applications: a repeatability analysis. Journal of Biomechanics 34, 1299–1307 (2001)

2. Eberly, D.: A Linear Algebraic Approach to Quaternions. Magic Software (2002)
3. Eberly, D.: Quaternion Algebra and Calculus. Magic Software Inc. (2002)
4. Hanson, A.J.: Visualizing Quaternions. Course Notes for SIGGRAPH 1999 (1999)
5. Hanson, A.J.: Visualizing Quaternions. Elsevier (2006)
6. Hanson, A.J., Ishkov, K., Ma, J.: Meshview: Visualizing the Fourth Dimension. Computer Science Department, Indiana University
7. Hanson, A.J.: Quaternion Rotation Demo, http://www.elsevierdirect.com/Companions/9780120884001/vq/QuatRot/
8. Hanson, A.J.: Quaternion-Maps, http://www.elsevierdirect.com/companions/9780120884001/vq/Quaternion-Maps/
9. Hart, J.C., Francis, G.K., Kauffman, L.H.: Visualizing Quaternion Rotation. ACM Transactions on Graphics (1994)
10. Hart, J.C.: Quaternion Demonstrator, http://graphics.stanford.edu/courses/cs348c-95-fall/software/quatdemo/
11. Johnson, M.P.: Exploiting Quaternions to Support Expressive Interactive Character Motion. PhD thesis, Massachusetts Institute of Technology (1995)
12. Vedenev, M.: Quaternions Visualizations in Matlab, http://quaternion.110mb.com
13. Mosseri, R., Dandoloff, R.: Geometry of entangled states, Bloch spheres and Hopf fibrations. Journal of Physics A: Mathematical and General 34, 10243–10252 (2001)
14. Gregory, R., Harvey, J.A., Moore, G.: Unwinding strings and T-duality of Kaluza-Klein and H-Monopoles. Advances in Theoretical and Mathematical Physics 1, 283–297 (1997)
15. Marsden, J., Ratiu, T.: Introduction to Mechanics and Symmetry. Springer, New York (1982)
16. Szczęsna, A., Słupik, J., Janiak, M.: Quaternion Lifting Scheme for Multi-resolution Wavelet-based Motion Analysis. In: Proceedings of the 1st International Workshop on Computer Vision and Computer Graphics, ThinkMind, ICONS 2012, pp. 223–228 (2012)
17. Świtoński, A., Stawarz, M., Boczarska-Jedynak, M., Sieroń, A., Polański, A., Wojciechowski, K.: The effectiveness of applied treatment in Parkinson disease based on feature selection of motion activities. Przegląd Elektrotechniczny 88(12b) (2012) ISSN 0033-2097
18. Stawarz, M., Polański, A., Kwiek, S., Boczarska-Jedynak, M., Janik, Ł., Przybyszewski, A., Wojciechowski, K.: A System for Analysis of Tremor in Patients with Parkinson's Disease Based on Motion Capture Technique. In: Bolc, L., Tadeusiewicz, R., Chmielewski, L.J., Wojciechowski, K. (eds.) ICCVG 2012. LNCS, vol. 7594, pp. 618–625. Springer, Heidelberg (2012)
19. Stawarz, M., Kwiek, S., Polański, A., Janik, Ł., Boczarska-Jedynak, M., Przybyszewski, A., Wojciechowski, K.: Algorithms for Computing Indexes of Neurological Gait Abnormalities in Patients After DBS Surgery for Parkinson Disease Based on Motion Capture Data. Machine Graphics & Vision, Special Issue (2012)
20. Szczęsna, A., Słupik, J., Janiak, M.: Quaternion Lifting Scheme for Multi-resolution Wavelet-based Motion Analysis. In: Proceedings of the First International Workshop on Computer Vision and Computer Graphics, VisGra 2012 (2012)
21. Szczęsna, A., Słupik, J., Janiak, M.: Denoising Motion Data Based on the Quaternion Lifting Scheme Multiresolution Transform. Machine Graphic & Vision, Special Issue (2012)

Heuristic Method of Feature Selection for Person Re-identification Based on Gait Motion Capture Data

Henryk Josiński[2], Agnieszka Michalczuk[1], Daniel Kostrzewa[3], Adam Świtoński[1], and Konrad Wojciechowski[1,2]

[1] Polish-Japanese Institute of Information Technology, Aleja Legionów 2, 41-902 Bytom, Poland
{amichalczuk,aswitonski,kwojciechowski}@pjwstk.edu.pl

[2] Silesian University of Technology, Institute of Computer Science, Akademicka 16, 44-100 Gliwice, Poland
{Henryk.Josinski,Konrad.Wojciechowski}@polsl.pl

[3] Silesian University of Technology, Department of Industrial Informatics, Krasińskiego 8, 40-019 Katowice, Poland
Daniel.Kostrzewa@polsl.pl

Abstract. The authors present a heuristic method of feature selection for gait mocap data, based on the exIWO metaheuristic which is characterized by both the hybrid strategy of the search space exploration and three variants of selection of individuals as candidates for next population. The proposed method was evaluated by the accuracy of person re-identification based on the selected feature subset. Because of the high-dimensional nature of motion data, feature selection was preceded by the data dimensionality reduction.

Keywords: feature selection, exIWO algorithm, gait analysis, person re-identification, dimensionality reduction, MPCA algorithm.

1 Introduction

Gait is defined as coordinated, cyclic combination of movements which results in human locomotion [1]. A unique advantage of gait as a biometric is that it offers potential for recognition at a distance or at low resolution or when other biometrics might not be perceivable [2]. Gait can be captured by two-dimensional video cameras of surveillance systems or by much accurate motion capture (*mocap*) systems which acquire motion data as a time sequence of poses. In the latter case the movement of an actor wearing a special suit with attached markers is recorded by NIR (Near Infrared) cameras. Positions of the markers in consecutive time instants constitute basis for reconstruction of their 3D coordinates.

Although the inconvenience of the acquisition process excludes direct application of the mocap system for human identification in a surveillance system, high precision of mocap recordings make them useful in the development stage

N.T. Nguyen et al. (Eds.): ACIIDS 2014, Part II, LNAI 8398, pp. 585–594, 2014.

of methods for solving the *person re-identification problem* which can be formulated as the question of when person detections in different views or at different time instants can be linked to the same individual [3].

Motion data lie in high-dimensional space, but the components of gait description are correlated, what allows to transform a high-dimensional data into a low-dimensional equivalent representation while retaining most of the information regarding the underlying structure or the actual physical phenomenon [4]. The dimensionality reduction problem can be solved, *inter alia*, by encoding an image object as a general tensor of second or higher order [5].

Feature (attribute, variable) selection is expected to simplify object description, discover most discriminative features and give a chance for more precise classification. Most methods involve searching the space of attributes for the subset that is most likely to predict the class best [6]. In the *filter* approach features are selected on the basis of statistical properties, i.e. by ranking them with correlation coefficients. *Wrappers* assess subsets of variables according to their usefulness to a given predictor. In practice, one needs to define: (i) how to search the space of all possible variable subsets; (ii) how to assess the prediction performance of a learning machine to guide the search and halt it; and (iii) which predictor to use [7]. Making use of the wrapper methodology, the authors applied the exIWO metaheuristic to choose a correct subset of predictive attributes. The exIWO constitutes an expanded version of the Invasive Weed Optimization (IWO) algorithm. The authors of the original method [8] from University of Tehran were inspired by observation of dynamic spreading of weeds and their quick adaptation to environmental conditions. The exIWO proposed by the authors of the present paper is characterized by both the hybrid strategy of the search space exploration and three variants of selection[1] of individuals as candidates for next population. According to the authors' knowledge, the algorithm has never been used for the purpose of feature selection.

Main goal of the research is an evaluation of the influence of a heuristic method of feature selection on accuracy of person re-identification based on gait mocap data.

The organization of this paper is as follows – section 2 contains a brief description of both human motion acquisition procedure and gait data tensor representation. Application of the MPCA algorithm for gait data dimensionality reduction is presented in section 3, whereas adaptation of the exIWO metaheuristic to the feature selection is discussed in section 4. Section 5 deals with procedure of the experimental research along with its results. The conclusions are formulated in section 6.

2 Gait Data Acquisition and Representation

Gait sequences were recorded in the Human Motion Laboratory (HML) of the Polish-Japanese Institute of Information Technology (`http://hm.pjwstk.edu.pl`) by means of the Vicon Motion Kinematics Acquisition and Analysis System

[1] It is necessary to mention that the term "selection" is used in the present paper in the following meanings: (i) feature selection, (ii) selection of individuals.

equipped with 10 NIR cameras with the acquisition speed of 100 to 2000 frames per second at full frame resolution of 4 megapixels and 8-bit grayscale (Fig. 1). The gait route was specified as a 5 meters long straight line. The acquiring process started and ended with a T-letter pose because of requirements of the Vicon calibration process. As a result of the acquisition procedure 353 sequences for 25 men aged 20-35 years were stored in a gait database.

Fig. 1. Recording session in the HML

Tensor object is a multidimensional object, the elements of which are to be addressed by indices. The number of indices determines the order of the tensor object, whereas each index defines one of the tensor modes. Gait silhouette sequences are naturally represented as third-order tensors with column, row, and time modes [9].

Description of each of the consecutive poses of a gait sequence depends on the assumed skeleton model. For a typical model containing 22 segments and a global skeleton rotation (Fig. 2), description of a single pose comprises values of 69 Euler angles. Three additional values are required for specification of a global translation.

The third-order tensor representation is based on modes indexed, respectively, by numbers of components of Euler angles ("angle mode"), numbers of skeleton components ("pose mode"), and numbers of sequence frames ("time mode"). Gait sequences consist of 128 frames. Euler angles increase the total number of features characterizing a single sequence to 8832.

3 Data Dimensionality Reduction – The MPCA Algorithm

Multilinear projection of tensor objects for the purpose of dimensionality reduction is the basis of the multilinear principal component analysis (MPCA) which is the multilinear extension of the PCA method. According to the authors of the MPCA algorithm: "Operating directly on the original tensorial data, the proposed MPCA is a multilinear algorithm performing dimensionality reduction in

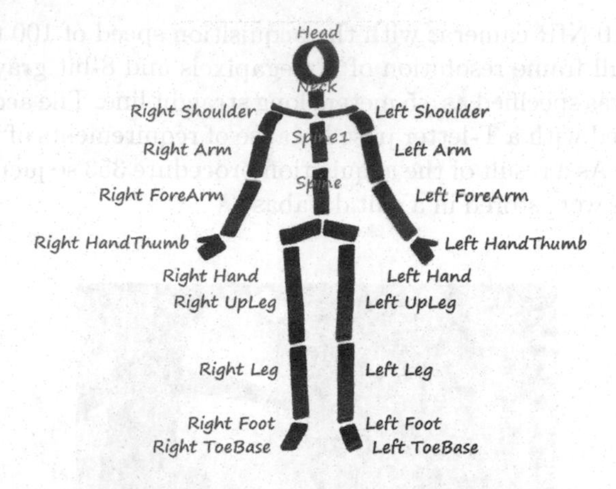

Fig. 2. Segments of the skeleton model

all tensor modes seeking those bases in each mode that allow projected tensors to capture most of the variation present in the original tensors" [9]. Input data are required to be tensor objects and the algorithm operates directly on tensors producing as output low-dimensional *feature tensors*.

In the MPCA an elementary matrix algebra is extended by two operations: tensor unfolding and the product of a tensor by a matrix (Fig. 3). The unfolding (matricization) transforms a tensor into a matrix along a specified mode. In other words, the tensor is decomposed into column vectors. The n-mode multiplication of a tensor x by a matrix U ($x \times_n U$) is realized by the product of the tensor x unfolded along the n-mode ($x_{(n)}$) by the matrix U, followed by the folding operation which creates a resultant tensor: $x \times_n U = \text{fold}_n(U\ \text{unfold}_n(x)) = \text{fold}_n(U x_{(n)})$. To put it another way, tensor x is projected in the n-mode vector subspace by the projection matrix U.

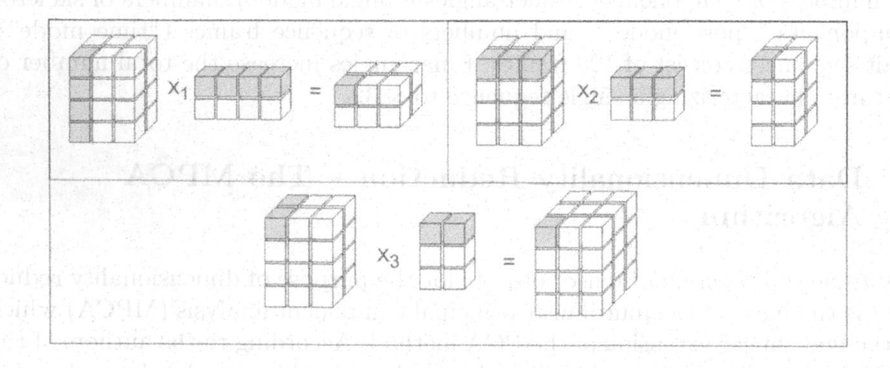

Fig. 3. Multiplication of a tensor by a matrix

The MPCA algorithm consists of the following steps:

1. Preprocessing – normalization of all M N-order input tensor samples $x_m \in \mathbb{R}^{I_1 \times I_2 \times \ldots \times I_N}$ $(m = 1..M)$ to zero mean value $(\tilde{x}_m)$. I_n denotes the n-mode dimension of the tensor $(n = 1..N)$.
2. Learning phase which has an iterative character:
   (a) Initialization – for each mode n:
   - Computation of the matrix $\Phi^{(n)*}$: $\Phi^{(n)*} = \sum_{m=1}^{M} \tilde{X}_{m(n)} \tilde{X}_{m(n)}^T$, where $\tilde{X}_{m(n)}$ is the n-mode unfolded matrix of $\tilde{x}_m$.
   - Eigendecomposition of the matrix $\Phi^{(n)*}$.
   - Selection of P_n most significant eigenvectors which form a projection matrix $\tilde{U}^{(n)}, \tilde{U}^{(n)} \in \mathbb{R}^{I_n \times P_n}, P_n \leq I_n$. Eigenvectors are evaluated on the basis of the corresponding eigenvalues. Variation coverage $Q^{(n)}$ in the n-mode after the truncation of the eigenvectors beyond the P_n-th could be calculated as follows: $Q^{(n)} = \sum_{i_n=1}^{P_n} \lambda_{i_n}^{(n)} / \sum_{i_n=1}^{I_n} \lambda_{i_n}^{(n)}$ where $\lambda_{i_n}^{(n)}$ is the i-th eigenvalue of the matrix $\Phi^{(n)*}$. In practice, only one user-defined value of variation coverage Q determines the number of selected eigenvectors for each mode separately.

   (b) Iterative local optimization of the projection matrices $\tilde{U}^{(n)}$ $(n = 1..N)$:
   - Computation of the matrix $\Phi^{(n)}$: $\Phi^{(n)} = \sum_{m=1}^{M} \left(X_{m(n)} - \bar{X}_{(n)}\right) \cdot \tilde{U}_{\Phi^{(n)}} \cdot \tilde{U}_{\Phi^{(n)}}^T \cdot \left(X_{m(n)} - \bar{X}_{(n)}\right)^T$, where $\bar{X}_{(n)} = \frac{1}{M} \sum_{m=1}^{M} X_{m(n)}$ and $\tilde{U}_{\Phi^{(n)}} = \left(\tilde{U}^{(n+1)} \otimes \tilde{U}^{(n+2)} \otimes \ldots \otimes \tilde{U}^{(N)} \otimes \tilde{U}^{(1)} \otimes \ldots \otimes \tilde{U}^{(n-1)}\right)$; $\otimes$ denotes the Kronecker product.
   - Eigendecomposition of the matrix $\Phi^{(n)}$.
   - Construction of the matrix $\tilde{U}^{(n)}$ which consists of the P_n eigenvectors corresponding to the largest P_n eigenvalues of the matrix $\Phi^{(n)}$.
   - In our version of the MPCA algorithm the local optimization stage is repeated until a user-specified number of iterations is reached.
3. Reduction phase – projection of the input samples using the matrices $\tilde{U}^{(n)}$ $(n = 1..N)$ results in construction of the low-dimensional representation of the input samples with Q % variation captured, in the form of feature tensors $y_m \in \mathbb{R}^{P_1 \times P_2 \times \ldots \times P_N}$ $(m = 1..M)$: $y_m = x_m \times_1 \tilde{U}^{(1)^T} \times_2 \tilde{U}^{(2)^T} \times_3 \ldots \times_N \tilde{U}^{(N)^T}$.

Our MPCA implementation is based on the Jama-1.0.2 library (`http://math.nist.gov/javanumerics/jama/`).

4 Feature Selection – The exIWO Algorithm

The simplified pseudocode describes the exIWO using terminological convention consistent with the "natural" inspiration of the authors of the original IWO version. Consequently, the words "*individual*", "*plant*", and "*weed*" are treated as synonyms.

```
Create the first population.
For each individual from the population:
  Compute the value of the fitness function.
While the stop criterion is not satisfied:
  For each individual from the population:
    Compute the number of seeds.
    For each seed:
      Draw the dissemination method.
      Create a new individual.
      Compute the value of its fitness function.
  Select individuals for a new population.
Return the best individual.
```

The optimization process starts with a random initialization of the first population. An individual is represented by a binary vector of a length equal to the entire number of features. "1" at position i in the vector denotes that the i-th feature belongs to the considered subset. Each weed, i.e. each subset of features constructed by the exIWO is evaluated by means of the 1NN classifier using five-fold cross-validation on a training set. Thus, the fitness function is equivalent to the classification accuracy expressed by means of the Correct Classification Rate (CCR) which indicates the percentage of correctly classified cases.

The number of seeds S_w produced by a single weed depends on the value of its fitness function f_w in the following way:

$$S_w = S_{min} + \left\lfloor (f_w - f_{min}) \left(\frac{S_{max} - S_{min}}{f_{max} - f_{min}} \right) \right\rfloor \tag{1}$$

where S_{max}, S_{min} denote maximum and minimum admissible number of seeds generated, respectively, by the best population member (fitness f_{max}) and by the worst one (fitness f_{min}).

According to the terminological convention the hybrid strategy of the search space exploration proposed by the authors of the paper can be called "dissemination of seeds". It consists of 3 methods randomly chosen for each seed: spreading, dispersing and rolling down. The draw procedure is based on the pseudorandom number generator of the uniform distribution on the interval $[0, 1)$. The sum of probabilities assigned to the particular methods should be equal to 1.

The *spreading* consists in random disseminating seeds over the whole of the search space, i.e. independently of the location of a parent plant.

The *dispersing* is based on the idea proposed in the original IWO version. The degree of difference between the individual and his offspring can be interpreted as the distance between the parent plant and the place where the seed falls on the ground. The distance is described by normal distribution with a mean equal to 0 and a standard deviation truncated to nonnegative values. The standard deviation is decreased in each algorithm iteration as follows:

$$\sigma_{iter} = \left(\frac{iter_{max} - iter}{iter_{max}} \right)^m (\sigma_{init} - \sigma_{fin}) + \sigma_{fin} \tag{2}$$

where *iter* denotes the current iteration ($iter \in [1, iter_{max}]$). Consequently, the distance is gradually reduced. The number of iterations $iter_{max}$ is used as stop criterion. The symbols σ_{init}, σ_{fin} represent, respectively, initial and final values of the standard deviation, whereas m is a nonlinear modulation factor. From the practical point of view, the distance between plants, rounded to the nearest integer value, is interpreted as the number of transformations of the parent individual. A single transformation is a simple binary mutation of a randomly chosen element of the feature vector.

The *rolling down* can be interpreted as a movement of a seed towards a "better" location with respect to the fitness function. The term "neighbours" stands for individuals located at the distance equal to 1 (transformation) from the current plant. The best adapted individual is chosen from among k neighbours (k is a parameter of the method), whereupon its neighbourhood is analyzed in search of the next best adapted individual. This procedure is repeated $k-1$ times giving the opportunity to select the best adapted individual found in the last iteration as a new one. Thus, the method enables exploration of the vicinity of the parent individual's location in the search space.

The term "selection" refers also to competitive exclusion of individuals. Candidates for next population are selected in a deterministic manner according to one of the following methods: global, offspring-based and family-based. Set of candidates for the *global* selection consists of all parent plants and all their newly created descendants. By contrast, the *offspring-based* selection is limited solely to the descendants and thus should decrease the risk of stagnation at non-optimal points in the search space [10]. If the best individual so far was grown in the current population, then despite the fact that it cannot be retained in the next population it will be stored with an eye to the final optimization result. According to the rules of the *family-based* selection [11], each plant from the first population is a protoplast of a separate family. A family consists of a parent weed and its direct descendants. Only the best individual of each family survives and becomes member of the next population. For all 3 aforementioned methods cardinality of a population remains constant in all algorithm iterations. The aspect of individuals' selection was also taken into account in the experimental research.

5 Experimental Research

The goal of the experiments was to evaluate the influence of feature selection for accuracy of person re-identification based on gait mocap data. Dimensionality of the recordings was previously reduced by means of the MPCA. Subsequently, gait sequences of each of the 25 actors were divided evenly into training set containing 180 samples and test set composed of 173 samples. The training set was subject to feature selection performed by the exIWO. Finally, samples from the test set were classified using the 1NN method in the space defined by the selected features. The findings of the initial experiments determined the most appropriate values of the exIWO parameters for the considered problem. They were collected in Table 1.

Table 1. Basic parameters of the exIWO used for feature selection

Description	Value
Population cardinality	$\{10, 50\}$
Number of iterations (stop criterion)	$\{20, 50, 100\}$
Number of seeds sowed by a weed ($S_{max} = S_{min}$)	2
Initial value of standard deviation σ_{init} (dispersing)	8.735
Final value of standard deviation σ_{fin} (dispersing)	0.01
Nonlinear modulation factor m (dispersing)	2.59
Number k of examined neighbours and neighbourhoods (rolling down)	3

The parameters from Table 1 were supplemented by different combinations of probabilities (p_{spr}, p_{disp}, p_{roll}) assigned to the particular methods of dissemination of seeds: spreading, dispersing and rolling down. The probabilities were taken from the range of $[0, 1]$ with a step value of 0.1 taking into account the fact that $p_{spr} + p_{disp} + p_{roll} = 1$. Only in case of the family-based selection values of p_{spr} were limited to $\{0, 0.1, 0.2, 0.3\}$, because this method is assumed to give a chance for the preservation of characteristic features of the plants family, what implicates a marginal importance or even absence of the "random oriented" spreading. Hence, in total, 170 combinations of probabilities were tested for the given set of tensor samples (i.e. for the given variation coverage Q). The number of trial runs in the presence of a single configuration of the probabilities was equal to 5. The workstation used for experiments is described by the following parameters: 2×Intel Xeon X5650 2.66GHz (×6) processor, RAM 12GB 1600MHz. Table 2 includes averaged results of the experiments conducted on several data sets previously reduced by the MCPA according to the given value of Q. The results comprise the classification accuracy (CCR) of the test set: (i) in the space defined by the selected features, (ii) in the space of all features, and total execution time of both selection and final classification. Mean values of the accuracy of the final classification test are illustrated in Fig. 4. The x-axis values denote entire number of features for the given Q (see Table 2).

Analysis of the results leads to the following remarks: (i) if number of features is too low (such as for example for $Q = 85$), the classification accuracy is not satisfying, (ii) subsets of high quality features are already constructed by the exIWO in early populations – extending the computational process through increase of numbers of individuals and iterations yielded very modest accuracy profit at best and turned out to be unrewarding ($Q \in \{91, 96\}$), (iii) increase of the number of features does not guarantee accuracy improvement ($Q = 89$ vs $Q \in \{91, 93, 99\}$), (iv) selection on the smaller feature subset gives accuracy similar to the larger subset in time which is several times shorter than that related to the larger subset ($Q = 89$ using 210 weeds vs $Q = 96$ using 510 plants), (v) feature selection leads to the improvement of accuracy, but not in all cases ($Q = 99$). Cardinality of the final selected subset of features oscillated in particular experiments from 42 to 64% of cardinality of the entire set.

Table 2. Results of the experiments

Variation coverage Q [%]	Entire number of features	Entire number of individuals	Mean CCR with feature selection [%]	Mean CCR without feature selection [%]	Mean execution time [s]
85	16	5050	97.68	97.17	584.61
89	30	510	99.74	99.15	52.29
89	30	210	99.69	99.15	27.87
91	33	5050	99.40	98.58	554.30
91	33	510	99.41	98.58	52.72
93	36	510	99.58	99.15	64.52
93	36	210	99.63	99.15	35.96
96	84	5050	99.84	99.72	948.01
96	84	510	99.71	99.72	97.03
99	270	510	97.85	99.15	298.60
99	270	210	97.51	99.15	109.68

Influence of the method of individuals' selection on the classification accuracy turned out to be not significant. Hence, the summary of this aspect of the research will be limited only to description of each method by means of a triple $a/b/c$, where a denotes number of cases (i.e. rows in Table 2) in which the mean CCR of the given method (not included in Table 2) was greater than the mean CCR for all 3 methods ($CCR_m > CCR_{all}$), b represents number of cases in which both mean values were equal, while c – number of cases in which $CCR_m < CCR_{all}$. Thus, descriptions of particular methods are as follows: global – 3/4/4, offspring-based – 8/2/1, family-based – 2/3/6. The outcome of the family-based method is supposed to result from the random initialization of the first population, instead of using any "well-considered" method.

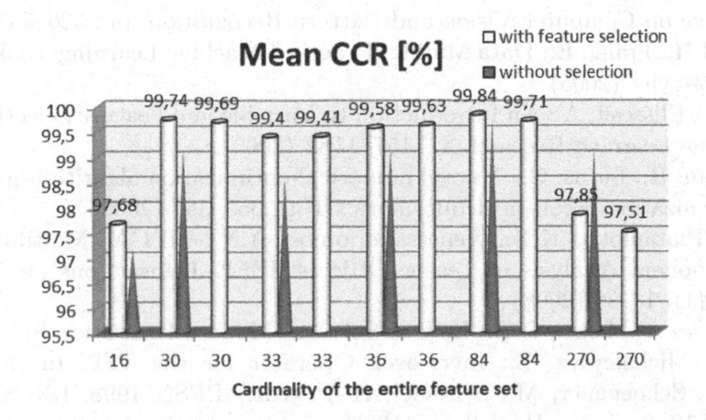

Fig. 4. Dependence between entire number of features and re-identification accuracy

6 Conclusion

The research revealed the usefulness of the exIWO for solving feature selection problem. However, the adaptation of the metaheuristic requires determination of the following components: a representation of a single solution, a method of initialization of the first population, admissible transformations of an individual, a formula of a fitness function, a stop criterion, and a thorough choice of appropriate values of algorithm parameters. Besides, evaluation of quality of the selected feature subset based on classification accuracy is classifier-dependent and rather time-consuming. With regard to this last aspect, the selection was preceded by the data dimensionality reduction performed by the MPCA algorithm. Future research will focus on alignment of parameters of both algorithms and feature selection for video data.

Acknowledgments. This work was supported by projects NN 516475740 and DEC-2011/01/B/ST6/06988 from the Polish National Science Centre.

References

1. Boyd, J.E., Little, J.J.: Biometric Gait Recognition. In: Tistarelli, M., Bigun, J., Grosso, E. (eds.) Advanced Studies in Biometrics. LNCS, vol. 3161, pp. 19–42. Springer, Heidelberg (2005)
2. Nixon, M.S., Tan, T.N., Chellappa, R.: Human Identification Based on Gait. Springer (2006)
3. Doretto, G., Sebastian, T., Tu, P., Rittscher, J.: Appearance-based person reidentification in camera networks: problem overview and current approaches. Journal of Ambient Intelligence and Humanized Computing 2, 127–151 (2011)
4. Law, M.H.C., Jain, A.K.: Incremental nonlinear dimensionality reduction by manifold learning. IEEE Transactions on Pattern Analysis and Machine Intelligence 28(3), 377–391 (2006)
5. Yan, S., Xu, D., Yang, Q., Zhang, L., Tang, X., Zhang, H.-J.: Discriminant Analysis with Tensor Representation. In: Proceedings of the 2005 IEEE Computer Society Conference on Computer Vision and Pattern Recognition, pp. 526–532 (2005)
6. Witten, I.H., Frank, E.: Data Mining. Practical Machine Learning Tools and Techniques. Elsevier (2005)
7. Guyon, I., Elisseeff, A.: An Introduction to Variable and Feature Selection. Journal of Machine Learning Research 3, 1157–1182 (2003)
8. Mehrabian, R., Lucas, C.: A novel numerical optimization algorithm inspired from weed colonization. Ecological Informatics 1(4), 355–366 (2006)
9. Lu, H., Plataniotis, K.N., Venetsanopoulos, A.N.: MPCA: Multilinear Principal Component Analysis of Tensor Objects. IEEE Transactions on Neural Networks 19(1), 18–39 (2008)
10. Michalewicz, Z., Fogel, D.B.: How to Solve It: Modern Heuristics. Springer (2004)
11. Tao, G., Michalewicz, Z.: Inver-over Operator for the TSP. In: Eiben, A.E., Bäck, T., Schoenauer, M., Schwefel, H.-P. (eds.) PPSN 1998. LNCS, vol. 1498, pp. 803–812. Springer, Heidelberg (1998)

3D Gait Recognition Using Spatio-Temporal Motion Descriptors

Bogdan Kwolek[1], Tomasz Krzeszowski[3], Agnieszka Michalczuk[2], and Henryk Josinski[2]

[1] AGH University of Science and Technology
30 Mickiewicza Av., 30-059 Krakow, Poland
bkw@agh.edu.pl

[2] Polish-Japanese Institute of Information Technology
Al. Legionów 2, 41-902 Bytom, Poland
{amichalczuk,hjosinski}@pjwstk.edu.pl

[3] Rzeszów University of Technology
Al. Powstańców Warszawy 12, 35-959 Rzeszów, Poland
tkrzeszo@prz.edu.pl

Abstract. We present a view independent algorithm for 3D human gait recognition. The identification of the person is achieved using motion data obtained by our markerless 3D motion tracking algorithm. We report its tracking accuracy using ground-truth data obtained by a marker-based motion capture system. The classification is done using SVM built on the proposed spatio-temporal motion descriptors. The identification performance was determined using 230 gait cycles performed by 22 persons. The correctly classified ratio achieved by SVM is equal to 93.5% for rank 1 and 99.6% for rank 3. We show that the recognition performance obtained with the spatio-temporal gait signatures is better in comparison to accuracy obtained with tensorial gait data and reduced by MPCA.

1 Introduction

Gait is an attractive biometric feature for human identification at a distance, and recently has gained much interest from academia. Vision-based gait recognition has attracted increased attention due to possible applications in intelligent biometrics and visual surveillance systems [2]. Compared with conventional biometric features, such as face or iris, gait has many unique advantages since the identification techniques are non-contact, non-invasive, perceivable at a larger distance and do not require a cooperation of the individual.

In general, there are two main approaches for gait-based person identification, namely appearance-based (model free) and model-based ones [10]. Appearance-based approaches focus on identifying persons using shape, silhouette, geometrical measures, etc. On the other hand, model-based approaches are focused on identifying persons taking into account the kinematic characteristics of the walking manner. The majority of the approaches are based on appearance and rely on analysis of image sequences acquired by a single camera. The main drawback

N.T. Nguyen et al. (Eds.): ACIIDS 2014, Part II, LNAI 8398, pp. 595–604, 2014.

of such approaches is that they can perform the recognition from a specific viewpoint. To achieve view-independent person identification, Jean et al. [4] proposed an approach to determine view-normalized body part trajectories of pedestrians walking on potentially non-linear paths. However, Yu et al. [15] reported that view changes cause a significant deterioration in gait recognition accuracy. Viewpoint dependence is still significant problem for many gait analysis techniques, since it is not easy to match between different orientations of the subject.

Model-based gait recognition is usually based on 2D fronto-parallel body models. In [14], a sequential set of 2D stick figures is utilized to extract gait patterns. Afterwards, a SVM classifier is used to classify gender using such gait signatures. The use of 3D technology for gait analysis [1] dates back to 1990. The 3D approaches for gait recognition model human body structure explicitly, often with support of the gait biomechanics [13]. They are far more resistant to view changes in comparison to 2D ones. In [11] 3D locations of markers were utilized, from which joint-angle trajectories measured from normal walks were derived. The recognition was performed using dynamic time warping on the normalized joint-angle information. It was done on two walking databases of 18 people and over 150 walk instances using nearest neighbor classifier with Euclidean distance. In [9] several ellipses are fitted to different parts of the binary silhouettes and the parameters of these ellipses (e.g., location and orientation) are used as gait features. In [12], an approach relying on matching 3D motion models to images, and then tracking and restoring the motion parameters is proposed. The evaluation was performed on datasets with four people, i.e. 2 women and 2 men walking at 9 different speeds ranging from 3 to 7 km/h by increments of 0.5 km/h. Motion models were constructed using Vicon motion capture system (moCap). To overcome the non-frontal pose problem, more recently a multi-camera based gait recognition method has been developed [3]. In the mentioned work, joint positions of the whole body are employed as a feature for gait recognition.

In controlled laboratory experiments, gait has been shown to be very effective biometric for distinguishing between individuals. However, in real world scenarios it has been found to be much harder to achieve good recognition accuracy. Most of gait analysis techniques, particularly neglecting 3D information, are unable to reliably match gait signatures from differing viewpoints. Moreover, they are also strongly dependant on the ability of the background segmentation and require accurate delineation between the subject and the background.

In this work we present an approach for 3D gait recognition. The motion parameters are estimated on the basis of markerless human motion tracking. They are inferred with the help of a 3D human model. The estimation takes place on video sequences acquired by four calibrated and synchronized cameras. We show the tracking performance of the motion tracking algorithm using ground-truth data acquired by a commercial motion capture (moCap) system from Vicon Nexus. The identification is done on the basis of the proposed spatio-temporal motion descriptors. We show that they allows us to achieve better results in comparison to an algorithm based on third order tensor and a reduction of the tensorial gait data by Multilinear Principal Components Analysis (MPCA) [7].

2 Markerless System for Articulated Motion Tracking

2.1 3D Human Body Model

The human body can be represented by a 3D articulated model formed by 11 rigid segments representing the key parts of the body. The 3D model specifies a kinematic chain, where the connections of body parts comprise a parent-child relationship, see Fig. 1. The pelvis is the root node in the kinematic chain and at the same time it is the parent of the upper legs, which are in turn the parents of the lower legs. In consequence, the position of a particular body limb is partially determined by the position of its parent body part and partially by its own pose parameters. In this way, the pose parameters of a body part are described with respect to the local coordinate frame determined by its parent. The 3D geometric model is utilized to simulate the human motion and to recover the persons's position, orientation and joint angles. To account for different body part sizes, limb lengths, and different ranges of motion we employ a set of pre-specified parameters, which express typical postures. For each degree of freedom there are constraints beyond which the movement is not allowed. The model is constructed from truncated cones and is used to generate contours, which are then matched with the image contours. The configuration of the body is parameterized by the position and the orientation of the pelvis in the global coordinate system and the angles between the connected limbs.

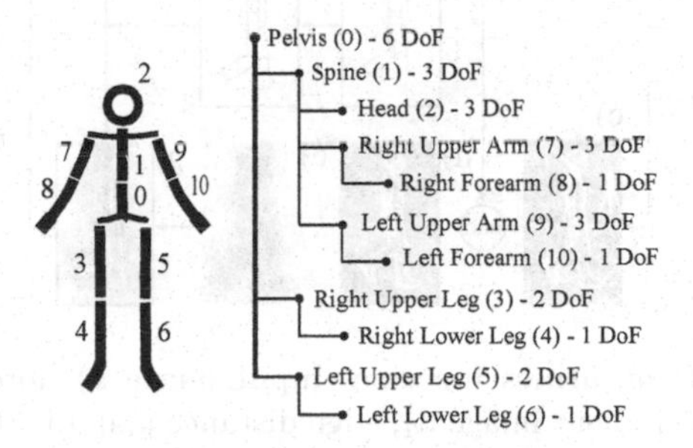

Fig. 1. 3D human body model consisting of 11 segments (left), hierarchical structure (right)

2.2 Articulated Motion Tracking

Estimating 3D motion can be cast as a non-linear optimization problem. The degree of similarity between the real and the estimated pose is evaluated using an objective function. Recently, particle swarm optimization (PSO) [5] has been successfully applied to full-body motion tracking [8]. In PSO each particle follows simple position and velocity update equations. Thanks to interaction between particles a collective behavior of the swarm arises. It leads to the emergence of global and collective search capabilities, which allow the particles to gravitate towards the global extremum. The motion tracking can be achieved by a sequence of static PSO-based optimizations, followed by re-diversification of the particles to cover the possible poses in the next frame. In this work the 3D motion tracking is achieved through the Annealed Particle Swarm Optimization (APSO) [8].

2.3 Fitness Function

The fitness function expresses the degree of similarity between the real and the estimated human pose. Figure 2 illustrates the calculation of the objective function. It is determined in the following manner: $f(x) = 1-(f_1(x)^{w_1} \cdot f_2(x)^{w_2})$, where x stands for the state (pose), whereas w denotes weighting coefficients that were determined experimentally. The function $f_1(x)$ reflects the degree of overlap between the extracted body and the projected 3D model into 2D image. The function $f_2(x)$ reflects the edge distance-based fitness. A background subtraction algorithm [8] is employed to extract the binary image of the person, see Fig. 2b. The binary image is then utilized as a mask to suppress edges not belonging to the person, see Fig. 2d. The projected model edges are then matched with the image edges using the edge distance map, see Fig. 2g.

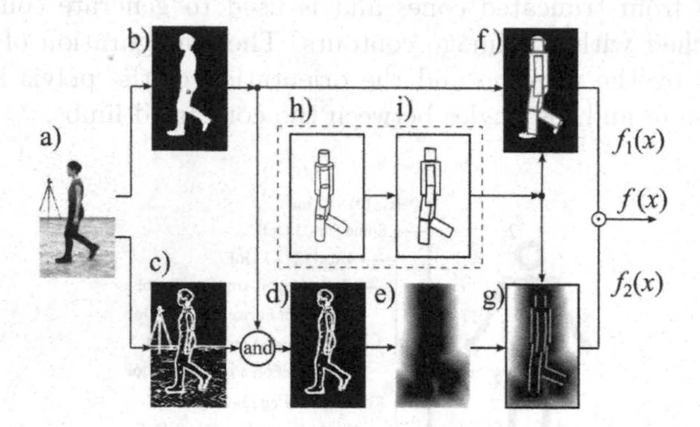

Fig. 2. Calculation of the fitness function. Input image a), foreground b), gradient magnitude c), masked gradient image d), edge distance map e), 3D model h) projected onto image 2D plane i) and overlaid on binary image f) and edge distance map g)

3 Gait Characterization and Recognition

The markerless motion tracking was achieved using color images of size 960×540, which were acquired at 25 fps by four synchronized and calibrated cameras. Each pair of the cameras is approximately perpendicular to the other camera pair. Figure 3 depicts the location of the cameras in the laboratory.

A commercial motion capture system from Vicon Nexus was employed to provide the ground truth data. The system uses reflective markers and ten cameras to recover the 3D location of such markers. The data are delivered with rate of 100 Hz and the synchronization between the moCap and multi-camera system is achieved using hardware from Vicon Giganet Lab.

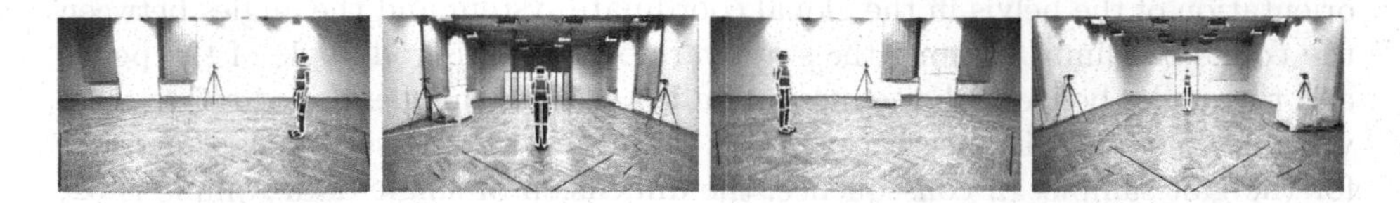

Fig. 3. Layout of the laboratory with four cameras. The images illustrate the initial model configuration, overlaid on the image in first frame and seen in view 1, 2, 3, 4.

A set of $M = 39$ markers attached to main body parts has been used. From the above set of markers, 4 markers were placed on the head, 7 markers on each arm, 12 on the legs, 5 on the torso and 4 markers were attached to the pelvis. Given such a placement of the markers on the human body and the estimated human pose, which has been determined by our APSO algorithm, the corresponding positions of virtual markers on the body model were determined. Figure 4 illustrates the distances between ankles, which were determined by our markerless motion tracking algorithm and the moCap system. High overlap between both curves formulates a rationale for the usage of the markerless motion tracking in view-independent gait recognition. In particular, as we can observe, the gait cycle and the stride length can be determined with high precision.

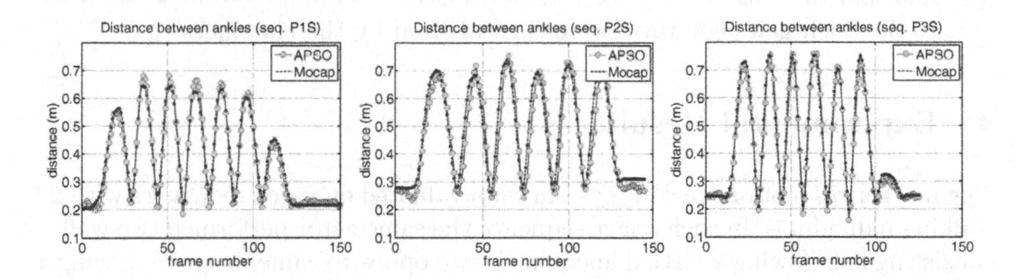

Fig. 4. Distance between ankles in sequences P1S, P2S and P3S (straight)

In a typical system for gait-based person identification a gait signature is extracted in advance. Given a gallery database consisting of gait patterns from a set of known subjects, the objective of the gait recognition system is to determine the identity of the probe samples. In this work we treat each gait cycle as a data sample. Thus, a gait sample consists of some attributes describing a person, like height, stride length and joint angles estimated by marker-less motion capture.

The data extracted by markerless motion tracking algorithm were stored in ASF/AMC data format. For a single gait cycle consisting of two strides a gait signature was determined. The number of frames in each gait sample has some variation and therefore the number of frames in each gait sample was subjected to normalization. The normalized time dimension was chosen to be 30, which was roughly the average number of frames in each gait cycle. As mentioned in Subsection 2.1, the body configuration is parameterized by the position and the orientation of the pelvis in the global coordinate system and the angles between the connected limbs. Among the state variables there is roll angle of the pelvis and the angles between the connected limbs. A set of the mentioned above state variables plus the distance between the ankles and the person's height account for the gait sample. In consequence, the dimension of single data sample is 32, and it is equal to the number of bones (excluding pelvis) times three angles plus two (i.e. distance between ankles and the person's height). From such a gait data sample we excluded the angles not used and the resulting dimension of the gait sample was equal to 22. Thus, a single gait pattern was of size 30×22.

In order to comprise the variation of the gait attributes over time we evaluated a number of spatio-temporal motion descriptors. The first group of gait descriptors was calculated on the basis of the normalized cumulative sums and the corresponding variances, which were calculated for each attribute in some time intervals. In the first gait signature g_cum5 they were calculated in every fifth frame, in the second signature g_cum10 they were calculated every tenth frame, whereas in third one g_cum15 they were calculated every fifteenth frame. Given the dimension of the gait sample, the lengths of the gait signatures were 264, 132 and 88. The second group of gait signatures was determined analogously, but instead of the normalized cumulative sums and variances, the averages and the corresponding variances were calculated for a specified range of adjacent frames, i.e. 5, 10 and 15. That means that the first gait signature g_ti5 consisted of the averages and the variances of the attributes corresponding to frames 1-5, 6-10, ..., 25-30. Such gait signatures were then utilized by the classifiers.

4 Experimental Results

The markerless motion tracking system was evaluated on video sequences with 22 walking individuals. In each image sequence the same actor performed two walks, consisting in following a virtual line joining two opposite cameras and following a virtual line joining two nonconsecutive laboratory corners. The first subsequence is referred to as 'straight', whereas the second one is called 'diagonal'. Given the estimated pose, the 3D model was projected to 2D plane and then overlaid on the images. Figure 5 depicts some results, which were obtained for person 1 in a straight walk. The degree of overlap of the projected 3D body model with the performer's silhouette reflects the accuracy of motion tracking.

Fig. 5. Articulated 3D human body tracking in sequence P1S. Shown are results in frames #0, 20, 40, 60, 80, 100, 120. The left sub-images are seen from view 1, whereas the right ones are seen from view 2.

The plots depicted in Fig. 6 illustrate the accuracy of motion estimation for some joints. As we can observe, the average tracking error of both legs is about 50 mm and the maximal error does not exceed 110 mm. The results presented above were obtained by APSO algorithm in 20 iterations using 300 particles.

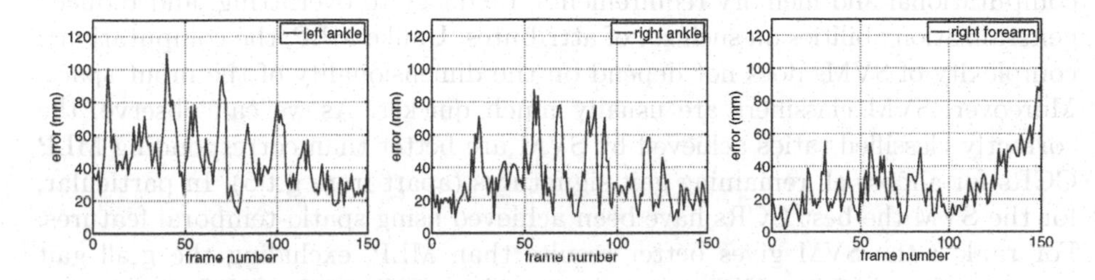

Fig. 6. Tracking errors [mm] versus frame number

In Table 1 are presented some quantitative results that were obtained using the discussed image sequences. The errors were calculated on the basis of 39 markers. For each frame they were computed as average Euclidean distance between individual markers and the recovered 3D joint locations. For each sequence they were then averaged over ten runs of the APSO with unlike initializations

Table 1. Average errors for $M = 39$ markers in three image sequences. The images from sequence P1S are depicted on Fig. 5.

			Seq. P1S	Seq. P2S	Seq. P3S
	#particles	it.	error [mm]	error [mm]	error [mm]
APSO	100	10	50.9±31.4	54.6±30.3	55.6±34.1
	100	20	46.9±26.5	49.0±27.1	49.7±26.5
	300	10	46.7±26.6	50.4±28.5	50.5±26.1
	300	20	44.4±25.9	46.6±24.8	48.2±25.2

Table 2 shows the recognition accuracy that was obtained on 230 gait cycles from both straight and diagonal walks. Each of 22 persons performed 2 walks in each direction, and each crossing consisted of 2 or 3 full gait cycles. 10-fold cross-validation was used to evaluate the performance of the proposed algorithm for view independent gait recognition. In the evaluation of the system we employed Naïve Bayes (NB), multilayer perceptron (MLP) and a linear SVM. The first column of the table contains acronym of the utilized gait signature, whereas the second one contains the number of the attributes of the gait signature. In order to show the usefulness of the spatio-temporal gait signatures we show also the correctly classified ratio (CCR), which was obtained by the use of all attributes, see signature g_all, and a signature g_avg consisting of the averages and the variances for each attribute. As we can observe, for rank 1 the best correctly classification ratio was obtained by MLP classifier operating on all attributes. However, taking into account that this result was obtained on large number of attributes, the usefulness of MLP classifier can be reduced in practice due to its computational and memory requirements, tendency to overfitting, and reduced generalization abilities on such set of attributes. Unlike MLP, the computational complexity of SVMs does not depend on the dimensionality of the input space. Moreover, SVM classifiers are usually much quicker. As we can observe, the correctly classified ratios achieved by SVM are better than corresponding MLP CCRs for almost all remaining gait signatures (apart from g_ti5). In particular, for the SVM the best CCRs have been achieved using spatio-temporal features. For rank 3 the SVM gives better results than MLP, excluding the g_all gait signature for which the CCR is equal to the best CCR of the SVM.

Table 2. Correctly classified ratio [%] using data from markerless motion capture

		rank 1			rank 3		
gait sig.	# att.	NB	MLP	SVM	NB	MLP	SVM
g_all	660	84.8	**96.1**	90.0	94.6	**99.6**	98.3
g_avg	44	87.0	90.9	91.6	95.2	97.0	**99.6**
g_cum5	264	82.6	91.3	**93.5**	92.6	97.8	99.1
g_cum10	132	83.0	90.0	**93.5**	92.6	97.8	98.7
g_cum15	88	83.9	88.7	91.7	93.9	97.0	99.1
g_ti5	264	81.7	91.3	90.4	91.7	98.3	98.3
g_ti10	132	80.4	90.9	92.2	93.6	98.3	98.3
g_ti15	88	83.0	91.3	93.0	95.2	98.7	**99.6**

The discussed experimental results were obtained using data, which were employed in [7] and were produced by our marker-less motion tracking system. For rank 1, the best identification performance obtained by SVM and operating on spatio-temporal features is better about 4% in comparison to results reported in [7], i.e. obtained by MLP operating on tensorial gait data that were reduced using Multilinear Principal Components Analysis (MPCA).

Figure 7 depicts the confusion matrix, which was obtained by SVM and g_cum10 signature. As we can see, eleven persons were classified with 100% probability. The lowest probability associated with true label is equal to 72.7%.

In Tab. 3 are shown results that were obtained using motion data from marker-based motion capture. As we can observe, the 3D data allows us to obtain very promising correctly classified ratios. In particular, the results demonstrate a potential of 3D techniques in view-independent gait analysis. The data obtained by motion capture systems are available at: `http://home.agh.edu.pl/~bkw/research/data/3DGaitData.7z`.

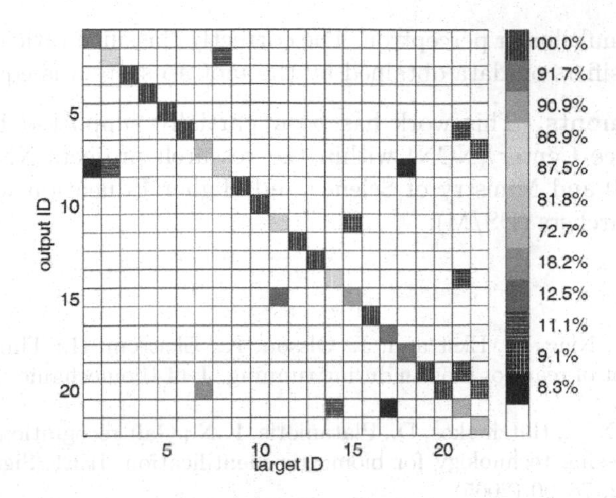

Fig. 7. Confusion matrix for SVM using g_cum10 (CCR=93.5%)

Table 3. Correctly classified ratio [%] using data from marker-based moCap (rank 1)

gait sig.	# att.	NB	MLP	SVM	1NN	3NN	1NN	3NN
					Euclidean dist.		Manhattan dist.	
g_all	8880	100	-	100	100	100	100	100
g_avg	148	99.1	100	100	100	100	100	100
g_cum20	888	99.6	100	100	100	100	100	100
g_cum40	444	99.6	100	100	100	100	100	100
g_cum60	296	99.6	100	100	100	100	100	100
g_ti20	888	99.6	100	100	99.6	100	100	100
g_ti40	444	99.6	100	100	100	100	100	100
g_ti60	296	99.6	100	100	99.6	100	100	100

The complete motion capture system was implemented in C/C++. The recognition performance was evaluated using WEKA software. The marker-less motion capture system runs on an ordinary PC. As demonstrated in [6], the full body motion tracking can be realized in real-time on modern GPUs.

5 Conclusions

We have presented an approach for view-independent gait recognition. The motion parameters were estimated using markerless human motion tracking. The person identification was done on the basis of the proposed spatio-temporal motion descriptors. The classification performance was determined on a dataset consisting of 230 gait cycles that were performed by 22 persons. The correctly classified ratio achieved by SVM is equal to 93.5% for rank 1 and 99.6% for rank 3. The identification accuracy is better than accuracy reported in [7], where tensorial gait data reduced by Multilinear Principal Components Analysis were classified by a multilinear perceptron. The correctly classified ratio of SVM, MLP and k-NN classifiers on data obtained by the moCap system is equal to 100%.

Acknowledgments. This work has been partially supported by the Polish National Science Center (NCN) within the research projects NN 516 475740, NN 516 483240 and Ministry of Science and Higher Education within a grant for young researchers (DS/M).

References

1. Areblad, M., Nigg, B., Ekstrand, J., Olsson, K., Ekström, H.: Three-dimensional measurement of rearfoot motion during running. J. of Biomechanics 23(9), 933–940 (1990)
2. Boulgouris, N.V., Hatzinakos, D., Plataniotis, K.N.: Gait recognition: a challenging signal processing technology for biometric identification. IEEE Signal Processing Magazine 22, 78–90 (2005)
3. Gu, J., Ding, X., Wang, S., Wu, Y.: Action and gait recognition from recovered 3-D human joints. IEEE Trans. Sys. Man Cyber. Part B 40(4), 1021–1033 (2010)
4. Jean, F., Albu, A.B., Bergevin, R.: Towards view-invariant gait modeling: Computing view-normalized body part trajectories. Pattern Recogn. 42(11) (November 2009)
5. Kennedy, J., Eberhart, R.: Particle swarm optimization. In: Proc. of IEEE Int. Conf. on Neural Networks, pp. 1942–1948. IEEE Press, Piscataway (1995)
6. Krzeszowski, T., Kwolek, B., Wojciechowski, K.: GPU-accelerated tracking of the motion of 3D articulated figure. In: Bolc, L., Tadeusiewicz, R., Chmielewski, L.J., Wojciechowski, K. (eds.) ICCVG 2010, Part I. LNCS, vol. 6374, pp. 155–162. Springer, Heidelberg (2010)
7. Krzeszowski, T., Michalczuk, A., Kwolek, B., Switonski, A., Josinski, H.: Gait recognition based on marker-less 3D motion capture. In: 10th IEEE Int. Conf. on Advanced Video and Signal Based Surveillance, pp. 232–237 (2013)
8. Kwolek, B., Krzeszowski, T., Wojciechowski, K.: Swarm intelligence based searching schemes for articulated 3D body motion tracking. In: Blanc-Talon, J., Kleihorst, R., Philips, W., Popescu, D., Scheunders, P. (eds.) ACIVS 2011. LNCS, vol. 6915, pp. 115–126. Springer, Heidelberg (2011)
9. Lee, L., Dalley, G., Tieu, K.: Learning pedestrian models for silhouette refinement. In: Proc. of the Ninth IEEE Int. Conf. on Computer Vision, pp. II:663–670 (2003)
10. Nixon, M.S., Carter, J.: Automatic recognition by gait. Proc. of the IEEE 94(11), 2013–2024 (2006)
11. Tanawongsuwan, R., Bobick, A.: Gait recognition from time-normalized joint-angle trajectories in the walking plane. IEEE Comp. Society Conf. on Computer Vision and Pattern Recognition 2, 726–731 (2001)
12. Urtasun, R., Fua, P.: 3D tracking for gait characterization and recognition. In: Proc. of IEEE Int. Conf. on Automatic Face and Gesture Rec., pp. 17–22 (2004)
13. Yam, C., Nixon, M.S., Carter, J.N.: Automated person recognition by walking and running via model-based approaches. Pattern Rec. 37(5), 1057–1072 (2004)
14. Yoo, J.-H., Hwang, D., Nixon, M.S.: Gender classification in human gait using support vector machine. In: Blanc-Talon, J., Philips, W., Popescu, D.C., Scheunders, P. (eds.) ACIVS 2005. LNCS, vol. 3708, pp. 138–145. Springer, Heidelberg (2005)
15. Yu, S., Tan, D., Tan, T.: Modelling the effect of view angle variation on appearance-based gait recognition. In: Narayanan, P.J., Nayar, S.K., Shum, H.-Y. (eds.) ACCV 2006. LNCS, vol. 3851, pp. 807–816. Springer, Heidelberg (2006)

Imaging and Evaluating Method as Part of Endoscopical Diagnostic Approaches

Martin Kuneš[1], Jaroslav Květina[2], Ilja Tacheci[2], Marcela Kopáčová[2], Jan Bureš[2], Milan Nobilis[3], Ondřej Krejcar[4], and Kamil Kuča[1]

[1] Biomedical Research Centre, University Hospital, Hradec Kralove, Czech Republic
{martin.kunes,kamil.kuca}@fnhk.cz

[2] 2nd Department of Internal Medicine – Gastroenterology, Charles University in Prague, Faculty of Medicine at Hradec Kralove, University Hospital, Hradec Kralove, Czech Republic
{tacheci,jan.bures,marcela.kopacova}@fnhk.cz

[3] Department of Pharmaceutical Chemistry and Drug Control, Faculty of Pharmacy, Charles University, Hradec Kralove, Czech Republic
nobilis@faf.cuni.cz

[4] Faculty of Informatics and Management, University of Hradec Kralove, Hradec Kralove, Czech Republic
ondrej.krejcar@uhk.cz

Abstract. Novel imaging diagnostic method - wireless capsule enteroscopy was used in experimental gastroenterology studies in pigs. We monitored the disintegration of two types of drug forms in gastrointestinal tract for next optimalisation of pharmaceutical technology process. In parallel, the plasma time profiles of the active component were detected. In second study, scanning of intestinal mucous lesions was done as diagnostic tool. The disintegration of tablets was perceptible in the duodenal-jejunal segment (until the 90th min) and culminated in the proximal jejunum (at the 3rd hour). The peak of maximal concentration was reached between the 3rd and 6th hour. The most common lesions induced by non-steroidal anti-inflammatory drugs were red spots and erosions. Sensitivity and specificity of capsule endoscopy were more than 80% and 95%, respectively. Wireless capsule endoscopy is a highly accurate non-invasive method for evaluation of experimental NSAID-induced enteropathy and evaluation of tablet disintegration proces in the intestine.

Keywords: wireless capsule enteroscopy, enteropathy, imaging method, drug forms.

1 Introduction

Diagnostic imaging methods in gastroenterology consists of various endoscopic techniques. Wireless video capsule endoscopy represents a novel, fundamental progress in non-invasive imaging of the gastrointestinal tract and allows direct visualization of the entire small bowell mucosa. The method was introduced into clinical practice relatively recently, 12 years [1–5]. At a very early stage, an initial

N.T. Nguyen et al. (Eds.): ACIIDS 2014, Part II, LNAI 8398, pp. 605–614, 2014.

study on experimental use of capsule endoscopy in dogs was published [6]. We used this diagnostic tool in several pre-clinical experiments [7, 8] however, other papers dealing with the use of capsule endoscopy in a different experimental setting are still missing. The first clinical studies on use of capsule endoscopy indiagnostics of NSAID-induced enteropathy in humans have already been published [9–12]. However, there are only limited data on experimental NSAID-induced enteropathy [13–17].

The aim of our experiments was to use the methodology of wireless video capsule endoscopy in experimental pigs in various experimenatal setups. Firstly, the confrontation of disintegration kinetics of drug formulations quantified by wireless capsule enteroscopy with levels of the active drug substance in the systemic circulation (plasma). Secondly, to diagnose the rate of gastrointestinal injury viewed in images scanned by capsule and to assess the diagnostic yield of this relatively novel method.

2 Methods

2.1 Capsule Enteroscopy

An EndoCapsule system (Olympus Optical Co, Tokyo, Japan) was used for capsule enteroscopies in pigs (Sus scrofa f. domestica), hybrids of Czech White and Landrace breeds, weighing 35.6 ± 3.3 kg (4–5 months old). Capsule enteroscopies in pigs were carried out under general anaesthesia - intramuscular injection of ketamine (Narkamon, Spofa, Prague, Czech Republic) and azaperone (Stresnil, Janssen-Pharmaceutica, Beerse, Belgium), continued by i.v. administration of 1% thiopental (Thiopental Valeant, Valeant Czech Pharma, Czech Republic) when appropriate to the lateral auricle vein. Syntostigmine (Hoechst-Biotika, Slovakia) is administrated intravenously, immediately after successful placement of the capsule endoscope into the duodenum (to eliminate the influence of general anesthesia on the gastrointestinal passage). Infusions of 0.9% saline solution was used to secure basal hydration.

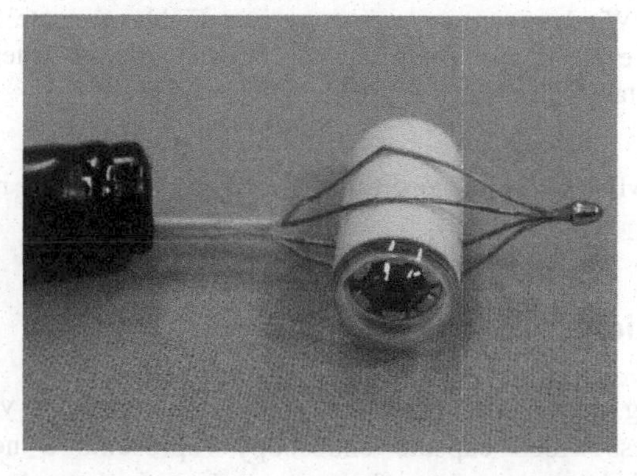

Fig. 1. Capsule endoscope is grasped by a special basket inserted through working channel of a flexible video-gastroscope and introduced into the porcine duodenum in this manner.

The wireless capsules (an outer diameter of 11 mm, length of 26 mm, and weighing 3.8 g, see Fig. 1) were in our studies introduced into the duodenum endoscopically. It moves through the small bowel, propelled by peristalsis and transmits data (2 images per second for 9 hours) to a portable data recorder. Endoscopy procedures were performed using video-gastroscopes GIF-Q130 (Olympus Optical Co, Tokyo, Japan). Data obtained during capsule enteroscopy are downloaded from the recorder unit (Fig. 2) to the workstation. It contains proprietary application software ‘‘Endo Capsule Software’’ (Olympus Optical Co, Tokyo, Japan), which enables the smooth management of capsule endoscope examination data. All video images are evaluated by a single physician, using usually auto speed adjustment (the review speed is automatically increased when there is little movement in consecutive images) with the lower limit for 12-frames/s of review speed in most recordings. Eight transmission antennas were fixed on the skin in the abdominal part of animal (Fig. 3).

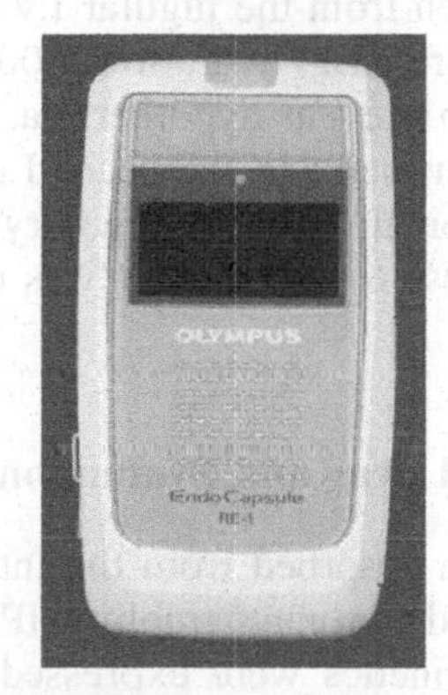

Fig. 2. Data recorder

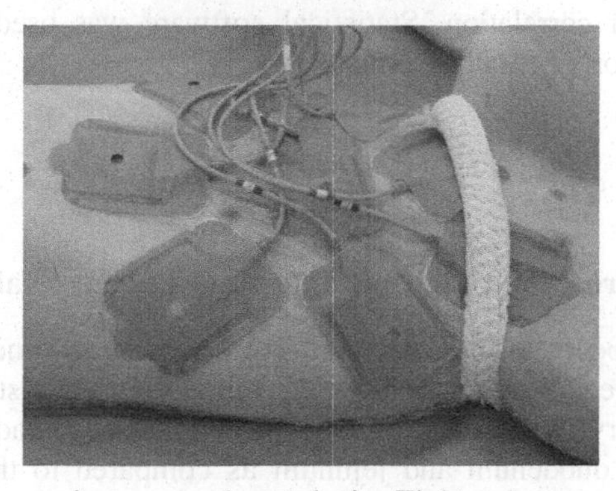

Fig. 3. Capsule endoscopy in an experimental pig. Eight antennae are fixed to skin of the abdomen.

2.2 Drugs

Two types of tablets prepared by compression of the technological intermediate – granulates (tablets “G”) and pellets (tablets “P”) were used in the study evaluated process of tablet disintegration and disaggregation in pig intestine.

2.3 Ethics

The study was approved by the Institutional Review Board of the Animal Care Committee at the Institute of Experimental Biopharmaceutics, the Czech Academy of Sciences (Protocol Number 149/2006). Animals were held and treated in accordance with the European Convention for the Protection of Vertebrate Animals Used for Experimental and Other Scientific Purposes (Council of Europe 1986).

2.4 Design of the Animal Study

Disintegration of the tablet was continuously (from administration to 8th hour) scanned by means of wireless capsule enteroscopy. In parallel, blood samples for the detection of active substance and comparison of the place of tablet disintegration and plasma time profiles were taken from the jugular I.v. infusion of a 0.9% saline solution was administrated to secure basal hydration (1000 mL per 8 hours). All animals were covered with blankets to prevent hypothermia. The jugular vein was incannulated one day before the experiment. In the second study, we induced intestinal lesions via 10-days of nonsteroidal anti-inflammatory drug (NSAID) administration. After NSAID medication, capsule enteroscopy was done and images scanned were evaluated.

2.5 Detection of the Model Drug and Evaluation

Active substance of drug form absorbed from the intestine into blood were detected using High-Performance Liquid Chromatography (HPLC). Plasma levels and numeric expression of disintegration kinetics were expressed as means and standard deviations. Data were analysed statistically with t-test, Mann-Whitney rank sum test and Spearman rank order correlation. Statistical software was used for these analyses (SigmaStat; Jandel Corp. Erkrath, Germany).

3 Results

3.1 Capsule Enteroscopy Usage and Intestinal Lesions Evaluation

The capsule endoscope reached the cecum during enteroscopy once (at 7 h 57 min), in the remaining cases, endoscopy recordings terminated in the distal or terminal ileum (life span of the battery). Movement of the capsule endoscope had less regional transit abnormalities in the duodenum and jejunum as compared to the ileum. The mean speed of the capsule endoscope in the proximal and middle jejunum was 4 cm/s. We estimated (from the pistures and transit time) the location of capsule in the intestine an in according to histological evaluation after next day autopsy. Due to the large volume of intestinal content in the ileum, the visibility of some ileal segments was a little bit worse. All capsule enteroscopies found a normal pattern of the small intestine in all experimental animals. Evaluating capsule endoscopy, visibility of the small intestinal mucosa was limited in parts of the ileum due to intestinal content (in 6/8 animals, comprising 27 ± 7% of images recorded from the ileum). Findings compatible with NSAID enteropathy were observed in 7/8 experimental pigs and were confirmed in 50.0% (6/12) on gross autopsy . Severe enteropathy represented by acute duodenal bleeding was classified on capsule endoscopy in one pig only. In six animals, we observed erosions in the jejunum and ileum (Fig. 4a) or multiple red spots in the duodenum, jejunum, and ileum (Fig. 4b).

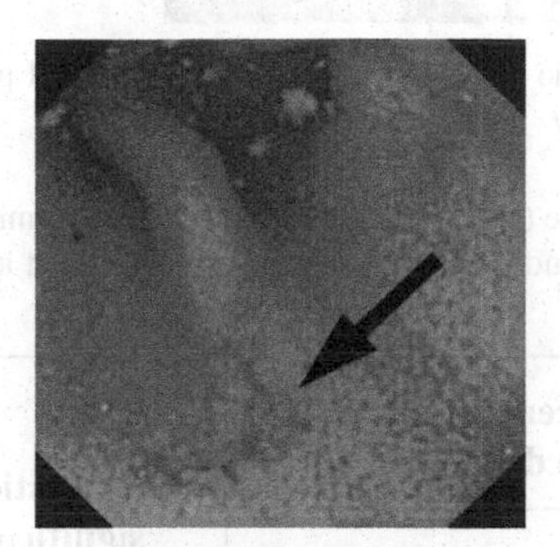

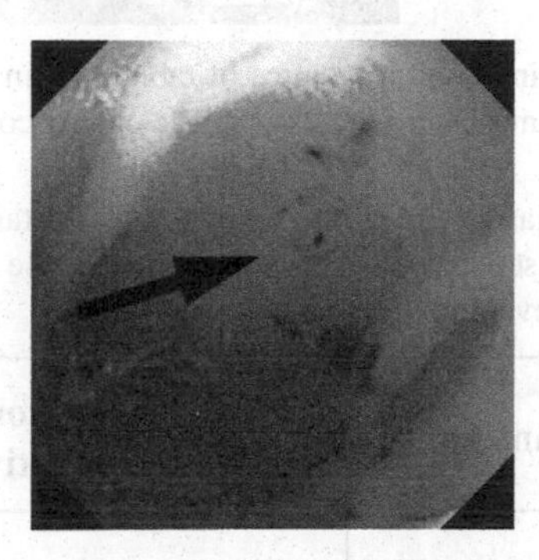

Fig. 4a,b. Erosions in proximal jejunum – mild enteropathy (Left image) and red spots in middle jejunum – mild enteropathy (Right image)

3.2 Disintegration of Tablets and Plasma Levels of Active Compound

Images were evaluated and representative pictures of findings sorted according to intestinal segments and specific time intervals are shown. In time 0 – immediately after tablet administration into the duodenum, 90 min. after administration (proximal jejunum), 3 h after administration (middle jejunum), 5 h after administration (distal jejunum) and 7 - 8 hours after administration (middle to distal ileum). The scale of number and size of fragments were evaluated. The statistically significant differences in the most of measured time intervals were found in tested drug formulations. The difference in behaviour of the studied tablet dosage forms is demonstrated by expression of the time relation between the disintegration of tablets in the intestine and systemic levels of released active compound. While in the case of tablets "G", plasma levels of active substance have a classical peak of c_{max} (3-4 hours) and subsequent gradual bioelimination phase. In tablets "P", a relatively low linear "steady state" phase occurred which is usually characteristic for tablets with slow-release of active ingredients (Figs 5-6 and Tab 1).

Fig. 5. Disintegration kinetics of tablet "G" in the intestinal lumen of experimental pigs documented by means of wireless capsule enteroscopy

Fig. 6. Disintegration kinetics of tablet "P" in the intestinal lumen of experimental pigs documented by means of wireless capsule enteroscopy

Table 1. Plasma concentration of active substance (pmol/μL) after intraduodenal administration of tablets (see text dealing with details of the study design). Results are expressed as mean ± standard deviation. NS = non-significant.

Parameter	**Plasma concentration of active drug**		**Statistical significance**
Time	**"G" (granulates)**	**"P" (pellets)**	
30 min.	248.5 ± 172.0	62.0 ± 79.1	p = 0.036
60 min.	429.0 ± 255.8	142.5 ± 134.1	p = 0.035
90 min.	3630.8 ± 2232.6	231.7 ± 159.8	p = 0.002
2 hours	3795.8 ± 2199.5	423.8 ± 268.1	p = 0.002
3 hours	7960.2 ± 2549.9	628.5 ± 483.7	p = 0.002
4 hours	515.7 ± 299.7	586.0 ± 400.7	NS (p = 0.738)
5 hours	225.8 ± 132.3	635.3 ± 480.0	NS (p = 0.180)
6 hours	219.8 ± 134.2	654.7 ± 464.9	NS (p = 0.180)
8 hours	207.8 ± 130.7	696.2 ± 468.4	p = 0.041

4 Discussion

Capsule enteroscopies are feasible in experimental design in pigs. In our study, the capsule endosopy was successfully accomplished in all animals. However, there were several difficulties during introducing the capsule into duodenum - gastric cardia is close to the pylorus, so the endoscopic approach to the duodenum is more hooked. To prevent a delay due to persistence of the video capsule in the stomach, we introduced all capsules directly into the duodenum by means of a standard flexible video-gastroscope. This study is continuation of our previously presented experiments pigs, including preliminary data on the use of capsule endoscopy [7, 8]. The movement of capsule endoscope was more rapid in the jejunum compared to the ileum. This fact reflects the natural motility character of the jejunum. The mean life span of the batteries used in the experiment was about 510 min. Capsule enteroscopy (similar to other methods) is not able to recognize the exact borderline between the jejunum and ileum. We have proposed a rough estimation of the approximate location of the capsule endoscope, being merely a conservative estimate. In comparison to clinical studies (the small bowel is twice shorter than a pig), the cecum is reached. In up to 25% of cases, the batteries also run out before the capsule passes the ileo-cecal valve in humans [18]. This paper concerns methodological preparation for further experimental pre-clinical studies such as the evaluation studies of the effect of different xenobiotics on the small intestine or disintegration of various type of drug forms. Our study provided several new data, differentiating capsule endoscopy in experimental pigs from those performed in humans. Small intestinal injury compatible with NSAID-induced enteropathy was observed in 7/8 animals. The most frequent findings on capsule endoscopy were mild mucosal injury (multiple red spots and erosions), in 6/8 animals. Its clinical impact is probably low. Although human capsule endoscopy studies reported such lesions in up to 40% of healthy volunteers or controls [14, 18, 19]. Prevalence of NSAID-induced enteropathy in humans treated with nonselective NSAIDs is 55–78% in the available literature (mostly for chronic users) [11, 13, 20, 21]. In four animals, red spots seen on capsule endoscopy were not confirmed on autopsy, probably due to the etiology of these changes (mucosal congestion could hardly be revealed on gross autopsy).

The use of capsule enteroscopy provided quite a new and original insight into the manner of disintegration techniques (dissolution techniques), regarding the description of disintegration kinetics of dosage form (in simulated digestive tract) and regarding effective release of a drug from the dosage form, this being in their economy/cost savings, in the relative speed of achieving results and in the possibility of flexibly changing the dissolution medium. During development of generic products, dissolution tests are therefore a suitable initial screening test to estimate the behaviour of the studied formulation in different segments of the GIT [20]. The composition of a gradually changed medium is a crucial input factor of dissolution methods [18]. Results might be less reliable if gastric, biliary and pancreatic juice is not added into the dissolution medium. These compounds may of course be involved in the metabolism both in the excipients and also active substances of the formulation. During dissolution tests, the eventual effect of intestinal passage (i.e. intestinal tonus and peristalsis) on disintegration tablet formulation cannot, of course, be registered and have a little predictive value from this point of view [23, 24] . Animal "in vivo" studies, which determine the pharmacokinetic parameters of released substances after gastro-intestinal administration of the dosage form, provide – in comparison with dissolution approaches – a broader basis for interpretation. The use of wireless capsule enteroscopy, with simultaneous drug administration to the most proximal intestinal segment, is thus one of the original and new additional technical approaches, which enable us to complete our knowledge of pharmacokinetic mechanisms of various dosage forms and the released substances. Future directions contain study of using novel localization techniques to determine exact position of capsule inside a body [25] and new imaging methods as well as possibilities in development of new type of capsule for wide spectrum of sensing [26,30].

5 Conclusions

The NSAID lesions localisation study revealed the presence of small intestinal mucosal lesions induced by oral drug administration on capsule endoscopy in the majority of experimental pigs (more than 85%).. We also confirmed the utility of capsule endoscopy in this experimental model for NSAID-induced enteropathy research. It was shown utilisation of capsule enteroscopy for showing the process of disintegration of drug form in vivo - diametrically different kinetics of tablet disintegration also different plasma levels of released. The results of this pre-clinical work showed the usefulness both from the aspects of more precise definition of pharmacokinetic mechanisms in the intestinal tract and from the ethical aspects (data transfer into clinical phases of research).

Acknowledgments. The study was supported by research grant IGA NT 13532 from the Ministry of Health (Czech Republic).

References

1. Eliakim, R.: Video capsule endoscopy of the small bowel. Curr. Opin. Gastroenterol. 24, 159–163 (2008)
2. El-Matary, W.: Wireless capsule endoscopy: indications, limitations, and future challenges. J. Pediatr. Gastroenterol. Nutr. 46, 4–12 (2008)
3. Keuchel, M., Hagenmüller, F., Fleischer, D.E. (eds.): Atlas of video capsule endoscopy. Springer, Heidelberg (2006)
4. Mishkin, D.S., Chuttani, R., Croffie, J., et al.: ASGE technology status evaluation report: wireless capsule endoscopy. Gastrointest. Endosc. 63, 539–545 (2006)
5. Sidhu, R., Sanders, D.S., Morris, A.J., McAlindon, M.E.: Guidelines on small bowel enteroscopy and capsule endoscopy in adults. Gut. 57, 125–136 (2008)
6. Appleyard, M., Fireman, Z., Glukhovsky, A., et al.: A randomized trial comparing wireless capsule endoscopy with push enteroscopy for the detection of small-bowel lesions. Gastroenterology 119, 1431–1438 (2000)
7. Kopacova, M., Tacheci, I., Kvetina, J., Bures, J., Kunes, M., Spelda, S., Tycova, V., Svoboda, Z., Rejchrt, S.: Wireless video capsule enteroscopy in preclinical studies: methodical design of its applicability in experimental pigs. Dig. Dis. Sci. 55, 626–630 (2010)

8. Kvetina, J., Kunes, M., Bures, J., Kopacova, M., Tacheci, I., Spelda, S., Herout, V., Rejchrt, S.: The use of wireless capsule enteroscopy in a preclinical study: a novel diagnostic tool for indomethacin-induced gastrointestinal injury in experimental pigs. Neuroendocrinol. Lett. 29, 763–769 (2008)
9. Goldstein, J.L., Eisen, G.M., Lewis, B., Gralnek, I.M., Zlotnick, S., Fort, J.G.: Video capsule endoscopy to prospectively assess small bowel injury with celecoxib, naproxen plus omeprazole, and placebo. Clin. Gastroenterol. Hepatol. 3, 133–141 (2005)
10. Graham, D.Y., Opekun, A.R., Willingham, F.F., Qureshi, W.A.: Visible small-intestinal mucosal injury in chronic NSAID users. Clin. Gastroenterol. Hepatol. 3, 55–59 (2005)
11. Maiden, L., Thjodleifsson, B., Seigal, A., Bjarnason, I.I., Scott, D., Birgisson, S., Bjarnason, I.: Long-term effects of nonsteroidal anti-inflammatory drugs and cyclooxygenase-2 selective agents on the small bowel: a cross-sectional capsule enteroscopy study. Clin. Gastroenterol. Hepatol. 5, 1040–1045 (2007)
12. Hawkey, C.J., Ell, C., Simon, B.: Less small-bowel injury with lumiracoxib compared with naproxen plus omeprazole. Clin. Gastroenterol. Hepatol. 5, 536–544 (2008)
13. Satoh, H., Hara, T., Murakawa, D., Matsuura, M., Takata, K.: Soluble dietary fiber protects against nonsteroidal anti-inflammatory drug-induced damage to the small intestine in cats. Dig. Dis. Sci. 16, 23–28 (2009)
14. Sidhu, R., Sanders, D.S., McAlindon, M.E., Kapur, K.: Capsule endoscopy for the evaluation of nonsteroidal anti-inflammatory drug-induced enteropathy: United Kingdom pilot data. Gastrointest. Endosc. 64, 1035 (2006)
15. Kunes, M., Kvetina, J., Malakova, J., Bures, J., Kopacova, M., Rejchrt, S.: Pharmacokinetics and organ distribution of fluorescein in experimental pigs: an input study for confocal laser endomicroscopy of the gastrointestinal tract. Neuroendocrinol. Lett. 31 (suppl. 2), 57–61 (2010)
16. Kunes, M., Kvetina, J., Bures, J.: Type and distribution of indomethacin-induced lesions in the gastrointestinal tract of rat. Neuroendocrinol. Lett. 30(suppl. 1), 96–100 (2009)
17. Kunes, M., Kvetina, J., Kubant, P., Nobilis, M., Smidova, I., Herout, V., Svoboda, Z.: The rat small intestine in situ perfusion technique as a tool for the bioequivalence estimation of suspension drug forms. Biomedical Papers 151(suppl. 1), 47–55 (2007)
18. Maiden, L., Thjodleifsson, B., Theodors, A., Gonzalez, J., Bjarnason, I.: A quantitative analysis of NSAID-induced small bowel pathology by capsule enteroscopy. Gastroenterology 128, 1172–1178 (2005)
19. Goldstein, J.L., Eisen, G.M., Lewis, B., Gralnek, I.M., Aisenberg, J., Bhadra, P., Berger, M.F.: Small bowel mucosal injury is reduced in healthy subjects treated with celecoxib compared with ibuprofen plus omeprazole, as assessed by video capsule endoscopy. Aliment. Pharmacol. Ther. 25, 1211–1222 (2007)
20. Amidon, K.S., Langguth, P., Lennernas, H., Yu, L., Amidon, G.L.: Bioequivalence of oral products and the biopharmaceutics classification system: science, regulation, and public policy. Clin. Pharmacol. Ther. 90, 467–470 (2011)
21. Kopylov, U., Seidman, E.G.: Clinical application of small bowel capsule endoscopy. Clin. Exp. Gastroenterol. 6, 129–137 (2013)
22. Lim, Y.J., Chun, H.J.: Recent advances in NSAIDs-Induced Enteropathy therapeutics: New options, new challenges. Gastroenterol. Res. Pract. 6, 1–7 (2013)
23. Amidon, G.L., Lennernas, H., Shah, V.P., Crison, J.R.: A theoretical basis for a biopharmaceutic drug classification: the correlation of in vitro drug product dissolution and in vivo bioavailaibility. Pharm. Res. 12, 413–420 (1995)

24. Kunes, M., Kvetina, J., Kholova, D., Tlaskalova-Hogenova, H., Pavlik, M.: Absorption kinetics of 5-aminosalicylic acid in rat: influence of indomethacin-induced gastrointestinal lesions and Escherichia Coli Nissle 1917 medication. Neuroendocrinol. Lett. 32 (suppl. 1), 46–52 (2011)
25. Benikovsky, J., Brida, P., Machaj, J.: Proposal of User Adaptive Modular Localization System for Ubiquitous Positioning. In: Pan, J.-S., Chen, S.-M., Nguyen, N.T. (eds.) ACIIDS 2012, Part II. LNCS, vol. 7197, pp. 391–400. Springer, Heidelberg (2012)
26. Cheng, W.C., Liou, J.W., Liou, C.Y.: Construct Adaptive Template Array for Magnetic Resonance Images. In: IEEE International Joint Conference on Neural Networks, Brisbane, June 10-15 (2012) doi: 10.1109/IJCNN.2012.6252560
27. Penhaker, M., Stankus, M., Prauzek, M., Adamec, O., Peterek, T., Cerny, M., Kasik, V.: Advanced Experimental Medical Diagnostic System Design and Realisation. Electronics and Electrical Engineering 1(117), 89–94 (2012)
28. Penhaker, M., Darebnikova, M., Cerny, M.: Sensor network for measurement and analysis on medical devices quality control. In: Yonazi, J.J., Sedoyeka, E., Ariwa, E., El-Qawasmeh, E. (eds.) ICeND 2011. CCIS, vol. 171, pp. 182–196. Springer, Heidelberg (2011)
29. Longo, L., Kane, B.: A Novel Methodology for Evaluating User Interfaces in Health Care. In: 24th IEEE International Symposium on Computer-Based Medical Systems, CBMS 2011, Bristol, England, June 27-30 (2011)
30. Liou, C.-Y., Cheng, W.-C.: Manifold Construction by Local Neighborhood Preservation. In: Ishikawa, M., Doya, K., Miyamoto, H., Yamakawa, T. (eds.) ICONIP 2007, Part II. LNCS, vol. 4985, pp. 683–692. Springer, Heidelberg (2008)

Author Index